Birkhäuser

Trends in Mathematics

Trends in Mathematics is a series devoted to the publication of volumes arising from conferences and lecture series focusing on a particular topic from any area of mathematics. Its aim is to make current developments available to the community as rapidly as possible without compromise to quality and to archive these for reference.

Proposals for volumes can be submitted using the Online Book Project Submission Form at our website www.birkhauser-science.com.

Material submitted for publication must be screened and prepared as follows:

All contributions should undergo a reviewing process similar to that carried out by journals and be checked for correct use of language which, as a rule, is English. Articles without proofs, or which do not contain any significantly new results, should be rejected. High quality survey papers, however, are welcome.

We expect the organizers to deliver manuscripts in a form that is essentially ready for direct reproduction. Any version of TEX is acceptable, but the entire collection of files must be in one particular dialect of TEX and unified according to simple instructions available from Birkhäuser.

Furthermore, in order to guarantee the timely appearance of the proceedings it is essential that the final version of the entire material be submitted no later than one year after the conference.

More information about this series at http://www.springer.com/series/4961

Ordered Structures and Applications

Marcel de Jeu • Ben de Pagter • Onno van Gaans
Mark Veraar
Editors

Positivity VII (Zaanen Centennial Conference),
22-26 July 2013, Leiden, the Netherlands

Birkhäuser

Editors
Marcel de Jeu
Mathematical Institute
Leiden University
Leiden, the Netherlands

Onno van Gaans
Leiden University
Leiden, the Netherlands

Ben de Pagter
Delft University of Technology
Delft, the Netherlands

Mark Veraar
Delft University of Technology
Delft, the Netherlands

ISSN 2297-0215 ISSN 2297-024X (electronic)
Trends in Mathematics
ISBN 978-3-319-27840-7 ISBN 978-3-319-27842-1 (eBook)
DOI 10.1007/978-3-319-27842-1

Library of Congress Control Number: 2016953686

Mathematics Subject Classification (2010): 46-06, 47-06, 28-06, 06-06

Printed on acid-free paper

This book is published under the trade name Birkhäuser, www.birkhauser-science.com
The registered company is Springer International Publishing AG
The registered company address is: Gewerbestrasse 11, 6330 Cham, Switzerland

Contents

Preface

The seventh Positivity conference was held from 22–26 July, 2013, at the Science campus of Leiden University, the Netherlands, jointly organized by Leiden University and Delft University of Technology. It was also the Zaanen Centennial Conference, on the occasion of the 100th birth year of Adriaan Cornelis Zaanen. The organising committee consisted of Marcel de Jeu (chair), Ben de Pagter, Miek Messerschmidt, Jan Rozendaal, Onno van Gaans and Mark Veraar.

The sixteen invited speakers Francesco Altomare (Bari, Italy), Wolfgang Arendt (Ulm, Germany), Karim Boulabiar (Tunis, Tunisia), Qingying Bu (Mississippi, USA), Guillermo Curbera (Sevilla, Spain), Julio Flores (Madrid, Spain), Yehoram Gordon (Haifa, Israel), Rien Kaashoek (Amsterdam, the Netherlands), Coenraad Labuschagne (Johannesburg, South Africa), Boris Mordukhovich (Detroit, Michigan, USA), Ioannis Polyrakis (Athens, Greece), Abdelaziz Rhandi (Salerno, Italy), Evgeny Semenov (Voronezh, Russia), Fedor Sukochev (Sydney, Australia), Jun Tomiyama (Tokyo, Japan) and Jan van Neerven (Delft, the Netherlands) covered a broad spectrum of topics in plenary lectures, ranging from order-theoretic approaches to stochastic processes via positive solutions of evolution equations to ordered structures in the context of algebras of operators on Hilbert spaces. This variety was also visible in the 122 contributed lectures in the various thematic special sessions, covering the field Positivity in virtually all its aspects. At the conference's website http://websites.math.leidenuniv.nl/positivity2013/ many of the PDF's have been posted.

During the Zaanen celebration, several of Zaanen's PhD students and his oldest son highlighted his life and works.

With 162 participants from 32 countries and 6 continents, this issue of the series of Positivity conferences was almost twice as large as its predecessor, illustrating the growth and vitality of the field. We are pleased that Birkhäuser has been found prepared to publish these proceedings, and we hope that this will stimulate a further development.

Positivity VII was made possible by the Delft Institute of Applied Mathematics, Foundation Compositio Mathematica, the Geometry and Quantum Theory research cluster (GQT), the Royal Netherlands Academy of Arts and Sciences (KNAW), the Royal Mathematical Society (KWG), the Leiden University Fund (LUF), the Mathematical Institute of Leiden University, the Nonlinear Dynamics in Natural Systems research cluster (NDNS+) and the Netherlands Organisation for Scientific Research (NWO). Their generous support is gratefully acknowledged.

Marcel de Jeu
Ben de Pagter
Onno van Gaans
Mark Veraar

Adriaan Cornelis Zaanen (1913–2003)

Ordered Structures and Applications: Positivity VII
Trends in Mathematics, xi–xxii

Adriaan Cornelis Zaanen

Koos Grobler, Rien Kaashoek, Anton Schep and Pieter Zaanen

Abstract. The Positivity VII conference in Leiden in July 2013 was the Zaanen Centennial Conference, on the occasion of the 100th birth year of Adriaan Cornelis Zaanen (1913–2003). Zaanen held the chair of analysis in Delft for five years and subsequently the chair of mathematical analysis in Leiden for more than twenty-five years, until his retirement in 1982. He played a prominent role in 'Positivity' throughout that period.

During the conference, Koos Grobler, Rien Kaashoek and Anton Schep, three of his former PhD students, shared their personal recollections of Zaanen with the audience. Pieter Zaanen, his oldest son, told about their family life. The written accounts of their speeches are included here, followed by a chronological list of material on the life and works of Zaanen that has appeared in print and on the internet over the years.

Mathematics Subject Classification (2010). 01A70.

Keywords. Adriaan Cornelis Zaanen.

1. Koos Grobler

"Adriaan C. Zaanen. Professor of Mathematics at the University of Leyden, Netherlands", it read on the title page of the book 'Introduction to the theory of integration'. This was my first encounter with his name as an honours student doing a course in measure and integration theory. Being a first encounter with 'abstract mathematics' the concise and exact style of the book was challenging and I admired the author, who, in my esteem, must have been a great mathematician to write a book like that. (Incidently this course was taught to me by a student of Zaanen, Ben Strydom, who sadly passed away on 3 December 2012, one day after his 80th birthday.) After completing my master's studies, Ben advised me to pursue my studies abroad and he recommended Zaanen. With some apprehension I wrote him a letter and promptly (6 weeks later) received a letter from Zaanen asking me to inform him in detail about my training: books studied with chapters and paragraphs in detail. Again 6 weeks later I received the good news that he would accept me as a student and he even suggested a topic.

I remember his letters: handwritten, extremely well-formulated and complete. I still keep them for this was the way we communicated – if you were lucky you had an answer to a letter after 6 weeks. This left no room for sloppy formulations or incompleteness!

I remember that Zaanen was not in Leiden when I arrived – he was with Wim Luxemburg in Pasadena, most probably putting the finishing touches to 'Riesz spaces I'. Thus I worked for the first 9 months without meeting him. However, I remember him as a man who invited a student to study with him at a time when South Africans were not welcomed with open arms – especially not in the Netherlands.

Then finally one day when I was in the Institute at the beginning of the new semester, with an inquiry at the secretary, we met. I remember his first words: "Welcome! We will get to know each other much better!" I remember his kindness and friendly way, and the way he struck me as a modest, unpretentious man. He invited me into his room and we started to work. We made appointments, I had to show him what I did and we went through it thoroughly. I remember the precise and complete way he demanded me to write mathematics. Nothing slipped by and he demanded full proofs.

After completion of my thesis, I was invited by him for a visit at home for dinner. There we (my wife and I) first met Ada. They immediately insisted to be called on their first names – something rather difficult for me at first. Ada was in many respects Aad's complement. She was very practical and outspoken, while Aad was, to say the least, not a very practical man and somewhat quiet. But in their kindness and friendliness they were on par. I will never open a bottle of wine without remembering the uncorking of the wine at this dinner: after fumbling with the corkscrew in vain for a while, Ada took the bottle from him (or I should say grabbed it from him) and uncorked it. According to her he was useless when it came to everyday chores and wanted to know from my wife if that was also the case with me. At this occasion our friendship started and we enjoyed many hours with them, inter alia at their holiday home on the lake of Geneva. Although not holding back her disgust at his impractical nature, one knew that she was in fact extremely proud of him.

I remember Aad to be a man who visited South Africa at a time this was frowned upon. As the official guest of the South African Mathematical Society he toured the country in 1973, visiting all universities. Perhaps, in his wisdom, he observed signs that the future of the country was not totally lost but that there were many signs that change may come to that society. I remember Aad as a friend who visited me on three occasions for longer periods. I also remember Aad as a loyal friend who insisted that I be invited to symposiums organized in his honour, even when it was official policy of Leiden University that its staff should have no contact with South Africans.

Through Aad I also met Pay Huijsmans, Ton Schep and Ben de Pagter, all of them people who enriched my life.

So this was the giant, the professor and the friend that Aad Zaanen was to me. I have only fond memories of him and salute him as the man who influenced my life profoundly.

2. Rien Kaashoek

In my contribution to this memorial session I would like to remember Aad Zaanen as my analysis professor and PhD supervisor at Leiden University.

My first year (1955–1956) as a student at Leiden University was a year without mathematics professors. The geometry professor Haantjes was seriously ill and passed away during the year, and the famous professor Kloosterman, chair of number theory and algebra, was on sabbatical leave. The analysis professor Droste had retired earlier and his chair was still vacant, and so were a few other positions. Zaanen, a former PhD student of Droste, was at that time professor at Delft. He was appointed as Droste's successor and came to Leiden in '56, in my second year.

In that year I followed Zaanen's one year analysis course, later known under the name 'Analysis B'. It was a great experience. The exposition was absolutely clear and transparent, systematic from the beginning to the end, like the Euclidean geometry I learned in my first year at high school, and no gaps in the proofs. Moreover, notes were easy to make: Zaanen presented his course on the blackboard from the beginning to the end, and not in telegram style. I admired Zaanen's course, and I got to know Zaanen as a great lecturer.

Later in my fourth year, when I was already Zaanen's assistant, I met one of his other attractive characteristics: generosity. One of my tasks was to present for approval to Zaanen the problems for the written examination for 'Analysis B', an examination that was feared by many students. Usually what I presented was accepted, but sometimes his reaction was: you make the problem too difficult, be more generous, and on the spot he would suggest a simpler version, which still required the knowledge the students were expected to have.

After attending Zaanen's functional analysis course I bought the 1956 edition of his book 'Linear analysis'. Zaanen had a broad view on functional analysis. The title of the book does not tell you that but it is immediately clear from the subtitle: 'Measure and integral, Banach and Hilbert space, linear integral equations'. Operators on the mentioned spaces, selfadjoint and non-selfadjoint, and non-singular linear integral equations are all covered by the book.

To understand the value of the book let me quote from the review Paul Halmos wrote in 1954 for the Bulletin of the American Mathematical Society. Halmos writes:

> "This book goes a long way toward filling the gap caused by the fact that Banach's book has been out of date for several years. The writing is clear and well organized; the author is an excellent expositor. A pleasing feature is the quality and quantity of examples. Not only is there an adequate supply of exercises at the end of each chapter, but throughout the book there are many detailed discussions of standard and non-standard examples: sequence spaces, the Orlicz generalization of L_p-spaces, sequential transformations, integral kernels, etc."

The book is still of great value today, to a large extent because of these examples and exercises.

In 1961 I became Zaanen's third PhD student. He left me a lot of freedom. He gave me Tosio Kato's 1958 paper 'Perturbation theory for nullity, defect and other quantities of linear operators'. Zaanen told me that he was going on sabbatical to the California Institute of Technology to work with Wim Luxemburg, and that he expected me to come up with a problem on his return. At that time I did not realize that the topic Zaanen had introduced me to was what we nowadays call the theory of Fredholm operators, for bounded as well as for unbounded operators. At that time a very modern topic with many ramifications to various mathematical areas, old and new, including Wiener–Hopf integral equations and singular integral equations, differential equations, evolution equations, equations with delay, Toeplitz operators, spectral theory of Banach algebras, index theory in geometry, etc., etc. I am very grateful to Zaanen for introducing me to this rich subject, and that he put me on such an exciting track in operator theory. Did I find a problem? Yes, I did, and it was accepted by Zaanen.

I mentioned the freedom Zaanen gave me. That changed when I started presenting him drafts of theorems and sections of the thesis for discussion. He read each part in great detail, made critical remarks, presented critical questions, offered suggestions for better proofs, for references or for other directions. It was extremely enriching. Zaanen also taught me what it means to write a good mathematical text. Sometimes he would return to me a draft of a section completely rewritten in his own handwriting.

Zaanen had several rules which I found very encouraging. One of them goes as follows: If you don't understand a paragraph in a paper, or an argument in a proof, don't blame it directly on yourself. It may very well be that the author made a mistake or was not careful enough. Another rule concerned lecturing at a conference: When you give a lecture for a math audience, of 30 minutes say, don't begin with your latest results, use the first 20 minutes for describing the area and the problems, and then in the last 10 minutes you can tell about your own achievements. Both rules I passed on to my own PhD students.

Zaanen had a great international standing. Being one of his students was a real advantage. When Zaanen wrote a letter for you, success was almost guaranteed. He also introduced me to many outstanding mathematicians, including Kato at Berkeley, Szökefalvi-Nagy in Szeged, Angus Taylor at UCLA, Smithies at Cambridge, and Köthe in Frankfurt, to mention just a few.

At that time assistants and PhD students did not call their supervisors by their first name. Immediately after my defense Zaanen told me that from now on I had to call him Aad. It took me a while to get used to that.

On this occasion I also want to remember Aad Zaanen's dear wife: Ada. She was for us, young PhD students and assistants, an inspiring lady who knew the world outside the university much better than we did. Ada's passing away was a great loss for Aad.

When Zaanen passed away in 2003 I informed my mathematical friends and co-authors. Israel Gohberg immediately replied. He wrote me:

> "It is sad news. It is such a pity that good people are dying also. My generation in functional analysis will remember professor Zaanen as a teacher, pioneer and innovator. His books helped us to learn the subject in times when the sources were very limited and the maps were not drawn yet. For you professor Zaanen played a very important role in the early stage of your career. Having him as an instructor you probably studied in one of the best schools of functional analysis of that time. I am sure that he will always have a place in your heart."

The latter is certainly true.

Albrecht Pietsch wrote me:

> "He was a really great mathematician who will stay alive through his works."

And so it is, as we are seeing at the present conference.

3. Anton Schep

3.1. Zaanen as teacher

I arrived as a student in the fall of 1969 at Leiden University, but it was not until the fourth semester, the spring of 2001, that I took my first course of professor Zaanen. The class was quite sizable and met in the Gorlaeus Laboratories, the exact same location as where the Positivity VII conference dedicated to the Zaanen centennial is now being held. The picture on the right is taken in that building and judging from the text on the blackboard it could well have been the class I took from him. At that time it was customary to address one's professor formally by 'professor', and it was not until I had defended my dissertation in 1977 that this changed. After that I was allowed to call him Aad, but it took me a few years to get used to that. My first course with Zaanen was 'Analysis IV', the first theoretical course in analysis and the recommended textbook was 'Principles of mathematical analysis' [7], also known as 'Baby Rudin'. There were no homework assignments in the course and at the end we had to take an oral exam. As there were a large number of students in the class the oral examinations were given by many people. My examiner wasn't Zaanen, but Rob Tijdeman. In the following years I took three more one year courses of Zaanen. The first one was a course in measure and integration. He did not follow his own book [8] in the course, but followed rather closely Bartle's 'The elements of integration' [1]. He did supplement the material of this book by a detailed treatment of Birkhoff's L_1 ergodic theorem. Notably missing from the course was a treatment of absolutely continuous functions and differentiation of monotone functions. The next course was a course

A.C. Zaanen

in functional analysis. By now Zaanen was aware of me as a student and asked me to write up notes for the course. If I did a good job at it, I would be exempt from the oral exam at the end. This was an offer I could not turn down. If my memory is correct it is during the second half of this course that Zaanen had to have a kidney removed and C.B. (Pay) Huijsmans finished the lectures for him. The final course I took from Zaanen was a course on integral equations. It included a detailed introduction to Banach function spaces and a detailed treatment of integral operators of finite double norm. I have still complete sets of notes of all three of these courses, but have only partial notes of the first course. Looking back it is quite remarkable that the notion of a Riesz space did not come up in any of these courses. Just a few years before he had published 'Riesz spaces I' [5] with Wim Luxemburg, but that did not yet trickle down in his lectures. I started learning about Riesz spaces in the seminars with Pay Huijsmans. I still remember the thrill of proving my first result, while bicycling to the Mathematical Institute. It solved an open problem posed by Huijsmans during one of these seminars.

Looking back at my notes of the courses I took from Zaanen, one sees the same attention to detail as in his books and papers. He never waives his hand in a argument and his proofs are crystal clear. This characteristic he passed on to his PhD students. I see the same style of lecturing by Luxemburg, Kaashoek, Grobler and de Pagter. I try to emulate his style in the classes I teach.

3.2. Zaanen as PhD advisor

In the spring of 1974, near the completion of undergraduate degree, I applied for and got one of three assistant positions. At that time I had not yet settled on Zaanen as a PhD advisor, but after exploring the alternatives, I decided quickly that my mathematical tastes aligned most with Zaanen's area of research. Initially I did not have a well-defined thesis topic. Zaanen had given me a set of notes on the lattice of Riesz seminorms on a Riesz space and said to take a look whether I could something more with it. After a month or two I returned these notes to him as I did not see any interesting remaining open questions connected to them. At that time he suggested that I take a look at his papers [3, 4] with Wim Luxemburg about the modulus of a linear integral operator. In particular, he mentioned that he did not know how to prove the ideal property for integral operators. In a few months I had found a proof of this property and completed in that way the proof that the kernel operators form a band. Zaanen was known for his attention to detail and it is therefore somewhat remarkable that he overlooked a quite subtle minor error I had in that proof. Fortunately when at some point I realized this, I was able to fix it, but this is the only case I know that he overlooked an error in an argument. As an advisor he left me rather free at that point, on how to continue my dissertation. I became intrigued by an earlier result of Nakano [6], which I felt to be relevant to the topic of my dissertation. Zaanen was always available for a discussion and from time to time he would mention a paper he had noticed while he had visited the library of the Royal Netherlands Academy of Arts and Sciences (KNAW) in Amsterdam. It was in that way he mentioned one Friday in the late fall of 1975 that he had seen a paper by Buhvalov [2]. As it turned out Buhvalov and I had independently of each other, by reading Nakano, arrived at the same criterion for an operator to be representable as a kernel operator. I had just finished writing my notes on this and was planning to give them to Zaanen the next week, which I did. As Buhvalov announced his version in print before me, he had the priority of the result. Zaanen nevertheless always stressed in his writings and presentations, that I had independently obtained the same result. I was always very grateful for his support in this matter, as it was naturally a disappointment to have somebody beat you in getting such an important theorem. He insisted that I include this material in my dissertation, even though I had in the meantime written another paper on a different topic, which could have been included in my dissertation. After handing in the draft of my dissertation, I was surprised to see how much he had marked it up with corrections, when he returned it to me. Each page was filled with corrections of my English or writing style! Zaanen clearly was instrumental in obtaining a post-doctoral position for me at Caltech, after I finished my PhD in 1977. We stayed in touch after that and every time I visited the Netherlands, I would either see him in his office, or I would visit his house at the Nassaulaan in Delft, where Ada would offer sometimes a sandwich for lunch and we would discuss mathematics at the kitchen table.

References

[1] R.G. Bartle, *The elements of integration*, John Wiley & Sons, Inc., New York-London-Sydney, 1966.

[2] A.V. Buhvalov, *A criterion for integral representability of linear operators*, Funkcional. Anal. i Priložen. **9** (1975), no. 1, 51.

[3] W.A.J. Luxemburg and A.C. Zaanen, *The linear modulus of an order bounded linear transformation. I*, Nederl. Akad. Wetensch. Proc. Ser. A **74**=Indag. Math. **33** (1971), 422–434.

[4] W.A.J. Luxemburg and A.C. Zaanen, *The linear modulus of an order bounded linear transformation. II*, Nederl. Akad. Wetensch. Proc. Ser. A **74**=Indag. Math. **33**(1971), 435–447.

[5] W.A.J. Luxemburg and A.C. Zaanen, *Riesz spaces. Vol. I*, North-Holland Publishing Co., Amsterdam, 1971.

[6] H. Nakano, *Product spaces of semi-ordered linear spaces*, J. Fac. Sci. Hokkaido Univ. Ser. I. **12** (1953), 163–210.

[7] W. Rudin, *Principles of mathematical analysis*, Second edition, McGraw-Hill Book Co., New York, 1964.

[8] A.C. Zaanen, *Integration*, Completely revised edition of *An introduction to the theory of integration*, North-Holland Publishing Co., Amsterdam, 1967.

4. Pieter Zaanen

First I would like to thank the organizers of this A.C. Zaanen Centennial Conference for the opportunity to say a few words about our father. It is indeed 100 years ago that our father was born, on 14th June 1913. I also speak on behalf of my three brothers, Herman, Gijs and Maarten.

We never saw our dad as famous when we were young. As children we just knew he was a mathematics professor. We also noticed that he spent a lot of time behind his desk, thinking and writing.

Our dad was raised in Rotterdam in a family with four sisters and his brother Cees. He never told us much about his youth, but we know he loved his bicycling trips to the Alblasserwaard, where he visited family in his teens.

His parents were 'Rotterdam working class' and there was certainly no academic history in the family. His teachers had to argue with his parents to let him go to high school (HBS) after primary school. He finished high school with the second best final exam of the Netherlands. The school principal convinced his parents again that, with his intellectual capabilities, he should study at Leiden University. He received a full government grant to pay for his mathematical study.

Our father was basically a serious person, dedicated to mathematics and science. He lived with his parents in the Delfshaven area in Rotterdam during his study years in Leiden. After his PhD in 1938 (promotor prof. Droste), he spent the WW-II years in Rotterdam. The Germans never deported our father during the war because he was a teacher at secondary school at the time, and the Germans wanted the schools to continue. From letters, we understand he undertook some adventurous trips, mainly to find food in the 'Hunger Winter' of 1944.

Amongst his hobbies were sailing (he had his own boat at the end of the 30s), water polo and he took part in the 'Vierdaagse' in Nijmegen.

In 1940 he met our mother, Ada. She was a young, pretty and extrovert girl of 18. He was 27. She was a pupil at the girls' school where he taught. Ada was the girl of his dreams. She was in many ways the opposite of our dad. Whilst our father was introvert, our mother was very extrovert. She was a good socialiser and made friends easily. Our parents married in 1943 and found a small room in Leiden. It was the first time he left home. In March 1946, I was born.

Our dad was young (34) when he was asked to become professor at Bandung Technical High School in Indonesia, the later Bandung Institute of Technology. He travelled by boat to Indonesia on his own in 1947. Our mother gave birth to Herman in March 1948 and she flew to Bandung soon afterwards with a two year old and a six week old baby. This took three days. They led an expatriate life. Parties, a clubhouse, swimming pool and a large house with live-in household help were unheard luxury after the dark days of WW-II. They made many friends in Indonesia. Our mother especially loved it and her appetite for travel and living abroad was born.

In 1951, our father was appointed professor at Delft Technical High School, the later Delft University of Technology. We moved into the new 'professor lane'

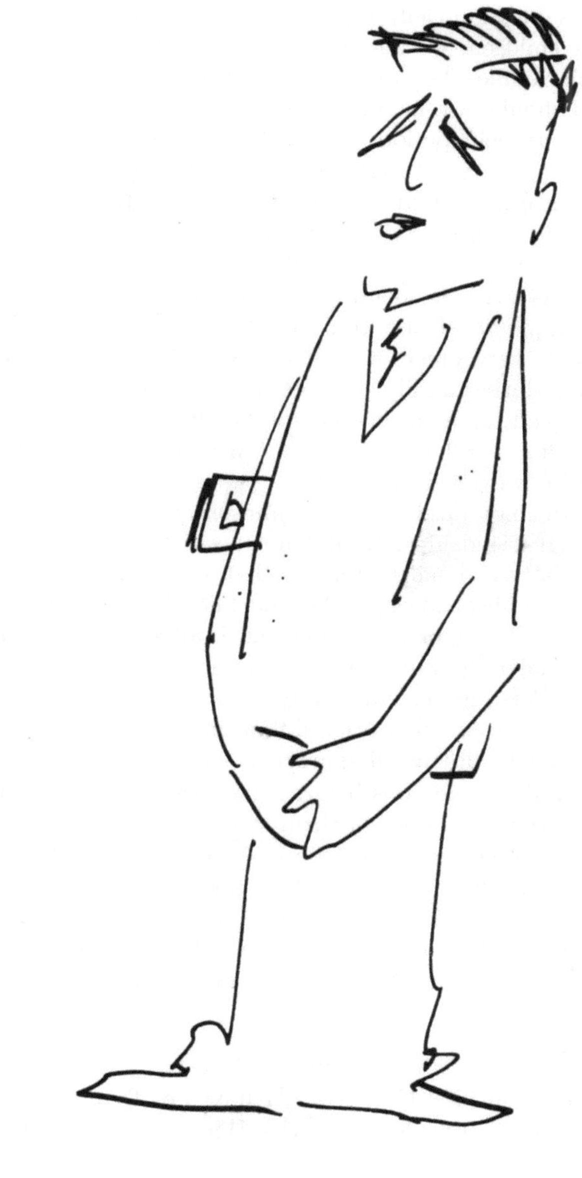

Zaanen as drawn in the 1960s by Ferry Rondagh

houses at the Nassaulaan in Delft. Our parents lived in that house until our mother died in 1996. As children, we got to know many of our father's colleagues and students and their families. We used to call them uncle and aunt. Those friendships last to this day. I should also mention that our younger brothers were born at the end of the 50s in our Delft period: my brother Gijs in 1957 and Maarten in 1959.

In 1956 he was appointed to the chair in Leiden of his promotor prof. Droste. He stayed there until he retired as Emeritus in 1982. He was very proud of that position. He was 69 when he retired and he told me later "That was too late – I should have retired earlier and enjoyed life when I was younger and healthier."

Our father was a true academic, a real analytical man. He was often in deep thought about complex abstracts concepts. Often he was absentminded. Even at parties he could be very quiet, with his mind on his mathematics. His lack of 'social networking' during family get-togethers was often a point of discussion at our home. I still recall the moments when he said: "I solved it" or "Wim is right." He would rush upstairs and start writing. A new, and as later transpired, unique mathematical proof was produced.

Our father did not need much for himself; he appreciated human and especially intellectual friendships. He did not need much material luxury. He saw intelligence and analytical skills as a gift that should be used for social benefit. The fact that none of his children (and grandchildren) pursued an academic career or worked in the public sector was somewhat disappointing to him.

As you may understand by now, the dominant force in the Zaanen household was our mother. The roles in our family were pretty clear. Our dad spent his time and effort in the academic world and our mother ran all family and financial matters. She also made all travelling arrangements.

Our dad was always nervous before a trip. He could see all kinds of terrible things happening. However, both of them loved to travel and it brought them to Indonesia, St Andrews in Scotland, Oberwolfach in Germany, many times to the USA – twice to work for a year with Wim Luxemburg at Caltech, in 1961 and 1969. My parents also went for longer periods to South Africa, Australia and even China. As sons we have had a unique life. We travelled to many places, met many colourful people. All four sons had a year in high school or junior high school in South Pasadena.

I would like to finish with some anecdotes:

I still remember our first USA trip to Knoxville (Tennessee) in the summer of 1953. Again our mother travelled later with Herman and me. We had no car and our parents had no driving licences at that time. These voyages were done with the famous boats of the Holland America Line and by train. I still remember my first large milkshake at the train station of Washington DC. A new world opened up for us children: colourful cars, toy stores, denim jeans, refrigerators, hamburgers and BBQs with roasted marshmallows.

When my parents bought their car in 1955 (a VW Beetle), my dad also got his driving license. In the beginning he drove the car – but he was not a great driver. Traffic was mild in those days but in the first couple of months he managed to have two minor accidents. His excuse was that mathematics had distracted him from the traffic around him. Our mother thereafter banned him from driving which he found quite OK. He loved taking the train – he found it relaxing to travel and think of things other than the traffic around him.

In conversation, our dad was very direct, with an occasional touch of irony.

When Herman started studying in Delft, he was partying and succeeded in passing only a single exam in his first year in Delft. Our dad then told him: "Your study in Delft is going as I had expected." This was the smartest thing he could have done. Herman's ego was dented so badly that he started to study very seriously. When Herman later graduated, there was a little celebration dinner. Our dad gave a speech and with a huge smile on his face stated: "I congratulate you but they must have lowered the standards in Delft after I left."

As children, we had a wonderful childhood and we are very grateful for all the things our parents did for us. Our dad certainly had a special personality and character. He gave us the encouragement and the means we needed in our youth and also later in life. For that we are very thankful.

5. Published material on Zaanen

1982 *Curriculum vitae, publications and PhD students of A.C. Zaanen*, in "From A to Z. Proceedings of a symposium in honour of A.C. Zaanen" (Leiden, 1982), Math. Centre Tracts, 149, Math. Centrum, Amsterdam, 1982, 123–130.

1984 J. Korevaar, *Adriaan Cornelis Zaanen*, in "Anniversary volume on approximation theory and functional analysis" (Oberwolfach, 1983), Internat. Schriftenreihe Numer. Math., 65, Birkhäuser, Basel, 1984, 41–45.

1988 *Gegevens over A.C. Zaanen* (Dutch), Nieuw Arch. Wisk. (4) **6** (1988), no. 1-2, 60–62. Erratum (Dutch): Nieuw Arch. Wisk. (4) **6** (1988), no. 3, i.

1995 C.B. Huijsmans, M.A. Kaashoek, W.A.J. Luxemburg and B. de Pagter, *Biographical notes, list of publications and curriculum vitae of A.C. Zaanen*, in "Operator theory in function spaces and Banach lattices", Oper. Theory Adv. Appl., 75, Birkhäuser, Basel, 1995, 1–14.

2004 W.A.J. Luxemburg, *Levensbericht A.C. Zaanen* (Dutch), in "Levensberichten en herdenkingen", Royal Netherlands Academy of Arts and Sciences (KNAW), Amsterdam, 2004, 108–114.

2004 M.A. Kaashoek, W. Luxemburg and B. de Pagter, *Adriaan Cornelis Zaanen* (Dutch), Nieuw Arch. Wiskd. (5) **5**, (2004), 21–25.

2005 C.D. Aliprantis, *In memoriam: Yuri Alexander Abramovich and Adriaan Cornelis Zaanen*, Positivity **9** (2005), no. 3, 269–272.

2014 M.A. Kaashoek, W. Luxemburg and B. de Pagter, *Adriaan Cornelis Zaanen*, Translated from the Dutch [Nieuw Arch. Wiskd. (5) **5**, (2004)], Indag. Math. (N.S.) **25** (2014), no. 2, 164–169.

Internet *Adriaan Cornelis Zaanen*, item in "MacTutor History of Mathematics archive", http://www-history.mcs.st-and.ac.uk/Biographies/Zaanen.html.

Internet *Adriaan Cornelis Zaanen*, item in "Wikipedia", http://en.wikipedia.org/wiki/Adriaan_Cornelis_Zaanen.

Koos Grobler
Unit for Business Mathematics and Informatics
North-West University
Potchefstroom Campus
Potchefstroom 2520, South Africa
e-mail: `Koos.Grobler@nwu.ac.za`

Rien Kaashoek
Department of Mathematics FEW
VU University
De Boelelaan 1081a
NL-1081 HV Amsterdam, The Netherlands
e-mail: `m.a.kaashoek@few.vu.nl`

Anton Schep
Department of Mathematics
University of South Carolina
Columbia, SC 29212, USA
e-mail: `schep@math.sc.edu`

Pieter Zaanen
e-mail: `pzaanen@ziggo.nl`

Ordered Structures and Applications: Positivity VII
Trends in Mathematics, 1–11

Minimal Projections with Respect to Numerical Radius

Asuman Güven Aksoy and Grzegorz Lewicki

Abstract. In this paper we survey some results on minimality of projections with respect to numerical radius. We note that in the cases L^p, $p = 1, 2, \infty$, there is no difference between the minimality of projections measured either with respect to operator norm or with respect to numerical radius. However, we give an example of a projection from l_3^p onto a two-dimensional subspace which is minimal with respect to norm, but not with respect to numerical radius for $p \neq 1, 2, \infty$. Furthermore, utilizing a theorem of Rudin and motivated by Fourier projections, we give a criterion for minimal projections, measured in numerical radius. Additionally, some results concerning strong unicity of minimal projections with respect to numerical radius are given.

Mathematics Subject Classification (2010). Primary 41A35,41A65, Secondary 47A12.

Keywords. Numerical radius, minimal projection, diagonal extremal pairs, Fourier projection.

1. Introduction

A projection from a normed linear space X onto a subspace V is a bounded linear operator $P : X \to V$ having the property that $P_{|V} = I$. P is called a *minimal projection* if $\|P\|$ is the least possible. Finding a minimal projection of the least norm has an obvious connection to approximation theory, since for any $P \in \mathcal{P}(X, V)$, the set of all projections from X onto V, and $x \in X$, from the inequality:

$$\|x - Px\| \leq (\|Id - P\|) \operatorname{dist}(x, V) \leq (1 + \|P\|) \operatorname{dist}(x, V), \tag{1}$$

one can deduce that Px is a good approximation to x if $\|P\|$ is small. Furthermore, any minimal projection P is an extension of Id_V to the space X of the smallest possible norm, which can be interpreted as a Hahn–Banach extension. In general, a given subspace will not be the range of a projection of norm 1, and the projection of least norm is difficult to discover even if its existence is known *a priori*. For example, the minimal projection of $C[0,1]$ onto the subspace Π_3 of polynomials of degree ≤ 3 is unknown. For an explicit determination of the projection of minimal norm from the subspace $C[-1,1]$ onto Π_2, see [8]. However, it is known that, see [10], for a Banach space X and a subspace $V \subset X$, $V = Z^*$ for some Banach space Z, if $\mathcal{P}(X,V) \neq \emptyset$, then there exists a minimal projection $P : X \to V$. A well-known example of a minimal projection, [13], is Fourier projection $F_m : C(2\pi) \to \Pi_M := \text{span}\{1, \sin x, \cos x, \ldots, \sin mx, \cos mx\}$ defined as

$$F_m(f) = \sum_{k=0}^{m} \alpha_k \cos kx + \sum_{k=0}^{m} \beta_k \sin kx \tag{2}$$

where α_k, β_k are Fourier coefficients and $C(2\pi)$ denotes 2π-periodic, real-valued functions equipped with the sup norm. For uniqueness of minimality of Fourier projection see [17]. Let X be a Banach space over $\mathbb{R}$ or $\mathbb{C}$. We write $B_X(r)$ for a closed ball with radius $r > 0$ and center at 0 (B_X if $r = 1$) and S_X for the unit sphere of X. The dual space of X is denoted by X^* and the Banach algebra of all continuous linear operators going from X into a Banach space Y is denoted by $B(X,Y)$ ($B(X)$ if $X = Y$).

The numerical range of a bounded linear operator T on X is a subset of a scalar field, constructed in such a way that it is related to both algebraic and norm structures of the operator. More precisely:

Definition 1.1. The *n*umerical range $T \in B(X)$ is defined by

$$W(T) = \{x^*(Tx) : x \in S_X, x^* \in S_{X^*}, x^*(x) = 1\}. \tag{3}$$

Notice that the condition $x^*(x) = 1$ gives us that x^* is a norm attaining functional.

The concept of a numerical range comes from Toeplitz' original definition of the *field of values* associated with a matrix, which is the image of the unit sphere under the quadratic form induced by the matrix A:

$$F(A) = \{x^*Ax : \|x\| = 1, x \in \mathbb{C}^n\}, \tag{4}$$

where x^* is the original conjugate transform and $\|x\|$ is the usual Euclidean norm. It is known that the classical numerical range of a matrix always contains the spectrum, and as a result the study of numerical range can help understand properties that depend on the location of the eigenvalues such as the stability and non-singularity of matrices. In case A is a normal matrix, then the numerical range is the polygon in the complex plane whose vertices are eigenvalues of A. In particular, if A is Hermitian, then the polygon reduces to the segment on the real axis bounded by the smallest and largest eigenvalue, which perhaps explains the name numerical range.

The *numerical radius* of T is given by

$$\|T\|_w = \sup\{|\lambda| : \lambda \in W(T)\}. \tag{5}$$

Clearly $\|T\|_w$ is a semi-norm on $B(X)$ and $\|T\|_w \leq \|T\|$ for all $T \in B(X)$. For example, if we consider $T : \mathbb{C}^n \to \mathbb{C}^n$ as a right shift operator

$$T(f_1, f_2, \dots, f_n) = (0, f_1, f_2, \dots, f_{n-1})$$

then $\langle Tf, f\rangle = f_1\overline{f}_2 + f_2\overline{f}_3 + \cdots f_{n-1}\overline{f}_n$ and consequently to find $\|T\|_w$ we must find $\sup\{|f_1||f_2| + \cdots + |f_{n-1}||f_n|\}$ subject to the condition $\sum_{i=1}^{n} |f_i|^2 = 1$. The solution [11] to this "Lagrange multiplier" problem is

$$\|T\|_w = \cos\left(\frac{\pi}{n+1}\right). \tag{6}$$

The *numerical index* of X is then given by

$$n(X) = \inf\left\{\|T\|_w : T \in S_{B(X)}\right\}. \tag{7}$$

Equivalently, the numerical index $n(X)$ is the greatest constant $k \geq 0$ such that $k\|T\| \leq \|T\|_w$ for every $T \in B(X)$. Note also that $0 \leq n(X) \leq 1$ and $n(X) > 0$ if and only if $\|\cdot\|_w$ and $\|\cdot\|$ are equivalent norms. The concept of numerical index was first introduced by Lumer [14] in 1968. Since then much attention has been paid to the constant of equivalence between the numerical radius and the usual norm of the Banach algebra of all bounded linear operators of a Banach space. Two classical books devoted to these concepts are [7] and [6]. For more recent results we refer the reader to [4], [15], [16] and [11].

In this paper, we study the minimality of projections with respect to numerical radius. Since the operator norm of T is defined as $\|T\| = \sup|\langle Tx, y\rangle|$ with $(x, y) \in B(X) \times B(X^*)$, while the numerical radius $\|T\|_w = \sup|\langle Tx, y\rangle|$ with $(x, y) \in B(X) \times B(X^*)$ and $\langle x, y\rangle = 1$, $\|T\|$ is bilinear and $\|T\|_w$ is quadratic in nature. However, $\|T\|_w \leq \|T\|$ implies that there are more spaces for which $\|T\| \geq 1$ but $\|T\|_w = 1$.

Furthermore, if T is a bounded linear operator on a Hilbert space H, then the numerical radius takes the form

$$\|T\|_w = \sup\{|\langle Tx, x\rangle| : \|x\| = 1\}. \tag{8}$$

This follows from the fact that for each linear functional x^* there is a unique $x_0 \in H$ such that $x^*(x) = \langle x, x_0\rangle$ for all $x \in H$. Moreover, if T is self-adjoint or a normal operator on a Hilbert space H, then

$$\|T\|_w = \|T\|. \tag{9}$$

Also, if a non-zero $T : H \to H$ is self-adjoint and compact, then T has an eigenvalue λ such that

$$\|T\|_w = \|T\| = \lambda. \tag{10}$$

These properties of numerical radius together with the desirable properties of diagonal projections from Hilbert spaces onto closed subspaces provides motivation to investigate minimal projections with respect to numerical radius.

2. Characterization of minimal numerical radius projections

In [1], a characterization of minimal numerical radius extension of operators from a Banach space X onto its finite-dimensional subspace $V = [v_1, v_2, \dots, v_n]$ is given. To express this theorem, we first set up our notation.

Notation 2.1. Let $T = \sum_{i=1}^{n} u_i \otimes v_i : V \to V$ where $u_i \in V^*$. Its extension to X is denoted by $\widetilde{T} : X \to V$ and defined as

$$\widetilde{T} = \sum_{i=1}^{n} \tilde{u}_i \otimes v_i, \tag{11}$$

where $\widetilde{u}_i \in X^*$.

Definition 2.2. Let X be a Banach space. If $x \in X$ and $x^* \in X^*$ are such that

$$|\langle x, x^* \rangle| = \|x\|\|x^*\| \neq 0, \tag{12}$$

then x^* is called an *extremal* of x and written as $x^* = \operatorname{ext} x$. Similarly, x is an extremal of x^*. We call $(\operatorname{ext} y, y) \in S_{X^{**}} \times S_{X^*}$ a *diagonal extremal pair* for $\widetilde{T} \in B(X, V)$ if

$$\langle \widetilde{T}^{**} x, y \rangle = \|\widetilde{T}\|_w, \tag{13}$$

where $\widetilde{T}^{**} : X^{**} \to V$ is the second adjoint extension of $\widetilde{T}$ are $V = [v_1, \dots, v_n] \subset X$. In other words, the map $\widetilde{T}$ has the expression $\widetilde{T} = \sum_{i=1}^{n} \tilde{u}_i \otimes v_i : X \to V$ and

$$\widetilde{T} x = \sum_{n=1}^{n} \langle x, \tilde{u}_i \rangle v_i \tag{14}$$

where $\tilde{u}_i \in X^*$, $v_i \in V$ and $\langle x, \tilde{u}_i \rangle$ denotes the functional $\tilde{u}_i$ is acting on x and

$$\widetilde{T}^{**} x = \sum_{i=1}^{n} \langle u_i, x \rangle v_i, \tag{15}$$

$u_i \in X^{***}$, $v_i \in V$, $x \in X^{**}$.

The set of all diagonal extremal pairs will be denoted by $\mathcal{E}_w(\widetilde{T})$ and defined as:

$$\mathcal{E}_w(\widetilde{T}) = \left\{ (\operatorname{ext} y, y) \in S_{X^{**}} \times S_{X^*} : \ \|\widetilde{T}\|_w = \sum_{i=1}^{n} \langle \operatorname{ext} y, u_i \rangle \cdot \langle v_i, y \rangle \right\}. \tag{16}$$

Note that to each $(x, y) \in X^{**} \times X^*$ we associate the rank-one operator $y \otimes x : X \to X^{**}$ given by

$$(y \otimes x)(z) = \langle z, y \rangle x \qquad \text{for } z \in X. \tag{17}$$

Accordingly, to each $(x, y) \in \mathcal{E}_w(\widetilde{T})$ we can associate the rank-one operator $y \otimes \operatorname{ext} y : X \to X^{**}$ given by

$$(y \otimes \operatorname{ext} y)(z) = \langle z, y \rangle \operatorname{ext} y. \tag{18}$$

By $\mathcal{E}(\widetilde{T})$ we denote the usual set of all extremal pairs for $\widetilde{T}$ and

$$\mathcal{E}(\widetilde{T}) = \left\{ (x,y) \in S_{X^{**}} \times S_{X^*} : \|\widetilde{T}\| = \sum_{i=1}^{n} \langle x, u_i \rangle \cdot \langle v_i, y \rangle \right\}. \tag{19}$$

In case of diagonal extremal pairs we require $|\langle \operatorname{ext} y, y \rangle| = 1$.

Definition 2.3. Let $T = \sum_{i=1}^{n} u_i \otimes v_i : V \to V = [v_1, v_2, \dots, v_n] \subset X$, where $u_i \in V^*$. Let $\widetilde{T} : \sum_{i=1}^{n} \tilde{u}_i \otimes v_i : X \to V$ be an extension of T to all of X. We say $\widetilde{T}$ is a *minimal numerical extension* of T if

$$\|\widetilde{T}\| = \inf \left\{ \|S\|_w : S : X \to V \; ; \; S_{|V} = T \right\}. \tag{20}$$

Clearly $\|T\|_w \leq \|\widetilde{T}\|_w$.

Theorem 2.4 ([1]). *$\widetilde{T}$ is a minimal radius-extension of T if an only if the closed convex hull of $\{y \otimes x\}$ where $(x,y) \in \mathcal{E}_w(\widetilde{T})$ contains an operator for which V is an invariant subspace.*

Theorem 2.5. *P is a minimal projection from X onto V if and only if the closed convex hull of $\{y \otimes x\}$, where $(x,y) \in \mathcal{E}_w(P)$ contains an operator for which V is an invariant subspace.*

Proof. By taking $T = I$ and $\widetilde{T} = P$ one can appropriately modify the proof given in [1] without much difficulty. The problem is equivalent to the best approximation in the numerical radius of a fixed operator from the space of operators

$$\mathcal{D} = \{\Delta \in \mathcal{B} : \Delta = 0 \text{ on } V\} = sp\{\delta \otimes v : \delta \in V^{\perp}; v \in V\}.$$

One of the main ingredients of the proof is Singer's identification theorem ([20], Theorem 1.1 (p. 18) and Theorem 1.3 (p. 29)) of finding the minimal operator as the error of best approximation in C(K) for K compact. In the case of numerical radius, one considers $K_w = K \cap \text{Diag} = \{(x,y) \in B(X^{**}) \times B(X^*) : x = \text{ext}(y) \text{ or } x = 0\}$ and shows K_w is compact. Thus the set $\mathcal{E}(P)$, being the set of points where a continuous (bilinear) function achieves its maximum on a compact set, is not empty. For further details see [1]. □

Theorem 2.6 (When minimal projections coincide). *In case $X = L^p$ for $p = 1, 2, \infty$, the minimal numerical radius projections and the minimal operator norm projections coincide with the same norms.*

Proof. In case of L^2, for any self-adjoint operator, we have

$$\|P\| = \|P\|_w = |\lambda|, \tag{21}$$

where λ is the maximum (in modulus) eigenvalue. In this case,

$$\|P\| = \|P\|_w = |\langle Px, x \rangle|, \tag{22}$$

where x is a norm-1 "maximum" eigenvector.

When $p = 1, \infty$, it is well known that $n(L^p) = 1$ ([7], Section 9) thus

$$\|P\| = \|P\|_w.$$

□

Example 2.7. The projection $P : l_3^p \to [v_1, v_2] = V$ where $v_1 = (1,1,1)$ and $v_2 = (-1,0,1)$ is minimal with respect to the operator norm, but not minimal with respect to numerical radius for $1 < p < \infty$ and $p \neq 2$. Let us denote by P_o, P_m projections minimal with respect to operator norm and numerical radius respectively. In other words

$$\|P_o\| = \inf\{\|P\| : P \in \mathcal{P}(X,V)\}$$
$$\|P_m\|_w = \inf\{\|P\|_w : P \in \mathcal{P}(X,V)\}.$$

Note that

$$P_o(f) = u_1(f)v_1 + u_2(f)v_2 \quad \text{and} \quad P_m(f) = z_1(f)v_1 + z_2(f)v_2. \tag{23}$$

Then it is easy to see that

$$u_1 = z_1 = \left(-\frac{1}{2}, 0, \frac{1}{2}\right), \qquad u_2 = \left(\frac{1-d}{2}, d, \frac{1-d}{2}\right),$$
$$z_2 = \left(\frac{1-g}{2}, g, \frac{1-g}{2}\right),$$

and for $p = 4/3$ it is possible to determine g and d to conclude $\|P_o\| = 1.05251$ while $\|P_m\|_w = 1.02751$, thus $\|P_o\| \neq \|P_m\|_w$.

V.P. Odinec in [19] (see also [18], [12]) proves that minimal projections of norm greater than one from a three-dimensional real Banach space onto any of its two-dimensional subspaces are unique. Thus in the above example, the projection from l_3^p onto a two-dimensional subspace not only proves the fact that $\|P_o\| \neq \|P_m\|_w$ for $p \neq 1, 2, \infty$, here once again we have the uniqueness of the minimal projections.

3. Rudin's projection and numerical radius

One of the key theorems on minimal projections is due to W. Rudin ([21] and [22]). The setting for his theorem is as follows. X is a Banach space and G is a compact topological group. Defined on X is a set $\mathcal{A}$ of all bounded linear bijective operators in a way that $\mathcal{A}$ is algebraically isomorphic to G. The image of $g \in G$ under this isomorphism will be denoted by T_g. We will assume that the map $G \times X \to X$ defined as $(g,x) \mapsto T_g x$ is continuous. A subspace V of X is called *G-invariant* if $T_g(V) \subset V$ for all $g \in G$ and a mapping $S : X \to X$ is said to *commute* with G if $S \circ T_g = T_g \circ S$ for all $g \in G$. In case $\|T_g\| = 1$ for all $g \in G$, we say g acts on G by *isometries*.

Theorem 3.1 ([22]). *Let G be a compact topological group acting by isomorphism on a Banach space X and let V be a complemented G-invariant subspace of X (there exists a bound projection P of X onto V). Then there exists a bounded linear projection Q of X onto V which commutes with G.*

The idea behind the proof of the above theorem is to obtain Q by averaging the operators $T_{g^{-1}}PT_g$ with respect to Haar measure μ on G, i.e.,

$$Q(x) := \int_G \left(T_{g^{-1}}PT_g\right)(x)\,d\mu(g). \tag{24}$$

Now assume X has a norm which contains the maps $\mathcal{A}$ to be *isometries* and that all of the hypotheses in Rudin's theorem are satisfied, then one can claim the following stronger version of Rudin's theorem:

Corollary 3.2. *If there is a* unique *projection* $Q : X \to V$ *which commutes with* G*, then for any* $P \in \mathcal{P}(X, V)$*, the projection*

$$Q(x) = \int_G \left(T_{g^{-1}}PT_g\right)(x)\,d\mu(g), \tag{25}$$

is a minimal projection *of* X *onto* V*.*

Theorem 3.3 ([3]). *Let* $\mathcal{A}$ *be a set of all bounded linear bijective operators on* X *such that* $\mathcal{A}$ *is algebraically isomorphic to* G*. Suppose that all of the hypotheses of Rudin's theorem above are satisfied and the maps in* $\mathcal{A}$ *are isometries. If* P *is any projection in the numerical radius of* X *onto* V*, then the projection* Q *defined as*

$$Q(x) = \int_G \left(T_{g^{-1}}PT_g\right)(x)\,d\mu(g) \tag{26}$$

satisfies $\|Q\|_w \leq \|P\|_w$*.*

Proof. Consider $\|Q\|_w = \sup\{|x^*(Qx)| : x^*(Qx) \in W(Q)\}$, where $W(Q)$ is the numerical range of Q. Notice that

$$\begin{aligned} |x^*(Qx)| &= \left| x^* \int_G \left(T_{g^{-1}}PT_g\right)(x)\,d\mu(g) \right| \\ &\leq \int_G \left| \left(x^* \circ T_{g^{-1}}\right) P(T_g x) \right|\,d\mu(g). \end{aligned} \tag{27}$$

But $\|x\| = 1$ and $\|x^*\| = 1$ which implies that $\|T_g x\| = 1$ and $\|x^* T_{g^{-1}}\| = 1$, moreover,

$$1 = x^*(x) = x^* T_{g^{-1}}(T_g x) \implies |x^*(Qx)| \leq \|P\|_w. \tag{28}$$

Consequently, $\|Q\|_w \leq \|P\|_w$ which proves Q is a minimal projection in numerical radius. □

Theorem 3.4 ([3]). *Suppose all hypotheses of the above theorem are satisfied and that there is exactly one projection* Q *which commutes with* G*. Then* Q *is a minimal projection with respect to numerical radius.*

Proof. Let $P \in \mathcal{P}(X, V)$. By the properties of Haar measure, Q_p given in the above theorem commutes with G. Since there is exactly one projection which commutes with G, $Q_p = Q$ and $\|Q\|_w \leq \|P\|_w$ as desired. □

Remark 3.5. In [3] it is shown that if G is a compact topological group acting by isometries on a Banach space X. If we let

$$\psi : B(X) \to [0, +\infty], \tag{29}$$

be a convex function which is lower semi-continuous in the strong operator topology and if one further assumes that

$$\psi\left(g^{-1} \circ P \circ g\right) \le \psi(P), \tag{30}$$

for some $P \in B(X)$ and $g \in G$, then $\psi(Q_P) \le \psi(P)$. This result leads to the calculation of minimal projections not only with respect to numerical radius but also with respect to p-summing, p-nuclear and p-integral norms. For details see [3].

4. An application

Let $C(2\pi)$ denote the set of all continuous 2π-periodic functions and Π_n be the space of all trigonometric polynomials of order $\le n$ (for $n \ge 1$).

The *Fourier projection* $F_n : C(2\pi) \to \Pi_n$ is defined by

$$F_n(f) = \sum_{k=0}^{2n} \left(\int_0^{2\pi} f(t) g_n(t) dt \right) g_k, \tag{31}$$

where $(g_k)_{k=0}^{2n}$ is an orthonormal basis in Π_n with respect to the scalar product

$$\langle f, g \rangle = \int_0^{2\pi} f(t) g(t) dt. \tag{32}$$

Lozinskii in [13] showed that F_n is a minimal projection in $\mathcal{P}(C(2\pi), \Pi_n)$. His proof is based on the equality which states that for any $f \in C(2\pi)$, $t \in [0, 2\pi]$ and $P \in \mathcal{P}(C(2\pi), \Pi_n)$, we have

$$F_n f(t) = \frac{1}{2\pi} \int_0^{2\pi} \left(T_{g^{-1}} P T_g f\right)(t)\, d\mu(g). \tag{33}$$

Here μ is a Lebesgue measure and $(T_g f)(t) = f(t+g)$ for any $g \in \mathbb{R}$. This equality is called Marcinkiewicz equality ([9] p. 233).

Notice that F_n is the only projection that commutes with G, where $G = [0, 2\pi]$ with addition mod 2π. In particular, F_n is a minimal projection with respect to numerical radius.

We know the upper and lower bounds on the operator norm of F_n satisfy ([9] p. 212):

$$(4/\pi^2) ln(n) \le \|F_n\| \le ln(n) + 3. \tag{34}$$

From the theorem (when minimal projections coincide) we know that in the cases of L^p, $p = 1, \infty$, the numerical radius projections and the operator norm projections are equal. Since $C(2\pi) \subset L^\infty$, we also have lower and upper bounds for the numerical radius of Fourier projections, i.e.,

$$(4/\pi^2)ln(n) \leq \|F_n\|_w \leq ln(n) + 3. \tag{35}$$

Remark 4.1. Lozinskii's proof of the minimality of F_n is based on the Marcinkiewicz equality. However, the Marcinkiewicz equality holds true if one replaces $C(2\pi)$ by $L^p[0, 2\pi]$ for $1 \leq p \leq \infty$ or Orlicz space $L^\phi[0, 2\pi]$ equipped with the Luxemburg or Orlicz norm provided ϕ satisfies the suitable Δ_2 condition. Hence, Theorem 3.3 can be applied equally well to numerical radius or norm in Banach operator ideals of p-summing, p-integral,p-nuclear operators generated by L^p-norm or the Luxemburg or Orlicz norm. For further examples see [3].

5. Strongly unique minimal extensions

In [19] (see also [18]) it is shown that a minimal projection of the operator norm greater than one from a three-dimensional real Banach space onto any of its two-dimensional subspace is the unique minimal projection with respect to the operator norm. Later in [12] this result is generalized as follows:
Let X be a three-dimensional real Banach space and V a two-dimensional subspace. Suppose $A \in B(V)$ is a fixed operator. Set

$$\mathcal{P}_A(X, V) = \{P \in B(X, V) : \ \ P\,|_V = A\ \}$$

and assume $\|\ P_0\ \| > \|\ A\ \|$, if $P_o \in \mathcal{P}_A(X, V)$ is an extension of minimal operator norm. Then P_o is *a strongly unique minimal extension with respect to operator norm.*
In other words there exists $r > 0$ such that for all $P \in \mathcal{P}_A(X, V)$ one has

$$\|P\| \geq \|P_o\| + r\,\|P - P_o\|.$$

Definition 5.1. We say an operator $A_o \in \mathcal{P}_A(X, V)$ is *a strongly unique minimal extension with respect to numerical radius* if there exists $r > 0$ such that

$$\|B\|_w \geq \|A_o\|_w + r\,\|B - A_o\|_w$$

for any $B \in \mathcal{P}_A(X, V)$.

A natural extension of the above-mentioned results to the case of numerical radius $\|\cdot\|_w$ was given in [2].

Theorem 5.2 ([2]). *Assume that X is a three-dimensional real Banach space and let V be its two-dimensional subspace. Fix $A \in B(V)$ with $\|A\|_w > 0$. Let*

$$\lambda_w^A = \lambda_w^A(V, X) = \inf\{\|B\|_w : B \in \mathcal{P}_A(X, V)\} > \|A\|,$$

where $\|A\|$ denotes the operator norm. Then there exist exactly one $A_o \in \mathcal{P}_A(X, V)$ such that

$$\lambda_w^A = \|A_o\|_w.$$

Moreover, A_o is the strongly minimal extension with respect to numerical radius.

Notice that if we take $A = id_V$ then $\|A\|_w = \|A\| = 1$. In this case Theorem (5.2) reduces to the following theorem:

Theorem 5.3 ([2]). *Assume that X is a three-dimensional real Banach space and let V be its two-dimensional subspace. Assume that*

$$\lambda_w^{id_V} > 1.$$

Then there exist exactly one $P_o \in \mathcal{P}(X,V)$ of minimal numerical radius. Moreover, P_o is a strongly unique minimal projection with respect to numerical radius. In particular P_o is the only minimal projection with respect to the numerical radius.

Remark 5.4 ([2]). Notice that in Theorem 5.2 the assumption that $\|A\| < \lambda_w^A$ is essential. Indeed, let $X = l_\infty^{(3)}$, $V = \{x \in X : x_1 + x_2 = 0\}$ and $A = id_V$. Define

$$P_1 x = x - (x_1 + x_2)(1,0,0) \quad \text{and} \quad P_2 x = x - (x_1 + x_2)(0,1,0).$$

It is clear that

$$\|P_1\| = \|P_1\|_w = \|P_2\| = \|P_2\|_w = 1$$

and $P_1 \neq P_2$. Hence, there is no strongly unique minimal projection with respect to numerical radius in this case.

Remark 5.5 ([2]). Theorem 5.3 cannot be generalized for real spaces X of dimension $n \geq 4$. Indeed let $X = l_\infty^{(n)}$, and let $V = \ker(f)$, where $f = (0, f_2, \dots, f_n) \in l_1^{(n)}$ satisfies $f_i > 0$ for $i = 2, \dots, n$, $\sum_{i=2}^n f_i = 1$ and $f_i < 1/2$ for $i = 1, \dots, n$. It is known (see, e.g., [5], [18]) that in this case

$$\lambda(V,X) = 1 + \left(\sum_{i=2}^n f_i/(1-2f_i) \right)^{-1} > 1, \quad \text{where } \lambda(V,X) = \inf\{\|P\| : P \in (X,V)\}.$$

By [1], $\lambda(V,X) = \lambda_w^{id_V}$. Define for $i = 1, \dots, n$ $y_i = (\lambda(V,X) - 1)(1 - 2f_i)$. Let $y = (y_1, \dots, y_n)$ and $z = (0, y_2, \dots, y_n)$. Consider mappings P_1, P_2 defined by

$$P_1 x = x - f(x) y \quad \text{and} \quad P_2 x = x - f(x) z$$

for $x \in l_\infty^{(n)}$. It is easy to see that $P_i \in \mathcal{P}(X,V)$, for $i = 1,2$, $P_1 \neq P_2$. By ([18] p. 104) $\|P_i\| = \|P_i\|_w = \lambda(V,X) = \lambda_w^{id_V}$. for $i = 1,2$.

Remark 5.6. Theorem 5.3 is not valid for complex three-dimensional spaces. For details see [2].

For Kolmogorov type criteria concerning approximation with respect to numerical radius, we refer the reader to [2].

References

[1] A.G. Aksoy and B.L. Chalmers, *Minimal numerical radius extensions of operators*, Proc. Amer. Math. Soc. **135** (2007), 1039–1050.

[2] A.G. Aksoy and G. Lewicki, *Best approximation in numerical radius*, Numer. Funct. Anal. Optim. **32** (2011), 593–609.

[3] A.G. Aksoy and G. Lewicki, *Minimal projections with respect to various norms*, Studia Math. **210** (2012), 1–16.

[4] A.G. Aksoy and G. Lewicki, *Limit theorems for the numerical index*, J. Math. Anal. Appl. **398** (2013), 296–302.

[5] J. Blatter and E.W. Cheney, *Minimal projections onto hyperplanes in sequence spaces*, Ann. Mat. Pura Appl., Vol. 101, (1974), 215–227.

[6] F.F. Bonsall and J. Duncan, *Numerical ranges II*, London Math. Soc. Lecture Note Ser. 10, Cambridge Univ. Press, 1971.

[7] F.F. Bonsall and J. Duncan, *Numerical ranges of operators on normed spaces and of elements of normed algebras*, London Math. Soc. Lecture Note Ser. 2, Cambridge Univ. Press, 1971.

[8] B.L. Chalmers and F.T. Metcalf, *Determination of minimal projection from* $C[-1,1]$ *onto the quadratics*, Numer. Funt. Anal. Optim. **11** (1990), 1–10.

[9] E.W. Cheney, *Introduction to approximation theory*, McGraw-Hill, 1966.

[10] E.W. Cheney and P.D. Morris, *On the existence and characterization of minimal projections*, J. Reine Angew. Math **279** (1974), 61–76.

[11] K.E. Gustafson and D.K.M. Rao, *Numerical range: The field of values of linear operators and matrices*, Springer, 1997.

[12] G. Lewicki, *Strong unicity criterion in some spaces of operators*, Comment. Math. Univ. Carolinae, Vol. 34, 1, (1993), 81–87.

[13] S.M. Lozinskii, *On a class of linear operators*, Dokl. Akad. Nauk SSSR **61** (1948), 193–196.

[14] G. Lumer, *Semi-inner product spaces*, Trans. Amer. Math. Soc. **100** (1961), 29–43.

[15] M. Martin, *A survey on the numerical index of Banach space*, Extracta Math. **15** (200), 265–276.

[16] M. Martin, J. Meri, and M. Popov, *On the numerical index of* $L^p(\mu)$*-spaces*, Israel J. Math. **184** (2011), 183–192.

[17] P.D. Morris, S.D. Fisher and D.E. Wulbert, *Unique minimal of Fourier projections*, Trans. Amer. Math. Soc. **265** (1981), 235–246.

[18] W. Odyniec, G. Lewicki *Minimal Projections in Banach spaces*, Lecture Notes in Mathematics, 1449, Springer-Verlag, 1990.

[19] V.P. Odinec, *The uniqueness of minimal projections in Banach spaces*, Dokl. Akad. Nauk SSSR **220** (1975), 779–781.

[20] I. Singer, *Best approximation in normed linear spaces by elements of linear subspaces*, Springer, 1970.

[21] W. Rudin, *Functional analysis*, McGraw-Hill, 1974

[22] W. Rudin, *Projections onto invariant subspaces*, Proc. Amer. Math. Soc. **13** (1962), 429–432.

Asuman Güven Aksoy
Department of Mathematics
Claremont McKenna College
Claremont, CA, 91711, USA
e-mail: `aaksoy@cmc.edu`

Grzegorz Lewicki
Dept. of Math and Comp. Sci.
Jagiellonian University
Lojasiewicza 6
PL-30-348 Kraków, Poland
e-mail: `Grzegorz.Lewicki@im.uj.edu.pl`

Ordered Structures and Applications: Positivity VII
Trends in Mathematics, 13–23

On Ordered C^*-algebras

Egor A. Alekhno

Abstract. An ordered Banach algebra A, which is also a C^*-algebra with positive involution, is called an ordered C^*-algebra. Elementary properties of such algebras are studied. In particular, the algebra $L(H)$, where H is an ordered Hilbert space with cone K satisfying $K^* = K$, is considered. An example is given of an ordered C^*-algebra A with an ideal $\mathcal{I}$ in A that contains elements $a, b \in A$ such that $0 \leq a \leq b \in \mathcal{I}$ and $a^n \notin \mathcal{I}$ for all $n \in \mathbb{N}$. Finally, ordered C^*-algebras with p-additive norms are studied.

Mathematics Subject Classification (2010). 46B40, 46L35.

Keywords. Ordered C^*-algebra, ordered Hilbert space.

1. Introduction

Let A be a (real or complex) Banach algebra with algebraic unit $\mathbf{e}$ and let A^+ be a (closed, convex) cone in A. If $\mathbf{e} \in A^+$ and the relations $a, b \in A^+$ imply $ab \in A^+$ then A is called an *ordered Banach algebra*. As usual, for elements $a, b \in A$, the symbol $a \geq b$ (or $b \leq a$) means $a - b \in A^+$. Throughout we will tacitly assume that $A \neq \{0\}$. The study of ordered Banach algebras was initiated in [11, 10]. In these papers and in a number of subsequent ones the main emphasis was on the study of spectral properties of positive elements.

However, in spite of the progress made in the development of the theory of ordered Banach algebras, several important aspects of the general theory of Banach algebras have received little or no attention. For example, up to now an important part of Banach algebras, namely, C^*-algebras or, more generally, algebras with an involution, has not been considered from the viewpoint of ordered Banach algebras and, in general, of the theory of cones. The main purpose of this note is to take a step in this direction.

The paper is organized as follows. In the second section examples are given and some simple properties of ordered C^*-algebras are considered. The third section is devoted to the domination problem. In the last section, a special class of ordered C^*-algebras, namely, algebras with p-additive norm, is studied.

For any unexplained terminology, notations, and elementary properties of ordered Banach spaces, we refer the reader to [4]. For information on the theory of normed Riesz spaces, we suggest [2, 3]. More detail on elementary properties of Banach algebras can be found in [12]. From another viewpoint, some results below were mentioned in [1].

2. Elementary properties of ordered C^*-algebras

An ordered Banach algebra A with a cone A^+ is called an *ordered C^*-algebra* if it is equipped with an involution $a \to a^*$ that maps A^+ into itself and under which A is a C^*-algebra (see [1]).

We recall that A is a C^*-algebra if $\|a^*a\| = \|a\|^2$ holds for all $a \in A$. If the latter relation is only valid for $a \in A^+$ then A is called an *almost ordered C^*-algebra.*

The condition that $a \mapsto a^*$ maps A^+ into itself, can also be expressed by saying that the involution is positive.

Example 2.1. (a) Let H be a (real or complex) Hilbert space with a (closed) cone K and let $L(H)$ be the algebra of all (bounded, linear) operators on H. In this case, to avoid ambiguity, $T \in L(H)$ will be called *positive* if and only if it maps K into itself, i.e., if and only if T is positive in the sense of the theory of ordered linear spaces. In this case we write $T \geq 0$.

A cone K is called *self-adjoint* whenever $K = K^*$, where K^* is the dual wedge of K. A Hilbert space H with a self -adjoint cone is called an *ordered Hilbert space.*

The next statement holds:

The algebra $L(H)$ is an ordered C^*-algebra with cone

$$(L(H))^+ = \{T \in L(H) : T(K) \subseteq K\}$$

if and only if K is a self-adjoint cone.

Indeed, to check the necessity, consider non-zero elements $x \in K^*$ and $y \in K$ and define an operator $x \otimes y$ on H via the formula $(x \otimes y)z = \langle z, x\rangle y$ for $z \in H$. Evidently, $x \otimes y \geq 0$ and, hence, $(x \otimes y)^* \geq 0$. Consequently, $0 \leq (x \otimes y)^* y = \|y\|^2 x$ and so $x \in K$. Next, for an arbitrary element $z \in K$, we have $0 \leq (x \otimes y)^* z = \langle z, y\rangle x$. Thus, $y \in K^*$.

For the converse, let $K = K^*$. This means that the dual wedge K^* is a cone and, consequently, the *linear space $K - K$ is dense in H*. Therefore, the set $(L(H))^+$ is a cone and $L(H)$ with this cone is an ordered Banach algebra. Let $0 \leq T \in L(H)$ and $x \in K$. For an arbitrary element $y \in K^*$, we have $\langle T^*x, y\rangle = \langle x, Ty\rangle \geq 0$ and, hence, $T^* \geq 0$. Thus, the involution is positive, as desired.

We will deduce some more properties of an ordered Hilbert space with a cone K. First of all, we mention that $\langle x, y\rangle \in \mathbb{R}$ for all $x, y \in H$ as $\langle x, y\rangle \geq 0$ holds for all $x, y \in K$. Therefore, H is a *real* Hilbert space. The norm on H is *strictly monotone*, i.e., the inequalities $0 \leq x < y$ imply $\|x\| < \|y\|$. In fact, we have the relations $\|x\|^2 = \langle x, x\rangle \leq \langle y, x\rangle \leq \langle y, y\rangle = \|y\|^2$. If $\|x\| = \|y\|$ then $\langle y - x, x\rangle = 0$ and $\langle y - x, y\rangle = 0$ and, hence, $y = x$, as required. In particular, the cone K is normal and so, according to the M.G. Krein theorem, (see [4, pp. 89, 90]), K is a generating cone and, moreover, K has the strong Levi property. By Andô's theorem (see [4, p. 92]), the ordered Hilbert space H is a Riesz space under the ordering induced by K if and only if K has the Riesz decomposition property; in this case, the Riesz space H is Dedekind complete (see [4, p. 97, Exercise 12(a)]).

Thus, every ordered Hilbert space is real. Nevertheless, the following approach to the complex case is possible. Let H be an arbitrary real Hilbert space with a cone K, and let the Hilbert space $H_{\mathbb{C}}$ be the complexification of H. The next statement holds:

The algebra $L(H_{\mathbb{C}})$ is an ordered C^*-algebra with cone

$$\{T \in L(H_{\mathbb{C}}) : T(H) \subseteq H \text{ and } T(K) \subseteq K\}$$

if and only if K is a self-adjoint cone in H.

In particular, the algebra $M_n(\mathbb{R})$ ($M_n(\mathbb{C})$) of all $n \times n$ matrices with real (complex) entries under the Euclidean norm and the natural ordering is an ordered C^*-algebra with an involution defined by the (complex) conjugate.

(b) In this part we consider an important example of a self-adjoint cone K in a Hilbert space H with $\dim H > 1$. Let $z \in H$ with $\|z\| = 1$ and let $\epsilon > 0$. Then, (see [4, § 2.6]), the *ice cream cone* is the cone

$$K_{z,\epsilon} = \{x \in H : \langle x, z\rangle \geq \epsilon\|x\|\}.$$

The next assertion holds:

In a real space H the ice cream cone $K_{z,\epsilon}$ is self-adjoint if and only if $\epsilon = \frac{1}{\sqrt{2}}$.

We begin with the check of the sufficiency, i.e., of the validity of the identity $K^*_{z,\frac{1}{\sqrt{2}}} = K_{z,\frac{1}{\sqrt{2}}}$. Let $x \in K^*_{z,\frac{1}{\sqrt{2}}}$ with $\|x\| = 1$. We firstly prove the inclusion $x \in K_{z,\frac{1}{\sqrt{2}}}$. Since H is real, z is an interior point of $K_{z,\frac{1}{\sqrt{2}}}$ and, hence, the inequality $\langle z, x\rangle > 0$ is valid. If $\langle z, x\rangle = 1$ then $x \in K_{z,\frac{1}{\sqrt{2}}}$. Consider the case in which $\langle z, x\rangle \in (0, 1)$. It is easy to see that for every scalar λ, satisfying $\lambda^2 \leq \frac{\langle z,x\rangle^2}{1-\langle z,x\rangle^2}$, the inclusion $(1-\lambda)\langle z, x\rangle z+\lambda x \in K_{z,\frac{1}{\sqrt{2}}}$ holds. In particular, for $\lambda_0 = -\frac{\langle z,x\rangle}{\sqrt{1-\langle z,x\rangle^2}}$, we get

$$\begin{aligned} 0 &\leq \langle (1 - \lambda_0)\langle z, x\rangle z + \lambda_0 x, x\rangle \\ &= \langle z, x\rangle^2 + \lambda_0(1 - \langle z, x\rangle^2) = \langle z, x\rangle^2 - \langle z, x\rangle\sqrt{1 - \langle z, x\rangle^2}, \end{aligned}$$

i.e., $\sqrt{1 - \langle z, x\rangle^2} \leq \langle z, x\rangle$. Finally, $\langle z, x\rangle^2 \geq \frac{1}{2}$ and so $x \in K_{z,\frac{1}{\sqrt{2}}}$. Let us verify the inclusion $K_{z,\frac{1}{\sqrt{2}}} \subseteq K^*_{z,\frac{1}{\sqrt{2}}}$. Consider two elements $y_1, y_2 \in K_{z,\frac{1}{\sqrt{2}}}$ satisfying

$\|y_i\| = 1$ and, hence,

$$\langle y_i, z \rangle \geq \frac{1}{\sqrt{2}} \tag{1}$$

for $i = 1, 2$. We must check the inequality $\langle y_1, y_2 \rangle \geq 0$. To this end, we define an orthogonal projection P_z on H via the formula $P_z y = \langle y, z \rangle z$. In view of (1),

$$\langle P_z y_1, P_z y_2 \rangle \geq \frac{1}{2}. \tag{2}$$

Obviously, we have $\|P_z y_i\| \geq \frac{1}{\sqrt{2}}$ for $i = 1, 2$, whence

$$1 = \|P_z y_i\|^2 + \|y_i - P_z y_i\|^2 \geq \frac{1}{2} + \|y_i - P_z y_i\|^2,$$

i.e., $\|y_i - P_z y_i\|^2 \leq \frac{1}{2}$. Taking into account the Cauchy–Schwarz inequality, we get $|\langle y_1 - P_z y_1, y_2 - P_z y_2 \rangle| \leq \frac{1}{2}$. Using the last inequality and (2), we infer that

$$\langle y_1, y_2 \rangle = \langle y_1 - P_z y_1, y_2 - P_z y_2 \rangle + \langle P_z y_1, P_z y_2 \rangle \geq -\frac{1}{2} + \frac{1}{2} = 0,$$

as required.

For the converse, let $K^*_{z,\epsilon} = K_{z,\epsilon}$. Clearly, $\epsilon \leq 1$. If $\epsilon = 1$ then for every $x \in K_{z,\epsilon}$, we have $\|x\| = \langle x, z \rangle$ and, hence, x and z are linearly dependent, i.e., $K_{z,\epsilon} = \{\lambda z : \lambda \geq 0\}$, a contradiction. Thus, $\epsilon \in (0, 1)$. Fix an arbitrary element w with $\|w\| = 1$ which is orthogonal to z. It is easy to check that for the scalar $\alpha_0 = \sqrt{\frac{1}{\epsilon^2} - 1}$ the inclusions $z \pm \alpha_0 w \in K_{z,\epsilon}$ are valid. Therefore, $0 \leq \langle z + \alpha_0 w, z - \alpha_0 w \rangle = 2 - \frac{1}{\epsilon^2}$, i.e., $\epsilon^2 \geq \frac{1}{2}$. In particular,

$$K_{z,\epsilon} \subseteq K_{z,\frac{1}{\sqrt{2}}}. \tag{3}$$

Next, $z + w \in K_{z,\frac{1}{\sqrt{2}}}$ and hence, as was shown above, $z + w \in K^*_{z,\frac{1}{\sqrt{2}}}$. In view of (3), $z + w \in K^*_{z,\epsilon} = K_{z,\epsilon}$. Thus, $1 = \langle z + w, z \rangle \geq \epsilon \|z + w\| = \epsilon\sqrt{2}$ and so $\frac{1}{\sqrt{2}} \geq \epsilon$. We have therefore proved that $\epsilon = \frac{1}{\sqrt{2}}$.

(c) Another important example of an ordered C^*-algebra is an algebra of all bounded, continuous functions $C(S)$ on some topological space S with an involution defined by the complex conjugate $x^*(s) = \overline{x(s)}$ with $s \in S$.

(d) Every ordered Banach algebra A such that $\|a^2\| = \|a\|^2$ for all $a \in A^+$ is an almost ordered C^*-algebra with the identity operator as the involution. In particular, such algebras are $\ell_1(\mathbb{Z})$ and $L_1(\mathbb{R}) \otimes \mathbb{R}$, where $\ell_1(\mathbb{Z})$ and $L_1(\mathbb{R})$ are considered with convolution as the product and $L_1(\mathbb{R}) \otimes \mathbb{R}$ is the unitization of $L_1(\mathbb{R})$. □

Let A be an ordered C^*-algebra. It is easy to see that the involution preserves lattice operations; e.g., if for $a, b \in A$ the least upper bound $a \vee b$ exists then the element $a^* \vee b^*$ is well defined and $(a \vee b)^* = a^* \vee b^*$. Thus, if an element $a \vee a^*$ exists then it is hermitian. In particular, if for an element $a \in A$ the modulus $|a|$

exists then $|a|^* = |a^*|$. Next, let us consider the order ideal $A_{\mathbf{e}}$ generated by $\mathbf{e}$ and defined by

$$A_{\mathbf{e}} = \{a \in A : -\lambda\mathbf{e} \leq a \leq \lambda\mathbf{e} \text{ for some scalar } \lambda \geq 0\}.$$

Clearly, $A_{\mathbf{e}}$ is a $*$-subalgebra. We note that if $A_{\mathbf{e}}$ is a (real) Riesz space under the ordering induced by A then $A_{\mathbf{e}}$ is an f-algebra and so that, by the Amemiya–Birkhoff–Pierce theorem (see [2, p. 117]), the algebra $A_{\mathbf{e}}$ is commutative.

The following result holds.

Theorem 2.2. *Let A be an almost ordered C^*-algebra with unit $\mathbf{e}$. Then every element $a \in A_{\mathbf{e}}$ such that $a \wedge a^*$ exists, is hermitian.*

Proof. As it is easy to see, we can assume $0 \leq a \leq \mathbf{e}$. Whence, $0 \leq a^* \leq \mathbf{e}$. Clearly,

$$0 \leq a - a \wedge a^* \leq \mathbf{e} \quad \text{and} \quad 0 \leq a^* - a \wedge a^* \leq \mathbf{e}.$$

Consequently,

$$0 \leq (a - a \wedge a^*)(a^* - a \wedge a^*) \leq (a - a \wedge a^*) \wedge (a^* - a \wedge a^*) = 0.$$

Thus, $(a^* - a \wedge a^*)^*(a^* - a \wedge a^*) = 0$. Since A is an almost ordered C^*-algebra, we have $a^* = a \wedge a^*$. It follows by symmetry that a is hermitian. □

Example 2.3. (a) First of all, we mention that it is not known if the preceding theorem holds if the condition that $a \wedge a^*$ exists is omitted, even when $A = L(H)$, where H is an ordered Hilbert space with a cone K (see Example 2.1 (a)).

However:

If I is the identity operator, then each of the following conditions implies that every operator T belonging to the order ideal $(L(H))_I$ is hermitian:

(i) *K coincides with the closed convex hull of its extremal rays;*

(ii) *K has a closed, bounded base (e.g., K is an ice cream cone or H is finite dimensional).*

Indeed, we can suppose $0 \leq T \leq I$. Let (i) hold and let x be an extremal vector of K (see [4, p. 37]). Clearly, $0 \leq Tx \leq x$, hence $Tx = \alpha x$ for some scalar $\alpha \geq 0$; analogously, $T^*x = \beta x$ with $\beta \geq 0$. On the other hand,

$$0 = \langle Tx, x\rangle - \langle x, T^*x\rangle = (\alpha - \beta)\|x\|^2.$$

Therefore, $\alpha = \beta$ or $Tx = T^*x$. Now, using our condition, we at once infer that $T = T^*$. Next, if K has a closed, bounded base $\mathcal{B}$ then (see [4, p. 41]) extreme points of $\mathcal{B}$ are exactly extremal vectors of K. Since H is reflexive, the set $\mathcal{B}$ is weakly compact and hence, by the Krein–Milman theorem (see [2, p. 137]), $\mathcal{B}$ coincides with a closed convex hull of its extreme points. Thus, the cone K satisfies the condition (i). It remains to observe that (see [4, pp. 100, 121]) every ice cream cone and every cone in a finite-dimensional space has a bounded, closed base.

The cone of non-negative functions in the ordered Hilbert space $L_2(\mu)$, with μ a σ-finite measure on a σ-algebra, does not have extremal rays if (and only if) the measure μ is diffuse (i.e., the σ-algebra does not contain any atoms). But, in this case Theorem 2.2 can be used. We mention another possible approach here. If $0 \le T \le I$ with $T \in L(L_2(\mu))$ then T is an orthomorphism and so it can be represented in the form $Tx = ux$ for some measurable function $u \in L_\infty(\mu)$ (see [2, p. 123]). The identity $T = T^*$ is then obvious.

Next, as was noted before in Theorem 2.2, in some cases the order ideal $A_{\mathbf{e}}$, where A is an arbitrary ordered Banach algebra, is a commutative subalgebra of A. However, in general it is not known if this assertion is valid even when $A = L(H)$. But, it is easy to see that the conditions (i) and (ii) above again guarantee an affirmative answer, i.e., in these cases, if $S, T \in (L(H))_I$ then $ST = TS$.

(b) The purpose of this part is to make the results of the preceding part more precise for the case of an ice cream cone. Namely, let H be a (real or complex) Hilbert space with $\dim H \neq 2$, let $z \in H$ with $\|z\| = 1$ and let $\epsilon \in (0,1)$. Let $K = K_{z,\epsilon}$ be the ice cream cone in H (see Example 2.1 (b)). Then *the identity* $(L(H))_I = \{\lambda I : \lambda \in \mathbb{R}\}$ *holds.* Actually, we mention first that if for a non-zero element $x \in H$ the relation $\langle x, z\rangle = \epsilon\|x\|$ is valid then x is an extremal vector of the cone K. In fact, the inequalities $0 \le y \le x$ imply

$$\epsilon\|x-y\| \le \langle x-y, z\rangle = \epsilon\|x\| - \langle y, z\rangle,$$

whence

$$\epsilon\|y\| \ge \epsilon(\|x\| - \|x-y\|) \ge \langle y, z\rangle \ge \epsilon\|y\|$$

and so $\|x\| = \|x-y\| + \|y\|$. Thus, x and y are linearly dependent. In particular, if an element w is orthogonal to z and $\|w\| = 1$ then for $\lambda = \sqrt{\frac{1}{\epsilon^2} - 1}$ the elements $z \pm \lambda w$ are extremal vectors of K. Consider an arbitrary orthonormal system $\{w_\alpha\}_{\alpha\in J}$ such that $\{z\} \cup \{w_\alpha\}_{\alpha\in J}$ is an orthonormal basis of H. Since for the case $\dim H < 2$ the assertion is trivial, we will assume $\operatorname{card} J \ge 2$. Let $T \in (L(H))_I$ and $0 \le T \le I$. In view of the remarks above, there exist scalars $\gamma_\beta^\pm \in [0,1]$ satisfying $T(z \pm \lambda w_\beta) = \gamma_\beta^\pm(z \pm \lambda w_\beta)$ for all $\beta \in J$. For some scalars $c_{\alpha\beta}$ with $\alpha, \beta \in \{0\} \cup J$ the decompositions $Tz = c_{00}z + \sum_{\alpha\in J} c_{\alpha 0}w_\alpha$ and $Tw_\beta = c_{0\beta}z + \sum_{\alpha\in J} c_{\alpha\beta}w_\alpha$ hold, whence

$$(c_{00} \pm \lambda c_{0\beta})z + \sum_{\alpha\in J}(c_{\alpha 0} \pm \lambda c_{\alpha\beta})w_\alpha = \gamma_\beta^\pm(z \pm \lambda w_\beta).$$

Using the uniqueness of a series representation, we get, on the one hand, $c_{\alpha 0} = c_{\alpha\beta} = 0$ for all $\alpha \in J$ and $\alpha \neq \beta$ (in particular, since β is arbitrary and $\operatorname{card} J \ge 2$, we have $c_{\alpha 0} = 0$ for all $\alpha \in J$). On the other hand,

$$\gamma_\beta^\pm = c_{00} \pm \lambda c_{0\beta} = \pm\frac{c_{\beta 0} \pm \lambda c_{\beta\beta}}{\lambda} = c_{\beta\beta}.$$

Thus, $c_{0\beta} = 0$ and $c_{00} = c_{\beta\beta}$ for all $\beta \in J$. Therefore, we finally have $Tz = cz$ and $Tw_\beta = cw_\beta$ for $\beta \in J$ with $c = c_{00}$, i.e., $T = cI$.

Obviously, the proven assertion does not hold for the case of $\dim H = 2$. Moreover, it implies that under the conditions that H is real, $\dim H > 2$, and $\epsilon \in (0,1)$, the ice cream cone $K_{z,\epsilon}$ does not have the Riesz decomposition property.

(c) Theorem 2.2 does not hold for the case of an arbitrary ordered Banach algebra with a positive involution. To see this, we consider the algebra $A = C[0, 1]$ under the natural norm, multiplication, and order. We define an involution via the formula $a^*(t) = \overline{a(1-t)}$. Evidently, this involution is positive. Next, the function $b(t) = t$ satisfies the inequalities $0 \leq b \leq \mathbf{e}$ while $b^*(t) = 1 - t$, i.e., $b^* \neq b$. We also mention that A is a $*$-normed algebra, i.e., $\|a\| = \|a^*\|$ for all $a \in A$. □

By the Gelfand–Naimark theorem (see [12, p. 244]), for an arbitrary complex C^*-algebra A there exists an isometric $*$-isomorphism φ from A to a closed $*$-subalgebra of $L(H)$, where H is some Hilbert space. For an ordered C^*-algebra A *it is an open question whether there exists a Hilbert space H with a cone K such that $\varphi(A^+) = \varphi(A) \cap (L(H))^+$.*

In conclusion of this section, the author should mention of the related concept of a "commutatively ordered C^*-algebra", which would be defined as a commutatively ordered Banach algebra (see [8]) which is also a C^*-algebra with positive involution.

3. The domination problem

In the theory of operators on a Banach lattice E the following *domination problem* plays a significant role:

(H_L) Let $\mathcal{I}$ be a closed (algebraical) ideal in the algebra $L(E)$. Does there exist a number $n \in \mathbb{N}$ (depending on $\mathcal{I}$) such that the relations $0 \leq S \leq T \in \mathcal{I}$ imply $S^n \in \mathcal{I}$?

Up to now *the validity of this conjecture in its general form remains open.* However, for various different operator ideals the answer is affirmative (see [1], where this problem is discussed in detail).

For the case of an ordered Banach algebra the hypothesis (H_L) should be formulated as follows:

(H_A) Let $\mathcal{I}$ be a closed ideal in an ordered Banach algebra A. Does there exist a number $n \in \mathbb{N}$ (depending on $\mathcal{I}$) such that the relations $0 \leq a \leq b \in \mathcal{I}$ imply $a^n \in \mathcal{I}$?

In its general form the hypothesis (H_A) is not valid (see [1]). Nevertheless, some versions of the domination problem have been studied in ordered Banach algebras in [5, 7, 9]. On the other hand, now it is well known that many results in the general theory of Banach algebras can be made more precise for the case of C^*-algebras. However, in the case of an ordered C^*-algebra the hypothesis (H_A) also has a negative answer. The following example clarifies the situation.

Example 3.1. (a) Let us consider the space $\mathbb{R}^k$ with $k \geq 2$ under addition and multiplication defined coordinatewise, with the norm $\|x\| = \max\limits_{1 \leq i \leq k} |x_i|$, where $x = (x_1, \ldots, x_k)$, and with the identity operator as involution. Let $b_1, \ldots, b_k$

be a collection of linearly independent elements in $\mathbb{R}^k$. We consider the cone $K = \left\{\sum_{i=1}^{k} \lambda_i b_i : \lambda_i \geq 0\right\}$ and will assume that

(i) $b_i b_j \in K$ for all $i, j = 1, \dots, k$ and

(ii) $\mathbf{e} = (1, \dots, 1) \in K$.

Then, under the ordering generated by the cone K, the space $\mathbb{R}^k$ is a real ordered C^*-algebra which is even a Riesz space. Now we assume that the following condition also holds:

(iii) there exist a regular element a and a singular one b satisfying the inequalities $0 \leq a \leq b$.

Then $a^n \notin b\mathbb{R}^k$ for all $n \in \mathbb{N}$ and, hence, the hypothesis (H_A) does not hold for $\mathbb{R}^k$. Thus we only need to see that a collection of elements $b_1, \dots, b_k$ in $\mathbb{R}^k$ with the properties (i)–(iii) exists. Indeed, depending on whether k is odd or is even, we can take $b_i = \sum_{j=1}^{i} \mathbf{e}_j$ or $b_i = \sum_{j=1}^{i+1} \mathbf{e}_j$ if i is odd and $b_i = -\mathbf{e}_{i+1} + \sum_{j=1}^{i} \mathbf{e}_j$ or $b_i = -\mathbf{e}_i + \sum_{j=1}^{i-1} \mathbf{e}_j$ if i is even, respectively, and put $a = b_{k-1} + 2b_k$ and $b = 2(b_{k-1} + b_k)$.

(b) Let us show that for an ordered C^*-algebra of the form $L(H)$, where H is an ordered Hilbert space, the hypothesis (H_A) also does not hold. We consider the real Hilbert space ℓ_2 and the ice cream cone (see Example 2.1 (b) with $z = (1, 0, 0, \dots)$ and $\epsilon = \frac{1}{\sqrt{2}}$)

$$K = \left\{x = (x_0, x_1, \dots) \in \ell_2 \quad : \quad x_0 \geq 0 \text{ and } x_0^2 \geq \sum_{i=1}^{\infty} x_i^2\right\}.$$

Then the algebra $A = L(\ell_2)$ with the normal cone $A^+ = \{T : T(K) \subseteq K\}$ is an ordered Banach algebra while there exists a rank-one operator T satisfying the inequalities $0 \leq I \leq T$ (see [6]). It remains only to observe that (see Example 2.1 (b)) A is an ordered C^*-algebra. □

4. C^*-algebras with p-additive norm

We recall that a norm on an ordered Banach space Z with a (not necessarily closed) cone K is called *p-additive* for $1 \leq p < \infty$, if $\|x + y\|^p = \|x\|^p + \|y\|^p$ holds for every $x, y \in Z$ with $x \wedge y = 0$.

Example 3.1′. In Example 3.1 (a) the norm is not p-additive if $p > 1$ (see Theorem 4.1 (a) below). If, however, $p = 1$ then under the choice of elements $b_1, \dots, b_k$ given in Example 3.1 (a), we do obtain a p-additive norm on the space $\mathbb{R}^k$ with the cone K. In fact, if $x = \sum_{i=1}^{k} \lambda_i b_i$ and $\lambda_i \geq 0$ then, it is easy to see, $\|x\| = \sum_{i=1}^{k} \lambda_i$ and, hence, the norm is p-additive and strictly monotone. However, if $k > 2$ then this norm is not a lattice norm (it is easy to see this immediately, but it is also implied by Theorem 4.1 (b) below). If $k = 2$ then $b_1 = \mathbf{e}$ and $b_2 = (1, -1)$ and it is not difficult to show that we now have a lattice norm. Moreover, $b_2^2 = \mathbf{e}$. In other words, in this case the algebra $\mathbb{R}^2$ is isomorphic to an algebra A_0 which is defined as the linear space $\mathbb{R}^2$ under the norm $\|x\| = |x_1| + |x_2|$, the natural order, the convolution as product, i.e., $x * y = (x_1y_1 + x_2y_2, x_1y_2 + x_2y_1)$, and the identity operator as the involution operator. It is easy to check that the algebra A_0 is an ordered C^*-algebra with unit $\mathbf{e}_0 = (1, 0)$. □

Our next objective is to show that, in fact, the algebra A_0 from the preceding example is the only type of ordered C^*-algebras with p-additive norm. To do this, we need to recall the following assertion: *in a normed Riesz space E with a strictly monotone norm the relation $x \perp y$ holds for two elements $x, y \geq 0$ if and only if $\|x - y\| = \|x + y\|$.* To see this, it is sufficient to observe the validity of the identity $x + y = |x - y| + 2(x \wedge y)$. Next, for every normed Riesz space E whose norm is p-additive there exists a norm-preserving Riesz isomorphism from E onto a Riesz subspace of some L_p (see [3, p. 92]). In particular, in this case, we obtain the strictly monotonicity of p-additive norm.

Theorem 4.1. *Let A be an almost C^*-algebra such that under the ordering and the norm induced by A, the space $A_r = A^+ - A^+$ is a Riesz space with p-additive norm. The following statements hold:*

(a) *If $p > 1$ then* $\dim A_r = 1$*;*

(b) *If $p = 1$ and A is an ordered C^*-algebra with a lattice norm on A_r then* $\dim A_r \leq 2$ *and in the case that* $\dim A_r = 2$*, the space A_r is isomorphic to the algebra A_0 defined in Example* 3.1′.

Proof. Let $p \geq 1$. We shall show first that the unit $\mathbf{e}$ is an atom in A_r. Actually, if $a + b \leq \mathbf{e}$ with $a \wedge b = 0$ then, it is easy to see, $ab = ba = 0$ and $a^2 \perp b^2$. In view of Theorem 2.2, the elements a and b are hermitian. Then

$$(\|a\|^p + \|b\|^p)^2 = \|a + b\|^{2p} = \|(a + b)^2\|^p = \|a^2 + b^2\|^p = \|a\|^{2p} + \|b\|^{2p}.$$

Thus, $a = 0$ or $b = 0$. Finally, $\mathbf{e}$ is an atom and, in particular, the band $B_{\mathbf{e}} = \{\lambda\mathbf{e} : \lambda \in \mathbb{R}\}$ generated by $\mathbf{e}$ is a projection band (see [3, p. 40]).

Next we show that if $p > 1$ then there exists no element $a > 0$ satisfying $a \perp \mathbf{e}$. The latter implies $\dim A_r = 1$ and the assertion (a) follows. Proceeding by contradiction and, in the case of the necessity, considering the element $\frac{a+a^*}{\|a+a^*\|}$, we can suppose $a = a^*$ and $\|a\| = 1$. There exist a scalar $\alpha_0 \geq 0$ and an element $a_0 \geq 0$ satisfying $a^2 = \alpha_0\mathbf{e} + a_0$ and $a_0 \perp \mathbf{e}$. For arbitrary scalars $\alpha, \beta > 0$, we have

$$\begin{aligned}(\alpha^p + \beta^p)^2 &= \|(\alpha\mathbf{e} + \beta a)\|^{2p} = \|\alpha^2\mathbf{e} + 2\alpha\beta a + \beta^2 a^2\|^p \\ &= (\alpha^2 + \alpha_0\beta^2)^p + \|2\alpha\beta a + \beta^2 a_0\|^p \geq \alpha^{2p} + \beta^p\|2\alpha a + \beta a_0\|^p,\end{aligned}$$

whence $2\alpha^p\beta^p + \beta^{2p} \geq \beta^p\|2\alpha a + \beta a_0\|^p$ or $2\alpha^p + \beta^p \geq \|2\alpha a + \beta a_0\|^p$. Letting in the last inequality $\beta \downarrow 0$, we obtain $2\alpha^p \geq 2^p\alpha^p$. Consequently, $2 \geq 2^p$ or $1 \geq p$, a contradiction.

(b) Again, we consider an arbitrary element $a > 0$ satisfying $a \perp \mathbf{e}$ and $\|a\| = 1$ and establish the equality

$$a^* a = \mathbf{e}. \tag{4}$$

To this end, let $d \geq 0$ be an arbitrary element with the property $d \perp \mathbf{e}+a$. We have

$$\|\mathbf{e} + a + d\|^2 = \|\mathbf{e} - a - d\|^2 = \|(\mathbf{e} - a^* - d^*)(\mathbf{e} - a - d)\| = \|d_1 - d_2\|$$

with $d_1 = \mathbf{e} + (a + d)^*(a + d)$ and $d_2 = a + d + a^* + d^*$. On the other hand, it is easy to see that $\|d_1 + d_2\| = \|d_1 - d_2\|$. In view of the remarks above, the last identity implies $d_1 \perp d_2$ and, in particular, $a^*a \perp a + d$. Since the element d is arbitrary, this yields the inclusion $a^*a \in B_{\mathbf{e}}$ and (4) is proved.

Now let $\mathbf{e}, a, b \in A^+$ be three pairwise disjoint elements (if such a and b do not exist then $\dim A_r \leq 2$). We have the equality

$$a^* b = b^* a = 0. \tag{5}$$

In order to see this, let $d \geq 0$ be an arbitrary element with the property $d \perp \mathbf{e}+a+b$. Using the identities

$$\|\mathbf{e} + a + b + d\|^2 = \|\mathbf{e} - a - b - d\|^2$$

and

$$\|\mathbf{e} + a + b + d\|^2 = \|\mathbf{e} + a - b - d\|^2$$

and arguing as above, we get from the first identity, the relation $a^*b \perp a+b+d$ and from the second one, the relation $a^*b \perp \mathbf{e}$. Therefore, $a^*b = 0$ and, hence, $b^*a = 0$.

If $\dim A_r > 2$, then we can find two elements $a, b > 0$ satisfying $a + b \perp \mathbf{e}$, $a \perp b$, and $\|a\| = \|b\| = 1$. Using (4) and (5), we have

$$4 = \|a + b\|^2 = \|(a^* + b^*)(a + b)\| = \|2\mathbf{e}\| = 2,$$

which is impossible. It follows that $\dim A_r \leq 2$.

If $\dim A_r = 2$, then every element $b \in A_r$ can be represented in the form of $b = \alpha \mathbf{e} + \beta b_0$ with $b_0 \in A^+$, $b_0 \perp \mathbf{e}$, and $\|b_0\| = 1$. Clearly, $b_0^* = b_0$, whence, in view of (4), $b_0^2 = \mathbf{e}$. Thus, A_r is algebraically isomorphic to the algebra A_0 with the *preservation of the ordering and the norm.* □

The statement (b) of the preceding theorem is not true for almost ordered C^*-algebras and for ordered C^*-algebras not having a lattice norm (see Example 2.1 (d) and Example 3.1′ above, respectively). Moreover, in the proof of Theorem 4.1 the completeness of the norm and the closedness of the cone A^+ in A were not used.

Corollary 4.2. *Let $E \neq \{0\}$ be a Riesz subspace of L_p with $1 \leq p < \infty$. Then there exists an isometry φ from E onto a $*$-subalgebra $\widehat{A}$ of some C^*-algebra with a unit $\mathbf{e}_0 \in \widehat{A}$ satisfying $\varphi^{-1}(\varphi(x)\varphi(y)) \geq 0$ and $\varphi^{-1}(\varphi(x)^*) \geq 0$ for $x, y \geq 0$ and $\varphi^{-1}(\mathbf{e}_0) \geq 0$ if and only if either* $\dim E = 1$ *or* $\dim E = 2$ *and* $p = 1$.

The next result gives the conditions under which in an ordered C^*-algebra the identity $|a^*a| = |a^*||a|$ holds.

Proposition 4.3. *Let A be an ordered C^*-algebra with a strictly monotone norm and suppose that, for $b \in A$, $\|b\| = \|\,|b|\,\|$ whenever the modulus $|b|$ exists. Then, for an element $a \in A$ such that the moduli $|a|$ and $|a^*a|$ ($|aa^*|$) exist, we have that $|a^*a| = |a^*||a|$ ($|aa^*| = |a||a^*|$).*

Proof. Obviously, if $|a^*a|$ exists then $|a^*a| \leq |a^*||a|$. Assuming $|a^*a| < |a^*||a|$, we have $\|a\|^2 = \||a^*a|\| < \||a^*||a|\| \leq \|a\|^2$, a contradiction. □

Acknowledgment

The author takes this opportunity to express his sincere gratitude to Prof. J.J. Grobler for reading an early draft of the paper and making numerous corrections which have been incorporated into the present paper. The author is also grateful to the referee for useful improvements and bringing some references into the author's attention.

References

[1] E.A. Alekhno, *Hereditary properties of dominated elements.* (Russian) Proceedings of the 6th Int. Conference "AMADE" (Minsk, 2011) **1** (2012), 23–27.

[2] C.D. Aliprantis, O. Burkinshaw, *Positive operators.* Academic Press, 1985.

[3] C.D. Aliprantis, O. Burkinshaw, *Locally solid Riesz Spaces with Applications to Economics.* Mathematical Surveys and Monographs **105** (2003).

[4] C.D. Aliprantis, R. Tourky, *Cones and duality.* Graduate Studies in Mathematics **84** (2007).

[5] D. Behrendt, H. Raubenheimer, *On domination of inessential elements in ordered Banach algebras.* Illinois J. Math. **51**(3) (2007), 927–936.

[6] L. Burlando, *Monotonicity of spectral radius for positive operators on ordered Banach spaces.* Arch. Math. **56**(1) (1991), 49–57.

[7] H. du T. Mouton, S. Mouton, *Domination properties in ordered Banach algebras.* Studia Math. **149**(1) (2002), 63–73.

[8] S. Mouton, K. Muzundu, *Commutatively ordered Banach algebras.* Quaest. Math. **36**(4) (2013), 559–587.

[9] S. Mouton, K. Muzundu, *Domination by ergodic elements in ordered Banach algebras.* Positivity **18**(1) (2014), 119–130.

[10] S. Mouton, H. Raubenheimer, *More spectral theory in ordered Banach algebras.* Positivity **1**(4) (1997), 305–317.

[11] H. Raubenheimer, S. Rode, *Cones in Banach algebras.* Indag. Math., N.S., **7**(4) (1996), 489–502.

[12] Ch.E. Rickart, *General theory of Banach algebras.* Van Nostrand, 1974.

Egor A. Alekhno
Faculty of Mechanics and Mathematics
Belarussian State University
Minsk, Belarus
e-mail: alekhno@bsu.by

Ordered Structures and Applications: Positivity VII
Trends in Mathematics, 25–33

On L-, M-Weakly Compact and Rank-One Operators

Şafak Alpay

Abstract. The finite rank, L- and M-weakly compact operators are studied. The inclusion of these classes with one another and with the weak Dunford–Pettis, almost Dunford–Pettis and b-weakly compact operators as well as the effect of these inclusions to the order structure of spaces serving as range or domain are examined.

Mathematics Subject Classification (2010). Primary 47B65; Secondary 46A42.

Keywords. L-weakly compact operator, M-weakly compact operator, weak Dunford–Pettis and almost Dunford–Pettis operators.

1. Introduction

In this note Banach lattices will be denoted by E, F and Banach spaces will be denoted by X, Y. All spaces are assumed to be nontrivial. $L_b(E, F)$ will denote order-bounded operators between E and F. A bounded linear operator between X and Y will be called an operator. The closed unit ball of X will be denoted by B_X. Positive elements of E will be denoted by E_+.

The difficulty of weak compactness in Banach lattices resulted in introduction of related notions of L- and M-weakly compactness. A nonempty bounded subset A of E is called L-weakly compact if for every disjoint sequence (x_n) in the solid hull of A, we have $\|x_n\| \to 0$. Every L-weakly compact subset of E is contained in E^a, the largest order ideal in E on which the norm is order continuous. Every relatively compact subset of E^a is L-weakly compact. Each L-weakly compact set is relatively weakly compact and a bounded subset of an AL-space is L-weakly compact if and only if it is relatively weakly compact[13, p. 212].

Definition 1.1. An operator $T : E \to X$ is called M-weakly compact if $\lim_n \|Tx_n\| = 0$ for every norm bounded disjoint sequence (x_n) in E. An operator $T : X \to E$ is called L-weakly compact if $T(B_X)$ is an L-weakly compact subset of E.

L- and M-weakly compact operators were introduced in [12] where it is also shown that $T : E \to X$ is M-weakly compact if and only if $T' : X' \to E'$ is L-weakly compact and $T : X \to E$ is L-weakly compact if and only if $T' : E' \to X'$ is M-weakly compact. L and M-weakly compact operators are weakly compact [13, 3.6.12]. If E is AL-space, then each operator $T : X \to E$ is weakly compact if and only if L-weakly compact and if E is an AM-space, then each operator $T : E \to X$ is weakly compact if and only if it is M-weakly compact [13, p. 214].

Example. l^∞ is an AM-space as well as a Grothendieck space. It is well known that each operator from a Grothendieck space into any separable Banach space is weakly compact [13, 5.3.10]. Hence every operator from l^∞ into c_0 is weakly and consequently, is M-weakly compact.

M- and L-weakly compact operators may be very large. If $1 \leq q < p \leq \infty$, then all regular operators from L^p into L^q are L- and M-weakly compact [13, 3.6.20]. On the other hand, they can also be very scarce. If $E = l^2[L^1[0,1]]$, then for each Banach space X the only M-weakly compact operator $T : E \to X$ is the zero operator [9]. L- and M-weakly compact operators were studied in [2, 6, 9, 10, 12] and [13].

We will also work with other operators which we recall now.

Definition 1.2. An operator $T : X \to Y$ is called a Dunford–Pettis operator if it maps weakly convergent sequences onto norm convergent sequences. An operator $T : E \to Y$ is called almost Dunford–Pettis if $\|Tx_n\| \to 0$ for each weakly null sequence (x_n) of disjoint elements in E_+. $T : X \to Y$ is called a weak Dunford–Pettis operator whenever $x_n \xrightarrow{w} 0$ in X and $y'_n \xrightarrow{w} 0$ in Y', we have $\lim_n (Tx_n, y'_n) = 0$.

Almost and weak Dunford–Pettis operators were studied in [5, 7, 8, 14, 15, 16] and the references therein.

An operator $T : X \to E$ is called semicompact if for each $\epsilon > 0$, there exists some $u \in E_+$ such that $T(B_X) \subseteq [-u, u] + \epsilon B_E$. An operator $T : E \to X$ is called b-weakly compact if $\|T(x_n)\| \to 0$ for each disjoint sequence (x_n) which is order bounded in E'' [3]. Each M-weakly compact operator is almost Dunford–Pettis and each almost Dunford–Pettis operator is b-weakly compact. Let us note that the identity of l^2 is weakly compact and therefore b-weakly compact but not an almost Dunford–Pettis operator. Let us note that there are almost Dunford–Pettis operators that are not L- or M-weakly compact. The identity operator $L^1[0,1] \to L^1[0,1]$ is almost Dunford–Pettis but is neither M-weakly nor L-weakly compact.

The main aim of this note is to study how the preceding spaces of operators accomodate one another and the effect of this to the properties of spaces serving as domain or range.

In all undefined terminology concerning Riesz spaces and Banach lattices, we will adhere to [2] and [13].

2. On L- and M-weakly compact operators

We can distinguish reflexive Banach lattices among spaces with order-continuous norm using M-weakly compact operators.

Proposition 2.1.

(a) *Let E be a Banach lattice with order-continuous norm. If each positive $T : E \to c_0$ is M-weakly compact, then E is reflexive.*
(b) *Suppose each positive operator $T : E \to l^\infty$ is M-weakly compact, then E is finite dimensional.*

Proof. (a) We first show E is a KB-space. If not, then E contains a copy of c_0 [13, 2.4.12]. There is a positive projection P onto this copy of c_0[13, 2.4.3]. By assumption P is an M-weakly compact operator. If i is the canonical embedding of c_0 in E, then $P \circ i$ is M-weakly compact whose restriction to c_0 is the identity of c_0. This is a contradiction. Now we show E' is a KB-space. If it is not, then E contains a copy of l^1 and there exists a positive projection $P : E \to l^1$ [13, 2.3.11]. Let $i : l^1 \to c_0$ be the canonical map of l^1 into c_0 and consider the operator $i \circ P$ from E into c_0. By the assumption it should be an M-weakly compact operator which in turn would imply that i is M-weakly compact. This contradiction shows E' also has order-continuous norm and therefore, E is reflexive by Theorem 14.22 in [2].

(b) The proof follows from Theorem 2 in [10] where it is proved that if F is Dedekind σ-complete, then each positive semicompact $T : E \to F$ is M-weakly compact if and only if one of: E' and F have order-continuous norms or E is finite dimensional. □

Order-continuity of E is essential in (a). If $E = l^\infty$, $F = c_0$, then each operator $T : l^\infty \to c_0$ is M-weakly compact but l^∞ is not reflexive. Considering the canonical embedding of l^2 into c_0, we see that the converse is not true.

If E' has order-continuous norm, then each Dunford–Pettis operator is M-weakly compact. The necessary conditions for the larger set of weak Dunford–Pettis operators to be M-weakly compact are given as follows:

Corollary 2.2. *Let F be σ-Dedekind complete. If each positive weak Dunford–Pettis operator $T : E \to F$ is an M-weakly compact operator, then one of the following holds:*

(a) *E is finite dimensional.*
(b) *F has order-continuous norm.*

Proof. We show that if the norm of F is not order continuous, then E is finite dimensional. If the norm of F is not order continuous, then F contains a copy of l^∞. Let $i : l^\infty \to F$ be this embedding. Since l^∞ has the Dunford–Pettis property, i is a weak Dunford–Pettis operator. Let $T : E \to l^\infty$ be an arbitrary positive operator, then $T = i \circ T$ is a weak Dunford–Pettis operator. Thus, T is M-weakly compact by the assumption. As T was arbitrary, the result follows from the proposition, part (b). □

The last corollary is similar to Theorem 2.3 in [14] where it is shown that if each positive weak Dunford–Pettis operator $T : E \to F$ is almost Dunford–Pettis then one of E has the Schur Property or F has order-continuous norm holds. Also see Remarks 2.2 and 2.3 in [14]. Dedekind σ-completeness of F is essential above. If $E = l^\infty$ and $F = c$, then each positive weak Dunford–Pettis operator T from E into F is M-weakly compact but E is not finite dimensional and c does not have order-continuous norm. The order-continuity of the norm of the range is not sufficient as the identity of c_0 is a weak Dunford–Pettis operator but not M-weakly compact.

In general L, M-weakly compact and Dunford–Pettis operators are distinct classes. The identity of l^1 is Dunford–Pettis but not an L- or M-weakly compact. On the other hand, the inclusion $i : L^2[0,1] \to L^1[0,1]$ is M-weakly compact but not a Dunford–Pettis operator. We now give sufficient condition for L-weakly compact operators to be Dunford–Pettis.

Proposition 2.3.

(a) *Let $0 \le T : E \to F$ be an L-weakly compact operator. If the lattice operations are weakly sequentially continuous in E, then T is a Dunford–Pettis operator.*

(b) *Let $T : E \to F$ be L-weakly compact. If the lattice operations in E' are weak* sequentially continuous, then $T' : F' \to E'$ is a Dunford–Pettis operator.*

Proof. (a) Let (x_n) be a weak null sequence in E. Since the lattice operations are weakly sequentially continuous, $|x_n| \to 0$ for $\sigma(E, E')$. Choose K such that $\|x_n\| \le K$ for all n. The adjoint $T' : F' \to E'$ is M-weakly compact. There exists $u \in E'_+$ such that $\|T'(|f| - u)^+\| \le \frac{\epsilon}{2K}$ for each $f \in B_{F'}$ by Theorem 18.9 in [2]. There exists some N such that $T'u(|x_n|) \le \frac{\epsilon}{2}$ for all $n \ge N$. Therefore, we have

$$|f(Tx_n)| \le \|T'(|f| - u)^+\| \cdot \|x_n\| + T'u(|x_n|) \le \epsilon$$

for all $n \ge N$ and $f \in B_{F'}$. Thus, $\|Tx_n\| \to 0$ and T is Dunford–Pettis.

(b) Let (g_n) be a weak null sequence in E'. Since the lattice operations are weak* sequentially continuous, $|g_n| \to 0$ in $\sigma(E', E)$. Let K be such that $\|g_n\| \le K$ for all n. Since T is L-weakly compact, given $\epsilon > 0$, there exists some $u \in E_+$ with $\|(|T(x)| - u)^+\| \le \frac{\epsilon}{2K}$ for all $x \in B_E$ by Theorem 18.9 in [2]. As $|g_n| \to 0$ for the topology $\sigma(E', E)$, there exists some N such that $|g_n|(u) < \epsilon/2$ for all $n \ge N$. Thus, for $n \ge N$ and $x \in B_E$, we have

$$|T'(g_n)(x)| \le |g_n|(|T(x)| - u)^+ + |g_n|(|T(x)| \wedge u) \le |g_n|(|T(x)| - u)^+ + |g_n|(u).$$

Hence,

$$|T'(g_n)(x)| \le \|g_n\| \cdot \|(|Tx| - u)^+\| + |g_n|(u) \le \epsilon$$

Thus, $\|T'g_n\| \to 0$ and T' is Dunford–Pettis. □

Dunford–Pettis operators are weakly Dunford–Pettis so if weak Dunford–Dunford operators are M-weakly compact, then, Dunford–Pettis operators are also weakly compact and E' has order-continuous norm by Theorem 19.23 in [2]. The next result shows M-weak compactness of almost Dunford–Pettis operators yields a similar result to M-weak compactness of weak Dunford–Pettis operators.

Proposition 2.4. *The following are equivalent:*

(a) *E' has order-continuous norm.*
(b) *Every Dunford–Pettis operator $T : E \to X$ is M-weakly compact.*
(c) *Each positive almost Dunford–Pettis operator $T : E \to X$ is M-weakly compact.*

Proof. The proof is the same as the proof of Theorem 3.7.10 in [13]. □

Sufficient conditions for a weak Dunford–Pettis operator to be an almost Dunford–Pettis were given in Theorem 5.3 in [8]. This result together with preceding observations yield the following.

Corollary 2.5. *Suppose E' is a KB-space. Let $T : E \to F$ be a positive weak Dunford–Pettis operator. Each of the following imply that T is M-weakly compact.*

(a) *F is a dual KB-space.*
(b) *F is a discrete KB-space.*
(c) *F'' has order-continuous norm.*
(d) *E has the positive Schur property.*
(e) *F has the positive Schur property.*

In (a–c), the range has order-continuous norm. Hence, each regular weak Dunford–Pettis operator is also L-weakly compact by 3.6.14 in [13].

None of the preceding conditions is sufficient. For example, take $E = l^\infty$ and $F = c_0$. Then, since E' has order-continuous norm, each operator $T : E \to F$ is Dunford-Pettis, and therefore, M-weakly compact but c_0 is not a KB-space, l^∞ does not have the Schur property, and $(c_0)'' = l^\infty$ does not have order-continuous norm.

Proposition 2.6. *Suppose E has order-continuous norm. If each positive weak Dunford–Pettis operator $T : E \to F$ is M-weakly compact, then one of E is a KB-space or F is a KB-space holds.*

Proof. Suppose E, F are not KB-spaces. Then, E contains a copy of c_0 [13, 2.4.12] and there exists a positive projection $P : E \to c_0$ [13, 2.4.3]. On the other hand, there is a lattice homomorphism i, mapping c_0 into F such that $K\|(\alpha_n)\|_\infty \leq \|i(\alpha_n)\|$ for all $(\alpha_n) \in c_0$. Clearly, the embedding i is not M-weakly compact. As the identity I_{c_0} of c_0 is a weak Dunford–Pettis operator, $i \circ P = i \circ I_{c_0} \circ P$ is a weak Dunford–Pettis operator but it is not M-weakly compact for otherwise I_{c_0} would be M-weakly compact. □

Observe that the assumption that E has order-continuous norm is essential as each operator $T : l^\infty \to c_0$ is M-weakly compact but l^∞ and c_0 are not KB-spaces.

Proposition 2.7. *Let E' be a KB-space and F be an AL-space. Let $T : E \to F$ be a regular operator such that $T[0, x]$ is $|\sigma|(F, F')$-totally bounded for each $x \in E_+$. Then, T is M-weakly compact.*

Proof. Let (e_n) be a norm bounded disjoint sequence in E. Then, (e_n) is a weak null sequence in E [13, 2.4.14]. Then, $|Te_n| \xrightarrow{w} 0$ in F by Theorem 19.17 in [2]. Let e' be the order unit in F'. Then, $\|Te_n\| = e'(|Te_n|) \xrightarrow{w} 0$. Therefore, T is M-weakly compact. □

An operator $T : E \to X$ is called AM-compact if $T[0, x]$ is relatively compact in X for each $x \in E_+$.

Corollary 2.8. *Let E, F be as above and $T : E \to F$ be a regular AM-compact operator. Then, T is M-weakly compact.*

Proposition 2.9. *Let E be a Banach lattice with order-continuous norm. Suppose each positive AM-compact operator $T : E \to F$ is M-weakly compact. Then, one of E or F is a KB-space.*

Proof. Suppose neither E nor F is a KB-space. Then, E contains c_0 and there exists a positive projection $P : E \to c_0$. On the other hand, there exists a lattice homomorphism i mapping c_0 into F such that $L\|x\| \leq \|i(x)\|$ for each $x \in c_0$. Consider the operator $i \circ P : E \to c_0 \to F$. Since c_0 has compact order intervals, it is AM-compact. However, if (e_n) is the canonical basis of c_0 then, $\|i \circ P(e_n)\| \geq L$, and therefore, $i \circ P$ is not M-weakly compact and this is a contradiction. □

Proposition 2.10. *Suppose F has order-continuous norm. If $T : E \to F$ is a disjointness preserving operator satisfying $|T(x)| \leq S(|x|)$ for each $x \in E$, and some positive compact operator S, then T is M-weakly compact.*

Proof. Let (e_n) be a norm bounded disjoint sequence in E and (f_n) be an arbitrary subsequence of (e_n). By Lemma 10.64 in [1], there exists a subsequence (g_n) of (f_n) such that $T(g_n)$ is order bounded and disjoint. Since F has order-continuous norm, $T(g_n)$ is norm convergent to zero. Thus, every subsequence of $T(e_n)$ has a subsequence that is norm convergent to zero. This shows that $T(e_n) \to 0$. □

An M-weakly compact operator need not have modulus. Even if it does, the modulus need not be M-weakly compact [9]. On the other hand, each order-bounded disjointness preserving operator T has a modulus which is given by $|T|(|x|) = |T(|x|)| = |Tx|$ by Theorem 8.6 in [2].

Corollary 2.11. *Let E, F and T, S be as above. Then, the modulus of the operator T is M-weakly compact.*

3. Rank-one operators may determine order

There are examples where rank-one operators solely determine the structure of the underlying space. Here is an example. An operator $T : X \to X$ satisfies the Daugavet property if $\|T + I\| = 1 + \|T\|$. If X satisfies the Daugavet property for rank-one operators, then X contains l^1 [1, 11.62].

A finite rank operator need not be an M- or L-weakly compact operator.

Example. Consider the operator $T : l^1 \to l^\infty$, defined by $T(\alpha_n) = (\sum_n \alpha_n)e$ for each $(\alpha_n) \in l^1$ where e is the constantly one sequence in l^∞. The operator T is rank-one but is neither L- nor M-weakly compact.

Recall that an operator $T : X \to Y$ is called an approximable operator if there exists a sequence (T_n) of finite rank operators such that $\|T - T_n\| \to 0$.

Proposition 3.1. *The following are equivalent:*

(a) *Each positive rank-one $T : X \to F$ is L-weakly compact.*
(b) *Each positive semicompact $T : X \to F$ is L-weakly compact.*
(c) *Each positive compact $T : X \to F$ is L-weakly compact.*
(d) *Every approximable operator into F is L-weakly compact.*
(e) *F has order-continuous norm.*

Proof. The equivalence of (b) and (e) is given in Theorem 3.1 [10]. The proof of equivalence of (a) and (e) is identical with this. (b) implies (c) follows from the semicompactness of positive compact operators. (c) implies L-weak compactness of finite rank operators and (d) follows from norm closedness of L-weakly compact operators. (d) implies (a) is clear. □

Using the fact that adjoint of a finite rank operator is of finite rank and the duality of L- and M-weakly compact operators, we have:

Corollary 3.2. *The dual E' of E has order-continuous norm if and only if every positive rank-one operator $T : E \to F$ is M-weakly compact.*

Corollary 3.3. *Each finite rank operator is L- and M-weakly compact if and only if E' and F have order-continuous norms.*

Suppose F is Dedekind complete. $T \in L_b(E, F)$ has order-continuous norm if for each sequence (T_n) of positive operators with $|T| \geq T_n \downarrow 0$ in $L_b(E, F)$, we have $\|T_n\| \to 0$. Positive operators that have order-continuous norms are precisely the operators which are simultaneously L- and M-weakly compact operators [2, 18.17].

Corollary 3.4. *Suppose E' and F have order-continuous norms. Then, every positive finite rank operator has order-continuous norm.*

The following was proved for positive semicompact operators in [6]. An inspection of the proof of Theorem 2.9 there yields the following.

Proposition 3.5. *For a Banach lattice E, the following are equivalent:*

(a) *For finite rank T and S with $0 \leq S \leq T$, S is M-weakly compact.*
(b) *Each positive finite rank operator T on E is M-weakly compact.*
(c) *For each positive finite rank T on E, T^2 is M-weakly compact.*
(d) *E' has order-continuous norm.*

It is possible to obtain a similar result for L-weakly compact operators by inspecting the proof of Theorem 2.8 in [6].

Theorem 4.1 in [9] shows that every regular M-weakly compact operator between Banach lattices E, F is L-weakly compact if and only if one of: F has order-continuous norm or $(E')^a = 0$ holds. The following, which is contained in the proof of Theorem 4.1 in [9], shows that under the assumption that each M-weakly compact, rank-one operator is L-weakly compact, Theorem 4.1 of [9] still holds.

Proposition 3.6. *The following are equivalent:*

(a) *At least one of, F has order-continuous norm or $(E')^a = 0$, holds.*
(b) *Every rank-one M-weakly compact $T : E \to F$ is L-weakly compact.*

There is a dual version of this as well. The proof of Theorem 4.2 in [9] shows that one of E' has order-continuous norm or $F^a = 0$ holds if and only if every L-weakly compact, rank-one $T : E \to F$ is M-weakly compact.

An operator $T : E \to X$ is called strong type B if T'' maps the band generated by E in E'' into X. If E' has the positive Schur property, then E' has order-continuous norm and operators of strong type B coincides with weakly compact operators [4]. In this case not only every weakly compact operator is M-weakly compact but the larger class of operators of strong type B, also are M-weakly compact and Theorem 3.3 in [9] takes the following form:

Proposition 3.7. *The following are equivalent for a Banach lattice E:*

(a) *E' has the positive Schur property.*
(b) *Every operator of strong type B, $T : E \to X$ is M-weakly compact.*
(c) *Every positive weakly compact $T : E \to c_0$ is M-weakly compact.*

We finish with a characterization of KB spaces in terms of weak Dunford–Pettis operators. We need to know the relation between weak Dunford–Pettis and b-weakly compact operators which was initiated in [7]. In general, they are distinct classes. The identity of c_0 is weak Dunford–Pettis but is not b-weakly compact. On the other hand, the identity of l^2 is weakly and therefore b-weakly compact but is not a weak Dunford–Pettis operator.

l^∞ is a Dedekind complete AM space with order unit and it has the b-property as each subset A in l^∞ which is order bounded in the bidual $(l^\infty)''$ is norm bounded there and therefore is order bounded in l^∞. On the other hand each operator $T : E \to l^\infty$ is order bounded and therefore is regular. Thus, taking $F = l^\infty$ in Corollary 2.6 in [7], we have the following.

Proposition 3.8. *Each weak Dunford–Pettis operator $T : E \to l^\infty$ is b-weakly compact if and only if E is a KB-space.*

Acknowledgment

The author wishes to thank the referee for numerous remarks, notes and suggestions.

References

[1] Y.A. Abramovich, C.D. Aliprantis, *An invitation to Operator Theory.* Amer. Math. Soc., Vol.**50**, 2002.

[2] C.D. Aliprantis, O. Burkinshaw, *Positive Operators.* Academic Press, 1985.

[3] S. Alpay, B. Altin, C. Tonyali, *On property* (b) *of vector lattices.* Positivity **7**(2003), 135–139.

[4] S. Alpay, *On operators of strong type B.* Hacet. J. Math. Stat. **41**(2012), No. 5, 627–633.

[5] B. Aqzzouz, A. Elbour, J. H'michane, *Some properties of the class of positive Dunford–Pettis Operators.* J. Math. Anal.Appl. **354**(2009), 295–300.

[6] B. Aqzzouz, A. Elbour, J. H'michane, *On some properties of the class of semicompact operators.* Bull. Belg. Math. Soc. Simon Stevin **18**(2011), 761–767.

[7] B. Aqzzouz, J. H'michane *The b-weak compactness of weak Dunford–Pettis operators.* Operators and Matrices **7** (2013), 219–224.

[8] B. Aqzzouz, K. Bouras, *Weak and almost Dunford–Pettis operators on Banach lattices.* Demonstratio Math. **XLVI** (2013), 165–179.

[9] Z.L. Chen, A.W. Wickstead, *L-weakly and M-weakly compact operators.* Indag. Math **10**(1999), 321–330.

[10] Z.L. Chen, Na Cheng, Y. Feng, *L- and M-weak compactness of positive semicompact operators.* Rend. Circolo Math. Palermo **59** (2010), 101–109.

[11] P.G. Dodds, D.H. Fremlin, *Compact operators on Banach lattices.* Israel J. Math. **34**(1979), 287–320.

[12] P. Meyer-Nieberg, *Über Klassen schwach kompakter Operatoren in Banachverbänden.* Math. Z. **138**(1974), 145–149.

[13] P. Meyer-Nieberg, *Banach Lattices.* Springer-Verlag, 1991.

[14] M. Moussa, K. Bouras, *About positive weak Dunford–Pettis Operators on Banach Lattices.* J. Math. Anal. Appl. **381**(2011), 891–896.

[15] W. Wnuk, *Banach lattices with the weak Dunford–Pettis property.* Atti. Sem. Mat. Fis. Univ. Modena **XLII** (1994), 227–236.

[16] W. Wnuk, *Banach lattices with properties of the Schur type – a survey.* Conf. Sem. Mat. Univ. Bari. **249**, (1993), 1–25.

Şafak Alpay
Middle East Technical University
Department of Mathematics
06800 Ankara, Turkey
e-mail: `safak@metu.edu.tr`

Ordered Structures and Applications: Positivity VII
Trends in Mathematics, 35–49

Characterization of Riesz Spaces with Topologically Full Center

Şafak Alpay and Mehmet Orhon

Abstract. Let E be a Riesz space and let $E^\sim$ denote its order dual. The orthomorphisms $\mathrm{Orth}(E)$ on E, and the ideal center $Z(E)$ of E, are naturally embedded in $\mathrm{Orth}(E^\sim)$ and $Z(E^\sim)$ respectively. We construct two unital algebra and order-continuous Riesz homomorphisms

$$\gamma : ((\mathrm{Orth}(E))^\sim)_n^\sim \to \mathrm{Orth}(E^\sim)$$

and

$$m : Z(E)'' \to Z(E^\sim)$$

that extend the above-mentioned natural inclusions respectively. Then, the range of γ is an order ideal in $\mathrm{Orth}(E^\sim)$ if and only if m is surjective. Furthermore, m is surjective if and only if E has a topologically full center. (That is, the $\sigma(E, E^\sim)$-closure of $Z(E)x$ contains the order ideal generated by x for each $x \in E_+$.) As a consequence, E has a topologically full center $Z(E)$ if and only if $Z(E^\sim) = \pi \cdot Z(E)''$ for some idempotent $\pi \in Z(E)''$.

Mathematics Subject Classification (2010). Primary 47B38, 46B42; Secondary 47B60, 46H25.

Keywords. Riesz space, ideal center, orthomorphism, Arens extension.

1. Introduction

Let E be a Banach lattice and let $Z(E)$ be its (ideal) center. In general $Z(E)$ is a subalgebra and sublattice of $Z(E')$, the center of the Banach dual E' of E. It is possible to extend this embedding to a contractive algebra and lattice homomorphism of $Z(E)''$ into $Z(E')$. It is natural to ask when the homomorphism would be onto $Z(E')$. It is clear that $Z(E)$ has to be large, since $Z(E')$ is always large. It turns out that the concept of largeness best suiting the center $Z(E)$ in this problem is that $Z(E)$ should be topologically full in the sense of Wickstead [17]. Namely, for each $x \in E_+$, the closure of $Z(E)x$ is the closed ideal generated by x. Then it is shown that the homomorphism is onto $Z(E')$ if and only if $Z(E)$ is topologically full [11, Corollary 2].

Our purpose in this paper is to consider the corresponding problem for a Riesz space E with point separating order dual $E^{\sim}$. In the case of Riesz spaces however, there are, in general, two different algebras that act on E which should be considered. Namely, $\mathrm{Orth}(E)$, the algebra of the orthomorphisms on E, and its subalgebra and (order) ideal $Z(E)$, the center of E. In the case of Banach lattices $\mathrm{Orth}(E) = Z(E)$, therefore this problem does not arise. Recall that an orthomorphism on E is an order-bounded operator on E that preserves bands, and $Z(E)$ is the ideal generated by the identity operator on E in the Riesz space $\mathrm{Orth}(E)$. In fact $\mathrm{Orth}(E)$ is an f-algebra. The ideal center $Z(E)$, on the other hand, is a normed AM-lattice where the identity operator is the order unit of the AM-lattice. Similar to the Banach lattice case, $\mathrm{Orth}(E)$ is embedded in $\mathrm{Orth}(E^{\sim})$ as an f-subalgebra and sublattice. Also as before, $Z(E)$ is embedded in $Z(E^{\sim})$ as a subalgebra and sublattice. We construct two unital algebra and order-continuous lattice homomorphisms

$$\gamma : (\mathrm{Orth}(E)^{\sim})^{\sim}_{n} \to \mathrm{Orth}(E^{\sim})$$

and

$$m : Z(E)'' \to Z(E^{\sim})$$

that extend the two embeddings mentioned above respectively. Then the range of γ is an order ideal in $\mathrm{Orth}(E^{\sim})$ if and only if m is onto $Z(E^{\sim})$ (Corollary 3.5). Also, m is onto $Z(E^{\sim})$ if and only if E has a topologically full center (Proposition 3.6). That is for each $x \in E_{+}$, the $\sigma(E, E^{\sim})$-closure of $Z(E)x$ in E contains the ideal generated by x. It follows that E has a topologically full center $Z(E)$ if and only if the ideal center of its order dual $E^{\sim}$ is given as $Z(E^{\sim}) = \pi \cdot Z(E)''$ for some idempotent $\pi \in Z(E)''$.

We point out that the proofs of the above-mentioned results differ from those used in the Banach lattice case [11]. The method used in this paper owes a lot to the work and results of Huijsmans and de Pagter [8] on the bidual of an f-algebra.

In the Banach lattice case it is shown that if $Z(E)$ is topologically full then it is maximal abelian [11, Corollary 3]. Wickstead [19] showed that the converse is not true. He constructed an interesting AM-lattice with a center that is maximal abelian but is not topologically full. In the case of Riesz spaces, even when $Z(E)$ is topologically full, it need not be maximal abelian. In fact, when $Z(E)$ is topologically full, its commutant in the order-bounded operators is $\mathrm{Orth}(E)$ (Corollary 3.13).

If E is a Riesz space, by $E^{\sim}$ we will denote the Riesz space of order-bounded linear functionals on E. All Riesz spaces considered in this paper are assumed to have separating order duals. $E^{\sim}_{n}$ will denote the order-continuous linear functionals in $E^{\sim}$. We let $L_b(E, F)$ denote the space of order-bounded linear operators from the Riesz space E into the Riesz space F. When $T : E \to F$ is an order-bounded operator between two Riesz spaces, the adjoint of T carries $F^{\sim}$ into $E^{\sim}$ and we will denote it by T'. The dual of a normed space will be denoted by E'. In all terminology concerning Riesz spaces we will adhere to the definitions in [1] and [20].

Let us recall that for any associative algebra A, a multiplication (called the Arens multiplication) can be introduced in the second algebraic dual A^{**} of A [2]. This is accomplished in three steps: given $a, b \in A$, $f \in A^*$ and $F, G \in A^{**}$, one defines $f \cdot a \in A^*$, $F \cdot f \in A^*$, and $F \cdot G \in A^{**}$ by the equations

$$(f \cdot a)(b) = f(ab)$$
$$(F \cdot f)(a) = F(f \cdot a)$$
$$(F \cdot G)(f) = F(G \cdot f).$$

For any Archimedean f-algebra A, the space $(A^\sim)_n^\sim$ is an Archimedean f-algebra with respect to the Arens multiplication [8].

We denote for any $F \in (A^\sim)_n^\sim$ the mapping $f \to F \cdot f$ by ν_F. The map ν_F is an orthomorphism on $A^\sim$ [8]. The mapping $\nu : (A^\sim)_n^\sim \to \mathrm{Orth}(A^\sim)$, defined by $\nu(F) = \nu_F$ is an algebra and Riesz homomorphism for any Archimedean f-algebra A. Moreover ν is onto $\mathrm{Orth}(A^\sim)$ if and only if $(A^\sim)_n^\sim$ has a unit element. In that case, υ is injective by Theorem 5.2 in [8]. The main purpose of this paper is to extend the latter result to an arbitrary Riesz space.

2. The Arens homomorphism

Let E be a Riesz space and consider the bilinear map

$$\mathrm{Orth}(E) \times E \to E \tag{1}$$

defined by $(\pi, x) \to \pi(x)$ for each $\pi \in \mathrm{Orth}(E)$ and $x \in E$. Related to (1), we define the following bilinear maps:

$$E \times E^\sim \to (\mathrm{Orth}(E))^\sim :: (x, f) \to \psi_{x,f} : \psi_{x,f}(\pi) = f(\pi x) \tag{2}$$

$$E^\sim \times (\mathrm{Orth}(E))^{\sim\sim} \to E^\sim :: (f, F) \to F \bullet f : F \bullet f(x) = F(\psi_{x,f}) \tag{3}$$

where $x \in E$, $f \in E^\sim$, $\pi \in \mathrm{Orth}(E)$ and $F \in \mathrm{Orth}(E)^{\sim\sim}$. We call the map defined in (3) the *Arens extension* of the map in (1).

For an Archimedean unital f-algebra A, we have $A^{\sim\sim} = (A^\sim)_n^\sim$ by Corollary 3.4 in [8]. Since $\mathrm{Orth}(E)$ is a unital f-algebra with point separating order dual, it is Archimedean. This enables us to use the identification $(\mathrm{Orth}(E))^{\sim\sim} = ((\mathrm{Orth}(E))^\sim)_n^\sim$ throughout this paper. It is straightforward to check that the Arens product on the f-algebra $((\mathrm{Orth}(E))^\sim)_n^\sim$ is compatible with the Arens extension defined in (3), that is, $E^\sim$ is a unital module over $((\mathrm{Orth}(E))^\sim)_n^\sim$.

We use (3) to define a linear operator

$$\gamma : ((\mathrm{Orth}(E))^\sim)_n^\sim \to L_b(E^\sim) \text{ by } \gamma(F)(f) = F \bullet f$$

for each $F \in ((\mathrm{Orth}(E))^\sim)_n^\sim$ and $f \in E^\sim$. We will call γ the Arens homomorphism of the order bidual of $\mathrm{Orth}(E)$.

Proposition 2.1. *γ is a unital algebra and order-continuous Riesz homomorphism such that $\gamma\left(((\mathrm{Orth}(E))^\sim)_n^\sim\right) \subset \mathrm{Orth}(E^\sim)$.*

Proof. It follows from the definition of γ that γ is a positive order-continuous unital algebra homomorphism. Also it is easily checked that $\gamma(\pi) = \pi'$ for each $\pi \in \mathrm{Orth}(E)$. Let $F \in ((\mathrm{Orth}(E))^\sim)^\sim_n$ such that $|F| \leq \pi$ for some $\pi \in \mathrm{Orth}(E)$. Then $|\gamma(F)| \leq \gamma(|F|) \leq \gamma(\pi) = \pi'$ by the positivity of γ. Since $\pi' \in \mathrm{Orth}(E^\sim)$, we see that $\gamma(F) \in \mathrm{Orth}(E^\sim)$. Then, since γ is order continuous and the ideal generated by $\mathrm{Orth}(E)$ is strongly order dense in $((\mathrm{Orth}(E))^\sim)^\sim_n$, the range of γ is contained in $\mathrm{Orth}(E^\sim)$. That γ is a Riesz homomorphism follows from the fact that it is an algebra homomorphism by Corollary 5.5 in [9]. □

If A is an f-algebra then the range of the homomorphism $\nu : (A^\sim)^\sim_n \to \mathrm{Orth}(A^\sim)$ is contained in the range of the Arens homomorphism γ.

Proposition 2.2. *Let A be an f-algebra with point separating order dual. The range of the homomorphism $\nu : (A^\sim)^\sim_n \to \mathrm{Orth}(A^\sim)$ is contained in the range of the Arens homomorphism γ.*

Proof. Let $P : A \to \mathrm{Orth}(A)$ be the canonical embedding of A into $\mathrm{Orth}(A)$ (i.e., $P(a)(b) = ab$ for all $a, b \in A$). It is well known that $P(A)$ is a sublattice and an algebra ideal in $\mathrm{Orth}(A)$. For each $\mu \in (\mathrm{Orth}(A))^\sim$, define $\widehat{\mu} \in A^*$ by $\widehat{\mu}(a) = \mu(P(a))$. Since positivity is preserved, it is clear that $\widehat{\mu} \in A^\sim$ for each $\mu \in (\mathrm{Orth}(A))^\sim$. Given $a \in A$, $f \in A^\sim$, we have $\psi_{a,f} \in (\mathrm{Orth}(A))^\sim$. Then

$$\widehat{\psi_{a,f}}(b) = \psi_{a,f}(P(b)) = f(P(b)(a)) = f(ab) = (f \cdot a)(b)$$

for each $b \in A$. That is $\widehat{\psi_{a,f}} = f \cdot a$ for all $a \in A$, $f \in A^\sim$. Now let $F \in (A^\sim)^\sim_n$. Define $\widehat{F} \in ((\mathrm{Orth}(A))^\sim)^*$ by $\widehat{F}(\mu) = F(\widehat{\mu})$ for each $\mu \in (\mathrm{Orth}(A))^\sim$. Since $0 \leq \mu$ implies $0 \leq \widehat{\mu}$, we have $0 \leq \widehat{F}$ whenever $0 \leq F$. That is $\widehat{F} \in ((\mathrm{Orth}(A))^\sim)^\sim = ((\mathrm{Orth}(A))^\sim)^\sim_n$. For each $F \in (A^\sim)^\sim_n$, $f \in A^\sim$ and $a \in A$, we have

$$\begin{aligned}\nu_F(f)(a) &= F \cdot f(a) = F(f \cdot a) = F(\widehat{\psi_{a,f}}) = \widehat{F}(\psi_{a,f})\\ &= \widehat{F} \bullet f(a) = \gamma(\widehat{F})(f)(a).\end{aligned}$$

Hence $\nu_F = \gamma(\widehat{F})$. □

Let us check the behavior of γ in some specific examples.

Example 1. Let ω denote all sequences and let $A = l^1$ the f-subalgebra of ω consisting of absolutely summable sequences. Then $A^\sim = l^\infty$ and $\mathrm{Orth}(A) = \mathrm{Orth}(A^\sim) = l^\infty$. Since $(A^\sim)^\sim_n = l^1$, ν is the inclusion map $l^1 \to l^\infty$ so that ν is one-to-one and not onto. On the other hand $((\mathrm{Orth}(A))^\sim)^\sim_n = (l^\infty)''$ and γ is the band projection of $(l^\infty)''$ onto l^∞. Thus γ is onto and not one-to-one.

Example 2. Let $A = c_0$ be the f-subalgebra of ω consisting of the sequences convergent to zero. Then $A^\sim = l^1$ and $\mathrm{Orth}(A) = \mathrm{Orth}(A^\sim) = l^\infty$. Since $(A^\sim)^\sim_n = l^\infty$, ν is the identity map on l^∞ so that ν is one-to-one and onto. On the other hand $((\mathrm{Orth}(A))^\sim)^\sim_n = (l^\infty)''$ and γ is the band projection of $(l^\infty)''$ onto l^∞. Thus γ is onto and not one-to-one.

Example 3. Consider $C[0,1]$ with the product $*$ defined by $a*b = iab$ with $i(x) = x$ for all $x \in [0,1]$. Then $A = (C[0,1],*)$ is an Archimedean f-algebra. As shown in [8], $(A^{\sim})_n^{\sim}$ is not semi-prime. Therefore υ is not one-to-one and not onto [8]. On the other hand $\mathrm{Orth}(A) = Z(C[0,1]) = C[0,1]$ and $\mathrm{Orth}(A^{\sim}) = Z(C[0,1]') = C[0,1]''$. Since $(C[0,1]^{\sim})_n^{\sim} = C[0,1]''$, γ is the identity map on $C[0,1]''$. Therefore γ is one-to-one and onto.

3. Riesz spaces with topologically full center

We start with the following definition that is due to Wickstead [17] in the case of Banach lattices.

Definition 3.1. Suppose E is a Riesz space. Then E is said to have a topologically full center if for each $x \in E_+$ the $\sigma(E, E^{\sim})$-closure of $Z(E)x$ contains the ideal generated by x.

Banach lattices with topologically full center were initiated in [17]. The class of Riesz spaces and the class Banach lattices with topologically full center are quite large. For example, in a σ-Dedekind complete Riesz space E each positive element generates a projection band. Therefore for each $x \in E_+$, $Z(E)x$ is an ideal and $Z(E)$ is topologically full. Also Banach lattices with a quasi-interior point or with a topological orthogonal system have topologically full center [19]. However not all Riesz spaces have topologically full center.

Example 4 (Zaanen [20, p. 664]). Let E be the Riesz space of piecewise affine continuous functions on $[0,1]$. Clearly the ideal generated by the constant 1 function equals E. But, as shown by Zaanen, $Z(E)$ is trivial, that is, it consists of the scalar multiples of the identity. Therefore E does not have a topologically full center.

The first example of an AM-space that has trivial center was given in [5]. A thorough study of Banach lattices with trivial center was undertaken in [18]. We refer the reader to [18] for further examples of Banach lattices with trivial center as well as a careful treatment of the following example of Goullet de Rugy mentioned above.

Example 5 ([5]). Let K be a compact Hausdorff space with a point $p \in K$ such that $\{p\}$ is not a G_δ-set in K (e.g., [12, Example 4, p. 170]). Let $C_0(K)$ denote the elements of $C(K)$ that vanish at p. Let H denote the positive unit ball of $C_0(K)'$ with the relative $\sigma(C_0(K)', C_0(K))$-topology. Let $E = \{f \in C(H) : f(r\mu) = rf(\mu)$ for all $r \in [0,1]$ and for each $\mu \in H$ with $\|\mu\| = 1\}$. Then E is an AM-space (without order unit). As a sublattice of $l^\infty(H \setminus \{0\})$, one has $E^d = \{0\}$. Therefore $Z(E)$ is embedded in $l^\infty(H \setminus \{0\})$ [15]. Then one may compute that $Z(E)$ consists of continuous bounded functions on $H \setminus \{0\}$ that are constant on the rays (i.e., $g(r\mu) = g(\mu)$ for all $r \in (0,1]$, for each $\mu \in H$ with $\|\mu\| = 1$). It follows by an argument in [5, p. 371] that if there is a non-constant function in $Z(E)$ then $\{0\}$ is a G_δ-set in H. That in turn implies that $\{p\}$ would be a G_δ-set in K. Therefore $Z(E)$ is trivial and E does not have a topologically full center.

$Z(E)$ is an Archimedean unital f-algebra with order unit. The order unit norm induced on $Z(E)$ is an algebra and lattice norm. $\widehat{Z(E)}$, the norm completion of $Z(E)$, is an AM-space and a partially ordered Banach algebra where the order unit and the algebra unit coincide. Therefore by the Stone algebra theorem $\widehat{Z(E)} \cong C(K)$ (isometric algebra and lattice homomorphism) for some compact Hausdorff space K. (Here $C(K)$ denotes the real-valued continuous functions on K.) Then $Z(E)' = \widehat{Z(E)}' = \widehat{Z(E)}^{\sim} = Z(E)^{\sim}$ and $Z(E)^{\sim\sim} = Z(E)''$. $Z(E)''$ is an AM-space and with the Arens product, it is a partially ordered Banach algebra (an Archimedean f-algebra with unit) where the order unit and the algebra unit coincide. Therefore $Z(E)'' \cong C(S)$ for some hyperstonian space S. That is the Arens product on $Z(E)''$ coincides with the pointwise product on $C(S)$ [2].

Given the bilinear map

$$Z(E) \times E \to E \tag{4}$$

defined by $(T, x) \to Tx$ for each $T \in Z(E)$ and $x \in E$, we define the following bilinear maps:

$$E \times E^{\sim} \to Z(E)' :: (x, f) \to \mu_{x.f} = \psi_{x,f}|_{Z(E)} \tag{5}$$

$$E^{\sim} \times Z(E)'' \to E^{\sim} :: (f, F) \to F \circ f : F \circ f(x) = F(\mu_{x,f}) \tag{6}$$

where $x \in E$, $f \in E^{\sim}$ and $F \in Z(E)''$. The Arens product on $Z(E)''$ is compatible with the bilinear map defined in (6). That is, $E^{\sim}$ is a unital module over $Z(E)''$. (6) allows us to define a linear operator $m : Z(E)'' \to L_b(E^{\sim})$ where $m(F)(f) = F \circ f$ for all $f \in E^{\sim}$ and $F \in Z(E)''$. It is easily checked that $m(T) = \gamma(T)$ whenever $T \in Z(E)$.

We have the following analogue of Proposition 2.1 for the map m.

Proposition 3.2. *m is a unital algebra and order-continuous lattice homomorphism such that $m(Z(E)'') \subset Z(E^{\sim})$.*

Proof. That m is a positive order-continuous algebra homomorphism is immediate from the definition of m. That it is a lattice homomorphism follows as in Proposition 2.1.

For each $F \in Z(E)''$, there is a net $\{T_\alpha\}$ in $Z(E)$ such that $\| T_\alpha \| \leq \| F \|$ and $T_\alpha \to F$ in $\sigma(Z(E)'', Z(E)')$-topology. Let $f \in E^{\sim}_{+}$ and $x \in E_{+}$. Then

$$-||F||f(x) \leq F \circ f(x) = \lim_{\alpha} f(T_\alpha x) \leq ||F||f(x).$$

So $F \in Z(E^{\sim})$. □

We will call the map $m : Z(E)'' \to Z(E^{\sim})$ the Arens homomorphism of the bidual of $Z(E)$ (into $Z(E^{\sim})$).

Proposition 3.3. *Let A be a unital f-algebra with point separating order dual. Then the Arens homomorphism of the bidual of $Z(A)$ is onto $Z(A^{\sim})$.*

Proof. Let $F \in Z(A^\sim)$ with $0 \leq F \leq 1$. Since A is unital, Theorem 5.2 in [8] implies that the algebra and lattice homomorphism $\nu : (A^\sim)_n^\sim \to \mathrm{Orth}(A^\sim)$ of [8] is one-to-one and onto. Therefore with a slight abuse of notation we will identify $(A^\sim)_n^\sim$ with $\mathrm{Orth}(A^\sim)$. In the duality $\langle (A^\sim)_n^\sim, A^\sim\rangle$, A is dense in $(A^\sim)_n^\sim$ with respect to the weak topology. Since the locally solid convex topology $|\sigma|((A^\sim)_n^\sim, A^\sim)$ is compatible with the duality $\langle (A^\sim)_n^\sim, A^\sim\rangle$ [1, Theorem 11.13, p.170], there is a net $\{a_\alpha\}$ in A such that $\{a_\alpha\}$ converges to F in the $|\sigma|((A^\sim)_n^\sim, A^\sim)$-topology. Lattice operations are continuous in the locally solid convex $|\sigma|((A^\sim)_n^\sim, A^\sim)$-topology and A is a sublattice of $(A^\sim)_n^\sim$ with $1 \in A$. Therefore we may suppose that $0 \leq a_\alpha \leq 1$. Hence, by the Alaoglu Theorem, there is $T \in Z(A)''$ with $0 \leq T \leq 1$ such that a subnet $\{a_\beta\}$ converges to T in the $\sigma(Z(A)'', Z(A)')$-topology. Then, when $f \in A^\sim$ and $b \in A$,

$$T \circ f(b) = T(\mu_{b,f}) = \lim_\beta \mu_{b,f}(a_\beta) = \lim_\beta \psi_{b,f}(a_\beta) = \lim_\beta f(a_\beta b).$$

On the other hand, since $\{a_\beta\}$ converges to F in the $|\sigma|((A^\sim)_n^\sim, A^\sim)$-topology implies convergence also in the $\sigma((A^\sim)_n^\sim, A^\sim)$-topology, we have

$$F \cdot f(b) = F(f \cdot b) = \lim_\beta f \cdot b(a_\beta) = \lim_\beta f(a_\beta b).$$

Therefore $m(T) = F$. □

Note that if E is a Riesz space and if we set $A = \mathrm{Orth}(E)$, then $Z(A) = Z(E)$. Also, since then A is a unital f-algebra with point separating order dual, Proposition 3.3 will be true for A. In what follows we will use these facts repeatedly.

Corollary 3.4. *Let E be a Riesz space. Let $A = \mathrm{Orth}(E)$ and m_A denote the Arens homomorphism of the bidual of $Z(A)$ onto $Z(A^\sim)$ (where $Z(A) = Z(E)$). Then*

(1) *the following diagram is commutative*

$$\begin{array}{ccccc} Z(E)'' & \overset{m}{\to} & Z(E^\sim) & \overset{i}{\to} & \mathrm{Orth}(E^\sim) \\ \| & & & & \uparrow \gamma \\ Z(A)'' & \overset{m_A}{\to} & Z(A^\sim) & \overset{i}{\to} & (A^\sim)_n^\sim \end{array}$$

where i denotes the natural inclusion map;

(2) $\gamma(Z((A^\sim)_n^\sim)) = m(Z(E)'')$.

Proof. (1) Suppose $F \in Z(A)''$ with $0 \leq F \leq 1$. Then, as in the proof of Proposition 3.3, let $0 \leq a_\alpha \leq 1$ be a net in $Z(A) \subset A$ such that $\{a_\alpha\}$ converges to F in the $\sigma(Z(A)'', Z(A)')$-topology and also to $m_A(F) \in (A^\sim)_n^\sim$ in the $\sigma((A^\sim)_n^\sim, A^\sim)$-topology. Since $(A^\sim)_n^\sim = (\mathrm{Orth}(E)^\sim)_n^\sim$, for each $x \in E$ and $f \in E^\sim$, we have

$$\begin{aligned}\gamma(m_A(F))(f)(x) &= m_A(F) \bullet f(x) = m_A(F)(\psi_{x,f}) = \lim_\alpha \psi_{x,f}(a_\alpha) \\ &= \lim_\alpha \mu_{x,f}(a_\alpha) = F(\mu_{x,f}) = m(F)(f)(x).\end{aligned}$$

(2) As in the proof of Proposition 3.3, we may identify $(A^\sim)_n^\sim$ with $\mathrm{Orth}(A^\sim)$ via the homomorphism ν. Then $Z(A^\sim) = Z((A^\sim)_n^\sim)$. Since, by Proposition 3.3, m_A is onto $Z(A^\sim)$, from part (1), we have $\gamma(Z((A^\sim)_n^\sim) = m(Z(E)'')$. □

Corollary 3.5. *Let E be a Riesz space. Then $\gamma((\mathrm{Orth}(E)^{\sim})_n^{\sim})$ is an order ideal in* $\mathrm{Orth}(E^{\sim})$ *if and only if* $m(Z(E)'') = Z(E^{\sim})$.

Proof. Let $A = \mathrm{Orth}(E)$. Suppose $\gamma((A^{\sim})_n^{\sim})$ is an order ideal in $\mathrm{Orth}(E^{\sim})$. Since γ is an order-continuous lattice homomorphism, the kernel $\mathrm{Ker}(\gamma)$ is a band in $(A^{\sim})_n^{\sim}$. Let $(1-e)$ be the band projection of $(A^{\sim})_n^{\sim}$ onto $\mathrm{Ker}(\gamma)$. Then $e \cdot (A^{\sim})_n^{\sim}$ is algebra and lattice isomorphic to the order ideal $\gamma((A^{\sim})_n^{\sim})$ with $\gamma(e) = 1$. Let $T \in Z(E^{\sim})$ with $0 \le T \le 1 = \gamma(e)$. So there is $F \in (A^{\sim})_n^{\sim}$ such that $\gamma(F) = T$. Moreover, since γ is an algebra and lattice homomorphism, we may choose F so that $0 \le F \le e \le 1$. That is we may suppose that $F \in Z((A^{\sim})_n^{\sim})$. Then Corollary 3.4 part (2) implies there is $G \in Z(E)''$ such that $m(G) = \gamma(F) = T$.

Conversely, suppose $m(Z(E)'') = Z(E^{\sim})$. Suppose $0 \le T \le \gamma(F)$ for some $0 \le F \in (A^{\sim})_n^{\sim}$ and for some $T \in \mathrm{Orth}(E^{\sim})$. Since $\mathrm{Orth}(E^{\sim})$ is Dedekind complete, there is $\widehat{T} \in Z(\mathrm{Orth}(E^{\sim})) = Z(E^{\sim})$ such that $0 \le \widehat{T} \le 1$ and $T = \widehat{T}\gamma(F)$. By Corollary 3.4 part (2), there is $G \in Z((A^{\sim})_n^{\sim})$ such that $\gamma(G) = \widehat{T}$. Then $T = \gamma(G)\gamma(F) = \gamma(G \cdot F)$ where $G \cdot F \in (A^{\sim})_n^{\sim}$. □

Corollary 3.5 gives the first part of the result stated in the abstract. We will complete the result by showing that the only Riesz spaces that satisfy Corollary 3.5 are necessarily those with a topologically full center. For Banach lattices a proof of this was given in [11]. Initially we will give the proof of the sufficiency.

Proposition 3.6. *Let E be a Riesz space. Then E has a topologically full center if and only if the Arens homomorphism $m : Z(E)'' \to Z(E^{\sim})$ is surjective. Then there exits an idempotent $\pi \in Z(E)''$ such that $Z(E^{\sim}) = \pi \cdot Z(E)''$ and* $\mathrm{Ker}(m) = (1-\pi) \cdot Z(E)''$.

Proof. (Sufficiency) Suppose m is surjective. To show that E has topologically full center it is sufficient to show that each $\sigma(E, E^{\sim})$-closed $Z(E)$ submodule of E is an ideal. This is equivalent to showing that each $\sigma(E^{\sim}, E)$-closed $Z(E)$ submodule of $E^{\sim}$ is an ideal. Let M be a $\sigma(E^{\sim}, E)$-closed $Z(E)$ submodule of $E^{\sim}$ and let $T \in Z(E)''$. There is a net $\{T_\alpha\}$ in $Z(E)$ that converges to T in the $\sigma(Z(E)'', Z(E)')$-topology. For each $x \in E$ and $f \in M$ we have

$$T_\alpha \circ f(x) = \mu_{x,f}(T_\alpha) \to T(\mu_{x,f}) = T \circ f(x).$$

Hence M is a $Z(E)''$-submodule of $E^{\sim}$. Since $Z(E^{\sim}) = m(Z(E)'')$, M is a $Z(E^{\sim})$-submodule. $E^{\sim}$ is Dedekind complete. It is well known that a subspace of $E^{\sim}$ is an ideal in $E^{\sim}$ if and only if it is a $Z(E^{\sim})$-submodule of $E^{\sim}$ (e.g., [19]). Hence M is an ideal in $E^{\sim}$. Since m is order continuous (Proposition 3.2), $\mathrm{Ker}(m)$ is a band in $Z(E)''$. Hence there exists a band projection $\pi \in Z(E)''$ such that $\mathrm{Ker}(m) = (1-\pi) \cdot Z(E)''$ and $Z(E^{\sim}) = \pi \cdot Z(E)''$. □

The proof of the converse requires some preparatory results.

Given a Riesz space E, let A be a unital subalgebra of $Z(E)$. Let A^0 denote the polar of A in $Z(E)'$ and let A^{00} denote the polar of A^0 in $Z(E)''$. By standard duality theory, we have that $A' = Z(E)'/A^0$ and $A'' = A^{00} \subset Z(E)''$. Since A is a normed algebra, A'' is a Banach algebra with the Arens product [2]. In fact A'' is a subalgebra of $Z(E)''$ when $Z(E)''$ has its Arens product.

Lemma 3.7. *Let A be a unital subalgebra of $Z(E)$. Then A^{00} is a subalgebra of $Z(E)''$ and the algebra product on A^{00} is identical with the Arens product on A'' under the canonical isomorphism of A^{00} with A''.*

Proof. Let $f \in A^0$ and $a \in A$. It is easily checked that $f \cdot a \in A^0$. Let $F, G \in A^{00}$. There is a net $\{a_\alpha\}$ in A that converges to F in the $\sigma(Z(E)'', Z(E)')$-topology. Let $f \in A^0$. Then

$$F \cdot G(f) = F(G \cdot f) = \lim_\alpha G \cdot f(a_\alpha) = \lim_\alpha G(f \cdot a_\alpha) = 0$$

and $F \cdot G \in A^{00}$.

On the other hand, denote the isomorphism of A'' onto A^{00} by $F \to \widehat{F}$. Given $f \in A'$, let $\widehat{f} \in Z(E)'$ denote any extension of f on $Z(E)$. When $F, G \in A''$, let $\{a_\alpha\}$ and $\{b_\beta\}$ be nets in A that converge to $\widehat{F}$ and $\widehat{G}$ respectively in the $\sigma(Z(E)'', Z(E)')$-topology. Then, evidently, the respective nets also converge to F and G respectively in the $\sigma(A'', A')$-topology. For any $f \in A'$, we have

$$F \cdot G(f) = \lim_\alpha \lim_\beta f(a_\alpha b_\beta) = \lim_\alpha \lim_\beta \widehat{f}(a_\alpha b_\beta) = \widehat{F} \cdot \widehat{G}(\widehat{f}). \qquad \square$$

Lemma 3.8. *Let I be a closed algebra ideal in $Z(E)$ and consider $A = Z(E)/I$ with the quotient norm. Then A is a normed algebra and with the Arens product A'' may be identified with the subalgebra of $Z(E)''$ given by*

$$(I^{00})^d = \{F \in Z(E)'' : |F| \wedge |G| = 0 \text{ for all } G \in I^{00}\}.$$

Proof. Let $\widehat{Z(E)} = C(K')$ for some compact Hausdorff space K' and let $\overline{I}$ denote the closure of I in $C(K')$. There is a closed subset K of K' such that $\overline{I} = \{a \in C(K') : a(K) = \{0\}\}$ and $C(K')/\overline{I} = C(K)$. Furthermore A is a subalgebra of $C(K)$. In fact $C(K)$ is the completion of A. Since $\overline{I}$ is an order ideal in $\widehat{Z(E)}$, $A' \cong I^0 = (\overline{I})^0$ is a band in $Z(E)'$. Then $A'' \cong Z(E)''/I^{00} = (I^{00})^d$, since I^{00} is a band in $Z(E)''$. It remains to check that the Arens product of A'' is identical with the product on the subalgebra $(I^{00})^d$. Let $F, G \in Z(E)''$ and $\widehat{F}, \widehat{G} \in A''$ such that $\widehat{F} = F|_{I^0}$, $\widehat{G} = G|_{I^0}$. Let $\{a_\alpha\}$ and $\{b_\beta\}$ be nets in $Z(E)$ that converge to F and G respectively in the $\sigma(Z(E)'', Z(E)'))$-topology. Let $[a] = a + I$ for each $a \in Z(E)$. Then, it follows that, $\{[a_\alpha]\}$ and $\{[b_\beta]\}$ converge to $\widehat{F}$ and $\widehat{G}$ respectively in the $\sigma(A'', A')$-topology. Then, for $f \in I^0$, we have

$$\widehat{F} \cdot \widehat{G}(f) = \lim_\alpha \lim_\beta f([a_\alpha][b_\beta]) = \lim_\alpha \lim_\beta f(a_\alpha b_\beta) = F \cdot G(f). \qquad \square$$

Let J be an ideal in $E^\sim$. For any $F \in Z(E)''$, $m(F)|_J \in Z(J)$. Each operator in $Z(J)$ has a unique extension to an operator in the ideal center of the band generated by J. Since bands in $E^\sim$ are projection bands, we have $Z(J) = Z(E^\sim)|_J$.

Suppose J is an ideal in $E^\sim$ that separates the points of E. Let A be a unital subalgebra of $Z(E)$.

Definition 3.9. E is called cyclic with respect to A for the dual pair $\langle E, J\rangle$ if there is $u \in E_+$ such that $(Au)^0 = \{0\}$ in J.

Lemma 3.10. *Let E be a Riesz space and J be an ideal in $E^\sim$ that separates the points of E. Suppose E is cyclic with respect to a unital f-subalgebra A of $Z(E)$ for the dual pair $\langle E, J\rangle$. Then $m(A'')|_J = Z(J)$.*

Proof. The norm completion $\widehat{A}$ of A is a unital closed subalgebra of $\widehat{Z(E)}$. Therefore $\widehat{A} = C(K)$ for some compact Hausdorff space K. Furthermore $A^\sim = A' = C(K)'$ and $A'' = C(K)'' = C(S)$ for some hyperstonian space S. (We again mention that the usual multiplication on $C(S)$ is the Arens extension of the product on $C(K)$ as shown in [2]. Also the usual $C(S)$ -module structure of $C(K)'$ via its ideal center is the Arens homomorphism of the bidual of $C(K)$ onto the ideal center of $C(K)'$ (e.g., Proposition 3.3).) In the rest of the proof, by Lemma 3.7, we consider A'' as a subalgebra of $Z(E)''$.

Let $u \in E_+$ be a cyclic vector and $f \in J_+$. Then $\mu_{u,f} \in Z(E)'_+$ and $\mu_{u,f}|_A = \widehat{\mu}_{u,f} \in C(K)'_+$. Let P_f be the band projection of $C(K)'$ onto the band $B(\widehat{\mu}_{u,f})$ generated by $\widehat{\mu}_{u,f}$. By the Lebesgue Decomposition Theorem and the Radon–Nikodym Theorem, we have

$$B(\widehat{\mu}_{u,f}) = P_f \cdot C(K)' = L^1(\widehat{\mu}_{u,f}) = \{\mu \in C(K)' : |\mu| << \widehat{\mu}_{u,f}\}.$$

The first equality above follows by Proposition 3.3.

Suppose $e \circ f = 0$ for some idempotent $e \in A'' = C(S)$. Let $\{a_\alpha\}$ be a net in A that converges to e in the $\sigma(A'', A')$-topology. Then for each $a \in A$,

$$\begin{aligned} e \cdot \widehat{\mu}_{u,f}(a) &= e(\widehat{\mu}_{u,f} \cdot a) = \lim_\alpha \widehat{\mu}_{u,f} \cdot a(a_\alpha) = \lim_\alpha \widehat{\mu}_{u,f}(a a_\alpha) = \lim_\alpha \mu_{u,f}(a_\alpha a) \\ &= \lim_\alpha f(a_\alpha a u) = \lim_\alpha \mu_{au,f}(a_\alpha) = e(\mu_{au,f}) = e \circ f(au) = \widehat{\mu}_{u,e\circ f}(a) = 0. \end{aligned}$$

Hence $0 \le e \le 1 - P_f$. Conversely, since u is a cyclic vector, $\widehat{\mu}_{u,(1-P_f)\circ f} = (1-P_f)\cdot\widehat{\mu}_{u,f} = 0$ implies that $(1-P_f)\circ f = 0$. Therefore $1 - P_f = \sup\{e \in C(S) : e \circ f = 0 \text{ and } e = e^2\}$. That is, $a \circ f = 0$ for some $a \in A''$ if and only if $P_f \cdot a = 0$.

Let $T \in Z(E^\sim)$ with $0 \le T \le 1$. Let $f \in J_+$. Then $0 \le \widehat{\mu}_{u,Tf} \le \widehat{\mu}_{u,f}$ in A'. Therefore, by the Radon–Nikodym Theorem, there is $a_f \in L^\infty(\widehat{\mu}_{u,f}) \subset C(S) = A''$ such that $\widehat{\mu}_{u,Tf} = a_f \cdot \widehat{\mu}_{u,f} = \widehat{\mu}_{u,a_f\circ f}$. Since u is a cyclic vector it follows that $Tf = a_f \circ f$. Suppose $g \in J$ such that $0 \le g \le f$. $E^\sim$ is Dedekind complete, there is $G \in Z(E^\sim)$ with $0 \le G \le 1$ such that $g = Gf$. Therefore

$$m(a_g)(g) = Tg = T(Gf) = G(Tf) = Gm(a_f)(f) = m(a_f)(Gf) = m(a_f)(g).$$

That is $(a_g - a_f)\circ g = 0$. Therefore $(a_g - a_f)\cdot P_g = 0$. Now suppose $f, g \in J_+$ and $h = f \vee g$. Then $(a_f - a_h)\cdot P_f = (a_g - a_h)\cdot P_g = 0$. Hence $a_g \cdot P_f = a_f \cdot P_g$ for all $f, g \in J_+$. Since S is Stonian and $0 \le a_f \le P_f \le 1$ for all $f \in J_+$, there is a unique $a \in A''$ such that $P_f \cdot a = a_f$ and $(1 - \sup\{P_f : f \in J_+\})\cdot a = 0$. Then

$$Tf = a_f \circ f = (a \cdot P_f)\circ f = a \circ (P_f \circ f) = a \circ f$$

for all $f \in J_+$. □

Now we are ready to complete the proof of Proposition 3.6.

Proof. (Necessity) Suppose $x \in E_+$. Let $I(x)$ denote the ideal generated by x in E and let $\operatorname{Ann}(x) = \{T \in Z(E) : Tx = 0\}$. $\operatorname{Ann}(x)$ is a closed order and algebra ideal in $Z(E)$. Consider the map from $Z(E)$ into $Z(I(x))$ defined by $T \to T|_{I(x)}$. Clearly the map is a norm reducing positive algebra homomorphism. Since the kernel of this map is equal to $\operatorname{Ann}(x)$, the map is also a lattice homomorphism. It induces a norm reducing lattice and algebra homomorphism of $Z(E)/\operatorname{Ann}(x)$ into $Z(I(x))$ where $T+\operatorname{Ann}(x) = [T] \to T|_{I(x)}$. The induced map is an isometry. For example, if $T \in Z(E)_+$ with norm $||T|_{I(x)}||$ in $Z(I(x))$, then $(T \wedge ||T|_{I(x)}||1)|_{I(x)} = T|_{I(x)}$ and $||[T]|| \leq ||T|_{I(x)}||$. Hence let $A = Z(E)|_{I(x)}$ be the normed unital f-subalgebra of $Z(I(x))$. Then since $A \cong Z(E)/\operatorname{Ann}(x)$ (isometric, lattice and algebra homomorphism), we may think of A'' as a subalgebra of $Z(E)''$ (cf., Lemma 3.8). Equivalently, we may think of A'' as a subalgebra of $Z(I(x))''$ (cf., Lemma 3.7).

Let $J = \{f|_{I(x)} : f \in E^\sim\}$. Clearly J is an ideal in $I(x)^\sim$ that separates the points of $I(x)$. Let $P_x \in Z(E^\sim)$ denote the band projection of $E^\sim$ onto $(I(x)^0)^d$. The map $f|_{I(x)} \to P_x(f)$ is a lattice homomorphism of J onto the band $(I(x)^0)^d$. Let m_x be the Arens homomorphism of $Z(I(x))''$ into $Z(I(x)^\sim)$ and let m be the same for $Z(E)''$ into $Z(E^\sim)$. We claim that on elements of A'', m_x restricted to J agrees with m restricted to $(I(x)^0)^d$. Namely, let $F \in A''$, $f \in E^\sim$ and $y \in I(x)$. Choose a net $\{a_\alpha\}$ in A that converges to F in the $\sigma(A'', A')$-topology. Then

$$\begin{aligned} m_x(F)(f|_{I(x)})(y) &= F \circ f|_{I(x)}(y) = F(\mu_{y,f|_{I(x)}}) = F([\mu_{y,f|_{I(x)}}]) \\ &= \lim_\alpha [\mu_{y,f|_{I(x)}}](a_\alpha) = \lim_\alpha \mu_{y,f|_{I(x)}}(a_\alpha) = \lim_\alpha f(a_\alpha y) \end{aligned}$$

where $[\mu_{y,f|_{I(x)}}] = \mu_{y,f|_{I(x)}} + A^0 \in A' = Z(I(x))'/A^0$. On the other hand

$$\begin{aligned} m(F)(P_x(f))(y) &= P_x(m(F)(f))(y) = m(F)(f)(y) = F \circ f(y) \\ &= F(\mu_{y,f}) = \lim_\alpha \mu_{y,f}(a_\alpha) = \lim_\alpha f(a_\alpha y). \end{aligned}$$

In the last string of equalities, the first equality follows because $Z(E^\sim)$ is commutative. The second equality follows because the range of $1 - P_x$ is $I(x)^0$ in $E^\sim$. Finally the fifth equality follows because $\mu_{y,f} \in \operatorname{Ann}(x)^0 = A'$ in $Z(E)'$. Hence, the claim is verified. In what follows we will keep the notation that we established in this initial part of the proof.

Suppose that E has topologically full center. This means that $I(x)$ is cyclic with respect to the unital f-subalgebra A of $Z(I(x))$ for the duality $\langle I(x), J\rangle$. Given $T \in Z(E^\sim)$ with $0 \leq T \leq 1$, let $\widehat{T} = T|_{(I(x)^0)^d} \in Z(J)$ where $\widehat{T}(f|_{I(x)})(y) = T(P_x(f))(y)$ for each $f \in E^\sim$ and $y \in I(x)$. Then by Lemma 3.10 there is $a_x \in A''$ with $0 \leq a_x \leq 1$ such that $m_x(a_x) = \widehat{T}$ on J. Then we have that

$$m(a_x)(P_x(f))(y) = m_x(a_x)(f|_{I(x)})(y) = \widehat{T}(f|_{I(x)})(y) = T(P_x(f))(y)$$

for all $f \in E^{\sim}$ and $y \in I(x)$. Since the range of $1 - P_x$ is $I(x)^0$ in $E^{\sim}$, we have that

$$m(a_x)(f)(y) = T(f)(y)$$

for all $f \in E^{\sim}$ and $y \in I(x)$.

Let $1 - e_x = \sup\{e \in Z(E)'' : e = e^2, m(e)(f)(x) = 0 \text{ for all } f \in E^{\sim}\}$. Since m is order continuous, we have $m(1 - e_x)(f)(x) = 0$ for all $f \in E^{\sim}$. Hence $F \circ f(x) = 0$ for all $f \in E^{\sim}$ for some $F \in Z(E)''$ if and only if $e_x \cdot F = 0$. (Note that $\{\mu_{x,f} : f \in E^{\sim}\}$ is an ideal in $Z(E)'$ and $1 - e_x$ is the band projection of $Z(E)''$ onto the band in $Z(E)''$ that annihilates this ideal.) Now repeating the argument in the proof of Lemma 3.10, we find a unique $a \in Z(E)''$ such that $0 \leq a \leq 1$, and $e_x \cdot a = a_x$ for each $x \in E_+$. Then, for each $f \in E^{\sim}$ and $x \in E_+$, we have

$$T(f)(x) = m(a_x)(f)(x) = m(a \cdot e_x)(f)(x) = m(e_x)m(a)(f)(x) = m(a)(f)(x).$$

Therefore $m(a) = T$. □

We will conclude this paper by stating some immediate consequences of Proposition 3.6.

Corollary 3.11. *Let A be an f-algebra with point separating order dual such that $(A^{\sim})_n^{\sim}$ has a unit. Then A has topologically full center.*

Proof. By Theorem 5.2 [8], the homomorphism υ of $(A^{\sim})_n^{\sim}$ is onto $\mathrm{Orth}(A^{\sim})$. Then Proposition 2.2 implies that the Arens homomorphism $\gamma : (\mathrm{Orth}(A)^{\sim})_n^{\sim} \to \mathrm{Orth}(A^{\sim})$ is also onto. Hence, by Corollary 3.5, m is onto $Z(A^{\sim})$. Therefore, by Proposition 3.6, A has topologically full center. □

Remark 3.12. Characterizations of f-algebras A such that $(A^{\sim})_n^{\sim}$ has unit are given in [8], [4], [10]. Related to Corollary 3.11, we mention that we do not know any examples of semi-prime f-algebras that do not have topologically full center.

Corollary 3.13. *Let E be a Riesz space with topologically full center. Suppose T is an order-bounded operator on E. Then T commutes with $Z(E)$ if and only if T is in* $\mathrm{Orth}(E)$.

Proof. T is order bounded implies $T' : E^{\sim} \to E^{\sim}$. If T commutes with $Z(E)$, then T' commutes with $m(Z(E)'')$. Since $Z(E)$ is topologically full, $m(Z(E)'') = Z(E^{\sim})$. Therefore T' commutes with $Z(E^{\sim})$. That is, T' commutes with the band projections on $E^{\sim}$. Since $E^{\sim}$ is Dedekind complete, each band in $E^{\sim}$ is a projection band. So T' is band preserving and therefore $T' \in \mathrm{Orth}(E^{\sim})$. By a result in [16, Theorem 3.3], $T \in \mathrm{Orth}(E)$. □

In view of Examples 4 and 5, the corollary may fail if $Z(E)$ is not topologically full. On the other hand, the result may be true even when $Z(E)$ is not topologically full. The example constructed by Wickstead in [19] shows this. We refer the reader to the introduction for more detailed information. Corollary 3.13 shows that Orth E) is maximal abelian when $Z(E)$ is topologically full. On the other hand, Wickstead's example in [19] shows that the converse is not true.

Corollary 3.14. *Let E be a Riesz space with topologically full center. Then the following are equivalent:*

(1) $E^{\sim\sim} = (E^{\sim})_n^{\sim}$.

(2) *m is continuous when its domain has the $\sigma(Z(E)'', Z(E)')$-topology and its range has the $\sigma(E^{\sim}, E^{\sim\sim})$-operator topology.*

We leave the straightforward proof of the corollary to the interested reader.

Before stating our final corollary, we want to discuss its content and fix some notation. Let K be a hyperstonian space. That is K is a Stonian compact Hausdorff space and $C(K)$ is a dual Banach space. Let $C(K)_*$ denote the predual of $C(K)$. Recall that $C(K)_* = C(K)'_n$, the order-continuous linear functionals on $C(K)$. Hence $C(K)_*$ is a band in the Dedekind complete Banach lattice $C(K)'$. Since $Z(C(K)') = C(K)''$, there is an idempotent $p \in C(K)''$ such that p is the band projection on $C(K)'$ with range $C(K)_*$. That is

$$p \cdot C(K)' = C(K)'_n = C(K)_*.$$

Let E be a Riesz space. Its order dual $E^{\sim}$ is a Dedekind complete Riesz space. Therefore $E^{\sim}$ has a topologically full center $Z(E^{\sim})$. Furthermore $Z(E^{\sim})$ is itself Dedekind complete as a Banach lattice. In fact, it is familiar that $Z(E^{\sim}) = C(K)$ for some hyperstonian space K.(This will become clear in the proof of the corollary.) Let $m : Z(E^{\sim})'' \to Z(E^{\sim\sim})$ be the Arens homomorphism of the bidual of $Z(E^{\sim})$. Since $Z(E^{\sim})$ is topologically full, we have $m(Z(E^{\sim})'') = Z(E^{\sim\sim})$ and $\mathrm{Ker}(m) = (1-\pi) \cdot Z(E^{\sim})''$ for some idempotent $\pi \in Z(E^{\sim})'' = C(K)''$ (Proposition 3.6). We will show that $p \circ E^{\sim\sim} = (E^{\sim})_n^{\sim}$.

Corollary 3.15. *Let E be a Riesz space with point separating order dual $E^{\sim}$. Let m be the Arens homomorphism of the bidual of $Z(E^{\sim})$ in $Z(E^{\sim\sim})$. Then*

(1) *$Z(E^{\sim})$ is topologically full and $Z(E^{\sim}) = C(K)$ for some hyperstonian space K.*

(2) *There is an idempotent $\pi \in C(K)''$ such that*

$$Z(E^{\sim\sim}) = \pi \cdot C(K)'' \text{ and } \mathrm{Ker}(m) = (1-\pi) \cdot C(K)''.$$

(3) *There is an idempotent $p \in C(K)''$ with $p \leq \pi$ such that*

$$p \cdot C(K)' = C(K)_* = C(K)'_n \text{ and } p \circ E^{\sim\sim} = (E^{\sim})_n^{\sim}.$$

(4) *$E^{\sim\sim} = (E^{\sim})_n^{\sim}$ if and only if $p = \pi$.*

Proof. (1) Since $E^{\sim}$ is Dedekind complete, $Z(E^{\sim}) = C(K)$ is topologically full and K is a Stonian compact Hausdorff space. (It is well known that K is hyperstonian. We include a proof for the sake of completeness.) To show that K is hyperstonian, it is sufficient to see that the order-continuous linear functionals on $C(K)$ separate the points of $C(K)$ [12]. Consider $E \subset (E^{\sim})_n^{\sim} \subset E^{\sim\sim}$. Take positive elements $x \in E$, $f \in E^{\sim}$ and $a_\tau, a \in C(K)$ such that $\{a_\tau\}$ is an increasing net with $\sup a_\tau = a$ in $C(K)$. Then, since $E^{\sim}$ is Dedekind complete, $\sup a_\tau f = af$ in $E^{\sim}$. Therefore $a_\tau f(x) \uparrow af(x)$, since x is an order-continuous linear functional on $E^{\sim}$. Consider $\mu_{f,x} \in C(K)'$ in the definition process of m, we have $\mu_{f,x}(b) = x(bf) = bf(x)$ for all

$b \in C(K)$. Hence it follows that $\mu_{f,x} \in C(K)'_n$ for all $x \in E$ and $f \in E^\sim$. Also it is clear that these linear functionals separate the points of the center $Z(E^\sim) = C(K)$. Therefore K is hyperstonian and $C(K)_* = C(K)'_n$ is the predual of $C(K)$.

(2) The existence of π is clear from Proposition 3.6. An equivalent means of defining $\pi \in C(K)''$ is by observing that π is the supremum of the band projections on $C(K)'$ obtained by considering the bands generated by each linear functional of the form $\mu_{f,x''} \in C(K)'$ when $f \in E^\sim$ and $x'' \in E^{\sim\sim}$.

(3) Let $p \in C(K)'' = Z(C(K)')$ be the band projection onto the band $C(K)'_n = C(K)_*$. An equivalent means of defining p would be to observe that p is the supremum of the band projections on $C(K)'$ obtained by considering the bands generated by each linear functional of the form $\mu_{f,x} \in C(K)'$ when $f \in E^\sim$ and $x \in E \subset E^{\sim\sim}$. Hence we have $p \leq \pi$. It remains to show that $p \circ E^{\sim\sim} = (E^\sim)^\sim_n$. Note that for each $f \in E^\sim$ and each $x'' \in E^{\sim\sim}$, we have $\mu_{f,p\circ x''} = p \cdot \mu_{f,x''}$. (Namely, let $\{a_\alpha\}$ be a net in $C(K)$ that converges to p in $\sigma(C(K)'', C(K)')$-topology. Then

$$\mu_{f,p\circ x''}(a) = p \circ x''(af) = p(\mu_{af,x''}) = \lim_\alpha \mu_{af,x''}(a_\alpha) = \lim_\alpha x''(a_\alpha af)$$

and

$$p \cdot \mu_{f,x''}(a) = p(\mu_{f,x''} \cdot a) = \lim_\alpha \mu_{f,x''} \cdot a(a_\alpha) = \lim_\alpha \mu_{f,x''}(aa_\alpha) = \lim_\alpha x''(a_\alpha af)$$

for each $a \in C(K)$. Here the second set of displayed equalities follows from the definition of the Arens product on the bidual of $C(K)$ when $C(K)$ is considered as a unital f-algebra [8].) But $p \cdot \mu_{f,x''} \in C(K)_* = C(K)'_n$ for each $f \in E^\sim$. By reversing the process we used in part (1), it follows that $p \circ x'' \in (E^\sim)^\sim_n$. Conversely if $x'' \in (E^\sim)^\sim_n$, the process we used in part (1) shows that $\mu_{f,x''} \in C(K)_*$. Therefore

$$\mu_{f,x''} = p \cdot \mu_{f,x''} = \mu_{f,p\circ x''}$$

for each $f \in E^\sim$. That is, $p \circ x'' = x''$ for all $x'' \in (E^\sim)^\sim_n$. So $p \circ E^{\sim\sim} = (E^\sim)^\sim_n$.

Now (4) is clear from parts (2) and (3). □

Remark 3.16. The article [8] has initiated considerable research on the Arens product on the biduals of lattice ordered algebras, we include a partial list [3, 6, 7, 13, 14].

References

[1] C.D. Aliprantis and O. Burkinsaw, *Positive Operators.* Academic Press, London, 1985

[2] R. Arens, *Operations induced in function classes.* Monatsh. Math. **55** (1951), 1–19.

[3] S.J. Bernau and C.B. Huijsmans, *The order bidual of almost f-algebras and d-algebras.* Trans. Amer. Math. Soc. **347** (1995), 4259–4275.

[4] K. Boulabiar, and J. Jaber, *The order bidual of f-algebras revisited.* Positivity **15** (2011), 271–279.

[5] A. Goullet de Rugy, *La structure idéale des M-espaces.* J. Math. Pures Appl. **55** (1972), 331–373.

[6] J.J. Grobler, *Commutativity of the Arens product in lattice ordered algebras.* Positivity **3** (1999), 357–364.

[7] C.B. Huijsmans, *The order bidual of lattice ordered algebras. II.* J. Operator Theory **22** (1989), 277–290.

[8] C.B. Huijsmans and B. de Pagter, *The order bidual of lattice ordered algebras.* J. Func. Anal. **59** (1984), 41–64.

[9] C.B. Huijsmans and B. de Pagter, *Subalgebras and Riesz subspaces of an f-algebra.* Proc. London Math. Soc. **48** (1984), 41–64.

[10] J. Jaber, *f-algebras with a σ-bounded approximate unit.* Positivity **18** (2014), 161–170.

[11] M. Orhon, *The ideal center of the dual of a Banach lattice.* Positivity **14** (2010), 841–847.

[12] H.H. Schaefer, *Banach lattices and positive operators.* Springer, Berlin, 1974.

[13] E. Scheffold, *Der Bidual von F-Banachverbandsalgebren.* Acta Sci. Math. (Szeged) **55** (1991), 167–179.

[14] E. Scheffold, *Über symmetrische Operatoren auf Banachverbänden und Arens-Regularität.* Czechoslovak Math. J. **48** (123) (1998), 747–753.

[15] A.W. Wickstead, *The structure space of a Banach lattice.* J. Math. Pures Appl. **56** (1977), 39–54.

[16] A.W. Wickstead, *Representation and duality of multiplication operators on archimedean Riesz spaces.* Comp. Math. **35** (1977), 225–238.

[17] A.W. Wickstead, *Extremal structures of cones of operators.* Quart. J. Math. Oxford (2). **32** (1981), 239–253.

[18] A.W. Wickstead, *Banach lattices with trivial centre.* Proc. R. Ir. Acad. **88A** (1988), 71–83.

[19] A.W. Wickstead, *Banach lattices with topologically full centre.* Vladikavkaz Mat. Zh. **11** (2009), 50–60.

[20] A.C. Zaanen, *Riesz Spaces II.* North-Holland, Amsterdam, 1983.

Şafak Alpay
Mathematics Department
Middle East Technical University
06531 Ankara, Turkiye
e-mail: `safak@metu.edu.tr`

Mehmet Orhon
Department of Mathematics and Statistics
University of New Hampshire
Durham, NH 03824, USA
e-mail: `mo@unh.edu`

Ordered Structures and Applications: Positivity VII
Trends in Mathematics, 51–75

Reproducing Kernels and Positivity of Vector Bundles in Infinite Dimensions

Daniel Beltiţă and José E. Galé

Abstract. We investigate the interaction between the existence of reproducing kernels on infinite-dimensional Hermitian vector bundles and the positivity properties of the corresponding bundles. The positivity refers to the curvature form of certain covariant derivatives associated to reproducing kernels on the vector bundles under consideration. The values of the curvature form are Hilbert space operators, and its positivity is thus understood in the usual sense from operator theory.

Mathematics Subject Classification (2010). Primary 46E22; Secondary 47B32, 46L05, 18A05, 58B12.

Keywords. Vector bundle, reproducing kernel, covariant derivative, Griffiths positivity.

1. Introduction

The idea of positivity plays a central role in Hilbert space operator theory, see for instance the whole panel of constructions of Gelfand–Naimark–Segal type, in particular dilation theory of completely positive maps or the theory of reproducing kernel Hilbert spaces.

On the other hand, ideas of positivity and order structures also hold a quite important place in branches of mathematics which might seem to be remote from operator theory, as it is the case with complex algebraic geometry; see for instance the impressive two-volume treatise [Lz04], which was devoted to a thorough discussion of that topic. We will show in this paper that these ideas of positivity

This research was partly supported by Project MTM2013-42105-P, DGI-FEDER, of the MEYC, Spain. Partial financial support is also acknowledged by the first-named author from the Grant of the Romanian National Authority for Scientific Research, CNCS-UEFISCDI, project number PN-II-ID-PCE-2011-3-0131, and by the second-named author from Project E-64, D.G. Aragón, Spain.

are actually rather close to each other, which sheds a fresh light on the relationship between the aforementioned theories and raises several interesting problems in both these research areas.

It is to be also mentioned at this point that the interaction between complex geometry and operator theory has already surfaced in the literature – as for instance in the famous theory of Cowen–Douglas operators ([CD78]), or in the method introduced in [Od88], [Od92] to approach the study of physical systems – without a particular emphasis on the concept of positivity. As regards the specific topic of the present paper, we will discuss the relationship between the Griffiths positivity of holomorphic vector bundles ([Gr69], [GH78]) and the reproducing kernels on infinite-dimensional vector bundles that we have studied recently ([BG08], [BG09], [BG11], [BG13]).

Reproducing kernels and covariant derivatives often occur simultaneously on the vector bundles involved in various problems in areas such as geometric quantization, geometric representation theory of Lie groups, theory of Cowen–Douglas operators, etc. This simple remark additionally motivated the present note, as we tried to provide an explanation for the aforementioned occurrence by using the universality properties of the reproducing kernels established in our previous paper [BG11]. A related issue is the infinite-dimensional extension of the Chern correspondence between the connections and the (almost) complex structures on the total space of a bundle, that we plan to consider in a forthcoming work; see Problem 3.17 below for some more details and references.

The simplest setting that illustrates that idea is provided by the *tautological vector bundle* $\Pi_{\mathcal{H}}$ corresponding to any complex Hilbert space $\mathcal{H}$. Let $\mathrm{Gr}(\mathcal{H})$ be the Grassmann manifold, whose points are the closed linear subspaces of $\mathcal{H}$, and define $\mathcal{T}(\mathcal{H}) = \{(\mathcal{S}, x) \in \mathrm{Gr}(\mathcal{H}) \times \mathcal{H} \mid x \in \mathcal{S}\}$ and $\Pi_{\mathcal{H}} \colon \mathcal{T}(\mathcal{H}) \to \mathrm{Gr}(\mathcal{H})$, $\Pi_{\mathcal{H}}(\mathcal{S}, x) = \mathcal{S}$. Both $\mathrm{Gr}(\mathcal{H})$ and $\mathcal{T}(\mathcal{H})$ are complex Banach manifolds and the mapping $\Pi_{\mathcal{H}}$ is a Hermitian holomorphic vector bundle. For the purposes of the present investigation it is important to single out two additional structures on this bundle:

(1) Let $p_1, p_2 \colon \mathrm{Gr}(\mathcal{H}) \times \mathrm{Gr}(\mathcal{H}) \to \mathrm{Gr}(\mathcal{H})$ the natural Cartesian projections. For any $\mathcal{S} \in \mathrm{Gr}(\mathcal{H})$ let us also denote by $p_{\mathcal{S}}$ the orthogonal projection of $\mathcal{H}$ onto $\mathcal{S}$. Then we can define

$$Q_{\mathcal{H}} \colon \mathrm{Gr}(\mathcal{H}) \times \mathrm{Gr}(\mathcal{H}) \to \mathrm{Hom}(p_2^*(\Pi_{\mathcal{H}}), p_1^*(\Pi_{\mathcal{H}})),$$
$$Q_{\mathcal{H}}(\mathcal{S}_1, \mathcal{S}_2) = (p_{\mathcal{S}_1})|_{\mathcal{S}_2} \colon \mathcal{S}_2 \to \mathcal{S}_1.$$

The mapping $Q_{\mathcal{H}}$ is the *universal reproducing kernel* associated with the Hilbert space $\mathcal{H}$ (see [BG11] and Example 2.5 below).

(2) Just as in the finite-dimensional case, there exists a unique *Chern connection* on $\Pi_{\mathcal{H}}$, that is, a linear connection which is compatible in the usual sense both with the holomorphic structure and with the Hermitian structure defined by the inner product on every fiber $\mathcal{S}$ (see Theorem 3.15 below).

Our present paper belongs to a project devoted to understanding the relationship between the above universal reproducing kernel and linear connections and how their interaction propagates from the tautological bundle, to a fairly wide class of reproducing kernels on Hermitian vector bundles, in a sense that will be made precise below. We already pointed out in [BG13] a certain functorial correspondence between reproducing kernels and linear connections on infinite-dimensional vector bundles, and in the present paper we take up the study of that circle of ideas from the perspective of the complex algebraic geometry, which eventually amounts to curvature positivity properties of the vector bundles involved in this discussion.

We will briefly survey some facts from our earlier papers [BG08], [BG09], [BG11], and [BG13], and will also raise some problems and announce a few partial results with sketchy proofs. Full details of these proofs will be published elsewhere.

The paper is structured as follows. In Section 2 we provide some background information on reproducing kernels on vector bundles along with some basic examples. Section 3 is devoted to a discussion of Chern covariant derivatives on Hermitian holomorphic vector bundles, which are covariant derivatives that are compatible both with the Hermitian and with the holomorphic structures of these bundles. In Section 4 we introduce the Griffith positivity condition for bundle-valued differential 2-forms in our setting of infinite-dimensional bundles. In Section 5 we finally relate the reproducing kernels to the Griffith positivity property of Hermitian vector bundles. The paper concludes by the Appendix which includes some basic observations on vector-valued differential forms, which are needed in the main body of our paper and were collected there for the reader's convenience.

2. Reproducing kernels on Hermitian vector bundles

Definition 2.1. Let Z be a Banach manifold. A *Hermitian structure* on a smooth Banach vector bundle $\Pi\colon D \to Z$ is a family $\{(\cdot \mid \cdot)_z\}_{z\in Z}$ with the following properties:

(a) For every $z \in Z$, $(\cdot \mid \cdot)_z\colon D_z \times D_z \to \mathbb{C}$ is a scalar product ($\mathbb{C}$-linear in the first variable) that turns the fiber D_z into a complex Hilbert space.

(b) If V is any open subset of Z, and $\Psi_V\colon V \times \mathcal{E} \to \Pi^{-1}(V)$ is a trivialization (whose typical fiber is the complex Hilbert space $\mathcal{E}$) of the vector bundle Π over V, then the function

$$(z,x,y) \mapsto (\Psi_V(z,x) \mid \Psi_V(z,y))_z, \quad V \times \mathcal{E} \times \mathcal{E} \to \mathbb{C}$$

is smooth.

A *Hermitian bundle* is a bundle endowed with a Hermitian structure as above.

Definition 2.2. Let $\Pi\colon D \to Z$ be a Hermitian bundle, and $p_1, p_2\colon Z \times Z \to Z$ be the Cartesian projections. A *reproducing kernel* on Π is a continuous section of the bundle $\mathrm{Hom}(\mathrm{p}_2^*\Pi, \mathrm{p}_1^*\Pi) \to \mathrm{Z} \times \mathrm{Z}$ such that the mappings $K(s,t)\colon D_t \to D_s$ ($s,t \in Z$) are bounded linear operators and such that K is positive definite in the following sense: For every $n \geq 1$ and $t_j \in Z$, $\eta_j \in D_{t_j}$ $(j = 1, \ldots, n)$,

$$\sum_{j,l=1}^{n} \left(K(t_l, t_j)\eta_j \mid \eta_l\right)_{t_l} \geq 0. \tag{2.1}$$

For every $\xi \in D$ we set

$$K_\xi := K(\cdot, \Pi(\xi))\xi \colon Z \to D$$

which is a section of the bundle Π. For $\xi, \eta \in D$, the prescriptions

$$(K_\xi \mid K_\eta)_{\mathcal{H}^K} := (K(\Pi(\eta), \Pi(\xi))\xi \mid \eta)_{\Pi(\eta)}, \tag{2.2}$$

define an inner product $(\cdot \mid \cdot)_{\mathcal{H}^K}$ on $\mathrm{span}\{K_\xi : \xi \in D\}$ whose completion gives rise to a Hilbert space denoted by $\mathcal{H}^K$, which consists of sections of the bundle Π. We also define the mappings

$$\widehat{K} \colon D \to \mathcal{H}^K, \quad \widehat{K}(\xi) = K_\xi,$$
$$\zeta_K \colon Z \to \mathrm{Gr}(\mathcal{H}^K), \quad \zeta_K(s) = \overline{\widehat{K}(D_s)},$$

where $\mathrm{Gr}(\mathcal{H}^K)$ is the Grassmann manifold of all closed subspaces of $\mathcal{H}^K$ and the bar over $\widehat{K}(D_s)$ indicates the topological closure. See [BG08] for details.

Example 2.3 (trivial bundles). Let Z be any Banach manifold (for instance any open subset of some real Banach space) and $\mathcal{E}$ be any complex Hilbert space and define the trivial bundle

$$\Pi \colon D = Z \times \mathcal{E} \to Z, \quad (z, x) \mapsto z.$$

For every $z \in Z$ we have $D_z = \{z\} \times \mathcal{E}$ and moreover there exists a one-to-one correspondence $\sigma \mapsto f_\sigma$ sections σ of Π and $\mathcal{E}$-valued functions F_σ on Z given by

$$(\forall z \in Z) \quad \sigma(z) = (z, F_\sigma(z)).$$

Denote by $\mathrm{GL}^+(\mathcal{E})$ the set of positive invertible operators on $\mathcal{E}$, which is an open subset of the C^*-algebra $\mathcal{B}(\mathcal{E})$. Then there exists a one-to-one correspondence between the Hermitian structures on Π and the smooth mappings $h \colon Z \to \mathrm{GL}^+(\mathcal{E})$ given by

$$(\forall z \in Z) \quad (\cdot \mid \cdot)_z \colon D_z \times D_z \to \mathbb{C}, \quad ((z, x_1) \mid (z, x_2))_z := (h(z)x_1 \mid x_2)_{\mathcal{E}}.$$

Also, there exists a one-to-one correspondence between the reproducing kernels K on Π and the $\mathcal{B}(\mathcal{E})$-valued reproducing kernels κ on Z (see [Ne00]) by

$$(\forall z_1, z_2 \in Z)(\forall x \in \mathcal{E}) \quad K(z_1, z_2) \colon D_{z_2} \to D_{z_1}, \quad (z_2, x) \mapsto (z_1, \kappa(z_1, z_2)x).$$

Example 2.4 (homogeneous bundles). Let G_A be a Banach–Lie group with a Banach–Lie subgroup G_B. Let $\rho_A \colon G_A \to \mathcal{B}(\mathcal{H}_A)$ and $\rho_B \colon G_B \to \mathcal{B}(\mathcal{H}_B)$ be uniformly continuous unitary representations with $\mathcal{H}_B \subseteq \mathcal{H}_A$, $\rho_B(u) = \rho_A(u)|_{\mathcal{H}_B}$ for $u \in G_B$ and $\mathcal{H}_A = \overline{\mathrm{span}}\rho_A(G_A)\mathcal{H}_B$.

Let us consider the homogeneous vector bundle

$$\Pi_\rho \colon G_A \times_{G_B} \mathcal{H}_B \to G_A/G_B$$

induced by the representation ρ_B. Recall that $\mathrm{G}_A \times_{G_B} \mathcal{H}_B$ is the Cartesian product $\mathrm{G}_A \times \mathcal{H}_B$ modulo the equivalence relation defined by

$$(u,h) \sim (u',h') \iff (\exists w \in G_B) \quad u' = uw, \; h' = \rho(w^{-1})h,$$

endowed with its canonical structure of a Banach manifold; see [KM97].

We provide Π_ρ with the Hermitian structure given by

$$([(u,f)],[(u,h)])_s := (f \mid h)_{\mathcal{H}}, \quad u \in G_A, s := uG_B, f,h \in \mathcal{H}_B.$$

Let $P\colon \mathcal{H}_A \to \mathcal{H}_B$ be the orthogonal projection.

We define the reproducing kernel K_ρ on the homogeneous Hermitian vector bundle $\Pi_\rho\colon D = G_A \times_{G_B} \mathcal{H}_B \to G_A/G_B$ by

$$K_\rho(uG_B, vG_B)[(v,f)] = [(u, P(\rho_A(u^{-1})\rho_A(v)f))], \tag{2.3}$$

for $uG_B, vG_B \in D$ and $f \in \mathcal{H}_B$ (see [BG08]).

There exists a unitary operator $W\colon \mathcal{H}^{K_\rho} \to \mathcal{H}_A$, $W(K_\eta) = \pi_A(v)f$ if $\eta = [(v,f)] \in D$; see the end of the proof of [BG08, Proposition 4.1].

Example 2.5 (tautological bundles). In Example 2.4 assume $G_A = U(\mathcal{H}_A)$ with the tautological representation ρ_A, and

$$G_B = \{u \in U(\mathcal{H}_A) \mid u(\mathcal{H}_B) \subseteq \mathcal{H}_B\} \simeq U(\mathcal{H}_B) \times U(\mathcal{H}_B^\perp).$$

Denote $\mathrm{Gr}_{\mathcal{H}_B}(\mathcal{H}_A) := \{u(\mathcal{H}_B) \mid u \in U(\mathcal{H}_A)\}$, and

$$\mathcal{T}_{\mathcal{H}_B}(\mathcal{H}_A) = \{(u(\mathcal{H}_B), x) \mid u \in U(\mathcal{H}_A), x \in u\mathcal{H}_B\} \subseteq \mathrm{Gr}_{\mathcal{H}_B}(\mathcal{H}_A) \times \mathcal{H}_A.$$

Then the pair of maps

$$\begin{aligned} G_A \times_{G_B} \mathcal{H}_B \to \mathcal{T}_{\mathcal{H}_B}(\mathcal{H}_A), &\quad [(u,x)] \mapsto (u(\mathcal{H}_B), u(x)),\\ G_A/G_B \to \mathrm{Gr}_{\mathcal{H}_B}(\mathcal{H}_A), &\quad uG_B \mapsto u(\mathcal{H}_B) \end{aligned}$$

defines an isomorphism of the vector bundle $\Pi_\rho\colon G_A \times_{G_B} \mathcal{H}_B \to G_A/G_B$ onto the *tautological bundle* $\Pi_{\mathcal{H}_A,\mathcal{H}_B}\colon \mathcal{T}_{\mathcal{H}_B}(\mathcal{H}_A) \to \mathrm{Gr}_{\mathcal{H}_B}(\mathcal{H}_A)$, $(\mathcal{S},x) \mapsto \mathcal{S}$. See [BG09] for some more details.

Now let $p_1, p_2\colon \mathrm{Gr}_{\mathcal{H}_B}(\mathcal{H}_A) \times \mathrm{Gr}_{\mathcal{H}_B}(\mathcal{H}_A) \to \mathrm{Gr}_{\mathcal{H}_B}(\mathcal{H}_A)$ be the natural Cartesian projections.

For any $\mathcal{S} \in \mathrm{Gr}_{\mathcal{H}_B}(\mathcal{H}_A)$ let us also denote by $p_{\mathcal{S}}\colon \mathcal{H}_A \to \mathcal{S}$ the corresponding orthogonal projection, whose adjoint operator is the inclusion map $p_{\mathcal{S}}^* = \iota_{\mathcal{S}}\colon \mathcal{S} \hookrightarrow \mathcal{H}_A$. Then we can define

$$\begin{aligned} &Q_{\mathcal{H}_A,\mathcal{H}_B}\colon \mathrm{Gr}_{\mathcal{H}_B}(\mathcal{H}_A) \times \mathrm{Gr}_{\mathcal{H}_B}(\mathcal{H}_A) \to \mathrm{Hom}(p_2^*(\Pi_{\mathcal{H}_A,\mathcal{H}_B}), p_1^*(\Pi_{\mathcal{H}_A,\mathcal{H}_B})),\\ &Q_{\mathcal{H}_A,\mathcal{H}_B}(\mathcal{S}_1,\mathcal{S}_2) = p_{\mathcal{S}_1}p_{\mathcal{S}_2}^* = (p_{\mathcal{S}_1})|_{\mathcal{S}_2}\colon \mathcal{S}_2 \to \mathcal{S}_1. \end{aligned}$$

The mapping $Q_{\mathcal{H}_A,\mathcal{H}_B}$ is the *universal reproducing kernel* corresponding to the Hilbert space $\mathcal{H}_A$ and its closed subspace $\mathcal{H}_B$; see [BG11]. Note that $Q_{\mathcal{H}_A,\mathcal{H}_B}$ actually depends on $\mathrm{Gr}_{\mathcal{H}_B}(\mathcal{H}_A)$ and not on $\mathcal{H}_B$.

3. Covariant derivatives on Hermitian vector bundles

This section contains the main tools that allow us to describe positivity properties of vector bundles. These tools are the covariant derivatives and their curvatures. As we will see, in the case of holomorphic vector bundles, the existence of nontrivial global holomorphic cross-sections entails a positive curvature property. The conclusion of all that will be that a certain *intrinsic positivity property* is necessary in order that a holomorphic vector bundle admits nontrivial reproducing kernels that give rise to Hilbert spaces of holomorphic cross-sections.

Covariant derivative and curvature

Definition 3.1. We first define the covariant derivatives on trivial vector bundles. So assume X is an open subset of any real Banach space $\mathcal{X}$, and let $\mathcal{V}$ be another real Banach space. A *linear connection form* on the trivial bundle $X \times \mathcal{V} \to X$, $(x, v) \mapsto x$, is any 1-form $A \in \Omega^1(X, \mathcal{B}(\mathcal{V}))$. The value of A at any point $x \in X$ is denoted by $A_x \in \mathcal{B}(\mathcal{X}, \mathcal{B}(\mathcal{V}))$ and we are going to use freely the natural topological isomorphisms

$$\mathcal{B}(\mathcal{X}, \mathcal{B}(\mathcal{V})) \simeq \mathcal{B}(\mathcal{X}, \mathcal{V}; \mathcal{V}) \simeq \mathcal{B}(\mathcal{X} \widehat{\otimes} \mathcal{V}, \mathcal{V})$$

where $\mathcal{B}(\mathcal{X}, \mathcal{V}; \mathcal{V})$ stands for the space of bounded bilinear maps from $\mathcal{X} \times \mathcal{V}$ into $\mathcal{V}$ and $\widehat{\otimes}$ denotes the projective tensor product of Banach spaces.

The *covariant derivative* corresponding to the above linear connection form is the sequence of linear operators $\nabla \colon \Omega^p(X, \mathcal{V}) \to \Omega^{p+1}(X, \mathcal{V})$ defined for $p = 0, 1, 2, \dots$ by

$$\nabla \sigma = \mathrm{d}\sigma + A \wedge \sigma$$

for every $\sigma \in \Omega^p(X, \mathcal{V})$, where the wedge product

$$\wedge \colon \Omega^1(X, \mathcal{B}(\mathcal{V})) \times \Omega^p(X, \mathcal{V}) \to \Omega^{p+1}(X, \mathcal{V})$$

is defined (Definition A.2) by using the natural bilinear map $\mathcal{B}(\mathcal{V}) \times \mathcal{V} \to \mathcal{V}$ given by the action of the operators in $\mathcal{B}(\mathcal{V})$ on $\mathcal{V}$.

If $\Pi \colon D \to Z$ is any (locally trivial) vector bundle, then for every $p = 0, 1, 2, \dots$ we define $\mathrm{Hom}(\wedge^p \tau_Z, \Pi)$ as the vector bundle over Z whose fiber over $z \in Z$ is the space $\mathcal{B}(\wedge T_z Z, D_z)$ of all bounded skew-symmetric p-linear maps $T_z Z \times \cdots \times T_z Z \to D_z$ (see the Appendix below). We denote by $\Omega^p(Z, D)$ the space of all locally defined smooth sections of $\mathrm{Hom}(\wedge^p \tau_Z, \Pi)$.

A *covariant derivative* on the vector bundle Π is any sequence of operators $\nabla \colon \Omega^p(Z, D) \to \Omega^{p+1}(Z, D)$ for $p = 0, 1, 2, \dots$ which can be expressed in terms of connection forms as above in any local trivialization of Π.

Remark 3.2. In Definition 3.1 we have

$$\begin{aligned}(\nabla \sigma)_x(x_1, \dots, x_{p+1}) =& (\mathrm{d}_x \sigma)(x_1, \dots, x_{p+1}) \\ &+ \sum_{j=1}^{p+1} \underbrace{A_x(x_j)}_{\in \mathcal{B}(\mathcal{V})} \underbrace{\sigma_x(x_1, \dots, x_{j-1}, x_{j+1}, \dots, x_{p+1})}_{\in \mathcal{V}}\end{aligned}$$

for every $\sigma \in \Omega^p(X, \mathcal{V})$, $x \in X$, and $x_1, \dots, x_{p+1} \in \mathcal{X}$. In particular, if $p = 0$, then $\sigma \in \mathcal{C}^\infty(X, \mathcal{V})$ and the 1-form $\nabla_\sigma \in \Omega^1(X, \mathcal{V})$ is given by

$$(\nabla\sigma)_x(x_1) = (\mathrm{d}_x\sigma)(x_1) + A_x(x_1)\sigma_x$$

for every $x \in X$ and $x_1 \in \mathcal{X}$.

Definition 3.3. Assume the setting of Definition 3.1. The *curvature form* corresponding to the linear connection form $A \in \Omega^1(X, \mathcal{B}(\mathcal{V}))$ is

$$\Theta := \mathrm{d}A + A \wedge A \in \Omega^2(X, \mathcal{B}(\mathcal{V}))$$

where the wedge product

$$\wedge \colon \Omega^1(X, \mathcal{B}(\mathcal{V})) \times \Omega^1(X, \mathcal{B}(\mathcal{V})) \to \Omega^2(X, \mathcal{B}(\mathcal{V}))$$

is defined (see Definition A.2) via the bilinear map $\mathcal{B}(\mathcal{V}) \times \mathcal{B}(\mathcal{V}) \to \mathcal{B}(\mathcal{V})$ given by product of operators in $\mathcal{B}(\mathcal{V})$.

If $\Pi \colon D \to Z$ is an arbitrary (locally trivial) vector bundle with covariant derivative $\nabla \colon \Omega^p(Z, D) \to \Omega^{p+1}(Z, D)$ defined for $p = 0, 1, 2, \dots$, then the curvature forms defined as above in local trivializations can be glued together into a global curvature form $\Theta \in \Omega^2(Z, \mathrm{End}\,(\Pi))$. The covariant derivative is said to be *flat* if its curvature is $\Theta = 0$.

Remark 3.4. In the notation of Definition 3.3, for every $z \in Z$ we have the skew-symmetric bilinear map $\Theta_z \colon T_zZ \times T_zZ \to \mathcal{B}(D_z)$. Moreover, for every $p \geq 0$ and $\sigma \in \Omega^p(Z, D)$ we have $\nabla(\nabla\sigma) = \Theta \wedge \sigma$. Hence the covariant derivative ∇ is flat if and only if $\nabla^2 = 0$.

Covariant derivatives compatible with Hermitian structures

Definition 3.5. Let $\Pi \colon D \to Z$ be any Hermitian vector bundle. We say that a covariant derivative ∇ on Π is *compatible with the Hermitian structure* if it satisfies the following condition: For any open set $W \subseteq Z$, if $\sigma_1, \sigma_2 \colon W \to D$ are smooth cross-sections of the bundle Π, then

$$\mathrm{d}(\sigma_1 \mid \sigma_2) = (\nabla\sigma_1 \mid \sigma_2) + (\sigma_1 \mid \nabla\sigma_2) \in \Omega^1(W, \mathbb{C})$$

that is, we have $(\mathrm{d}_z(\sigma_1 \mid \sigma_2))(x) = ((\nabla\sigma_1)(x) \mid \sigma_2(z))_z + (\sigma_1(z) \mid (\nabla\sigma_2)(x))_z$ for all $z \in W$ and $x \in T_zZ$. This condition has a local character, so it suffices to check it in local trivializations of the bundle Π.

Remark 3.6. Assume that X is an open set in the real Banach space $\mathcal{X}$ and $\mathcal{E}$ is any complex Hilbert space. A Hermitian structure on $\mathcal{E}$ is then the same thing as a smooth mapping $h \colon X \to \mathrm{GL}^+(\mathcal{E})$. If we define the mapping

$$(\cdot \mid \cdot)_x \colon \mathcal{E} \times \mathcal{E} \to \mathbb{C}, \quad (v_1 \mid v_2)_x = (h(x)v_1 \mid v_2)$$

for every $x \in X$, then we can note the following:

1. For every $x \in X$ the mapping $(\cdot \mid \cdot)_x$ is a scalar product compatible with the topology of $\mathcal{E}$ since $h_x \in \mathrm{GL}^+(\mathcal{E})$.

2. Since the Hermitian structure h is smooth, we can define a natural sesquilinear map
$$h(\cdot,\cdot)\colon \mathcal{C}^\infty(X,\mathcal{E})\times\mathcal{C}^\infty(X,\mathcal{E})\to\mathcal{C}^\infty(X,\mathbb{C})$$
such that $(h(\phi,\psi))_x=(\phi_x\mid\psi_x)_x$ for all $\phi,\psi\in\mathcal{C}^\infty(X,\mathcal{E})$ and $x\in X$.
3. We can also define a sesquilinear map denoted in the same way,
$$h(\cdot,\cdot)\colon \Omega^1(X,\mathcal{E})\times\mathcal{C}^\infty(X,\mathcal{E})\to\Omega^1(X,\mathbb{C})$$
such that for all $\sigma\in\Omega^1(X,\mathcal{E})$, $\psi\in\mathcal{C}^\infty(X,\mathcal{E})$, and $x\in X$ we have
$$(h(\sigma,\psi))_x=(\sigma_x(\cdot)\mid\psi_x)_x=(h_x\sigma_x(\cdot)\mid\psi_x)\colon \mathcal{X}\to\mathbb{C},$$
where we recall that $\sigma_x\in\mathcal{B}(\mathcal{X},\mathcal{E})$. Similarly, one defines a sesquilinear map
$$h(\cdot,\cdot)\colon \mathcal{C}^\infty(X,\mathcal{E})\times\Omega^1(X,\mathcal{E})\to\Omega^1(X,\mathbb{C})$$
such that
$$\begin{aligned}(h(\psi,\sigma))_x(v)&=\overline{h(\sigma,\psi)_x(y)}=\overline{(\sigma_x(y)\mid\psi_x)_x}=\overline{(h_x\sigma_x(y)\mid\psi_x)}\\&=(\psi_x\mid h_x\sigma_x(y))\end{aligned}$$
for all $\sigma\in\Omega^1(X,\mathcal{E})$, $\psi\in\mathcal{C}^\infty(X,\mathcal{E})$, $y\in\mathcal{X}$, and $x\in X$.

The following result is suggested by the classical situation of finite-dimensional bundles; see for instance the computations prior to [We08, Ch. III, Prop. 1.11].

Proposition 3.7. *In the setting of Remark* 3.6, *assume that a linear connection form $A\in\Omega^1(X,\mathcal{B}(\mathcal{E}))$ is also given. Then the following assertions are equivalent:*

1. *We have*
$$\mathrm{d}(h(\phi,\psi))=h(\nabla\phi,\psi)+h(\phi,\nabla\psi)$$
for all $\phi,\psi\in\mathcal{C}^\infty(X,\mathcal{E})$.
2. *The equation*
$$\mathrm{d}_x h=h_xA_x(\cdot)+A_x(\cdot)^*h_x\in\mathcal{B}(\mathcal{X},\mathcal{B}(\mathcal{E}))$$
is satisfied for every $x\in X$.

Proof. Since the mapping $\mathcal{B}(\mathcal{E})\times\mathcal{E}\times\mathcal{E}\to\mathbb{C}$, $(T,v,w)\mapsto(Tv\mid w)$ is trilinear and continuous, it follows by the product rule of differentiation (see for instance [Nl69, Ch. 1, Th. 1]) that for all $\phi,\psi\in\mathcal{C}^\infty(X,\mathcal{E})$ and $x\in X$ we have the following equalities in $\mathcal{B}(\mathcal{X},\mathbb{C})$:
$$\mathrm{d}_x(h(\phi,\psi))=\mathrm{d}_x(h\phi\mid\psi)=(\mathrm{d}_xh(\cdot)\phi_x\mid\psi_x)+(h_x\mathrm{d}_x\phi(\cdot)\mid\psi_x)+(h_x\phi_x\mid\mathrm{d}_x\psi(\cdot)),$$
hence by using the fact that $h_x^*=h_x$ in $\mathcal{B}(\mathcal{E})$ we get
$$\mathrm{d}_x(h(\phi,\psi))=(\mathrm{d}_xh(\cdot)\phi_x\mid\psi_x)+(h(\mathrm{d}\phi,\psi))_x+(h(\phi,\mathrm{d}\psi))_x.$$
Since $\nabla=\mathrm{d}+A$, we get further
$$\begin{aligned}&\mathrm{d}_x((h(\phi,\psi)))-(h(\nabla\phi,\psi))_x-(h(\phi,\nabla\psi))_x\\&=(\mathrm{d}_xh(\cdot)\phi_x\mid\psi_x)-(h(A\wedge\phi,\psi))_x+(h(\phi,A\wedge\psi))_x\\&=(\mathrm{d}_xh(\cdot)\phi_x\mid\psi_x)-(h_xA_x(\cdot)\phi_x\mid\psi_x)-(h_x\phi_x\mid A_x(\cdot)\psi_x)\\&=(\mathrm{d}_xh(\cdot)\phi_x-h_xA_x(\cdot)\phi_x-A_x(\cdot)^*h_x\phi_x\mid\psi_x).\end{aligned}$$

With this equality at hand, it follows at once that the assertions in the statement are equivalent to each other. □

Linear connections compatible with complex structures

Definition 3.8. Assume X is any open subset of some complex Banach space $\mathcal{X}$ and $\mathcal{E}$ is another complex Banach space. Let $A \in \Omega^1(X, \mathcal{B}(\mathcal{E}))$ be any connection form, hence $A\colon TX = X \times \mathcal{X} \to \mathcal{B}(\mathcal{E})$ is smooth and $\mathbb{R}$-linear in the second variable. Since both $\mathcal{X}$ and $\mathcal{E}$ are complex vector spaces, we can use the direct sum decomposition (A.3) to define the linear operators

$$\nabla'\colon \mathcal{C}^\infty(X, \mathcal{E}) \to \Omega^{(1,0)}(X, \mathcal{E}) \text{ and } \nabla''\colon \mathcal{C}^\infty(X, \mathcal{E}) \to \Omega^{(0,1)}(X, \mathcal{E})$$

such that

$$\nabla = \nabla' + \nabla''$$

where $\nabla\colon \mathcal{C}^\infty(X, \mathcal{E}) \to \Omega^1(X, \mathcal{E})$ is the covariant derivative corresponding to A. So for every $\sigma \in \mathcal{C}^\infty(X, \mathcal{E})$ and $x \in X$ we have

$$(\nabla\sigma)(x) = (\nabla'\sigma)(x) + (\nabla''\sigma)(x),$$

the unique decomposition for which the operator $(\nabla'\sigma)(x)\colon \mathcal{X} \to \mathcal{B}(\mathcal{E})$ is $\mathbb{C}$-linear while $(\nabla''\sigma)(x)\colon \mathcal{X} \to \mathcal{B}(\mathcal{E})$ is conjugate linear.

The following result is suggested by the beginning remark in the proof of [We08, Ch. III, Th. 2.1].

Proposition 3.9. *In the setting of Definition* 3.8, *the following assertions are equivalent:*

1. *For every* $\sigma \in \mathcal{O}(X, \mathcal{E})$ *we have* $\nabla''\sigma = 0$.
2. *We have* $A \in \Omega^{(1,0)}(X, \mathcal{B}(\mathcal{E}))$.

Proof. First note that the 1-form A takes values in the complex vector space $\mathcal{B}(\mathcal{E})$, hence we get a decomposition $A = A^{(1,0)} + A^{(0,1)}$ (see the Appendix), and then condition (2) is equivalent to $A^{(0,1)} = 0$.

If $\sigma \in \mathcal{O}(X, \mathcal{E})$, then

$$\nabla\sigma = \mathrm{d}\sigma + A \wedge \sigma = \partial\sigma + \bar{\partial}\sigma + A \wedge \sigma = \partial\sigma + A \wedge \sigma$$

hence, by Remark A.3(1),

$$\nabla'\sigma = \partial\sigma + A^{(1,0)} \wedge \sigma$$

and

$$\nabla''\sigma = A^{(0,1)} \wedge \sigma.$$

By considering constant $\mathcal{E}$-valued functions on X, we see that $\mathcal{E}$ is generated by the values of functions in $\mathcal{O}(X, \mathcal{E})$. Then the above equality implies that Assertion (1) is equivalent to $A^{(0,1)} = 0$, and this concludes the proof. □

Definition 3.10. If the assertions in Proposition 3.9 are satisfied, then we say that the linear connection corresponding to A is compatible with the complex structures of X and $\mathcal{E}$. More generally, a linear connection on a holomorphic Banach vector bundle is *compatible with the complex structure* if its local connection form in any local holomorphic trivialization is compatible (in the above sense) with the complex structures of the base and the fiber. It is easily seen that this property has a local character and does not depend on the choice of a local holomorphic trivialization.

Chern covariant derivatives

Definition 3.11. A *Hermitian holomorphic vector bundle* is any holomorphic vector bundle $\Pi\colon D \to Z$ with smoothly paracompact base, endowed with a Hermitian structure. In this framework, a *Chern covariant derivative* on Π is any covariant derivative which is compatible both with the complex structure and with the Hermitian structure of the vector bundle Π.

We are now able to prove an infinite-dimensional version of [We08, Ch. III, Th. 2.1] for trivial bundles.

Lemma 3.12. *Let X be any open subset of some complex Banach space $\mathcal{X}$, $\mathcal{E}$ be any complex Hilbert space, and $h\colon X \to \mathrm{GL}^+(\mathcal{E})$ be any smooth mapping. Then there exists a unique connection form $A \in \Omega^1(X, \mathcal{B}(\mathcal{E}))$ that is compatible both with the Hermitian structure given by h and with the complex structures of X and $\mathcal{E}$, and it is given by*

$$A_x = h_x^{-1}(\partial h)_x \tag{3.1}$$

for every $x \in X$.

Proof. We first prove the uniqueness assertion. If A is a connection form that satisfies the compatibility conditions mentioned in the statement, then by Proposition 3.7(2) we obtain

$$(\partial h)_x + (\bar{\partial} h)_x = \mathrm{d}_x h = h_x A_x + A_x^* h_x$$

for every $x \in X$. On the other hand $A \in \Omega^{(1,0)}(X, \mathcal{B}(\mathcal{E}))$ by Proposition 3.9(2), hence the above equation is equivalent to

$$(\partial h)_x = h_x A_x \text{ and } (\bar{\partial} h)_x = A_x^* h_x. \tag{3.2}$$

The first of these equations is clearly equivalent to (3.1).

To prove the existence, just note that the connection form defined by (3.1) belongs to $\Omega^{(1,0)}(X, \mathcal{B}(\mathcal{E}))$, hence it is compatible with the complex structures by Proposition 3.9. On the other hand, if we define A by the formula in the statement and we use the above formulas we see that Assertion (2) of Proposition 3.9 holds true, and then by that proposition we see that the covariant derivative corresponding to the connection form A is compatible with the Hermitian structure as well. □

Before we go further, let us establish the infinite-dimensional version of [We08, Ch. III, Prop. 2.2].

Proposition 3.13. *Assume the setting of Lemma* 3.12 *and let* $\Theta \in \Omega^2(X, \mathcal{B}(\mathcal{E}))$ *be the curvature form corresponding to* A. *Then the following assertions hold:*

1. *We have* $A \in \Omega^{(1,0)}(X, \mathcal{B}(\mathcal{E}))$ *and* $\partial A = -A \wedge A$.
2. *We have* $\Theta = \bar{\partial} A \in \Omega^{(1,1)}(X, \mathcal{B}(\mathcal{E}))$.

Proof. We have $h \cdot h^{-1} = \mathbf{1}$, hence $\partial h \cdot h^{-1} + h\partial(h^{-1}) = 0$. Therefore

$$\partial(h^{-1}) = -h^{-1} \cdot \partial h \cdot h^{-1}$$

and now by (3.1) we get

$$\partial A = \partial(h^{-1} \cdot \partial h) = \partial(h^{-1}) \wedge \partial h + h^{-1} \cdot \partial^2 h = -h^{-1} \cdot \partial h \cdot h^{-1} \wedge \partial h = -A \wedge A$$

We have already seen in Proposition 3.9 that $A \in \Omega^{(1,0)}(X, \mathcal{B}(\mathcal{E}))$. Hence

$$\Theta = \mathrm{d}A + A \wedge A = \bar{\partial} A + \partial A + A \wedge A = \bar{\partial} A \in \Omega^{(1,1)}(X, \mathcal{B}(\mathcal{E}))$$

and this concludes the proof. □

Remark 3.14. As above, let X be any open subset of the complex Banach space $\mathcal{X}$ and $\mathcal{E}$ be any complex Hilbert space. In Proposition 3.13(2), recall that the curvature property $\Theta \in \Omega^{(1,1)}(X, \mathcal{B}(\mathcal{E}))$ means that for every $x \in X$ the map $\Theta_z \colon \mathcal{X} \times \mathcal{X} \to \mathcal{B}(\mathcal{E})$ is sesquilinear (more precisely, is $\mathbb{C}$-linear in the first variable and conjugate linear in the second).

Theorem 3.15. *Every Hermitian holomorphic vector bundle has a unique Chern covariant derivative.*

Proof. The existence in the case of the trivial bundles, as well as the uniqueness in the general case follow by Lemma 3.12, by using a family of local holomorphic trivializations. See for instance [We08] or [De12] for the proof of the existence in the classical situation of finite-dimensional vector bundles. The full details of the proof in the general case will be included in a forthcoming paper. □

Example 3.16. In Example 2.5, the bundle $\Pi_{\mathcal{H}_A,\mathcal{H}_B} \colon \mathcal{T}_{\mathcal{H}_B}(\mathcal{H}_A) \to \mathrm{Gr}_{\mathcal{H}_B}(\mathcal{H}_A)$ is a Hermitian holomorphic vector bundle if $\dim \mathcal{H}_B < \infty$, hence it carries a unique Chern covariant derivative $\nabla_{\mathcal{H}_A,\mathcal{H}_B}$ by Theorem 3.15. See for instance [We08, Ch. III, Ex. 2.4] for more details on that covariant derivative in the case when $\dim \mathcal{H}_A < \infty$.

Problem 3.17. It would be interesting to establish a version of the Koszul–Malgrange integrability theorem of [KM58] (see also [AHS78, Th. 5.1]) for Banach vector bundles (with infinite-dimensional base). Some results in this direction were recently obtained in [DP12] and [Ne13].

Some computations of Chern covariant derivatives. In the following proposition we denote by $\mathfrak{S}_2(\mathcal{H}_1, \mathcal{H}_2)$ the complex Hilbert space consisting of the Hilbert–Schmidt operators from any complex Hilbert space $\mathcal{H}_1$ into another complex Hilbert space $\mathcal{H}_2$, with the usual scalar product on $\mathfrak{S}_2(\mathcal{H}_1, \mathcal{H}_2)$ defined in terms of the operator trace. If $\Pi_j \colon D_j \to Z$ is any Hermitian vector bundle for $j = 1, 2$, then $\mathfrak{S}_2(\Pi_1, \Pi_2) \colon D \to Z$ denotes the Hermitian vector bundle whose fiber over any $z \in Z$ is the space of Hilbert–Schmidt operators $\mathfrak{S}_2(\Pi_1^{-1}(z), \Pi_2^{-1}(z))$. If $\mathcal{E}_j$ is the typical fiber of Π_j for $j = 1, 2$ and $\Phi_j \colon V \to U(\mathcal{E}_j)$ gives a local change of coordinates in Π_j over some open set $V \subseteq Z$, then $\Phi \colon V \to U(\mathfrak{S}_2(\mathcal{E}_1, \mathcal{E}_2))$, $\Phi(z)T = \Phi_2(z)T\Phi_1(z)^{-1}$ for $z \in V$ and $T \in \mathfrak{S}_2(\mathcal{E}_1, \mathcal{E}_2)$ gives a local change of coordinates in $\mathfrak{S}_2(\Pi_1, \Pi_2)$.

Proposition 3.18. *Let $\Pi_j \colon D_j \to Z$ be any Hermitian holomorphic vector bundle with the Chern covariant derivative ∇_j for $j = 1, 2$. Then $\mathfrak{S}_2(\Pi_1, \Pi_2)$ is a Hermitian holomorphic vector bundle with the Chern covariant derivative satisfying $\nabla\Gamma = \nabla_2\Gamma - \Gamma\nabla_1$ for every $\Gamma \in \Omega^0(Z, \mathfrak{S}_2(\Pi_1, \Pi_2))$.*

Proof. The conclusion has a local character hence we may assume for $j = 1, 2$ that $\Pi_j \colon Z \times \mathcal{E}_j \to Z$ is a trivial vector bundle, where $\mathcal{E}_j$ is some complex Hilbert space. Let $h_j \colon Z \to \mathrm{GL}^+(\mathcal{E}_j)$ be the Hermitian structure of Π_j. Then it is easily checked that the Hermitian structure of the vector bundle

$$\Pi := \mathfrak{S}_2(\Pi_1, \Pi_2) \colon Z \times \mathfrak{S}_2(\mathcal{E}_1, \mathcal{E}_2) \to Z$$

is given by

$$H \colon Z \to \mathrm{GL}^+(\mathfrak{S}_2(\mathcal{E}_1, \mathcal{E}_2)), \quad H(z)S = h_2(z)Sh_1(z)^{-1}$$

where $S \in \mathfrak{S}_2(\mathcal{E}_1, \mathcal{E}_2)$ and $z \in Z$.

It follows that

$$\begin{aligned} H'(z)(\cdot)S &= h_2'(z)(\cdot)Sh_1(z)^{-1} - h_2(z)Sh_1(z)^{-1}h_1'(z)(\cdot)h_1(z)^{-1} \\ &\in \mathcal{B}_{\mathbb{R}}(T_zZ, \mathfrak{S}_2(\mathcal{E}_1, \mathcal{E}_2)) \end{aligned}$$

hence

$$H(z)^{-1}H'(z)S = h_2(z)^{-1}h_2'(z)S - Sh_1(z)^{-1}h_1'(z).$$

Therefore, if we denote by A, A_1, and A_2 the linear connection forms of ∇, ∇_1, and ∇_2, respectively, then by using Lemma 3.12 we obtain

$$A(z)S = A_2(z)S - SA_1(z) \in \mathcal{B}(T_zZ, \mathfrak{S}_2(\mathcal{E}_1, \mathcal{E}_2)) \text{ if } S \in \mathfrak{S}_2(\mathcal{E}_1, \mathcal{E}_2) \text{ and } z \in Z.$$

Now the assertion follows easily since $\nabla = \mathrm{d} + A$ and $\nabla_j = \mathrm{d} + A_j$ for $j = 1, 2$. □

We now compute the Chern covariant derivatives of holomorphic subbundles of Hermitian holomorphic vector bundles.

Proposition 3.19. *Let $\Pi \colon D \to Z$ be any Hermitian holomorphic vector bundle with a holomorphic vector subbundle $\Pi_1 \colon D_1 \to Z$ and its fiberwise orthogonal complement $\Pi_2 \colon D_2 \to Z$. For $j = 1, 2$ we regard Π_j as a Hermitian holomorphic bundle with respect to the Hermitian structure induced from Π. Denote by ∇, ∇_1, and ∇_2 the Chern covariant derivatives of Π, Π_1, and Π_2, respectively. Also let Θ, Θ_1, and Θ_2 be the corresponding curvatures. Then with respect to the fiberwise orthogonal direct sum decomposition $D = D_1 \oplus D_2$ we have*

$$\nabla = \begin{pmatrix} \nabla_1 & -\beta^* \\ \beta & \nabla_2 \end{pmatrix}$$

and

$$\Theta = \begin{pmatrix} \Theta_1 - \beta^* \wedge \beta & * \\ * & \Theta_2 - \beta \wedge \beta^* \end{pmatrix}$$

for some $\beta \in \Omega^{(1,0)}(Z, \mathrm{Hom}(D_1, D_2))$ *where* $\beta^* \in \Omega^{(0,1)}(Z, \mathrm{Hom}(D_2, D_1))$ *is its pointwise adjoint 1-form.*

Proof. With Proposition 3.18 at hand, one can use the method of proof of [De12, Ch. V, Th. 14.3 and 14.5]; see also [GH78, Ch. 0, Sect. 5, pages 73 and 78]. □

4. Positivity and global sections of holomorphic vector bundles

For the sake of completeness, we include in this section a brief discussion on the properties of Griffiths positivity of holomorphic vector bundles. We refer to [Gr69], [GH78], [Lz04], and particularly to the elegant exposition in [De12] for further details.

Quotient tautological bundles. We will give here some straightforward infinite-dimensional versions of certain constructions from [De12, Ch. V, §16]. For any Hilbert space $\mathcal{H}$ and any integer $k \geq 1$ we denote by $\mathrm{Gr}^{(k)}(\mathcal{H})$ the set of all k-codimensional subspaces of $\mathcal{H}$, which has the natural structure of a complex $U(\mathcal{H})$-homogeneous Banach manifold. Recall that the tautological bundle over $\mathrm{Gr}^{(k)}(\mathcal{H})$ is

$$\Pi^{(k)} : \mathcal{T}^{(k)}(\mathcal{H}) \to \mathrm{Gr}^{(k)}(\mathcal{H}), \quad (\mathcal{S}, v) \mapsto \mathcal{S},$$

where

$$\mathcal{T}^{(k)}(\mathcal{H}) = \{(\mathcal{S}, v) \in \mathrm{Gr}^{(k)}(\mathcal{H}) \times \mathcal{H} \mid v \in \mathcal{S}\} \subseteq \mathrm{Gr}^{(k)}(\mathcal{H}) \times \mathcal{H}.$$

On the other hand, the *quotient tautological bundle* over $\mathrm{Gr}^{(k)}(\mathcal{H})$ is

$$\mathcal{Q}^{(k)}(\mathcal{H}) \to \mathrm{Gr}^{(k)}(\mathcal{H}), \quad (\mathcal{S}, v + \mathcal{S}) \mapsto \mathcal{S},$$

where

$$\mathcal{Q}^{(k)}(\mathcal{H}) = \{(\mathcal{S}, v + \mathcal{S}) \in \mathrm{Gr}^{(k)}(\mathcal{H}) \times (\mathcal{H}/\mathcal{S}) \mid v \in \mathcal{H}\}.$$

Note that there is the short exact sequence of holomorphic vector bundles over $\mathrm{Gr}^{(k)}(\mathcal{H})$

$$\begin{array}{ccccccccc} 0 & \longrightarrow & \mathcal{T}^{(k)}(\mathcal{H}) & \longrightarrow & \mathrm{Gr}^{(k)}(\mathcal{H}) \times \mathcal{H} & \longrightarrow & \mathcal{Q}^{(k)}(\mathcal{H}) & \longrightarrow & 0 \\ & & & \searrow & \downarrow & \swarrow & & & \\ & & & & \mathrm{Gr}^{(k)}(\mathcal{H}) & & & & \end{array}$$

where the vertical arrow in the middle is the projection of the trivial bundle with the typical fiber $\mathcal{H}$.

Globally generated holomorphic vector bundles. Unless otherwise specified, we let $\Pi\colon D \to Z$ be any holomorphic vector bundle whose fibers have finite dimension k and whose base is a complex Banach manifold. Moreover, $\mathcal{O}(Z,D)$ stands for the space of global holomorphic sections endowed with the topology of uniform convergence on compact sets, and we define the evaluation maps

$$(\forall z \in Z) \quad \mathrm{ev}_z\colon \mathcal{O}(Z,D) \to D_z, \ \mathrm{ev}_z(\sigma) = \sigma(z).$$

Definition 4.1. The bundle $\Pi\colon D \to Z$ is *globally generated* by the complex Hilbert space $\mathcal{H}$ if we have a continuous inclusion map $\mathcal{H} \hookrightarrow \mathcal{O}(Z,D)$ and for which for arbitrary $z \in Z$ we have $\mathrm{ev}_z(\mathcal{H}) = D_z$.

Remark 4.2. It follows by [De12, Ch. VII, Prop. 11.2] that if Z is a finite-dimensional manifold, then the above notion of globally generated holomorphic vector bundle agrees with the one introduced in [De12, Ch. VII, Def. 11.1(a)].

Remark 4.3. If the bundle $\Pi\colon D \to Z$ is globally generated by the complex Hilbert space $\mathcal{H}$, then we define

$$\begin{aligned}(\forall z \in Z) \quad N_z &:= \{(z,\sigma) \in Z \times \mathcal{H} \mid \sigma(z) = 0\} \\ &= \{z\} \times \mathrm{Ker}\,(\mathrm{ev}_z\,|_{\mathcal{H}}) \\ &\subseteq Z \times \mathcal{H}\end{aligned}$$

and $N := \bigcup\limits_{z \in Z} N_z$.

Now *assume the fibers of* Π *are finite dimensional.* Then N is the total space of a subbundle of the trivial Hermitian bundle $Z \times \mathcal{H} \to Z$. We have the fiberwise exact sequence of Hermitian bundles

$$0 \to N \hookrightarrow Z \times \mathcal{H} \xrightarrow{\mathrm{ev}} D \to 0 \tag{4.1}$$

and the commutative diagram

$$\begin{array}{ccc} D & \xrightarrow{\Psi_{\mathcal{H}}} & \mathcal{Q}^{(k)}(\mathcal{H}) \\ {\scriptstyle \Pi}\downarrow & & \downarrow \\ Z & \xrightarrow{\psi_{\mathcal{H}}} & \mathrm{Gr}^{(k)}(\mathcal{H}) \end{array}$$

where for every $z \in Z$ we have

$$(\forall \xi \in D_z) \quad \Psi_{\mathcal{H}}(\xi) = \{\sigma \in \mathcal{H} \mid \sigma(z) = \xi\} \in \mathcal{H}/\mathrm{Ker}\,(\mathrm{ev}_z\,|_{\mathcal{H}})$$

where we performed the identification

$$\mathcal{H}/\mathrm{Ker}\,(\mathrm{ev}_z\,|_{\mathcal{H}}) \simeq D_z, \quad \sigma + \mathrm{Ker}\,(\mathrm{ev}_z\,|_{\mathcal{H}}) \mapsto \mathrm{ev}_z(\sigma).$$

Positivity curvature condition. In order to introduce the positivity curvature condition on the covariant derivatives, we need the following remark, which is well known at least in the case of the scalar-valued bilinear maps.

Remark 4.4. Let $\mathcal{V}$ be any complex Banach space and $\mathcal{A}$ be a complex associative Banach $*$-algebra, and denote $\mathcal{A}^{\rm sa} := \{a \in \mathcal{A} \mid a^* = a\}$. We define the following spaces of bounded $\mathbb{R}$-bilinear maps:

- the space $\mathrm{Herm}(\mathcal{V}, \mathcal{A})$ of all $\mathbb{R}$-bilinear maps $\Psi\colon \mathcal{V} \times \mathcal{V} \to \mathcal{A}$ satisfying

$$(\forall v_1, v_2 \in \mathcal{V}) \quad \Psi(v_1, v_2)^* = \Psi(v_2, v_1) = \mathrm{i}\Psi(v_2, \mathrm{i}v_1)$$

- the space $\mathrm{Symm}(\mathcal{V}, \mathcal{A})$ of all $\mathbb{R}$-bilinear maps $\psi\colon \mathcal{V} \times \mathcal{V} \to \mathcal{A}$ satisfying

$$(\forall v_1, v_2 \in \mathcal{V}) \quad \psi(v_1, v_2) = \psi(v_2, v_1) = \psi(\mathrm{i}v_1, \mathrm{i}v_2)$$

- the space $\mathrm{Skew}(\mathcal{V}, \mathcal{A})$ of all $\mathbb{R}$-bilinear maps $\omega\colon \mathcal{V} \times \mathcal{V} \to \mathcal{A}^{\rm sa}$ satisfying

$$(\forall v_1, v_2 \in \mathcal{V}) \quad \omega(v_1, v_2) = -\omega(v_2, v_1) = \omega(\mathrm{i}v_1, \mathrm{i}v_2).$$

If $\Psi \in \mathrm{Herm}(\mathcal{V}, \mathcal{A})$, $\psi \in \mathrm{Symm}(\mathcal{V}, \mathcal{A})$, and $\omega \in \mathrm{Skew}(\mathcal{V}, \mathcal{A})$, then any of these three bilinear maps determines the other two maps in a unique manner such that the equation

$$(\forall v_1, v_2 \in \mathcal{V}) \quad \Psi(v_1, v_2) = \psi(v_1, v_2) + \mathrm{i}\omega(v_1, v_2)$$

be satisfied, and *canonical* $\mathbb{R}$*-linear isomorphisms* are thus defined between the spaces $\mathrm{Herm}(\mathcal{V}, \mathcal{A})$, $\mathrm{Symm}(\mathcal{V}, \mathcal{A})$, and $\mathrm{Skew}(\mathcal{V}, \mathcal{A})$, respectively.

More precisely, the $\mathbb{R}$-bilinear maps involved in the above equation are related by the formulas

$$\begin{aligned}
\omega(v_1, v_2) &= \psi(v_1, \mathrm{i}v_2) \\
\psi(v_1, v_2) &= \frac{1}{2}(\Psi(v_1, v_2) + \Psi(v_2, v_1)) \\
\omega(v_1, v_2) &= \frac{1}{2\mathrm{i}}(\Psi(v_1, v_2) - \Psi(v_2, v_1))
\end{aligned}$$

for all $v_1, v_2 \in \mathcal{V}$.

We now introduce the notion of Griffiths positivity of bundle-valued differential 2-forms, which goes back to [Gr69]; see also [GH78], [Lz04], [De12]. In the case of infinite-rank vector bundles, a version of this notion was also used in [Ber09].

Definition 4.5. Let $\Pi\colon D \to Z$ be any Hermitian holomorphic bundle. A bundle-valued differential form $\omega \in \Omega^2(Z, \mathrm{End}\,(\Pi))$ is *Griffiths nonnegative* if for every $z \in Z$ the bounded $\mathbb{R}$-bilinear map $\omega_z\colon T_zZ \times T_zZ \to \mathcal{B}(D_z)$ satisfies the conditions $\omega_z \in \mathrm{Skew}(T_zZ, \mathcal{B}(D_z))$ and $\Psi_z(x, x) \geq 0$ in $\mathcal{B}(D_z)$ for all $x \in T_zZ$, where $\Psi_z \in \mathrm{Herm}(T_zZ, \mathcal{B}(D_z))$ is the sesquilinear map which canonically corresponds to ω_z via Remark 4.4. If moreover for every $x \in T_zZ \setminus \{0\}$ we have $\Psi_z(x, x) \neq 0$, then we say that ω is *Griffiths positive*.

In the case of the bundles with a finite-dimensional base, the following result can be found in [GH78, Ch. 0, Sect. 5] or [De12, Ch. VII, Cor. 11.5].

Theorem 4.6. *Let* $\Pi\colon D \to Z$ *be any holomorphic vector bundle which is globally generated by the complex Hilbert space* $\mathcal{H} \hookrightarrow \mathcal{O}(Z, D)$ *and has finite-dimensional fibers. Then there exists a unique Hermitian structure on* Π *such that for every* $z \in \mathcal{H}$ *the adjoint of the evaluation map* $\mathrm{ev}_z\colon \mathcal{H} \to D_z$ *is an isometry, and the curvature of the corresponding Chern covariant derivative is Griffiths nonnegative.*

Proof. One can use the method of proof from [GH78, Ch. 0, Sect. 5], by relying on the above Proposition 3.18 and Remark 4.3. □

5. Reproducing kernels and Griffiths positivity

This section contains one of our main results, which is a necessary condition for existence of reproducing kernels on vector bundles (Theorem 5.4). It relies on Griffith positivity properties of certain covariant derivatives associated with reproducing kernels on Hermitian holomorphic bundles which satisfy a certain admissibility condition. In order to introduce the latter notion, we need the following lemma.

Lemma 5.1. *In the setting of Definition 2.2, consider the following assertions at an arbitrary point* $s \in Z$*:*

1. *The operator* $\widehat{K}|_{D_s}\colon D_s \to \mathcal{H}^K$ *is injective and has closed range.*
2. *The operator* $K(s,s) \in \mathcal{B}(D_s)$ *is invertible.*
3. *The operator* $K(s,s) \in \mathcal{B}(D_s)$ *is surjective.*
4. *The evaluation map* $\mathrm{ev}_s\colon \mathcal{H}^K \to D_s$ *is surjective.*

Then we have (1) $\iff$ (2) $\iff$ (3) $\implies$ (4)*, and all the above four assertions are equivalent if moreover* $\dim D_s < \infty$*.*

Proof. The equivalence (1) $\iff$ (2) was established in [BG13, Lemma 3.4]. Moreover we have (2) $\iff$ (3) since $K(s,s)$ is always a bounded (nonnegative) self-adjoint operator on the complex Hilbert space D_s, as a consequence of (2.1) in Definition 2.2 for $n = 1$, hence $\mathrm{Ker}\,(K(s,s)) = (\mathrm{Im}(K(s,s)))^\perp$.

Next, for every $\xi \in D_s$ we have $\widehat{K}(\xi) = K_\xi = K(\cdot, s)\xi$ hence

$$(\mathrm{ev}_s \circ \widehat{K}|_{D_s})(\xi) = K(s,s)\xi$$

and this shows that (3) $\implies$ (4).

Now note that for all $t \in Z$, $\eta \in D_t$, and $\xi \in D_s$ we have

$$\begin{aligned}((\widehat{K}|_{D_s})(\xi) \mid K_\eta)_{\mathcal{H}^K} &= (K_\xi \mid K_\eta)_{\mathcal{H}^K} = (K(t,s)\xi \mid \eta)_{D_t} = (\xi \mid K(s,t)\eta)_{D_s} \\ &= (\xi \mid \mathrm{ev}_s(K_\eta))_{D_s}\end{aligned}$$

hence the operators $\mathrm{ev}_s\colon \mathcal{H}^K \to D_s$ and $\widehat{K}|_{D_s}\colon D_s \to \mathcal{H}^K$ are adjoint to each other. This implies $\mathrm{Ker}\,(\widehat{K}|_{D_s}) = (\mathrm{Im}\,\mathrm{ev}_s)^\perp$. Therefore, if (4) holds true, then $\widehat{K}|_{D_s}\colon D_s \to \mathcal{H}^K$ is injective, and if moreover $\dim D_s < \infty$, then the range of $\widehat{K}|_{D_s}$ is in turn finite dimensional hence is a closed subspace of $\mathcal{H}^K$, and thus (1) also holds true. This concludes the proof. □

The following is a special case of [BG13, Def. 3.5].

Definition 5.2. Assume $\Pi\colon D \to Z$ is a Hermitian bundle whose fibers are finite dimensional (for instance, Π is a line bundle). A reproducing kernel K on Π is called *admissible* if it has the following properties:

(a) The kernel K is smooth as a section of the bundle $\mathrm{Hom}(\mathrm{p}_2^*\Pi, \mathrm{p}_1^*\Pi)$.
(b) For every $s \in Z$ the operator $K(s,s) \in \mathcal{B}(D_s)$ is invertible.

Remark 5.3. In the setting of Definition 5.2, the admissible reproducing kernel K has the additional property that the mapping $\zeta_K\colon Z \to \mathrm{Gr}(\mathcal{H}^K)$ is smooth. See [BG13, Ex. 3.6] for details.

If $\Pi\colon D \to Z$ is any Hermitian holomorphic bundle with the space of holomorphic sections denoted by $\mathcal{O}(Z,D)$ and K is any reproducing kernel on Π, then we say that K is *holomorphic* if for every $\xi \in D$ we have $K_\xi \in \mathcal{O}(Z,D)$.

Theorem 5.4. *Let $\Pi\colon D \to Z$ be any Hermitian holomorphic vector bundle with finite-dimensional fibers and with its Hermitian structure denoted by $\{(\cdot \mid \cdot)_z\}_{z\in Z}$. If K is a holomorphic admissible reproducing kernel on the vector bundle Π, then $\{(K(z,z)\cdot \mid \cdot)_z\}_{z\in Z}$ is a new Hermitian structure on Π, for which the curvature of its Chern covariant derivative is Griffiths positive.*

Proof. Let $\mathcal{H}^K$ be the reproducing kernel Hilbert space associated to K. By hypothesis K is a holomorphic reproducing kernel on Π, hence we have a continuous inclusion map $\mathcal{H}^K \hookrightarrow \mathcal{O}(Z,D)$. Since K is admissible, Lemma 5.1 implies that the evaluation map $\mathrm{ev}_s\colon \mathcal{H}^K \to D_s$ is surjective for arbitrary $s \in Z$. Thus the bundle Π is globally generated in the sense of Definition 4.1, and then the conclusion follows by Theorem 4.6. □

Example 5.5. Consider the special case of Example 2.3 with $\mathcal{E} = \mathbb{C}$,

$$Z = \mathbb{D} = \{z \in \mathbb{C} \mid |z| < 1\}$$

and for every $\nu \geq 1$ define

$$K_{\mathbb{D}}^{(\nu)}\colon \mathbb{D}\times\mathbb{D} \to \mathbb{C}, \quad K_{\mathbb{D}}^{(\nu)}(z_1,z_2) = \frac{1}{(1 - z_1\bar{z}_2)^\nu}$$

which is the reproducing kernel of the Bergman space on the unit disc if $\nu > 1$ and of the Hardy space if $\nu = 1$; see also [BG13, subsect. 5.1, (b.1)].

The new Hermitian structure on the trivial bundle $\Pi\colon D = Z \times \mathbb{C} \to Z$ (referred to in Theorem 5.4) is in this case given by

$$h^{(\nu)}\colon \mathbb{D} \to \mathrm{GL}^+(\mathcal{E}) \simeq (0,\infty), \quad h^{(\nu)}(z) = \frac{1}{(1-|z|^2)^\nu} = \frac{1}{(1-z\bar{z})^\nu}.$$

We have

$$\frac{1}{h^{(\nu)}(z)} \cdot \partial h^{(\nu)}(z) = (1-z\bar{z})^\nu \cdot \frac{\nu\bar{z}}{(1-z\bar{z})^{\nu+1}} = \frac{\nu\bar{z}}{1-z\bar{z}}$$

hence by using Lemma 3.12 we obtain the following expression for the linear connection form of the Chern covariant derivative corresponding to the Hermitian structure determined by h:

$$A^{(\nu)}\colon \mathbb{D} \to \mathcal{B}(\mathbb{C}), \quad A^{(\nu)}_z = \frac{\nu\bar{z}}{1 - z\bar{z}}\mathrm{d}z.$$

Since

$$\frac{\partial}{\partial \bar{z}}\Big(\frac{\bar{z}}{1 - z\bar{z}}\Big) = \frac{(1 - z\bar{z}) - \bar{z}\cdot(-z)}{(1 - z\bar{z})^2} = \frac{1}{(1 - z\bar{z})^2}$$

it then follows by Proposition 3.13(2) that the curvature of the aforementioned Chern covariant derivative is the 2-form

$$\Theta^{(\nu)}\colon \mathbb{D} \to \mathcal{B}_{\mathbb{R}}(\mathbb{C}\wedge\mathbb{C}, \mathbb{C}), \quad \Theta^{(\nu)}_z = \frac{\nu}{(1 - |z|^2)^2}\,\mathrm{d}\bar{z}\wedge\mathrm{d}z$$

which clearly is Griffiths positive. It also follows by the above formula that the curvature $\Theta^{(\nu)}$ depends linearly on ν, and for all $\nu \geq 1$ we have $\Theta^{(\nu)} = \nu\Theta^{(1)}$.

Example 5.6. Recall that if $\mathcal{H}$ is any complex Hilbert space, then the mapping $h \mapsto (\cdot \mid h)_{\mathcal{H}}$ is an antilinear isometric isomorphism from $\mathcal{H}$ onto its topological dual $\mathcal{H}^*$. For this reason $\mathcal{H}^*$ will be alternatively described as the complex Hilbert space whose underlying structure of real Hilbert space is that of $\mathcal{H}$, while the complex structure is the opposite to the complex structure of $\mathcal{H}$; that is, we may assume $\mathcal{H}^* = \mathcal{H}$ as real vector spaces, with the complex scalar products related by the equality $(h_1 \mid h_2)_{\mathcal{H}^*} = (h_2 \mid h_1)_{\mathcal{H}}$ for all $h_1, h_2 \in \mathcal{H} = \mathcal{H}^*$. So $\mathcal{H}$ and $\mathcal{H}^*$ have the same closed complex subspaces and the identity map is an antiholomorphic diffeomorphism between their Grassmann manifolds $\mathrm{Gr}(\mathcal{H})$ and $\mathrm{Gr}(\mathcal{H}^*)$.

With this convention, if $\Pi\colon D \to Z$ is a holomorphic Hermitian bundle and $D^* := \bigsqcup_{s\in Z} D^*_s = \bigsqcup_{s\in Z} D_s = D$ as real manifolds, then the dual bundle $\Pi^*\colon D^* \to Z$ is again a holomorphic Hermitian bundle, whose complex structure is fiberwise the opposite to the complex structure of D, while both mappings Π and Π^* are holomorphic (i.e., they are smooth and the differentials are $\mathbb{C}$-linear) onto the same complex manifold Z. In particular, if K is a reproducing kernel on Π, then it is also a reproducing kernel on Π^*, to be denoted by K^*, and it follows by (2.2) that the corresponding reproducing kernel Hilbert spaces are related by $\mathcal{H}^{K^*} = (\mathcal{H}^K)^*$. In addition, one can also check that K is admissible if and only if K^* is. In this case, if $\Theta \in \Omega^2(Z, \Pi)$ and $\Theta^* \in \Omega^2(Z, \Pi^*)$ are the curvatures of the Chern connections associated to K and K^* as in Theorem 5.4, respectively, then by using Proposition 3.13(2) along with equations (3.1)–(3.2) one can show that $\Theta^* = -\Theta$.

By using a suitable method of localization of reproducing kernel Hilbert spaces on vector bundles, one can obtain infinite-dimensional versions of the properties of Bergman kernels established in [MP97]. Put $\delta_K := (\zeta_K \circ \Pi, \widehat{K})$, where ζ_K and $\widehat{K}$ are as in Definition 2.2.

Theorem 5.7. *Let $\Pi\colon D \to Z$ be a holomorphic Hermitian bundle with finite-dimensional fibers and K be a holomorphic admissible reproducing kernel on Π. Then the following assertions hold:*

1. *The mapping $\widehat{K}\colon D^* \to (\mathcal{H}^K)^*$ is holomorphic.*

2. *The pair* $\Delta_K = (\delta_K, \zeta_K)$ *is a holomorphic morphism of vector bundles from* $\Pi^* \colon D^* \to Z$ *to* $\Pi_{(\mathcal{H}^K)^*} \colon \mathcal{T}((\mathcal{H}^K)^*) \to \mathrm{Gr}((\mathcal{H}^K)^*)$.

Proof. By the note prior to the theorem, it suffices to prove the assertions under the assumption that $\Pi \colon D \to Z$ is a trivial vector bundle, say $D = Z \times \mathcal{V}$, where the complex Hilbert space $\mathcal{V}$ is the typical fiber. Recall from [BG13, Lemma 3.3] that $\widehat{K} \colon D \to \mathcal{H}^K$ is smooth. On the other hand, since K is an admissible holomorphic reproducing kernel, it is given by an operator-valued reproducing kernel $\kappa \colon Z \times Z \to \mathcal{B}(\mathcal{V})$ (see for instance [BG13, subsect. 5.1]) such that $\kappa(\cdot, t) \in \mathcal{O}(Z, \mathcal{V})$ and $\kappa(t,t) \in \mathcal{B}(\mathcal{V})$ is invertible for every $t \in Z$. Then for every $\eta = (t, v) \in Z \times \mathcal{V} = D$ we have

$$\widehat{K}(\eta) = (\cdot, \kappa(\cdot, t)v) \in \mathcal{O}(Z, D)$$

which implies at once that the differential of $\widehat{K} \colon D^* = Z \times \mathcal{V}^* \to (\mathcal{H}^K)^*$ at every point is $\mathbb{C}$-linear, hence the mapping $\widehat{K} \colon D^* \to (\mathcal{H}^K)^*$ is holomorphic. Since $\kappa(t, \cdot)^* = \kappa(\cdot, t) \in \mathcal{O}(Z, \mathcal{V})$ and $\zeta_K(t) = \widehat{K}(\{t\} \times \mathcal{V})$, it also follows by the above formula for $\widehat{K}(\eta)$ that $\zeta_K \colon Z \to \mathrm{Gr}((\mathcal{H}^K)^*)$ is holomorphic, since it is smooth by the assumption that K is admissible, and its tangent map at any point is $\mathbb{C}$-linear. □

In connection with Theorem 5.7, we note that a certain procedure to associate linear connections Φ_K to reproducing kernels K on infinite-dimensional vector bundles Π was established in [BG13]. That method relies on canonical pullback operations by starting from tautological bundles on Grassmann manifolds. Then one can also prove that the linear connection Φ_{K^*} associated with K^* is compatible both with the complex structure of the dual bundle $\Pi^* \colon D^* \to Z$ and with the Hermitian structure $\{(K^*(s,s)\cdot \mid \cdot)^*_s\}_{s \in Z}$ where $(K^*(s,s)\cdot \mid \cdot)^*_s := \overline{(K(s,s)\cdot \mid \cdot)_s}$ for all $s \in Z$. That is, on dual vector bundles of vector bundles with reproducing kernel (and finite-dimensional fibers), the covariant derivatives associated with the linear connections defined in [BG13] are also examples of the Griffiths-positive Chern derivatives of Theorem 5.4 above. As the details of these results are beyond the scope of the present work, we defer them to a forthcoming paper.

Remark 5.8. It is well known that there also exist holomorphic vector bundles with finite-dimensional fibers which do not admit any nontrivial global holomorphic cross-section, so they do not carry any reproducing kernel satisfying the hypothesis of Theorem 5.4. An example in this sense (with one-dimensional fibers) is provided by the tautological vector bundle over the projective space, in the above notation

$$\Pi^{(n)} \colon \mathcal{T}^{(n)}(\mathbb{C}^{n+1}) \to \mathrm{Gr}^{(n)}(\mathbb{C}^{n+1})$$

see for instance [De12, Ch. V, Cor. 15.6], where this line bundle was denoted by $\mathcal{O}(-1)$.

Remark 5.9. In connection with Theorem 5.4, we recall that there exist several open problems on sufficient conditions for Griffiths' positivity. There is for instance

Griffiths' problem ([Gr69]) which asks whether or not every ample holomorphic vector bundle with compact base is Griffiths positive.

It is also unknown whether any holomorphic vector bundle is Griffiths positive if there exists some integer $k_0 \geq 1$ such that the symmetric kth tensor powers of that bundle are globally generated for all $k \geq k_0$. See however [De12, Ch. VII, Cor. 11.6] for an affirmative answer to that problem in the case of the line bundles. From the perspective of the above Theorem 5.4, the problem would be to construct reproducing kernels on some Hermitian vector bundle by using some reproducing kernels on symmetric tensor powers of that bundle.

Appendix: Complements on vector-valued differential forms

Definition A.1. Assume X is any open subset of some real Banach space $\mathcal{X}$ and $\mathcal{V}$ is another real Banach space. The space of *$\mathcal{V}$-valued differential forms of degree* $p \geq 0$ on X is defined by

$$\Omega^p(X,\mathcal{V}) = \begin{cases} \mathcal{C}^\infty(X,\mathcal{V}) & \text{if } p = 0 \\ \mathcal{C}^\infty(X,\mathcal{B}(\wedge^p\mathcal{X},\mathcal{V})) & \text{if } p \geq 1, \end{cases}$$

where $\mathcal{B}(\wedge^p\mathcal{X},\mathcal{V})$ denotes the space of all bounded p-linear skew-symmetric maps $\mathcal{X} \times \cdots \times \mathcal{X} \to \mathcal{V}$. For every $\sigma \in \Omega^p(X,\mathcal{V})$ and $x \in X$ we denote $\sigma_x := \sigma(x)$.

The *exterior derivative* $d\colon \Omega^p(X,\mathcal{V}) \to \Omega^{p+1}(X,\mathcal{V})$ is defined for $\sigma \in \Omega^p(X,\mathcal{V})$ as follows:

1. If $p = 0$, then for every $x \in X$ we set $(\mathrm{d}\sigma)_x = \sigma'_x \in \mathcal{B}(\mathcal{X},\mathcal{V})$.
2. If $p \geq 1$, then for every $x \in X$ we define $(\mathrm{d}\sigma)_x = \mathrm{d}_x\sigma \in \mathcal{B}(\wedge^p\mathcal{X},\mathcal{V})$ as a bounded skew-symmetric $(p+1)$-linear mapping $\mathcal{X} \times \cdots \times \mathcal{X} \to \mathcal{V}$ by the formula
$$(\mathrm{d}\sigma)_x(x_1,\dots,x_{p+1}) = \sum_{j=1}^{p+1}(-1)^{j-1}(\sigma'_x(x_j))(x_1,\dots,x_{j-1},x_{j+1},\dots,x_{p+1})$$
for every $x_1,\dots,x_{p+1} \in \mathcal{X}$. Note that $\sigma\colon X \to \mathcal{B}(\wedge^p\mathcal{X},\mathcal{V})$ is a smooth mapping, hence $\sigma'_x \in \mathcal{B}(\mathcal{X},\mathcal{B}(\wedge^p\mathcal{X},\mathcal{V}))$.

We have $d^2 = 0$ (see for instance [Lg01]) as an operator from $\Omega^p(X,\mathcal{V})$ into $\Omega^{p+2}(X,\mathcal{V})$ for every $p \geq 0$.

Definition A.2. Now assume the following setting:

- X is an open set in the real Banach space $\mathcal{X}$;
- $\mathcal{V}_1$, $\mathcal{V}_2$, and $\mathcal{V}$ are real Banach spaces endowed with a continuous bilinear mapping $\mathcal{V}_1 \times \mathcal{V}_2 \to \mathcal{V}$ denoted simply by $(v_1,v_2) \mapsto v_1 \cdot v_2$.

Then for $p_1,p_2 \geq 0$ we define the *exterior product* (cf. [Ne06, Subsect. I.4])

$$\wedge\colon \Omega^{p_1}(X,\mathcal{V}_1) \times \Omega^{p_2}(X,\mathcal{V}_2) \to \Omega^{p_1+p_2}(X,\mathcal{V})$$

in the following way. Let $\sigma_j \in \Omega^{p_j}(X,\mathcal{V}_j)$, $j = 1,2$, and $x \in X$.

1. If $p_1 = 0$, then we set

$$(\sigma_1 \wedge \sigma_2)_x(x_1, \dots, x_{p_2}) = \underbrace{(\sigma_1)_x}_{\in \mathcal{V}_1} \cdot \underbrace{(\sigma_2)_x(x_1, \dots, x_{p_2})}_{\in \mathcal{V}_2} \in \mathcal{V}$$

for $x_1, \dots, x_{p_2} \in \mathcal{X}$. We proceed in a similar manner if $p_2 = 0$.

2. If $p_1, p_2 \geq 1$, then

$$(\sigma_1 \wedge \sigma_2)_x(x_1, \dots, x_{p_1+p_2}) = \frac{1}{p_1! p_2!} \sum_\tau \epsilon(\tau) \underbrace{(\sigma_1)_x(x_{\tau(1)}, \dots, x_{\tau(p_1)})}_{\in \mathcal{V}_1} \cdot \underbrace{(\sigma_2)_x(x_{\tau(p_1+1)}, \dots, x_{\tau(p_1+p_2)})}_{\in \mathcal{V}_2}$$

where the sum is taken for every permutation τ of the set $\{1, \dots, p_1 + p_2\}$ and $\epsilon(\tau) \in \{\pm 1\}$ denotes the signature of τ (compare [Lg01, Ch. V, §3]).

Just as in the scalar-valued case, one can check the formula

$$\mathrm{d}(\sigma_1 \wedge \sigma_2) = \mathrm{d}\sigma_1 \wedge \mathrm{d}\sigma_2 + (-1)^{p_1} \sigma_1 \wedge \mathrm{d}\sigma_2$$

for $\sigma_j \in \Omega^{p_j}(X, \mathcal{V}_j)$ and $j = 1, 2$.

Remark A.3. Let $\mathcal{Y}$ and $\mathcal{V}$ be complex Banach spaces and denote by $\mathcal{B}_{\mathbb{R}}(\mathcal{Y}, \mathcal{V})$ the Banach space of bounded $\mathbb{R}$-linear operators from $\mathcal{Y}$ into $\mathcal{V}$. Also denote by

$$\mathcal{B}_{\mathbb{R}}^{(0,1)}(\mathcal{Y}, \mathcal{V}) = \{T \in \mathcal{B}_{\mathbb{R}}(\mathcal{Y}, \mathcal{V}) \mid (\forall x \in \mathcal{Y}) \quad T(\mathrm{i}x) = -\mathrm{i}Tx\}$$

the space of bounded conjugate-linear operators from $\mathcal{Y}$ into $\mathcal{V}$, and use for the moment the notation $\mathcal{B}_{\mathbb{R}}^{(1,0)}(\mathcal{Y}, \mathcal{V}) := \mathcal{B}(\mathcal{Y}, \mathcal{V})$ for the space of $\mathbb{C}$-linear operators.

Then we note the following facts:

1. The mapping

$$\mathcal{B}_{\mathbb{R}}^{(1,0)}(\mathcal{Y}, \mathcal{V}) \times \mathcal{B}_{\mathbb{R}}^{(0,1)}(\mathcal{Y}, \mathcal{V}) \to \mathcal{B}_{\mathbb{R}}(\mathcal{Y}, \mathcal{V}), \quad (R, S) \mapsto R + S \tag{A.1}$$

is a linear topological isomorphism. Indeed, it is clear that this mapping is linear and continuous, hence it suffices to prove that it is bijective. In fact it is easily checked that for every $T \in \mathcal{B}_{\mathbb{R}}(\mathcal{Y}, \mathcal{V})$ there exist uniquely determined operators $T^{(1,0)} \in \mathcal{B}_{\mathbb{R}}^{(1,0)}(\mathcal{Y}, \mathcal{V})$ and $T^{(0,1)} \in \mathcal{B}_{\mathbb{R}}^{(0,1)}(\mathcal{Y}, \mathcal{V})$ such that $T = T^{(1,0)} + T^{(0,1)}$, namely

$$T^{(1,0)}x = \frac{1}{2}(Tx - \mathrm{i}T(\mathrm{i}x)) \text{ and } T^{(0,1)}x = \frac{1}{2}(Tx + \mathrm{i}T(\mathrm{i}x))$$

for every $x \in \mathcal{Y}$.

2. Each of the spaces $\mathcal{B}_{\mathbb{R}}^{(1,0)}(\mathcal{Y}, \mathcal{V})$ and $\mathcal{B}_{\mathbb{R}}^{(0,1)}(\mathcal{Y}, \mathcal{V})$ has a natural structure of *complex* Banach space, defined by multiplying the values of any operator by complex numbers (This is possible since $\mathcal{V}$ is a complex Banach space.)

3. By using the above items (1) and (2), one can obtain direct sum decompositions similar to (A.1) for spaces of $\mathbb{R}$-multilinear mappings in a higher number of variables. For instance, for bilinear mappings we have

$$\begin{aligned}\mathcal{B}_{\mathbb{R}}(\mathcal{Y}\widehat{\otimes}_{\mathbb{R}}\mathcal{Y},\mathcal{V}) &\simeq \mathcal{B}_{\mathbb{R}}(\mathcal{Y},\mathcal{B}_{\mathbb{R}}(\mathcal{Y},\mathcal{V}))\\ &\simeq \mathcal{B}_{\mathbb{R}}(\mathcal{Y},\mathcal{B}_{\mathbb{R}}^{(1,0)}(\mathcal{Y},\mathcal{V}) \dotplus \mathcal{B}_{\mathbb{R}}^{(0,1)}(\mathcal{Y},\mathcal{V}))\\ &\simeq \mathcal{B}_{\mathbb{R}}(\mathcal{Y},\mathcal{B}_{\mathbb{R}}^{(1,0)}(\mathcal{Y},\mathcal{V})) \dotplus \mathcal{B}(\mathcal{Y},\mathcal{B}_{\mathbb{R}}^{(0,1)}(\mathcal{Y},\mathcal{V}))\\ &\simeq \mathcal{B}_{\mathbb{R}}^{(1,0)}(\mathcal{Y},\mathcal{B}_{\mathbb{R}}^{(1,0)}(\mathcal{Y},\mathcal{V})) \dotplus \mathcal{B}_{\mathbb{R}}^{(1,0)}(\mathcal{Y},\mathcal{B}_{\mathbb{R}}^{(0,1)}(\mathcal{Y},\mathcal{V}))\\ &\quad \dotplus \mathcal{B}_{\mathbb{R}}^{(0,1)}(\mathcal{Y},\mathcal{B}_{\mathbb{R}}^{(1,0)}(\mathcal{Y},\mathcal{V})) \dotplus \mathcal{B}_{\mathbb{R}}^{(0,1)}(\mathcal{Y},\mathcal{B}_{\mathbb{R}}^{(0,1)}(\mathcal{Y},\mathcal{V})).\end{aligned}$$

4. We now use the above remarks to obtain a direct sum decomposition for the space of skew-symmetric $\mathbb{R}$-bilinear maps from $\mathcal{Y}\times\mathcal{Y}$ into $\mathcal{V}$. Let us consider the bounded $\mathbb{R}$-linear operator

$$A\colon \mathcal{Y}\widehat{\otimes}_{\mathbb{R}}\mathcal{Y}\to\mathcal{Y}\widehat{\otimes}_{\mathbb{R}}\mathcal{Y},\quad A(y_1\otimes y_2)=\frac{1}{2}(y_1\otimes y_2-y_2\otimes y_1).$$

Then we have $A^2=A$, and we define $\mathcal{Y}\wedge\mathcal{Y}:=\operatorname{Ran}A$. We also define

$$\mathcal{A}\colon \mathcal{B}_{\mathbb{R}}(\mathcal{Y}\widehat{\otimes}_{\mathbb{R}}\mathcal{Y},\mathcal{V})\to\mathcal{B}_{\mathbb{R}}(\mathcal{Y}\widehat{\otimes}_{\mathbb{R}}\mathcal{Y},\mathcal{V}),\quad \mathcal{A}(\Phi):=\Phi\circ A$$

and then $\mathcal{A}^2=\mathcal{A}$ and it is easily seen that

$$\operatorname{Ran}\mathcal{A}=\{\Phi\in\mathcal{B}_{\mathbb{R}}(\mathcal{Y}\widehat{\otimes}_{\mathbb{R}}\mathcal{Y},\mathcal{V})\mid(\forall y_1,y_2\in\mathcal{Y})\quad \Phi(y_1\otimes y_2)=-\Phi(y_2\otimes y_1)\}.$$

In order to study the behavior of $\mathcal{A}$ with respect to the direct sum decomposition established in (3) above, we introduce the operator

$$\mathcal{Q}\colon \mathcal{B}_{\mathbb{R}}(\mathcal{Y}\widehat{\otimes}_{\mathbb{R}}\mathcal{Y},\mathcal{V})\to\mathcal{B}_{\mathbb{R}}(\mathcal{Y}\widehat{\otimes}_{\mathbb{R}}\mathcal{Y},\mathcal{V}),\quad (\mathcal{Q}(\Phi))(y_1\otimes y_2)=\Phi((\mathrm{i}y_1)\otimes(\mathrm{i}y_2)).$$

Then it is easily seen that $\mathcal{Q}^2=\mathrm{id}$ and $\mathcal{Q}\mathcal{A}=\mathcal{A}\mathcal{Q}$, hence we have the direct sum decomposition

$$\mathcal{B}_{\mathbb{R}}(\mathcal{Y}\widehat{\otimes}_{\mathbb{R}}\mathcal{Y},\mathcal{V})=\operatorname{Ker}(\mathcal{Q}-\mathrm{id})\dotplus\operatorname{Ker}(\mathcal{Q}+\mathrm{id}) \tag{A.2}$$

and both subspaces involved in this decomposition are invariant under $\mathcal{A}$. On the other hand, it is easily seen that

$$\begin{aligned}\mathcal{B}_{\mathbb{R}}^{(1,0)}(\mathcal{Y},\mathcal{B}_{\mathbb{R}}^{(1,0)}(\mathcal{Y},\mathcal{V})) \dotplus \mathcal{B}_{\mathbb{R}}^{(0,1)}(\mathcal{Y},\mathcal{B}_{\mathbb{R}}^{(0,1)}(\mathcal{Y},\mathcal{V})) &\subseteq \operatorname{Ker}(\mathcal{Q}+\mathrm{id}),\\ \mathcal{B}_{\mathbb{R}}^{(1,0)}(\mathcal{Y},\mathcal{B}_{\mathbb{R}}^{(0,1)}(\mathcal{Y},\mathcal{V})) \dotplus \mathcal{B}_{\mathbb{R}}^{(0,1)}(\mathcal{Y},\mathcal{B}_{\mathbb{R}}^{(1,0)}(\mathcal{Y},\mathcal{V})) &\subseteq \operatorname{Ker}(\mathcal{Q}-\mathrm{id}).\end{aligned}$$

It then follows by (A.2) and the decomposition established above in (3) that the above inclusions are actually equalities. In particular, the space

$$\mathcal{B}_{\mathbb{R}}^{(1,0)}(\mathcal{Y},\mathcal{B}_{\mathbb{R}}^{(0,1)}(\mathcal{Y},\mathcal{V})) \dotplus \mathcal{B}_{\mathbb{R}}^{(0,1)}(\mathcal{Y},\mathcal{B}_{\mathbb{R}}^{(1,0)}(\mathcal{Y},\mathcal{V}))$$

is invariant under $\mathcal{A}$, and we denote by $\mathcal{B}_{\mathbb{R}}^{(1,1)}(\mathcal{Y}\widehat{\otimes}\mathcal{Y},\mathcal{V})$ the image of the corresponding restriction of $\mathcal{A}$. On the other hand, it is easily checked that each of the spaces $\mathcal{B}_{\mathbb{R}}^{(1,0)}(\mathcal{Y},\mathcal{B}_{\mathbb{R}}^{(1,0)}(\mathcal{Y},\mathcal{V}))$ and $\mathcal{B}_{\mathbb{R}}^{(0,1)}(\mathcal{Y},\mathcal{B}_{\mathbb{R}}^{(0,1)}(\mathcal{Y},\mathcal{V}))$ is invariant under $\mathcal{A}$, and $\mathcal{B}_{\mathbb{R}}^{(2,0)}(\mathcal{Y}\widehat{\otimes}\mathcal{Y},\mathcal{V})$ and $\mathcal{B}_{\mathbb{R}}^{(0,2)}(\mathcal{Y}\widehat{\otimes}\mathcal{Y},\mathcal{V})$ will denote the images of the corresponding restrictions of $\mathcal{A}$, respectively. We thus get a direct sum decomposition

$$\operatorname{Ran}\mathcal{A} = \mathcal{B}_{\mathbb{R}}^{(2,0)}(\mathcal{Y}\widehat{\otimes}\mathcal{Y},\mathcal{V}) \dotplus \mathcal{B}_{\mathbb{R}}^{(1,1)}(\mathcal{Y}\widehat{\otimes}\mathcal{Y},\mathcal{V}) \dotplus \mathcal{B}_{\mathbb{R}}^{(0,2)}(\mathcal{Y}\widehat{\otimes}\mathcal{Y},\mathcal{V}).$$

Note the natural isomorphism $\operatorname{Ran}\mathcal{A} \simeq \mathcal{B}_{\mathbb{R}}(\mathcal{Y}\wedge\mathcal{Y},\mathcal{V})$, $\Phi \mapsto \Phi|_{\operatorname{Ran} A}$, and thus the above decomposition gives rise to a direct sum decomposition

$$\mathcal{B}_{\mathbb{R}}(\mathcal{Y}\wedge\mathcal{Y},\mathcal{V}) = \mathcal{B}_{\mathbb{R}}^{(2,0)}(\mathcal{Y}\wedge\mathcal{Y},\mathcal{V}) \dotplus \mathcal{B}_{\mathbb{R}}^{(1,1)}(\mathcal{Y}\wedge\mathcal{Y},\mathcal{V}) \dotplus \mathcal{B}_{\mathbb{R}}^{(0,2)}(\mathcal{Y}\wedge\mathcal{Y},\mathcal{V}).$$

For every $\Phi \in \mathcal{B}_{\mathbb{R}}(\mathcal{Y}\wedge\mathcal{Y},\mathcal{V})$ we denote by $\Phi = \Phi^{(2,0)} + \Phi^{(1,1)} + \Phi^{(0,2)}$ the corresponding decomposition. This is the bilinear version of the decomposition established above in (1).

Definition A.4. Let $\mathcal{X}$ and $\mathcal{V}$ be complex Banach spaces. If $p \in \{1,2\}$, then by using Remark A.3 for the values of the differential forms in $\Omega^p(X,\mathcal{V})$, we get the direct sum decompositions

$$\Omega^1(X,\mathcal{V}) = \Omega^{(1,0)}(X,\mathcal{V}) \dotplus \Omega^{(0,1)}(X,\mathcal{V}) \tag{A.3}$$

and

$$\Omega^2(X,\mathcal{V}) = \Omega^{(2,0)}(X,\mathcal{V}) \dotplus \Omega^{(1,1)}(X,\mathcal{V}) \dotplus \Omega^{(0,2)}(X,\mathcal{V}).$$

By using these decompositions we can define the operators

$$\bar\partial\colon \mathcal{C}^\infty(X,\mathcal{V}) \to \Omega^{(0,1)}(X,\mathcal{V}) \text{ and } \bar\partial\colon \Omega^{(r,1-r)}(X,\mathcal{V}) \to \Omega^{(r,2-r)}(X,\mathcal{V})$$

for $r \in \{0,1\}$ as the corresponding projections of the exterior derivatives

$$\mathrm{d}\colon \mathcal{C}^\infty(X,\mathcal{V}) \to \Omega^1(X,\mathcal{V}) \text{ and } \mathrm{d}\colon \Omega^1(X,\mathcal{V}) \to \Omega^2(X,\mathcal{V}),$$

respectively. We also define $\partial := \mathrm{d} - \bar\partial$ on any of the above spaces where $\bar\partial$ is defined.

Remark A.5. In the setting of Definition A.4, we have $\bar\partial^2 = 0$, $\partial^2 = 0$, and

$$\mathrm{d}\colon \Omega^{(r,s)}(X,\mathcal{V}) \to \Omega^{(r+1,s)}(X,\mathcal{V}) \dotplus \Omega^{(r,s+1)}(X,\mathcal{V}),$$

hence $\partial\colon \Omega^{(r,s)}(X,\mathcal{V}) \to \Omega^{(r+1,s)}(X,\mathcal{V})$ for $r,s \in \{0,1\}$ with $r+s=1$.

Remark A.6. In Definition A.4, a function $\sigma \in \mathcal{C}^\infty(X,\mathcal{V})$ is holomorphic if and only if its differential is $\mathbb{C}$-linear at every point of X, which is equivalent to the Cauchy–Riemann equation $\bar\partial\sigma = 0$.

We refer to [Ll98, Sect. 2] for a definition of the Dolbeault operator $\bar\partial$ in a more general setting.

Acknowledgment

We wish to thank the referee for several remarks which improved our presentation.

References

[AHS78] M.F. Atiyah, N.J. Hitchin, I.M. Singer, *Self-duality in four-dimensional Riemannian geometry.* Proc. Roy. Soc. London Ser. A **362** (1978), no. 1711, 425–461.

[BG08] D. Beltiţă, J.E. Galé, *Holomorphic geometric models for representations of C^*-algebras.* J. Funct. Anal. **255** (2008), no. 10, 2888–2932.

[BG09] D. Beltiţă, J.E. Galé, *On complex infinite-dimensional Grassmann manifolds.* Complex Anal. Oper. Theory **3** (2009), no. 4, 739–758.

[BG11] D. Beltiţă, J.E. Galé, *Universal objects in categories of reproducing kernels.* Rev. Mat. Iberoamericana **27** (2011), no. 1, 123–179.

[BG13] D. Beltiţă, J.E. Galé, *Linear connections for reproducing kernels on vector bundles.* Math. Z. (to appear; http://dx.doi.org/10.1007/s00209-013-1243-9).

[Ber09] B. Berndtsson, *Curvature of vector bundles associated to holomorphic fibrations.* Ann. of Math. (2) **169** (2009), no. 2, 531–560.

[BH98] W. Bertram, J. Hilgert, *Reproducing kernels on vector bundles.* in: *Lie Theory and Its Applications in Physics III*, World Scientific, Singapore, 1998, pp. 43–58.

[CG99] G. Corach, J.E. Galé, *On amenability and geometry of spaces of bounded representations.* J. London Math. Soc. **59** (1999), no. 2, 311–329.

[CD78] M.J. Cowen, R.G. Douglas, *Complex geometry and operator theory.* Acta Math. **141** (1978), no. 3-4, 187–261.

[DP12] Th. Delzant, P. Py, *Kähler groups, real hyperbolic spaces and the Cremona group.* Compos. Math. **148** (2012), no. 1, 153–184.

[De12] J.-P. Demailly, *Complex Analytic and Differential Geometry.* Open Content Book, version of June 21, 2012.
(http://www-fourier.ujf-grenoble.fr/ demailly/documents.html)

[DG01] M.J. Dupré, J.F. Glazebrook, *The Stiefel bundle of a Banach algebra.* Integral Equations Operator Theory **41** (2001), no. 3, 264–287.

[DG10] M.J. Dupré, J.F. Glazebrook, E. Previato, *Curvature of universal bundles of Banach algebras.* In: J.A. Ball, V. Bolotnikov, J.W. Helton, L. Rodman and I.M. Spitkovsky (eds.), *Topics in operator theory. Volume* 1. *Operators, matrices and analytic functions*, Oper. Theory Adv. Appl., 202, Birkhäuser Verlag, Basel, 2010, pp. 195–222.

[GH78] Ph. Griffiths, J. Harris, *Principles of algebraic geometry.* Pure and Applied Mathematics. Wiley-Interscience, New York, 1978.

[Gr69] Ph. Griffiths, *Hermitian differential geometry, Chern classes, and positive vector bundles.* in: D.C. Spencer, S. Iyanaga (eds.), *Global Analysis* (Papers in Honor of K. Kodaira), Univ. Tokyo Press, Tokyo; Princeton University Press, Princeton, N.J., pp. 185–251.

[KM58] J.-L. Koszul, B. Malgrange, *Sur certaines structures fibrées complexes.* Arch. Math. (Basel) **9** (1958), 102–109.

[KM97] A. Kriegl, P.W. Michor, *The Convenient Setting of Global Analysis.* Mathematical Surveys and Monographs, 53. American Mathematical Society, Providence, RI, 1997.

[Lg01] S. Lang, *Fundamentals of Differential Geometry* (corrected second printing), Graduate Texts in Mathematics, 191. Springer-Verlag, New York, 2001.

[Lz04] R. Lazarsfeld, *Positivity in algebraic geometry.* I–II. Ergebnisse der Mathematik und ihrer Grenzgebiete. 3. Folge. A Series of Modern Surveys in Mathematics, 48. Springer-Verlag, Berlin, 2004.

[Ll98] L. Lempert, *The Dolbeault complex in infinite dimensions. I.* J. Amer. Math. Soc. **11** (1998), no. 3, 485–520.

[MP97] M. Monastyrski, Z. Pasternak-Winiarski, *Maps on complex manifolds into Grassmann spaces defined by reproducing kernels of Bergman type.* Demonstratio Math. **30** (1997), no. 2, 465–474.

[Ne00] K.-H. Neeb, *Holomorphy and Convexity in Lie Theory.* de Gruyter Expositions in Mathematics 28, Walter de Gruyter & Co., Berlin, 2000.

[Ne06] K.-H. Neeb, *Towards a Lie theory of locally convex groups.* Japanese J. Math. **1** (2006), no. 2, 291–468.

[Ne13] K.-H. Neeb, *Holomorphic realization of unitary representations of Banach–Lie groups*, in: A. Huckleberry, I. Penkov, G. Zuckerman (eds.), *Lie Groups: Structure, Actions, and Representations* (in Honor of Joseph A. Wolf on the Occasion of his 75th Birthday) Progress in Mathematics 306, Birkhäuser, Springer, 2013, pp. 185–223.

[Nl69] E. Nelson, *Topics in Dynamics. I: Flows.* Mathematical Notes. Princeton University Press, Princeton, N.J.; University of Tokyo Press, Tokyo, 1969.

[Od88] A. Odzijewicz, On reproducing kernels and quantization of states. *Comm. Math. Phys.* **114** (1988), no. 4, 577–597.

[Od92] A. Odzijewicz, Coherent states and geometric quantization. *Comm. Math. Phys.* **150** (1992), no. 2, 385–413.

[PR86] H. Porta, L. Recht, *Classification of linear connections.* J. Math. Anal. Appl. **118** (1986), no. 2, 547–560.

[We08] R.O. Wells, Jr., *Differential Analysis on Complex Manifolds.* Third edition. With a new appendix by O. Garcia-Prada. Graduate Texts in Mathematics, 65. Springer, New York, 2008.

Daniel Beltiţă
Institute of Mathematics "Simion Stoilow"
of the Romanian Academy
P.O. Box 1-764
Bucharest, Romania
e-mail: `beltita@gmail.com`
`Daniel.Beltita@imar.ro`

José E. Galé
Departamento de matemáticas and I.U.M.A.
Universidad de Zaragoza
50009 Zaragoza, Spain
e-mail: `gale@unizar.es`

Ordered Structures and Applications: Positivity VII
Trends in Mathematics, 77–90

Algebraic Absolutely Invertible Elements in Archimedean Riesz Algebras

Fethi Ben Amor and Karim Boulabiar

Abstract. Let $\mathfrak{A}$ be an Archimedean Riesz algebra with a positive unit element e. An element $f \in \mathfrak{A}$ is said to be algebraic if $P(f) = 0$ for some non-zero polynomial P with real coefficients. Moreover, f is called an e-step function in $\mathfrak{A}$ if there exist pairwise disjoint components $p_1, \ldots, p_n$ of e and real numbers $\alpha_1, \ldots, \alpha_n$ such that

$$e = p_1 + \cdots + p_n \quad \text{and} \quad f = \alpha_1 p_1 + \cdots + \alpha_n p_n.$$

First, we shall prove that if $\mathfrak{A}$ is an f-algebra, then f is algebraic if and only if f is an e-step function. This leads to the main result of this paper, which asserts that if f is an absolutely invertible element (i.e., $|f|$ is invertible and its inverse $|f|^{-1}$ is positive) in an arbitrary Archimedean Riesz algebra with positive identity, then f is algebraic if and only if f has an e-step function power in $\mathfrak{A}$. As a consequence, we obtain all previous results by Boulabiar, Buskes, and Sirotkin who investigated the special case of disjointness preserving operators on Archimedean Riesz spaces.

Mathematics Subject Classification (2010). Primary 06F25; Secondary 47B25.

Keywords. Absolutely invertible, algebraic, disjointness preserving, lattice-ordered algebra, order-bounded, step function.

1. Introduction

Let n be a positive integer and $\mathrm{M}_n(\mathbb{R})$ indicate the Archimedean Riesz algebra of all $n \times n$-matrices with entries in the field $\mathbb{R}$ of all real numbers. Choose an invertible positive matrix M in $\mathrm{M}_n(\mathbb{R})$ and assume that its inverse is again positive. Hence, M defines a lattice isomorphism on the Riesz space $\mathbb{R}^n$. Thus, the replacement of each nonzero entry by 1 transforms M into a permutation matrix (see [16]). It readily follows that M has a diagonal power. In what follows, the

This reaserch is supported by the LATAO-LR11ES12 grant of the Tunis El Manar University.

extent to which such a result can be generalized to a wider class of Riesz algebras is investigated. In the general setting of a Riesz algebra with a positive unit element e, diagonals are naturally replaced by e-step functions. Moreover, when leaving the realm of finite-dimensional spaces, some reasonable constraints have to be imposed. In this respect, we have been largely motivated by the recent papers [5, 6] (see also [4]) which deal with operators under polynomial constraints, viz., algebraic order-bounded operators on Riesz spaces (see [10]). A more detailed overview seems to be in order.

Let $\mathfrak{A}$ be an Archimedean Riesz algebra (i.e., an Archimedean lattice-ordered algebra) with a positive unit element e. Our first result asserts that if f is an e-step function in $\mathfrak{A}$ (see [1] or [18] for the definition and elementary properties), then f is an algebraic element in $\mathfrak{A}$. Then, we prove that the converse does not hold, unless $\mathfrak{A}$ is in addition a function algebra (briefly, f-algebra) in the sense of Birkhoff and Pierce [7]. However, our main result is certainly the following. Let f be an absolutely invertible element in $\mathfrak{A}$, that is to say, $|f|$ is invertible and its inverse $|f|^{-1}$ is positive. Then, f is algebraic if and only if f has an e-step function power. This result is based upon the fact that the principal band generated by e in a uniformly complete Riesz algebra is a projection band. Using duality arguments, we show that all results obtained in [5, 6] for disjointness preserving operators turn out to be special cases of our study.

The paper is organized as follows. After this introductory section, we provide a complete description of algebraic elements in Archimedean f-algebra with identity as step functions. The third section contains the main result of this paper. Namely, if $\mathfrak{A}$ is an Archimedean Riesz algebra with a positive identity e, then an element $f \in \mathfrak{A}$ is algebraic if and only if f^p is an e-step function for some nonzero natural number. In the last section, we prove that characterizations of algebraic order-bounded disjointness preserving operators on Archimedean Riesz spaces obtained in [5, 6] can be derived from our main result.

Finally, we point out that our references on Riesz spaces and disjointness preserving (linear) operators are the great texts [1] by Aliprantis and Burkinshaw, and [14] by Meyer-Nieberg. However, the standard monograph [19] by Zaanen seems to be the most complete reference on Riesz algebras, f-algebras and orthomorphisms.

2. Step functions in Riesz algebras

As usual, the symbol $\mathbb{R}[X]$ is used to indicate the principal ideal ring of all polynomials with coefficients in $\mathbb{R}$ [13]. Let $\mathfrak{A}$ be an arbitrary associative real algebra $\mathfrak{A}$ with a unit element e. An element $f \in \mathfrak{A}$ is said to be *algebraic* if $P(f) = 0$ for some nonzero $P \in \mathbb{R}[X]$. In this case, the unique monic generator of the ring ideal $\{P \in \mathbb{R}[X] : P(f) = 0\}$ in $\mathbb{R}[X]$ is denoted by π_f and is called the *minimal polynomial* of f. Let f be an algebraic element in $\mathfrak{A}$. It is not hard to see that the set

$$\mathbb{R}[f] = \{P(f) : P \in \mathbb{R}[X]\}$$

is a finite-dimensional commutative subalgebra of $\mathfrak{A}$. By the way, the dimension of $\mathbb{R}[f]$ equals the degree of the minimal polynomial of f. It follows that all powers of f are again algebraic.

Beginning with this sentence, $\mathfrak{A}$ stands for an Archimedean Riesz algebra with a positive unit element e.

A subalgebra $\mathfrak{B}$ of $\mathfrak{A}$ is called a *Riesz subalgebra* of $\mathfrak{A}$ if $\mathfrak{B}$ is a Riesz subspace (i.e., a vector sublattice) of the underlying Riesz space of $\mathfrak{A}$. An element $p \in \mathfrak{A}$ is called an *e-component* in $\mathfrak{A}$ if $p \wedge (e-p) = 0$. The set of all e-components in $\mathfrak{A}$ is denoted by $\mathcal{C}_e(\mathfrak{A})$. Clearly, $0 \leq p \leq e$ and $e - p \in \mathcal{C}_e(\mathfrak{A})$ for all $p \in \mathcal{C}_e(\mathfrak{A})$. Moreover,

$$pq = 0 \quad \text{for all } p, q \in \mathcal{C}_e(\mathfrak{A}) \text{ with } p \wedge q = 0. \tag{1}$$

Indeed,

$$0 \leq pq = (pq) \wedge (pq) \leq (pe) \wedge (eq) = p \wedge q = 0.$$

It follows in particular that

$$p^2 = p \quad \text{for all } p \in \mathcal{C}_e(\mathfrak{A}). \tag{2}$$

By an *e-step function* in $\mathfrak{A}$ we mean an element $s \in \mathfrak{A}$ for which there exist pairwise disjoint $p_1, \dots, p_n \in \mathcal{C}_e(\mathfrak{A})$ and $\alpha_1, \dots, \alpha_n \in \mathbb{R}$ such that

$$e = p_1 + \dots + p_n \quad \text{and} \quad s = \alpha_1 p_1 + \dots + \alpha_n p_n.$$

This last expression is referred to as an *e-representation* of s. It turns out that the set of all e-step functions in $\mathfrak{A}$ is a Riesz subspace of the underlying Riesz space of $\mathfrak{A}$. (see [1] or [18]). Moreover, from (1) and (2) it follows quickly that the set of all e-step functions in $\mathfrak{A}$ is a Riesz subalgebra of $\mathfrak{A}$. The following simple lemma will be useful for later purposes.

Lemma 1. *Let $\mathfrak{A}$ be an Archimedean Riesz algebra with a positive unit element e and s be an e-step function in $\mathfrak{A}$ with an e-representation*

$$s = \alpha_1 p_1 + \dots + \alpha_n p_n.$$

Then

$$P(s) = P(\alpha_1) p_1 + \dots + P(\alpha_n) p_n \quad \textit{for all } P \in \mathbb{R}[X].$$

Proof. Since the equality $e = p_1 + \dots + p_n$ holds, a simple induction via (1) and (2) reveals that

$$s^k = \alpha_1^k p_1 + \dots + \alpha_n^k p_n \quad \text{for all } k \in \{0, 1, \dots\}.$$

By linearity, we derive that

$$P(s) = P(\alpha_1) p_1 + \dots + P(\alpha_n) p_n \quad \text{for all } P \in \mathbb{R}[X]$$

and we are done. □

Another lemma is needed to prove the main result of this section.

Lemma 2. *Let $\mathfrak{A}$ be an Archimedean Riesz algebra with a positive unit element e and $u, v \in \mathfrak{A}^+$ for which there exist $f \in \mathfrak{A}$ and $\alpha, \beta \in \mathbb{R}$ such that $\alpha \neq \beta$ and*

$$|f - \alpha e|\, u = |f - \beta e|\, v = 0.$$

Then $u \wedge v = 0$.

Proof. If such f and α, β exist then

$$\begin{aligned} 0 &\leq |f\,(u \wedge v) - \alpha\,(u \wedge v)| \\ &\leq |f - \alpha e|\,(u \wedge v) \leq |f - \alpha e|\, u = 0. \end{aligned}$$

Whence,

$$f\,(u \wedge v) = \alpha\,(u \wedge v)\,.$$

Analogously,

$$f\,(u \wedge v) = \beta\,(u \wedge v)\,.$$

Therefore,

$$\alpha\,(u \wedge v) = \beta\,(u \wedge v)\,.$$

Since $\alpha \neq \beta$, we get $u \wedge v = 0$, as desired. □

The connection between algebraic elements and e-step functions in $\mathfrak{A}$ is established next.

Theorem 3. *Let $\mathfrak{A}$ be an Archimedean Riesz algebra with a positive unit element e. Then, any e-step function in $\mathfrak{A}$ is algebraic. The converse is true if $\mathfrak{A}$ is in addition an f-algebra.*

Proof. Let s be an e-step function in $\mathfrak{A}$ with e-representation

$$s = \alpha_1 p_1 + \cdots + \alpha_n p_n.$$

Define the nonzero polynomial

$$P\,(X) = (X - \alpha_1) \cdots (X - \alpha_n) \in \mathbb{R}\,[X]\,.$$

By Lemma 1, we have

$$P\,(s) = P\,(\alpha_1)\, p_1 + \cdots + P\,(\alpha_n)\, p_n = 0.$$

This means that s is algebraic.

Conversely, suppose that $\mathfrak{A}$ is an f-algebra and let f be an algebraic element in $\mathfrak{A}$. Assume that there exist $a, b \in \mathbb{R}$ and $P \in \mathbb{R}\,[X]$ such that $b^2 - a < 0$ and

$$\pi_f\,(X) = \left(X^2 + 2bX + a\right) P\,(X)\,.$$

Observe that

$$0 < \left(a - b^2\right) e \leq \left(a - b^2\right) e + (f + be)^2 = f^2 + 2bf + ae.$$

Hence,

$$\begin{aligned} 0 &\leq \left(a - b^2\right) |P\,(f)| \leq \left(f^2 + 2bf + ae\right) |P\,(f)| \\ &= \left|\left(f^2 + 2bf + ae\right) P\,(f)\right| = |\pi_f\,(f)| = 0. \end{aligned}$$

This yields that $P(f) = 0$, contradicting the minimality of π_f. It follows that there are no quadratic polynomials in the factorization of π_f into irreductible elements in $\mathbb{R}[X]$. Consequently, there exist $k_1, \ldots, k_n \in \{1, 2, \ldots\}$ and pairwise different $\alpha_1, \ldots, \alpha_n \in \mathbb{R}$ such that

$$\pi_f(X) = (X - \alpha_1)^{k_1} \cdots (X - \alpha_n)^{k_n}.$$

Therefore,

$$(f - \alpha_1 e)^{k_1} \cdots (f - \alpha_n e)^{k_n} = \pi_f(f) = 0.$$

Hence,

$$[(f - \alpha_1 e) \cdots (f - \alpha_n e)]^{k_1 \cdots k_n} = 0.$$

But then

$$(f - \alpha_1 e) \cdots (f - \alpha_n e) = 0$$

because 0 is the only nilpotent element in an f-algebra with unit element (see [19]). We derive that

$$\pi_f(X) = (X - \alpha_1) \cdots (X - \alpha_n).$$

Define

$$P_k(X) = \frac{\pi_f(X)}{X - \alpha_k} \in \mathbb{R}[X] \quad \text{for all } k \in \{1, \ldots, n\}.$$

By the classical Bézout Theorem, there exist $Q_1, \ldots, Q_n \in \mathbb{R}[X]$ such that

$$1 = P_1 Q_1 + \cdots + P_n Q_n.$$

This yields that

$$P_1(f) Q_1(f) + \cdots + P_n(f) Q_n(f) = e.$$

Put

$$u_k = P_k(f) Q_k(f) \text{ and } p_k = |u_k| \quad \text{for all } k \in \{1, \ldots, n\}.$$

Hence,

$$u_1 + \cdots + u_n = e.$$

Moreover,

$$(f - \alpha_k e) u_k = \pi_f(f) Q_k(f) = 0 \quad \text{for all } k \in \{1, \ldots, n\}.$$

Hence,

$$|f - \alpha_k e| p_k = |f - \alpha_k e| |u_k| = |(f - \alpha_k e) u_k| = 0 \quad \text{for all } k \in \{1, \ldots, n\}.$$

By Lemma 2, we derive that $p_1, \ldots, p_n$ are pairwise disjoint and so are $u_1, \ldots, u_n$. Then,

$$e = |e| = |u_1 + \cdots + u_n| = |u_1| + \cdots + |u_n| = p_1 + \cdots + p_n.$$

This yields quickly that $p_1, \ldots, p_n \in \mathcal{C}_e(\mathfrak{A})$. Furthermore,

$$|f p_k - \alpha_k p_k| = |f - \alpha_k e| p_k = 0 \quad \text{for all } k \in \{1, \ldots, n\}.$$

So,

$$\begin{aligned} f &= fe = f(p_1 + \cdots + p_n) \\ &= fp_1 + \cdots + fp_n = \alpha_1 p_1 + \cdots + \alpha_n p_n. \end{aligned}$$

This means that f is an e-step function in $\mathfrak{A}$ and completes the proof. □

We notice finally that, without any extra condition on the Riesz algebra $\mathfrak{A}$, an algebraic element in $\mathfrak{A}$ need not be an e-step function in $\mathfrak{A}$. This can be seen directly with the Riesz algebra of all 2×2 matrices with entries in $\mathbb{R}$.

3. Algebraic absolutely invertible elements

Again in this section, $\mathfrak{A}$ *is an Archimedean* Riesz algebra *with a positive unit element* e. The principal band generated by an element f in $\mathfrak{A}$ is denoted by $\mathcal{B}(f)$. Furthermore, a Riesz subalgebra of $\mathfrak{A}$ which is an f-algebra is called an *f-subalgebra* of $\mathfrak{A}$. For instance, it is shown in [11] that $\mathcal{B}(e)$ is an f-subalgebra of $\mathfrak{A}$ with e as a unit element. Moreover, if $\mathfrak{A}$ in addition uniformly complete [14], then $\mathcal{B}(e)$ is a projection band in $\mathfrak{A}$, that is, the direct sum

$$\mathcal{B}(e)^{\mathrm{d}} \oplus \mathcal{B}(e) = \mathfrak{A}$$

holds, where

$$\mathcal{B}(e)^{\mathrm{d}} = \{f \in \mathfrak{A} : |f| \wedge e = 0\}$$

(this result is proved in [8]). The band projection on $\mathcal{B}(e)$ is denoted by $\mathcal{P}_e$. In particular, $\mathcal{P}_e$ is an element in the f-algebra $\mathrm{Orth}(\mathfrak{A})$ of all orthomorphisms on $\mathfrak{A}$, and the inequalities

$$0 \leq \mathcal{P}_e \leq I,$$

hold, where I is the identity operator on $\mathfrak{A}$ [1]. Another property of $\mathcal{P}_e$ is labeled and proved next.

Lemma 4. *Let* $\mathfrak{A}$ *be a uniformly complete Riesz algebra with a positive unit element* e. *Then,*

$$\mathcal{P}_e(f)\,\mathcal{P}_e(g) \leq \mathcal{P}_e(fg) \quad \text{for all } f, g \in \mathfrak{A}^+.$$

In particular,

$$\mathcal{P}_e(f)^k \leq \mathcal{P}_e\left(f^k\right) \quad \text{for all } f \in \mathfrak{A}^+ \text{ and } k \in \{1, 2, \dots\}.$$

Proof. Let $f, g \in \mathfrak{A}^+$ and notice that $0 \leq \mathcal{P}_e(f) \leq f$ and $0 \leq \mathcal{P}_e(g) \leq g$. It follows that

$$0 \leq \mathcal{P}_e(f)\,\mathcal{P}_e(g) \leq fg.$$

Since $\mathcal{P}_e(f)\,\mathcal{P}_e(g) \in \mathcal{B}(e)$, we get

$$0 \leq \mathcal{P}_e(\mathcal{P}_e(f)\,\mathcal{P}_e(g)) = \mathcal{P}_e(f)\,\mathcal{P}_e(g) \leq \mathcal{P}_e(fg),$$

as desired. The second inequality follows from the first one via an obvious induction process. □

For every $f \in \mathfrak{A}$ we define the multiplication operators L_f and R_f on $\mathfrak{A}$ by

$$L_f(g) = fg \text{ and } R_f(g) = gf \quad \text{for all } g \in \mathfrak{A}.$$

Clearly, if $f \in \mathfrak{A}$ then L_f and R_f are order bounded. Moreover, L_f and R_f are positive whenever f is positive in $\mathfrak{A}$. The following lemmas play key roles in the proof of the main result of this section.

Lemma 5. *Let $\mathfrak{A}$ be an Archimedean Riesz algebra with a positive unit element e. If $f \in \mathcal{B}(e)$ and $g \in \mathfrak{A}$ such that $f^n g = 0$ for some $n \in \{2, 3, \dots\}$, then $fg = 0$.*

Proof. Since $f \in \mathcal{B}(e)$, Proposition 1 in [11] implies that $L_f \in \mathrm{Orth}(\mathfrak{A})$. Then,

$$L_{f^n} = (L_f)^n \in \mathrm{Orth}(\mathfrak{A}).$$

Moreover, since $n \geq 2$, Theorem 2.52 in [1] and the equality $f^n g = 0$ imply that

$$\mathrm{Im}(L_f) \ni (L_f)^{n-1} g \in \ker(L_f) = [\mathrm{Im}(L_f)]^{\mathrm{d}}.$$

Therefore $(L_f)^{n-1}(g) = 0$. Repeating this argument again $n-2$ times we finally find $L_f(g) = 0$ and the lemma follows. □

Lemma 6. *Let $\mathfrak{A}$ be a uniformly complete Riesz algebra with a positive unit element e and $f \in \mathfrak{A}$ such that $L_{|f|}$ is a lattice homomorphism on $\mathfrak{A}$. Then*

$$|\mathcal{P}_e(f)| \, |f - \mathcal{P}_e(f)| = 0.$$

Proof. Put

$$u = \mathcal{P}_e(f) \in \mathcal{B}(e) \quad \text{and} \quad v = f - \mathcal{P}_e(f) \in \mathcal{B}(e)^{\mathrm{d}}.$$

Hence, $f = u + v$ and $|f| = |u| + |v|$. Since $L_{|f|}$ is a lattice homomorphism and $e \wedge |v| = 0$, we obtain

$$L_{|f|}(e) \wedge L_{|f|}(|v|) = |f| \wedge (|f| \, |v|) = 0.$$

These equalities together with the inequalities $|v| \leq |f|$ and $|u| \, |v| \leq |f| \, |v|$ yield that $|v| \wedge |u| \, |v| = 0$. This means that

$$|u| \, |v| \in \mathcal{B}(v)^{\mathrm{d}}. \tag{3}$$

On the other hand, $L_{|u|}$ is an orthomorphism on $\mathfrak{A}$ because $u \in \mathcal{B}(e)$ (where we use Theorem 1 in [8] or Proposition 1 in [11]). But then

$$|u| \, |v| = L_{|u|}(|v|) \in \mathcal{B}(v) \tag{4}$$

because orthomorphisms preserve bands. Combining (3) and (4) we obtain $|u| \, |v| = 0$ and the proof is complete. □

An element f in $\mathfrak{A}$ is said to be *absolutely invertible* in $\mathfrak{A}$ if $|f|$ has positive inverse $|f|^{-1}$ in $\mathfrak{A}$. Next, we collect some facts on the multiplication operators on $\mathfrak{A}$ in connection with absolutely invertible elements in $\mathfrak{A}$.

Lemma 7. *Let f be an absolutely invertible element in an Archimedean Riesz algebra $\mathfrak{A}$ with a positive unit element. Then the following hold.*

(i) $L_{|f|}$ *and* $R_{|f|}$ *are lattice isomorphisms.*
(ii) L_f *and* R_f *are order-bounded disjointness preserving operators.*
(iii) $|f|^k = |f^k|$ *for all* $k \in \{1, 2, \dots\}$.

Proof. (i) As previously observed, $L_{|f|}$ is a positive operator. Moreover, it is readily checked that $L_{|f|}$ is bijective and that $(L_{|f|})^{-1} = L_{|f|^{-1}}$. It follows that $(L_{|f|})^{-1}$ is again positive. By Theorem 2.15 in [1], we derive that $L_{|f|}$ is a lattice isomorphism and so is $R_{|f|}$.

(ii) We already observed that L_f is order bounded. Hence, choose $g, h \in \mathfrak{A}$ with $|g| \wedge |h| = 0$. Using (i), we can write

$$\begin{aligned}0 &\leq |L_f(g)| \wedge |L_f(h)| = |fg| \wedge |fh| \\ &\leq (|f|\,|g|) \wedge (|f|\,|h|) = L_{|f|}(|g|) \wedge L_{|f|}(|h|) \\ &= L_{|f|}(|g| \wedge |h|) = 0.\end{aligned}$$

It follows that L_f preserves disjointness. The same proof works for R_f.

(iii) We argue by induction. Let $k \in \{1, 2, \dots\}$ and assume that $|f|^k = |f^k|$. By (ii) and Theorem 3.1.4 in [14], we get

$$|f^{k+1}| = |ff^k| = |L_f(f^k)| = \left|L_f\left(|f|^k\right)\right| = \left|R_{|f|^k}(f)\right|.$$

Now, observe that $|f|^k$ is again absolutely invertible in $\mathfrak{A}$. Applying (i) to $|f|^k$, we obtain

$$|f^{k+1}| = \left|R_{|f|^k}(f)\right| = R_{|f|^k}(|f|) = |f|^{k+1},$$

completing the induction process. □

The order ideal in a Riesz space generated by $f_1, \dots, f_n$ is denoted by $\mathcal{I}(f_1, \dots, f_n)$. The following lemma is the last one we need before proving the central theorem of this section.

Lemma 8. *Let f be an absolutely invertible element in a uniformly complete Riesz algebra $\mathfrak{A}$ with a positive unit element e. If $e \in \mathcal{I}(f, f^2, \dots, f^n)$ for some $n \in \{1, 2, \dots\}$, then $f^{n!} \in \mathcal{B}(e)$.*

Proof. Lemma 7 (i) asserts that $L_{|f|}$ is a lattice isomorphism and so is $(L_{|f|})^{n!}$. Since $(L_{|f|})^{n!} = L_{|f|^{n!}}$ and $|f^{n!}| = |f|^{n!}$ (where we use Lemma 7 (iii)), we derive that $L_{|f^{n!}|}$ is again a lattice homomorphism. By Lemma 6, we get that

$$|P_e(f^{n!})|\,|f^{n!} - \mathcal{P}_e(f^{n!})| = 0.$$

Moreover, if $k \in \{1, \dots, n\}$ then Lemma 4 and Lemma 7 (iii) yield that

$$|\mathcal{P}_e(f^{n!})| = \mathcal{P}_e(|f^{n!}|) = \mathcal{P}_e\left(|f|^{n!}\right) = \mathcal{P}_e\left(\left(|f|^k\right)^{n!/k}\right) \geq \mathcal{P}_e\left(|f|^k\right)^{n!/k}$$

(the first equality comes from Theorem 2.40 in [1]). Hence,

$$0 \leq \mathcal{P}_e\left(|f|^k\right)^{n!/k}\left|f^{n!} - \mathcal{P}_e\left(f^{n!}\right)\right| \leq \left|\mathcal{P}_e\left(f^{n!}\right)\right|\left|f^{n!} - \mathcal{P}_e\left(f^{n!}\right)\right| = 0$$

Therefore,

$$\mathcal{P}_e\left(|f|^k\right)^{n!/k}\left|f^{n!} - \mathcal{P}_e\left(f^{n!}\right)\right| = 0.$$

By Lemma 5, we obtain

$$\mathcal{P}_e\left(|f|^k\right)\left|f^{n!} - \mathcal{P}_e\left(f^{n!}\right)\right| = 0 \quad \text{for all } k \in \{1, \ldots, n\}. \tag{5}$$

Since $e \in \mathcal{I}\left(f, f^2, \ldots, f^n\right)$, there is $a \in \mathbb{R}$ for which

$$e \leq a\left(|f| + \left|f^2\right| + \cdots + |f^n|\right).$$

Therefore,

$$e = \mathcal{P}_e(e) \leq a\left(\mathcal{P}_e(|f|) + \mathcal{P}_e\left(|f|^2\right) + \cdots + \mathcal{P}_e(|f|^n)\right). \tag{6}$$

Combining (5) and (6), we derive that

$$\begin{aligned}\left|f^{n!} - \mathcal{P}_e\left(f^{n!}\right)\right| &= e\left|f^{n!} - \mathcal{P}_e\left(f^{n!}\right)\right| = \mathcal{P}_e(e)\left|f^{n!} - \mathcal{P}_e\left(f^{n!}\right)\right| \\ &\leq a\left(\mathcal{P}_e(|f|) + \mathcal{P}_e\left(|f|^2\right) + \cdots + \mathcal{P}_e(|f|^n)\right)\left|f^{n!} - \mathcal{P}_e\left(f^{n!}\right)\right| = 0\end{aligned}$$

Finally, $f^{n!} = \mathcal{P}_e\left(f^{n!}\right) \in \mathcal{B}(e)$. This completes the proof. □

An element $f \in \mathfrak{A}$ is said to have *an e-step function power* in $\mathfrak{A}$ if f^n is an e-step function in $\mathfrak{A}$ for some $n \in \{2, 3, \ldots\}$. We are in position now to prove the main (and the last) result of this section.

Theorem 9. *Let f be an absolutely invertible element in an Archimedean Riesz algebra $\mathfrak{A}$ with a positive unit element e. Then, f is algebraic if and only if f has an e-step function power in $\mathfrak{A}$.*

Proof. If f^n is an e-step function in $\mathfrak{A}$ for some $n \in \{2, 3, \ldots\}$, then Theorem 3 implies that f^n is algebraic and so is f.

Conversely, suppose that f is algebraic in $\mathfrak{A}$. We have to prove that f has an e-step function power. We start with the case where $\mathfrak{A}$ is uniformly complete. Put

$$\pi_f(X) = a_0 + a_1 X + \cdots + a_n X^n$$

with $a_n \neq 0$. Let m denote the multiplicity of 0 as a root of π_f. Therefore,

$$a_m \neq 0 \quad \text{and} \quad a_m f^m + \cdots + a_n f^n = 0.$$

Thus,

$$|f|^m \leq \left|\frac{a_{m+1}}{a_m}\right| |f|^{m+1} + \cdots + \left|\frac{a_n}{a_m}\right| |f|^n.$$

Observe that $m < n$. Otherwise, f is nilpotent and, by Lemma 7 (iii), so is $|f|$. This contradicts the fact that f is absolutely invertible in $\mathfrak{A}$. Thus, we may write

$$e \leq \left|\frac{a_{m+1}}{a_m}\right| |f| + \cdots + \left|\frac{a_n}{a_m}\right| |f|^{n-m}.$$

Therefore, $e \in \mathcal{I}\left(f, f^2, \ldots, f^n\right)$ and so $f^{n!} \in \mathcal{B}(e)$ (where we use Lemma 8). Furthermore, $f^{n!}$ is algebraic because f is algebraic. In summary, $f^{n!}$ is an algebraic element of $\mathcal{B}(e)$, which is f-subalgebra of $\mathfrak{A}$. From Theorem 3 it follows that $f^{n!}$ is an e-step function in $\mathfrak{A}$.

Now, we focus on the general case. It is well known that the multiplication in $\mathfrak{A}$ can be extended to its uniformly completion $\mathfrak{A}^{\mathrm{ru}}$ so that $\mathfrak{A}^{\mathrm{ru}}$ becomes a uniformly complete Riesz algebra with the same unit element e (see [17]). It is obvious that f is again algebraic in $\mathfrak{A}^{\mathrm{ru}}$. By the first case, there exists $n \in \{2, 3, \}$ such that f^n is an e-step function in $\mathfrak{A}^{\mathrm{ru}}$ with e-representation

$$f^n = \alpha_1 p_1 + \cdots + \alpha_s p_s.$$

Here, $p_k \in \mathcal{C}_e\left(\mathfrak{A}^{\mathrm{ru}}\right)$ for all $k \in \{1, \ldots, s\}$. We claim that f^n is an e-step function in $\mathfrak{A}$. To this end, it suffices to prove that $p_k \in \mathfrak{A}$ for all $k \in \{1, \ldots, s\}$. First, observe that if p, q are disjoint e-components then $p+q$ is again an e-component. It follows readily that we can assume without lose of generality that $\alpha_1, \ldots, \alpha_s$ are pairwise different. Then, put

$$Q(X) = (X - \alpha_1) \cdots (X - \alpha_s)$$

and

$$Q_k(X) = \frac{Q(X)}{X - \alpha_k} \quad \text{for all } k \in \{1, \ldots, n\}.$$

Pick $k \in \{1, \ldots, s\}$ and observe that $Q_k(\alpha_k) \neq 0$ and $Q_k(\alpha_i) = 0$ if $i \neq k$. These observations together with Lemma 1 yield that

$$Q_k(f^n) = Q_k(\alpha_1) p_1 + \cdots + Q_k(\alpha_s) p_s = Q_k(\alpha_k) p_k,$$

so

$$p_k = \frac{1}{Q_k(\alpha_k)} Q_k(f^n) \in \mathfrak{A}.$$

This completes the proof of the theorem. □

4. Algebraic disjointness preserving operators revisited

The main purpose of this section is to apply Theorem 9 to order-bounded disjointness preserving operators on an Archimedean Riesz space. Our approach relies heavily on the notion of duality in Riesz spaces (see [1, 14]).

Throughout this last section, E stands for an Archimedean Riesz space.

If T is an order-bounded operator on E, then its adjoint $T^{\sim}$ is defined on the order dual $E^{\sim}$ of E by

$$T^{\sim}(\varphi) = \varphi \circ T \quad \text{for all } \varphi \in E^{\sim}.$$

Moreover, recall that any order-bounded disjointness preserving operator (called *Lamperti operator* by Arendt in [2]) T on E has an absolute value $|T|$ in the ordered real algebra $\mathfrak{L}^{\mathrm{b}}(E)$ of all order-bounded operators on E (see [14]). By the way, the first result of this section concerns the adjoint (and its absolute value) of a bijective order-bounded disjointness preserving operator on E. We omit the proof because it is just a slight modification of Proposition 2.7 in [2] by Arendt, who obtained the result for Banach lattices.

Proposition 10. *Let E be an Archimedean Riesz space and T be a bijective order-bounded disjointness preserving operator on E. Then $T^{\sim}$ is again an order-bounded disjointness preserving operator and $|T|^{\sim} = |T^{\sim}|$.*

As usual, a subset D of the Archimedean Riesz space is said to be *T-invariant*, where T is an order-bounded operator on E, if T sends D to itself. Furthermore, recall that E^{ru} indicates the uniform completion of E (see [14]). The next lemma is a crucial step in the proof of the central result of this section.

Lemma 11. *Let E be an Archimedean Riesz space and T be an algebraic order-bounded disjointness preserving operator on E. Then the following holds.*

(i) *T has a unique extension T^{ru} to E^{ru} such that T^{ru} is an algebraic order-bounded disjointness preserving operator on E^{ru} and $\pi_{T^{\mathrm{ru}}} = \pi_T$.*

(ii) *Every $f \in E$ is contained in a T-invariant principal order ideal of E.*

Proof. (i) See Lemma 5.1 in [5].

(ii) This follows directly from Lemma 5.2 in [5]. □

To proceed our investigation, we have to give further notations and remarks. First, Let I denote the identity operator on any given space. Furthermore, let $m, n \in \{1, 2, \dots\}$ and T be an order-bounded operator on E. The range of T^m on E is denoted by $\mathrm{Im}(T^m)$. Next, we shall obtain as consequences of Theorem 9 the main results of [5, 6]. To hit this objective, we need a result from [3], viz., if T is an algebraic operator on an arbitrary vector space then T is injective if and only if T is surjective.

Theorem 12. *Let E be a Riesz space and T be an order-bounded disjointness preserving operator on E. If T is injective or surjective, then T is algebraic if and only if T has an I-step function power in* $\mathrm{Orth}(E)$.

Proof. The 'if' part is obvious. The 'only if' part needs a detailed proof. Suppose that T is algebraic and put $n = \deg(\pi_T)$. First, we claim that $T^{n!} \in \mathrm{Orth}(E)$. By Lemma 11 (i), we can assume without loss of generality that E is uniformly complete. Choose $f, g \in E$ such that $|f| \wedge |g| = 0$. Lemma 11 (ii) implies quickly that there exists a T-invariant principal order-ideal $\mathcal{I}(h)$ of E such that $f, g \in \mathcal{I}(h)$. The restriction of T to $\mathcal{I}(h)$, denoted by R, can be seen as an order-bounded disjointness preserving operator on $\mathcal{I}(h)$. Moreover, it is easily seen that R is algebraic and bijective. By Proposition 10, we derive that $R^{\sim}$ is an order-bounded disjointness preserving operator on $\mathcal{I}(h)^{\sim}$. Besides, it is routine to show that $R^{\sim}$

is algebraic and injective (so bijective). In particular, $R^{\sim}$ is an algebraic element in the Riesz space $\mathfrak{L}^{\mathrm{b}}\left(\mathcal{I}\left(h\right)^{\sim}\right)$ which is a Dedekind complete (and then Archimedean) Riesz algebra with I as a positive unit element. Furthermore, Proposition 10 yields that $R^{\sim}$ is an absolutely invertible element in $\mathfrak{L}^{\mathrm{b}}\left(\mathcal{I}\left(h\right)^{\sim}\right)$. By Theorem 9, we get

$$\left(R^{n}\right)^{\sim}=\left(R^{\sim}\right)^{n}\in\operatorname{Orth}\left(\mathcal{I}\left(h\right)^{\sim}\right)\quad\text{for some } n\in\{1,2,\dots\}.$$

On the other hand, since E is uniformly complete, $\mathcal{I}\left(h\right)$ is a Banach lattice and so $\mathcal{I}\left(h\right)^{\sim}$ separates the points of $\mathcal{I}\left(h\right)$ (see Proposition 1.2.13 in [14]). This implies, via Theorem 2.60 in [1], that $R^{n}\in\operatorname{Orth}\left(\mathcal{I}\left(h\right)\right)$. Consequently,

$$\left|T^{n}\left(f\right)\right|\wedge\left|g\right|=\left|R^{n}\left(f\right)\right|\wedge\left|g\right|=0.$$

Thus, $T^{n}\in\operatorname{Orth}\left(E\right)$. As T is algebraic, so is T^{n}. Hence, since $\operatorname{Orth}\left(E\right)$ is an Archimedean f-algebra with I as a unit element (see Theorem 3.1.10 in [14]), from Theorem 3 it follows that T^{n} is an I-step function in $\operatorname{Orth}\left(E\right)$. This completes the proof of the theorem. □

A disjointness preserving operator T on E is positive if and only if T is a lattice homomorphism on E. In this situation, each power T^{k} of T with $k\in\{1,2,\dots\}$ is again a lattice homomorphism and hence its range $\operatorname{Im}\left(T^{k}\right)$ is a T-invariant Riesz subspace of E. These observations lead to the last result of this paper (see [5, 6]).

Corollary 13. *Let E be an Archimedean Riesz space and T be a lattice homomorphism on E. Then T is algebraic if and only if there exist $m\in\{1,2,\dots\}$ such that the restriction of T to* $\operatorname{Im}\left(T^{m}\right)$ *has an I-step function power in* $\operatorname{Orth}\left(\operatorname{Im}\left(T^{m}\right)\right)$.

Proof. Assume that there is $m\in\{1,2,\dots\}$ for which the restriction R of T to $\operatorname{Im}\left(T^{m}\right)$ has an I-step function power in $\operatorname{Orth}\left(\operatorname{Im}\left(T^{m}\right)\right)$. It follows from Theorem 3 that R is algebraic. Hence, if $P\in\mathbb{R}\left[X\right]$ is a non zero polynomial such that $P\left(R\right)=0$, then $P\left(T\right)\circ T^{m}=0$. This means that T is algebraic, as required.

Conversely, suppose that T is algebraic and put

$$\pi_{T}(X)=X^{n}+a_{n-1}X^{n-1}+\cdots+a_{m}X^{m}$$

with $a_{m}\neq 0$ the minimal polynomial of T. Clearly, T induces a lattice homomorphism R on $\operatorname{Im}\left(T^{m}\right)$. Since T is algebraic, then R is algebraic as well. Moreover, R is injective. Indeed, let $g\in\operatorname{Im}\left(T^{m}\right)$ with $R\left(g\right)=0$ and pick $f\in E$ such that $g=T^{m}\left(f\right)$. Thus

$$T^{m+1}\left(f\right)=T\left(T^{m}\left(f\right)\right)=T\left(g\right)=R\left(g\right)=0.$$

Since $\pi_{T}\left(T\right)=0$ we get $a_{m}T^{m}\left(g\right)=0$. Hence, $f=T^{m}\left(g\right)=0$ because $a_{m}\neq 0$. By Theorem 12, there is $n\in\{1,2,\dots\}$ for which R^{n} is an I-step function in $\operatorname{Orth}\left(\operatorname{Im}\left(T^{m}\right)\right)$ and the proof is complete. □

We end the paper with the following remark (see [6]). Let T be an order-bounded disjointness preserving operator on the Archimedean Riesz space E. If T is either injective, surjective, or positive, then T is algebraic if and only if T^{p} is an I-step function in $\operatorname{Orth}\left(E\right)$ for some nonzero positive integer p. It turns out that, without extra condition, this equivalence fails as illustrated by the following simple example. Put $E = \mathbb{R}^2$ and define $T \in \mathfrak{L}_b(E)$ by

$$T(\alpha, \beta) = (\alpha, -\alpha) \quad \text{for all } \alpha, \beta \in \mathbb{R}.$$

Obviously, T is not an I-step function in $\mathrm{Orth}(E)$. However, it is readily verified that T is an order-bounded disjointness preserving operator on E. Also, one may check easily that

$$T^k = T \quad \text{for all } k \in \{1, 2, \dots\}.$$

Therefore, T is algebraic and has no I-step function powers.

References

[1] C.D. Aliprantis and O. Burkinshaw, *Positive Operators*, Springer-Verlag, Dordrecht, 2006.

[2] W. Arendt, Spectral properties of Lamperti operators, *Indiana Univ. Math. J.*, 32 (1983), 199–215.

[3] K. Boulabiar and G. Buskes, After the determinants are down: a criterion for invertibility, *Amer. Math. Monthly* 110 (2003), 737–741.

[4] K. Boulabiar, G. Buskes and G. Sirotkin, A strongly diagonal power of algebraic order-bounded disjointness preserving operators, *Electron. Res. Announc. Amer. Math. Soc.*, 9 (2003), 94–98.

[5] K. Boulabiar, G. Buskes, and G. Sirotkin, Algebraic order bounded disjointness preserving operators and strongly diagonal operators, *Integr. Equ. Oper. Theory*, 54 (2006), 9–31.

[6] K. Boulabiar, G. Buskes, and G. Sirotkin, Erratum to: Algebraic order bounded disjointness preserving operators and strongly diagonal operators, *Integr. Equ. Oper. Theory*, 75 (2013), 443–444.

[7] G. Birkhoff and R.S. Pierce, Lattice-ordered rings, *An. Acad. Brasil Ciênc.*, 28 (1956), 41–69.

[8] C.B. Huijsmans, Elements with unit spectrum in a Banach lattice algebra, *Indag. Math.*, 91 (1988), 43–51.

[9] C.B. Huijsmans and B. de Pagter, Invertible disjointness preserving operators, *Proc. Edinburgh Math. Soc.*, 37 (1994), 125–132.

[10] I. Kaplansky, *Infinite Abelian Groups*, University of Michigan Press, Ann Arbor, 1954.

[11] B. Lavrič, A note on Archimedean $\mathfrak{L}^b$gebras, *An. Stiint. Univ. Al. I. Cuza Iaşi Sect. I a Mat.*, 39 (1993), 397–400.

[12] W.A.J. Luxemburg and A.C. Zaanen, *Riesz spaces* I, North-Holland, Amsterdam, 1971.

[13] S. Mc Lane and G. Birkhoff, *Algebra*, Third edition, Chelsea Publishing Co., New York, 1988.

[14] P. Meyer-Nieberg, *Banach Lattices*, Springer-Verlag, Berlin, 1991.

[15] J. Quinn, Intermediate Riesz spaces, *Pacific J. Math.*, 56 (1975), 225–263.

[16] H.H. Schaefer, *Banach Lattices and Positive Operators*, Springer-Verlag, Berlin-Heidelberg-New York,

[17] A. Triki, On algebra homomorphisms in complex almost f-algebras, *Comment. Math. Univ. Carolinae* 43 (2002), 23–31.

[18] A.C. Zaanen, *Introduction to Operator Theory in Riesz Spaces*, Springer-Verlag, Berlin-Heidelberg-New York, 1997.

[19] A.C. Zaanen, *Riesz Spaces* II, North-Holland, Amsterdam, 1971.

Fethi Ben Amor and Karim Boulabiar
Research Laboratory of
Algebra-Topology-Arithmetic-Order
Department of Mathematics
Faculty of Sciences of Tunis
Tunis-El Manar University
2092 Tunis, Tunisia
e-mail: fethi.benamor@ipest.rnu.tn
karim.boulabiar@ipest.rnu.tn

Ordered Structures and Applications: Positivity VII
Trends in Mathematics, 91–98

σ-Weak Orthomorphisms

Elmiloud Chil and Mohamed Mokaddem

Abstract. In this paper we introduce a new class of weak orthomorphisms, so-called, σ-weak orthomorphisms. We prove that for a uniformly complete vector lattice, σ-weak orthomorphisms and σ-extended orthomorphisms coincide. As application we study some important structural properties of such operators.

Mathematics Subject Classification (2010). Primary 06F25; Secondary 46A40.

Keywords. Extended orthomorphism, weak orthomorphism, vector lattice.

The study of orthomorphisms on an Archimedean vector lattice E, that is, of band preserving order-bounded linear operator $T : E \to E$, has been greatly developed these last years. It is well known that the collection $\mathrm{Orth}(E)$ of all orthomorphisms on E is an Archimedean Riesz space, and even an f-algebra, with the identity I on E as algebra and weak order unit; the richness of this structure gives a particular interest to $\mathrm{Orth}(E)$. But are there always enough non trivial orthomorphisms on E? This problem was studied in [8], where it was shown that the answer may be negative; in this case it is thus difficult to compare the properties of E with those of $\mathrm{Orth}(E)$. To avoid this difficulty, it is useful to consider the space $\mathrm{Orth}^{\infty}(E)$ (resp. $\mathrm{Orth}^{w}(E)$) of equivalence classes of extended orthomorphisms (resp. weak orthomorphisms) on E.

Buskes and van Rooij in [2] called an order-bounded linear operator $T : D_T \to E$, where D_T is a vector sublattice of E, an *orthomorphism* if $|x| \wedge |y| = 0$ in D_T implies $|Tx| \wedge |y| = 0$ in E. If in addition D_T is order dense in E, we call the linear operator $T : D_T \to E$ a *weak orthomorphism* after Wickstead in [11]. A natural equivalence relation can be introduced in the set of all weak orthomorphisms on E as follows. Two weak orthomorphisms on E are equivalent whenever they agree on an order-dense vector sublattice of E. Amongst those extensions of weak orthomorphisms on E, which are again weak orthomorphisms of E, there is one which has a largest domain. The set of all weak orthomorphisms of E which have maximal domain is denoted by $\mathrm{Orth}^{w}(E)$ (see [11] for more details). We denote by $M(T)$ the largest domain of a weak orthomorphism extension of a weak orthomorphism T.

In [6], Luxemburg and Schep defined an *extended orthomorphism* to be a weak orthomorphism $T : D_T \to E$ such that D_T is an order-dense ideal in E. For extended orthomorphisms equivalence reduces to equality on the intersection of domains. The set of all equivalence classes of extended orthomorphisms of E is denoted by $\mathrm{Orth}^{\infty}(E)$. Clearly, $\mathrm{Orth}^{\infty}(E)$ is the subset of $\mathrm{Orth}^{w}(E)$ consisting of those operators T for which $M(T)$ contains an order-dense ideal of E. Observing that, for every band B in E, the band projection $\pi_B : B \oplus B^d \to E$ is a member of $\mathrm{Orth}^{\infty}(E)$, we see there are always many extended so weak orthomorphisms on E. For more details about extended and weak orthomorphisms we refer to [4, 5, 6, 9, 11].

Since weak orthomorphisms seem to be well-behaved and they form a natural generalization of extended orthomorphisms, why have they received no previous attention? The reason is that they do not, in general, have an additive structure see ([11], example 4.4). It seems natural therefore to ask what is missing for a vector lattice to obtain an addive structure in the set of all weak orthomorphisms. In [11], Wickstead showed that this kind of behavior is avoided in a special case where E is an Archimedean semiprime f-algebra. It turns out that pointwise operations and ordering make $\mathrm{Orth}^{w}(E)$ into an Archimedean f-algebra with unit element (see [11, Theorem 4.8]). Moreover, if in addition E, is uniformly complete then $\mathrm{Orth}^{w}(E) = \mathrm{Orth}^{\infty}(E)$ (see [11,Theorem 4.10]).

It will be noted that the proofs of the last results are heavily based on the algebra structure of E and hence it can not be expected that it can be adapted to the vector lattice case. Therefore, it seems to be an interesting question to characterizes the vector lattice E such that $\mathrm{Orth}^{w}(E) = \mathrm{Orth}^{\infty}(E)$. In this direction, recently the first author has shown in [3] that if the vector lattice E is uniformly complete then $\mathrm{Orth}^{w}(E)$ is a vector lattice since, in this case, we have $\mathrm{Orth}^{w}(E) = \mathrm{Orth}^{\infty}(E)$. The question of course arises whether the converse is true. However, as has been recently shown by the author in [3], this is not true in general. In [3], the author asked if the result above can be generalized to a larger class of vector lattice. Therefore, it seems natural to describe various conditions under which $\mathrm{Orth}^{w}(E)$ is a vector lattice. In [4] the first author proved that the condition "the uniform completeness of E" can be weakened as follows: if the Archimedean vector lattice E has a uniformly complete order-dense ideal then $\mathrm{Orth}^{w}(E) = \mathrm{Orth}^{\infty}(E)$ and therefore $\mathrm{Orth}^{w}(E)$ is a vector lattice.

Duhoux and Meyer in [5] defined a *σ-extended orthomorphism* to be an extended orthomorphism which can be defined on a super order-dense ideal in E (or, equivalently, such that its maximal domain is super order dense). It is well known that the collection $\mathrm{Orth}^{\sigma\infty}(E)$ of all σ-extended orthomorphism on E is an f-subalgebra of $\mathrm{Orth}^{\infty}(E)$. It is obvious that $\mathrm{Orth}(E) \subset \mathrm{Orth}^{\sigma\infty}(E) \subset \mathrm{Orth}^{\infty}(E)$, and it is shown in [5] that, in general, both inclusions may be proper. If B is a band in E, then the band projection $\pi_B : B \oplus B^d \to E$ is a member of $\mathrm{Orth}^{\sigma\infty}(E)$ if and only if B is an s-band, that is, $B \oplus B^d$ super order dense. That shows that if $\mathrm{Orth}^{\sigma\infty}(E) = \mathrm{Orth}^{\infty}(E)$ then every band in E must be an s-band. In the other direction, if E is order separable, every order-dense ideal of E is super order dense and so $\mathrm{Orth}^{\sigma\infty}(E) = \mathrm{Orth}^{\infty}(E)$. For more details about σ-extended orthomorphisms we refer to [5].

In view of this context we define a *σ-weak orthomorphism* to be a weak orthomorphism which can be defined on a super order-dense vector sublattice of E. The set of all σ-weak orthomorphism of E which have maximal domain will be denote by $\mathrm{Orth}^{\sigma_w}(E)$. We always have

$$\mathrm{Orth}(E) \subset \mathrm{Orth}^{\sigma_\infty}(E) \subset \mathrm{Orth}^{\sigma_w}(E) \subset \mathrm{Orth}^{w}(E)$$

and

$$\mathrm{Orth}(E) \subset \mathrm{Orth}^{\sigma_\infty}(E) \subset \mathrm{Orth}^{\infty}(E) \subset \mathrm{Orth}^{w}(E).$$

Note that all inclusions may be proper (see [5]). It follows easily from [3] that if E is uniformly complete then

$$\mathrm{Orth}(E) \subset \mathrm{Orth}^{\sigma_\infty}(E) \subset \mathrm{Orth}^{\sigma_w}(E) \subset \mathrm{Orth}^{w}(E) = \mathrm{Orth}^{\infty}(E).$$

It is worth recording that by [4] we can do a little better. We have the previous inclusions if E has a uniformly complete order-dense ideal. We next look at the algebraic structure of $\mathrm{Orth}^{\sigma_w}(E)$. Note that σ-weak orthomorphisms do not, in general, have an additive structure as it is shown in the following example.

Example. Let $C(X)$ be the vector lattice of continuous functions on a topological space X and E be a vector subspace of $C(X)$, then we shall say that a function on X is locally in E if it is defined on a dense open subset of X and if it coincides, on some neighborhood of each point of its domain, with some element of E. Moreover, $L_E(X)$ denotes the vector lattice of equivalence classes of such functions, under the relation of coinciding on a dense open subset of X and with vector and lattice operations defined modulo dense open sets. A moment's reflection shows that if X has the Baire property then $L_E(X)$ will be Archimedean and will be an f-algebra if E is an algebra. See [10] pages 90 and 91 for more details of this construction. Now, let F be the vector space of all polynomials functions on $[0, 1]$ which vanish at 0, and G the space of the continuous piecewise linear functions vanishing at 1. Now, let E be the vector space of continuous functions on $[0, 1]$ generated by F and G. By a classical Weierstrass Theorem it is quite easy to show that $L_F([0, 1])$ is a super order-dense vector sublattice of $L_E([0, 1])$. Now, by using a classical approximation Theorem, it is easy to check that $L_G([0, 1])$ is a super order-dense vector sublattice of $L_E([0, 1])$. Now define $T_1 : L_F([0, 1]) \to L_E([0, 1])$ by $T_1(f)(x) = \frac{f(x)}{x}$ and $T_2 : L_G([0, 1]) \to L_E([0, 1])$ by $T_2(f) = \frac{f(x)}{x-1}$. Clearly T_1 and T_2 are σ-weak orthomorphisms and are defined on their largest possible domain, yet $L_F([0, 1]) \cap L_G([0, 1]) = \{0\}$, so there is no hope of defining $T_1 + T_2$.

We will look at the following problem: what about weaker conditions on the order structure of E to obtain an additive structure in $\mathrm{Orth}^{\sigma_w}(E)$. The above results tell us what is happening in some cases, but there are certainly gaps in the general case. If we limit our attention to some of the classical vector lattices then we would hope to establish more. Even here our knowledge remains incomplete. The most notable gap in our knowledge is because the intersection of two super order-dense vector sublattice may be equal to zero.

It is our main purpose in this paper to describe a generalization of parts of the theory of extended and weak orthomorphisms. This generalization is perhaps not quite in final form yet, but it is already good enough to give new results which seem deep for the orthomorphisms defined on a super order-dense vector sublattice. We have not been able to characterize the vector lattice for which we obtain an additive structure in $\mathrm{Orth}^{\sigma_w}(E)$. Nonetheless, we have been able to show that if we assume that the vector lattice E is uniformly complete then $\mathrm{Orth}^{\sigma_w}(E)$ is a vector lattice since, in this case, we have $\mathrm{Orth}^{\sigma_w}(E)$=$\mathrm{Orth}^{\sigma_\infty}(E)$.

It is convenient to use the monograph [1] for basic information concerning the general theory of vector lattices, f-algebras and orthomorphisms. Properties and definitions of the uniform topology can be found in [7].

Theorem 1. *Let E be a uniformly complete vector lattice. Then*

$$\mathrm{Orth}^{\sigma_w}(E) = \mathrm{Orth}^{\sigma_\infty}(E).$$

Proof. We have already mentioned that every σ-extended orthomorphism is a σ-weak orthomorphism so we have the inclusion $\mathrm{Orth}^{\sigma_\infty}(E) \subset \mathrm{Orth}^{\sigma_w}(E)$. To establish the converse inclusion we propose to show that every σ-weak orthomorphism can be extended to a σ-extended orthomorphism. To this end denote E^δ the Dedekind completion of E and let $\pi : D \to E$ a σ-weak orthomorphism where D is a super order-dense Riesz subspace of E, using the Veksler theorem ([1], Theorem 1.65) π can be extended to E_D^δ (the ideal generated by D in E^δ), we denote this extension by:

$$\pi^\delta : E_D^\delta \to E^\delta \in \mathrm{Orth}^\infty(E^\delta)$$

To end the proof is suffices to show that the restriction of π^δ on E_D (the ideal generated by D in E) has its range in E. To see this denote $\pi_n^\delta \in \mathrm{Orth}(E_D^\delta)$ defined by:

$$\pi_n^\delta = \pi^\delta \wedge nI : E_D^\delta \to E_D^\delta$$

and denote π_n the restriction of π_n^δ on D:

$$\pi_n = \pi_{n|D}^\delta : D \to E_D$$

by ([2], Theorem 4.1) π_n can be extended to an orthomorphism $\pi_n^* : E_D \to E_D$.

Now since $\mathrm{Orth}^\infty(E^\delta)$ is an f-algebra with multiplicative unit we have by ([1], Theorem 2.57) that

$$0 \leq \pi^\delta - \pi_n^\delta \leq \frac{1}{n}(\pi^\delta)^2.$$

So

$$0 \leq \pi^\delta(x) - \pi_n^\delta(x) \leq \frac{1}{n}(\pi^\delta)^2(x) \quad \forall x \in E_D^+$$

since $\pi_n^* = \pi_{n|E_D}^\delta$ and E majorizes E_D^δ there is $u \in E^+$ such that

$$0 \leq \pi^\delta(x) - \pi_n^*(x) \leq \frac{1}{n}(\pi^\delta)^2(x) \leq \frac{1}{n}u \quad \forall x \in E_D^+$$

Therefore the sequence $(\pi_n^*(x)) \in E$ converges u-uniformly to $\pi^\delta(x)$ for all $x \in E_D^+$ then it follows from the uniform completeness of E that $\pi^\delta(x) \in E$ for all $x \in E_D^+$, thus $\pi^\delta(x) \in E$ for all $x \in E_D$, so that

$$\pi^\delta_{|E_D} : E_D \to E \quad \in \mathrm{Orth}^{\sigma_\infty}(E)$$

and the proof is finished. □

The next proposition and lemma are immediate from ([4], Proposition 1 and Lemma 1).

Proposition 2. *Let E be an Archimedean vector lattice and $T : D \to E$ be a σ-weak orthomorphism on E. Then* $\mathrm{Ker}\, T = D \cap T(D)^d$. *In particular* $\mathrm{Ker}\, T$ *is a band in D.*

Lemma 3. *Under the same notation of the previous proposition, we have:*

(i) $Tx \neq 0$ *for all* $0 \neq x \in D \cap T(D)$.
(ii) $A^{dd} \cap T(D)^{dd} = T(A)^{dd}$ *for all subset A of D.*

At this level we have to recall the notions of s-band and c-band originally introduced by M. Meyer and M. Duhoux in [5]: Given an Archimedean Riesz space E, an ideal A of E is called an s-ideal if $A \oplus A^d$ is super order dense. More generally A is called a c-ideal if for every $x \in E^+$ there exist a sequence $(x_n) \in A^+$ such that

$$x = \sup \{x_n + y \;;\; n = 1, 2, \ldots, \text{ and } y \in A^d \cap [0, x]\}$$

An s-band (resp. c-band) is an s-ideal (resp. c-ideal) which is also a band.

Lemma 4. *Let E be an Archimedean Riesz space and let* $T \in \mathrm{Orth}^{\sigma_w}(E)$ *then* $T(D)^{dd}$ *is a c-Band.*

Proof. According to ([5], Section 3) it suffices to show that for all $x \in E^+$, $T(D)^{dd} \cap \{x\}$ is countably generated. So let $x \in E^+$, since D is a super order-dense Riesz subspace of E there exist a sequence (x_n) in D^+ such that $x_n \uparrow x$, thus $\{x_n, n \in \mathbb{N}\}^{dd} = \{x\}^{dd}$ and then $T(D)^{dd} \cap \{x\}^{dd} = T(D)^{dd} \cap \{x_n, n \in \mathbb{N}\}^{dd} = \{T(x_n), n \in \mathbb{N}\}^{dd}$. Now by (ii) of the above lemma the following equality holds $T(D)^{dd} \cap \{x_n, n \in \mathbb{N}\} = \{T(x_n), n \in \mathbb{N}\}^{dd}$ so $T(D)^{dd}$ is a c-band and the proof is finished. □

It is shown in ([4], Theorem 2 and Corollary 1) that when E is just an Archimedean Riesz space then for every $T \in \mathrm{Orth}^w(E)$ there exists $0 \leq S \in \mathrm{Orth}^w(E)$ such that $TST = T$. Under an extra condition on E, namely the almost σ-Dedekind completeness, the same result still true for $\mathrm{Orth}^{\sigma_w}(E)$. Recall that a Riesz space E is said to be almost σ-Dedekind completeness if it can be embedded as a super order-dense Riesz subspace of a σ-Dedekind complete Riesz space. M. Meyer and M. Dhoux give an interesting characterisation of such spaces as follows: E is an almost σ-Dedekind complete Riesz if and only if every c-band is an s-band ([5], Theorem 3.3). Note that order-separable Riesz spaces are a particular almost-σ-Dedekind complete Riesz spaces.

Proposition 5. *Let E be an almost σ-Dedekind complete Riesz space, then for every $T \in \mathrm{Orth}^{\sigma_w}(E)$ there exists $S \in \mathrm{Orth}^{\sigma_w}(E)$ such that $TST = T$.*

Proof. Let $T : D \to E \in \mathrm{Orth}^{\sigma_w}(E)$ where D is a super order-dense Riesz subspace of E. Denote $S : T(D \cap T(D)^{dd}) \oplus \mathrm{Ker}\, T \to E$ the mapping defined as follows:

$$\begin{cases} ST(x) = x & \text{if } x \in D \cap T(D)^{dd} \\ S(x) = 0 & \text{if } x \in \mathrm{Ker}\, T. \end{cases}$$

S is well defined indeed, let $x, y \in D \cap T(D)^{dd}$ such that $Tx = Ty$, then $x - y \in \mathrm{Ker}\, T = D \cap T(D)^d$ so $x - y \in T(D)^d$, on the other hand $x - y \in T(D)^{dd}$ hence $x = y$.

Now let us prove that $(D \cap T(D)^{dd}) \oplus \mathrm{Ker}\, T$ is super order-dense Riesz subspace of E. It is trivial that $D \cap T(D)^{dd}$ and $\mathrm{Ker}\, T = D \cap T(D)^d$ are respectively super order dense in $T(D)^{dd}$ and $T(D)^d$ then it follows that $(D \cap T(D)^{dd}) \oplus \mathrm{Ker}\, T$ is super order dense in $T(D)^{dd} \oplus T(D)^d$. To show that $(D \cap T(D)^{dd}) \oplus \mathrm{Ker}\, T$ is super order dense in E it remains to prove that $T(D)^{dd}$ is an s-band. By the above lemma we have that $T(D)^{dd}$ is a c-band and since E is almost σ-Dedekind complete it follows immediately that $T(D)^{dd}$ is an s-band. Clearly S is a σ-weak orthomorphism and $TST = T$. □

At present, we are not able to say whether the almost σ-Dedekind completeness condition is necessary or not. However a natural and interesting question can be raised: Let E be an Archimedean Riesz space, is there equivalence between the two following statements:

a) Every c-band in E is an s-band.
b) The c-band $T(D)^{dd}$ is an s-band for very $T \in \mathrm{Orth}^{\sigma_w}(E)$.

If the question is negatively answered, the almost σ-Dedekind completeness condition can be weakened.

Let E be a Riesz space and F a Riesz subspace of E, unlike weak orthomorphisms and extended orthomorphisms the inclusion $\mathrm{Orth}^{\sigma_w}(F) \subset \mathrm{Orth}^{\sigma_w}(E)$ does not holds in general as shown by the following example.

Example. Let X be a non countable set, $F(X)$ be the Riesz space of all real functions on X. Define

$$E = \{f \in F(X); \{x \in X : f(x) \neq a\} \text{ is at most countable for some } a \in \mathbb{R}\}$$

and

$$F = C_0(X) \text{ (functions in } E \text{ vanishing at infinity).}$$

Then we have

$$\mathrm{Orth}^{\sigma_w}(F) = \mathrm{Orth}^{\sigma}(F) = F(X), \;\; \textit{but} \;\; \mathrm{Orth}^{\sigma_w}(E) = E$$

But in the case where F is super order dense we have $\mathrm{Orth}^{\sigma_w}(F) \subset \mathrm{Orth}^{\sigma_w}(E)$ and we have more when F is a super order-dense ideal in E as shown in the following proposition.

Proposition 6. *Let E be an Archimedean Riesz space, F be a super order-dense ideal in E then* $\mathrm{Orth}^{\sigma_w}(F) = \mathrm{Orth}^{\sigma_w}(E)$.

Proof. It is easy to verify that $\mathrm{Orth}^{\sigma_w}(F) \subset \mathrm{Orth}^{\sigma_w}(E)$. For the converse inclusion, take $T \in \mathrm{Orth}^{\sigma_w}(E)$ with domain D, it is easily seen that $D \cap F$ is a super order-dense Riesz subspace of F and then a super order-dense Riesz subspace of E. Using almost the same techniques used in ([4], Theorem 1) we can prove that $F_1 = \{x \in D \cap F, T(x) \in E_x\}$ is super order dense Riesz subspace of F. Now the restriction $T_{|F_1}$ can be considered as a σ-weak orthomorphism of F since $T(F_1) \subset F$, so $\mathrm{Orth}^{\sigma_w}(E) \subset \mathrm{Orth}^{\sigma_w}(F)$ and thus $\mathrm{Orth}^{\sigma_w}(E) = \mathrm{Orth}^{\sigma_w}(F)$. □

As an immediate application

Theorem 7. *Let E be an Archimedean Riesz space having a uniformly complete super order-dense ideal then*

$$\mathrm{Orth}^{\sigma_w}(E) = \mathrm{Orth}^{\sigma\infty}(E).$$

Proof. Let F be a uniformly complete super order-dense ideal of E. Then by Theorem 2 we have $\mathrm{Orth}^{\sigma_w}(F) = \mathrm{Orth}^{\sigma\infty}(F)$. In other hand by Proposition 8 we have $\mathrm{Orth}^{\sigma_w}(E) = \mathrm{Orth}^{\sigma_w}(F)$. And by ([5], Theorem 3.15) it holds that $\mathrm{Orth}^{\sigma\infty}(E) = \mathrm{Orth}^{\sigma\infty}(F)$. So we conclude that

$$\mathrm{Orth}^{\sigma_w}(E) = \mathrm{Orth}^{\sigma\infty}(E).$$ □

It follows from the above theorem that if E is an Archimedean Riesz space having a uniformly complete super order-dense ideal (particularly when E is uniformly complete) then $\mathrm{Orth}^{\sigma_w}(E)$ is a Riesz space, even more an f-algebra with the identity on E as unit element.

We devote the remaining of this paper to the relation between $\mathrm{Orth}(E)$, $Z(E)$ and $\mathrm{Orth}^{\sigma_w}(E)$. The following example shows that neither $\mathrm{Orth}(E)$ nor $Z(E)$ are ideals in $\mathrm{Orth}^{\sigma_w}(E)$.

Example. Let $E = C([0,1])$; it is clear that E is uniformly complete; by ([5], Lemma 4.1) and the fact that E is order separable $\mathrm{Orth}^{\sigma_w}(E) = \mathrm{Orth}^{\infty}(E) = C(D,[0,1])$ where D is the filter of all dense open subsets of $[0,1]$. Since $\mathrm{Orth}(E) = Z(E) = E = C([0,1])$ we see that neither $\mathrm{Orth}(E)$ nor $Z(E)$ are ideals in $\mathrm{Orth}^{\sigma_w}(E)$.

Now if E is an order Cauchy complete Riesz space, then $\mathrm{Orth}(E)$ and $Z(E)$ are not only ideals in $\mathrm{Orth}^{\sigma_w}(E)$ but even more, as shown in the following theorem.

Theorem 8. *Let E be an order Cauchy complete Riesz space; then* $\mathrm{Orth}(E)$ *and* $Z(E)$ *are super order-dense ideals in* $\mathrm{Orth}^{\sigma_w}(E)$.

Proof. Given $S : D_S \to E$ and $T : D_T \to E$ in $\mathrm{Orth}^{\sigma_w}(E)$ such that $0 \leq S \leq T$, we can define S on D_T. Indeed let $(0 \leq x \in D_T)$ and (x_n) be any sequence in D_S such that $0 \leq x_n \uparrow x$. Note first of all that since T is a positive order-continuous operator we have $T(x_n) \uparrow T(x)$ and then from the inequality $0 \leq S(x_{n+p} - x_n) \leq$

$T(x_{n+p} - x_n)$ we deduce that $(S(x_n))_n$ is an increasing order Cauchy sequence, thus $\sup S(x_n)$ exists. So we can define S on D_T by $S(x) = supS(x_n)$ for all $x \in D_T^+$. In particular Orth(E) and $Z(E)$ are ideals in Orth$^{\sigma_w}(E)$. □

We end this paper by some open problems:

- What is the necessary and sufficient condition on E for which Orth(E) and $Z(E)$ are ideals in Orth$^{\sigma_w}(E)$?
- Can Orth$^{\sigma_w}(E)$ have a structure of Riesz space even if it does not coincide with Orth$^{\sigma_\infty}(E)$?

References

[1] Aliprantis, C.D and O. Burkinshaw, "Positive Operators", Springer, Dordrecht, 2006. Reprint of the 1985 original.

[2] Buskes, G. and A. van Rooij, Small Riesz spaces, Math. Proc. Camb. Phil. Soc., 105(1989), 523–536.

[3] E. Chil, A note on weak orthomorphisms. Arch. Math. 92 (2009), 266–269.

[4] E. Chil, On weak orthomorphisms, Mathematical Proceedings of the Royal Irish Academy 111A (2011), 13–20.

[5] M. Duhoux and M. Meyer, Extended orthomorphisms on Archimedean Riesz spaces. Ann. Mat. pura et appl. (IV) 33 (1983) 193–226.

[6] W.A.J. Luxemburg and A.R. Schep, A Radon–Nikodym type theorem for positive operators and a dual. Indag. Math. 40 (1978) 357–375.

[7] W.A.J. Luxembourg and A.C. Zaanen, "Riesz spaces I". North-Holland, Amsterdam, 1971.

[8] M. Meyer, Richesses du centre d'un espace réticulé, Math. Ann., 236 (1978), pp. 147–169.

[9] B. De Pagter, The space of extended orthomorphisms in a Riesz space, *Pacific J. Math.*, 112 (1) (1984) 193–210.

[10] A.W. Wickstead, Extensions of orthomorphisms. J. Austral. Math. Soc. 29 (1980) 87–98.

[11] A.W. Wickstead, The injective hull of an Archimedean f-algebra, Compos. Math., 62 (1987) 329–342.

Elmiloud Chil
2 Rue jawaher lel Nehrou Monflery 1008 Tunisia
e-mail: `Elmiloud.chil@ipeit.rnu.tn`

Mohamed Mokaddem
Research Laboratory of Algebra, Topology, Arithmetic, and Order
Departement of Mathematics
Faculty of Mathematical and Physical Sciences of Tunis
Tunis-El Manar University
2092-El Manar, Tunisia
e-mail: `m.mokaddem1@yahoo.fr`

Ordered Structures and Applications: Positivity VII
Trends in Mathematics, 99–110

Ergodic Theorems for L_1-L_∞ Contractions in Banach–Kantorovich L_p-lattices

V.I. Chilin and I.G. Ganiev

Abstract. We present versions of ergodic theorems for L_1-L_∞ contractions in Banach–Kantorovich L_p-lattices associated with the disjunctive decomposable measure taking values in the algebra of measurable functions.

Mathematics Subject Classification (2010). Primary 37A30; Secondary 46G10, 47A35.

Keywords. Banach–Kantorovich lattice; measurable Banach bundle; disjunctive decomposable vector-valued measure; L_1-L_∞ contraction; ergodic theorem.

1. Introduction

Let (Ω, Σ, μ) be a measure space with σ-finite measure μ and let $L_p(\Omega, \Sigma, \mu)$ be the Banach space of all measurable functions f on (Ω, Σ, μ) such that

$$\|f\|_p = \left(\int_\Omega |f|^p d\mu \right)^{1/p} < \infty, \quad (1 \leq p < \infty).$$

The well-known M.A. Akcoglu theorem [1] asserts that for any positive contraction T in the space $L_p(\Omega, \Sigma, \mu)$, $1 < p < \infty$, the averages

$$s_n(T)(f) = \frac{1}{n} \sum_{i=0}^{n-1} T^i(f)$$

converge almost everywhere for every $f \in L_p(\Omega, \Sigma, \mu)$, in addition

$$f^* = \sup_{n \geq 1} s_n(T)(|f|) \in L_p(\Omega, \Sigma, \mu) \quad \text{and} \quad \|f^*\|_p \leq \frac{p}{p-1} \|f\|_p.$$

The same ergodic theorem holds for non positive $L_1 - L_\infty$ contraction T in $L_p(\Omega, \Sigma, \mu)$ [7, Ch. VIII, §6]. Further development extends this ergodic theorem to the space of Banach-valued functions as follows (see, e.g., [12, Ch. 4, §4.2]): if T is an L_1-L_∞ contraction in the space $L_p(\Omega, X)$ of Bochner maps from (Ω, Σ, μ) into a reflexive Banach space $(X, \|\cdot\|_X)$, $1 < p < \infty$, then there exists $\widetilde{f} \in L_p(\Omega, X)$ such that $\|s_n(T)(f)(\omega) - \widetilde{f}(\omega)\|_X \to 0$ almost everywhere on (Ω, Σ, μ) for any $f \in L_p(\Omega, X)$.

The following natural advance in extension of this area of applications of ergodic theorems is a description of ergodic properties of L_1-L_∞ contractions in Banach–Kantorovich L_p-lattices $L_p(\nabla, m)$ associated with a vector-valued measure m, defined on a complete Boolean algebra ∇.

In [2] the (o)-convergence of averages $s_n(T)(f)$ in the vector lattice $L_p(\nabla, m)$ has been established for positive contractions $T : L_p(\nabla, m) \to L_p(\nabla, m)$, satisfying the condition $T\mathbf{1} \leq \mathbf{1}$, where $1 < p < \infty$. In [4] a similar ergodic theorem has been extended for positive contractions of the Orlicz–Kantorovich lattices $L_M(\nabla, m)$ in case when N-function M satisfies the condition

$$\sup_{s\geq 1}\{\frac{1}{M(s)} \int_1^s M(t^{-1}s)dt\} < \infty.$$

In the present paper we establish ergodic theorems for L_1-L_∞ contractions in Banach–Kantorovich lattices $L_p(\nabla, m)$ and $L_M(\nabla, m)$. Moreover, we present "vector" versions of weighted ergodic theorems obtained in [10]. In the proof of these theorems we use a representation of a Banach–Kantorovich lattice as a space of measurable sections of a measurable Banach bundle [8], [9]. This fact allows us to obtain the required properties of the Banach–Kantorovich lattice by means of the corresponding stalkwise verification of these properties. Using this approach, a version of the dominated ergodic theorem was obtained for positive contractions in the $L_p(\nabla, m)$ [2]. This approach we use in the present article. Of a particular assistance for us is the representation theorem of the Banach–Kantorovich lattice $L_p(\nabla, m)$ as a measurable bundle of classical L_p spaces associated with scalar measures [8]. We apply this representation and the corresponding ergodic theorems for L_1-L_∞ contractions in classical L_p-spaces in order to obtain versions of ergodic theorems for L_1-L_∞ contractions in Banach–Kantorovich lattices $L_p(\nabla, m)$ and $L_M(\nabla, m)$.

We use the terminology and notation of the theory of Riesz spaces, vector integration, and lattice-normed spaces [14], as well as the terminology of measurable bundles of Banach lattices [8], [9].

2. Preliminaries

Let (Ω, Σ, μ) be a measure space with the direct sum property [14, 1.1.8]. Denote by $\mathcal{L}(\Omega)$ (respectively, $\mathcal{L}_\infty(\Omega)$) the set of all (respectively, essentially bounded) measurable real functions defined a.e. on Ω. Denote by $L_0(\Omega)$ the algebra of all classes of functions from $\mathcal{L}(\Omega)$ equal almost everywhere, and by B the complete Boolean algebra of all idempotents from $L_0(\Omega)$. The set $L_\infty(\Omega)$ of all bounded functions from $L_0(\Omega)$ is a subalgebra in $L_0(\Omega)$ with the unity $\mathbf{1}(\omega) = 1$ a.e., in addition, $L_\infty(\Omega)$ is a Banach algebra with respect to the norm $\|x\|_\infty = \sup\{|x(\omega)| : \omega \in \Omega\}$.

Let ∇ be an arbitrary complete Boolean algebra and let $m : \nabla \to L_0(\Omega)$ be an $L_0(\Omega)$-valued measure on ∇, i.e., (i). $m(e) \geq 0$ for all $e \in \nabla$ and $m(e) = 0 \Leftrightarrow e = 0$; (ii). $m(e \vee g) = m(e) + m(g)$ if $e \wedge g = 0, e, g \in \nabla$; (iii). $m(e_\alpha) \downarrow 0$ for every net $e_\alpha \downarrow 0$.

We assume that the measure m is a disjunctive decomposable, i.e., for every $e \in \nabla$ and a decomposition $m(e) = a_1 + a_2$, $a_1 \wedge a_2 = 0$, $a_i \in L_0(\Omega)$ there exists $e_i \in \nabla$ such that $e = e_1 \vee e_2$ and $m(e_i) = a_i, i = 1, 2$. In this case, as it is shown in [3], there exist a regular Boolean subalgebra ∇_0 in ∇ and an isomorphism $\varphi : B \to \nabla_0$ such that $m(\varphi(q)e) = qm(e)$ for all $q \in B, e \in \nabla$. Further we identify the Boolean algebras ∇_0 and B and assume that the measure m is a B-modular, i.e., $m(qe) = qm(e)$ for all $q \in B = \nabla_0, e \in \nabla$. In this case, the algebra $L_0(\Omega)$ is a subalgebra of $L_0(\nabla)$ and the element $m(\mathbf{1})$ is invertible in $L_0(\Omega)$. It is clear that $m_1(e) = m(e)m(\mathbf{1})^{-1}$ is an $L_0(\Omega)$-valued measure on ∇, for which we have $m_1(\mathbf{1}) = \mathbf{1}$. As a result, in what follows we assume that $m(\mathbf{1}) = \mathbf{1}$.

Denote by $L_0(\nabla)$ the universally complete vector lattice $C_\infty(X(\nabla))$, where $X(\nabla)$ is the Stone space of ∇ [14, 1.4.2]. Let $L_p(\nabla, m)$ be the Banach–Kantorovich lattice of all elements x from $L_0(\nabla)$, for which there exists the $L_0(\Omega)$-valued norm $\|x\|_p = \left(\int |x|^p dm\right)^{1/p}$, $p \geq 1$ (see for example [14, 6.1]). We need the representation of $L_p(\nabla, m)$ as a measurable bundle of classical L_p spaces associated with scalar measures [8]. Let $l : L_\infty(\Omega) \to \mathcal{L}_\infty(\Omega)$ be a lifting (since (Ω, Σ, μ) has a direct sum property then lifting l there always exists [14, 1.4.8]).

For each $\omega \in \Omega$, $e \in \nabla$ we set $m^0_\omega(e) = l(m(e))(\omega)$ and $I^0_\omega = \{e \in \nabla : m^0_\omega(e) = 0\}$. Denote by ∇_ω the completion of the quotient Boolean algebra ∇/I^0_ω by the metric $\rho_\omega([e], [g]) = m^0_\omega(e \Delta g)$, where $[e]$ is the coset in ∇/I^0_ω of $e \in \nabla$. Let m_ω be a measure on the complete Boolean algebra ∇_ω extending m^0_ω and let $L_p(\nabla_\omega, m_\omega)$ be the classical L_p-space associated with ∇_ω and scalar measure m_ω. Consider the homomorphism $\gamma_\omega = i_\omega \circ \pi_\omega : \nabla \to \nabla_\omega$, where $\pi_\omega : \nabla \to \nabla^0_\omega$ is the quotient homomorphism and $i_\omega : \nabla^0_\omega \to \nabla_\omega$ is the natural embedding. Denote by $(Y_p, \mathcal{E})$ the Banach bundle over Ω, where $Y_p(\omega) = L_p(\nabla_\omega, m_\omega)$ for each $\omega \in \Omega$ and

$$\mathcal{E} = \left\{ \sum_{i=1}^{n} \lambda_i \gamma_\omega(e_i) : \lambda_i \in \mathbb{R},\ e_i \in \nabla,\ i = 1, \ldots, n,\ n \in \mathbb{N} \right\}$$

is a measurable structure in Y_p. Let $\mathcal{M}(\Omega, Y_p)$ be a set of all $\mathcal{E}$-measurable sections of bundle $(Y_p, \mathcal{E})$ and let $L_0(\Omega, Y_p)$ be the Banach–Kantorovich lattice, which is the factorization of $\mathcal{M}(\Omega, Y_p)$ by the equality almost everywhere. In [8] it is shown that there exists an isometric isomorphism Φ from $L_p(\nabla, m)$ onto $L_0(\Omega, Y_p)$ such that

$$\Phi\left(\sum_{i=1}^{n} \lambda_i e_i\right) = \sum_{i=1}^{n} \lambda_i \gamma_\omega(e_i)$$

for all $\lambda_i \in \mathbb{R}$, $e_i \in \nabla, i = 1, \ldots, n$, $n \in \mathbb{N}$. Thus, every element $x \in L_p(\nabla, m)$ is identified with the bundle $\Phi(x) \in L_0(\Omega, Y_p)$, where $\Phi(x)(\omega) \in L_p(\nabla_\omega, m_\omega)$ a.e.

Set

$$\mathcal{L}_\infty(\Omega, Y_p) = \{u \in \mathcal{M}(\Omega, Y_p) : ||u(\omega)||_p \in \mathcal{L}_\infty(\Omega)\}$$

and

$$L_\infty(\Omega, Y_p) = \{u^\sim : u \in \mathcal{L}_\infty(\Omega, Y_p)\},$$

where $u^\sim$ is the coset of u in $L_0(\Omega, Y_p)$.

Denote by $L_\infty(\nabla)$ the subalgebra in $L_0(\nabla)$ of all bounded elements, i.e., those $x \in L_0(\nabla)$ such that $|x| \leq \lambda\mathbf{1}$ for some $\lambda > 0$. The algebra $L_\infty(\nabla)$ is a Banach algebra with respect to the norm $\|x\|_\infty = \inf\{\lambda > 0 : |x| \leq \lambda\mathbf{1}\}$, in addition, $L_\infty(\nabla) \subset L_p(\nabla, m)$ and $L_\infty(\nabla)$ is *bo*-dense in $L_p(\nabla, m)$ for all $p \geq 1$.

Let $l : L_\infty(\Omega) \to \mathcal{L}_\infty(\Omega)$ be a lifting. In [8] it is proved that there exists a linear mapping $\ell_\nabla : L_\infty(\nabla) \to \mathcal{L}_\infty(\Omega, Y_p)$ such that for all $x, y \in L_\infty(\nabla), h \in L_\infty(\Omega) \subset L_\infty(\nabla)$ the following properties hold:

1) $\ell_\nabla(x) \in \Phi(x)$ *and* $\mathrm{dom}(\ell_\nabla(x)) = \Omega$, *where* $\mathrm{dom}(\ell_\nabla(x))$ *is the domain of* $\ell_\nabla(x)$*;*
2) $\|\ell_\nabla(x)(\omega)\|_{L_p(\nabla_\omega, m_\omega)} = l(\|x\|_p)(\omega)$ *(we see that the equality* $m(\mathbf{1}) = \mathbf{1}$ *implies that* $\|x\|_p \in L_\infty(\Omega)$ *for all* $x \in L_\infty(\nabla)$*);*
3) $\ell_\nabla(x)(\omega) \geq 0$ *if* $x \geq 0$*;*
4) $\ell_\nabla(hx) = l(h)\ell_\nabla(x)$*;*
5) $\{\ell_\nabla(x)(\omega) : x \in L_\infty(\nabla)\}$ *is dense in* $L_p(\nabla_\omega, m_\omega), \omega \in \Omega$*;*
6) $\ell_\nabla(x \vee y) = \ell_\nabla(x) \vee \ell_\nabla(y)$.

A mapping $\ell_\nabla : L_\infty(\nabla) \to \mathcal{L}_\infty(\Omega, Y_p)$ is called *a vector-valued lifting* associated with the lifting $l : L_\infty(\Omega) \to \mathcal{L}_\infty(\Omega)$.

Using a vector-valued lifting ℓ_∇, we can obtain a stalkwise representation of the linear operator $T : L_p(\nabla, m) \to L_p(\nabla, m)$, in case when T is $L_0(\Omega)$-bounded, i.e., if there exists $0 \leq c \in L_0(\Omega)$ such that $\|Tx\|_p \leq c\|x\|_p$ for all $x \in L_p(\nabla, m)$. For a positive $L_0(\Omega)$-contraction $T : L_p(\nabla, m) \to L_p(\nabla, m)$, satisfying the condition $T(\mathbf{1}) \leq \mathbf{1}$, this stalkwise representation of T is given in [8]. The following theorem establishes similar result without the assumption of positivity of the operator T.

Theorem 2.1. *Let* $T : L_1(\nabla, m) \to L_1(\nabla, m)$ *be a linear operator such that*

$$T(L_\infty(\nabla)) \subset L_\infty(\nabla) \quad \textit{and} \quad \|T(x)\|_1 \leq \|x\|_1$$

for all $x \in L_1(\nabla, m)$. *Then for every* $\omega \in \Omega$ *there exists a linear contraction* $T_\omega : L_1(\nabla_\omega, m_\omega) \to L_1(\nabla_\omega, m_\omega)$ *such that* $\Phi(Tx)(\omega) = T_\omega(\Phi(x)(\omega))$ *a.e. for every* $x \in L_1(\nabla, m)$. *In addition, if the operator* T *is positive, then the operator* T_ω *is also positive for all* $\omega \in \Omega$.

Proof. We define a mapping $\varphi(\omega)$ from $\{\ell_\nabla(x)(\omega) : x \in L_\infty(\nabla, m)\}$ into $L_1(\nabla_\omega, m_\omega)$ by the equality $\varphi(\omega)(\ell_\nabla(x)(\omega)) = \ell_\nabla(Tx)(\omega), \omega \in \Omega$. By $\|Tx\|_1 \leq \|x\|_1$ we have that

$$\|\ell_\nabla(Tx)(\omega)\|_{L_1(\nabla_\omega, m_\omega)} = l(\|Tx\|_1)(\omega) \leq l(\|x\|_1)(\omega) = \|\ell_\nabla(x)(\omega)\|_{L_1(\nabla_\omega, m_\omega)},$$

and therefore the operator $\varphi(\omega)$ is well defined and bounded with respect to the norm $\|\cdot\|_{L_1(\nabla_\omega, m_\omega)}$. Since $\{\ell_\nabla(x)(\omega) : x \in L_\infty(\nabla, m)\}$ is dense in $L_1(\nabla_\omega, m_\omega)$, then the linear operator $\varphi(\omega)$ can be extended to the contraction $T_\omega : L_1(\nabla_\omega, m_\omega) \to L_1(\nabla_\omega, m_\omega)$.

We shall show that $\Phi(Tx)(\omega) = T_\omega(\Phi(x)(\omega))$ for a.e. $\omega \in \Omega$, where $x \in L_1(\nabla, m)$. Choose $\{x_n\} \in L_\infty(\nabla, m)$ such that the sequence $\|x_n - x\|_1$ (o)-converges to zero. Then $\|\Phi(x_n)(\omega) - \Phi(x)(\omega)\|_{L_1(\nabla_\omega, m_\omega)} \to 0$ for a.e. $\omega \in \Omega$. Since $\|Tx_n - Tx\|_1 \xrightarrow{(o)} 0$, it follows that $\|\ell_\nabla(Tx_n)(\omega) - \Phi(Tx)(\omega)\|_{L_1(\nabla_\omega, m_\omega)} \to 0$ for a.e. $\omega \in \Omega$. Moreover, the continuity of the operator T_ω implies that $\|\ell_\nabla(Tx_n)(\omega) - T_\omega(\Phi(x))(\omega)\|_{L_1(\nabla_\omega, m_\omega)} \to 0$ for a.e. $\omega \in \Omega$. Hence $\Phi(Tx)(\omega) = T_\omega(\Phi(x)(\omega))$ for a.e. $\omega \in \Omega$. It is clear that for the positive operator T, by property 3) of the vector-valued lifting ℓ_∇, the operator T_ω is also positive. □

A linear operator $T : L_1(\nabla, m) \to L_1(\nabla, m)$ is said to be regular if it can be represented as a difference of two positive operators. The set of all regular operators on $L_1(\nabla, m)$ is denoted by $H_r(L_1(\nabla, m))$. It is known that $H_r(L_1(\nabla, m))$ forms a complete vector lattice, in addition for every $T \in H_r(L_1(\nabla, m))$ the module $|T|$ is a positive linear operator and

$$|T|(x) = \sup\{\, |Ty| : y \in L_1(\nabla, m), |y| \le x\,\}$$

where $0 \le x \in L_1(\nabla, m)$ [14, 3.1.2]. In addition

$$|Tx| \le |T||x|$$

for all $x \in L_1(\nabla, m)$.

We denote the set of all $L_0(\Omega)$-bounded linear operators acting in the Banach–Kantorovich lattice $L_1(\nabla, m)$ by $B(L_1(\nabla, m))$. With respect to the $L_0(\Omega)$-valued norm $\|T\| := \|T\|_{L_1(\nabla,m)\to L_1(\nabla,m)} = \sup\{\|Tx\|_1 : \|x\|_1 \le \mathbf{1}\}$ this space is a Banach–Kantorovich space [14, 4.2.6]. We need the following property of regularity for operators $T \in B(L_1(\nabla, m))$.

Proposition 2.2. $B(L_1(\nabla, m)) \subset H_r(L_1(\nabla, m))$.

Proof. Let $T \in B(L_1(\nabla, m))$, $0 \le x \in L_1(\nabla, m)$. The set of all elements from $L_1(\nabla, m)$ of the form $y = |T(x_1)| + \cdots + |T(x_n)|$ is denoted by $E(x)$, where $x = x_1 + \cdots + x_n, x_i \ge 0, i = 1, 2, \ldots, n$. It is clear that for $y \in E(x)$ the following inequalities hold

$$\|y\|_1 \le \sum_{i=1}^n \|Tx_i\|_1 \le \|T\| \sum_{i=1}^n \|x_i\|_1 = \|T\|\|x\|_1.$$

Repeating the proof of [15, Theorem VIII.7.2] we obtain that for any $y_1, y_2, \ldots, y_k$ from $E(x)$ there exists $y \in E(x)$ such that $\sup\limits_{1\le i\le k} y_i \le y$. Since $\|y\|_1 \le \|T\|\|x\|_1$ we have that $\|\sup\limits_{1\le i\le k} y_i\|_1 \le \|T\|\|x\|_1$.

We denote by A the direction of finite subsets of $E(x)$, ordered by inclusion and for every $\alpha \in A$ we set $y_\alpha = \sup\{y : y \in \alpha\}$. It is clear that $\{y_\alpha\}_{\alpha\in A}$ is an increasing net of positive elements from $L_1(\nabla, m)$, in addition $\|y_\alpha\|_1 \le \|T\|\|x\|_1$ for all $\alpha \in A$. By the theorem of monotone convergence [5] there exists $z \in L_1(\nabla, m)$ such that $y_\alpha \uparrow z$. Hence $E(x)$ is an order-bounded set in $L_1(\nabla, m)$.

Repeating again the proof of [15, Theorem VIII.7.2] we have that $T(F)$ is an order-bounded set in $L_1(\nabla, m)$ for any order-bounded set $F \subset L_1(\nabla, m)$. Therefore, by [15, Theorem VIII 2.2], $T \in H_r(L_1(\nabla, m))$. □

By Proposition 2.2 for every $T \in B(L_1(\nabla, m))$ there exists its module $|T| \in B(L_1(\nabla, m))$. Repeating the proof of [15, Theorem VIII.6.3] we have that $\|T\| = \| \, |T| \, \|$.

A linear operator $T \in B(L_1(\nabla, m))$ is called an L_1-L_∞ contraction if

$$T(L_\infty(\nabla)) \subset L_\infty(\nabla)$$

and

$$\|T\|_{L_1(\nabla,m)\to L_1(\nabla,m)} \leq \mathbf{1}, \quad \|T\|_{L_\infty(\nabla)\to L_\infty(\nabla)} \leq 1.$$

We denote the set of all L_1-L_∞ contractions by $C_{1,\infty}(\nabla, m)$.

Proposition 2.3. *If $T \in C_{1,\infty}(\nabla, m)$ then $|T| \in C_{1,\infty}(\nabla, m)$.*

Proof. It is sufficient to show that $|T|(L_\infty(\nabla)) \subset L_\infty(\nabla)$ and that the inequality $\| \, |T| \, \|_{L_\infty(\nabla)\to L_\infty(\nabla)} \leq 1$ holds. If $y \in L_\infty(\nabla)$ and $|y| \leq \mathbf{1}$ then $\|y\|_\infty \leq 1$ and

$$|Ty| \leq \|Ty\|_\infty \mathbf{1} \leq \|T\|_{L_\infty(\nabla)\to L_\infty(\nabla)} \|y\|_\infty \mathbf{1} \leq \mathbf{1}.$$

Consequently,

$$| \, |T|(y) \, | \leq |T|(|y|) = \sup\{|Tz| : z \in L_1(\nabla, m), |z| \leq |y|\} \leq \mathbf{1}. \qquad \square$$

Theorem 2.1 and Proposition 2.3 imply the following

Corollary 2.4. *If $T \in C_{1,\infty}(\nabla, m)$, $p > 1$, then $T(L_p(\nabla, m)) \subset L_p(\nabla, m)$ and $\|T\|_{L_p(\nabla,m)\to L_p(\nabla,m)} \leq \mathbf{1}$.*

Proof. By Theorem 2.1 for every $\omega \in \Omega$ there exists a positive linear contraction $S_\omega : L_1(\nabla_\omega, m_\omega) \to L_1(\nabla_\omega, m_\omega)$ such that $S_\omega(\Phi(x)(\omega)) = \Phi(|T|x)(\omega)$ a.e. for every $x \in L_1(\nabla, m)$. Since $|T| \in C_{1,\infty}(\nabla, m)$ (see Proposition 2.3), it follows that $|T|(\mathbf{1}) \leq \mathbf{1}$, and therefore

$$S_\omega \mathbf{1}_\omega = \Phi(|T|\mathbf{1})(\omega) \leq \Phi(\mathbf{1})(\omega) = \mathbf{1}_\omega$$

for every $\omega \in \Omega$, where $\mathbf{1}_\omega$ is the unit element of the Boolean algebra ∇_ω. Hence S_ω is a positive linear contraction in $L_\infty(\nabla_\omega)$ for a.e. $\omega \in \Omega$.

Since $L_p(\nabla_\omega, m_\omega)$ is an interpolation space between $L_1(\nabla_\omega, m_\omega)$ and $L_\infty(\nabla_\omega)$ [13, Ch. II, §4], we have that

$$S_\omega(L_p(\nabla_\omega, m_\omega)) \subset L_p(\nabla_\omega, m_\omega) \quad \text{and} \quad \|S_\omega\|_{L_p(\nabla_\omega,m_\omega)\to L_p(\nabla_\omega,m_\omega)} \leq 1.$$

Hence $|T|(L_p(\nabla, m)) \subset L_p(\nabla, m)$. Using the equality $\Phi(|x|^p)(\omega) = (\Phi(|x|)(\omega))^p$ a.e., where $x \in L_p(\nabla, m)$ [8], we obtain that

$$\| \, |T|(|x|)\|_p^p(\omega) = \|S_\omega(\Phi(|x|)(\omega))\|^p_{L_p(\nabla_\omega,m_\omega)} \leq \|\Phi(|x|)(\omega)\|^p_{L_p(\nabla_\omega,m_\omega)} = \|x\|_p^p(\omega)$$

for a.e. $\omega \in \Omega$. Hence $\||T|\|_{L_p(\nabla,m)\to L_p(\nabla,m)} \leq \mathbf{1}$. The inequality $|Tx| \leq |T||x|$ implies that $T(L_p(\nabla,m)) \subset L_p(\nabla,m)$, in addition

$$\|T\|_{L_p(\nabla,m)\to L_p(\nabla,m)} \leq \| \, |T| \, \|_{L_p(\nabla,m)\to L_p(\nabla,m)} \leq \mathbf{1}. \qquad \square$$

The following theorem is a version of Theorem 2.1 for an operator $T \in C_{1,\infty}(\nabla,m)$.

Theorem 2.5. *If $T \in C_{1,\infty}(\nabla,m)$, then for every $\omega \in \Omega$ there exists*

$$T_\omega \in C_{1,\infty}(\nabla_\omega, m_\omega) \quad \textit{such that} \quad T_\omega(\Phi(x)(\omega)) = \Phi(Tx)(\omega)$$

a.e. for every $x \in L_1(\nabla,m)$.

Proof. Theorem 2.1 provides the existence of a linear operator T_ω in $L_1(\nabla_\omega,m_\omega)$ satisfying $T_\omega(\Phi(x)(\omega)) = \Phi(Tx)(\omega)$, in addition $T_\omega(\ell_\nabla(x)(\omega)) = \ell_\nabla(Tx)(\omega)$ for all $\omega \in \Omega$, $x \in L_\infty(\nabla)$. Let S_ω be the positive linear contractions in $L_1(\nabla_\omega,m_\omega)$ as in the proof of Corollary 2.4. Since $|Tx| \leq |T||x|$ and $S_\omega(\ell_\nabla(x)(\omega)) = \ell_\nabla(|T|x)(\omega)$ for all $\omega \in \Omega$, $x \in L_\infty(\nabla)$ we have that $|T_\omega(\ell_\nabla(x)(\omega))| = |\ell_\nabla(T(x))(\omega)| \leq \ell_\nabla(|T||x|)(\omega) = S_\omega(\ell_\nabla(|x|)(\omega)) = S_\omega(|\ell_\nabla(x)(\omega)|)$. Using density of the linear space $\{\ell_\nabla(x)(\omega) : x \in L_\infty(\nabla)\}$ in $L_1(\nabla_\omega,m_\omega)$ (see property 5) of ℓ_∇) we obtain that $|T_\omega g| \leq S_\omega|g|$ for all $g \in L_1(\nabla_\omega,m_\omega)$. Since every bounded linear operator in $L_1(\nabla_\omega,m_\omega)$ is regular, the module $|T_\omega|$ is defined, which is a positive contraction in $L_1(\nabla_\omega,m_\omega)$, in addition

$$|T_\omega|h = \sup\{|T_\omega g| : g \in L_1(\nabla_\omega,m_\omega), |g| \leq h \} \leq S_\omega h$$

for all $0 \leq h \in L_1(\nabla_\omega,m_\omega)$. In particular $|T_\omega|(\mathbf{1}_\omega) \leq S_\omega(\mathbf{1}_\omega) \leq \mathbf{1}_\omega$, that implies $|T_\omega|(L_\infty(\nabla_\omega)) \subset L_\infty(\nabla_\omega)$ and $\| \, |T_\omega| \, \|_{L_\infty(\nabla_\omega)\to L_\infty(\nabla_\omega)} \leq 1$. Since $|T_\omega g| \leq |T_\omega||g|$ for all $g \in L_1(\nabla_\omega,m_\omega)$, it follows that $T_\omega(L_\infty(\nabla_\omega)) \subset L_\infty(\nabla_\omega)$ and $\|T_\omega\|_{L_\infty(\nabla_\omega)\to L_\infty(\nabla_\omega)} \leq 1$. $\square$

3. Ergodic theorems for L_1-L_∞ contractions in $L_p(\nabla, m)$

Consider on the vector lattice $L_0(\nabla)$ the metric $\rho(x,y) = \int |x-y|(\mathbf{1}+|x-y|)^{-1}dm$ with values in $L_0(\Omega)$. Let $Z : \omega \to Z(\omega) = (L_0(\nabla_\omega),\rho_\omega)$ be a bundle over Ω of metric spaces $(L_0(\nabla_\omega),\rho_\omega)$, where

$$\rho_\omega(u(\omega),v(\omega)) = \int |u(\omega)-v(\omega)|(\mathbf{1}_\omega + |u(\omega)-v(\omega)|)^{-1}dm_\omega, \omega \in \Omega.$$

In [8] it is established that there exists an isometric isomorphism Ψ from $(L_0(\nabla),\rho)$ onto the complete $L_0(\Omega)$-metrisable vector lattice $L_0(\Omega,(Z,\mathcal{E}))$ such that

$$\Psi\left(\sum\nolimits_{i=1}^{n} \lambda_i e_i\right) = \sum\nolimits_{i=1}^{n} \lambda_i \gamma_\omega(e_i) = \Phi\left(\sum\nolimits_{i=1}^{n} \lambda_i e_i\right),$$

for all $\lambda_i \in \mathbb{R}$, $e_i \in \nabla, i = 1,\ldots,n$, $n \in \mathbb{N}$, in addition, $L_0(\Omega,(Y_1,\mathcal{E}))$ can be identified with a vector sublattice in $L_0(\Omega,(Z,\mathcal{E}))$ and $\Psi(x) = \Phi(x)$ for all $x \in L_1(\nabla,m)$, where $Y_1(\omega) = L_1(\nabla_\omega,m_\omega)$, $\omega \in \Omega$.

In order to obtain various results of ergodic theorems related to L_1-L_∞ contractions in $L_p(\nabla,m)$, we need to employ the following connection between (o)-

convergence of a sequence $\{x_n\} \subset L_0(\nabla)$ and (o)-convergence of the sequences $\{\Psi(x_n)(\omega)\} \subset L_0(\nabla_\omega)$, $\omega \in \Omega$.

Theorem 3.1. [2] *If $x_n, x \in L_0(\nabla)$ and $x_n \stackrel{(o)}{\to} x$, then $\Psi(x_n)(\omega) \stackrel{(o)}{\to} \Psi(x)(\omega)$ in $L_0(\nabla_\omega)$ for a.e. $\omega \in \Omega$. Conversely, if $x_n \in L_0(\nabla)$ and $\Psi(x_n)(\omega) \stackrel{(o)}{\to} v(\omega)$ a.e. for some $v(\omega) \in L_0(\nabla_\omega)$, then there exists $x \in L_0(\nabla)$ such that $\Psi(x)(\omega) = v(\omega)$ a.e. and $x_n \stackrel{(o)}{\to} x$ in $L_0(\nabla)$.*

The following theorem is a vector version of the well-known N. Danford and J.T. Schward's ergodic theorems for a L_1-L_∞ contraction in the Banach–Kantorovich lattice $L_p(\nabla, m)$ associated with an $L_0(\Omega)$-valued measure.

Theorem 3.2. *If $T \in C_{1,\infty}(\nabla, m)$, $1 < p < \infty$, $x \in L_p(\nabla, m)$ then the sequence $s_n(T)(x) = \frac{1}{n}\sum_{i=0}^{n-1} T^i(x)$ is order bounded in the Banach–Kantorovich lattice $L_p(\nabla, m)$ and*

$$\| \sup_{n\geq 1} |s_n(T)(x)| \; \|_p \leq \left(\frac{p}{p-1}\right) \|x\|_p,$$

in addition, there exists $\tilde{x} \in L_p(\nabla, m)$ such that the sequence $s_n(T)(x)$ is (o)-convergent to $\tilde{x}$ in $L_p(\nabla, m)$.

Proof. From the proof of Corollary 2.4 it follows that $|T|$ is a positive contraction in $L_p(\nabla, m)$ and $|T|(\mathbf{1}) \leq \mathbf{1}$. From [2] follows that the sequence $s_n(|T|)(|x|)$ is order bounded in $L_p(\nabla, m)$ and

$$\| \sup_{n\geq 1} |s_n(|T|)(|x|)| \; \|_p \leq \left(\frac{p}{p-1}\right) \|x\|_p.$$

Since $|T^i x| \leq |T|^i(|x|), i = 1, 2, \dots$ it follows that

$$|s_n(T)(x)| \leq \frac{1}{n}\sum_{i=0}^{n-1} |T^i(x)| \leq \frac{1}{n}\sum_{i=0}^{n-1} |T|^i |x| = s_n(|T|)(|x|)$$

and

$$\| \sup_{n\geq 1} |s_n(T)(x)| \|_p \leq \| \sup_{n\geq 1} s_n(|T|)(|x|) \|_p \leq \left(\frac{p}{p-1}\right) \|x\|_p.$$

According to Theorem 2.5 and Corollary 2.4 we have that $s_n(T_\omega)(\Phi(x)(\omega)) = \Phi(s_n(T)(x))(\omega)$ a.e. Since $T_\omega \in C_{1,\infty}(\nabla_\omega, m_\omega)$ (Theorem 2.5), Theorem 6 [7, Ch. VIII, §6] implies that there exists $v(\omega) \in L_0(\nabla_\omega)$ such that

$$s_n(T_\omega)(\Phi(x)(\omega)) \stackrel{(o)}{\to} v(\omega)$$

in $L_0(\nabla_\omega, m_\omega)$ for a.e. $\omega \in \Omega$. Since $\Phi(s_n(T)) \in L_0(\Omega, Y_p)$, it follows that $v \in L_0(\Omega, Y_p)$ and there exists $\tilde{x} \in L_0(\nabla, m)$ such that $\Psi(\tilde{x}) = v$.

Theorem 3.1 implies that $s_n(T)(x) \stackrel{(o)}{\to} \tilde{x}$ in $L_0(\nabla)$. Using this (o)-convergence and order-boundedness in $L_p(\nabla, m)$ of the sequence $\{s_n(T)(x)\}$ we have that $\tilde{x} \in L_p(\nabla, m)$ and $s_n(T)(x)$ is (o)-convergent to $\tilde{x}$ in $L_p(\nabla, m)$. □

Remark 3.3. Repeating the proof of Theorem 3.2 and using Corollaries 4 and 5 [7, Ch. VIII, §5] we get that for every $T \in C_{1,\infty}(\nabla, m)$, $x \in L_1(\nabla, m)$ there exists $\tilde{x} \in L_1(\nabla, m)$ such that the sequence $s_n(T)(x)$ is (o)-convergent to $\tilde{x}$ in $L_0(\nabla)$ and $\|s_n(T)(x) - \tilde{x}\|_1 \xrightarrow{(o)} 0$.

Now, we shall present a version of Theorem 3.2 for Orlicz–Kantorovich lattices $L_M(\nabla, m)$.

Let $M : R \to [0, \infty)$ be an N-function and let M^* be the complementary N-function to M [11, Ch. I, §§1–2].

In the same way as in [4], we consider the following subsets in $L_1(\nabla, m)$:

$$L_M^0(\nabla, m) = \{x \in L_0(\nabla) : M(x) \in L_1(\nabla, m)\},$$

$$L_M(\nabla, m) = \{x \in L_0(\nabla) : xy \in L_1(\nabla, m), \forall y \in L_{M^*}^0(\nabla, m)\}$$

for which the inclusions

$$L_\infty(\nabla) \subset L_M^0(\nabla, m) \subset L_M(\nabla, m) \subset L_1(\nabla, m)$$

hold.

The set $L_M(\nabla, m)$ is a vector sublattice in $L_1(\nabla, m)$ and with respect to the $L_0(\Omega)$-valued norm

$$\|x\|_M := \sup\left\{\left|\int xy\, dm\right| : y \in L_{M^*}^0(\nabla, m), \int M^*(y) dm \leq \mathbf{1}\right\}$$

the pair $(L_M(\nabla, m), \|\cdot\|_M)$ is a Banach–Kantorovich lattice, which is called an Orlicz–Kantorovich lattice [4].

Proposition 3.4. *If $T \in C_{1,\infty}(\nabla, m)$, then*

$$T(L_M(\nabla, m)) \subset L_M(\nabla, m) \quad \textit{and} \quad \|T\|_{L_M(\nabla,m)\to L_M(\nabla,m)} \leq \mathbf{1}.$$

Proof. By [4, Proposition 2.3] we have that an element $x \in L_1(\nabla, m)$ belongs to $L_M(\nabla, m)$ if and only if $\Phi(x)(\omega) \in L_M(\nabla_\omega, m_\omega)$ a.e., moreover $\|x\|_M(\omega) = \|\Phi(x)(\omega)\|_{L_M(\nabla_\omega, m_\omega)}$ a.e. Since $L_M(\nabla_\omega, m_\omega)$ is an interpolation space between $L_1(\nabla_\omega, m_\omega)$ and $L_\infty(\nabla_\omega, m_\omega)$ [13, Ch. II, §4], repeating the proof of Corollary 2.4 we obtain that $T(L_M(\nabla, m)) \subset L_M(\nabla, m)$ and $\|T\|_{L_M(\nabla,m)\to L_M(\nabla,m)} \leq \mathbf{1}$. □

For establishing the statistic ergodic theorem for L_1-L_∞ contractions in $L_M(\nabla, m)$ we need the next properties of the classical Orlicz spaces $L_M(\nabla_\omega, m_\omega)$ which immediately follow from [6, Proposition 2.1].

Proposition 3.5. *Let N-function M meets $\triangle_2$-condition and let K be a norm bounded set in $L_M(\nabla_\omega, m_\omega)$. Then K is relatively weak compact if and only if for each $f \in L_M^*(\nabla_\omega, m_\omega) = L_{M^*}(\nabla_\omega, m_\omega)$ and a sequence $q_n \in \nabla_\omega$ with $q_n \downarrow 0$ the convergence*

$$\sup\left\{\left|\int (q_n f h) dm_\omega\right| : h \in K\right\} \to 0$$

holds.

Now let us give a version of the statistic ergodic theorem for L_1-L_∞ contraction in the Banach–Kantorovich lattice $L_M(\nabla, m)$.

Theorem 3.6. *If $T \in C_{1,\infty}(\nabla, m)$, $x \in L_M(\nabla, m)$ and the N-function M meets the $\triangle_2$-condition, then there exists $\tilde{x} \in L_M(\nabla, m)$ such that*

$$\|s_n(T)(x) - \tilde{x}\|_M \overset{(o)}{\to} 0.$$

Proof. Since $m_\omega(\mathbf{1}_\omega) = 1$, Proposition 3.5 implies that ∇_ω is relatively weak compact in $L_M(\nabla_\omega, m_\omega)$.

Let T_ω be an L_1-L_∞ contraction in $L_1(\nabla_\omega, m_\omega)$ such that $T_\omega(\Phi(x)(\omega)) = \Phi(Tx)(\omega)$ a.e. for every $x \in L_1(\nabla, m)$ (see Theorem 2.5). It is clear that

$$\|(1/n)s_n(T_\omega)(h)\|_{L_M(\nabla_\omega, m_\omega)} \to 0 \quad \text{as} \quad n \to \infty$$

for all $h \in L_M(\nabla_\omega, m_\omega)$.

Since the N-function M meets $\triangle_2$-condition, the linear subspace

$$\left\{\sum\nolimits_{i=1}^{n} \lambda_i e_i : \lambda_i \in \mathbb{R},\ e_i \in \nabla_\omega,\ i = 1, \ldots, n,\ n \in \mathbb{N}\right\}$$

is dense in $L_M(\nabla_\omega, m_\omega)$, in addition, ∇_ω is a relatively weak compact set in $L_M(\nabla_\omega, m_\omega)$ (see Proposition 3.5). Hence by Corollary 3 [7, Ch. VIII, §5] there exists $v(\omega) \in L_M(\nabla_\omega, m_\omega)$ such that $\|s_n(T_\omega)(\Phi(x)(\omega)) - v(\omega)\|_{L_M(\nabla_\omega, m_\omega)} \to 0$ as $n \to \infty$ for a.e. $\omega \in \Omega$. Since $\Phi(s_n(T)) \in L_0(\Omega, Y_1)$, it follows that $v \in L_0(\Omega, Y_1)$. By [4, Proposition 2.3] we have that there exists $\tilde{x} \in L_M(\nabla, m)$ such that $\Phi(\tilde{x})(\omega) = v(\omega)$ a.e., in addition, $\|s_n(T)(x) - \tilde{x}\|_M \overset{(o)}{\to} 0$. □

Repeating the proof of Theorems 3.2, 3.6 and using [4, Theorem 3.3], we obtain the following version of the individual ergodic theorem for L_1-L_∞ contraction in the Orlicz–Kantorovich lattice $L_M(\nabla, m)$.

Theorem 3.7. *If $T \in C_{1,\infty}(\nabla, m)$, $x \in L_M(\nabla, m)$ and the N-function M has the property $\sup\limits_{s \geq 1}\{\frac{1}{M(s)} \int\limits_1^s M(t^{-1}s)dt\} < \infty$, then the sequence $s_n(T)(x)$ is order bounded in the Orlicz–Kantorovich lattice $L_M(\nabla, m)$ and $s_n(T)(x)$ is (o)-convergent in $L_M(\nabla, m)$.*

4. Weighted ergodic theorems in Banach–Kantorovich lattice $L_p(\nabla, m)$

Let $\mathbb{S}$ be the unit circle in the field $\mathbb{C}$ of complex numbers and let $\mathbb{Z}$ be the ring of integer numbers. A function $P_s : \mathbb{Z} \to \mathbb{C}$ is called a trigonometric polynomial if

$$P_s(k) = \sum\nolimits_{j=1}^{s} r_j \lambda_j^k, k \in \mathbb{Z}, \quad \text{for some} \quad \{r_j\}_{j=1}^{s} \subset \mathbb{C}$$

and $\{\lambda_j\}_{j=1}^{s} \subset \mathbb{S}$. A sequence $\{\alpha(k)\}$ of complex numbers is called a bounded Besicovich sequence (BB-sequence) if $\sup\{|\alpha(k)| : k \in \mathbb{Z}\} < \infty$ and for every $\varepsilon > 0$

there exists a sequence of trigonometric polynomials P_s, such that

$$\limsup_n \frac{1}{n} \sum_{k=0}^{n-1} |\alpha(k) - P_s(k)| < \varepsilon.$$

The following theorem is a vector version of the weighted ergodic theorem for L_1-L_∞ contractions in the Banach–Kantorovich lattice $L_p(\nabla, m)$.

Theorem 4.1. *Let $\{\alpha(k)\}$ be a BB-sequence of real numbers and $T \in C_{1,\infty}(\nabla, m)$, $x \in L_p(\nabla, m)$. If $1 < p < \infty$ ($p = 1$) then there exists $\tilde{x} \in L_p(\nabla, m)$ such that $s_n(\alpha, T)(x) = \frac{1}{n} \sum_{k=0}^{n-1} \alpha(k) T^k(x) \xrightarrow{(o)} \tilde{x}$ in $L_p(\nabla, m)$ (respectively, $s_n(\alpha, T)(x) \xrightarrow{(o)} \tilde{x}$ in $L_0(\nabla, m)$).*

Proof. Let T_ω be an L_1-L_∞ contraction in $L_1(\nabla_\omega, m_\omega)$ such that $T_\omega(\Phi(x)(\omega)) = \Phi(Tx)(\omega)$ a.e. (see Theorem 2.5). Since $T_\omega \in C_{1,\infty}(\nabla_\omega, m_\omega)$, Theorem 1.4 [10] implies that there exists $v(\omega) \in L_1(\nabla_\omega, m_\omega)$ such that

$$s_n(\alpha, T_\omega)(\Phi(x)(\omega)) \xrightarrow{(o)} v(\omega) \text{ in } L_0(\nabla_\omega) \quad \text{for a.e. } \omega \in \Omega.$$

By Theorem 3.1 there exists $\tilde{x} \in L_1(\nabla, m)$ such that $s_n(\alpha, T)(x) \xrightarrow{(o)} \tilde{x}$ in $L_0(\nabla)$.

If $1 < p < \infty$, $x \in L_p(\nabla, m)$, then using Theorem 3.2 and the inequality

$$|s_n(\alpha, T)(x)| \leq \frac{1}{n} \sum_{k=0}^{n-1} |\alpha(k)||T^k(x)| \leq \sup_k |\alpha(k)| s_n(|T|)(|x|),$$

we have that the sequence $s_n(\alpha, T)(x)$ is order bounded in the Banach–Kantorovich lattice $L_p(\nabla, m)$. Consequently, $\tilde{x} \in L_p(\nabla, m)$ and the convergence $s_n(\alpha, T)(x) \xrightarrow{(o)} \tilde{x}$ in $L_0(\nabla)$ imply the (o)-convergence of the sequence $s_n(\alpha, T)(x)$ to $\tilde{x}$ in $L_p(\nabla, m)$. □

Remark 4.2. Since the norm $\|\cdot\|_p$ is order continuous in $L_p(\nabla, m)$ [5], Theorem 4.1 implies that under the conditions of Theorem 4.1 we have that the convergence $\|s_n(\alpha, T)(x) - \tilde{x}\|_p \xrightarrow{(o)} 0$ is provided by the condition $1 < p < \infty$, $x \in L_p(\nabla, m)$, in addition (see Theorem 3.2),

$$\| \sup_{n \geq 1} |s_n(\alpha, T)(x)| \|_p \leq \sup_k |\alpha(k)| \cdot \| \sup_{n \geq 1} s_n(|T|)(|x|) \|_p \leq \left(\frac{p}{p-1} \right) \sup_k |\alpha(k)| \cdot \|x\|_p.$$

Acknowledgment

The second author (I.G) acknowledges the MOHE Grant FRGS13-071-0312.

References

[1] M.A. Akcoglu, *A pointwise ergodic theorems in L_p-spaces.* Can. J. Math. **27** (1975), 1075–1082.

[2] V.I. Chilin, I.G. Ganiev, *An individual ergodic theorem for contractions in the Banach–Kantorovich lattice $L^p(\nabla, m)$.* Russian Math. (Izv. VUZ). **44** (2000), No. 7, 77–79.

[3] V.I. Chilin, B.S. Zakirov, *Decomposition of the measure with the values in order complete vector lattices.* Vladikavkaz. Math. J. **4(10)** (2008), 31–38, (Russian).

[4] V.I. Chilin, B.S. Zakirov, *Ergodic theorems for contractions in Orlicz–Kantorovich lattices.* Sib. Math. J. **50** (2009), 1027–1037.

[5] V.I. Chilin, B.S. Zakirov, *Noncommutative integration for traces with values in Kantorovich–Pinsker spaces.* Russian Math. (Izv. VUZ) **54** (2010), No. 10, 15–26.

[6] P.G. Doodds, F.A. Sukochev, G. Schluchtermann, *Weak compactness criteria in symmetric spaces of measurable operators.* Math. Proc. Camb. Phil. Soc. **131** (2001), 363–384.

[7] N. Dunford and J. Schwartz, *Linear Operators. Part I*, Interscience, New York (1958).

[8] I.G. Ganiev, *Measurable bundles of lattices nd their applications.* In: Studies on Functional Analysis and its Applications, Nauka, Moscow (2006), pp. 9–49, (Russian).

[9] A.E. Gutman, *Banach bundles in the theory of lattice-normed spaces. Order-compatible linear operators*, Novosibirsk, Trudy Inst. Mat. **29** (1995), 63–211, (Russian).

[10] R.L. Jones, J. Olsen, *Multiparameter weighted ergodic theorems.* Can. J. Math. **46** (1994), 343–356.

[11] M.A. Krasnoselskii and Ja.B. Rutickii, *Convex functions and Orlicz spaces.* Translated from the first Russian edition. Groningen, 1961.

[12] U. Krengel, *Ergodic Theorems,* Walter de Grugwer, Berlin, New York. 1985.

[13] S.G. Kreyin, Yu.I. Petunin, Y.M. Semyenov, *Interpolation of linear operators*, Translations of Mathematical Monographs, **54**, American Mathematical Society, Providence (1982).

[14] A.G. Kusraev, *Dominanted operators*, Mathematics and its Applications, **519**. Kluwer Academic Publishers, Dordrecht, 2000.

[15] B.Z. Vulich, *Introduction to the theory of partially ordered spaces.* Wolters-Noordhoff. Groningen. 1967.

V.I. Chilin
Department of Mathematics
National University of Uzbekistan
100174, Tashkent, Uzbekistan
e-mail: `chilin@ucd.uz`

I.G. Ganiev
Department of Science in Engineering
International Islamic University Malaysia
P.O. Box 10, 50728
Kuala Lumpur, Malaysia
e-mail: `inam@iium.edu.my`

Ordered Structures and Applications: Positivity VII
Trends in Mathematics, 111–126

The Dedekind Completion of $C(X)$ with Pointwise Discontinuous Functions

Nicolae Dăneţ

Abstract. In this paper we show that whenever X is a topological space, which is completely regular and Baire, then the Dedekind completion of $C(X)$, the space of all real continuous functions on X, is the Dedekind complete Riesz space of all pointwise discontinuous functions, where two functions that coincides on a dense set are identified.

Mathematics Subject Classification (2010). Primary 46A40; Secondary 06B23, 26A15.

Keywords. Dedekind order completion, Riesz spaces, vector lattices, pointwise discontinuous functions, semicontinuous functions.

1. Introduction

In the theory of linear operators between Riesz spaces almost all good results use the fact that the range space is Dedekind complete. Two of the most important spaces in functional analysis, $C(X)$ and $C_b(X)$, the spaces of all real continuous functions or real bounded continuous functions on a topological space X, are not Dedekind complete. The problem of construction of the Dedekind completion of these spaces has been addressed by many authors in the last sixty five years, but a universally accepted solution has not been found yet. In the last years, after Rosinger and Anguelov showed that the solutions of some partial differential equations can be assimilated with functions in the Dedekind completion of the above two spaces [3, 4], the interest in this problem has become even more.

The terminology used in this paper for Riesz spaces is that of [20]. We recall however the definition of the Dedekind completion of a Riesz space ([20], Definition 32.1).

Definition 1.1. The Dedekind complete Riesz space L^δ is called a Dedekind completion of the Riesz space L if the following conditions hold.

(i) There exists a Riesz subspace $\widehat{L}$ of L^δ such that L and $\widehat{L}$ are Riesz isomorphic.
(ii) $\widehat{f} = \bigvee\{\widehat{g} : \widehat{g} \in \widehat{L},\, \widehat{g} \le \widehat{f}\} = \bigwedge\{\widehat{g} : \widehat{g} \in \widehat{L},\, \widehat{g} \ge \widehat{f}\}$, for all $\widehat{f} \in L^\delta$.

Every Archimedean Riesz space L has a Dedekind completion L^δ ([20], Theorem 32.5), which is unique up to a Riesz isomorphism. The Riesz space $C(X)$ is Archimedean, therefore its Dedekind completion $C(X)^\delta$ exists.

The aim of this paper is to describe a construction of $C(X)^\delta$ using functions (more precisely, equivalent classes of functions) defined on the space X. We are not interested in obtaining a Riesz isomorphism with another space $C(Y)$, with Y a topological space different from X.

In 1937 MacNeille [21] proved his famous result about the order completion by cuts of a partially ordered set. Let P be a nonempty partially ordered set containing no smallest or largest element. If A and B are subsets of P we define the subsets A^u and B^l of P by putting $A^u = \{p \in P : p \geq a \text{ for all } a \in A\}$ and $B^l = \{p \in P : p \leq a \text{ for all } a \in A\}$, respectively. (For the properties of the sets A^u and B^l see [20], pp. 187–188.) A subset A of P is called a cut if $A = A^{ul}$. Let $\widetilde{P}$ be the set of all cuts of P. MacNeille showed that the set $\widetilde{P}$, partially ordered by inclusion, is an order-complete lattice and the map $\phi : P \longrightarrow \widetilde{P}$, given by $\phi(x) = \{x\}^{ul}$, is a one-to-one map which preserves suprema and infima. In addition, every cut A satisfies

$$A = \bigvee \{\{x\}^{ul} : \{x\}^{ul} \subset A\} = \bigwedge \{\{x\}^{ul} : \{x\}^{ul} \supset A\}$$

([20], Theorem 32.3). The order-complete lattice $\widetilde{P}$ is called the *completion by cuts* or the *normal completion* of the partially ordered set P.

In 1950 Dilworth used MacNeille's result to construct the completion by cuts of the lattice $C_b(X)$. In his seminal paper [11] Dilworth introduced the notion of *normal upper semicontinuous* functions, that is, the functions f that satisfy the equality $S(I(f)) = f$ (see (2.1) for the definitions of the operators I and S) and proved that, if X is a completely regular space, the completion by cuts of $C_b(X)$ is isomorphic with the complete lattice $\mathcal{NU}^b_{\mathrm{sc}}(X)$ of all bounded normal upper semicontinuous functions ([11], Theorem 4.1). The isomorphism is only for the lattice structure and not for the vector structure. It is worth to note that the sum and the pointwise infimum of two normal upper semicontinuous functions are semicontinuous but not normal. The author showed in [10] how the set $\mathcal{NU}^b_{\mathrm{sc}}(X)$ can be organized as a Riesz space.

In 1953 Horn [14] proved a similar result to that of Dilworth, but for unbounded continuous functions. First he developed a general theory for the completion by cuts of a partially ordered set C, which is a subset of a complete lattice B. Applying this construction to the lattice $C(X)$ Horn proved that, for X a completely regular space, the completion by cuts of $C(X)$ is isomorphic with $\mathcal{NL}^{\mathrm{cb}}_{\mathrm{sc}}(X)$, the complete lattice of all *normal lower semicontinuous* functions (see Definition 2.2), which are C-bounded ([14], Theorem 11). (A function $f : X \longrightarrow \mathbb{R}$ is called C-bounded if there exist $g_1, g_2 \in C(X)$ such that $g_1 \leq f \leq g_2$.)

The results of Dilworth and Horn concern only the completion by cuts of the lattices $C_b(X)$ and $C(X)$, respectively. In 1962 Kuzumi Nakano and Shimogaki [22] constructed the Dedekind completion of the Riesz space $C(X)$, for X a compact space, using *quasicontinuous* functions. They used the name of quasicontinuous function for a real bounded function, which satisfies the equality $IS(f) = ISI(f)$. This definition was introduced by Hidegorô Nakano in his book "Measure Theory" published in 1948 in Japanese [23]. We will call this type of functions *N-quasicontinuous*, that is, quasicontinuous in the sense of Nakano. Nowadays a real function f on a topological space X is called *quasicontinuous* on X if for every $x \in X$ and for every $\varepsilon > 0$ and for every neighborhood U of x there exists a nonempty open set $G \subset U$ such that $|f(y) - f(x)| < \varepsilon$, for all $y \in G$. (Note that it is not necessary that $x \in G$.) If f is quasicontinuous then $IS(f) = I(f)$ and $SI(f) = S(f)$ ([7], Proposition 3.1) and, in consequence, $IS(f) = ISI(f)$, that is, f is N-quasicontinuous. Lemma 1 in [22] (with reference for proof to the inaccessible book [23]) asserts that a function f is N-quasicontinuous if and only if it is pointwise discontinuous. (A function $f : X \longrightarrow \mathbb{R}$ is called *pointwise discontinuous* on X if the set of points of continuity of f is dense in X.) We will give a proof of this lemma in Proposition 4.6.

In a series of four papers [15, 16, 17, 18] published between 1957 and 1964, and later in a book [19], Kaplan studied extensively $C(X)$, its dual $C(X)^*$ and its second dual $C(X)^{**}$. In this context he gave several descriptions for $C(X)^\delta$ using $C(X)^{**}$ [18, 19]. In these papers Kaplan developed the arithmetic of the operators ℓ and u (see definitions in (2.7) and (2.8)), which are the basic tools for the construction of $C(X)^\delta$. These operators can be also defined on every Dedekind complete Riesz space and they have almost all the properties highlighted by Kaplan.

In 2004 Anguelov [1] constructed the completion by cuts of $C(X)$ using *Hausdorff continuous interval-valued functions* (in the sense of Sendov [24], H-continuous functions for short), that is, the functions $\overline{\underline{f}} : X \longrightarrow \mathbb{IR}$, which associate to every point $x \in X$ the real closed interval $[\underline{f}(x), \overline{f}(x)]$ whose components $(\underline{f}, \overline{f})$ form a regular pair (see Definition 5.6). Anguelov showed that the completion by cuts of $C(X)$ is isomorphic with $\mathbb{H}_{\mathrm{cb}}(X)$, the complete lattice of all H-continuous functions that are C-bounded on X ([1], Theorems 9 and 10). The algebraic structure of the set $\mathbb{H}(\Omega)$ of all H-continuous functions defined on an open set $\Omega \subset \mathbb{R}^n$ was studied later in [2]. The interest in study of the space $\mathbb{H}(\Omega)$, and hence of $C(\Omega)^\delta$, comes from the fact that the solutions of some partial differential equations can be assimilated with H-continuous functions [3, 4].

In 2010 Becker [6] gave a short description of how can be constructed the Dedekind completion of the Riesz space $C_b(X)$, when X is a Baire space, using *upper semicontinuous* functions.

In 2010–2011 the author had the following contributions to this topic: (a) every H-continuous function on a completely regular topological space X corresponds uniquely to a Dedekind cut in $C(X)$ [9]; (b) if X is a compact space or a complete metric space, then the completion by cuts of $C(X)$ is isomorphic with $\mathcal{Q}(X, \mathbb{R})$, the complete lattice of all equivalence classes of quasicontinuous functions, where $f \sim g$ if and only if $f = g$ on the dense set of all common points of continuity of f and g [7]; (c) Anguelov's construction can be deduced via an order isomorphism from Horn's construction [8].

The short list from above, certainly incomplete, shows that in all constructions of the completion by cuts of $C(X)$ or of the Dedekind completion $C(X)^\delta$ the authors have used functions with certain types of discontinuity. A careful analysis shows that all these functions have in common one property: they are pointwise discontinuous.

In this paper we show that the Dedekind completion of $C(X)$, for X a completely regular Baire space, is the Dedekind complete Riesz space of all pointwise discontinuous functions, where two functions that coincides on a dense set are identified. Our second goal is to give proofs based on Riesz space techniques, which can be used to any Riesz space.

In Section 2 we establish the terminology and recall the definitions and the basic properties of some nonlinear operators I and S, called Baire operators [5], which are the principal tools in our proofs. Because the Dedekind completion is an order process and not a topological one, we characterize all topological properties of the functions with the aid of the Baire operators. In Section 3 we show that a function is pointwise discontinuous on a Baire space if and only if $I(S(f)-I(f)) = 0$, and that the set of all pointwise discontinuous functions $C_d(X)$ is a Riesz space. In Section 4, following Kaplan [18], we call a function f rare if $IS(|f|) = 0$ and show that f is rare if and only if $f = 0$ on a dense set in a Baire space X. Since the set $\mathcal{R}a(X)$ of all rare functions is an ideal in the Riesz space $\mathcal{B}_{\mathrm{loc}}(X)$ of all real locally bounded functions, we consider the quotient space $\mathcal{B}_{\mathrm{loc}}(X)/\mathcal{R}a(X)$ and study its properties in Section 5. The most important result in this section is Theorem 5.7, which shows that in every equivalence class $\widehat{f} \in C_d(X)/\mathcal{R}a(X)$ there exists a unique regular pair $(\underline{f}, \overline{f})$ such that $\widehat{f} = \widehat{\underline{f}} = \widehat{\overline{f}}$. Finally, Section 6 contains the construction of $C(X)^\delta$.

2. Preliminaries and notation

2.1. Baire operators and their properties

Let X be a topological space. A function $f : X \longrightarrow \mathbb{R}$ is called *locally bounded* on X if for every $x \in X$ there exists a neighborhood V of x such that f is bounded on V, that is, $a \le f(y) \le b$, for some real numbers a and b and all $y \in V$. We denote by $\mathcal{B}_{\mathrm{loc}}(X)$ the Dedekind complete Riesz space of all locally bounded functions on X. To every function $f \in \mathcal{B}_{\mathrm{loc}}(X)$ we associate two new functions, the lower limit function and the upper limit function of f, $I(f)$ and $S(f)$, respectively, defined as follows,

$$I(f)(x) = \sup_{V\in\mathcal{V}_x} \inf_{y\in V} f(y) \quad \text{and} \quad S(f)(x) = \inf_{V\in\mathcal{V}_x} \sup_{y\in V} f(y), \tag{2.1}$$

where $\mathcal{V}_x$ denotes the set of all neighborhoods of the point $x \in X$. So we obtain two nonlinear operators $I, S : \mathcal{B}_{\mathrm{loc}}(X) \longrightarrow \mathcal{B}_{\mathrm{loc}}(X)$ called the *lower Baire operator* and the *upper Baire operator*, respectively. The definitions of these operators are pointwise and use explicitly the topology of the space X by using the neighborhoods of x, unlike the operators ℓ and u (see (2.7) and (2.8)), which are defined using the order relation between functions.

The Baire operators have the following properties:

(B1) $I(f) \leq f \leq S(f)$, for all $f \in \mathcal{B}_{\text{loc}}(X)$.

(B2) I and S are *monotone*, that is, $I(f) \leq I(g)$ and $S(f) \leq S(g)$, whenever $f \leq g$.

(B3) I and S are *idempotent*, that is, $I \circ I = I$ and $S \circ S = S$.

(B4) $I \circ S$ and $S \circ I$ are also *monotone* and *idempotent*.

(B5) If $\lambda \geq 0$, then $I(\lambda f) = \lambda I(f)$ and $S(\lambda f) = \lambda S(f)$.

(B6) $S(-f) = -I(f)$.

(B7) The operator I is *supra-additive*, the operator S is *sub-additive* and for any $f, g \in \mathcal{B}_{\text{loc}}(X)$ we have ([12], p. 22)

$$I(f) + I(g) \leq I(f+g) \leq I(f) + S(g) \leq S(f+g) \leq S(f) + S(g).$$

(B8) In consequence,

$$I(f) - S(g) \leq I(f-g) \leq \begin{matrix} S(f) - S(g) \\ I(f) - I(g) \end{matrix} \leq S(f-g) \leq S(f) - I(g).$$

(B9) $I(f \wedge g) = I(f) \wedge I(g)$.

(B10) $S(f \vee g) = S(f) \vee S(g)$.

(B11) $f \wedge g = 0 \Rightarrow I(f) \wedge S(g) = 0$.

(B12) $S(f^+) = S(f)^+$, $I(f^+) = I(f)^+$, $S(f^-) = I(f)^-$, $I(f^-) = S(f)^-$.

2.2. Baire operators and semicontinuous functions

A function $f \in \mathcal{B}_{\text{loc}}(X)$ is lower semicontinuous if and only if $I(f) = f$, and upper semicontinuous if and only if $S(f) = f$. We denote by $\mathcal{L}_{\text{sc}}(X)$ the set of all locally bounded lower semicontinuous functions, and by $\mathcal{U}_{\text{sc}}(X)$ the set of all locally bounded upper semicontinuous functions. The sets $\mathcal{L}_{\text{sc}}(X)$ and $\mathcal{U}_{\text{sc}}(X)$ are Dedekind complete lattices in which the supremum and the infimum of any nonempty order-bounded subset $\{f_\gamma\}_{\gamma \in \Gamma}$ are given by the formulae:

$$\bigvee_{\mathcal{L}} f_\gamma = \bigvee f_\gamma, \qquad \bigwedge_{\mathcal{L}} f_\gamma = I\left(\bigwedge f_\gamma\right), \tag{2.2}$$

$$\bigvee_{\mathcal{U}} f_\gamma = S\left(\bigvee f_\gamma\right), \qquad \bigwedge_{\mathcal{U}} f_\gamma = \bigwedge f_\gamma. \tag{2.3}$$

The linear subspace of $\mathcal{B}_{\text{loc}}(X)$ generated by lower or upper semicontinuous functions will be denoted by $\mathcal{S}(X)$. Therefore a function $f \in \mathcal{S}(X)$ can have one of the following descriptions: $f = g - h$, with $g, h \in \mathcal{L}_{\text{sc}}(X)$ or $g, h \in \mathcal{U}_{\text{sc}}(X)$, or $f = g + h$, with $g \in \mathcal{L}_{\text{sc}}(X)$ and $h \in \mathcal{U}_{\text{sc}}(X)$.

Proposition 2.1. *$\mathcal{S}(X)$ is a Riesz subspace of $\mathcal{B}_{\text{loc}}(X)$.*

Proof. Let $f = g - h \in \mathcal{S}(X)$, with $g, h \in \mathcal{L}_{\rm sc}(X)$, and $\lambda \in \mathbb{R}$. Then $\lambda f = \lambda g - \lambda h$, if $\lambda \geq 0$, and $\lambda f = (-\lambda)\, h - (-\lambda)\, g$, if $\lambda < 0$. Hence $\lambda f \in \mathcal{S}(X)$.

If $f_1 = g_1 - h_1$, $f_2 = g_2 - h_2 \in \mathcal{S}(X)$, with $g_1, g_2, h_1, h_2 \in \mathcal{L}_{\rm sc}(X)$, then

$$f_1 + f_2 = (g_1 + g_2) - (h_1 + h_2) \in \mathcal{S}(X),$$

$$f_1 \vee f_2 = (g_1 - h_1) \vee (g_2 - h_2) = (g_1 + h_2) \vee (g_2 + h_1) - (h_1 + h_2) \in \mathcal{S}(X),$$

and a similar formula holds for $f_1 \wedge f_2$. Therefore $\mathcal{S}(X)$ is a Riesz subspace of $\mathcal{B}_{\rm loc}(X)$. □

The compositions $I \circ S$ and $S \circ I$ of Baire operators are also monotone and idempotent. The functions that are fixed points for these operators were introduced and studied by Dilworth [11] under the name of *normal semicontinuous* functions. More precisely, we have the following definition.

Definition 2.2. A function $f \in \mathcal{L}_{\rm sc}(X)$ is called normal lower semicontinuous if $I(S(f)) = f$, and a function $f \in \mathcal{U}_{\rm sc}(X)$ is called normal upper semicontinuous if $S(I(f)) = f$.

The set of all normal lower semicontinuous functions is denoted by $\mathcal{NL}_{\rm sc}(X)$ and the set of all normal upper semicontinuous functions is denoted by $\mathcal{NU}_{\rm sc}(X)$. These sets are Dedekind complete lattices in which the supremum and the infimum of any nonempty order-bounded subset $\{f_\gamma\}_{\gamma\in\Gamma}$ are given by the formulae ([11], Theorem 4.2):

$$\bigvee\nolimits_{\mathcal{NL}} f_\gamma = IS\left(\bigvee f_\gamma\right), \qquad \bigwedge\nolimits_{\mathcal{NL}} f_\gamma = I\left(\bigwedge f_\gamma\right), \tag{2.4}$$

$$\bigvee\nolimits_{\mathcal{NU}} f_\gamma = S\left(\bigvee f_\gamma\right), \qquad \bigwedge\nolimits_{\mathcal{NU}} f_\gamma = SI\left(\bigwedge f_\gamma\right). \tag{2.5}$$

2.3. Saltus operator and its properties

For $f \in \mathcal{B}_{\rm loc}(X)$ we define the saltus of f by putting $\omega(f) = S(f) - I(f)$. The saltus $\omega(f)$ has the following properties:

(S1) $\omega(f)$ is an upper semicontinuous function.

(S2) $\omega(f) \geq 0$ and $\omega(f) = 0 \Leftrightarrow f \in C(X)$. If we denote by C_f the set of points of continuity of f, then we have

$$f \in C(X) \Leftrightarrow C_f = X \Leftrightarrow \omega(f) = 0. \tag{2.6}$$

(S3) $\omega(\lambda f) = |\lambda|\, \omega(f)$, for all $\lambda \in \mathbb{R}$. In particular, $\omega(-f) = \omega(f)$.

(S4) $\begin{matrix}\omega(f \vee g)\\ \omega(f \wedge g)\end{matrix} \leq \omega(f) \vee \omega(g)$.

(S5) $|\omega(f) - \omega(g)| \leq \begin{matrix}\omega(f + g)\\ \omega(f - g)\end{matrix} \leq \omega(f) + \omega(g)$.

(S6) $\omega(f + g) \leq \omega(f \vee g) + \omega(f \wedge g) \leq \omega(f) + \omega(g)$.

(S7) $\omega(f) = \omega(f^+) + \omega(f^-)$.

(S8) $\omega(|f|) \leq \omega(f) \leq 2S(|f|)$.

For the proof of these properties see [19], pp. 256–260.

2.4. Kaplan operators

We denote by $\mathcal{B}_c(X)$ the set of all real functions that are C-bounded on X. (We recall that a function $f : X \longrightarrow \mathbb{R}$ is called C-bounded if there exist $g_1, g_2 \in C(X)$ such that $g_1 \leq f \leq g_2$.) Obviously, $\mathcal{B}_c(X) \subset \mathcal{B}_{\text{loc}}(X)$, and since $\mathcal{B}_c(X)$ is the ideal generated by $C(X)$ in $\mathcal{B}_{\text{loc}}(X)$, $\mathcal{B}_c(X)$ is a Dedekind complete Riesz space. For $f \in \mathcal{B}_c(X)$ the sets $\{g \in C(X) : g \leq f\}$ and $\{g \in C(X) : g \geq f\}$ are nonempty and we can define two new functions by putting

$$\ell(f)(x) = \bigvee\{g(x) : g \in C(X),\ g \leq f\}, \quad x \in X, \tag{2.7}$$

$$u(f)(x) = \bigwedge\{g(x) : g \in C(X),\ g \geq f\}, \quad x \in X, \tag{2.8}$$

where $\bigvee$ and $\bigwedge$ denote the pointwise supremum and the pointwise infimum, respectively. So we obtain two nonlinear operators $\ell, u : \mathcal{B}_c(X) \longrightarrow \mathcal{B}_c(X)$, which are monotone and idempotent. For every $f \in \mathcal{B}_c(X)$, we have $\ell(f) \in \mathcal{L}_{\text{sc}}(X)$, $u(f) \in \mathcal{U}_{\text{sc}}(X)$ and $\ell(f) \leq f \leq u(f)$.

I called the operators ℓ and u Kaplan operators because Samuel Kaplan studied in details the properties of these operators in [18] and [19].

2.5. Relations between Baire and Kaplan operators and the characterization of a completely regular space

For any $f \in \mathcal{B}_c(X)$ all Baire operators and Kaplan operators are well defined and the following inequalities hold:

$$\ell(f) \leq I(f) \leq f \leq S(f) \leq u(f). \tag{2.9}$$

The equalities between ℓ and I or u and S characterize the completely regular topological spaces.

Theorem 2.3. *Let X be topological space. The following statements are equivalent:*

(i) *X is completely regular.*
(ii) *$\ell(f) = I(f)$, for every $f \in \mathcal{B}_c(X)$.*
(iii) *$u(f) = S(f)$, for every $f \in \mathcal{B}_c(X)$.*

Proof. For (i) $\Rightarrow$ (ii) and (i) $\Rightarrow$ (iii) see Proposition 6 in [8].

(ii) $\Rightarrow$ (i) Assume that $\ell(f) = I(f)$, for every $f \in \mathcal{B}_c(X)$. Let x be any point in X and let U be an open subset of X such that $x \in U$. Put $f = \chi_U$. Since U is open, f is lower semicontinuous and then we have $f = I(f)$. Therefore $f = \ell(f)$. Since

$$1 = f(x) = \ell(f)(x) = \sup\{g(x) : g \in C(X),\ g \leq f\},$$

there exists a function $g \in C(X)$, $g \leq f$, such that $g(x) > 0$. We can assume that $g \geq 0$. (Otherwise, we can replace g with g^+.) Then the inequalities $0 \leq g \leq f = \chi_U$ implies that $g(y) = 0$, for all $y \notin U$. If we define $h(z) = \inf\{g(z)/g(x), 1\}$, for all $z \in X$, we have a continuous function on X with values in $[0, 1]$, $h(x) = 1$ and $h(y) = 0$, for all $y \notin U$. Hence X is a completely regular topological space. $\square$

The implication (i) $\Rightarrow$ (iii) was proved by Dilworth ([11], Lemma 4.1) for real bounded functions. The equivalence (i) $\Leftrightarrow$ (iii) was proved by Horn ([14], Theorem 8) for functions $f : X \longrightarrow \overline{\mathbb{R}}$. My proof of Theorem 2.3 is for real-valued functions that are C-bounded.

For every $f \in \mathcal{B}_c(X)$ we have $\ell(f) \leq I(f) \leq f$ (see (2.9)). These inequalities become equalities in the following conditions: (a) $I(f) = f \Leftrightarrow f$ is lower semicontinuous; (b) $\ell(f) = I(f) \Leftrightarrow X$ is completely regular; (c) $\ell(f) = f \Leftrightarrow f$ is lower semicontinuous and X is completely regular. Since we also have $f \leq S(f) \leq u(f)$, similar statements hold in this case.

3. Pointwise discontinuous functions

A function $f : X \longrightarrow \mathbb{R}$ is called *pointwise discontinuous* or *densely continuous* on X if C_f, the set of points of continuity of f, is dense in X, or, equivalently, every nonempty open subset of X contains a point of continuity of f. We denote by $C_d(X)$ the set of all locally bounded functions on X that are pointwise discontinuous.

The following theorem gives a characterization of a function $f \in C_d(X)$. The theorem shows that $\overline{C_f} = X$ if and only if $I(\omega(f)) = 0$, that is, a topological property of the function f can be characterized with the aid of the Baire operators. This characterization appears in the thesis of Lester Ford in 1912 ([12], Theorem 16).

Theorem 3.1. *Let X be a Baire space and $f \in \mathcal{B}_{\mathrm{loc}}(X)$. The following assertions are equivalent.*

(i) *f is pointwise discontinuous, that is, $\overline{C_f} = X$.*
(ii) *$I(\omega(f)) = 0$.*
(iii) *For every real number $\lambda > 0$ the sets $A_\lambda(f) = \{x \in X : \omega(f)(x) \geq \lambda\}$ are closed and nowhere dense.*
(iv) *The set of points of discontinuity of f is a set of the first category.*

Proof. The implications (i) $\Rightarrow$ (ii) $\Rightarrow$ (iii) $\Rightarrow$ (iv) hold for any topological space X. For (iv) $\Rightarrow$ (i) we need X to be a Baire space.

(i) $\Rightarrow$ (ii) If f is pointwise discontinuous, then, for every $x \in X$, every neighborhood V of x contains a point y_0 where f is continuous, that is, $\omega(f)(y_0) = 0$. In consequence, $I(\omega(f))(x) = \sup_{V \in \mathcal{V}_x} \inf_{y \in V} \omega(f)(y) = 0$.

(ii) $\Rightarrow$ (iii) Since $\omega(f)$ is an upper semicontinuous function, all the sets $A_\lambda(f)$ are closed. We must show that $\mathrm{int} A_\lambda(f) = \emptyset$, for all $\lambda > 0$. By way of contradiction let us assume that $\mathrm{int} A_\lambda(f) \neq \emptyset$, for some λ. Let x be a point in $\mathrm{int} A_\lambda(f)$. Then we obtain the following contradiction

$$0 = I(\omega(f))(x) = \sup_{V \in \mathcal{V}_x} \inf_{y \in V} \omega(f)(y) \geq \inf_{y \in int A_\lambda(f)} \omega(f)(y) \geq \lambda > 0.$$

(iii) $\Rightarrow$ (iv) Since the set of points of continuity can be written in the form

$$C_f = \bigcap_{n=1}^{\infty} \{x \in X : \omega(f)(x) < 1/n\},$$

the set of points of discontinuity is

$$X \setminus C_f = \bigcup_{n=1}^{\infty} A_{1/n}(f),$$

where each $A_{1/n}(f)$ is nowhere dense. Hence $X \setminus C_f$ is of the first category.

(iv) $\Rightarrow$ (i) If we assume that (iv) holds, then

$$C_f = \bigcap_{n=1}^{\infty} \left(A_{1/n}(f)\right)^c,$$

with $\left(A_{1/n}(f)\right)^c = X \setminus A_{1/n}(f)$ open and dense. Since X is a Baire space, C_f is dense in X. □

The above theorem characterizes a function $f \in C_d(X)$ just as (2.6) characterizes a function $f \in C(X)$. Analogously, we can write

$$f \in C_d(X) \Leftrightarrow \overline{C_f} = X \Leftrightarrow I(\omega(f)) = 0. \tag{3.1}$$

Theorem 3.2. *If X is a Baire space, then the set $C_d(X)$ is a Riesz subspace of the Dedekind complete Riesz space $\mathcal{B}_{\mathrm{loc}}(X)$.*

Proof. Let f and g be two functions in $C_d(X)$. Theorem 3.1 shows that $I(\omega(f)) = 0$ and $I(\omega(g)) = 0$. Using the property (S6) of the saltus operator and the properties (B2), (B3) and (B7) of the Baire operator I, we have

$$\begin{aligned} 0 &\le I(\omega(f+g)) \le II(\omega(f \vee g) + \omega(f \wedge g)) \le I\left(I(\omega(f) + \omega(g))\right) \\ &\le I(\underbrace{I(\omega(f))}_{=0} + \underbrace{S(\omega(g))}_{=\omega(g)}) = I(\omega(g)) = 0. \end{aligned}$$

Hence $I(\omega(f+g)) = 0$ and $I(\omega(f \vee g) + \omega(f \wedge g)) = 0$. By Theorem 3.1 the first equality shows that $f + g \in C_d(X)$. Since I is supra-additive, $0 \le I(\omega(f \vee g)) + I(\omega(f \wedge g)) \le I(\omega(f \vee g) + \omega(f \wedge g)) = 0$. Therefore, $I(\omega(f \vee g)) = 0$ and $I(\omega(f \wedge g)) = 0$, that is, $f \vee g$, $f \wedge g \in C_d(X)$.

For any real number λ we have $I(\omega(\lambda f)) = I(|\lambda|\, \omega(f)) = |\lambda|\, I(\omega(f)) = 0$. Hence, $\lambda f \in C_d(X)$. □

4. The rare functions

A function $f \in \mathcal{B}_{\mathrm{loc}}(X)$ will be called *rare* if $IS\,(|f|) = 0$ ([18, 19]). The set of all rare functions f is denoted by $\mathcal{R}a(X)$. This subset of $\mathcal{B}_{\mathrm{loc}}(X)$ has a good structure since it is an ideal.

Proposition 4.1. ([18], (3.2)) *$\mathcal{R}a(X)$ is an ideal of $\mathcal{B}_{\mathrm{loc}}(X)$.*

A useful characterization of a rare function is given in the following proposition.

Proposition 4.2. *For $f \in \mathcal{B}_{\text{loc}}(X)$ the following conditions are equivalent:*

(i) $f \in \mathcal{R}a(X)$.
(ii) $f^+, f^- \in \mathcal{R}a(X)$.
(iii) $IS(f) = 0 = SI(f)$.
(iv) $IS(f) \leq 0 \leq SI(f)$.

The next proposition gives another characterization of a rare function, which is the first step in obtaining a topological characterization. In particular, the proposition shows that a rare function is pointwise discontinuous on a Baire space.

Proposition 4.3. $f \in \mathcal{R}a(X)$ *if and only if* $I(|f|) = 0$ *and* $I(\omega(f)) = 0$.

Proof. Let f be rare. Then we have $0 \leq I(|f|) \leq IS(|f|) = 0$, and therefore $I(|f|) = 0$. Moreover, since $\omega(f) \leq 2S(|f|)$ (see (S8)), we also have $0 \leq I(\omega(f)) \leq 2IS(|f|) = 0$, and so $I(\omega(f)) = 0$.

Conversely, assume that $I(|f|) = 0$ and $I(\omega(f)) = 0$. In order to prove that $f \in \mathcal{R}a(X)$, by Proposition 4.2 it is sufficient to show that $f^+, f^- \in \mathcal{R}a(X)$. Therefore we can prove the statement for f positive. If $f \in \mathcal{B}_{\text{loc}}(X)_+$ and $I(f) = 0$ and $I(\omega(f)) = 0$, then $IS(f) = I(S(f) - I(f)) = I(\omega(f)) = 0$, and so $f \in \mathcal{R}a(X)$. □

Since $\omega(f)$ is upper semicontinuous and then $S(\omega(f)) = \omega(f)$, the condition $I(\omega(f)) = 0$ means that $\omega(f)$ is rare. Proposition 4.3 shows that if f is rare, then $\omega(f)$ is also rare. We denote by $\omega^{-1}(\mathcal{R}a(X))$ the set of all locally bounded functions for which $\omega(f) \in \mathcal{R}a(X)$, that is, $I(\omega(f)) = 0$. With this notation, by Proposition 4.3 we have $\mathcal{R}a(X) \subset \omega^{-1}(\mathcal{R}a(X))$. Theorem 3.1 shows that $\omega^{-1}(\mathcal{R}a(X)) = C_d(X)$, for X a Baire space. Then the following corollary holds.

Corollary 4.4. *If X is a Baire space, then $\mathcal{R}a(X)$ is an ideal in $C_d(X)$.*

In the next proposition we give a topological characterization of a rare function.

Proposition 4.5. *Let X be a Baire space. A function $f \in \mathcal{B}_{\text{loc}}(X)$ is rare if and only if $f(x) = 0$, for all $x \in C_f$, the dense G_δ set of points of continuity of f.*

Proof. If f is rare, then $I(|f|) = 0$ and $I(\omega(f)) = 0$ (Proposition 4.3). By Theorem 3.1, $I(\omega(f)) = 0$ implies that f is pointwise discontinuous, that is, the set C_f of points of continuity of f is a dense G_δ subset of X. At every point $x \in C_f$ we have $|f|(x) = I(|f|)(x) = 0$, since f is continuous at x. So $f = 0$ on C_f.

Conversely, if $f(x) = 0$, for all $x \in C_f$, where C_f is a dense G_δ subset of X, then C_f is the set of points of continuity of f, and therefore f is pointwise discontinuous. By Theorem 3.1 we have $I(\omega(f)) = 0$. Moreover, since every open set contains at least one point where f is equal to zero, we obtain $I(|f|)(x) = \sup_{V \in \mathcal{V}_x} \inf_{y \in V} |f|(y) = 0$, for all $x \in X$. Thus, by Proposition 4.3, f is a rare function. □

Proposition 4.6. *The following assertions are equivalent.*

(i) $f \in \omega^{-1}(\mathcal{R}a(X))$, *that is,* $I(\omega(f)) = 0$.
(ii) $SI(f) \geq IS(f)$.
(iii) $IS(f) = ISI(f)$.
(iv) $SI(f) = SIS(f)$.

Proof. (i) $\Rightarrow$ (ii) results immediately by using property (B8).

$$0 = I(\omega(f)) = I(S(f) - I(f)) \geq IS(f) - SI(f).$$

(ii) $\Rightarrow$ (iii) $IS(f) = IIS(f) \leq ISI(f) \leq IS(f)$.

(iii) $\Rightarrow$ (i) First we show that $S(f) - SI(f) \in \mathcal{R}a(X)$. Indeed,

$$0 \leq IS(S(f) - SI(f)) \leq I(SS(f) - ISI(f)) \leq IS(f) - IISI(f) = 0.$$

Then $I(\omega(f)) = I(S(f) - I(f)) \leq S(f) - SI(f)$, from where we have

$$I(\omega(f)) = ISIS(\omega(f)) \leq IS(S(f) - SI(f)) = 0.$$

The proof of the equivalence (iii) $\Leftrightarrow$ (iv) uses the fact that IS and SI are idempotent operators. □

If X is a Baire space, the equivalence (i) $\Leftrightarrow$ (iii) in Proposition 4.6 shows that a function is *pointwise discontinuous* if and only if it is *N-quasicontinuous*. So we gave a proof for Lemma 1 in [22].

Proposition 4.7. *The Riesz space* $\mathcal{S}(X)$ *has the properties:*

(i) $\mathcal{S}(X) \subset \omega^{-1}(\mathcal{R}a(X))$.
(ii) *If* X *is a Baire space, then* $\mathcal{S}(X) \subset C_d(X)$.

Proof. (i) If $f \in \mathcal{L}_{\mathrm{sc}}(X)$, then $f = I(f)$ and $0 \leq I(\omega(f)) = I(S(f) - f) \leq SS(f) - S(f) = 0$.

For $f \in \mathcal{S}(X)$ there exist $g, h \in \mathcal{L}_{\mathrm{sc}}(X)$ such that $f = g - h$ with $\omega(g), \omega(h) \in \mathcal{R}a(X)$. Using the inequality (S5) we have $\omega(f) = \omega(g - h) \leq \omega(g) + \omega(h) \in \mathcal{R}a(X)$ and, since $\mathcal{R}a(X)$ is an ideal, it results $\omega(f) \in \mathcal{R}a(X)$, that is, $f \in \omega^{-1}(\mathcal{R}a(X))$. □

5. The space $\mathcal{B}_{\mathrm{loc}}(X)/\mathcal{R}a(X)$

In the Dedekind complete Riesz space $\mathcal{B}_{\mathrm{loc}}(X)$ we have the ideal $\mathcal{R}a(X)$ of all rare functions. The quotient space $\mathcal{B}_{\mathrm{loc}}(X)/\mathcal{R}a(X)$ is a Riesz space and the quotient mapping $\pi : \mathcal{B}_{\mathrm{loc}}(X) \longrightarrow \mathcal{B}_{\mathrm{loc}}(X)/\mathcal{R}a(X)$, $\pi(f) = \widehat{f}$, is a Riesz homomorphism ([20], Theorem 18.9).

The equivalence relation on $\mathcal{B}_{\mathrm{loc}}(X)$ is

$$f \sim g \Leftrightarrow f - g \in \mathcal{R}a(X) \Leftrightarrow IS(|f - g|) = 0, \tag{5.1}$$

and using Proposition 4.5 we can say that $f \sim g$ if and only if $f = g$ on some dense G_δ subset of X (the set of all common points of continuity of f and g).

The order relation on $\mathcal{B}_{\mathrm{loc}}(X)/\mathcal{R}a(X)$ is

$$\widehat{f} \leq \widehat{g} \Leftrightarrow (f-g)^{+} \in \mathcal{R}a(X) \Leftrightarrow IS(f-g) \leq 0. \tag{5.2}$$

For the first equivalence see [13], p. 14. The second equivalence results from the equality $0 = IS((f-g)^{+}) = (IS(f-g))^{+}$.

Proposition 5.1. *Let $f, g \in \mathcal{B}_{\mathrm{loc}}(X)$ such that $\widehat{f} \leq \widehat{g}$. Then:*

(i) $\widehat{f} \leq \widehat{g} \Rightarrow IS(f) \leq S(g)$. *The converse implication holds for $g \in \mathcal{S}(X)$.*
(ii) $\widehat{f} \leq \widehat{g} \Rightarrow I(f) \leq SI(g)$. *The converse implication holds for $f \in \mathcal{S}(X)$.*

Proof. (i) Let $\widehat{f} \leq \widehat{g}$. Then $IS(f-g) \leq 0$. Using (B8), we have

$$IS(f) - S(g) = IS(f) - SS(g) \leq I(S(f) - S(g)) \leq IS(f-g) \leq 0.$$

Now assume that $g \in \mathcal{S}(X)$ and $IS(f) \leq S(g)$. Using (B8) we obtain

$$IS(f-g) \leq I\left(S(f) - I(g)\right) \leq IS(f) - II(g) \leq S(g) - I(g) = \omega(g),$$

from where it results $IS(f-g) \leq IS\left(\omega(g)\right)$. Since $g \in \mathcal{S}(X)$, then $IS\left(\omega(g)\right) = 0$ (Proposition 4.7). Therefore the converse implication holds. □

Corollary 5.2. *Let $f, g \in \mathcal{B}_{\mathrm{loc}}(X)$. If $f - g \in \mathcal{R}a$, then $IS(f) = IS(g)$ and $SI(f) = SI(g)$.*

Proposition 5.3. *The quotient mapping $\pi : \mathcal{B}_{\mathrm{loc}}(X) \longrightarrow \mathcal{B}_{\mathrm{loc}}(X)/\mathcal{R}a(X)$ has the following properties of order-continuity.*

(i) *If $g = \bigwedge_{\gamma} g_{\gamma}$, where $\{g_{\gamma}\}_{\gamma\in\Gamma}$ is a subset of $\mathcal{U}_{\mathrm{sc}}(X)$ (hence g is also in $\mathcal{U}_{\mathrm{sc}}(X)$), then $\pi(g) = \bigwedge_{\gamma} \pi(g_{\gamma})$.*
(ii) *If $g = \bigvee \gamma g_{\gamma}$, where $\{g_{\gamma}\}_{\gamma\in\Gamma}$ is a subset of $\mathcal{L}_{\mathrm{sc}}(X)$ (hence g is also in $\mathcal{L}_{\mathrm{sc}}(X)$), then $\pi(g) = \bigvee_{\gamma} \pi(g_{\gamma})$.*

Proof. See [19], p. 384. □

The following proposition shows that, when X is a Baire space, every pointwise continuous function on X is the sum of a semicontinuous function and a rare function.

Proposition 5.4. *Let X be a Baire space. The following conditions are equivalent:*

(i) $f \in C_d(X)$
(ii) $f = I(f) + r_1$
(iii) $f = S(f) + r_2$
(iv) $f = IS(f) + r_3$
(v) $f = SI(f) + r_4$

where r_1, r_2, r_3, r_4 are some rare functions.

Proof. If $f \in C_d(X)$, $\omega(f)$ is in the Riesz ideal $\mathcal{R}a(X)$ (Theorem 3.1). Then the inequalities $0 \le S(f) - f \le \omega(f)$ and $0 \le f - I(f) \le \omega(f)$ show that $S(f) - f$ and $f - I(f)$ are in $\mathcal{R}a(X)$. Using that $SI(f) \ge IS(f)$ (Proposition 4.6) we have

$$-\omega(f) = I(f) - S(f) \le f - SI(f) \le f - IS(f) \le S(f) - I(f) = \omega(f),$$

from where it results $|f - SI(f)| \le \omega(f)$, and $|f - IS(f)| \le \omega(f)$. Hence $f - SI(f)$ and $f - IS(f)$ are in $\mathcal{R}a(X)$. So we just proved that (i) implies all the other conditions.

(ii) $\Rightarrow$ (i) If $f = I(f) + r_1$, then $\omega(f) \le \omega(I(f)) + \omega(r_1) = \omega(r_1)$, since ω is null on semicontinuous functions (Proposition 4.7). Using the monotonicity of I we obtain $0 \le I(\omega(f)) \le I(\omega(r_1)) = 0$. Hence $f \in C_d(X)$.

Similar proofs can be given for the rest of the converse implications. □

Corollary 5.5. *For X a Baire space the following subsets of $\mathcal{B}_{\mathrm{loc}}(X)/\mathcal{R}a(X)$ coincide:*

$$\pi(C_d(X)),\ \pi(\mathcal{S}(X)),\ \pi(\mathcal{U}_{\mathrm{sc}}(X)),\ \pi(\mathcal{L}_{\mathrm{sc}}(X)),\ \pi(\mathcal{N}\mathcal{U}_{\mathrm{sc}}(X)),\ \pi(\mathcal{N}\mathcal{L}_{\mathrm{sc}}(X)).$$

Definition 5.6. A pair of functions $(\underline{f}, \overline{f})$ is called regular if $\underline{f} \in \mathcal{L}_{\mathrm{sc}}(X)$, $\overline{f} \in \mathcal{U}_{\mathrm{sc}}(X)$, $\underline{f} \le \overline{f}$, and

$$S(\underline{f}) = \overline{f}, \quad I(\overline{f}) = \underline{f}.$$

If $(\underline{f}, \overline{f})$ is a regular pair, then the lower component $\underline{f} \in \mathcal{N}\mathcal{L}_{\mathrm{sc}}(X)$, the upper component $\overline{f} \in \mathcal{N}\mathcal{U}_{\mathrm{sc}}(X)$ and $\overline{f} - \underline{f} \in \mathcal{R}a(X)$. A regular pair $(\underline{f}, \overline{f})$ defines a H-continuous function $\overline{\underline{f}} = [\underline{f}, \overline{f}]$ and conversely ([1], Theorem 1). For the relations between regular pairs, H-continuous functions and cuts in $C(X)$ see [9], Theorem 4.6.

Theorem 5.7. *Let X be a Baire space. If $\widehat{f} \in \mathcal{C}_d(X)/\mathcal{R}a(X)$, then:*

(i) *$\widehat{f}$ contains exactly one regular pair $(\underline{f}, \overline{f})$ and $\widehat{f} = \widehat{\underline{f}} = \widehat{\overline{f}}$.*

(ii) *If $g, h \in \widehat{f}$, $g \in \mathcal{L}_{\mathrm{sc}}(X)$ and $h \in \mathcal{U}_{\mathrm{sc}}(X)$, then $g \le \underline{f} \le \overline{f} \le h$, that is, $\underline{f}$ is the largest lower semicontinuous function in $\widehat{f}$, and $\overline{f}$ is the smallest upper semicontinuous function in $\widehat{f}$.*

Proof. (i) Let $\widehat{f} \in \mathcal{C}_d(X)/\mathcal{R}a(X)$. Since $f \in \mathcal{C}_d(X)$, then $IS(f)$ and $SI(f)$ belong to $\widehat{f}$ (Proposition 5.4), and $IS(f) \le SI(f)$ (Proposition 4.6). Define

$$\underline{f} = IS(f), \quad \overline{f} = SI(f). \tag{5.3}$$

The pair $(\underline{f}, \overline{f})$ is regular because $\underline{f} \in \mathcal{L}_{\mathrm{sc}}(X)$, $\overline{f} \in \mathcal{U}_{\mathrm{sc}}(X)$, and, by using Proposition 4.6, we have

$$S(\underline{f}) = SIS(f) = SI(f) = \overline{f}, \quad I(\overline{f}) = ISI(f) = IS(f) = \underline{f}$$

The regular pair defined by (5.3) is unique, because for every other function $g \in \widehat{f}$, by Corollary 5.2, we have $IS(g) = IS(f) = \underline{f}$ and $SI(g) = SI(f) = \overline{f}$.

(ii) Let $g \in \widehat{f}$ be a lower semicontinuous function. Then $g = I(g) \le IS(g) = \underline{f}$. If $h \in \widehat{f}$ is upper semicontinuous, then $h = S(h) \ge SI(h) = \overline{f}$. □

Proposition 5.8. *The quotient mapping* $\pi : \mathcal{B}_{\rm loc}(X) \longrightarrow \mathcal{B}_{\rm loc}(X)/\mathcal{R}a(X)$ *restricted to the ordered set* $\mathcal{NL}_{\rm sc}(X)$ *is an order embedding, that is,* $f \le g \Leftrightarrow \pi(f) \le \pi(g)$, *for all* f, g *in* $\mathcal{NL}_{\rm sc}(X)$. *The assertion remains valid if we replace* $\mathcal{NL}_{\rm sc}(X)$ *by* $\mathcal{NU}_{\rm sc}(X)$.

Proof. Let f, g be in $\mathcal{NL}_{\rm sc}(X)$. We have to prove the implication $\pi(f) \le \pi(g) \Rightarrow f \le g$. By Proposition 5.1, if $\pi(f) \le \pi(g)$, then $IS(f) \le S(g)$, and so $IS(f) \le IS(g)$. Since $f, g \in \mathcal{NL}_{\rm sc}(X)$, we obtain $f \le g$. □

6. Dedekind completion of $C(X)$

In this section we assume that X is a completely regular Baire space. The most useful topological spaces in functional analysis are the compact spaces and the complete metric spaces, which are completely regular Baire spaces. We denote by $C_d^{\rm cb}(X)$ the Riesz space of all pointwise discontinuous functions on X that are C-bounded.

Theorem 6.1. *Let* X *be a completely regular Baire space. Then*

$$C(X)^\delta = C_d^{\rm cb}(X)/\mathcal{R}a(X),$$

that is, the Dedekind completion of $C(X)$ *is* $C_d^{\rm cb}(X)/\mathcal{R}a(X)$.

Proof. We must show that the Riesz space $C(X)^\delta$ satisfies the conditions of Definition 1.1:

(a) $C(X)^\delta$ is a Dedekind complete Riesz space.
(b) $\pi(C(X))$ is a Riesz subspace of $C(X)^\delta$ such that $C(X)$ and $\pi(C(X))$ are Riesz isomorphic.
(c) $\widehat{f} = \bigvee\{\widehat{g} : \widehat{g} \in \pi(C(X)),\ \widehat{g} \le \widehat{f}\} = \bigwedge\{\widehat{g} : \widehat{g} \in \pi(C(X)),\ \widehat{g} \ge \widehat{f}\}$, for all $\widehat{f} \in C(X)^\delta$.

(a) Since $C_d^{\rm cb}(X)$ is a Riesz space (Theorem 3.2) and $\mathcal{R}a(X)$ is an ideal in $C_d^{\rm cb}(X)$ (Corollary 4.4), then $C(X)^\delta = C_d^{\rm cb}(X)/\mathcal{R}a(X)$ is a Riesz space ([20], Theorem 18.9).

In order to prove that $C(X)^\delta$ is Dedekind complete, let $\{\widehat{f_\gamma}\}_{\gamma\in\Gamma}$ be a subset of $C(X)^\delta$, which is bounded above by $\widehat{h}$. By Theorem 5.7 we can assume that $f_\gamma, h \in \mathcal{NL}_{\rm sc}(X)$, and also that they are C-bounded. Therefore $f_\gamma = IS(f_\gamma)$ and $h = IS(h)$. The inequality $\widehat{f_\gamma} \le \widehat{h}$ implies $f_\gamma = IS(f_\gamma) \le S(h)$, for all γ (Proposition 5.1). In consequence, there exists $\bigvee f_\gamma$ in $\mathcal{B}_{\rm loc}(X)$ with $\bigvee f_\gamma \le S(h)$, from where we obtain $IS(\bigvee f_\gamma) \le ISS(h) = h$. Define $f = IS\left(\bigvee f_\gamma\right)$. It is worth to note that f is the supremum of the set $\{f_\gamma\}$ in the lattice $\mathcal{NL}_{\rm sc}^{\rm cb}(X)$ (see formulae (2.4)). Since $f \in \mathcal{NL}_{\rm sc}(X)$, then $f \in C_d(X)$ (Proposition 4.7), and it is easy to see that f is C-bounded. Then $f \le h$ and $f_\gamma = IS(f_\gamma) \le IS\left(\bigvee f_\gamma\right) = f$ implies that $\widehat{f} \le \widehat{h}$ and $\widehat{f_\gamma} \le \widehat{f}$, for all $\gamma \in \Gamma$. This shows that $\bigvee \widehat{f_\gamma} = \widehat{f}$.

(b) We have the diagram $C(X) \hookrightarrow C_d^{cb}(X) \xrightarrow{\pi} C(X)^\delta$. The quotient mapping π is a Riesz isomorphism and π restricted to $C(X)$ is a one-to-one mapping. Indeed, if $g_1, g_2 \in C(X)$ and $\pi(g_1) = \pi(g_2)$, then $g_1 - g_2 \in \mathcal{R}a(X)$. By Corollary 5.2 we have $IS(g_1) = IS(g_2)$. Since $g_1, g_2 \in C(X)$, we obtain $g_1 = g_2$.

(c) Let $\widehat{f} \in C(X)^\delta$. By Theorem 5.7, $\widehat{f} = \widehat{\overline{f}} = \widehat{\underline{f}}$, where $(\underline{f}, \overline{f})$ is a regular pair, and it is easy to see that the two components are C-bounded. Since X is completely regular, Theorem 2.3 gives the equalities $I(\underline{f}) = \ell(\underline{f})$, $S(\overline{f}) = u(\overline{f})$. Using the properties of order-continuity of the quotient mapping (Proposition 5.3), we have

$$\widehat{f} = \widehat{\underline{f}} = \pi(\underline{f}) = \pi(I(\underline{f})) = \pi(\ell(\underline{f})) = \pi\left(\bigvee\{g : g \in C(X),\ g \le \underline{f}\}\right)$$
$$= \bigvee\{\pi(g) : \pi(g) \in \pi\left(C(X)\right),\ g \le \underline{f}\}.$$

Since π restricted to $\mathcal{NU}_{sc}(X)$ is an order embedding (Proposition 5.8), $g \le \underline{f}$ if and only if $\widehat{g} \le \widehat{\underline{f}} = \widehat{f}$. Therefore $\widehat{f} = \bigvee\{\widehat{g} : \widehat{g} \in \pi\left(C(X)\right),\ \widehat{g} \le \widehat{f}\}$.

A similar proof can be given for the second equality $\widehat{f} = \bigwedge\{\widehat{g} : \widehat{g} \in \pi\left(C(X)\right),\ \widehat{g} \ge \widehat{f}\}$. □

Corollary 5.5 shows that for the construction of $C(X)^\delta$ we can use any type of semicontinuous functions, because all those sets coincide.

The Riesz ideal of the rare functions $\mathcal{R}a(X)$ coincides with the Riesz ideal N in the paper of K. Nakano and Shimogaki (see Proposition 4.5). Our Theorem 6.1 coincides with Theorem 1 in [22] for X a compact space, but the proof is different.

References

[1] Anguelov, R.: *Dedekind order completion of C(X) by Hausdorff continuous functions.* Quaest. Math. **27** (2004), 153–169.

[2] Anguelov, R., Markov, S., Sendov, B.: *The set of Hausdorff continuous functions – The largest linear space of interval functions.* Reliab. Comput. **12** (2006), 337–363.

[3] Anguelov, R., Rosinger, E.E.: *Hausdorff continuous solutions of nonlinear partial differential equations through the order completion method.* Quaest. Math. **28** (2005), 271–285.

[4] Anguelov, R., Rosinger, E.E.: *Solving large classes of nonlinear systems of PDE's.* Comput. Math. Appl. **53**, no. 3-4 (2007), 491–507.

[5] Baire, R.: *Leçon sur les fonctions discontinues.* Gauthier-Villars, Paris, 1905.

[6] Becker, R.: *Vector lattices associated with ordered vector spaces.* Mediterr. J. Math. **7** (2010), 313–322.

[7] Dăneţ, N.: *Hausdorff continuous interval-valued functions and quasicontinuous functions.* Positivity **14** (2010), 655–663.

[8] Dăneţ, N.: *The Dedekind completion of $C(X)$: an interval-valued functions approach.* Quaest. Math. **34**, no.2 (2011), 213–223.

[9] Dăneţ, N.: *Dedekind cuts in C(X).* Marcinkiewicz Centenary Volume, Polish Academy of Sciences, Banach Center Publications, Vol. 95, pp. 287–297, Warszawa, 2011.

[10] Dăneţ,N.: *Riesz spaces of normal semicontinous functions.* Mediterr. J. Math., **12** (2015), 1345–1355.

[11] Dilworth, R.P.: *The normal completion of the lattice of continuous functions.* Trans. Amer. Math. Soc. **68** (1950), 427–438.

[12] Ford, L.R.: *Pointwise discontinuous functions.* M.A. Thesis, Graduate School of the University of Missouri, 1912.

[13] Fremlin, D.H.: *Topological Riesz Spaces and Measure Theory*, Cambridge University Press, 1974.

[14] Horn, A.: *The normal completion of a subset of a complete lattice and lattices of continuous functions.* Pacific J. Math. **3** (1953), 137–152.

[15] Kaplan, S.: *On the second dual of the space of continuous functions.* Trans. Amer. Math. Soc. **86**, no.1 (1957), 70–90.

[16] Kaplan, S.: *The second dual of the space of continuous functions*, II. Trans. Amer. Math. Soc. **93**, no.2 (1959), 329–350.

[17] Kaplan, S.: *The second dual of the space of continuous functions*, III. Trans. Amer. Math. Soc. **101**, no.1 (1961), 34–51.

[18] Kaplan, S.: *The second dual of the space of continuous functions*, IV. Trans. Amer. Math. Soc. **113**, no.3 (1964), 512–546.

[19] Kaplan, S.: *The Bidual of C(X)* I, North-Holland Mathematics Studies 101, Amsterdam, 1985.

[20] Luxemburg, W.A.J., Zaanen, A.C.: *Riesz Spaces*, vol. I, North-Holland Publishing Company, Amsterdam-London, 1971.

[21] MacNeille, H.M.: *Partially Ordered Sets*, Trans. Amer. Math. Soc. **42** (1937), 416–460.

[22] Nakano, K. and Shimogaki, T.: *A note on the cut extensions of C-spaces.* Proc. Japan Acad. **38**, no.8 (1962), 473–477.

[23] Nakano, H.: *Measure Theory* II (in Japanese), Tokyo, 1948.

[24] Sendov, B.: *Hausdorff Approximations.* Kluwer Academic Publishers, Dordrecht, The Netherlands, 1990.

Nicolae Dăneţ
Technical University of Civil Engineering of Bucharest
Department of Mathematics and Computer Science
124, Lacul Tei Blvd.
Bucharest, Romania
e-mail: ndanet@cfdp.utcb.ro

Ordered Structures and Applications: Positivity VII
Trends in Mathematics, 127–159

A Mazur–Orlicz Type Theorem in Interval Analysis and Its Consequences

Rodica-Mihaela Dăneţ

Abstract. Classical extension theorems for linear functionals or, more generally, for linear operators in the setting of vector spaces are well known. For example, the Hahn–Banach Theorem and the Mazur–Orlicz Theorem extend linear functionals (operators) dominated in a certain sense by sublinear functionals (operators). It is also known that these theorems have many applications. To get more applications we intend to give some versions of these theorems in interval analysis. In the literature of this field, intervals are viewed as an extension of any value that they contain, motivated by the fact that in many practical situations some values are known with interval uncertainty. We will work with intervals in ordered vector spaces. It is known that the set of all closed intervals in such spaces is not a vector space. Indeed, for example, there is no additive inverse element for each closed interval. Therefore certain difficulties arise in the proofs of the extension results.

Mathematics Subject Classification (2010). Primary 46A22; Secondary 06F05, 46S99.

Keywords. Mazur–Orlicz Theorem, Hahn–Banach Theorem, interval analysis, interval-spaces, interval-linear functionals, interval-sublinear functionals.

1. Introduction

According to R. Moore [26], who can be considered the father of *modern interval analysis*, this branch of mathematics arose from the observation that if we compute a number a and a rigorous bound b on the total error in a, as an approximation to some unknown number x, such that $|x - a| \leq b$, then no matter how we compute a and b, we certainly know that x lies in the interval $[a - b, a + b]$. This idea naturally lead to investigation of computations with intervals.

An *interval* in the real line has a dual nature. Intervals are *sets* of real numbers, to which the usual operations on sets may be applied. *Intervals* can also be interpreted as "numbers", each represented by a pair of real numbers with suitable defined arithmetic operations – see the name of *interval arithmetic* given initial to the *interval analysis.* As sets, intervals are used, for example, for computer algorithms; see, for example, [5]. As numbers, intervals are used, to calculate upper and lower endpoints for the range of a function; see [1, p. 422].

Defining addition of intervals as the usual Minkowski addition of sets

$$A + B = \{a + b \mid a \in A, b \in B\},$$

we follow the logic of the principle of containing: the sum of two intervals certainly contains the sums of all pairs of real numbers, one from each of two intervals. The major obstacle in applications is the fact that for this addition, the additive inverse (that is, the opposite) generally does not exist.

In [2], R. Anguelov considered that this "defect" of the addition of algebraic structure of the space of intervals is "possibly one of the most important challenges associated with the development of the theory" of this space. We meet this difficulty, for example, when we try to extend classical results of functional analysis in the interval analysis, more precisely, when we try to formulate in this analysis extension results in the line of the Mazur–Orlicz and Hahn–Banach Theorems (as they appear in [6]). In addition, if we want to extend a type of *positive* (linear) functionals, we need to consider order relations in the interval-spaces. Consequently, a question arises: "*What is the best ordering*?"

Many *order relations* were considered in the interval analysis, most of them on $I\mathbb{R}$, the space of all real closed intervals (see, for example, [8]). The fundamental idea in defining such relations was that these orders would have to extend the order of $\mathbb{R}$, which, of course, is a total order. But the known order relations on $I\mathbb{R}$ are partial orders. This is consistent with the fact that in some practical situations, we are interested in quantities which are only partially ordered. An example is given in [36].

Thus, in *space-time geometry*, we do not have the exact location of an event in space-time; we usually only know the event $\underline{x}$ that can cause the given event x ($\underline{x} \leq x$) and the event $\bar{x}$ that can causally be affected by x ($x \leq \bar{x}$). In this case, the only information that we have about the event x is that it belongs to the interval $[\underline{x}, \bar{x}] = \{x \mid \underline{x} \leq x \leq \bar{x}\}$. This description looks similar to the interval $[\underline{a}, \bar{a}]$ of real numbers but the important difference is that the causality relation in space-time is only a partial order: there are two events x and y, for which $x \not\leq y$ and $y \not\leq x$. Such events are called *incompatible* and are sometimes denoted by $x \parallel y$; see, for example [12], [24], [7], [35].

Moving from the space-time geometry in an *economic* framework, the order relation introduced in the set of the *preferences* of agents is most often a partial ordering (here, if $[a] \neq [b]$ are two goods, then $[a] \prec [b]$ means that $[b]$ is strictly preferred to $[a]$); see [10], [31], [33].

Depending on the problem to be solved, in the interval analysis are used one or other of the order relations that can be entered – see, for example in the context of the *decision-making* problems, the *maximization* (of interval *profits*) or *minimization* (of interval *costs*).

A chronological study for the order relations in interval analysis reveals that at the beginning were considered *transitive relations.* The pioneer of this study was R. Moore [25]. He introduced two transitive relations on the space $I\mathbb{R}$ of all closed real intervals, later called the *strong relation* and the *containment relation*, respectively:

1) $[a] \prec [b] \Leftrightarrow \bar{a} \leq \underline{b}$,
2) $[a] \subseteq [b] \Leftrightarrow \underline{b} \leq \underline{a} \leq \bar{a} \leq \bar{b}$,

where $[a] = [\underline{a}, \bar{a}]$ and $[b] = [\underline{b}, \bar{b}]$ are in $I\mathbb{R}$.

Referring to "1)", it should be noted that it seems to be a natural relation in terms of time-reasoning. Indeed, "1)" is very convenient to describe a causal relation. Thus if the two events $[a] = [\underline{a}, \bar{a}]$ and $[b] = [\underline{b}, \bar{b}]$ are such that $[a] \prec [b]$, then certainly $[a]$ has happened before $[b]$, possibly even that $[a]$ causes $[b]$. Due to the obvious property of relation "$\prec$" to be *transitive*, it follows that $[a] \prec [b] \prec [c]$ means that we can deduce that $[a]$ causes $[c]$ if we know that $[a]$ causes $[b]$ and $[b]$ causes $[c]$.

In this paper we will use the *closed intervals* in an arbitrary ordered vector space E, to prove some *extension theorems* in the line of Mazur–Orlicz Theorem and Hahn–Banach Theorem. As we will see, a function $g : A \longrightarrow IE$ will appear in the context of our first result (a Mazur–Orlicz type theorem). Here A is an arbitrary nonempty set and IE is the (interval-) space of all closed intervals in E. Note that such functions appear in the literature of the interval analysis. As an example, we cite [32], a paper by W.T. Trotter; in this paper, an ordered vector space $(A, \leq)$ is called an *interval-function* if there exists a function g assigning to each element $a \in A$, a closed interval of a totally ordered set E. Usually E is the real line $\mathbb{R}$. The function g is called an *interval representation* of A. Note that the *endpoints* of the intervals used in the above representation may be identical.

Also note that in this paper we will consider the space IE endowed with the following ordering, known as the *weak order*:

$$[a] \leq [b] \Leftrightarrow \underline{a} \leq \underline{b} \text{ and } \bar{a} \leq \bar{b}, \text{ if } [a] = [\underline{a}, \bar{a}] \text{ and } [b] = [\underline{b}, \bar{b}] \, .$$

This (partial) order was introduced and studied by S. Markov, in [13] and [14]. According to [36], this order relation is a "very natural sense of an interval order, for example, saying that one event extended in time can be prior to another event if it is still underway when the subsequent event initiates". We will use this ordering in order to define *positive interval-linear functionals* and then to study the extension problem for such functionals.

2. Interval-spaces

An interval-space will be associated to an arbitrary real ordered vector space.

Firstly, if $E\,(= (E, \leq))$ is a real ordered vector space, and $\underline{a}, \bar{a} \in E$ are such that $\underline{a} \leq \bar{a}$, we will denote by $[a] = [\underline{a}, \bar{a}]$ the *order interval* $\{x \in E \mid \underline{a} \leq x \leq \bar{a}\}$. In the literature, the interval $[\underline{a}, \bar{a}]$ is sometimes denoted by $[a_L, a_R]$, or, in short, by $\bar{\underline{a}}$.

Of course, the notation $[a]$ for an interval $[\underline{a}, \bar{a}]$ enable us to write formulas and proofs in a closed form without using the endpoints of the interval.

If $[a] = [\underline{a}, \bar{a}]$ with $\underline{a} = \bar{a}$ denoted by a (hence $[a] = [a, a]$) we call this order interval, a *degenerate interval*. We can identify the degenerate interval $[a, a] = \{a\}$ with the element $a \in E$. We will call an order interval $[\underline{a}, \bar{a}]$ with $\underline{a} < \bar{a}$ in E, a *nondegenerate interval.*

We will also consider *symmetric intervals* in IE, denoted, for example, by $[-b, b]$, with $b \geq 0$ in E and *symmetric nondegenerate intervals* $[-b, b]$, with $b > 0$ in E.

The *interval-set* (in short, *i-set*) associated to the real ordered vector space E, is the set $IE = \{[a] = [\underline{a}, \bar{a}] \mid \underline{a}, \bar{a} \in E\}$.

Definition 1. We say that an interval-set (i-set) IE is an *interval-space* (in short, *i-space*) if it is endowed with the following operations, called the *usual algebraic operations*:

1. the *addition*, defined by:
$$[a] \oplus [b] = \{x + y \mid x \in [a], y\} \in [b], \text{that is,}$$
$$[a] \oplus [b] = \left[\underline{a} + \underline{b}, \bar{a} + \bar{b}\right] \text{if} \, [a] = [\underline{a}, \bar{a}] \in IE \text{ and } [b] = \left[\underline{b}, \bar{b}\right] \in IE;$$
2. the *scalar multiplication* with reals defined by:
$$\alpha \cdot [a] = \{\alpha x \mid x \in [\underline{a}, \bar{a}]\}, \text{that is,}$$
$$\alpha \cdot [a] = \begin{cases} [\alpha\underline{a}, \alpha\bar{a}], & \text{if } \alpha \in \mathbb{R}, \alpha \geq 0 \\ [\alpha\bar{a}, \alpha\underline{a}], & \text{if } \alpha \in \mathbb{R}, \alpha < 0 \end{cases}, \text{ where } [a] = [\underline{a}, \bar{a}] \in IE.$$

Sometimes we will denote $\alpha \cdot [a]$ by $\alpha [a]$.

Endowed with this algebraic operations, IE is **not** a real vector space. More precisely, $(IE, \oplus)$ is a commutative monoid with the neutral (or identity) element $\mathbf{0}$ ($\mathbf{0} = [0, 0]$, sometimes also denoted by $[0]$), but it is not a group, because a nondegenerate interval has no inverse with respect to addition, that is, has no *opposite.* Indeed, by way of contradiction, suppose that for the order interval $[a] = [\underline{a}, \bar{a}]$, with $\underline{a} < \bar{a}$ in E, there exists an inverse $[b] = \left[\underline{b}, \bar{b}\right]$. Hence $[\underline{a}, \bar{a}] \oplus \left[\underline{b}, \bar{b}\right] = \mathbf{0}$, that is, $\underline{a} + \underline{b} = 0$ and $\bar{a} + \bar{b} = 0$. Therefore, $\underline{b} = -\underline{a}$ and $\bar{b} = -\bar{a}$. But $\underline{b} \leq \bar{b}$ implies that $-\underline{a} \leq -\bar{a}$ or, equivalently $\bar{a} \leq \underline{a}$, which contradicts that $\underline{a} < \bar{a}$. The scalar multiplication in IE has the following *properties*:

$$(\alpha + \beta) [a] = \alpha [a] \oplus \beta [a], \text{if } [a] \in IE \text{ and } \alpha, \beta \in \mathbb{R}, \text{with } \alpha\beta > 0, \tag{1}$$
$$\alpha (\beta [a]) = (\alpha\beta) [a], \text{if } [a] \in IE \text{ and } \alpha, \beta \in \mathbb{R}, \tag{2}$$
$$1 \cdot [a] = [a] \text{ for each } [a] \in IE, \tag{3}$$
$$\alpha ([a] \oplus [b]) = \alpha [a] \oplus \alpha [b], \text{if} \, [a], [b] \in IE \text{ and } \alpha \in \mathbb{R}. \tag{4}$$

We can also consider the *subtraction* in IE:
$$[a] \ominus [b] = [a] \oplus (-[b]), \text{ where } -[b] \text{ means } (-1)[b].$$
If $[a] = [\underline{a}, \bar{a}]$ and $[b] = \left[\underline{b}, \bar{b}\right]$, then $[a] \ominus [b] = [\underline{a}, \bar{a}] \ominus \left[\underline{b}, \bar{b}\right] = \left[\underline{a} - \bar{b}, \bar{a} - \underline{b}\right]$.

In particular, if $[a] = [\underline{a}, \bar{a}]$ it follows that

$$[a] \ominus [a] = [\underline{a} - \bar{a}, \bar{a} - \underline{a}] = [-(\bar{a} - \underline{a}), \bar{a} - \underline{a}].$$

Thus, if $[a]$ is a nondegenerate interval ($\underline{a} < \bar{a}$) then $[a] \ominus [a] \neq \mathbf{0}$, and again we conclude that IE is not a real vector space.

Note that $[a] \ominus [a]$ is a symmetric interval. Let us denote such an interval by $[o]$. Therefore it is justified to consider the set $\mathcal{O}$ of all symmetric intervals of IE, hence $\mathcal{O} = \{[-b, b] \mid b \geq 0, b \in E\}$. Because $[a] \ominus [a] \in \mathcal{O}$ for each $[a] \in IE$, and obviously, $[a] \ominus [a] = \mathbf{0} \in \mathcal{O}$ for a degenerate interval $[a]$, we call the set $\mathcal{O}$, the *null set* of IE. In what follows we will denote by $[o] \in IE$, the *generic element* of $\mathcal{O}$.

Remark 1. The null set $\mathcal{O}$ of the i-space IE is closed under the algebraic operations on IE, that is, $[a] \oplus [b] \in \mathcal{O}$ and $\alpha [a] \in \mathcal{O}$ for all $[a], [b] \in \mathcal{O}$ and $\alpha \in \mathbb{R}$.

In the particular case $E = \mathbb{R}$ of reals, let us denote by $\mathbf{1}$ the symmetric interval $[-1, 1]$ from the null set $\mathcal{O}$ of $I\mathbb{R}$. Then, every interval $[o] = [-b, b] \in \mathcal{O}$ ($b \geq 0$) can be written as $b \cdot \mathbf{1}$. Let us call $\mathbf{1} \in I\mathbb{R}$ the *generator* of $\mathcal{O}$. With this "generator" we can write the proof of Remark 1 (for $E = \mathbb{R}$) without using the left and right bounds of the intervals $[a]$ and $[b]$. Indeed, for $[a] = a \cdot \mathbf{1}$ and $[b] = b \cdot \mathbf{1}$ in the null set $\mathcal{O}$ of $I\mathbb{R}$ ($a \geq 0, b \geq 0$) and $\alpha \in \mathbb{R}$, it follows that:

1. $[a] \oplus [b] = (a + b)\, \mathbf{1} \in \mathcal{O}$;
2. $\alpha [a] = \alpha (a \cdot \mathbf{1}) = \begin{cases} (\alpha a) \cdot \mathbf{1}, & \text{if } \alpha \geq 0 \\ (\alpha(-a)) \cdot \mathbf{1}, & \text{if } \alpha < 0 \end{cases}$, that is, $\alpha [a] \in \mathcal{O}$, too.

To **summarize** this section, let us mention that the i-set IE endowed with the usual algebraic operations is **not** a vector space, because:

a) $(IE, \oplus)$ is a commutative monoid with the neutral element $\mathbf{0}\,(= [0, 0])$, but it is not a group, since a nondegenerate interval has no inverse with respect to the addition;
b) the axiom $(\alpha + \beta)[a] = \alpha [a] \oplus \beta [a]$ is certainly true only when $\alpha, \beta \in \mathbb{R}$ are such that $\alpha\beta > 0$ (see (1)).

Historical remarks. Firstly, we notice that, just like in [20, p. 2] or in [18, p. 272], we will use the terms of "linear space" and "vector space" as synonyms. By looking in the literature for earlier contributions, we find that the notion of *interval-space* is a particular case of the notion of *quasilinear space*. There are two important cases of so-called quasilinear spaces.

1. Quasilinear spaces in the sense of Aseev

In [4] (see also [34]), a *quasilinear space* is a set E together two algebraic operations (an addition and a multiplication with real scalars) and a partial order on E (reflexive, antisymmetric and transitive) such that:

A1) $(E, +)$ is an abelian monoid (that is, "+" is associative, commutative and has a neutral element);

A2) $\alpha \cdot (\beta \cdot x) = (\alpha\beta) \cdot x$;

A3) $\alpha \cdot (x + y) = \alpha \cdot x + \alpha \cdot y$;
A4) $1 \cdot x = x$ (unit element);
A5) $0 \cdot x = 0$ (neutral or identity element);
A6) $(\alpha + \beta) \cdot x \leq \alpha \cdot x + \beta \cdot x$ (sub-distributivity);
A7) $x \leq y$ and $z \leq v$ imply $x + z \leq y + v$;
A8) $x \leq y$ implies $\alpha \cdot x \leq \alpha \cdot y$ $(x, y, z, v \in E, \alpha, \beta \in \mathbb{R})$.

In a quasilinear space, the neutral element 0 is minimal, that is, $x = 0$ if $x \leq 0$; see [34, Lemma 2.1] and, in fact, [4]. An element x' in a quasilinear space E is called an *opposite* (an *inverse*) of an element $x \in E$ if $x + x' = 0$. If an opposite element exists, then it is unique. Suppose that any element x in the *quasilinear space* E has an opposite element $x' \in E$. Then the partial order in E is determined by equality, the distributivity type conditions A3) and A6) hold and, consequently, E is a linear space; see [34, Lemma 2.3] and, in fact, [4]. Hence, a linear space (a vector space) is a quasilinear space with the partial ordering relation:

"$x \leq y$ if and only if $x = y$".

Moreover, according to [4] (see also [34, Corollary 2.4]), in a linear space, equality is the only way to define a partial ordering such that the conditions A1)–A8) hold.

Now it is easy to prove that an interval-space (i-space) IE is a quasilinear space with the following order relation:

"$[x] \leq [y]$ if and only if $[x] \subseteq [y]$"

(see, for example, [34, pp. 2–3]).

2. Quasilinear spaces in the sense of Markov

A definition of the notion of quasilinear spaces of monoid structure, can be found in [20]. To give this definition first we recall that a *linear space* (over $\mathbb{R}$) is a set E endowed with an addition and a multiplication with (real) scalars such that:

M1) $(x + y) + z = x + (y + z)$, for all $x, y, z, \in E$;
M2) $\exists\, 0 \in E$ with $x + 0 = x$, for each $x \in E$;
M3) $x + y = y + x$, for all $x, y \in E$;
M4) $\forall\, x \in E$, $\exists - x \in E$ with $x + (-x) = 0$;
M5) $\alpha \cdot (\beta \cdot x) = (\alpha\beta) \cdot x$, for all $x \in E$ and $\alpha, \beta \in \mathbb{R}$;
M6) $1 \cdot x = x$, for all $x \in E$;
M7) $\alpha \cdot (x + y) = \alpha \cdot x + \alpha \cdot y$, for all $x, y \in E$ and $\alpha \in \mathbb{R}$
(the so-called "*first distribution law*");
M8) $(\alpha + \beta) \cdot x = \alpha \cdot x + \beta \cdot x$, for all $x \in E$ and $\alpha \in \mathbb{R}$
(the so-called *second distributive law*).

Obviously, the axioms M1)–M4) show that E is an abelian additive group. In [20], Markov notes that a linear space can be defined also by relaxing the group axiom M4), replacing it by the weaker *cancelation law*:

M4′) $x + z = y + z \Rightarrow x = y$ (for all $x, y, z \in E$).

Hence, the axioms M1)–M3), M4), M5)–M8) are equivalent with the axioms M1)–M3), M4′), M5)–M8). Indeed, in [20] Markov noted that: "If in M8) we put $\alpha =$

$1, \beta = -1$ we obtain $0 = x + (-1) \cdot x$, that is, the element $(-1) \cdot x$ is the opposite to x, symbolically $-x = (-1) \cdot x$. One can also observe that the condition for the existence of a neutral element for addition is redundant, as such is the element $0 \cdot x$. Indeed, using M8) with $\alpha = 1, \beta = 0$, we obtain

$$x + 0 \cdot x \overset{\text{M6)}}{=} 1 \cdot x + 0 \cdot x \overset{\text{M8)}}{=} (1 + 0) \cdot x = 1 \cdot x \overset{\text{M6)}}{=} x,$$

implying $0 \cdot x = 0$.”. Also in [20] Markov defined the notion of *quasilinear space* that is obtained by relaxing the so-called second distributive law axiom of the linear space.

More precisely, a *quasilinear space* (*of monoid structure*) *in the sense of Markov*, is a set E endowed with an addition and a multiplication with (real) scalars such that the axioms M1)–M3), M4′), M5)–M7) and M8′) are satisfied, where M8’) is the so-called *quasidistributive law*:

M8′) $(\alpha + \beta) \cdot x = \alpha \cdot x + \beta \cdot x$, for all $x \in E$ and $\alpha, \beta \in \mathbb{R}$ with $\alpha \cdot \beta \geq 0$.

Following the proofs (found in [20]) for M8) $\Rightarrow$ M2) (that is, the existence of the opposite element to $x \in E$), and M8) $\Rightarrow$ M3) (that is, the existence of the neutral element in E), we remark that now we have only M8′) $\Rightarrow$ M2), that is, M8′) does not imply the existence of the opposite element for each element $x \in E$. In other words, the restriction $\alpha\beta \geq 0$ in M8′) does not permit us to conclude that a quasilinear space is an abelian group but, obviously, is an abelian (and cancelative) monoid.

The notion of *quasilinear space* in the sense of Markov, more precisely the notion of *quasilinear system*, appeared in [15], see also [17], [19], [18]. In [15], Markov gave a brief history of the term of “*quasilinear space*”. Thus, this term was introduced in [22] and [23]. In [15] and [16], S. Markov stated that the definition of quasilinear spaces given in [23] does not require the cancelation law and is thus more general. Markov considered (see [16, p. 134]) that the cancelative quasilinear spaces are useful for the study of the algebraic properties of convex sets and intervals. Such spaces have been considered in [27] under the name “*R-semigroups with cancelation law*”.

Of course, a *linear space* is a special case of a *quasilinear space* in the sense of Markov. Also an *interval-space* (i-space) IE is a *quasilinear space in the sense of Markov.*

3. Interval-subspaces

The following definition is very natural.

Definition 2. Let IE be an interval-space. We say that a nonempty subset IS of IE is an *interval-subspace* (in short, *i-subspace*) of IE, if it is closed under the algebraic operations, that is, for any $[a], [b] \in IS$ and $\alpha \in \mathbb{R}$, it follows that $[a] \oplus [b] \in IS$, and $\alpha [a] \in IS$.

Obviously $\mathbf{0} = (0,0) \in IS$ (because for any $[a] \in IS$, taking $\alpha = 0$, it follows that $\mathbf{0} = 0 \cdot [a] \in IS$). In addition, the null set $\mathcal{O}$ of IE, is an interval-subspace of IE (see Remark 1).

Moreover, for all $[a] \in IS$, $[a] \ominus [a] \in IS \cap \mathcal{O}$, that is, $IS \cap \mathcal{O}$ is a nonempty set. So, using the null set $\mathcal{O}$ of IE, we can define the *null part* $\mathcal{O}_{IS}$ of IS.

Remark 2. It follows that:

$$\mathcal{O}_{IS} = \{[a] \ominus [a] \mid [a] \in IS\}. \tag{5}$$

(To justify, we need to prove only the inclusion "$\subseteq$". For this, assume that $[b] \in \mathcal{O}_{IS}$. It follows that $[b] \in IS$ and $[b] = [-b, b]$. Then we have: $[b] \ominus [b] = [-b, b] \oplus (-[-b, b]) = [-b, b] \oplus [-b, b] = 2[-b, b] = 2[b] \Rightarrow [b] = \frac{1}{2}([b] \ominus [b]) = \left[\frac{1}{2}b\right] \ominus \left[\frac{1}{2}b\right]$ and $\left[\frac{1}{2}b\right] \in IS$, because $[b] \in IS$ and IS is an i-subspace of IE.)

Taking into account that $\mathcal{O} \subset IE$, sometimes we will be interested in the i-subspaces IS of IE for which the following condition is fulfilled:

$$\mathcal{O} \subseteq IS \tag{6}$$

We will call such an i-subspace, a *standard i-subspace* of IE.

Remark 3. Obviously, the condition (6) is equivalent to the following condition:

$$\mathcal{O} = \mathcal{O}_{IS} \tag{7}$$

To better understand the meaning of (6) and (7) we will analyze what happens when $E = \mathbb{R}$, and hence $IE = I\mathbb{R}$. In this case we can prove the following result.

Proposition 1. *Let IS be an i-subspace of $I\mathbb{R}$, $\mathcal{O}$ the null part of $I\mathbb{R}$ and $\mathcal{O}_{IS} = \mathcal{O} \cap IS$. Then,* (i) $\Leftrightarrow$ (ii) $\Leftrightarrow$ (iii), (iv) $\Rightarrow$ (i) *and* (iv) $\Rightarrow$ (v), *where:*

(i) $\mathcal{O} \subseteq IS$;
(ii) $\mathcal{O} = \mathcal{O}_{IS}$;
(iii) *IS contains at least one nondegenerate order interval*;
(iv) *$IS \oplus [o] = IS$ for all $[o] \in \mathcal{O}$*;
(v) *$IS \oplus [u] = IS$ for all $[u] \in IS$.*

(Note that, for example, $IS \oplus [o]$ is defined as the set $\{[a] \oplus [o] \mid [a] \in IS\}$.)

Proof. (i) $\Leftrightarrow$ (ii) and (i) $\Rightarrow$ (iii) are obvious.

(iii) $\Rightarrow$ (i). Let $[a] \in IS$ be a nondegenerate order interval. (Hence, $[a] = [\underline{a}, \bar{a}]$ with $\underline{a} < \bar{a}$ in E.) Then taking $[o] = [a] \ominus [a]$, it follows that $[o] \in IS$ (because IS is an i-subspace of $I\mathbb{R}$). Put $[o] = [-v, v] = v \cdot \mathbf{1}$ with $v > 0$, (recall that $\mathbf{1} = [-1, 1] \subset \mathbb{R}$). We have to prove that $\mathcal{O} \subseteq IS$. But taking $[u] = [-u, u] = u \cdot \mathbf{1} \in \mathcal{O}$, it follows that $u \cdot \mathbf{1} = \frac{u}{v}(v \cdot \mathbf{1}) \in IS$.

(iv) $\Rightarrow$ (i) is immediate, because if $[o] \in \mathcal{O}$, then taking $\mathbf{0} \in IS$, it follows that: $[o] = \mathbf{0} + [o] \in IS \oplus [o] \overset{\text{(iv)}}{=} IS$.

To prove (iv) $\Rightarrow$ (v), we observe that if we take $[o] = [u] \ominus [u]$ with $[u] \in IS$, then it follows:

$$IS \oplus [o] \subseteq IS \oplus [u] \subseteq IS. \tag{8}$$

(The *second inclusion* in (8) is true since IS is an i-subspace of $I\mathbb{R}$. The *first inclusion* in (8): if $[v] \in IS$, then $[v] \oplus [o] = [v] \oplus ([u] \ominus [u]) = ([v] \ominus [u]) \oplus [u] \in IS \oplus [u]$, because $[v]$ and $[u] \in IS$, and IS is an i-subspace of $I\mathbb{R}$.) From (iv), it follows that $IS \oplus [o] = IS$. But then, from (8), it follows that $IS \oplus [u] = IS$, too. So (iv) $\Rightarrow$ (v) is true. □

It is known that if V is a real vector space and $S \subseteq V$ is any vector subspace, then for all $u \in S$,

$$S + u = S. \tag{9}$$

The above Proposition 1 shows us that there is a big difference between the notion of *subspace* of a vector space and the notion of *i-subspace* of the i-space IE, at least for the case $E = \mathbb{R}$. Indeed, the corresponding equality to (9) for the last notion, that is,

$$IS \oplus [u] = IS \tag{10}$$

for all $[u] \in IS$ (and more precisely the inclusion "$\supseteq$" in (10)) is not always valid. Obviously, a necessary condition for (10) is that $IS \oplus [o] = IS$ for all $[o] \in \mathcal{O}_{IS}$, which is also a consequence of the statement (iv).

In [16, p. 134], S. Markov introduced the notion of a *subspace of a quasilinear space* E, that a subset $S \subset E$ endowed with the induced algebraic operations of E, hence, $(S, +)$ is a submonoid of $(E, +)$. Obviously, an *interval-subspace* IS (i-subspace) of an interval-space IE is a subspace of the quasilinear space IE.

4. Interval-linear functionals on an interval-subspace

Let IE be an i-space, IS an arbitrary i-subspace of IE, and $f : IS \to \mathbb{R}$ a map.

Definition 3. We say that f is an *interval-linear functional* on IS (in short, *i-linear functional*) if:

1. $f([a] \oplus [b]) = f([a]) + f([b])$ for all $[a], [b] \in IS$;
2. $f(\alpha [a]) = \alpha f([a])$ for all $[a] \in IS$ and $\alpha \in \mathbb{R}$.

Definition 4. The *kernel* of an i-linear functional $f : IS \to \mathbb{R}$ is the set: $\ker f = \{[x] \in IS \mid f([x]) = 0\}$.

It is known that, in the setting of vector spaces, the kernel of any linear functional f is nonempty, because it certainly contains the null element 0 of the domain of f.

The kernel of an i-linear functional f on an interval-subspace IS is nonempty, too. Indeed, the null part $\mathcal{O}_{IS}$ of IS is included in $\ker f$ and $\mathcal{O}_{IS}$ is a nonempty set. To prove that $\mathcal{O}_{IS} \subseteq \ker f$, choose $[o] \in \mathcal{O}_{IS} = \mathcal{O} \cap IS$; it follows that, there exists $a \geq 0$ such that $[o] = [-a, a]$, and therefore, $(-1)[o] = [-a, a]$, that is, $[o] = (-1)[o]$. But f is i-linear on IS, and thus $f([o]) = f((-1) \cdot [o]) =$

$(-1)\cdot f([o]) = -f([o])$. Then, $2f([o]) = 0$ and hence $f([o]) = 0$, that is, $[o] \in \ker f$. As a consequence of the inclusion

$$\mathcal{O}_{IS} \subseteq \ker f \tag{11}$$

the following observation is true.

Remark 4. For any i-linear functional $f : IS \to \mathbb{R}$, it follows that

$$f([a] \oplus [o]) = f([a]) \tag{12}$$

for each $[o] \in \mathcal{O}_{IS}$. (Indeed, we have: $f([a] \oplus [o]) = f([a]) + f([o]) = f([a])$, because $f([o]) = 0$, according to (11).) In particular, if $[u] \in \ker f$, then for all $[o] \in \mathcal{O}_{IS}$, $[u] \oplus [o] \in \ker f$.

5. Interval-sublinear functionals on an interval-subspace

A) *What are they?*

We recall that if V is a real vector space, and $S \subseteq V$ is a vector subspace, a real *sublinear functional* on S is a map $p : S \to \mathbb{R}$ such that:

1. $p(x+y) \leq p(x) + p(y)$ for all $x, y \in S$;
2. $p(\alpha x) = \alpha p(x)$ for all $x \in S$ and $\alpha > 0$.

Now we will introduce an equivalent of this notion in the interval-spaces setting.

Definition 5. A real-valued map p defined on an i-subspace IS of an i-space IE is called an *interval-sublinear functional* (in short, *i-sublinear functional*) on IS, if:

1. $p([x] \oplus [y]) \leq p([x]) + p([y])$ for all $[x], [y] \in IS$;
2. $p(\alpha [x]) = \alpha p([x])$ for all $[x] \in IS$ and $\alpha > 0$;
3. $p([x] \oplus [o]) = p([x])$ for all $[x] \in IS$ and $[o] \in \mathcal{O}_{IS}$.

Using the same terminology as in the theory of vector spaces we will say that in the Definition 5,

"1." means that p is *interval-subadditive* (in short, *i-subadditive*) on IS;
"2." means that p is *interval-positively homogeneous* (in short, *i-positively homogeneous*) on IS.

We also mention that the hypothesis "3." in Definition 5 is related to the remark that any i-linear functional on IS (that obviously has to be i-sublinear on IS) satisfies the equality $f([x] \oplus [o]) = f([x])$ for all $[x] \in IS$ and $[o] \in \mathcal{O}_{IS}$.

B) *Properties of interval-sublinear functionals.*

Let IS be an i-subspace of the i-space IE, and $p : IS \to \mathbb{R}$ an i-sublinear functional. Then the list of properties of p may be completed with the following:

4. $p([o]) = 0$ for all $[o] \in \mathcal{O}_{IS}$.

(Indeed, we have: $2p([o]) \stackrel{2.}{=} p(2[o]) = p([o] \oplus [o]) \stackrel{3.}{=} p([o])$.)

5. $p(0 \cdot [u]) = 0$ for all $[u] \in IS$.

(Recall that $\mathbf{0} = [0,0] \in \mathcal{O}_{IS}$. Then $p(0 \cdot [u]) = p(\mathbf{0}) \stackrel{4.}{=} 0$.)

6. If IS is a standard i-subspace of IE (and $\mathcal{O}$ is the null set of IE), then $p([o]) = 0$, for all $[o] \in \mathcal{O}$. To prove this, apply "4." and Remark 3.

C) *Pointwise ordering on a set of i-sublinear functionals.*

In what follows we will fix an i-sublinear functional p on an i-subspace IS of the i-space IE (with E an arbitrary ordered vector space). We consider the collection $\mathcal{S}$ $(= \mathcal{S}_p)$ of all i-sublinear functionals q on IS such that $q([v]) \leq p([v])$ for all $[v] \in IS$. We remark that $\mathcal{S}$ is a nonempty set, because, obviously, $p \in \mathcal{S}$. Consider in $\mathcal{S}$ the *pointwise ordering* "$\leq$", that is, for $q_1, q_2 \in \mathcal{S}$,

$$q_1 \leq q_2 \Leftrightarrow q_1([v]) \leq q_2([v]) \text{ for all } [v] \in IS. \tag{13}$$

Proposition 2. *The set $\mathcal{S}$ is inductively ordered from below (that is, each totally ordered subset of $\mathcal{S}$ has a lower bound).*

Proof. Suppose that $\mathcal{T} = \{q_j\}_{j\in J}$ is a totally ordered subset of $\mathcal{S}$ and define $q : IS \to \mathbb{R}$ by

$$q([v]) = \inf_{j\in J} q_j([v]) \quad \text{for all } [v] \in IS.$$

1. Firstly, we remark that q is *well defined*. Indeed, by the way of contradiction, suppose that there exists $[u] \in IS$ such that $q([u]) = -\infty$. Then it follows that

$$q([u] \oplus [v]) = -\infty \tag{14}$$

for all $[v] \in IS$. Indeed, let ε be any real number. Then, there exists $j_\varepsilon \in J$ such that

$$q_{j_\varepsilon}([u]) < \varepsilon. \tag{15}$$

Since q_{j_ε} is an i-subadditive functional, it follows that

$$q([u] \oplus [v]) \leq q_{j_\varepsilon}([u] \oplus [v]) \overset{(15)}{<} \varepsilon + q_{j_\varepsilon}([v]) \quad \text{for all; } [v] \in IS.$$

By choosing ε sufficiently small, it follows (14). But (14) implies that for $[o] = [u] \ominus [u] \in \mathcal{O}_{IS}$

$$q([o]) = -\infty. \tag{16}$$

(Indeed, because for any $[v] \in IS$, IS also contains $[v] \ominus [u]$, and for all $j \in J$, q_j is an i-sublinear functional, it follows that

$$\begin{aligned} q([v]) &= \inf_{j\in J} q_j([v]) = \inf_{j\in J} q_j([v] \oplus [o]) \\ &= \inf_{j\in J} q_j([v] \oplus ([u] \ominus [u])) \\ &= \inf_{j\in J} q_j([u] \oplus ([v] \ominus [u])) \overset{(14)}{=} -\infty. \end{aligned}$$

Now put $[o]$ instead of $[v]$ and obtain (16).)

But (16) contradicts with $q([o]) = \inf_{j\in J} q_j([o]) = 0$. (Here we used again that for all $j \in J$, q_j is an i-sublinear functional and apply property "4." of such functionals.) The above-mentioned contradiction shows us that $q([v]) > -\infty$ for all $[v] \in IS$, that is, q is well defined. Obviously, q will be a lower bound of $\mathcal{T}$ in $\mathcal{S}$, if $q \in \mathcal{S}$.

2. Now we will prove that $q \in \mathcal{S}$. Because $q([v])$ is defined as an infimum of $q_j([v]), j \in J$ and for all $j \in J$, $q_j \in \mathcal{S}$ it follows that:

a) $q([v]) \leq p([v])$ for all $[v] \in IS$;
b) $q([v] \oplus [o]) = q([v])$ for all $[o] \in \mathcal{O}_{IS}$;
c) q is an *i-positively homogeneous functional.*

It remains to prove only that q is an *i-subadditive functional*, that is:

$$q([v_1] \oplus [v_2]) \leq q([v_1]) + q([v_2]) \quad \text{for all } [v_1], [v_2] \in IS.$$

But

$$q([v_1] \oplus [v_2]) \leq \inf_{j \in J} (q_j([v_1]) + q_j([v_2])). \tag{17}$$

Now we will prove that

$$\inf_{j \in J} (q_j([v_1]) + q_j([v_2])) \leq q([v_1]) + q([v_2]). \tag{18}$$

We know that $q(v_1) > -\infty$ and $q(v_2) > -\infty$. Given any $\varepsilon > 0$, there exists $j_1, j_2 \in J$ such that $\inf_{j \in J} q_j([v_1]) + \varepsilon > q_{j_1}([v_1])$, and $\inf_{j \in J} q_j([v_2]) + \varepsilon > q_{j_2}([v_2])$. Therefore, since $\mathcal{T}$ is a totally ordered subset of $\mathcal{S}$, it follows that:

$$q([v_1]) + q([v_2]) + 2\varepsilon = \inf_{j \in J} q_j([v_1]) + \inf_{j \in J} q_j([v_2]) + 2\varepsilon > q_{j_1}([v_1]) + q_{j_2}([v_2])$$

$$\geq \begin{cases} q_{j_1}([v_1]) + q_{j_1}([v_2]), & \text{if } q_{j_1} \leq q_{j_2} \text{ in } \mathcal{T} \\ q_{j_2}([v_1]) + q_{j_2}([v_2]), & \text{if } q_{j_2} \leq q_{j_1} \text{ in } \mathcal{T} \end{cases} \geq \inf_{j \in J} (q_j([v_1]) + q_j([v_2])).$$

Since $\varepsilon > 0$ is arbitrary, it follows that the inequality (18) is true. From (17) and (18) it follows that q is an i-subadditive functional. Therefore, according to the above-mentioned properties a), b), c) of q, we infer that $q \in \mathcal{S}$. This shows us that q is a lower bound of $\mathcal{T} = (q_j)_{j \in J}$ in $\mathcal{S}$. □

6. Hahn–Banach existence type theorem for i-linear functionals on interval-subspaces

It is well known that the following result is called the (classical) *Hahn–Banach existence theorem* in the setting of vector spaces:

"*If X is a real vector space and $s : X \to \mathbb{R}$ is a sublinear functional, then there exists a linear functional $\ell : X \to \mathbb{R}$ such that $\ell \leq s$ (that is, $\ell(x) \leq s(x)$, for all $x \in X$).*"

We know that (see, for example, [30]) this theorem was proved in three ways, by:

1. Kelley–Namioka [9, 3.4, p. 21], using *cones*;
2. Rudin [28, 3.2, pp. 56–57], using an *extension by subspaces*;
3. König [11] and Simons [29], using the pointwise *order on sublinear functionals.*

The following result – see Theorem 3 – extends the classical Hahn–Banach existence Theorem in interval analysis, for i-sublinear functionals on an i-subspace IS of any i-space IE (E is an arbitrary ordered vector space). Our proof will be in the line of "3." ([11], [29]).

Theorem 3 (Hahn–Banach existence type theorem in the setting of interval-spaces). *Let IE be an arbitrary i-space and $IS \subseteq IE$ an i-subspace. Let also $s : IS \to \mathbb{R}$ be an i-sublinear functional. Then there exists an i-linear functional $\ell : IS \to \mathbb{R}$ such that*

$$\ell([v]) \leq s([v]) \tag{19}$$

for all $[v] \in IS$.

Proof. Denote by $\mathcal{S}$ the (nonempty) set of all i-sublinear functionals $q : IS \to \mathbb{R}$ such that $q \leq s$. Using Proposition 2 it follows that $\mathcal{S}$ endowed with the pointwise ordering is inductively ordered from below. Therefore, by Zorn's lemma, there exists at least one *minimal* element $\ell \in \mathcal{S}$. We will prove that ℓ is an *i-linear functional* on IS. Firstly, we will prove that $\ell = h$, where $h : IS \to \mathbb{R}$ is defined by

$$h([v]) = \inf_{\substack{\alpha>0 \\ [u]\in IS}} (\ell([v] \oplus \alpha [u]) - \alpha\ell([u])) \quad \text{for all } [v] \in IS.$$

Step 1. h is *well defined*, that is, $h([v]) > -\infty$ for all $[v] \in IS$. To prove this, consider $[o] = [v] \ominus [v] \in \mathcal{O}_{IS}$, and let $[u] \in IS$ and $\alpha > 0$. Because ℓ is an i-sublinear functional on IS, it follows that for all $[u] \in IS$ the following hold:

$$\begin{aligned} \alpha\ell([u]) &= \ell(\alpha[u]) = \ell(\alpha[u] \oplus [o]) = \ell(\alpha[u] \oplus ([v] \ominus [v])) \\ &= \ell(\alpha[u] \oplus [v] \oplus (-[v])) \leq \ell(\alpha[u] \oplus [v]) + \ell(-[v]). \end{aligned}$$

Hence, $-\ell(-[v]) \leq \ell(\alpha[u] \oplus [v]) - \alpha\ell([u])$ for all $\alpha > 0$ and $[u] \in IS$. By taking the infimum, it follows: $-\ell(-[v]) \leq h([v])$ and then $h([v]) > -\infty$.

Step 2. h is an *i-sublinear functional.*

a) Firstly, we will prove that $\lambda h([v]) = h(\lambda[v])$ for all $\lambda > 0$ and $[v] \in IS$. But:

$$\begin{aligned} \lambda h([v]) &= \lambda \inf_{\substack{\alpha>0 \\ [u]\in IS}} (\ell([v] \oplus \alpha[u]) - \alpha\ell([u])) \\ &\overset{\lambda\geq 0}{=} \inf_{\substack{\alpha>0 \\ [u]\in IS}} (\ell(\lambda[v] \oplus \lambda\alpha[u]) - \lambda\alpha\ell([u])) \\ &= \inf_{\substack{\mu>0 \\ [u]\in IS}} (\ell(\lambda[v] \oplus \mu[u]) - \mu\ell([u])) \\ &= h(\lambda[v]). \end{aligned}$$

(We used that ℓ is an i-sublinear functional and we have denoted $\lambda\alpha$ by μ.)

b) Now we will prove that $h([v] \oplus [o]) = h([v])$ for all $[v] \in IS$ and $[o] \in \mathcal{O}_{IS}$. Using the definition of h and that ℓ is an i-sublinear functional, it follows:

$$\begin{aligned} h([v] \oplus [o]) &= \inf_{\substack{\alpha>0 \\ [u]\in IS}} \left(\ell([v] \oplus [o] \oplus \alpha[u]) - \alpha\ell([u])\right) \\ &= \inf_{\substack{\alpha>0 \\ [u]\in IS}} \left(\ell([v] \oplus \alpha[u]) - \alpha\ell([u])\right) = h([v]). \end{aligned}$$

As a consequence of "a)" and "b)", $h(0 \cdot [v]) = 0$, for all $[v] \in IS$. (Indeed, for all $[o] \in IS$, it follows: $2h([o]) \overset{a)}{=} h(2[o]) = h([o] \oplus [o]) \overset{b)}{=} h([o])$, hence $h([o]) = 0$. Taking $[o] = \mathbf{0} = [0,0]$, it follows $h(\mathbf{0}) = 0$, too.) Hence, $h(\lambda[v]) = \lambda h[v]$, for all $\lambda \geq 0$.

c) Finally we will prove that h is an *i-subadditive functional*. Let $[v_1]$ and $[v_2] \in IS$. Then, using that ℓ is an i-subadditive functional, it follows:

$$\begin{aligned} h([v_1] \oplus [v_2]) &= \inf_{\substack{\alpha>0 \\ [u]\in IS}} \left(\ell([v_1] \oplus [v_2] \oplus \alpha[u]) - \alpha\ell([u])\right) \\ &\leq \ell([v_1] \oplus [v_2] \oplus (\alpha_1+\alpha_2)[u]) - (\alpha_1+\alpha_2)\ell([u])) \\ &\leq \left(\ell([v_1] \oplus \alpha_1[u]) - \alpha_1\ell([u])\right) + \left(\ell([v_2] \oplus \alpha_2[u]) - \alpha_2\ell([u])\right) \end{aligned}$$

where for a fixed $\alpha > 0$, we took arbitrary $\alpha_1 > 0, \alpha_2 > 0$ such that $\alpha_1 + \alpha_2 = \alpha$.

Taking the infimum and denoting

$$\begin{aligned} a_1(\alpha_1, [u_1]) &= \ell([v_1] \oplus \alpha_1[u_1]) - \alpha_1\ell([u_1]) \in \mathbb{R} \text{ and} \\ a_2(\alpha_2, [u_2]) &= \ell([v_2] \oplus \alpha_2[u_2]) - \alpha_2\ell([u_2]) \in \mathbb{R} \end{aligned}$$

it follows:

$$\begin{aligned} h([v_1] \oplus [v_2]) &\leq \inf_{\substack{\alpha_1>0 \\ [u_1]\in IS}} \inf_{\substack{\alpha_2>0 \\ [u_2]\in IS}} \left(a_1(\alpha_1,[u_1]) + a_2(\alpha_2,[u_2])\right) \\ &= \inf_{\substack{\alpha_1>0 \\ [u_1]\in IS}} a_1(\alpha_1,[u_1]) + \inf_{\substack{\alpha_2>0 \\ [u_2]\in IS}} a_2(\alpha_2,[u_2]) = h([v_1]) + h([v_2]). \end{aligned}$$

Hence:

$$h([v_1] \oplus [v_2]) \leq h([v_1]) + h([v_2]) \quad \text{for all } [v_1], [v_2] \in IS,$$

that is, h is an i-subadditive functional.

Step 3. We will prove that $h \in \mathcal{S}$. Since $\ell \in \mathcal{S}$ is an i-sublinear functional, we have for $[v] \in IS : \ell([v] \oplus \alpha[u]) - \alpha\ell([u]) \leq \ell([v]) \leq s([v])$ for all $\alpha > 0$ and $[u] \in IS$. Hence $h([v]) \leq s([v])$ for all $[v] \in IS$, that is, $h \in \mathcal{S}$. Moreover

$$h \leq \ell. \tag{20}$$

Step 4. Because $h \in \mathcal{S}$ and ℓ is a *minimal* element in $\mathcal{S}$, it follows that $\ell \leq h$ on IS. Then, according to (20), it follows that $h = \ell$ on IS.

Step 5. Now we will prove that, actually, ℓ is even an *i-additive* functional. But ℓ is an i-subadditive functional, hence it remains to prove that $\ell([v_1]) + \ell([v_2]) \leq \ell([v_1] \oplus [v_2])$ for all $[v_1], [v_2] \in IS$.

But: $\ell([v_1]) = h([v_1]) \leq \ell([v_1] \oplus \alpha[v_2]) - \alpha\ell([v_2])$ for all $\alpha > 0$. Taking $\alpha = 1$, it follows that $\ell([v_1]) \leq \ell([v_1] \oplus [v_2]) - \ell([v_2])$. Therefore, ℓ is an i-additive functional.

Step 6. Now we can prove that ℓ is an *i-linear* functional. (Obviously, because $\ell \in \mathcal{S}$, it follows that $\ell \leq s$ on IS.) But ℓ has the following properties:

1. ℓ is an i-additive functional on IS,
2. ℓ is an i-sublinear functional and, consequently, $\ell(\mathbf{0}) = \ell(0 \cdot [v]) = 0$ and, moreover $\ell([o]) = 0$ for all $[o] \in \mathcal{O}_{IS}$, equality which can be similarly demonstrated with the equality $h(\mathbf{0}) = 0$ – see *Step* 2.

It remains only to prove that $\ell(\lambda[v]) = \lambda\ell([v])$ for all $[v] \in IS$ and $\lambda < 0$. Put $\mu = -\lambda$. Then $\mu > 0$ and hence, $\mu\ell([v]) = \ell(\mu[v])$. Therefore it follows:

$$-\lambda\ell([v]) = \ell(-\lambda[v]). \tag{21}$$

Taking $[o] = [v] \ominus [v] \in \mathcal{O}_{IS}$, and using the properties of ℓ, it follows according to *Step* 5: $0 = \ell([o]) = \ell([v] \ominus [v]) = \ell([v] \oplus (-[v])) = \ell([v]) + \ell(-[v])$.

Then $\ell(-[v]) = -\ell([v])$ for all $[v] \in IS$. It follows that $\ell(-\lambda[v]) = -\ell(\lambda[v])$ and hence, according to (21), $-\lambda\ell([v]) = -\ell(\lambda[v])$, that is, $\lambda\ell([v]) = \ell(\lambda[v])$. This completes the proof.

Hence, there exists an i-linear functional $\ell : IS \to \mathbb{R}$ with $\ell \leq s$ on IS, that is, $\ell([v]) \leq s([v])$ for all $v \in [IS]$. □

If $\ell \leq s$ on IS as in the previous theorem we will say that ℓ is *dominated by* s on IS.

7. Mazur–Orlicz type theorem in the setting of interval-spaces

The following result is in the line of the classical Mazur–Orlicz Theorem (see [21] and [6, Theorem 2.1]). Notice that we will denote by $\mathbb{N}^*$ the set of all natural numbers n, with $n \geq 1$.

Theorem 4. *Let IS be an i-subspace of the i-space IE and $s : IS \to \mathbb{R}$ an i-sublinear functional. Let also A be a nonempty arbitrary set and $f : A \to \mathbb{R}$, $g : A \to IS$ two arbitrary maps. Then the following are equivalent:*

(i) *there exists an i-linear functional $\ell : IS \to \mathbb{R}$, such that*
   a) *$\ell \leq s$ on IS;*
   b) *$f(a) \leq \ell([g(a)])$ for each $a \in A$;*

(ii) *the inequality $\sum\limits_{j=1}^{n} \lambda_j f(a_j) \leq s\left(\bigoplus\limits_{j=1}^{n} \lambda_j [g(a_j)]\right)$ holds for all finite subsets $\{a_1, \ldots, a_n\}$ in A and $\lambda_1 \geq 0, \ldots, \lambda_n \geq 0$ in $\mathbb{R}$.*

Notice that we denoted $[g(a)]$ and $[g(a_j)]$ instead of $g(a)$ and $g(a_j)$, respectively, only to remind that, actually, $g(a)$ and $g(a_j)$ are order intervals from IS.

Proof. (i) $\Rightarrow$ (ii). It is obvious. Indeed, because ℓ is an i-linear functional and $\ell \leq s$, it follows:

$$\sum_{j=1}^{n}\lambda_j f(a_j) \leq \sum_{j=1}^{n}\lambda_j \ell([g(a_j)]) = \ell\left(\overset{n}{\underset{j=1}{\oplus}}\lambda_j [g(a_j)]\right) \leq s\left(\overset{n}{\underset{j=1}{\oplus}}\lambda_j [g(a_j)]\right).$$

(ii) $\Rightarrow$ (i). Firstly we will prove that (ii) $\Rightarrow$ (i) a), by using the *technique of the auxiliary i-sublinear functional* that is similar to the technique of the auxiliary sublinear operator, used, for example, in [6].

For every $[v] \in IS$, define $p([v])$ by the following infimum:

$$p([v]) = \inf_{\substack{n\in\mathbb{N}^* \\ \lambda_j\geq 0 \\ a_j\in A \\ j=\overline{1,n}}} \left(s\left([v] \oplus \left(\overset{n}{\underset{j=1}{\oplus}}\lambda_j [g(a_j)]\right)\right) - \sum_{j=1}^{n}\lambda_j f(a_j)\right). \tag{22}$$

The functional p (that will be "the auxiliary i-sublinear functional") has the following properties:

1. p is *well defined.* Using (ii) and that s is an i-sublinear functional, it follows:

$$\begin{aligned}\sum_{j=1}^{n}\lambda_j f(a_j) &\overset{\text{(ii)}}{\leq} s\left(\overset{n}{\underset{j=1}{\oplus}}\lambda_j [g(a_j)]\right) = s\left(\overset{n}{\underset{j=1}{\oplus}}\lambda_j [g(a_j)] \oplus [o]\right) \\ &= s\left(\overset{n}{\underset{j=1}{\oplus}}\lambda_j [g(a_j)] \oplus [v] \ominus [v]\right) \\ &\leq s\left([v] \oplus \left(\overset{n}{\underset{j=1}{\oplus}}\lambda_j [g(a_j)]\right)\right) + s(-[v]).\end{aligned}$$

Consequently it follows:

$$-s(-[v]) \leq s\left([v] \oplus \left(\overset{n}{\underset{j=1}{\oplus}}\lambda_j [g(a_j)]\right)\right) - \sum_{j=1}^{n}\lambda_j f(a_j).$$

(In the above we have chosen $[o] \in \mathcal{O}_{IS}$ given by $[o] = [v] \ominus [v]$.)

Then, the set considered in the right side of the equality (22) is lower bounded in $\mathbb{R}$, and therefore there exists its infimum.

2. p is an *i-sublinear* functional.

2. a) Firstly, p is an *i-subadditive* functional. Indeed,

$$p([v_1] \oplus [v_2]) = \inf_{\substack{n\in\mathbb{N}^* \\ \lambda_j\geq 0 \\ a_j\in A \\ j=\overline{1,n}}} \left(s\left([v_1] \oplus [v_2] \oplus \left(\overset{n}{\underset{j=1}{\oplus}}\lambda_j [g(a_j)]\right)\right) - \sum_{j=1}^{n}\lambda_j f(a_j)\right)$$

hence

$$p([v_1] \oplus [v_2]) \leq \left(s\left([v_1] \oplus \left(\overset{n}{\underset{j=1}{\oplus}} \mu_j [g(a_j)] \right) \right) - \sum_{j=1}^{n} \mu_j f(a_j) \right)$$
$$+ \left(s\left([v_2] \oplus \left(\overset{n}{\underset{j=1}{\oplus}} \nu_j [g(a_j)] \right) \right) - \sum_{j=1}^{n} \nu_j f(a_j) \right)$$

(where for any $j = \overline{1,n}$ we took $\mu_j \geq 0, \nu_j \geq 0$ such that $\lambda_j = \mu_j + \nu_j$ and we used that s is an i-sublinear functional). By taking the infimum in the right side of the last inequality, it follows:

$$p([v_1] \oplus [v_2]) \leq p([v_1]) + p([v_2]).$$

2. b) Now we will prove that $p(\alpha [v]) = \alpha p[v]$ for all $\alpha > 0$ and $[v] \in IS$. Using that s is an i-sublinear functional, we have:

$$p(\alpha [v]) = \inf_{\substack{n \in \mathbb{N}^* \\ \lambda_j \geq 0 \\ a_j \in A \\ j=\overline{1,n}}} \left(s\left(\alpha [v] \oplus \left(\overset{n}{\underset{j=1}{\oplus}} \lambda_j [g(a_j)] \right) \right) - \sum_{j=1}^{n} \lambda_j f(a_j) \right)$$

$$= \alpha \inf_{\substack{n \in \mathbb{N}^* \\ \lambda_j \geq 0 \\ a_j \in A \\ j=\overline{1,n}}} \left(s\left([v] \oplus \left(\overset{n}{\underset{j=1}{\oplus}} \frac{\lambda_j}{\alpha} [g(a_j)] \right) \right) - \sum_{j=1}^{n} \frac{\lambda_j}{\alpha} f(a_j) \right)$$

$$= \alpha \inf_{\substack{n \in \mathbb{N}^* \\ \mu_j \geq 0 \\ a_j \in A \\ j=\overline{1,n}}} \left(s\left([v] \oplus \left(\overset{n}{\underset{j=1}{\oplus}} \mu_j [g(a_j)] \right) \right) - \sum_{j=1}^{n} \mu_j f(a_j) \right) = \alpha p([v]).$$

(We denoted $\mu_j = \frac{\lambda_j}{a_j}$ for all $j = \overline{1,n}$.)

2. c) It remains to prove that $p([v] \oplus [o]) = p([v])$ for all $[v] \in IS$ and $[o] \in \mathcal{O}_{IS}$. But using again that s is an i-sublinear functional, it follows:

$$p([v] \oplus [o]) = \inf_{\substack{n \in \mathbb{N}^* \\ \lambda_j \geq 0 \\ a_j \in A \\ j=\overline{1,n}}} \left(s\left([v] \oplus [o] \oplus \left(\overset{n}{\underset{j=1}{\oplus}} \lambda_j [g(a_j)] \right) \right) - \sum_{j=1}^{n} \lambda_j f(a_j) \right)$$

$$= \inf_{\substack{n \in \mathbb{N}^* \\ \lambda_j \geq 0 \\ a_j \in A \\ j=\overline{1,n}}} \left(s\left([v] \oplus \left(\overset{n}{\underset{j=1}{\oplus}} \lambda_j [g(a_j)] \right) \right) - \sum_{j=1}^{n} \lambda_j f(a_j) \right) = p([v]).$$

Thus we have shown that p is an i-sublinear functional. Now we will use the Hahn–Banach existence type theorem for i-linear functionals on interval subspaces (see Theorem 3 above). By using this theorem, it follows that there exists an *i-linear functional* $\ell : IS \to \mathbb{R}$ such that $\ell([v]) \leq p([v])$ for all $[v] \in IS$. Using the definition of p and taking $\lambda_j = 0$ for all $j = \overline{1,n}\,(n \in \mathbb{N}^*)$ it follows that $p([v]) \leq s([v])$ for all $[v] \in IS$, and then, $\ell \leq s$ on IS.

Now we will prove that (ii) $\Rightarrow$ (i) b).

We have to prove that $f(a) \leq \ell([g(a)])$ for all $a \in A$. We will use the inequality $\ell \leq p$ and the fact that ℓ is an i-linear functional. Then, for all $a \in A$, it follows:

$$\begin{aligned}-\ell([g(a)]) &= \ell(-[g(a)]) \leq p(-[g(a)]) \\ &\leq s([g(a)] \oplus (-[g(a)])) - f(a) \\ &= s([g(a)] \ominus [g(a)]) - f(a) \\ &= s([o]) - f(a) = -f(a)\end{aligned}$$

hence

$$f(a) \leq \ell([g(a)]) \text{ for all } a \in A \qquad \square$$

Remark 5. Note that in the previous theorem, A is a nonempty arbitrary set and thus we can include the case where A is a *set of intervals.* (Of course, in this case we have to make a slight correction in this theorem. So, we will put $[a_j]$ instead of a_j, and if more, $A \subseteq IS$ and g is this inclusion, then $[g(a_j)]$ and $f(a_j)$ becomes $[a_j]$ and $f([a_j])$, respectively.) Then the resulted Mazur–Orlicz type theorem can be viewed as a generalization of the following Hahn–Banach type theorem.

Corollary 5 (Hahn–Banach extension type theorem in the setting of interval-spaces). *Let IE be an i-space and $IS \subseteq IE$ an i-subspace. Let also $s : IE \to \mathbb{R}$ be an i-sublinear functional and $t : IS \to \mathbb{R}$ an i-linear functional. Then the following are equivalent:*

(i) *There exists an i-linear functional $\ell : IE \to \mathbb{R}$ such that:*

  a) *$\ell \leq s$ on IE, and*

  b) *$\ell = t$ on IS, that is, ℓ is an i-linear extension of t;*

(ii) *$t \leq s$ on IS.*

Proof. Take in the version of Theorem 4, mentioned in Remark 5, IE instead of IS, IS instead of A, g the inclusion of IS in IE, and $f = t$. Then, the inequality

$$\sum_{j=1}^{n} \lambda_j f([a_j]) \leq s\left(\underset{j=1}{\overset{n}{\oplus}} \lambda_j [g([a_j])]\right),$$

that is,

$$\sum_{j=1}^{n} \lambda_j t([a_j]) \leq s\left(\underset{j=1}{\overset{n}{\oplus}} \lambda_j [a_j]\right)$$

becomes (using the i-linearity of t):

$$t\left(\bigoplus_{j=1}^{n}\lambda_j\,[a_j]\right)\leq s\left(\bigoplus_{j=1}^{n}\lambda_j\,[a_j]\right)$$

for all $n\in\mathbb{N}^*,\lambda_1\geq 0,\ldots,\lambda_n\geq 0$ and $[a_1],\ldots,[a_n]\in IS$ or, equivalently, $t([a])\leq s([a])$ for all $[a]\in IS$, that is, $t\leq s$ on IS. □

8. Extension results with convexity assumptions in the setting of interval-spaces

In this section we will give some consequences of the Hahn–Banach existence Theorem in the setting of interval-spaces. Using the ideas of S. Simons (see [30]), formulated in the setting of vector spaces, we introduce certain convexity hypotheses.

Firstly, we consider an i-space IE, an i-sublinear functional $s: IE\to\mathbb{R}$, a nonempty convex subset K of an arbitrary interval-space IF (with F an arbitrary ordered vector space) and two arbitrary maps $g: K\to IE$ and $f: K\to\mathbb{R}$. (A subset $K\subseteq IF$ is called a *convex subset* if $\alpha\,[a]+(1-\alpha)\,[b]\in K$, for all $[a],[b]\in K$ and $\alpha\in(0,1)$.)

a) We say that g is *s-convex*, if

$$s\left([x]\oplus[g(\alpha[a]\oplus(1-\alpha)\,[b])]\right)\leq s\left([x]\oplus\alpha\,[g([a])]\oplus(1-\alpha)\,[g([b])]\right)$$

for all $[x]\in IE$, $[a],[b]\in K$ and $\alpha\in(0,1)$.
(Here we wrote, for example, $[g(\alpha[a]\oplus(1-\alpha)\,[b])]$ instead of $g(\alpha[a]\oplus(1-\alpha)\,[b])$ only to remind that $g(\alpha[a]\oplus(1-\alpha)\,[b])$ is an ordered interval in E.)

Note that the previous inequality can also be written in a simpler form if we introduce in IE the following order:

$$[u]\leq_{IE}[v]\Leftrightarrow s([x]\oplus[u])\leq s([x]\oplus[v])$$

for all $[x]\in IE$. Indeed, g is s-convex if and only if

$$[g(\alpha[a]\oplus(1-\alpha)\,[b])]\leq_{IE}\alpha\,[g([a])]\oplus(1-\alpha)\,[g([b])]\,.$$

Any i-linear functional is clearly s-convex.

b) We say that f is an *interval*-convex functional (in short, an *i-convex functional*) if $f(\alpha[a]\oplus(1-\alpha)\,[b])\leq\alpha f([a])+(1-\alpha)f([b])$, for all $[a],[b]\in K$ and $\alpha\in(0,1)$. We also say that f is an *interval-concave functional* (in short, *i-concave functional*), if $-f$ is an i-convex functional.

Proposition 6 (in the line of [30, Lemma 1.4). *] Let IE be an i-space and $s: IE\to\mathbb{R}$ an i-sublinear functional. Let also K be a nonempty convex subset of an interval-space, $g: K\to IE$ a s-convex map and $f: K\to\mathbb{R}$ an arbitrary i-convex functional. Suppose that the following infimum $\lambda=\inf_{a\in K}(f([a])+s([g([a])]))$ exists in $\mathbb{R}$. For all $[x]\in IE$, define*

$$t([x])=\inf_{\substack{[a]\in K\\ \alpha>0}}\left(s([x]\oplus\alpha\,[g([a])])+\alpha f([a])-\alpha\lambda\right).\tag{23}$$

Then:

1. t *is well defined;*
2. t *is an i-sublinear functional;*
3. $t([x]) \leq s([x])$ *for all* $[x] \in IE$;
4. $-t(-[g([a])]) + f([a]) \geq \lambda$ *for all* $[a] \in K$.

(Recall that we denoted $[g([a])]$ instead of $g([a])$, to remind that $g([a])$ is an order interval in E.)

Proof. 1. We will prove that t is *well defined*. Since s is an i-sublinear functional it follows that

$$s(\alpha[g([a])]) = s(\alpha[g([a])] \oplus [o]) \tag{24}$$

for all $[a] \in K, \alpha > 0$ and $[o] \in \mathcal{O}$. (Recall that $\mathcal{O}$ is the null set of IE.)

Take $[o] = [x] \ominus [x]$, where $[x] \in IE$ is a fixed interval. According to (24) and using the i-sublinearity of s, it follows:

$$\begin{aligned} s(\alpha[g([a])]) &= s(\alpha[g([a])] \oplus ([x] \ominus [x])) \\ &= s(\alpha[g([a])] \oplus [x]) \oplus (-[x]) \\ &\leq s(\alpha[g([a])] \oplus [x]) + s(-[x]). \end{aligned}$$

Then:

$$-s(-[x]) \leq s([x] \oplus \alpha[g([a])]) - s(\alpha[g([a])]). \tag{25}$$

Because s is an i-positively homogeneous functional, $\alpha > 0$ and $\lambda \leq f([a]) + s([g([a])])$ for all $[a] \in K$, it follows that: $\alpha\lambda \leq \alpha f([a]) + s(\alpha[g([a])])$ and hence $-s(\alpha[g([a])]) \leq \alpha f([a]) - \alpha\lambda$. Therefore, from (25) we obtain: $-s(-[x]) \leq s([x] \oplus \alpha[g([a])]) + \alpha f([a]) - \alpha\lambda$ for all $[a] \in K$ and $\alpha > 0$.

Hence, the set $\{s([x] \oplus \alpha[g([a])]) + \alpha f([a]) - \alpha\lambda \mid [a] \in K, \alpha > 0\}$ is minorized in $\mathbb{R}$ by $-s(-[x])$, and so, there exists its infimum, that is, $t([x])$ is well defined.

2. We will prove that t is an *i-sublinear functional*, that is,

2. a) t is i-subadditive on IE;

2. b) t is i-positively homogeneous on IE;

2. c) $t([x] \oplus [o]) = t([x])$ for all $[x] \in IE$ and $[o] \in \mathcal{O}$.

So: 2. a) Let us prove that $t([x_1] \oplus [x_2]) \leq t([x_1]) + t([x_2])$ for all $[x_1], [x_2] \in IE$. We have:

$$\begin{aligned} t([x_1] \oplus [x_2]) &= \inf_{\substack{[a]\in k \\ \alpha>0}} (s([x_1] \oplus [x_2] \oplus \alpha[g([a])]) + \alpha f([a]) - \alpha\lambda) \\ &\leq s([x_1] \oplus [x_2] \oplus \alpha[g([a])]) + \alpha f([a]) - \alpha\lambda \end{aligned}$$

for all $[a] \in K$ and $\alpha > 0$. We take $[a_1], [a_2] \in K$ and $\alpha_1 > 0, \alpha_2 > 0$ with $\alpha = \alpha_1 + \alpha_2$, and $[a] = \beta_1[a_1] \oplus \beta_2[a_2]$ were $\beta_1 = \frac{\alpha_1}{\alpha}, \beta_2 = \frac{\alpha_2}{\alpha}$.

It follows that $\beta_1 > 0, \beta_2 > 0$ and $\beta_1 + \beta_2 = 1$, hence $[a] \in K$ (K being a convex set). Because s is i-positively homogeneous, g is s-convex, f is i-convex and s is i-subadditive, it follows that:

$$\begin{aligned}
&s\left([x_1] \oplus [x_2] \oplus \alpha\left[g\left([a]\right)\right]\right) + \alpha f\left([a]\right) - \alpha\lambda \\
&= \alpha\left(s\left(\frac{1}{\alpha}\left([x_1] \oplus [x_2]\right) \oplus \left[g\left([a]\right)\right]\right) + f\left([a]\right)\right) - \alpha\lambda \\
&\leq \alpha\left(s\left(\frac{1}{\alpha}\left([x_1] \oplus [x_2]\right) \oplus \beta_1\left[g\left([a_1]\right)\right] \oplus \beta_2\left[g\left([a_2]\right)\right]\right)\right. \\
&\quad \left. + \beta_1 f\left([a_1]\right) + \beta_2 f\left([a_2]\right)\right) - \alpha\lambda \\
&\leq \alpha s\left(\frac{1}{\alpha}\left([x_1] \oplus \beta_1\left[g\left([a_1]\right)\right]\right)\right) + \beta_1 f\left([a_1]\right) \\
&\quad + \alpha s\left(\frac{1}{\alpha}\left([x_2] \oplus \beta_2\left[g\left([a_2]\right)\right]\right)\right) + \beta_2 f\left([a_2]\right) - \left(\alpha_1 + \alpha_2\right)\lambda \\
&= \left(s\left([x_1] \oplus \alpha\beta_1\left[g\left([a_1]\right)\right]\right) + \alpha\beta_1 f\left([a_1]\right) - \alpha_1\lambda\right) \\
&\quad + \left(s\left([x_2] \oplus \alpha\beta_2\left[g\left([a_2]\right)\right]\right) + \alpha\beta_2 f\left([a_2]\right) - \alpha_2\lambda\right).
\end{aligned}$$

Hence we proved that:

$$\begin{aligned}
&s\left([x_1] \oplus [x_2] \oplus \alpha\left[g\left([a]\right)\right]\right) + \alpha f\left([a]\right) - \alpha\lambda \\
&\quad \leq \left(s\left([x_1] \oplus \alpha_1\left[g\left([a_1]\right)\right]\right) + \alpha_1 f\left([a_1]\right) - \alpha_1\lambda\right) \\
&\qquad + \left(s\left([x_2] \oplus \alpha_2\left[g\left([a_2]\right)\right]\right) + \alpha_2 f\left([a_2]\right) - \alpha_2\lambda\right).
\end{aligned}$$

Taking the infimum over $[a_1], [a_2] \in K$ and $\alpha_1 > 0, \alpha_2 > 0$ it follows that:

$$t\left([x_1] \oplus [x_2]\right) \leq t\left([x_1]\right) + t\left([x_2]\right).$$

2. b) It is immediate that t is *i-positively homogeneous* , that is, $t\left(\lambda[x]\right) = \lambda t\left([x]\right)$ for all $[x] \in IE$ and $\lambda > 0$ (because s is i-positively homogeneous).

2. c) Let us prove that $t\left([x] \oplus [o]\right) = t\left([x]\right)$ for all $[x] \in IE$ and $[o] \in \mathcal{O}$. But

$$\begin{aligned}
t\left([x] \oplus [o]\right) &= \inf_{\substack{[a]\in K \\ \alpha>0}} \left(s\left([x] \oplus [o] \oplus \alpha\left[g\left([a]\right)\right]\right) + \alpha f\left([a]\right) - \alpha\lambda\right) \\
&= \inf_{\substack{[a]\in K \\ \alpha>0}} \left(s\left([x] \oplus \alpha\left[g\left([a]\right)\right]\right) + \alpha f\left([a]\right) - \alpha\lambda\right) = t\left([x]\right).
\end{aligned}$$

(We used that s is an i-sublinear functional.)

3. For all $[x] \in IE$, according to the properties of s it follows that

$$t\left([x]\right) \leq s\left([x]\right) + \alpha\left(s\left(\left[g\left([a]\right)\right]\right) + f\left([a]\right) - \lambda\right).$$

Letting $\alpha \to 0$ in $\mathbb{R}$, it follows that $t\left([x]\right) \leq s\left([x]\right)$.

4. Now we will prove that $-t(-[g([a])]) + f([a]) \geq \lambda$ for all $[a] \in K$. Recall that

$$t([x]) = \inf_{\substack{[a]\in K \\ \alpha>0}} (s([x] \oplus \alpha [g([a])]) + \alpha f([a]) - \alpha\lambda).$$

Taking $[x] = -[g([a])]$ and $\alpha = 1$, it follows that

$$t(-[g([a])]) \leq s(-[g([a])] \oplus [g([a])]) + f([a]) - \lambda.$$

But s is an i-sublinear functional and then, $s([o]) = 0$, for all $[o] \in \mathcal{O}$ (see property "6." of the i-sublinear functionals). So, it follows that

$$t(-[g([a])]) \leq f([a]) - \lambda \Rightarrow -t(-[g([a])]) + f([a]) \geq \lambda$$

for all $[a] \in K$. □

Remark 6. In the previous result, K is a (nonempty) *convex* subset of an interval-space IF. Remark that the proof also works for K *a convex subset of an arbitrary vector space.* Of course, in this case we will say that $g : K \to IE$ is *s-convex* ($s : IE \to \mathbb{R}$ being an i-sublinear functional) if:

$$\begin{aligned} &s([x] \oplus [g(\alpha a + (1-\alpha) b)]) \\ &\quad \leq s([x] \oplus \alpha [g(a)] \oplus (1-\alpha) [g(b)]) \end{aligned}$$

for all $[x] \in IE$, $a, b \in K$ and $\alpha \in (0,1)$. Recall that $f : K \to \mathbb{R}$ is *convex*, if for all $a, b \in K$ and $\alpha \in (0,1)$ it follows that

$$f(\alpha a + (1-\alpha) b) \leq \alpha f(a) + (1-\alpha) f(b).$$

Remark 7. Then Proposition 6 is valid with a slight modification: put a instead of $[a]$ and therefore $\lambda = \inf_{a\in K} (f(a) + s([g(a)]))$ and (23) becomes:

$$t([x]) = \inf_{\substack{a\in K \\ \alpha>0}} (s([x] \oplus \alpha [g(a)]) + \alpha f(a) - \alpha\lambda) \quad \text{for all } [x] \in IE.$$

Theorem 7 (in the line of [30, Theorem 1.5]). *Let IE be an i-space and $s : IE \to \mathbb{R}$ an i-sublinear functional. Let also K be a nonempty convex subset of an i-space, $g : K \to IE$ a s-convex map and $f : K \to \mathbb{R}$ an i-convex functional. Then there exists an i-linear functional $\ell : IE \to \mathbb{R}$ such that:*

a) *$\ell \leq s$ on IE;*

b) $$\inf_{[a]\in K} (f([a]) + \ell([g([a])])) = \inf_{[a]\in K} (f([a]) + s([g([a])])). \qquad ((26))$$

(Note that we wrote $[g([a])]$ instead of $g([a])$, to remind that $g([a])$ is an order interval.)

Proof. We will use the *technique of the auxiliary i-sublinear functional.* Denote $\lambda = \inf_{[a]\in K} (f([a]) + s([g([a])])) \in \bar{\mathbb{R}}$.

Case 1. $\lambda = -\infty$.

Step 1. Apply Theorem 3 (Hahn–Banach existence type theorem in the setting of interval-spaces), for $IS = IE$. Then take any i-linear functional

$$\ell : IE \to \mathbb{R} \text{ such that } \ell \leq s \text{ on } IE.$$

Step 2. For all $[a] \in K$ it follows that

$$f([a]) + \ell([g([a])]) \leq f([a]) + s([g([a])])$$

and hence,

$$\inf_{[a]\in K} (f([a]) + \ell([g([a])])) \leq \lambda = -\infty,$$

that is, the equality (26) holds.

Case 2. $\lambda \in \mathbb{R}$.

Step 1. Apply Theorem 3 for $IS = IE$ and t instead of s, with t from Proposition 6, that is,

$$t([x]) = \inf_{\substack{[a]\in K \\ \alpha>0}} (s([x] \oplus \alpha [g([a])]) + \alpha f([a]) - \alpha\lambda)$$

for all $[x] \in IE$ (see (23)).

Then take any i-linear functional ℓ on IE such that $\ell \leq t$ on IE. (There exists at least one such ℓ according to Theorem 3.) But $t \leq s$ on IE and therefore $\ell \leq s$ on IE.

Step 2. Taking any $[a] \in K$, using the i-linearity of ℓ, the inequality $\ell \leq t$ on IE and "4." from Proposition 6, it follows:

$$\ell([g([a])]) + f([a]) = -\ell(-[g([a])]) + f([a]) \geq -t(-[g([a])]) + f([a]) \geq \lambda.$$

Taking the infimum over $[a] \in K$, we obtain:

$$\inf_{[a]\in K} (f([a]) + \ell([g([a])])) \geq \lambda.$$

The *converse inequality* is also true, since $\ell \leq s$ on IE, and hence:

$$\inf_{[a]\in K} (f([a]) + \ell([g([a])])) \leq \inf_{[a]\in K} (f([a]) + s([g([a])])) = \lambda$$

for all $[a] \in K$. □

Remark 8. We can make the same observation as in the Remark 6 after Proposition 6, about the fact that Theorem 7 still working when K is an arbitrary *convex* subset of a *vector space*, g is s-convex, f is *a convex functional*, and the interval $[a]$ is replaced by an element $a \in K$. More precisely, (26) becomes:

$$\inf_{a\in K} (f(a) + \ell([g(a)])) = \inf_{a\in K} (f(a) + s([g(a)])).$$

(Note again that we wrote $[g(a)]$ instead of $g(a)$ to recall that $g(a)$ is an interval in E.)

Remark 9. If in Theorem 7 we take $K = IS$ (an i-subspace of IE), g the inclusion of IS in IE and $f : K \to \mathbb{R}$ the null function (that is, $f([a]) = 0$, for all $[a] \in IS$), then we obtain the following result.

Corollary 8. *Let IE be an arbitrary i-space, $s : IE \to \mathbb{R}$ an i-sublinear functional, $K = IS \subseteq IE$ an i-subspace and g the inclusion of K in IE. Then there exists an i-linear functional $\ell : IE \to \mathbb{R}$ such that:*

a) *$\ell \leq s$ on IE;*
b) $\inf\limits_{[a]\in K} \ell([a]) = \inf\limits_{[a]\in K} s([a])$.

Proof. We will apply the previous Remark, mentioning that, obviously, $f = 0$ is i-convex and the inclusion g is s-convex. Indeed, the inequality

$$s([x] \oplus [g(\alpha[a] \oplus (1-\alpha)[b])]) \leq s([x] \oplus \alpha[g([a])] \oplus (1-\alpha)[g([b])])$$

is satisfied with equality for all $[x] \in IE, [a], [b] \in K$ and $\alpha \in (0,1)$. □

9. Ordered interval-spaces. Monotone interval-sublinear functionals. Positive interval-linear functionals

Let IE be an i-space (E an arbitrary ordered vector space), endowed with the following *order relation* (the *weak order*):

$$[a] \leq [b] \Leftrightarrow \underline{a} \leq \underline{b} \text{ and } \bar{a} \leq \bar{b}, \text{ where } [a] = [\underline{a}, \bar{a}] \text{ and } [b] = [\underline{b}, \bar{b}].$$

It is immediate that this order is compatible with the algebraic structure on IE, that is:

1. $[a] \leq [b]$ in $IE \Rightarrow [a] \oplus [c] \leq [b] \oplus [c]$, where $[a], [b], [c] \in IE$;
2. $[a] \leq [b]$ in IE and $\alpha > 0$ in $\mathbb{R} \Rightarrow \alpha[a] \leq \alpha[b]$, where $[a], [b] \in IE$.

We call IE endowed with this ordering, an *ordered interval-space* (in short, *ordered i-space*). Let IS be an i-subspace of the ordered i-space IE.

An i-sublinear functional $s : IS \to \mathbb{R}$ is called *monotone* if $[a] \leq [b]$ in IS implies $s([a]) \leq s([b])$ in $\mathbb{R}$. Now let $f : IS \to \mathbb{R}$ be an i-linear functional. It is called *positive* if $[a] \geq \mathbf{0}$ in IS implies $f([a]) \geq 0$ in $\mathbb{R}$. (Recall that $\mathbf{0} = [0,0]$ and observe that $[a] = [\underline{a}, \bar{a}] \geq \mathbf{0}$ means that $\underline{a} \geq 0$.)

Proposition 9. *Let $s : IS \to \mathbb{R}$ be a monotone i-sublinear functional and $\ell : IS \to \mathbb{R}$ an i-linear functional such that $\ell \leq s$ on IS. Then ℓ is positive.*

Proof. It is known (see property "5." of the i-sublinear functionals) that $s(\mathbf{0}) = 0$. Now let $[x] \geq \mathbf{0}$ in IS. Using that $\ell \leq s$, ℓ is i-linear, and s is i-sublinear and monotone, it follows:

$$-\ell([x]) = \ell(-[x]) \leq s(-[x]) \leq s(\mathbf{0}) = 0 \Rightarrow \ell([x]) \geq 0.$$ □

In the following section, firstly the set K will be a nonempty convex subset (of an arbitrary interval-space). We say that a map $p : K \to IS$ is an *interval-convex* (in short, *i-convex-*) *valued* map if

$$[p(\alpha[x] + (1-\alpha)[y])] \leq \alpha[p([x])] \oplus (1-\alpha)[p([y])]$$

for all $[x], [y] \in K$ and $\alpha \in (0,1)$. (Note that we denoted, for example, $[p([x])]$ instead of $p([x])$, only to remind that $p([x])$ is an order interval.)

10. Extension results in the setting of ordered interval-spaces. Mazur–Orlicz type theorem with other convexity assumptions

The following theorem is in the line of [6, Theorem 2.4].

Theorem 10. *Let IS be an i-subspace of an ordered i-space and $s : IS \to \mathbb{R}$ a monotone i-sublinear functional. Let also K be a nonempty convex set in an arbitrary i-space, $p : K \to IS$ be an i-convex-valued map and $q : K \to \mathbb{R}$ an i-concave map. Then the following are equivalent:*

(i) *There exists a positive i-linear functional $\ell : IS \to \mathbb{R}$ such that:*
  a) *$\ell \leq s$ on IS, and*
  b) *$q([a]) \leq \ell([p([a])])$ for all $[a] \in K$;*

(ii) *The following inequality holds: $q([a]) \leq s([p([a])])$ for all $[a] \in K$.*

(We mention again that we wrote $[p([a])]$ instead of $p([a])$ only to recall that $p([a])$ is an order interval.)

Proof. This result is a consequence of our Mazur–Orlicz type theorem in the setting of interval-spaces (see Theorem 4). For this is suffices to prove that "(ii)" from Theorem 10 implies "(ii)" from Theorem 4, for $A = K$ (actually, these conditions are equivalent). Let $\{[a_1], \ldots, [a_n]\}$ be a finite subset of K and $\lambda_1, \ldots, \lambda_n \in \mathbb{R}_+$ such that $\lambda := \lambda_1 + \cdots + \lambda_n > 0$. Denote $\mu_j = \frac{\lambda_j}{\lambda}$, $j = \overline{1,n}$. It follows that $\mu_j > 0$ and $\sum_{j=1}^{n} \mu_j = 1$. Because q is i-concave and p is i-convex-valued, it follows that

$$\sum_{j=1}^{n} \mu_j q([a_j]) \leq q\left(\bigoplus_{j=1}^{n} \mu_j [a_j]\right), \quad \text{and} \quad \left[p\left(\bigoplus_{j=1}^{n} \mu_j [a_j]\right)\right] \leq \bigoplus_{j=1}^{n} \mu_j [p([a_j])].$$

Since $\lambda > 0$ and s is a monotone i-sublinear functional, it follows that:

$$\begin{aligned}\sum_{j=1}^{n} \lambda_j q([a_j]) &= \lambda \sum_{j=1}^{n} \mu_j q([a_j]) \leq \lambda q\left(\bigoplus_{j=1}^{n} \mu_j [a_j]\right) \overset{\text{(ii)}}{\leq} \lambda s\left(\left[p\left(\bigoplus_{j=1}^{n} \mu_j [a_j]\right)\right]\right) \\ &\leq \lambda s\left(\bigoplus_{j=1}^{n} \mu_j [p([a_j])]\right) = s\left(\bigoplus_{j=1}^{n} \lambda_j [p([a_j])]\right).\end{aligned}$$

Apply now Theorem 4, for $A = K$. It follows that there exists an i-linear functional $\ell : IS \to \mathbb{R}$, such that $\ell \leq s$ on IS, and $q([a]) \leq \ell([p([a])])$ for all $[a] \in K$. Using Proposition 9, because $\ell \leq s$ and s is monotone, it follows that ℓ is a positive i-linear functional. □

Remark 10. As in Remark 5 after Theorem 4, note again that in the previous result the hypothesis that the set K be any convex subset of an arbitrary i-space, can be replaced by the hypothesis that the set K can be any convex set in an arbitrary vector space.

Therefore:

a) $p : K \to IS$ "i-convex-valued" means that:

$$[p(\alpha a + (1-\alpha) b)] \leq \alpha [p(a)] \oplus (1-\alpha) [p(b)]$$

for all $a, b \in K$ and $\alpha \in (0,1)$;

b) $q : K \to \mathbb{R}$ "interval-concave" (in short, "i-concave") will be replaced by "concave" meaning that:

$$q(\alpha a + (1-\alpha) b) \geq \alpha q(a) + (1-\alpha) q(b)$$

for all $a, b \in K$ and $\alpha \in (0,1)$.

In this case, the proof of Theorem 10 can be slightly modified. So, its part containing some inequalities, becomes for $\{a_1, \ldots, a_n\}$ a finite subset in K and $\lambda_1, \ldots, \lambda_n \in \mathbb{R}_+$ such that $\lambda = \lambda_1 + \cdots + \lambda_n > 0$ and $\mu_j = \frac{\lambda_j}{\lambda}, j = \overline{1,n}$ ($\mu_j > 0$ for all $j = \overline{1,n}$ and $\sum_{j=1}^{n} \mu_j = 1$):

$$\sum_{j=1}^{n} \mu_j q(a_j) \leq q\left(\sum_{j=1}^{n} \mu_j a_j\right), \quad \text{and} \quad \left[p\left(\sum_{j=1}^{n} \mu_j a_j\right)\right] \leq \bigoplus_{j=1}^{n} \mu_j [p(a_j)].$$

It follows that:

$$\begin{aligned}\sum_{j=1}^{n} \lambda_j q(a_j) &= \lambda \sum_{j=1}^{n} \mu_j q(a_j) \leq \lambda q\left(\sum_{j=1}^{n} \mu_j a_j\right) \\ &\leq \lambda s\left(\left[p\left(\bigoplus_{j=1}^{n} \mu_j a_j\right)\right]\right) \leq s\left(\bigoplus_{j=1}^{n} \lambda_j [p(a_j)]\right).\end{aligned}$$

(We used "(ii)" modified in the following manner: $q(a) \leq s([p(a)])$ for all $a \in K$.)

The following consequence of Theorem 10 is in the line of [6, Proposition 5.1].

Corollary 11 (Hahn–Banach extension type theorem for positive i-linear functionals on ordered interval-spaces). *Let IE be an i-space and $IS \subseteq IE$ an i-subspace. Let also $s : IE \to \mathbb{R}$ be a monotone i-sublinear functional and $f : IS \to \mathbb{R}$ a positive i-linear functional. Then, the following are equivalent:*

(i) *There exists a positive i-linear functional $\ell : IE \to \mathbb{R}$ such that $\ell \leq s$ on IE, and $\ell = f$ on IS;*

(ii) *$f([v]) \leq s([v])$ for all $[v] \in IS$.*

Proof. In Theorem 10, take $K = IS$, $q = f$ and p the inclusion of IS in IE. Apply also Proposition 9 to prove that ℓ is positive. □

The following result is inspired by [6, Proposition 5.2]. (Note that the paper [6] contains extension results for positive linear operators in the setting of ordered vector spaces). The point is that in the following theorem, the hypothesis (used in Corollary 11 above) that the i-sublinear functional s is *monotone* can be dropped if the condition $f([a]) \leq s([a])$ for all $[a] \in IS$ is replaced by $f([a]) \leq s([a] \oplus [u])$ for all $[a] \in IS$ and $[u] \geq \mathbf{0}$ in IE.

Proposition 12. *Let IE be an i-space and $IS \subseteq IE$ an i-subspace. Let also $s : IE \to \mathbb{R}$ be an i-sublinear functional and $f : IS \to \mathbb{R}$ a positive i-linear functional. Then the following are equivalent:*

(i) *there exists $\ell : IE \to \mathbb{R}$ a positive i-linear extension of f such that $\ell \leq s$ on IE;*
(ii) *$f([a]) \leq s([a] \oplus [u])$ for all $[a] \in IS$ and $\mathbf{0} \leq [u] \in IE$;*
(iii) *$f([a]) \leq s([x])$ for all $[a] \in IS$ and $[x] \in IE$ such that $[a] \leq [x]$.*

Proof. (i) $\Rightarrow$ (ii). If ℓ is a positive i-linear extension of f such that $\ell \leq s$ on IE, then for all $[a] \in IS$ and $[u] \in IE$, $[u] \geq \mathbf{0}$, it follows that

$$f([a]) \leq \ell([a]) \leq \ell([a] \oplus [u]).$$

(ii) $\Rightarrow$ (i). We will use the *technique of the auxiliary i-sublinear functional.* Define $s_1 : IE \to \mathbb{R}$ by the formula $s_1([y]) = \inf\limits_{\substack{[x]\in IE \\ [x]\geq \mathbf{0}}} s([y] \oplus [x])$.

Then s_1 has the following properties:

1. s_1 is *well defined.* To prove this, fix $[y] \in IE$ and take $\mathbf{0} \leq [x] \in IE$. From (ii), since s is i-sublinear, it follows that $0 = f(\mathbf{0}) \leq s(\mathbf{0} \oplus [x])$, that is, $0 \leq s([x])$. Now, for all $[x] \in IE, [x] \geq \mathbf{0}$, because s is an i-sublinear functional, we have $0 \leq s([x]) = s([x] \oplus [o])$ for all $[o] \in \mathcal{O}$. Taking $[o] = [y] \ominus [y]$, we obtain: $0 \leq s([x] \oplus [y] \ominus [y]) \leq s([x] \oplus [y]) + s([-y])$. Hence, for all $[x] \in IE[x] \geq \mathbf{0}$ we have: $-s(-[y]) \leq s([y] \oplus [x])$. Then there exists the infimum of the set that defines $s_1([y])$ for all $[y] \in IE$.
2. s_1 is a *monotone i-sublinear functional.* Let us prove only that s_1 is *monotone.* Let $[y], [x] \in IE$ and $[x] \geq \mathbf{0}$. Then: $[y] \leq [y] \oplus [x]$ and $s_1([y] \oplus [x]) = \inf\limits_{\substack{[z]\in IE \\ [z]\geq \mathbf{0}}} s([y] \oplus [x] \oplus [z])$. Denoting $[a] = [x] \oplus [z]$, it follows that $[a] \geq \mathbf{0}$.

   Therefore:

$$\inf_{\substack{[z]\in IE \\ [z]\geq \mathbf{0}}} s([y] \oplus [x] \oplus [z]) \geq \inf_{\substack{[a]\in IE \\ [a]\geq \mathbf{0}}} s([y] \oplus [a]) = s_1([y]).$$

   Hence $s_1([y]) \leq s_1([y] \oplus [x])$.
3. $f([a]) \leq s_1([a])$ for all $[a] \in IS$. Indeed, for all $[u] \in IE, [u] \geq \mathbf{0}$, it follows from (ii): $f([a]) \leq s([a] \oplus [u])$ and hence

$$f([a]) \leq \inf_{\substack{[u]\in IE \\ [u]\geq \mathbf{0}}} s([a] \oplus [u]) = s_1([a]).$$

4. $s_1([a]) \leq s([a])$ for all $[a] \in IE$. Now using Corollary 11, it follows that there exists an *i-linear functional* $\ell : IE \to \mathbb{R}$ such that $\ell = f$ on IS and $\ell \leq s_1$ and therefore $\ell \leq s$ on IE.

To end the proof, let us remark that (ii) $\Leftrightarrow$ (iii) is obvious. □

11. Some examples

Firstly, notice that *an interesting example* of an i-subspace IS of an i-space IE, with $E=\mathbb{R}^{[0,1]}$ can be found in [3]:

$$IS=\left\{f=[\underline{f},\overline{f}]\in IE \,\middle|\, \begin{array}{l}\overline{f}\text{ is upper semi-continuous}\\ \underline{f}\text{ is lower semi-continuous}\end{array}\right\}.$$

Thus, IS is the set of all "S-continuous functions", an important class of functions in interval analysis; see for instance [3] and the references cited there.

Next, we will give *other examples* of i-spaces, i-subspaces, i-linear functionals, i-sublinear functionals. There are simple examples that we have constructed ad hoc.

Example 1. We consider the vector space $E=\mathbb{R}^2$ endowed with the usual algebraic operations and the *usual ordering.* Then E is an *ordered vector space* having the positive cone:

$$E_+=\{(x,y)\mid x\geq 0, y\geq 0\}.$$

- We also consider the *interval-space (i-space)* IE:

  $$IE=\left\{[a]=[\underline{a},\bar{a}]\mid \underline{a},\bar{a}\in\mathbb{R}^2\right\},$$

  the set of all "boxes" in plan. If $\underline{a}=(\underline{a}_1,\underline{a}_2)$ and $\bar{a}=(\bar{a}_1,\bar{a}_2)$, then $[a]=[\underline{a},\bar{a}]$ is the following "box":

  $$\left\{(x,y)\in\mathbb{R}^2\mid \underline{a}_1\leq x\leq\bar{a}_1, \underline{a}_2\leq y\leq\bar{a}_2\right\}=[\underline{a}_1,\bar{a}_1]\times[\underline{a}_2,\bar{a}_2].$$

- IE is endowed with the *usual algebraic operations*:
  1. If $[a]=[\underline{a},\bar{a}]=[(\underline{a}_1,\underline{a}_2),(\bar{a}_1,\bar{a}_2)]\in IE$ and
     $[b]=\left[\underline{b},\bar{b}\right]=\left[(\underline{b}_1,\underline{b}_2),(\bar{b}_1,\bar{b}_2)\right]\in IE$, then
     $[a]\oplus[b]=\left[\underline{a}+\underline{b},\bar{a}+\bar{b}\right]=\left[(\underline{a}_1+\underline{b}_1,\underline{a}_2+\underline{b}_2),(\bar{a}_1+\bar{b}_1,\bar{a}_2+\bar{b}_2)\right].$
  2. If $[a]$ is like in "1." and $\alpha\in\mathbb{R}$, then

     $$\alpha[a]=\begin{cases}\left[\alpha(\underline{a}+\underline{b}),\alpha\left(\bar{a}+\bar{b}\right)\right], & \text{if }\alpha\geq 0\\ \left[\alpha\left(\bar{a}+\bar{b}\right),\alpha(\underline{a}+\underline{b})\right], & \text{if }\alpha<0\end{cases}.$$

- As an *interval-subspace (i-subspace)* of IE, we consider the set IS of all *quadratic boxes* in $\mathbb{R}^2$. Hence, if $[a]=[\underline{a},\bar{a}]=[(\underline{a}_1,\underline{a}_2),(\bar{a}_1,\bar{a}_2)]$, then: $IS=\{[a]\in IE\mid \bar{a}_1-\underline{a}_1=\bar{a}_2-\underline{a}_2\}$.
- The *null set* of IE is the following set: $\mathcal{O}=\left\{[o]=[-a,a]\mid a\in\mathbb{R}^2_+\right\}$, that is, the set of all *symmetric rectangular boxes* in $\mathbb{R}^2$. Then, the *null part* $\mathcal{O}_{IS}$ of IS is the set of all symmetric quadratic boxes in $\mathbb{R}^2$.
- Now, we consider the following two real maps defined on IS:

  $$\ell([a])=\underline{a}_1+\bar{a}_1+\underline{a}_2+\bar{a}_2,\text{ and}$$
  $$p([a])=|\underline{a}_1+\bar{a}_1|+|\underline{a}_2+\bar{a}_2|$$

  where $[a]=[(\underline{a}_1,\underline{a}_2),(\bar{a}_1,\bar{a}_2)]$.

It is easy to prove that:

1. ℓ is an *interval-linear* (*i-linear*) functional on IS, that is,
   a) $\ell([a] \oplus [b]) = \ell([a]) + \ell([b])$, and
   b) $\ell(\alpha[a]) = \alpha\ell([a])$, for all $[a], [b] \in IS$ and $\alpha \in \mathbb{R}$;
2. p is an *interval-sublinear* (*i-sublinear*) functional on IS, that is,
   a) $p([a] \oplus [b]) \leq p([a]) + p([b])$,
   b) $p(\alpha[a]) = \alpha p([a])$ for all $[a] \in IS$ and $\alpha > 0$, and
   c) $p([a] \oplus [o]) = p([a])$, for all $[a] \in IS$ and $[o] \in \mathcal{O}_{IS}$.

- Also it is immediate that ℓ is *dominated* by p on IS, that is, $\ell([a]) \leq p([a])$ for all $[a] \in IS$ or, in short, $\ell \leq p$. Indeed, for any $[a] = [\underline{a}, \bar{a}]$, with $\underline{a} = (\underline{a}_1, \underline{a}_2)$ and $\bar{a} = (\bar{a}_1, \bar{a}_2)$ in $\mathbb{R}^2$, we have: $\ell([a]) = \underline{a}_1 + \underline{a}_2 + \bar{a}_1 + \bar{a}_2 \leq |\underline{a}_1 + \underline{a}_2 + \bar{a}_1 + \bar{a}_2| \leq |\underline{a}_1 + \bar{a}_1| + |\underline{a}_2 + \bar{a}_2| = p([a])$.

Example 2 (First example for the framework of our Mazur–Orlicz type theorem in the setting of interval-spaces – see Theorem 4).

- Let $E = \mathbb{R}^2$ endowed with the *usual algebraic* and *order structures* . As an i-subspace IS, consider even the i-space IE. Let also A be the following set: $A = \{(x_1, y_1, x_2, y_2) \in \mathbb{R}^4 \mid x_2 - x_1 = y_2 - y_1\}$.
- We consider the following maps:
  1. $s : IS \to \mathbb{R}$, $s([a]) = |\underline{a}_1 + \bar{a}_1| + |\underline{a}_2 + \bar{a}_2|$, where $[a] = [\underline{a}, \bar{a}] = [(\underline{a}_1, \underline{a}_2), (\bar{a}_1, \bar{a}_2)] \in IS$;
  2. $g : A \to IS$, $a \mapsto [g(a)]$, where $a = (x_1, y_1, x_2, y_2)$ and $[g(a)] = [(x_1, y_1), (x_2, y_2)]$;
  3. $f : A \to \mathbb{R}$, $f(x_1, y_1, x_2, y_2) = x_1 + y_1 + x_2 + y_2$.

  Then, s is an *i-sublinear functional* on IS – see the previous example.
- The statement (ii) in our Mazur–Orlicz type theorem (see Theorem 4 above) is satisfied, that is:

$$\sum_{j=1}^{n} \lambda_j f(a_j) \leq s\left(\bigoplus_{j=1}^{n} \lambda_j [g(a_j)]\right)$$

for all $n \in \mathbb{N}^*, a_1, \ldots, a_n \in A$, and $\lambda_1 \geq 0, \ldots, \lambda_n \geq 0$.

Indeed, we have for $a_j = (x_{j_1}, y_{j_1}, x_{j_2}, y_{j_2}) \in A, j = \overline{1, n}$:

$$\sum_{j=1}^{n} \lambda_j f(a_j) = \sum_{j=1}^{n} \lambda_j (x_{j_1} + y_{j_1} + x_{j_2} + y_{j_2}) \leq \left|\sum_{j=1}^{n} \lambda_j (x_{j_1} + y_{j_1} + x_{j_2} + y_{j_2})\right|$$

$$\leq \left|\sum_{j=1}^{n} \lambda_j (x_{j_1} + x_{j_2})\right| + \left|\sum_{j=1}^{n} \lambda_j (y_{j_1} + y_{j_2})\right| = s\left(\left[\sum_{j=1}^{n} \lambda_j (x_{j_1}, y_{j_1}), \sum_{j=1}^{n} \lambda_j (x_{j_2}, y_{j_2})\right]\right)$$

$$= s\left(\bigoplus_{j=1}^{n} \lambda_j [(x_{j_1}, y_{j_1}), (x_{j_2}, y_{j_2})]\right) = s\left(\bigoplus_{j=1}^{n} \lambda_j [g(a_j)]\right).$$

Example 3 (Second example for the framework of our Mazur–Orlicz type theorem in the setting of interval-spaces – see Theorem 4).

- Let $E = \mathbb{R}^{[0,1]}$ be the space of all real-valued functions on $[0,1]$, endowed with the *pointwise algebraic* and *order* structures. (Hence, $f \geq 0$ in E if and only if $f(t) \geq 0$ for each $t \in [0,1]$). $IE = \{[\underline{u}, \bar{u}] \mid \underline{u} \leq \bar{u} \text{ in } E\}$ is the set of all closed intervals in E.
- Define:

$$A = \mathbb{R}^{[0,1]}; \quad g : A \to IE,\ g(u) = [0, |u|];$$
$$f : A \to \mathbb{R},\ f(u) = u(0); \quad s : IE \to \mathbb{R},\ s([\underline{u}, \bar{u}]) = |\underline{u}(0) + \bar{u}(0)|.$$

Obviously, s is an i-sublinear functional on IE. Indeed, it is i-subadditive and i-positively homogeneous and, in addition, if $\mathcal{O}$ is the null set of IE and $[-v, v] \in \mathcal{O}$, then

$$\begin{aligned} s([\underline{u}, \bar{u}] \oplus [o]) &= s([[\underline{u} - v, \bar{u} + v]]) = |\underline{u}(0) - v(0) + \bar{u}(0) + v(0)| \\ &= |\underline{u}(0) + \bar{u}(0)| = s([\underline{u}, \bar{u}]) \text{ for al } [\underline{u}, \bar{u}] \in IE. \end{aligned}$$

Now we remark that for each $u \in A$, $f(u) \leq s([g(u)])$ (since $f(u) = u(0)$ and $s([g(u)]) = s([0, |u|]) = |u(0)|$).

- The statement (ii) in our Mazur–Orlicz type theorem (see Theorem 4 above) is satisfied, that is:

$$\sum_{j=1}^{n} \lambda_j f(u_j) \leq s\left(\bigoplus_{j=1}^{n} \lambda_j (g[u_j])\right)$$

for all $n \in \mathbb{N}^*, u_1, \ldots, u_n \in A$ and $\lambda_1 \geq 0, \ldots, \lambda_n \geq 0$.

Indeed:

$$\sum_{j=1}^{n} \lambda_j f(a_j) = \sum_{j=1}^{n} \lambda_j u_j(0),$$

and

$$s\left(\bigoplus_{j=1}^{n} \lambda_j [g(u_j)]\right) = s\left(\bigoplus_{j=1}^{n} \lambda_j [0, |u_j|]\right) = \sum_{j=1}^{n} \lambda_j |u_j(0)|.$$

Example 4 (for the framework of our Mazur–Orlicz type theorem in the setting of interval-spaces – see Theorem 10).

- We consider $E = \mathbb{R}^{[0,1]}$ and the convex set $K = E$.
- Define: $s : IE \to \mathbb{R}$ by $s([\underline{u}, \bar{u}]) = |\underline{u}(0) + \bar{u}(0)|$, $p : K \to IE$ by $p(u) = [0, |u|]$, and $q : K \to \mathbb{R}$ by $q(u) = -|u(0)|$. It follows that:
  1. s is an *i-sublinear functional* – see the previous example.

2. the map p is *i-convex-valued*. Indeed, let $u, v \in K$ and $\alpha \in (0,1)$. Then
$$[p(\alpha u + (1-\alpha)v)] = [0, |\alpha u + (1-\alpha)v|]$$
and
$$\begin{aligned}\alpha [p(u)] \oplus (1-\alpha)[p(v)] &= \alpha [0, |u|] \oplus (1-\alpha)[0, |v|] \\ &= [0, \alpha |u| + (1-\alpha)|v|].\end{aligned}$$
Now it is obvious that $|\alpha u + (1-\alpha)v| \le \alpha |u| + (1-\alpha)|v|$ and hence
$$[p(\alpha u + (1-\alpha)v)] \le \alpha [p(u)] \oplus (1-\alpha)[p(u)].$$
3. The map q is *concave*. Indeed, for $u, v \in K$ and $\alpha \in (0,1)$, it follows:
$$\begin{aligned}q(\alpha u + (1-\alpha)v) &= -|\alpha u(0) + (1-\alpha)v(0)| \\ \ge -\alpha |u(0)| - (1-\alpha)|v(0)| &= \alpha q(u) + (1-\alpha) q(v).\end{aligned}$$

- Statement (ii) in our Theorem 10, that is $q(u) \le s([p(u)])$ is satisfied, for all $u \in K$. Indeed, we have:
$$q(u) = -|u(0)| \quad \text{and} \quad s([p(u)]) = s([0, |u|]) = |u|(0) = |u(0)|.$$

References

[1] Alefeld, G.; Mayer, G.: *Interval analysis: Theory and applications*, Journal of Computational and Applied Mathematics **121**, 421–464 (2000).

[2] Anguelov, R.: *The algebraic structure of spaces of intervals-Contributions of Svetoslav Markov to interval analysis and its applications*, International Conference on Mathematical Methods and Models in Bioscience, 16–21 June, Sofia, Bulgaria, Biomath **2013**, Conference Book, 22–24 (2013).

[3] Anguelov, R.; Markov, S.; Sendov, B.: *The set of Hausdorff continuous functions – the largest linear space of interval functions*, Reliable Computing **12**, 337–363 (2006).

[4] Aseev, S.M.: *Quasilinear operators and their application in the theory of multivalued mappings*, Proceedings of the Steklov Institute of Mathematics **2**, 23–52 (1986).

[5] Casado, L.G.; Garcia, I.; Sergeyev, Ya.D.: *Interval algorithms for finding the minimal root in a set of multiextremal one-dimensional non-differentiable functions*, SIAM Journal of Scientific Computing **24** (2), 359–376 (2002).

[6] Danet, N., Danet, R.-M.: *Existence and extensions of positive linear operators*, Positivity **13**, 89–106 (2009).

[7] Feynman, R.R.; Leighton, R.R.; Sands, M.: *The Feynman Lectures on Physics*, Addison Wesley, Boston, Massachusetts 2005.

[8] Karmakar, S.; Bhunia, A.K.: *A comparative study of different order relations of intervals*, Reliable Computing **16**, 38–72 (2012).

[9] Kelley, J.L.; Namioka, I. and co-authors: *Linear Topological Spaces*, Princeton-Toronto-London-Melbourne, Van Nostrand 1963.

[10] Kosheleva, O.M.; Kreinovich, V.; Nguyen, H.T.; Bouchon-Meunier, B.: *How to describe partially ordered preferences mathematical foundations*, Proceedings of the Vietnam-Japan Bilateral Symposium on Fuzzy Systems and Applications VJ-FUZZYY **98**, Ha Long Bay, Vietnam, 269–278 (1998).

[11] König, H.: *On some basic theorems in convex analysis*, B. Korte Ed., North-Holland, Amsterdam-New York-Oxford, 107–144 (1982).

[12] Kronheimer, C.H.; Penrose, S.: *On the structure of causal spaces*, Mathematical Proceedings of the Cambridge Philosophical Society **63** (2), 481–501 (1967).

[13] Markov, S.: *Calculus for interval functions of real variable*, Computing **22**, 325–327 (1979).

[14] Markov, S.: *Extended interval arithmetic involving infinite intervals*, Mathematica Balkanica, New Series, Vol. **6**, Fasc. 3, 269–304 (1992).

[15] Markov, S.: *On the algebra of intervals and convex bodies*, Journal of Universal Computer Science **4**, 34–47 (1998).

[16] Markov, S.: *On the algebraic properties of convex bodies and some applications*, Journal of Convex Analysis **7**, 129–166 (2000).

[17] Markov, S.: *On the algebraic properties of intervals and some applications*, Reliable Computing **7**, 113–127 (2001).

[18] Markov, S.: *On quasivector spaces of convex bodies and zonotopes*, Numerical Algorithms **37**, 263–274 (2004).

[19] Markov, S.: *On quasilinear spaces of convex bodies and intervals*, Journal of Computational and Applied Mathematics **162**, 93–112 (2004).

[20] Markov, S.: *Quasilinear spaces and their relation to vector spaces*, Electronic Journal on Mathematics of Computation, 2-1 (2005).

[21] Mazur, S.; Orlicz, W.: *Sur les espaces métrique linéaires II*, Studia Mathematica **13**, 137–179 (1953).

[22] Mayer, O.: *Über Eigenschaften von Intervallprozessen*, ZAAM **48**, T86-T87 (1968).

[23] Mayer, O.: *Algebraische und metrische Strukturen in der Intervallrechnung* und einige Auwendugen, Computing **5**, 144–162 (1970).

[24] Misner, C.W.; Thorne, K.S.; Wheeler, J.A.: *Gravitation*, W.H. Freeman and company, San Francisco 1973.

[25] Moore, R.E.: *Methods and Applications of Interval Analysis*, SIAM studies in Applied Mathematics, SIAM, Philadelphia, Pennsylvania 1979.

[26] Moore, R.E.; Lodwich, W.: *Interval analysis and fuzzy set theory*, Fuzzy sets and systems **135** **1**, 5–9 (2003).

[27] Ratschek, H.; Schröder, G.: *Representation semigroups as systems of compact convex sets*, Proceedings of the American Mathematical Society **65**, 24–28 (1977).

[28] Rudin, W.: *Functional Analysis*, McGraw-Hill Book Company New York 1973.

[29] Simons, S.: *Minimal sublinear functionals*, Studia Mathematica **37**, 37–56 (1970).

[30] Simons, S.: *A new version of the Hahn–Banach theorem*, Archiv der Mathematik **80**, 630–646 (2003).

[31] Tanenbaum, P.J.; de la Mora, C.; Wojciechowski, P.; Kosheleva, O.; Kreinovich, V.; Starks, S.A.; Kuzminykh, A.: *Robust methodology for characterizing system response to damage; approach based on partial order*, in: Lirkov, I.; Margenov, S.; Wasniewski,

J.; Yalamov, P. (eds.), Proceedings of the 4th International Conference (LSSC), Sozopol, Bulgaria, June 4–8, 2003. Large-Scale scientific Computing, Springer Lecture Notes in Computer Science **2004**, 276–283 (2004).

[32] Trotter, W.T.: *New perspectives on interval orders and interval graphs*, in Surveys in Combinatorises, (16th British Combinatorial Conference), London Mathematical Society, Lecture Notes series 241, Cambridge University Press 1997.

[33] Xu, Y.; Liu, J.; Zhong, X.; Chen, S.: *Lattice-valued matrix game with mixed strategies for intelligent decision support*, Journal of Knowledge-Based Systems **32**, 56–64 (2012).

[34] Yilmaz, Y.; Çakan, S.; Aytekin, S.: *Topological quasilinear spaces*, Abstract and Applied Analysis **2012**, Article ID 951374, 10 pages (2012).

[35] Zapata, F.; Kreinovich, V.: *Reconstructing an open order from its closure, with applications to space-time physics and to logic*, Studia Logica **100** (1-2), 419–435 (2012).

[36] Zapata, F.; Kreinovich, V.; Joslyn, C.; Hogan, E.: *Orders on intervals over partially ordered sets. Extending Allen's Algebra and Interval Graf Results*, Soft Computing **17**, 137–139 (2013).

Rodica-Mihaela Dăneţ
Department of Mathematics and Computer Science
Technical University of Civil Engineering of Bucharest
124, Lacul Tei Blvd.
036296 Bucharest, Romania

Ordered Structures and Applications: Positivity VII
Trends in Mathematics, 161–170

Narrow Operators on Lattice-normed Spaces and Vector Measures

Dina Dzadzaeva and Marat Pliev

Dedicated to the memory of Adriaan Cornelis Zaanen,
one of the great founders of the theory of vector lattices

Abstract. We consider linear narrow operators on lattice-normed spaces. We prove that, under mild assumptions, every finite rank linear operator is strictly narrow (before it was known that such operators are narrow). Then we show that every dominated, order-continuous linear operator from a lattice-normed space over atomless vector lattice to an atomic lattice-normed space is order narrow.

Mathematics Subject Classification (2010). Primary 46B99; Secondary 46G12.

Keywords. Narrow operator, lattice-normed space, Banach space with mixed norm, vector lattice, vector measure.

1. Introduction

Today the theory of narrow operators is a growing and active field of Functional Analysis (see the recent monograph [18]). Plichko and Popov were the first [15] who systematically studied this class of operators. Later many authors have studied linear and nonlinear narrow operators in functional spaces and vector lattices [3, 4, 12, 14, 17]. In the article [16] the second named author have considered a general lattice-normed space approach to narrow operators. Recently it became clear that a technique of vector measures is relevant to narrow operators [13]. The aim of this article is to use this new technique for investigation of linear narrow operators on lattice-normed spaces.

This work was completed with the support of the Russian Foundation for Basic Research, grant number 14-01-91339.

2. Preliminaries

The goal of this section is to introduce some basic definitions and facts. General information on vector lattices, Banach spaces and lattice-normed spaces can be found in the books [1, 2, 7, 11].

Consider a vector space V and a real archimedean vector lattice E. A map $|\cdot| : V \to E$ is a *vector norm* if it satisfies the following axioms:

1) $|v| \geq 0$; $|v| = 0 \Leftrightarrow v = 0$; $(\forall v \in V)$.
2) $|v_1 + v_2| \leq |v_1| + |v_2|$; $(v_1, v_2 \in V)$.
3) $|\lambda v| = |\lambda| \, |v|$; $(\lambda \in \mathbb{R}, v \in V)$.

A vector norm is called *decomposable* if

4) for all $e_1, e_2 \in E_+$ and $x \in V$ with $|x| = e_1 + e_2$ there exist $x_1, x_2 \in V$ such that $x = x_1 + x_2$ and $|x_k| = e_k$, $(k := 1, 2)$.

A triple $(V, |\cdot|, E)$ (in brief $(V, E), (V, |\cdot|)$ or V with default parameters omitted) is a *lattice-normed space* if $|\cdot|$ is a E-valued vector norm in the vector space V. If the norm $|\cdot|$ is decomposable then the space V itself is called decomposable. A subspace V_0 of V is called a (bo)-ideal of V if for $v \in V$ and $u \in V_0$, from $|v| \leq |u|$ it follows that $v \in V_0$. We say that a net $(v_\alpha)_{\alpha \in \Delta}$ *(bo)-converges* to an element $v \in V$ and write $v = bo\text{-}\lim v_\alpha$ if there exists a decreasing net $(e_\gamma)_{\gamma \in \Gamma}$ in E^+ such that $\inf_{\gamma \in \Gamma}(e_\gamma) = 0$ and for every $\gamma \in \Gamma$ there is an index $\alpha(\gamma) \in \Delta$ such that $|v - v_{\alpha(\gamma)}| \leq e_\gamma$ for all $\alpha \geq \alpha(\gamma)$. A net $(v_\alpha)_{\alpha \in \Delta}$ is called *(bo)-fundamental* if the net $(v_\alpha - v_\beta)_{(\alpha, \beta) \in \Delta \times \Delta}$ (bo)-converges to zero. A lattice-normed space is called *(bo)-complete* if every (bo)-fundamental net (bo)-converges to an element of this space. Every decomposable (bo)-complete lattice-normed space is called a *Banach–Kantorovich space* (a BKS for short).

Let V be a lattice-normed space and $y, x \in V$. If $|x| \wedge |y| = 0$ then we call the elements x, y *disjoint* and write $x \perp y$. As in the case of a vector lattice, a set of the form $M^\perp = \{v \in V : (\forall u \in M) u \perp v\}$, with $\emptyset \neq M \subset V$, is called a *band.* The equality $x = \coprod_{i=1}^n x_i$ means that $x = \sum_{i=1}^n x_i$ and $x_i \perp x_j$ if $i \neq j$. An element $z \in V$ is called a *component* or a *fragment* of $x \in V$ if $z \perp (x - z)$. Two fragments x_1, x_2 of x are called *mutually complemented* or *MC*, in short, if $x = x_1 + x_2$. The notations $z \sqsubseteq x$ means that z is a fragment of x. The set of all fragments of the element $v \in V$ is denoted by $\mathfrak{F}_v$. Following ([2], p. 111) an element $e > 0$ of a vector lattice E is called an *atom*, whenever $0 \leq f_1 \leq e$, $0 \leq f_2 \leq e$ and $f_1 \perp f_2$ imply that either $f_1 = 0$ or $f_2 = 0$. A vector lattice E is atomless if there is no atom $e \in E$.

Remark that vector lattices and normed spaces are simple examples of lattice normed spaces. Indeed, if $V = E$ then the modules of an element can be taken as its lattice norm: $|v| := |v| = v \vee (-v)$; $v \in E$. Decomposability of this norm easily follows from the Riesz Decomposition Property holding in every vector lattice. If $E = \mathbb{R}$ then V is a normed space. A reader can find many nontrivial examples of lattice-normed spaces in the book [7].

Let E be a Banach lattice and let (V, E) be a lattice-normed space. Since $|x| \in E_+$ for every $x \in V$ we can define a *mixed norm* in V by the formula

$$|||x||| := \| \, |x| \, \| \quad (\forall x \in V).$$

The normed space $(V, ||| \cdot |||)$ is called a *space with a mixed norm.* In view of the inequality $| \, |x| - |y| \, | \leq |x - y|$ and monotonicity of the norm in E, we have

$$\| \, |x| - |y| \, \| \leq |||x - y||| \quad (\forall x, y \in V),$$

so a vector norm is a norm continuous operator from $(V, ||| \cdot |||)$ to E. A lattice-normed space (V, E) is called a *Banach space with a mixed norm* if the normed space $(V, ||| \cdot |||)$ is complete with respect to the norm convergence.

Consider lattice-normed spaces (V, E) and (W, F), a linear operator $T : V \to W$ and a positive operator $S \in L_+(E, F)$. If the condition

$$|Tv| \leq S \, |v| \, ; \, (\forall v \in V)$$

is satisfied then we say that S *dominates* or *majorizes* T or that S is *dominant* or majorant for T. In this case T is called a *dominated* or *majorizable* operator. The set of all dominants of the operator T is denoted by maj(T). If there is the least element in maj(T) with respect to the order induced by $L_+(E, F)$ then it is called the *least* or the *exact dominant* of T and it is denoted by $|T|$.

We follow [16] in the next definition.

Definition 2.1. Let (V, E) be an lattice-normed space over an atomless vector lattice E and X a vector space. A linear operator $T : V \to X$ is called:

- *strictly narrow* if for every $v \in V$ there exists a decomposition $v = v_1 \sqcup v_2$ of v such that $T(v_1) = T(v_2)$;
- *narrow* if X is a normed space, and for every $v \in V$ and every $\varepsilon > 0$ there exists a decomposition $v = v_1 \sqcup v_2$ of v such that $\|T(v_1 - v_2)\| < \varepsilon$;
- *order-narrow* if X is a Banach space with a mixed norm, and for every $v \in V$ there exists a net of decompositions $v = v_\alpha^1 \sqcup v_\alpha^2$ such that $T(v_\alpha^1 - v_\alpha^2) \xrightarrow{(bo)} 0$.

A linear operator T from a lattice-normed space V to a Banach space X is called:

- *order-to-norm* σ-continuous if T sends (bo)-convergent sequences in V to norm convergent sequences in X;
- *order-to-norm* continuous provided T sends (bo)-convergent nets in V to norm convergent nets in X.

Necessary information on Boolean algebras can be found, for instance, in [5], [7], [11]. The most common example of a Boolean algebra is an algebra $\mathcal{A}$ of subsets of a set Ω, that is, a subset of the set $\mathcal{P}(\Omega)$ of all subsets of Ω, closed under the union, intersection and complementation and containing $\emptyset$ and Ω. The Boolean operations on $\mathcal{A}$ are $A \vee B = A \cup B$; $A \wedge B = A \cap B$ and $\neg A = \Omega \setminus A$, and the constants are $\mathbf{0} = \emptyset$, $\mathbf{1} = \Omega$.

A map $h : \mathcal{A} \to \mathcal{B}$ between two Boolean algebras is called a *Boolean homomorphism* if the following conditions hold for all $x, y \in \mathcal{A}$:

1. $h(\mathbf{0}) = \mathbf{0}$;
2. $h(\mathbf{1}) = \mathbf{1}$;
3. $h(x \vee y) = h(x) \vee h(y)$;
4. $h(x \wedge y) = h(x) \wedge h(y)$;
5. $h(\neg x) = \neg h(x)$.

A bijective Boolean homomorphism which is called a *Boolean isomorphism.* Two Boolean algebras $\mathcal{A}$ and $\mathcal{B}$ are called *Boolean isomorphic* if there is a Boolean isomorphism $h : \mathcal{A} \to \mathcal{B}$. The following remarkable result is known as the Stone representation theorem.

Theorem 2.2 ([5, **Theorem 7.11**]). *Every Boolean algebra is Boolean isomorphic to an algebra of subsets of some set.*

Every Boolean algebra $\mathcal{A}$ is a partially ordered set with respect to the partial order "$x \leq y$ if and only if $x \wedge y = x$", with respect to which $\mathbf{0}$ is the least element, $\mathbf{1}$ is the greatest element, $x \wedge y$ is the infimum and $x \vee y$ the supremum of the two-point set $\{x, y\}$ in $\mathcal{A}$. A Boolean algebra $\mathcal{A}$ is called Dedekind complete (resp., σ-Dedekind complete) if so is $\mathcal{A}$ as a partially ordered set, that is, if every (resp., countable) order-bounded nonempty subset of $\mathcal{A}$ has the least upper and the greatest lower bounds in $\mathcal{A}$. Obviously, a Boolean algebra is σ-Dedekind complete if and only if it is a σ-algebra. A Boolean algebra A can be viewed as an algebra over field $\mathbb{Z}_2 := \{\mathbf{0}, \mathbf{1}\}$ with respect of an algebraic operations:

$$x + y := x \triangle y,\ xy := x \wedge y\ (x, y \in A),$$

where $x \triangle y := (x \wedge y^\star) \vee (x^\star \wedge y)$ is a *symmetric difference* of x and y.

There is a natural connection between Boolean algebras and lattice-normed spaces. Let (V, E) be a lattice-normed space. Given $L \subset E$ and $M \subset V$, we let by definition $h(L) = \{v \in V : |v| \in L\}$ and $|M| = \{|v| : v \in M\}$. It is clear that $|h(L)| \subset L \bigcap |V|$.

Proposition 2.3 ([7, **Proposition 2.1.2**]). *Suppose that every band of vector lattice $E_0 = |V|^{\perp\perp}$ contains the norm of some nonzero element. Then the set of all bands of the lattice-normed space V is a complete Boolean algebra and the map $L \mapsto h(L)$ is an isomorphism of the Boolean algebras of bands of E_0 and V.*

All lattice-normed spaces considered below are decomposable and satisfy Proposition 2.3.

3. Vector measures on Boolean algebras

By a *measure* on a Boolean algebra $\mathcal{A}$ we mean a finitely additive function $\mu : \mathcal{A} \to X$ of $\mathcal{A}$ to a vector space X, that is, a map satisfying

$$(\forall x, y \in \mathcal{A})\Big(\big(x \wedge y = \mathbf{0}\big) \Rightarrow \big(\mu(x + y) = \mu(x) + \mu(y)\big)\Big).$$

If, moreover, $\mathcal{A}$ is a Boolean σ-algebra and X is a topological vector space then a *σ-additive measure* is a measure $\mu : \mathcal{A} \to X$ possessing the property that if $(x_n)_{n=1}^{\infty}$ is a sequence in $\mathcal{A}$ with $x_n \uparrow x \in \mathcal{A}$ then $\lim_{n\to\infty} \mu(x_n) = \mu(x)$.

3.1. Definitions and simple properties

We follow [13] in the definitions below.

Definition 3.1. Let $\mathcal{A}$ be a Boolean algebra and X a normed space. A measure $\mu : \mathcal{A} \to X$ is called *almost dividing* if for every $x \in \mathcal{A}$ and every $\varepsilon > 0$ there is a decomposition $x = y \sqcup z$ with $\|\mu(y) - \mu(z)\| < \varepsilon$.

Definition 3.2. Let $\mathcal{A}$ be a Boolean algebra and V a lattice-normed space. A measure $\mu : \mathcal{A} \to V$ is called *order dividing* if for every $x \in \mathcal{A}$ there is a net of decompositions $x = y_\alpha \sqcup z_\alpha$ with $\big(\mu(y_\alpha) - \mu(z_\alpha)\big) \xrightarrow{(bo)} 0$.

Definition 3.3. Let $\mathcal{A}$ be a Boolean algebra and X a vector space. A measure $\mu : \mathcal{A} \to X$ is called *dividing* if for every $x \in \mathcal{A}$ there is a decomposition $x = y \sqcup z$ with $\mu(y) = \mu(z)$.

Definition 3.4. Let V be a lattice-normed space, X a vector space. To every linear operator $T : V \to X$ we associate a family of measures $(\mu_v^T)_{v\in V}$ as follows. Given any $v \in V$, we define a measure $\mu_v^T : \mathfrak{F}_v \to X$ on the Boolean algebra $\mathfrak{F}_v$ of fragments of v by setting $\mu_v^T u = T(u)$, μ_v^T is called the *associated measure* of T at v.

The next proposition directly follows from the definitions.

Proposition 3.5. *Let (V, E) be a lattice-normed space, E be an atomless vector lattice, X a vector space and $T : V \to X$ a linear operator. Then the following assertions hold.*

1. *T is strictly narrow if and only if for every $v \in V$ the measure μ_v^T is dividing.*
2. *Let X be a normed space. Then T is narrow if and only if for every $v \in V$ the measure μ_v^T is almost dividing.*
3. *Let X be a lattice-normed space. Then T is order narrow if and only if for every $v \in V$ the measure μ_v^T is order dividing.*

Obviously, a dividing measure is both almost dividing and order dividing, for an appropriate range space. The following three propositions are close to propositions 10.7 and 10.9, and Example 10.8 from [18].

Proposition 3.6. *Let $\mathcal{A}$ be a Boolean algebra and W a Banach space with a mixed norm. Then every almost dividing measure $\mu : \mathcal{A} \to W$ is order dividing.*

Proof. Let $\mu : \mathcal{A} \to W$ be an almost dividing measure and $x \in \mathcal{A}$. Choose a sequence of decompositions $x = y_n \sqcup z_n$ with $|||\mu(y_n) - \mu(z_n)||| \le 2^{-n}$. Then for $u_n = \sum_{k=n}^{\infty} \big|\mu(y_n) - \mu(z_n)\big|$ one has $\big|\mu(y_n) - \mu(z_n)\big| \le u_n \downarrow 0$. Hence, $\big(\mu(y_n) - \mu(z_n)\big) \xrightarrow{(bo)} 0$. □

Proposition 3.7. *Let Σ be the Boolean σ-algebra of Lebesgue measurable subsets of $[0,1]$. Then there exists an order-dividing measure $\mu : \Sigma \to L_\infty$ which is not dividing.*

Proof. Use Proposition 3.5 and [18, Example 10.8]. □

Proposition 3.8. *Let $\mathcal{A}$ be a Boolean algebra and (W,F) a Banach space with a mixed norm, where F is an order-continuous Banach lattice. Then a measure $\mu : \mathcal{A} \to W$ is order dividing if and only if it is almost dividing.*

Proof. Let $\mu : \mathcal{A} \to W$ be order dividing. Given any $x \in \mathcal{A}$, let $x = y_\alpha \sqcup z_\alpha$ be a net of decompositions with $\big(\mu(y_\alpha) - \mu(z_\alpha)\big) \xrightarrow{(bo)} 0$. By the order-continuity of F, $\big\| \big| \mu(y_\alpha) - \mu(z_\alpha) \big| \big\| \to 0$, and hence, μ is almost dividing by arbitrariness of $x \in \mathcal{A}$. By Proposition 3.6, the proof is completed. □

A nonzero element u of a Boolean algebra $\mathcal{A}$ is called an *atom* if for every $x \in \mathcal{A}$ the condition $0 < x \leq u$ implies that $x = u$. Every dividing (of any type) measure sends atoms to zero.

Proposition 3.9. *Let $\mathcal{A}$ be a Boolean algebra and V a vector space (a normed space, or a lattice-normed space) and $\mu : \mathcal{A} \to V$ a dividing (an almost dividing or an order-dividing, respectively) measure. If $a \in \mathcal{A}$ is an atom then $\mu(a) = 0$.*

The proof is an easy exercise.

3.2. The range convexity of vector measures

We need the following remarkable result known as the Lyapunov[1] convexity theorem.

Theorem 3.10 ([10, Theorem 2, p.9]). *Let (Ω, Σ) be a measurable space, X a finite-dimensional normed space and $\mu : \Sigma \to X$ an atomless σ-additive measure. Then the range $\mu(\Sigma) = \{\mu(A) : A \in \Sigma\}$ of μ is a compact convex subset of X.*

The following theorem was proven in [13], but for sake of completeness we include the proof here.

Theorem 3.11. *Let $\mathcal{A}$ be a Boolean σ-algebra and X a finite-dimensional vector space. Then every atomless σ-additive measure $\mu : \mathcal{A} \to X$ is dividing.*

For a Boolean σ-algebra $\mathcal{A}$ and $x \in \mathcal{A} \setminus \{0\}$ by $\mathcal{A}_x$ we denote the Boolean σ-algebra $\{y \in \mathcal{A} : y \leq x\}$ with the unit $\mathbf{1}_{\mathcal{A}_x} = x$ and the operations induced by $\mathcal{A}$.

Proof. Let $\mu : \mathcal{A} \to X$ be an atomless σ-additive measure and $x \in \mathcal{A}$. If $x = 0$ then there is nothing to prove. Let $x \neq 0$. Then the restriction $\mu_x = \mu|_{\mathcal{A}_x} : \mathcal{A}_x \to X$ is an atomless σ-additive measure. By Theorem 2.2, $\mathcal{A}_x$ is Boolean isomorphic to some measurable space (Ω, Σ) by means of some Boolean isomorphism $J : \mathcal{A}_x \to \Sigma$. Since $\mathcal{A}_x$ is a Boolean σ-algebra, Σ is a σ-algebra. Then the map $\nu : \Sigma \to X$ given by $\nu(A) = \mu\big(J^{-1}(A)\big)$ for all $A \in \Sigma$, is an atomless σ-additive measure.

[1]= Lyapounoff, the old spelling

By Theorem 3.10, the range $\nu(\Sigma)$ of ν is a convex subset of X. In particular, since $0, \nu(J(x)) \in \nu(\Sigma)$, we have that $\nu(J(x))/2 \in \nu(\Sigma)$. Let $B \in \Sigma$ be such that $\nu(B) = \nu(J(x))/2 = \mu(x)/2$. Then for $y = J^{-1}(B)$ one has that $y \leq x$ and $\mu(y) = \nu(B) = \mu(x)/2$. Thus, for $z = x \wedge \neg y$ one has $x = y \sqcup z$ and $\mu(z) = \mu(x) - \mu(y) = \mu(x)/2 = \mu(y)$. □

3.3. Strict narrowness of order-continuous finite rank operators

The following theorem is the main result of this subsection. This assertion strengthens Theorem 4.12 from [16]. Using a method based on the Lyapunov theorem, we prove the strict narrowness of an operator.

Theorem 3.12. *Let (V, E) be a lattice-normed space, E an atomless vector lattice with the principal projection property, X a finite-dimensional normed space (resp., a lattice-normed space). Then every σ-order-to-norm continuous (resp., σ-order-continuous) linear operator $T : V \to X$ is strictly narrow.*

To use the technique of dividing vector measures, we preliminarily need the σ-additivity of a measure.

Lemma 3.13. *Let (V, E) be a lattice-normed space, E an atomless vector lattice with the principal projection property, X a normed space (resp., a lattice-normed space), $T : V \to X$ an order-to-norm continuous (resp., an order-continuous) linear operator. Then for every $v \in V$ the associated measure μ_v^T is atomless and σ-additive.*

Proof of Lemma 3.13. Fix any $v \in V$. The σ-additivity of μ_v^T directly follows from the order-continuity. We show that μ_v^T is atomless. Assume $v_0 \in \mathfrak{F}_v$ and $\mu_v^T(v_0) \neq 0$, that is, $T(v_0) \neq 0$. Set $Z = \{u \in \mathfrak{F}_{x_0} : T(u) = 0\}$. By the order-continuity and Zorn's lemma, Z has a maximal element $z \in Z$. Since $T(z) = 0$, one has that $T(v_0 - z) = T(z) + T(v_0 - z) = T(v_0) \neq 0$. Since E is atomless, we split $v_0 - z = w_1 \sqcup w_2$ with $w_1, w_2 \in \mathfrak{F}_{v_0} \setminus \{0\}$. By maximality of z, $T(w_1) \neq 0$ and $T(w_2) \neq 0$. Thus, $v_0 = (z + w_1) \sqcup w_2$ is a decomposition with $\mu_v^T(z + w_1) = \mu_v^T(w_2) \neq 0$ and $\mu_e^T(w_1) \neq 0$. □

Proof of Theorem 3.12. Let $v \in V$. By Lemma 3.13, the associated measure $\mu_v^T : \mathfrak{F}_v \to X$ is atomless and σ-additive. By Theorem 3.11, μ_v^T is dividing. So, we split $v = v_1 \sqcup v_2$ with $\mu_v^T(v_1) = \mu_v^T(v_2)$, that is, $T(v_1) = T(v_2)$. □

4. Operators from arbitrary to atomic lattice-normed spaces are order narrow

Definition 4.1. An element u of a lattice-normed space (V, E) is called an *atom*, whenever $0 \leq |v_1| \leq |u|$, $0 \leq |v_2| \leq |u|$ and $v_1 \perp v_2$ imply that either $v_1 = 0$ or $v_2 = 0$.

It is clear that $u \perp v$ for every different atoms $u, v \in V$.

Definition 4.2. A lattice-normed space V is said to be *atomic* if there is a collection $(u_i)_{i\in I}$ of atoms in V, called a *generating collection of atoms*, such that $u_i \bot u_j$ for $i \neq j$ and for every $v \in V$ if $|v| \wedge |u_i| = 0$ for each $i \in I$ then $v = 0$.

It follows from the definition that the (bo)-ideal V_0 generated by generating collection of atoms coincides with V.

Example. Let X be a Banach space and consider the lattice-normed space (V, E), where $E = c_0$ or $E = l^p$; $0 < p \leq \infty$ and

$$V = \{(x_n)_{n=1}^{\infty};\ x_n \in X : (\|x_n\|)_{n=1}^{\infty} \in E\}.$$

Elements $\{(0, \dots, x_i, \dots, 0, \dots) : x_i \in X;\ i \in \mathbb{N}\}$ are atoms and the space V is an atomic lattice-normed space.

Proposition 4.3. *Let (V, E) be a lattice-normed space such that every band of the vector lattice E contains the vector norm of some nonzero element. Then V is atomic if and only if the vector lattice E is atomic.*

Proof. Let V be atomic. We shall prove that E is atomic too. First we prove that a norm of an arbitrary atom in V is also an atom on E. Assume that $v \in V$ is an atom, $e = |v| \in E^+$ and there are elements f_1, f_2, such that $0 < f_1 \leq e$, $0 < f_2 \leq e$ and $f_1 \bot f_2$. Then the band $F_1 = \{f_1\}^{\bot\bot}$ is disjoint to the band $F_2 = \{f_2\}^{\bot\bot}$ and by our assumption there exist two nonzero elements $v_1, v_2 \in V$, such that $|v_1| \in F_1$ and $|v_2| \in F_2$. Then we may write

$$|v_i| = |v_i| \wedge f_i + e_i;\ e_i \geq 0;\ i \in \{1, 2\}.$$

By decomposability of the vector norm there exist $w_1, w_2 \in V$, so that $0 \leq |w_i| = |v_i| \wedge f_i \leq e = |v|$, $i \in \{1, 2\}$ and $w_1 \bot w_2$. But this is a contradiction and therefore $|v|$ is an atom in E. Let $e \in E^+$ and $e \bot |v|$ for every atom $v \in V$. Then there exists a nonzero element $h \in V$, such that $|h| \in \{e\}^{\bot\bot}$. Hence $e = 0$ and E is an atomic vector lattice. The converse assertion is obvious. □

The following theorem is the second main result of the our paper.

Theorem 4.4. *Let $(V, E), (W, F)$ be lattice-normed spaces, where W is atomic, E, F are vector lattices with the principal projection property and E is atomless. Then every dominated, order-continuous linear operator $T : V \to W$ is order narrow.*

For the proof we need an auxiliary result that gives a representation of an element of an atomic lattice-normed space via atoms. Let (W, F) be an atomic lattice-normed space with a generating collection of atoms $(u_i)_{i\in I}$ and F be a vector lattice with the principal projection property. Let Λ denote the directed set of all finite subsets of I ordered by inclusion, that is, $\alpha \leq \beta$ for $\alpha, \beta \in \Lambda$ if and only if $\alpha \subseteq \beta$. For every $\alpha \in \Lambda$ we set

$$\mathbf{P}_\alpha = \sum_{i\in\alpha} P_{u_i}, \tag{4.1}$$

where P_{u_i} is the band projection of W onto the band generated by the element u_i. It is immediate that P_{u_i} is a band projection of F onto the band generated by the element $|u_i|$ and $\mathbf{P}_\alpha$ is the band projection of W onto the band generated by $\{u_i : i \in \alpha\}$.

Proposition 4.5. *Let (W, F) be an atomic lattice-normed space, F be a vector lattice with the principal projection property and $(u_i)_{i\in I} \subset W$ be a generating collection of atoms. If $g \in W$ then $P_{u_i} g = a_i$ for every $i \in I$ and some $a_i \in \mathbb{R}$.*

Proof of Proposition 4.5. (1) Let $g \in W$. Then there exists a finite collection of atoms $u_1, \dots, u_n$, such that $|g| \leq \sum_{i=1}^{n} \lambda_i |u_i|$, $\lambda_i \in \mathbb{R}_+$, $i \in \{1, \dots, n\}$. By decomposability of the vector norm there exist $g_1, \dots, g_n$ in W, such that

$$g = \sum_{i=1}^{n} g_i; \ |g_i| \leq \lambda_i |u_i| ; \ i \in \{1, \dots, n\}.$$

Moreover $g_i = a_i u_i$ for some $a_i \in \mathbb{R}$ and every $i \in \{1, \dots, n\}$, $|g_i| \perp |g_j|$, $j \neq i$. Therefore $P_{u_i} g = a_i u_i$. □

Proof of Theorem 4.4. Let $T : V \to W$ be a dominated operator. Fix any $v \in V$ and $e = |v| \in E^+$. Since the set $\mathfrak{F}_e$ of all fragments of e is order bounded in E, its image $|T|(\mathfrak{F}_e)$ is order bounded in F, say, $|Tx| \leq |T| \, |x| \leq f$ for some $f \in F^+$ and all $x \sqsubseteq v$. Let $(u_i)_{i\in I}$ be a generating collection of atoms of W, Λ the directed set of all finite subsets of I ordered by inclusion, and $(\mathbf{P}_\alpha)_{\alpha\in\Lambda}$ the net of band projections of F defined by (4.1). By Proposition 4.5, $\mathbf{P}_\alpha$ is a finite rank operator for every $\alpha \in \Lambda$. Being a band projection, $\mathbf{P}_\alpha$ is order continuous. Then for each $\alpha \in \Lambda$ the composition operator $S_\alpha = \mathbf{P}_\alpha \circ T$ is a finite rank dominated, order-continuous operator which is strictly narrow by Theorem 3.12. So, for each $\alpha \in \Lambda$ we choose a decomposition $v = v'_\alpha \sqcup v''_\alpha$ with $S_\alpha(v'_\alpha) = S_\alpha(v''_\alpha)$. Then

$$\begin{aligned} |T(v'_\alpha) - T(v''_\alpha)| &= |(I - \mathbf{P}_\alpha) \circ T(v'_\alpha) - (I - \mathbf{P}_\alpha) \circ T(v''_\alpha)| \\ &\leq |(I - \mathbf{P}_\alpha) \circ T(v'_\alpha)| + |(I - \mathbf{P}_\alpha) \circ T(v''_\alpha)| \\ &\leq (I - \mathbf{P}_\alpha) \, |T(v'_\alpha)| + (I - \mathbf{P}_\alpha) \, |T(v''_\alpha)| \\ &\leq 2 \, (I - \mathbf{P}_\alpha)(f) \xrightarrow{(o)} 0. \end{aligned}$$

□

Acknowledgement

The authors are very grateful to the referee for his (her) valuable remarks and suggestions and Helen Basaeva for applying her wonderful expertise of TEX to the final preparation of the text.

References

[1] Yu.A. Abramovich, C.D. Aliprantis, *An invitation to operator theory*, Graduate Studies in Math., Amer. Math. Soc., Providence, Rhode Island. 50, 2002.

[2] C.D. Aliprantis, O. Burkinshaw, *Positive Operators*, Springer, Dordrecht, 2006.

[3] D. Bilik, V. Kadets, R. Shvidkov, D. Werner, *Narrow operators and the Daugavet property for ultraproducts*, Positivity, (9), 2005. 45–62.

[4] J. Flores C. Ruiz, *Domination by Positive Narrow Operators*, Positivity, 7, (17), 2003. 303–321.

[5] Th. Jech, *Set Theory*, Springer, Berlin, 2003.

[6] V.M. Kadets, M.M. Popov, *On the Lyapunov convexity theorem with applications to sign-embeddings*, Ukraïn. Mat. Zh., 44, No 9 (1992), 1192–1200.

[7] A.G. Kusraev, *Dominated Operators*, Kluwer Acad. Publ., Dordrecht–Boston–London, 2000.

[8] H.E. Lacey, *The Isometrical Theory of Classical Banach Spaces*, Springer-Verlag, Berlin–Heidelberg–New York, 1974.

[9] J. Lindenstrauss, L. Tzafriri, *Classical Banach spaces, Vol.* 1, *Sequence spaces*, Springer-Verlag, Berlin–Heidelberg–New York, 1977.

[10] J. Lindenstrauss, L. Tzafriri, *Classical Banach spaces, Vol.* 2, *Function spaces*, Springer-Verlag, Berlin–Heidelberg–New York, 1979.

[11] W.A.J. Luxemburg, A.C. Zaanen, *Riesz Spaces. Vol.* 1, North-Holland Publ. Comp., Amsterdam–London, 1971.

[12] O.V. Maslyuchenko, V.V. Mykhaylyuk, M.M. Popov, *A lattice approach to narrow operators*, Positivity 13 (2009), 459–495.

[13] V. Mykhaylyuk, M. Pliev, M. Popov, O. Sobchuk, *Dividing measures and narrow operators*, Studia Math. 231 (2015), 97–116.

[14] V. Mykhaylyuk, M. Popov, B. Randrianantoanina, G. Schechtman, *On narrow and ℓ_2-strictly singular operators on L_p*, Isr. J. Math. 203 (2014), 81–108.

[15] A.M. Plichko, M.M. Popov, *Symmetric function spaces on atomless probability spaces*, Dissertationes Math. (Rozprawy Mat.) 306 (1990), pp. 1–85.

[16] M. Pliev, *Narrow operators on lattice-normed spaces*, Cent. Eur. J. Math. 9, No 6 (2011), pp. 1276–1287.

[17] M. Pliev, M. Popov, *Narrow orthogonally additive operators*, Positivity v. 18, 4, (2014), 641–667.

[18] M. Popov, B. Randrianantoanina, *Narrow Operators on Function Spaces and Vector Lattices*, De Gruyter Studies in Mathematics 45, De Gruyter, 2013.

Dina Dzadzaeva
Department of Mathematics and Informatics
North-Osetian State University
Vatutina str. 44
Vladikavkaz 362025, Russia

Marat Pliev
Southern Mathematical Institute
of the Russian Academy of Sciences
Markusa str. 22
Vladikavkaz 362027, Russia

Ordered Structures and Applications: Positivity VII
Trends in Mathematics, 171–178

Quasi-compactness and Uniform Convergence of Markov Operator Nets on KB-spaces

Nazife Erkurşun Özcan

Abstract. We investigate quasi-compactness of operator nets on KB-spaces which is equivalent to uniform convergence. Moreover we study the uniform ergodicity condition of a single operator or operator semigroups.

Mathematics Subject Classification (2010). 47A35, 47B65, 46B42, 47D03, 47L05.

Keywords. KB-space, quasi-compactness, Markov operator net, uniform mean ergodic.

1. Introduction

Kryloff and Bogoliouboff [16, 17] in 1937 introduced an important class of quasi-compact operators for which the uniform mean ergodic theorem can be obtained. This result was generalized by Yosida and Kakutani [29] showing that any quasi-compact operator with bounded powers on a Banach space is uniformly ergodic. Since this result, many contributions have been made in probability and ergodic theory using the spectral properties of quasi-compact operators. Quasi-compactness for strongly convergent semigroups and its relation to uniform ergodicity and to the asymptotic behaviour of the solutions of abstract Cauchy problems is studied in [25].

2. Terminology and notations

Let E be a Banach lattice. Then $E_+ := \{x \in E : x \geq 0\}$ denotes the positive cone of E. On $\mathcal{L}(E)$ there is a canonical order given by $S \leq T$ if $Sx \leq Tx$ for all $x \in E_+$. If $0 \leq T$, then T is called positive. The dual space E' equipped with the canonical order is again a Banach lattice. Instead of the operations sup and inf on E we often write $\vee$ and $\wedge$, respectively. For $x \in E_+$ we denote by $[-x, x] := \{y \in E : |y| \leq x\}$ the order interval generated by x. A linear subspace of E is an ideal if $[-|x|, |x|] \subseteq I$ for all $x \in I$. An ideal I in E is called a band if for every subset $M \subseteq I$ such that $\sup M$ exists in E one has $\sup M \in I$. An ideal I in E is called a projection band if there is an ideal J in E such that $E = I \oplus J$ is the topological sum of I and J. In this case J is uniquely determined and I and J are bands. The projection P from E onto I with kernel $kerP = J$ is called the band projection corresponding to I and satisfies $0 \leq P \leq I_E$. The norm on E is order continuous if every order interval is weakly compact. A Banach lattice E is called a KB-space whenever every increasing norm bounded sequence of E_+ is norm convergent. In particular, it follows that every KB-space has order-continuous norm. Reflexive Banach lattices and AL-spaces are examples of KB-spaces. The following theorem combines some of the properties (see [1, 24, 28]).

Theorem 2.1. *For a Banach lattice E the following statements are equivalent:*

- *E is a KB-space*
- *E is a band of E''*
- *E is weakly sequentially complete*
- *c_0 is not embeddable in E.*
- *c_0 is not lattice embeddable in E*

The concept of a classical Markov operator was generalized to Banach lattices in [15]. Both the properties of Markov operators on L^1-spaces and the properties of their adjoints motivated the following definition.

Definition 2.2. Let E be a Banach lattice. A positive linear contraction $T : E \to E$ is called a Markov operator if there exists $0 < e' \in E'_+$ such that $T'e' = e'$.

It is well known that a positive linear operator T defined on a Banach lattice E, is continuous. It is also well known that if the Banach lattice E has order-continuous norm, then the positive operator T is moreover order continuous. According to the definition, we note that the Markov operators are again contained in the class of all positive contractions and that the adjoint T' is also a positive contraction. For more details, we refer to [15].

3. LR-nets

The class of nets considered in this paper was introduced by Lotz in [23] and then generalized by Räbiger [26]. We use the slightly modified terminology following the paper [8] and call them LR-nets.

Definition 3.1. A net $\Theta = (T_\lambda)_{\lambda \in \Lambda}$ is called an LR-net if

LR1 Θ is uniformly bounded;
LR2 $\lim_{\lambda\to\infty} \|T_\lambda \circ (T_\mu - I)x\| = 0$ for every $\mu \in \Lambda$ and for every $x \in X$;
LR3 $\lim_{\lambda\to\infty} \|(T_\mu - I) \circ T_\lambda x\| = 0$ for every $\mu \in \Lambda$ and for every $x \in X$.

If the limit conditions hold in the uniform operator topology, then the operator net is called uniform LR-net, or ULR-net.

Many examples of LR-nets appear in the investigation of operator semigroups. Thus every strongly convergent uniformly bounded Abelian operator semigroup itself is an LR-net with respect to the natural partial order $\prec$ defined by $T \prec S$ if there exists an R with $S = R \circ T$. If a semigroup $\mathcal{T} \subseteq \mathcal{L}(X)$ admits a $\mathcal{T}$-ergodic net Λ, then it is an LR-net. In particular, the Cesaro averages of a power bounded operator form an LR-net and moreover encompasses Cesaro averages of higher orders for both discrete and continuous semigroups. See [4, 7, 8, 10].

The convergence question is probably the most important one in applications of the concept of an LR-net. The net Θ is strongly convergent if the norm-limit $\|\cdot\| - \lim_{\lambda\to\infty} T_\lambda x$ exists for each $x \in X$. The following theorem states several equivalent conditions to convergence. For the proof, see [8].

Theorem 3.2. *Let $\Theta = (T_\lambda)_{\lambda\in\Lambda}$ be a Lotz–Räbiger net in a Banach space X. The following conditions are equivalent:*

a) *Θ converges strongly;*
b) *for every $x \in X$, the net $(T_\lambda x)_{\lambda\in\Lambda}$ has a weak cluster point;*
c) $X = \mathrm{Fix}(\Theta) \bigoplus \overline{\bigcup_{\lambda\in\Lambda}(I - T_\lambda)X}$;
d) $\mathrm{Fix}(\Theta)$ *separates* $\mathrm{Fix}(\Theta^*) = \{y \in X^* : T_\lambda^* y = y \text{ for all } \lambda \in \Lambda\}$.

If any of these conditions is satisfied, then the strong limit of $(T_\lambda)_\lambda$ is a projection onto the fixed space of Θ, $\mathrm{Fix}(\Theta) = \{x \in X : T_\lambda x = x, \forall \lambda \in \Lambda\}$.

4. Ergodicity of Markov LR-nets on KB-spaces

The concept of an attractor or constrictor was used by several mathematicians to characterize the asymptotic behavior of operators. In this section we show that a positive LR-net on KB-spaces is strongly convergent if an LR-net has an attractor of the form $W + \eta B_E$ where W is weakly compact. Moreover if the weakly compact part of an attractor is an order interval, then a Markovian LR-net converges strongly to a finite-rank projection, see [14].

If T is a Markov operator on L^1-space then the Cesaro averages $A_n^T = \frac{1}{n}\sum_{k=0}^{n-1} T^k$ are strongly convergent with $dimFix(T) < \infty$ whenever there exists a function $h \in L^1_+$ and a real $0 \le \eta < 1$ such that $\lim_{n\to\infty} \left\|(h - \frac{1}{n}\sum_{k=0}^{n-1} T^k f)_+\right\| \le \eta$ for every density f, see [13]. In [7] and [3] some generalizations of this results are given. Firstly we prove the above result for positive operator nets. The principal trick in the proof of the main results of [7] was using the additivity of the norm on the positive part of the L^1-space. Since this is no longer the case for a general KB-space, we use different ideas in this part, inspired by [27].

Theorem 4.1. *Let E be a KB-space with a quasi-interior point e and $\Theta = (T_\lambda)_{\lambda\in\Lambda}$ be a positive LR-net in E which has a cofinal subsequence, W be a weakly compact subset of E, and $\eta \in \mathbb{R}$, $0 \le \eta < 1$ such that*

$$\lim_{\lambda\to\infty} \mathrm{dist}(T_\lambda x, W + \eta B_E) = 0$$

for any $x \in B_E$. Then Θ converges strongly.

The theorem is also true if we replace a weakly compact subset W of E by an order interval $[-g, g]$ for any $g \in E_+$ because in KB-spaces, every order interval is weakly compact. In this case we moreover get that the dimension of the fixed space is finite.

Theorem 4.2. *Let E be a KB-space with a quasi-interior point e and $\Theta = (T_\lambda)_{\lambda\in\Lambda}$ be a Markov LR-net in E which has a cofinal subsequence. Then the following are equivalent*

(i) *there exists a function $g \in E_+$ and $\eta \in \mathbb{R}$, $0 \leq \eta < 1$ such that*

$$\lim_{\lambda\to\infty} \operatorname{dist}(T_\lambda x, [-g, g] + \eta B_E) = 0 \ \forall x \in B_E$$

(ii) *the net Θ is strongly convergent and* $\dim \operatorname{Fix}(\Theta) < \infty$.

For the proofs and details, we refer the reader to [14].

5. Quasi-compact Markov operator net on KB-spaces

The systematic study of a quasi-compact linear operator T on a Banach space was initiated by Yoshida and Kakutani [29], in order to obtain some of the limit theorems of Doeblin [6]. We recall that an operator $T \in \mathcal{L}(X)$ is called quasi-compact if there exist a positive integer n and a compact operator K with $\|T^n - K\| < 1$, see [16, 17]. It is well known that an operator T is quasi-compact if and only if there exists a sequence $(K_n)_{n\in\mathbb{N}} \subseteq \mathcal{L}(X)$ of compact operators with $\lim_{n\to\infty} \|T^n - K_n\| = 0$, see [18]. This motivates the following definition in [11].

Definition 5.1. An operator net $(T_\lambda)_{\lambda\in\Lambda}$ is called quasi-compact if for every $\lambda \in \Lambda$ there is a compact operator K_λ with $\lim_{\lambda\to\infty} \|T_\lambda - K_\lambda\| = 0$.

The quasi-compactness in the sense of Definition 5.1 is a property of an operator net, not of operators belonging to this net. Nevertheless the following theorem characterizes quasi-compactness of a ULR-net through quasi-compactness of its operators.

Theorem 5.2 ([11], Thm. 7). *Let $\Theta = (T_\lambda)_\lambda$ be a ULR-net then the following conditions are equivalent:*

(i) *Θ is quasi-compact;*

(ii) *there exists $\lambda_0 \in \Lambda$ such that T_λ are quasi-compact for all $\lambda \succeq \lambda_0$;*

(iii) *there exists $\lambda_0 \in \Lambda$ such that T_{λ_0} is quasi-compact;*

(iv) *there exists $S \in \overline{\operatorname{co}(\operatorname{sem}(\Theta))}$ such that S is quasi-compact, where $\overline{\operatorname{co}(\operatorname{sem}(\Theta))}$ is the closure in the operator norm of the convex hull* $\operatorname{co}(\operatorname{sem}(\Theta))$ *of the semigroup* $\operatorname{sem}(\Theta) \subseteq L(X)$ *generated by* $\{T_\lambda : \lambda \in \Lambda\}$.

Emel'yanov proved also the existence of the uniform limit in terms of quasi-compactness. A strongly convergent ULR-net on a Banach space is quasi-compact if and only if it converges uniformly to a finite-rank projection. It is worth adding that ULR-net condition is essential, otherwise there exists a quasi-compact LR-nets, they converge strongly but not uniformly, see [11].

Our goal is to prove that for a Markov ULR-net on KB-space then the net is quasi-compact if and only if it converges uniformly.

Theorem 5.3. *Let $\Theta = (T_\lambda)_{\lambda\in\Lambda}$ be a Markov ULR-net which has a cofinal subsequence on the KB-space E with a quasi-interior point e. Then Θ is quasi-compact if and only if it converges uniformly to a finite-rank projection.*

Proof. If Θ converges uniformly to a finite rank projection then it is automatically quasi-compact by Definition 5.1.

Let Θ be a quasi-compact Markov ULR-net on a KB-space. By Theorem 5.2 for $\epsilon > 0$ there exists $\lambda_0 \in \Lambda$ such that T_{λ_0} is quasi-compact and so $K_{\lambda_0}(B_E)+\epsilon B_E$ is an attractor of Θ with compact $\overline{K_{\lambda_0}(B_E)}$. Hence by Theorem 4.1 Θ converges strongly, say to P. By the compactness of $K_{\lambda_0} = K$, we can take $x_1, \ldots, x_k \in B_E$ such that for every $x \in B_E$, there exists x_j with $\|Kx - Kx_j\| \le \epsilon$. Then the interval $[-\sum Kx_k, \sum Kx_k] + 2\epsilon B_E$ is an attractor and Θ is a Markov ULR-net, so it converges to a finite-rank projection by Theorem 4.2.

Emel'yanov proved that if Θ is a strongly convergent ULR-net satisfying quasi-compactness then it converges uniformly in [11]. Hence the proof is completed. □

6. Quasi-compactness of single operator and operator semigroups

Yoshida and Kakutani proved firstly that if the contraction T is quasi-compact then it is uniformly ergodic, i.e., the Cesaro averages A_n^T converges uniformly, to a finite-rank projection. Later Lin showed that quasi-compactness is both sufficient and necessary for uniform ergodicity of positive operators on L^1-spaces or $C(X)$, [21, 22]. An important tool for proving is the following. Let $\|T\| \le 1$, then $\frac{1}{n}\sum_{k=0}^{n-1} T^k$ converges uniformly if and only if $I - T$ has a closed range. It is the statement of uniform ergodic theorem and now we generalize it to Markov operators on KB-spaces.

Theorem 6.1. *Let T be a Markov operator on KB-space E. Then the following conditions are equivalent.*

(i) *T is quasi-compact*
(ii) *T is uniformly ergodic to a finite-rank projection*
(iii) *$(I - T)E$ is closed and $\mathrm{Fix}(T) = \{x : Tx = x\}$ is finite dimensional.*
(iv) *$(I - T')E'$ is closed and $\mathrm{Fix}(T')$ is finite dimensional.*

Proof. (i) $\Rightarrow$ (ii) is due to Yosida and Kakutani [29].

(ii) $\Leftrightarrow$ (iii) By the uniform ergodic theorem.

(iii) $\Rightarrow$ (iv) By the uniform ergodic theorem $\frac{1}{n}\sum_{i=0}^{n-1} T^i$ converges uniformly to a projection P on $\mathrm{Fix}(T)$. Therefore $\dim \mathrm{Fix}(T') = \dim \mathrm{Fix}(T) < \infty$. Moreover $\frac{1}{n}\sum_{i=0}^{n-1} T'^i$ converges to P' uniformly and so $(I - T')E'$ is closed.

(iv) $\Rightarrow$ (iii) By the uniform ergodic theorem $\frac{1}{n}\sum_{i=0}^{n-1} T'^i$ converges uniformly and hence T is uniformly ergodic which implies $(I - T)E$ is closed.

(ii) $\Rightarrow$ (i) Since T is Markov on KB-space, the Cesaro averages $\frac{1}{n}\sum_{i=0}^{n-1} T^i$ is an ULR-net. For the limit condition of Definition 3.1,

$$\begin{aligned}\lim_{n\to\infty} \left\|A_n^T(A_m^T - I)\right\| &= \lim_{n\to\infty} \left\|A_n^T \left(\frac{1}{m}\sum_{k=0}^{m-1} T^k - \frac{1}{m}\sum_{k=0}^{m-1} I^k\right)\right\| \\ &= \lim_{n\to\infty} \frac{1}{m}\left\|A_n^T \sum_{k=0}^{m-1}(T^k - I^k)\right\| \\ &\leq \lim_{n\to\infty} \frac{1}{m}\frac{1}{n}\left\|T^n\right\| \left\|T^{m-1} + \cdots + I\right\|.\end{aligned}$$

Since T is Markov, so a positive contraction, limit is 0 and it satisfies the ULR-net conditions. By Theorem 5.3, A_n^T is quasi-compact. By Theorem 5.2, T is quasi-compact. □

Remark that for the Cesaro averages of a single operator without loss of generality we may assume that E has a quasi-interior point e, [3].

Uniform LR-net assumption is necessary because if we replace the condition ULR-net by the condition LR-net, the implication $(ii) \Rightarrow (i)$ is not true, for the example refer to [11].

Quasi-compactness for strongly-continuous semigroups and its relation to uniform ergodicity is studied in [25]. Quasi-compactness means that $T(t)$ approaches the compact operators as $t \to \infty$, to be precise, a strongly continuous semigroup $(T(t))_{t\geq 0}$ is quasi-compact if $\lim_{t\to\infty} \operatorname{dist}(T(t), \mathcal{K}(E)) = 0$ where $\mathcal{K}(E)$ stands for the ideal of compact operators on E and

$$\operatorname{dist}(T(t), \mathcal{K}(E)) = \inf_{K\in\mathcal{K}(E)} \left\|T(t) - K\right\|.$$

Quasi-compactness can be characterized in different ways and in [25] the notion of the essential growth bound of a semigroup $(T(t))_{t\geq 0}$ is used. Lotz gave a criterion for quasi-compactness of positive semigroups on $C(X)$. It is based on a criterion given by Doeblin for operators on spaces of bounded measurable functions. In [4] quasi-compactness is investigated for ULR-nets in $C(X)$.

Now, unfortunately, we could not consider the Yoshida ad Kakutani result with Lin's characterization for one-parameter semigroups on KB-spaces. The Cesaro averages of uniformly continuous one-parameter semigroup $A_t = \frac{1}{t}\int_0^t T(s)ds$ form an ULR-net. Its uniform convergence to a finite-dimensional projection does not imply quasi-compactness of $(T(t))_{t\geq 0}$ in general even on L^1-spaces. For an concrete example, see [4].

We could only state the following theorem.

Theorem 6.2. *Let $(T(t))_{t\geq 0}$ be uniformly continuous (norm-continuous) one-parameter Markov semigroup on KB-spaces with $T(0) = I$ satisfying $T(t)/t \to 0$ in norm. Then the Cesaro averages of $(T(t))_{t\geq 0}$ form a quasi-compact ULR-net if and only if it is uniformly ergodic onto the finite-dimensional fixed space*

Proof. Since $(T(t))_{t\geq 0}$ is bounded and uniformly continuous semigroup satisfying $\|T(t)/t\| \to 0$, the Cesaro averages of it is a ULR-net. By Theorems 5.3 and Theorem 5.2, the result follows. □

Acknowledgment

The author would like to thank the referee for helpful remarks and comments.

References

[1] Aliprantis C.D., Burkinshaw O. Positive Operators, New York, 1985

[2] Aliprantis C.D., Burkinshaw O. Locally Solid Riesz Spaces with Applications to Economics, Amer. Math. Soc., 2003

[3] Alpay S., Binhadjah A., Emelyanov Yu.E., Ercan Z. Mean ergodicity of positive operators in KB-spaces, J. Math. Anal. Appl. 2006; Vol. 323; no. 1; pp. 371–378

[4] Bartoszek W., Erkurşun N. On quasi compact Markov nets. Ergodic Theory Dynam. Systems 2011; Vol. 31; no. 4; pp.1081–1094

[5] Bartoszek W. Asymptotic periodicity of the iterates of positive contractions on Banach lattices. Studia Math. 1988; Vol. 91; no. 3; pp. 179–188.

[6] Doeblin W., Sur les propriétés asymptotiques de mouvements régis par certains types de chaînes simples. Bull. Math. Soc. Roum. Sci. 1937; Vol. 39; no. 1; pp. 57–115; no.2; pp.3–61

[7] Emel'yanov E., Erkursun N. Lotz–Räbiger's nets of Markov operators in L^1-spaces. J. Math. Anal. Appl. 2010; Vol. 371; no. 2; pp. 777–783.

[8] Emel'yanov E., Erkursun N. Generalization of Eberlein's and Sine's ergodic theorems to LR-nets. Vladikavkaz. Mat. Zh. 2007; Vol. 9; no. 3; pp. 22–26.

[9] Emel'yanov E., Erkursun N. Asymptotically absorbing nets of positive operators. Siberian Adv. Math. 2012; Vol. 22; no. 4; pp. 243–260.

[10] Emel'yanov E. Asymptotic Behavior of Lotz–Räbiger and Martingale Nets. Russian, with Russian summary, Sibirsk. Mat. Zh. 2010; Vol. 51; no. 5; pp. 1017–1026, [translation in Siberian Math. J. 2010; Vol. 51; no. 5; pp. 810–817.

[11] Emel'yanov E. On Quasi-compactness of Operator Nets on Banach Spaces, Studia Math. 2011; Vol. 203; no. 2; pp. 163–170.

[12] Emel'yanov E., Wolff MPH. Mean ergodicity on Banach lattices and Banach spaces. Arch. Math. (Basel) 1999; Vol. 72; no. 3; pp. 214–218.

[13] Emel'yanov E., Wolff MPH. Mean lower bounds for Markov operators. Ann. Polon. Math. 2004; Vol. 83; pp. 11–19

[14] Erkurşun Özcan, N. Asymptotic behavior of operator sequences on KB-spaces, http://arxiv.org/abs/1403.2114.

[15] Hawke P. Markov Operators on Banach Lattices, Thesis for the degree of Master of Science. 2006.

[16] Kryloff N.M., Bogoliouboff N.N. Les propriétés ergodiques des suites des probabilités en chaîne. Comptes Rendus Mathématique 1937; Vol. 204; pp. 1386–1388.

[17] Kryloff N.M., Bogoliouboff N.N. Sur les propriétés en chaîne. Comptes Rendus Mathématique 1937; Vol. 204; pp. 1454–1456.

[18] Krengel U. Ergodic theorems. de Gruyter Studies in Mathematics; Vol. 6, pp. viii+357, with a supplement by Antoine Brunel, Berlin: Walter de Gruyter & Co., (1985).

[19] Lasota A., Mackey M.C. Chaos, Fractals and Noise. Stochastics Aspect of Dynamics, second edition, Appl. Math. Sci., Vol. 97, Springer-Verlag, New York, 1994, xiv + 472 pp.

[20] Lasota A., Socala J. Asmptotic properties of constrictive Markov operators. Bull. Pol. Acad. Math. 1987; Vol. 35; pp. 71–76

[21] Lin M. Quasi-compactness and uniform ergodicity of positive operators. Israel J. Math. 1978; Vol. 29; no. 2-3; pp. 309–311.

[22] Lin M. Quasi-compactness and uniform ergodicity of Markov operators. Ann. Inst. H. Poincaré Sect. B (N.S.) 1975; Vol. 11; no. 4; pp. 345–354 (1976).

[23] Lotz HP. Tauberian theorems for operators on L_∞ and similar spaces, in: Functional Analysis: Surveys and Recent Results, III, Paderborn, 1983, in: North-Holland Math. Stu., North-Holland, Amsterdam, Vol. 90; pp. 117–133

[24] Meyer-Nieberg P. Banach Lattices, Universitext, Springer-Verlag, Berlin, 1991.

[25] Nagel R. (ed.), One-Parameter Semigroups of Positive Operators. (Lect. Notes Math. Vol. 1184), Berlin-Heidelberg-New York: Springer 1986

[26] Räbiger F. Stability and ergodicity of dominated semigroups: II. The strong case. Math. Ann. 1993; Vol. 297; pp. 103–116.

[27] Räbiger F. Attractors and asymptotic periodicity of positive operators on Banach lattices. Forum Math. 1995; Vol. 7; no. 6; pp. 665–683.

[28] Schaefer H.H. Banach Lattices and Positive Operators, Grundlehren Math. Wiss., Band 215, Springer-Verlag, New York, 1974.

[29] Yosida K., Kakutani S., Operator theoretical treatment of Markoff's process and mean ergodic theorem. Ann. of Math. 1941; Vol. 42; no. 2; pp. 188–228.

Nazife Erkurşun Özcan
Department of Mathematics
Hacettepe University
Ankara, Turkey, 06800
e-mail: `erkursun.ozcan@hacettepe.edu.tr`

Ordered Structures and Applications: Positivity VII
Trends in Mathematics, 179–201

Disjointly Homogeneous Banach Lattices and Applications

Julio Flores, Francisco L. Hernández and Pedro Tradacete

Abstract. This is a survey on disjointly homogeneous Banach lattices and their applications. Several structural properties of this class are analyzed. In addition we show how these spaces provide a natural framework for studying the compactness of powers of operators allowing for a unified treatment of well-known results.

Mathematics Subject Classification (2010). Primary 47B38, 46E30; Secondary 46B42, 47B07.

Keywords. Disjointly homogeneous Banach lattices.

Contents

The research has been supported by the Spanish Ministerio de Economía y Competitividad through grant MTM2012-31286 and Grupo UCM 910346. P. Tradacete has also been partially supported by MTM2010-14946.

1. Introduction

This paper is a survey on the properties of the recently introduced class of disjointly homogeneous Banach lattices, as well as the compactness properties of the operators defined on these spaces. It collects information previously given in ([15, 16, 17, 21]) and some new facts.

Recall that a Banach lattice E is *disjointly homogeneous* if two arbitrary sequences of normalized pairwise disjoint elements in E always have equivalent subsequences. The motivation which gave rise to the definition of a disjointly homogeneous space was to decide the compactness of the iterations of a given operator which already enjoyed nice close-to-compactness properties. Thus the first and third authors together with V.G. Troitsky considered this notion for the first time in [15].

Later on E.M. Semenov and the authors ([16]) analyzed the general problem of obtaining compactness of the iterations of a strictly singular operator on a Banach lattice, extending the classical result by V.D. Milman ([35]) which states that strictly singular operators in $L_p(\mu)$, $1 \le p \le \infty$, have compact square. In fact, one of the purposes of this survey is to offer compelling evidence that the class of disjointly homogeneous Banach lattices constitutes a proper setting for treating these questions. It is particularly evident in connection with the Kato property, i.e., when the class of compact and strictly singular operators coincide ([21]). From the point of view of the structural properties of the class of disjointly homogeneous Banach lattices, several aspects have been explored by V.G. Troitsky and E. Spinu jointly with the authors in [17]; particularly two of these aspects have been studied in detail, namely the problem of self-duality and the problem of obtaining complemented copies of the span of disjoint sequences.

The paper is organized in two clearly differentiated parts. The first one includes sections one through five which focus on disjointly homogeneous Banach lattices themselves. Definitions and examples are given, and structure properties such as the self-duality of this class and the existence of complemented copies of disjoint sequences are addressed. The second part, including the remaining sections, focuses on the operators defined on disjointly homogeneous Banach lattices and the properties they have; particularly, attention is given to the compactness properties of the iterations of endomorphisms as well as the relation between disjointly homogeneous Banach lattices and the Kato property. The paper concludes with a list of some open questions.

We follow the standard terminology concerning Banach spaces and Banach lattices as in the monographs [2, 32, 33, 34]. In the sequel by an operator we always mean a bounded linear operator. Given a sequence (x_n) in a Banach space, we write $[x_n]$ for the closed linear span of the sequence. Given basic sequences (x_n), (y_n), and $C > 0$, the notation $(x_n) \overset{C}{\sim} (y_n)$ means that for every scalars $(a_n)_{n=1}^{\infty}$

$$C^{-1}\Big\|\sum_{n=1}^{\infty} a_n y_n\Big\| \le \Big\|\sum_{n=1}^{\infty} a_n x_n\Big\| \le C\Big\|\sum_{n=1}^{\infty} a_n y_n\Big\|.$$

2. Disjointly homogeneous Banach lattices: definition and examples

The notion of disjointly homogeneous Banach lattices was first introduced in [15]; let us recall its definition.

Definition 2.1. A Banach lattice E is *disjointly homogeneous* (*DH*) if for every pair (x_n), (y_n) of normalized disjoint sequences in E, there exist $C > 0$ and a subsequence (n_k) such that $(x_{n_k}) \overset{C}{\sim} (y_{n_k})$.

Our interest will focus on those Banach lattices for which there is $1 \leq p \leq \infty$ such that every normalized disjoint sequence (x_n) has a subsequence (x_{n_k}) equivalent to the unit vector basis of ℓ_p (or c_0 for $p = \infty$), i.e.,

$$C^{-1}\Big(\sum_{k=1}^{\infty}|a_k|^p\Big)^{1/p} \leq \Big\|\sum_{k=1}^{\infty}a_k x_{n_k}\Big\| \leq C\Big(\sum_{k=1}^{\infty}|a_k|^p\Big)^{1/p},$$

for some $C > 0$. These form an important class of DH spaces, which will be denoted *p-disjointly homogeneous*, in short p-DH (resp. ∞-disjointly homogeneous, in short ∞-DH). Clearly, L_p-spaces are p-DH.

Note that 1-DH Banach lattices have been considered previously under a different approach. Recall that a Banach lattice E has the *positive Schur property* if every weakly null sequence (x_n) of positive vectors is norm convergent, see [25, 42, 43, 44]. It follows from, e.g., [34, Corollary 2.3.5], that it suffices to verify this condition for disjoint sequences. Using Rosenthal's ℓ_1-theorem, it was proved in [17] that a Banach lattice E is 1-DH if and only if E has the positive Schur property.

Observe that, in the definition of a DH Banach lattice, it is important to allow for the possibility of passing to subsequences in order to get the required equivalence. Otherwise, the class reduces to the spaces $L_p(\mu)$ or $c_0(\Gamma)$ ([17, Proposition 2.2]).

Thus $L_p(\mu)$-spaces exhibit a particularly strong version of this definition. But these are not the only examples. For instance, in the context of function spaces, Lorentz spaces $\Lambda(W, q)$ and $L_{p,q}$ on $[0, 1]$ are q-DH.

Recall that given $1 \leq q < \infty$ and W a positive, non-increasing function in $[0, 1]$, such that $\lim_{t\to 0} W(t) = \infty$, $W(1) > 0$ and $\int_0^1 W(t)dt = 1$, the *Lorentz function space* $\Lambda(W, q)[0, 1]$ is the space of all measurable functions f on $[0, 1]$ such that

$$\|f\| = \Big(\int_0^1 f^*(t)^q W(t)dt\Big)^{1/q} < \infty,$$

where f^* denotes the decreasing rearrangement of the function f (cf. [33, Chapter 2]). Let us also recall that for $1 < p < \infty$ and $1 \leq q \leq \infty$, the Lorentz space

$L_{p,q}[0,1]$ is the space of all measurable functions f in $[0,1]$ such that

$$\|f\|_{p,q} = \begin{cases} \left(\int_0^\infty (t^{1/p} f^*(t))^q \frac{dt}{t} \right)^{1/q} < \infty & \text{for } 1 \le q < \infty, \\ \sup_{t>0} t^{1/p} f^*(t) < \infty, & \text{if } q = \infty. \end{cases}$$

The following (see [7], [14, Proposition 5.1]) shows that these belong to the class of DH spaces.

Proposition 2.2. *Let $1 \le q < \infty$. Let $(f_n)_n$ be a disjoint normalized sequence in $\Lambda(W,q)[0,1]$ (resp. $L_{p,q}[0,1]$). For each $\varepsilon > 0$, there exists a subsequence (f_{n_k}) which is $(1+\varepsilon)$-equivalent to the unit vector basis of ℓ_q, whose span is a complemented subspace of $\Lambda(W,q)[0,1]$ (resp. $L_{p,q}[0,1]$).*

For the maximal Lorentz spaces $L_{p,\infty}[0,1]$, $1 < p < \infty$, the situation is different. Indeed, the space $L_{p,\infty}[0,1]$ satisfies that every disjoint sequence in its order-continuous part $(L_{p,\infty}(0,1))^0$ (the closed linear span of the characteristic functions in $L_{p,\infty}[0,1]$) has a subsequence equivalent to the unit vector basis of c_0 (see [37]). But there exists a disjoint normalized sequence (f_n) in $L_{p,\infty}[0,1]$ equivalent to the unit basis of ℓ_p (which generates a complemented subspace) (see [16]). Therefore, $L_{p,\infty}[0,1]$ is not DH.

In the class of Orlicz spaces we have further examples of DH spaces. Recall that given an Orlicz function $\varphi : \mathbb{R}_+ \to \mathbb{R}_+$, the *Orlicz function space* $L_\varphi(\Omega, \Sigma, \mu)$ is the space of all Σ-measurable functions f on Ω such that $\int_\Omega \varphi\left(\frac{|f|}{r}\right) d\mu < \infty$ for some $r > 0$. This is a Banach lattice endowed with the Luxemburg norm

$$\|f\|_{L_\varphi} = \inf\left\{ r > 0 : \int_\Omega \varphi\left(\frac{|f|}{r}\right) d\mu \le 1 \right\}.$$

A characterization of DH Orlicz spaces, over finite [16] and infinite [17] measure spaces, is known. In order to state this, let us first recall the definition of certain subsets of the space of continuous functions $C[0,1]$ associated to the Orlicz function φ (see [31]):

$$E_{\varphi,s}^\infty = \overline{\left\{ \frac{\varphi(r\cdot)}{\varphi(r)} : r \ge s \right\}}, \quad E_\varphi^\infty = \bigcap_{s>1} E_{\varphi,s}^\infty, \quad \text{and } C_\varphi^\infty = \overline{\operatorname{conv}}(E_\varphi^\infty).$$

Similarly, let

$$E_\varphi(0,\infty) = \overline{\left\{ F \in C[0,1] \; : \; F(\cdot) = \frac{\varphi(s\cdot)}{\varphi(s)}, \text{ for some } s \in (0,\infty) \right\}},$$

and $C_\varphi(0,\infty) = \overline{\operatorname{conv}}\, E_\varphi(0,\infty)$, in the space $C[0,1]$.

As usual, for a subset $A \subset C[0,1]$ and a function h, we will write $A \cong \{h\}$ whenever every function in A is equivalent to the function h at 0.

Theorem 2.3.

(i) *A separable Orlicz space* $L_\varphi[0,1]$ *is DH if and only if* $E_\varphi^\infty \cong \{t^p\}$ *(for some* $1 \le p < \infty$*).*

(ii) *A separable Orlicz space* $L_\varphi(0,\infty)$ *is DH if and only if* $C_\varphi(0,\infty) \cong \{t^p\}$ *(for some* $1 \le p < \infty$*).*

Moreover, in each case the space is p-DH for the corresponding p.

The proof of the finite measure case is based on techniques from [31]. For instance, if $\varphi(x) = x^p log(1+x)$, for $1 \le p < \infty$, then $L_\varphi[0,1]$ is p-DH.

For the infinite measure case, among other things, the proof makes use of [36, Theorem 1.1], which asserts that if an Orlicz function F is equivalent to a function in $C_\varphi(0,\infty)$ then $L_\varphi(0,\infty)$ contains a lattice copy of the Orlicz sequence space ℓ_F and, conversely, every normalized disjoint sequence in $L_\varphi(0,\infty)$ contains a subsequence equivalent to the unit vector basis of ℓ_F for some $F \in C_\varphi(0,\infty)$.

In the discrete setting, the class of DH Banach lattices is considerable smaller. Clearly, it contains the spaces c_0 and l_p, $1 \le p \le \infty$; by contrast, Orlicz and Lorentz sequence spaces other than ℓ_p cannot be DH. This follows from the well-known fact that these are stable spaces ([29]). Indeed, if one starts with a given pairwise disjoint sequence (x_n) in an Orlicz or Lorentz sequence space E, then the stability implies that there is some block sequence (w_n) of (x_n) equivalent to the unit vector basis of some ℓ_p for $1 \le p < \infty$. If E is assumed to be DH, then for some subsequence (n_k), (x_{n_k}) and (w_{n_k}) are equivalent to the unit vector basis of ℓ_p. But the unit basis in E is symmetric; thus, it must be equivalent to the unit vector basis of ℓ_p.

Tsirelson space also falls within the category of DH Banach lattices, as shown in [15]. As a consequence, we deduce that DH Banach lattices need not be p-DH for any $1 \le p \le \infty$. Some modifications of Tsirelson space (cf. [9]) are easily seen to be also DH.

Observe that in the definition of a DH Banach lattice it is enough to consider only positive disjoint normalized (or even semi-normalized) sequences. A formally weaker version of DH has been considered also in [17]: namely, a Banach lattice is *quasi-DH* if any two sequences of disjoint elements (x_n) and (y_n) have equivalent subsequences. This means that $(x_{n_k}) \sim (y_{m_k})$ for some, non necessarily equal, subsequences (n_k) and (m_k). The following result follows from a standard application, based on [39], of the infinite Ramsey theorem, and solves a natural question posed in [17].

Proposition 2.4. *A Banach lattice is DH if and only if it is quasi-DH.*

Proof. We prove that a quasi-DH Banach lattice X is DH. Let (x_n) be a disjoint sequence in X. For an infinite set A, by $\mathcal{P}_\infty(A)$ we denote the family of infinite subsets of A. We claim that $\mathcal{P}_\infty(\mathbb{N})$ contains some set $\mathbb{M} = \{m_k : k \in \mathbb{N}\}$ with $m_1 < m_2 < \cdots$ such that for every infinite subset $\mathbb{P} = \{p_j : j \in \mathbb{N}\} \subset \mathbb{M}$, the equivalence $(x_{p_{2j}}) \sim (x_{p_{2j+1}})$ holds.

Indeed, let

$$\mathcal{S} = \Big\{\{m_k : k \in \mathbb{N}\} \in \mathcal{P}_\infty(\mathbb{N}) : \forall k,\, m_k < m_{k+1}, \text{ and } (x_{m_{2k}}) \sim (x_{m_{2k+1}})\Big\}.$$

It is easy to check that $\mathcal{S}$ is a Borel subset of $\mathcal{P}_\infty(\mathbb{N})$. By the Galvin–Prikry Theorem (cf. [11]), there is $\mathbb{M} \in \mathcal{P}_\infty(\mathbb{N})$ such that either $\mathcal{P}_\infty(\mathbb{M}) \subset \mathcal{S}$ or $\mathcal{P}_\infty(\mathbb{M}) \cap \mathcal{S} = \emptyset$. Now, suppose that $\mathcal{P}_\infty(\mathbb{M}) \cap \mathcal{S} = \emptyset$. Since X is quasi-DH the disjoint sequences $(x_{m_{2k}})$ and $(x_{m_{2k+1}})$ have equivalent subsequences, that is (j_k), (l_k) such that

$$(x_{m_{2j_k}}) \sim (x_{m_{2l_k+1}}).$$

Passing to further subsequences we have that either $2j_1 < 2l_1+1 < 2j_2 < 2l_2+1 < \cdots$ or $2l_1+1 < 2j_1 < 2l_2+1 < 2j_2 < \cdots$. In both cases we have that

$$I = \{m_{2j_1}, m_{2l_1+1}, m_{2j_2}, m_{2l_2+1}, \ldots\} \in \mathcal{P}_\infty(\mathbb{M}) \cap \mathcal{S}.$$

This contradiction implies that $\mathcal{P}_\infty(\mathbb{M}) \subset \mathcal{S}$, and the claim follows.

To finish the proof, let (x_n) and (y_n) be two sequences of normalized disjoint elements in X. By the claim, we can assume, passing to some subsequence, that both (x_n) and (y_n) run on $\mathbb{M}$ and also that $(x_{m_{2k}}) \sim (x_{m_{2k+1}})$ for every (m_k) with $m_1 < m_2 < \cdots$ Since X is quasi-DH, there exist (n_k) and (p_k) such that $(x_{n_k}) \sim (y_{p_k})$. Passing to a further subsequence, we can assume that $n_1 < p_1 < n_2 < p_2 < \cdots$ or $p_1 < n_1 < p_2 < n_2 < \cdots$ By the properties of the sequence $(x_n)_{n\in\mathbb{M}}$, it follows that

$$(y_{p_k}) \sim (x_{n_k}) \sim (x_{p_k}).$$

□

3. Duality for disjointly homogeneous Banach lattices

It is natural to inquire about the stability by duality of the class of DH Banach lattices. Note that by Proposition 2.2, for $1 < p < \infty$, the Lorentz space $L_{p,1}[0,1]$ is 1-DH. However, as mentioned above, its dual $L_{p',\infty}[0,1]$ is not DH (here $\frac{1}{p}+\frac{1}{p'} = 1$). Thus, in the non-reflexive case, the class of DH Banach lattices is not stable under duality.

By contrast, in the reflexive case, all the examples of DH Banach lattices mentioned above have DH duals. This is obviously true for $L_p(\mu)$ spaces with $1 < p < \infty$ as well as for Lorentz spaces since Proposition 2.2 can also be applied to their duals. In addition, as shown in Theorem 2.3, an Orlicz space $L_\varphi[0,1]$ is DH if and only if every function in the set E_φ^∞ is equivalent to the function t^p for some fixed $1 \le p < \infty$. Note however that if φ' denotes the conjugate Orlicz function of φ, then every function in $E_{\varphi'}^\infty$ is easily seen to be equivalent to the function $t^{p'}$, which again is tantamount to the space $L_\varphi[0,1]^* = L_{\varphi'}[0,1]$ being DH.

Therefore, based on these examples, one might reasonably conjecture that among reflexive spaces being DH is indeed a self dual property. As it will be shown this turns out to be false. Still it holds true under additional assumptions which are of interest. The rest of the section is devoted to clarifying this.

A natural approach to proving a positive result of stability by duality should look more or less like this: start with two arbitrarily chosen disjoint normalized sequences (x_n) and (y_n) in a reflexive Banach lattice E whose dual E^* is DH. We would like to prove that, up to passing to some subsequence, (x_n) and (y_n) are equivalent. Certainly, two disjoint normalized sequences (x_n^*) and (y_n^*) in E^* can be found in E^* such that $x_n^*(x_m) = y_n^*(y_m) = \delta_{nm}$ for each $n, m \in \mathbb{N}$. Since E^* is DH, after passing to subsequences we may assume that (x_n^*) and (y_n^*) are equivalent in E^*. On the other hand, for each m, we can consider x_m^* as a functional on $[x_n]$ (formally speaking, we are taking the restriction of x_m^* to $[x_n]$); moreover, since E is reflexive, (x_m^*) is a basis of $[x_n]^*$. Then for any coefficients $\alpha_1, \ldots, \alpha_m$ we have

$$\begin{aligned}\Big\|\sum_{i=1}^m \alpha_i x_i\Big\| &= \sup\Big\{\Big|\Big\langle \sum_{i=1}^m \alpha_i x_i, \sum_{i=1}^m \beta_i x_i^*\Big\rangle\Big| \ :\ \Big\|\sum_{i=1}^m \beta_i x_i^*\Big\|_{[x_n]^*} \leq 1\Big\}\\ &= \sup\Big\{\Big|\sum_{i=1}^m \alpha_i\beta_i\Big| \ :\ \Big\|\sum_{i=1}^m \beta_i x_i^*\Big\|_{[x_n]^*} \leq 1\Big\}.\end{aligned}$$

In general, clearly $\|\sum_{i=1}^m \beta_i x_i^*\|_{[x_n]^*} \leq \|\sum_{i=1}^m \beta_i x_i^*\|_{E^*}$. However, if we could somehow control the converse estimate, we could continue, using the equivalence of (x_n^*) and (y_n^*) in E^* as follows

$$\begin{aligned}\Big\|\sum_{i=1}^m \alpha_i x_i\Big\| &\sim \sup\Big\{\Big|\sum_{i=1}^m \alpha_i\beta_i\Big| \ :\ \Big\|\sum_{i=1}^m \beta_i x_i^*\Big\|_{E^*} \leq 1\Big\}\\ &\sim \sup\Big\{\Big|\sum_{i=1}^m \alpha_i\beta_i\Big| \ :\ \Big\|\sum_{i=1}^m \beta_i y_i^*\Big\|_{E^*} \leq 1\Big\} \sim \Big\|\sum_{i=1}^m \alpha_i y_i\Big\|,\end{aligned}$$

which would imply that (x_n) and (y_n) are equivalent. In particular, such an argument would work if we could find a bounded operator $S\colon [x_n]^* \to E^*$ such that $Sx_m^* = x_m^*$ for each m and a similar operator for (y_n). The previous discourse is collected in the following

Definition 3.1. A Banach lattice E has the $\mathfrak{P}$ *property* if for every disjoint positive normalized sequence $(f_n) \subset E$ there exists an operator $T : E \to [f_n]$, such that some subsequence $(T^* f_{n_k}^*)$ is equivalent to a seminormalized disjoint sequence in E^* (here (f_n^*) denote the corresponding biorthogonal functionals in $[f_n]^*$).

Given a disjoint sequence (f_n) as in the above definition, we can consider $Px = \sum_{k=1}^\infty f_{n_k}^*(x) f_{n_k}$, the canonical projection from $[f_n]$ onto $[f_{n_k}]$ (which has $\|P\| = 1$ because (f_n) is 1-unconditional). If E has the $\mathfrak{P}$ property, then we can now view

$$PTx = \sum_{k=1}^\infty f_{n_k}^*(Tx) f_{n_k} = \sum_{k=1}^\infty (T^* f_{n_k}^*)(x) f_{n_k}$$

as a bounded operator on E.

The $\mathfrak{P}$ property can be characterized as follows ([17, Proposition 3.3]):

Proposition 3.2. *Let E be a reflexive Banach lattice. The following are equivalent:*

(i) *For every disjoint positive normalized sequence $(f_n) \subset E$ there exists a positive operator $T : E \to [f_n]$, with $\liminf_n \operatorname{dist}(f_n, T(B_E)) < 1$.*

(ii) *For every disjoint positive normalized sequence $(f_n) \subset E$ there exists a positive operator $T : E \to [f_n]$, such that $\|T^* f_n^*\| \nrightarrow 0$.*

(iii) *E has the $\mathfrak{P}$ property.*

Notice that Banach lattices in which every disjoint positive sequence has a subsequence whose span is complemented by a positive projection satisfy the $\mathfrak{P}$ property. Examples of these include L_p spaces, Lorentz function spaces $\Lambda(W, p)$, Tsirelson's space, etc.

As intended, the assumption of the $\mathfrak{P}$ property yields a partial positive answer to the problem of stability by duality of DH Banach lattices.

Theorem 3.3. *Let E be a reflexive Banach lattice with the $\mathfrak{P}$ property. If E^* is DH, then E is DH. Moreover, in the particular case when E^* is p-DH, for some $1 < p < \infty$, then E is q-DH with $\frac{1}{p} + \frac{1}{q} = 1$.*

This fact, which was given in [17], can be used in particular to show that if a reflexive Banach lattice E is p-DH and satisfies a lower p-estimate, for some $1 < p < \infty$, then E^* is q-DH (with $\frac{1}{p} + \frac{1}{q} = 1$).

We focus now our attention on some examples of DH Banach lattices with non-DH duals. The existence of these examples shows that the $\mathfrak{P}$ property cannot be removed from Theorem 3.3.

Theorem 3.4. *Let $1 < p < \infty$ and φ an Orlicz function such that $\varphi(t) \simeq t^p$ on $[0,1]$ and $\varphi(t) \simeq t^p \log(1+t)$ on $[1,\infty)$. Then the Orlicz space $L_\varphi(0,\infty)$ is a reflexive p-DH Banach lattice whose dual is not DH.*

The proof of the fact that $L_\varphi(0,\infty)^*$ is not DH is based on a representation of functions in the set $C_\varphi(0,\infty)$ given in [36, p. 242] and Theorem 2.3. In particular, one can see that this dual Orlicz space contains sublattices isomorphic to the Orlicz sequence space ℓ_{ψ_α}, for $\psi_\alpha(t) = t^q |\log t|^\alpha$, where $\frac{1}{p} + \frac{1}{q} = 1$ and every $\alpha \in (0, \min\{1, q-1\})$.

This example can be used to construct another one within the category of *atomic* reflexive p-DH Banach lattices, more precisely, a weighted Orlicz sequence space.

Recall that given a sequence of positive numbers $w = (w_n)$ and an Orlicz function φ, the *weighted Orlicz sequence space* $\ell_\varphi(w)$ is the space of all sequences (x_n) such that $\sum_{n=1}^\infty \varphi(\frac{|x_n|}{s}) w_n < \infty$ for some $s > 0$, endowed with the Luxemburg norm. Notice that the unit vectors form an unconditional basis of $\ell_\varphi(w)$ when φ satisfies the Δ_2-condition.

Theorem 3.5. *Let $w = (w_n)$ be a sequence of positive numbers such that there is a subsequence (w_{n_k}) with $w_{n_k} \to 0$ and $\sum_{k=1}^\infty w_{n_k} = \infty$. If φ is an Orlicz function* *as in the previous theorem then the weighted Orlicz sequence space $\ell_\varphi(w)$ is p-DH but its dual is not DH.*

The proof is based on the space constructed in Theorem 3.4 together with an identification theorem for weighted Orlicz sequence spaces [18] and a universal property of these spaces given in to [12].

It should be noted that this kind of examples cannot be adapted to Orlicz spaces over a probability space (see Theorem 2.3 and the comments at the beginning of this section). But more generally, one might wonder whether a reflexive p-DH rearrangement invariant function space ([33, Chapter 2]) on the interval $[0,1]$ whose dual is not DH may exist.

4. Complemented disjoint sequences

It was mentioned earlier that Banach lattices in which every positive disjoint sequence has some subsequence whose span is complemented by a positive projection necessarily satisfy the $\mathfrak{P}$ property. We take now a closer look at this situation. We will say that a sequence (x_n) is said to be complemented in E if there is a projection P on E with Range $P = [x_n]$.

Notice that given a positive projection P onto the span of a disjoint sequence $(x_n) \subset E$, if (x_n^*) denote the biorthogonal functionals, then the sequence $(P^*x_n^*)$ need not be disjoint in E^*:

Example. Take $E = \mathbb{R}^3$ and let

$$x_1 = \begin{bmatrix} 1 \\ 0 \\ 0 \end{bmatrix}, \quad x_2 = \begin{bmatrix} 0 \\ 1 \\ 0 \end{bmatrix}, \quad \text{and} \quad P = \begin{bmatrix} 1 & 0 & 1 \\ 0 & 1 & 1 \\ 0 & 0 & 0 \end{bmatrix}.$$

Note that $Pe_1 = x_1$, $Pe_2 = x_2$, and $Pe_3 = x_1 + x_2$. It follows from $(P^*x_n^*)_i = \langle P^*x_n^*, e_i \rangle = \langle x_n^*, Pe_i \rangle$ that

$$P^*x_1^* = \begin{bmatrix} 1 \\ 0 \\ 1 \end{bmatrix} \quad \text{and} \quad P^*x_2^* = \begin{bmatrix} 0 \\ 1 \\ 1 \end{bmatrix},$$

so that $P^*x_1^*$ and $P^*x_2^*$ are not disjoint.

Interestingly enough, the following result proved in [17] shows that if a disjoint positive sequence spans a complemented subspace, then a positive projection whose adjoint sends the biorthogonal functionals to a disjoint sequence can be found.

Proposition 4.1. *Let E be a reflexive Banach lattice, (f_n) a positive disjoint sequence, and $R \in \mathcal{L}(E)$ a projection onto $[f_n]$. Then there exists a positive disjoint sequence (g_n^*) in E^* with $\langle g_n^*, f_m \rangle = \delta_{n,m}$ such that the operator $Px = \sum_{n=1}^{\infty} g_n^*(x) f_n$ defines a positive projection onto $[f_n]$ with $\|P\| \le \|R\|$.*

This fact gains relevance in connection with the following problem: *Does every reflexive Banach lattice contain a complemented positive disjoint sequence?*

We don't know the answer to this question. However, the following result, which is derived from Proposition 4.1, provides a useful reformulation.

Corollary 4.2. *Given a positive disjoint sequence (e_n) in a reflexive Banach lattice E, the following are equivalent:*

(i) *The subspace $[e_n]$ is complemented in E.*

(ii) *There exists a disjoint positive sequence (e_n^*) in E^* with $\langle e_n^*, e_m\rangle = \delta_{mn}$ such that $\sum_{n=1}^{\infty} e_n^*(x)e_n$ converges for each $x \in E$.*

Note that if $\sum_{n=1}^{\infty} e_n^*(x)e_n$ converges for every $x \in E$, then the map $P\colon x \mapsto \sum_{n=1}^{\infty} e_n^*(x)e_n$ defines a positive projection from E onto $[e_n]$. In particular, the above result yields that a reflexive Banach lattice E contains a complemented positive disjoint sequence if and only if E^* does.

The question about the existence of complemented disjoint sequences has a positive answer for most examples of Banach lattices considered in the literature. For instance, if a Banach lattice is atomic (or has an infinite atomic part), that means that E has an unconditional basis inducing the order, and trivially this provides a positive disjoint complemented sequence.

On the other hand, it is well known that in a non-atomic order-continuous Banach lattice E, every unconditional basic sequence (u_n) spanning a complemented subspace is equivalent to a disjoint sequence (f_n) spanning also a complemented subspace provided that $[u_n]$ is lattice anti-Euclidean (that is, $[u_n]$ does not contain uniformly complemented lattice copies of ℓ_2^n for every n, see [8, Theorem 3.4]).

Another family of spaces which always contain complemented disjoint sequences is that of rearrangement invariant spaces. Using the averaging projection, every sequence of normalized characteristic functions over a family of disjoint sets is complemented in any r.i. space (cf. [33, Theorem 2.a.4]).

For DH Banach lattices, the existence of complemented disjoint sequences turns out to be equivalent to the $\mathfrak{P}$ property studied above. This was proved in [17, Theorem 4.4]:

Theorem 4.3. *Let E be a DH Banach lattice. E has the $\mathfrak{P}$ property if and only if E contains a complemented positive disjoint sequence.*

In fact, in most instances of DH Banach lattices such as L_p spaces, Lorentz spaces and some Orlicz spaces, every disjoint sequence has a complemented subsequence. This motivates the following.

Definition 4.4. A Banach lattice E is called *disjointly complemented* (DC) if every disjoint sequence (x_n) has a subsequence whose span is complemented in E.

The study of the relation between DC and DH Banach lattices appears now natural. More specifically, we are interested in deciding whether DH Banach lattices must be DC.

Let us consider first the non-reflexive case. Recall that if E is non-reflexive, then E either contains a lattice copy of c_0 or of ℓ_1 (cf. [34, Theorem 2.4.15]). Therefore, if E is DH and non-reflexive, it follows that it is either 1-DH or ∞-DH.

From Sobczyk's Theorem ([1, Theorem 2.5.9]), it easily follows that if E is a separable Banach lattice which is ∞-DH then it is DC. For the 1-DH case, we will use the following fact ([34, Lemma 2.3.11]): If a positive disjoint sequence in a Banach lattice is equivalent to the unit vector basis of ℓ_1 then its closed span is complemented. It follows that if E is 1-DH, then every positive disjoint sequence has a complemented subsequence. Splitting a sequence into its positive and negative parts it can be seen that 1-DH Banach lattices are in fact DC.

Hence, *if E is a separable non-reflexive Banach lattice which is DH, then E is DC.* Clearly, the separability of E is essential here: ℓ_∞ is non-reflexive and DH, however it is not DC. In fact, every normalized disjoint sequence is equivalent to the unit vector basis of c_0 and by Phillips–Sobczyk's theorem (cf. [1, Theorem 2.5.5], [32, Theorem 2.a.7]), the space ℓ_∞ does not contain any complemented subspace isomorphic to c_0.

Let us consider now the case of reflexive Banach lattices. Does DH imply DC in this context? Recall that in Theorem 3.3 it was proved that a reflexive Banach lattice E with the $\mathfrak{P}$ property is DH provided so is E^*. The following theorem gives a partial answer to this question. It summarizes the work done in [17].

Theorem 4.5. *Let E be a reflexive Banach lattice which contains a complemented positive disjoint sequence. If E is DH, then the following are equivalent:*

(a) *E^* is DH,*
(b) *E^* has the $\mathfrak{P}$ property,*
(c) *E^* is DC.*

Furthermore, if E and E^* are DH, then E is DC.

Using ideas from [10] it can also be shown that if E is p-DH and p-convex Banach lattice for some $1 \le p < \infty$, then E is DC.

5. Uniformly DH Banach lattices

Until now no attention has been given to the equivalence constants involved in the definition of a DH Banach lattice. The purpose of this section is to illustrate the role played by these.

In general an ℓ_p-sum of p-DH spaces need not be DH. Indeed, given $n \in \mathbb{N}$, let X_n denote the completion of the space of all eventually zero sequences c_{00} with respect to the norm

$$\|(a_k)\|_{X_n} = \sup\Big\{\sum_{i=1}^{n}|a_{k_i}| + \Big(\sum_{i>n}|a_{k_i}|^p\Big)^{1/p} : k_1 < k_2 < \cdots < k_i < \cdots\Big\}.$$

It is easy to see that $\|\cdot\|_{X_n}$ is equivalent to the ℓ_p norm. In fact, we have

$$\|(a_k)\|_{\ell_p} \le \|(a_k)\|_{X_n} \le (n^{\frac{1}{q}} + 1)\|(a_k)\|_{\ell_p}.$$

In [17, Example 6.4] it was proved that the space $\left(\bigoplus_{n=1}^{\infty} X_n\right)_{\ell_p}$ endowed with the ℓ_p-sum of the corresponding norms $\|\cdot\|_{X_n}$ is not DH.

Note that in the above example, the equivalence constant of disjoint sequences in different X_n-summands grows without bound. After this example it seems only natural to introduce the following:

Definition 5.1. A Banach lattice E is *uniformly disjointly homogeneous* if there is a constant $C > 0$ such that every two disjoint normalized sequences (x_n) and (y_n) in E, have subsequences such that $(x_{n_k}) \overset{C}{\sim} (y_{n_k})$.

Clearly every uniformly DH Banach lattice E is DH. The converse is nevertheless not true. In fact, stemming from deep results of W.B. Johnson and E. Odell in [23], and H. Knaust and E. Odell in [27] the following result is given in [17].

Theorem 5.2. *For every $1 < p < \infty$, there exists a super-reflexive atomic Banach lattice E_p which is p-DH but not uniformly DH.*

As a by-product, another example of a reflexive DH Banach lattice whose dual is not DH is obtained.

In Theorem 3.5 we have constructed examples of reflexive atomic Banach lattices (with the order induced by a 1-unconditional basis), which are DH, but whose dual spaces are not. The case of atomic Banach lattices with the order induced by a subsymmetric basis deserves some attention. Recall that a basis (x_n) is called *subsymmetric* if it is unconditional and every subsequence (x_{n_i}) is equivalent to (x_n) (cf. [32, Chapter 3]).

Also, recall that a normalized basis (e_n) in a Banach space X is said to be a *Rosenthal basis* if every normalized block-sequence of (e_n) contains a subsequence equivalent to (e_n). It is an open question whether such a basis is necessarily equivalent to the unit basis of ℓ_p or c_0, see [13] for further details and partial results in this direction. In particular, it was observed in [13, p. 397] that a Rosenthal basis (x_n), always satisfies that every subsequence (x_{n_i}) is equivalent to (x_n).

Proposition 5.3. *Let E be a reflexive atomic Banach lattice with the order induced by a subsymmetric basis (e_n). Then E is DH if and only if (e_n) is a Rosenthal basis.*

Let X be a Banach space with a Rosenthal basis (e_n). It was proved in [13, Theorem 1, Proposition 7] that (e_n) is equivalent to the unit basis of ℓ_p or c_0 if (e_n) is "uniformly" Rosenthal or if E^* also has a Rosenthal basis. In view of Proposition 5.3, we can now restate these statements in terms of disjoint homogeneity as follows.

Proposition 5.4. *Let E be a reflexive atomic Banach lattice with the order induced by a subsymmetric normalized basis (e_n). Then (e_n) is equivalent to the unit basis of ℓ_p for some $1 < p < \infty$ if any of the following conditions is satisfied:*

(i) *E is uniformly DH, or*

(ii) *E and E^* are both DH.*

In particular, if (e_n) is symmetric, then Proposition 5.4 also follows from [32, Theorem 3.a.10] due to Z. Altshuler. Indeed, if E is DH and (v_n) is a sequence generated by one vector (which is automatically symmetric), then (v_n) and (e_n) have equivalent subsequences, hence are themselves equivalent. Now apply the same argument to (e_n^*) in E^*.

We do not know whether every atomic reflexive Banach lattice with the order induced by a subsymmetric basis which is DH must be isomorphic to ℓ_p for some $1 < p < \infty$. In this direction, if we consider the symmetric version of Tsirelson space (see [9, Chapter X, B]), which does not contain ℓ_p subspaces, then it is not hard to see that this space fails being DH. However, let us suppose that E is a reflexive Banach lattice with the $\mathfrak{P}$ property containing a disjoint subsymmetric sequence, if E^* is *DH*, then E must be *p-DH* for some $1 < p < \infty$ (see [17, Corollary 6.10]).

6. Compact powers of strictly singular operators

The purpose of this section is to show how DH Banach lattices can be applied to the theory of strictly singular operators. In particular, we are interested in the extension of a result by V. Milman [35], which asserts that every strictly singular endomorphism on L_p has compact square. This kind of results have also been studied in [3] in the context of Banach spaces.

Recall that an operator between Banach spaces is *strictly singular* if it is not invertible on any infinite-dimensional subspace. This is an important class of operators which was first introduced in connection with the perturbation of Fredholm operators [26], and has later proved relevant in the modern theory of Banach spaces (see [4]).

Given a Banach space X, we will denote by $\mathcal{K}(X)$ (respectively $\mathcal{S}(X)$) the space of all compact (resp. strictly singular) endomorphisms on X.

A close notion to strict singularity was introduced in the setting of Banach lattices ([20]): given a Banach lattice E and a Banach space X, an operator $T : E \to X$ is *disjointly strictly singular* (DSS) if for any sequence of pairwise disjoint elements (x_n) in E, the restriction of T to the span $[x_n]$ is not invertible. Recall also that an operator $T : E \to X$ is *AM-compact* whenever $T([-x, x])$ is a relatively compact set in X for every $x \in E_+$ (recall that the order interval $[-x, x]$ is the set $\{y \in E : |y| \leq x\}$).

In [15] several results about compactness of operators belonging to the singular classes given above were proved in the context of regular operators. Recall that an operator between Banach lattices is positive when it maps positive elements to positive elements, and a regular operator is a difference of two positive ones.

Theorem 6.1. *Suppose that E is a DH Banach lattice with order-continuous norm and a weak unit. Suppose that S and T are two regular operators on E such that S is disjointly strictly singular and T is AM-compact.*

(i) *If E^* is order continuous then ST is compact.*
(ii) *If E^* is not order continuous then TS is compact.*

In particular, if R is disjointly strictly singular and regular, then STR is compact.

Observe that Theorem 6.1 remains valid in the case that S is not regular. Also, it remains valid if, instead of being disjointly strictly singular, S is only assumed to be weakly compact. In particular, the above result yields that if E is DH and $T : E \to E$ is regular, disjointly strictly singular, and AM-compact, then T^2 is compact.

A Banach lattice E has finite cotype (or equivalently finite concavity) if and only if E does not contain copies of ℓ_∞^n uniformly (cf. [33]). Moreover, every Banach lattice E with finite concavity satisfies the *subsequence splitting property* ([41]). This means that every bounded sequence (x_n) in E has a subsequence that can be written as $x_{n_k} = g_k + h_k$, with $|g_k| \wedge |h_k| = 0$, the sequence (g_k) being equi-integrable and (h_k) disjoint. Recall that a bounded sequence (g_n) in a Banach lattice of measurable functions over a measure space (Ω, Σ, μ) is equi-integrable if $\sup_n \|g_n \chi_A\| \to 0$ as $\mu(A) \to 0$. Note that every Banach lattice with finite cotype is order continuous.

Theorem 6.2. *Let E be a DH Banach lattice with the subsequence splitting property, such that E^* is order continuous. If $T : E \to E$ is a regular operator which is disjointly strictly singular and AM-compact, then T is compact.*

In [16] several results in similar spirit were given without the restriction of regularity. An important technique that was exploited in these arguments is the well-known Kadec–Pełczyński's dichotomy (see [14], [33]): given a normalized sequence (x_n) in an order-continuous Banach lattice E

(i) either $(\|x_n\|_{L_1})$ is bounded away from zero,
(ii) or there exist a subsequence (x_{n_k}) and a disjoint sequence (z_k) in E such that $\|z_k - x_{n_k}\| \longrightarrow 0$ as $k \to \infty$.

An operator $T : E \to X$ is called M-*weakly compact* if it maps disjoint sequences in B_E to sequences converging to zero. Notice that an operator is compact if and only if it is AM-compact and M-weakly compact ([34, Proposition 3.7.4]). There is a notion dual to M-weak compactness, namely, an operator $T : X \to E$ is L-*weakly compact* if every disjoint sequence in the solid hull of $T(B_X)$ tends to zero in norm. The following fact was given in [40] for endomorphism on L_p spaces.

Proposition 6.3. *Let E be a reflexive DH Banach lattice and $T : E \to E$ be a positive operator. The following are equivalent:*

(i) *T is disjointly strictly singular.*
(ii) *T is M-weakly compact.*
(iii) *T is L-weakly compact.*

Proof. An M-weakly compact operator is clearly disjointly strictly singular. For the converse implication, suppose that T is not M-weakly compact. Thus, there is a disjoint normalized sequence (x_n) in E such that $\|Tx_n\|_E \geq \alpha > 0$.

Observe that $(|x_n|)$ is also a disjoint normalized sequence, and since E is reflexive it must be weakly null. Hence, so is $(T|x_n|)$, and since T is positive it follows that $\|T|x_n|\|_{L_1} \to 0$. Note that

$$\|T|x_n|\|_E \geq \|Tx_n\|_E \geq \alpha > 0,$$

so by Kadec–Pełczyński's dichotomy, $(T|x_n|)$ has a subsequence equivalent to a disjoint sequence in E. Since E is DH, $(|x_n|)$ and $(T|x_n|)$ have an equivalent subsequence and T is not DSS. This proves the equivalence of the first two statements. The equivalence with the third one follows from [34, Theorem 3.6.17]. □

Recall that an operator between Banach spaces $T : X \to Y$ is *Dunford–Pettis* if it maps weakly null sequences to sequences converging to zero. The following result from [16] can be seen as an extension of the classical result stating that weakly compact operators on L_1 are Dunford–Pettis.

Theorem 6.4. *Let E be a 1-DH Banach lattice with finite cotype. Every operator $T \in \mathcal{S}(E)$ is Dunford–Pettis.*

Using that the composition of a weakly compact with a Dunford–Pettis operator is a compact operator, it follows that every strictly singular operator on a 1-DH Banach lattice with finite cotype has compact square.

Before we present the extension of Milman's result ([35]) on compactness of the square of strictly singular endomorphisms on L_p spaces we need the following.

Definition 6.5. A Banach lattice E has property (C) if it is order continuous, and there exist $q < \infty$ and a probability space (Ω, Σ, μ) such that the inclusions $L_q(\mu) \hookrightarrow E \hookrightarrow L_1(\mu)$ hold.

Note that condition (C) is a very mild assumption. Indeed, every Banach lattice with a weak order unit (for instance separable) and finite cotype satisfies property (C) (see [22, p. 14]). Moreover, every order-continuous rearrangement invariant function space on $[0, 1]$ with upper Boyd index $q_X < \infty$ also has property (C) (though it may have trivial cotype, [33, Proposition 2.b.3]). Recall that for an r.i. function space X the *Boyd indices* are given by

$$p_X = \lim_{s\to\infty} \frac{\log s}{\log \|D_s\|} \qquad\qquad q_X = \lim_{s\to 0^+} \frac{\log s}{\log \|D_s\|},$$

where $D_s : X \to X$ is the dilation operator given by $(D_s f)(t) = f(t/s)$ for $t \leq \min(1, s)$ and zero otherwise (see [33, Section 2.b]).

In [16], the following was proved.

Proposition 6.6. *Let E be a DH Banach lattice with property (C). If $T \in \mathcal{S}(E)$ then T^2 is AM-compact.*

This is a first step in the proof of the next theorem also from [16].

Theorem 6.7. *Let E be a DH Banach lattice with finite cotype and an unconditional basis. Every operator $T \in \mathcal{S}(E)$ satisfies that the square T^2 is compact.*

The existence of an unconditional basis is a technical condition in the previous proof and indeed not a restriction for many spaces. We conjecture that the result is still true without it. In fact, there are some situations in which there is no need to impose it:

Theorem 6.8. *If E is a p-DH Banach lattice ($2 \leq p \leq \infty$) with property (C), then every operator $T \in \mathcal{S}(E)$ has compact square.*

A classical result of J. Calkin [6] states that the only non-trivial closed ideal of operators in Hilbert space is the ideal of compact operators. In particular, as pointed out by T. Kato [26], on Hilbert spaces the ideals of strictly singular and compact operators coincide. In fact, this is a particular case of a more general result involving 2-DH Banach lattices ([16, Theorem 2.12]):

Theorem 6.9. *If E is a 2-DH Banach lattice with property (C), then $\mathcal{S}(E) = \mathcal{K}(E)$.*

To finish this section, the case of atomic Banach lattices deserves its proper space. Note that in the atomic case the class of DH Banach lattices E with a basis of disjoint vectors is a rather small class, since "most" basic sequences in E are equivalent to disjoint sequences. As mentioned earlier the examples include ℓ_p spaces and c_0, and also Tsirelson spaces and their generalizations (cf. [9]), as well as ℓ_p-sums of finite-dimensional Banach lattices $\ell_p(X_n)$. Also mentioned earlier was that Lorentz and Orlicz sequence spaces (distinct from spaces ℓ_p) are not disjointly homogeneous.

In the atomic setting, Theorem 6.7 is improved, similarly to the case of ℓ_p spaces where strictly singular and compact endomorphisms coincide (cf. [32, p. 76]). This is shown in the following result ([16]):

Theorem 6.10. *Let E be an atomic Banach lattice with a basis. If E is DH then every operator $T \in \mathcal{S}(E)$ is compact.*

7. Applications to operators on rearrangement invariant spaces

In the class of rearrangement invariant spaces it happens that the behaviour of powers of endomorphisms determines the behaviour of the composition of (different) operators. This is the content of the next result (see [16] for details).

Proposition 7.1. *Given a rearrangement invariant space E on $[0,1]$ and $n \in \mathbb{N}$ the following statements are equivalent:*

(i) *If an operator $T \in \mathcal{S}(E)$, then the power T^n is compact.*
(ii) *If $T_1, \ldots, T_n$ belong to $\mathcal{S}(E)$, then the composition $T_n \cdots T_1$ is compact.*

As a consequence of Theorems 6.4, 6.7 and 6.9 we get the following:

Proposition 7.2. *Given $1 < p < \infty$ and $1 \leq q < \infty$, every operator $T \in \mathcal{S}(\Lambda(W,q)[0,1])$ or $T \in \mathcal{S}(L_{p,q}[0,1])$ has a compact square. Moreover, if $q = 2$ then T is already compact, while if $q = 1$, then T is Dunford–Pettis.*

By contrast, if $q \neq 2$, then strictly singular non-compact operators on $L_{p,q}$ can be found. Take for instance a complemented subspace isomorphic to ℓ_q and the span of the Rademacher functions which is isomorphic to ℓ_2. Denote $P_1 : L_{p,q} \to \ell_q$ and $P_2 : L_{p,q} \to \ell_2$ the corresponding projections, $i_{s,t} : \ell_s \to \ell_t$ the canonical inclusion, and $Q : \ell_q \hookrightarrow L_{p,q}$ and $R : \ell_2 \hookrightarrow L_{p,q}$ the corresponding embeddings. When $q < 2$ consider $T = Ri_{q,2}P_1 \in \mathcal{S}(L_{p,q})\backslash\mathcal{K}(L_{p,q})$, and when $q > 2$ take $S = Qi_{2,q}P_2 \in \mathcal{S}(L_{p,q})\backslash\mathcal{K}(L_{p,q})$.

The behavior of the maximal Lorentz spaces $L_{p,\infty}$ is quite different. In fact, there exists an operator $T \in \mathcal{S}(L_{p,\infty})$, for $p \neq 2$, whose cube T^3 is not compact. The proof is based on a particular way of embedding ℓ_p as a complemented subspace into $L_{p,\infty}$ (notice that for $p < 2$ even L_p can be embedded as a complemented subspace of $L_{p,\infty}$, see [24]), and the fact that $\ell_{p,\infty}$ embeds as a complemented sublattice into $L_{p,\infty}$ [30]. See [16, Proposition 3.3] for details.

Similarly strictly singular operators on $L_{2,\infty}$ with non-compact squares can be defined. However, we do not know whether there might exist some $n \in \mathbb{N}$ such that T^n is compact whenever $T \in \mathcal{S}(L_{p,\infty})$, or even whether every operator $T \in \mathcal{S}(L_{p,\infty})$ is power-compact.

Observe also that if $L^0_{p,\infty}$ denotes the order-continuous part of $L_{p,\infty}$, then every strictly singular operator on $L^0_{p,\infty}$ has compact square. This follows from Theorem 6.8, since $L^0_{p,\infty}$ is ∞-DH and his upper Boyd index equals p. A similar statement also holds for order-continuous Marcinkiewicz spaces $M(\varphi)$ with finite upper Boyd index (since they are also ∞-DH Banach lattices, cf. [38]).

Note however that these results do not hold for Lorentz spaces $L_{p,q}(0,\infty)$ (for $p \neq q$) as they contain complemented lattice copies of the non-DH spaces $\ell_{p,q}$.

In the case of Orlicz spaces over a probability measure space L_φ is DH if and only if $E^\infty_\varphi \cong \{t^p\}$ (Theorem 2.3). This condition implies the equality of the indices $s(L_\varphi) = \sigma(L_\varphi) = p$, or equivalently the equality of the associated Boyd indices p_{L_φ} and q_{L_φ} as it follows from the identities $s(L_\varphi) = p_{L_\varphi}$ and $\sigma(L_\varphi) = q_{L_\varphi}$ (cf. [33, p. 139]). Thus, from Theorems 6.4 and 6.7 the following result is obtained.

Proposition 7.3. *Let φ be an Orlicz function such that $E^\infty_\varphi \cong \{t^p\}$, for some $1 \leq p < \infty$. If an operator $T \in \mathcal{S}(L_\varphi[0,1])$ then the square T^2 is compact. Furthermore for $p = 2$, the operator T is already compact, while for $p = 1$, T is Dunford–Pettis.*

Many Orlicz functions satisfy the condition $E^\infty_\varphi \cong \{t^p\}$, for example the class of all Orlicz functions of regular variation, i.e.,

$$\lim_{t\to\infty} \frac{t\varphi'(t)}{\varphi(t)} = p.$$

In general, we cannot weaken this condition on E^∞_φ, as there exist Orlicz spaces L_φ with indices $s(L_\varphi) = \sigma(L_\varphi) = p$, and an operator $T \in \mathcal{S}(L_\varphi)$ whose square T^2 is not compact (while for $p = 2$, we have a strictly singular non-compact operator, see [16, Proposition 4.3]).

Clearly these Orlicz spaces are not disjointly homogeneous (this follows from Theorem 6.7). More generally, every *minimal* Orlicz function space L_φ (different from L_p) is not disjointly homogeneous. Indeed, recall that in general for each $\psi \in C_\varphi^\infty$ there exists a sequence of normalized disjoint functions in L_φ equivalent to the symmetric canonical basis of ℓ_ψ ([31, Proposition 4]). Now, since φ is minimal, we have, by [19, Proposition 1], that $E_{\varphi,1}^\infty = E_\varphi^\infty = E_\varphi = E_{\varphi,1}$ and the set $E_{\varphi,1}^\infty$ contains uncountably many mutually non-equivalent Orlicz functions (see the proof of [32, Theorem 4.b.9]). Hence, using the symmetry, we deduce that in L_φ there are uncountable many sequences of normalized disjoint functions with no equivalent subsequence.

Notice also that in the class of Orlicz spaces L_φ with different indices ($s(L_\varphi) \neq \sigma(L_\varphi)$) there are no DH spaces. This follows from the fact that for each $p \in [s(L_\varphi), \sigma(L_\varphi)]$ we have $t^p \in C_\varphi^\infty$ and there exist sequences of normalized disjoint functions in L_φ that are equivalent to the canonical basis of ℓ_p ([31, Proposition 4]).

New examples of DH r.i. spaces in connection with interpolation theory can be found also in the recent paper [5].

8. The Kato property in rearrangement invariant spaces

As mentioned above the ideals of strictly singular and compact operators coincide on Hilbert spaces as well as on 2-DH function spaces (under some mild assumptions).

In this section we consider the natural converse question: Assume that E is an r.i. function space such that every strictly singular operator $T \in \mathcal{L}(E)$ is compact. Must E be 2-DH?

This question has been addressed in [21]. First, let us introduce the following

Definition 8.1. A Banach space X has the *Kato property* whenever $\mathcal{S}(X) = \mathcal{K}(X)$.

Examples of function spaces with the Kato property clearly include Hilbert spaces, Lorentz spaces of the form $L_{p,2}[0,1]$ and $\Lambda(W,2)[0,1]$ as well as certain Orlicz spaces $L_\varphi[0,1]$ like $\varphi(t) = t^2 \log^\alpha(1+t)$ for arbitrary α. The Kato property is also enjoyed by the sequence spaces ℓ_p ($1 \leq p < \infty$), c_0, Tsirelson space (and some of its modifications) and also some not so classical spaces such as the space $\mathcal{X}_{AH}$ (constructed in [4] as a solution to the scalar-plus-compact problem). Notice that the Kato property is an isomorphic property. Moreover, we have the following:

Proposition 8.2. *Let X be a Banach space with the Kato property. Suppose that for some subspace $Y \subset X$, there is $Z \subset X$ such that $Y \simeq X/Z$, then Y also has the Kato property. In particular, every complemented subspace of a space with the Kato property also has the Kato property.*

A weaker version of the 2-DH property is the following:

Definition 8.3. An r.i. space on $[0,1]$ is *restricted 2-DH* if for every sequence of disjoint sets $(A_n)_{n=1}^\infty$ in $[0,1]$ there is a subsequence such that $(\frac{1}{\|\chi_{A_{n_k}}\|}\chi_{A_{n_k}})_{k=1}^\infty$ is equivalent to the unit vector basis of ℓ_2.

Proposition 8.4. *Let E be an r.i. space on $[0,1]$. The space E is restricted 2-DH if and only if every subsequence of disjoint elements of the normalized Haar basis in E has a further subsequence equivalent to the unit vector basis of ℓ_2.*

This implies the following: *If E is an r.i. space on $[0,1]$ which is isomorphic to a 2-DH r.i. space F, then E is restricted 2-DH.* Indeed, this follows from [22, Theorem 6.1], since either $E = F$ up to equivalence of norms, or $E = L_2[0,1]$, or the Haar basis in E is equivalent to a sequence of disjoint elements in F and, in this case, the result is a consequence of Proposition 8.4.

As mentioned in Section 3, the 2-DH property is not stable in general by duality. However, restricted 2-DH r.i. spaces on $[0,1]$ are stable under duality. We do not know if in general an r.i. space E on $[0,1]$ which is restricted 2-DH, must be 2-DH. In the class of Orlicz spaces $L_\varphi[0,1]$ this is the case:

Proposition 8.5. *For an Orlicz space $L_\varphi[0,1]$, the following are equivalent:*

(i) *$L_\varphi[0,1]$ is 2-DH.*
(ii) *$L_\varphi[0,1]$ is restricted 2-DH.*
(iii) *Every function in E_φ^∞ is equivalent to the function t^2 at 0.*

For non-reflexive r.i. spaces we can use Lozanovski's theorem and the factorization through ℓ_1 and c_0 to obtain the following:

Proposition 8.6. *If E is a non-reflexive r.i. space on $[0,1]$, then E fails to have the Kato property.*

In the class of Lorentz function spaces we have that *$L_{p,q}[0,1]$ and $\Lambda(W,p)[0,1]$ have the Kato property if and only if they are 2-DH.*

On the other hand, for Orlicz spaces the study of the Kato property is more involved. Clearly every 2-DH Orlicz space on $[0,1]$ has the Kato property. Remarkably *if $L_\varphi[0,1]$ is a reflexive 2-convex (or 2-concave) Orlicz space with the Kato property then it is 2-DH.* The proof uses the fact that the associated Orlicz sequence space ℓ_ψ is 2-convex (or 2-concave), so that the inclusion $\ell_2 \hookrightarrow \ell_\psi$ (or $\ell_\psi \hookrightarrow \ell_2$) is strictly singular. Finally, by composing with canonic projections we get a non-compact strictly singular operator on $L_\varphi[0,1]$.

The condition (iii) in Proposition 8.5 which characterizes 2-DH Orlicz spaces can be rewritten as the formula

$$\sup_{0<t<\infty} \limsup_{u\to\infty} \frac{\varphi(tu)}{t^2\varphi(u)} < \infty.$$

Strengthening slightly this condition we get further necessary conditions for the Kato property in Orlicz spaces.

Theorem 8.7. *Let $L_\varphi[0,1]$ be a reflexive Orlicz space. If*

$$\lim_{t\to 0} \lim_{u\to\infty} \frac{\varphi(tu)}{t^2\varphi(u)} \in \{0,\infty\},$$

then $L_\varphi[0,1]$ fails to have the Kato property.

The proof relies on factorization results. Thus, in the case that φ satisfies

$$\lim_{t\to 0}\lim_{u\to\infty}\frac{\varphi(tu)}{t^2\varphi(u)}=\infty,$$

it can be shown, using N. Kalton's characterization for strictly singular inclusions between Orlicz sequence spaces (cf. [32]), that there exist a sequence of disjoint measurable sets (A_k) in $[0,1]$, and an Orlicz function ψ such that the sequence $(\chi_{A_k}/\|\chi_{A_k}\|)$ is equivalent to the unit vector basis of the sequence space ℓ_ψ; in addition, the inclusion $\ell_\psi\subset\ell_2$ holds and it is strictly singular.

We do not know whether every Orlicz space $L_\varphi[0,1]$ with the Kato property must be 2-DH. Notice that for infinite measures the answer is negative as shown with the following:

Example. Consider

$$\varphi(t)=\begin{cases}\frac{1}{\log 2}t^2 & t\in[0,1],\\ \frac{t^2}{\log(1+t)} & t\in[1,\infty).\end{cases}$$

Then the reflexive space $L_\varphi(0,\infty)$ has the Kato property but is not 2-DH.

Indeed, note that $L_\varphi(0,\infty)$ is isomorphic to $L_\varphi[0,1]$ ([22]). But $L_\varphi[0,1]$ is 2-DH and thus it has the Kato property. On the other hand, it can be shown that the function $t^2\log(|\log t|)$ belongs to the set $C_\varphi(0,\infty)$. Hence, using Theorem 2.3 we conclude that the space $L_\varphi(0,\infty)$ is not 2-DH.

9. Some open questions

In this final section we collect several questions that arose along the preceding sections.

Question 1. Does every reflexive Banach lattice contain a disjoint sequence whose span is complemented?

Question 2. Is every separable DH Banach lattice DC?

Question 3. Is there a reflexive p-DH r.i. space on $[0,1]$ whose dual is not DH?

Question 4. Is every DH atomic Banach lattice with a symmetric basis isomorphic to $l_p,(1\le p<\infty)$ or c_0?

Question 5. If X is an r.i. space on $[0,1]$ with the Kato property, must X be 2-DH? In particular, is every Orlicz space $L_\varphi[0,1]$ with the Kato property necessarily 2-DH?

Acknowledgment

We want to thank E.M. Semenov, E. Spinu and V.G. Troitsky for all the stimulating discussions that motivated many of the results included in this survey.

References

[1] F. Albiac and N.J. Kalton, *Topics in Banach space theory.* Graduate Texts in Mathematics 233, Springer, 2006.

[2] C.D. Aliprantis and O. Burkinshaw, *Positive Operators.* Springer, Dordrecht, 2006.

[3] G. Androulakis, P. Dodos, G. Sirotkin, and V.G. Troitsky, *Classes of strictly singular operators and their products.* Israel J. Math **169** (2009), 221–250.

[4] S.A. Argyros and R.G. Haydon, *A hereditarily indecomposable $\mathcal{L}_\infty$-space that solves the scalar-plus-compact problem.* Acta Math. **206** (2011), no. 1, 1–54.

[5] S. Astashkin *Disjointly homogeneous rearrangement invariant spaces via interpolation.* J. Math. Anal. Appl. **421** (2015), no. 1, 338–361.

[6] J.W. Calkin, *Abstract symmetric boundary conditions.* Trans. Amer. Math. Soc. **45** (1939), no. 3, 369–442.

[7] N.L. Carothers and S.J. Dilworth, *Subspaces of $L_{p,q}$.* Proc. Amer. Math. Soc. **104** (1988), no. 2, 537–545.

[8] P.G. Casazza and N.J. Kalton, *Uniqueness of unconditional bases in Banach spaces.* Israel J. Math. **103** (1998), 141–175.

[9] P.G. Casazza and T.J. Shura, *Tsirelson's space.* Lecture Notes in Mathematics, 1363, 1989.

[10] D. Chen, W.B. Johnson, and B. Zheng, *Commutators on $(\sum \ell_q)_p$.* Studia Math. **206** (2011), no. 2, 175–190.

[11] J. Diestel, *Sequences and series in Banach spaces.* Graduate Texts in Mathematics, 92. Springer-Verlag, New York, 1984.

[12] L. Drewnowski, *F-spaces with a basis which is shrinking but not hyper-shrinking.* Studia Math. **64** (1979), no. 2, 97–104.

[13] V. Ferenczi, A.M. Pelczar, and C. Rosendal, *On a question of Haskell P. Rosenthal concerning a characterization of c_0 and ℓ_p.* Bull. London Math. Soc. **36** (2004), no. 3, 396–406.

[14] T. Figiel, W.B. Johnson, and L. Tzafriri, *On Banach lattices and spaces having local unconditional structure with applications to Lorentz funtion spaces.* J. Approximation Theory **13** (1975), 395–412.

[15] J. Flores, P. Tradacete, and V.G. Troitsky, *Disjointly homogeneous Banach lattices and compact product of operators.* J. Math. Anal. Appl. **354** (2009), 657–663.

[16] J. Flores, F.L. Hernández, E.M. Semenov, and P. Tradacete, *Strictly singular and power-compact operators on Banach lattices.* Israel J. Math. **188** (2012), 323–352.

[17] J. Flores, F.L. Hernández, E. Spinu, P. Tradacete, and V.G. Troitsky, *Disjointly homogeneous Banach lattices: duality and complementation.* J. Funct. Anal. **266**, (2014), no. 9, 5858–5885.

[18] F. Fuentes and F.L. Hernández, *On weighted Orlicz sequence spaces and their subspaces.* Rocky Mountain J. Math. **18** (1988), no. 3, 585–599.

[19] F.L. Hernández and V. Peirats, *Orlicz function spaces without complemented copies of ℓ_p.* Israel J. Math. **56** (1986), no. 3, 355–360.

[20] F.L. Hernández and B. Rodríguez-Salinas, *On l_p complemented copies in Orlicz spaces II.* Israel J. Math. **68**, (1989) 27–55.

[21] F.L. Hernández, E.M. Semenov, and P. Tradacete, *Rearrangement invariant spaces with Kato property.* Special Issue dedicated to L. Drewnowski. Funct. Approx. Comment. Math. **50** (2014), no. 2, 215–232.

[22] W.B. Johnson, B. Maurey, G. Schechtman, and L. Tzafriri, Symmetric structures in Banach spaces. *Mem. Amer. Math. Soc.* 217, (1979).

[23] W.B. Johnson and E. Odell, *Subspaces of L_p which embed into ℓ_p.* Compositio Math. **28** (1974), 37–49.

[24] N.J. Kalton, *Banach spaces embedding into L_0.* Israel J. Math. **52** (1985), 305–319.

[25] A. Kamińska and M. Mastylo, *The Schur and (weak) Dunford–Pettis properties in Banach lattices.* J. Aust. Math. Soc. **73** (2002), no. 2, 251–278.

[26] T. Kato, *Perturbation theory for nullity deficiency and other quantities of linear operators.* J. Analyse Math. **6** (1958) 273–322.

[27] H. Knaust and E. Odell, *Weakly null sequences with upper ℓ_p-estimates.* Functional analysis (Austin, TX, 1987/1989), 85–107, Lecture Notes in Math., 1470, Springer, Berlin, 1991.

[28] M.A. Krasnosel'skiĭ and Ja.N. Rutickiĭ, *Convex functions and Orlicz spaces.* Izdat. Fiz.-Mat. Lit., Moscow, 1958. English translation Noordhoff Ltd., Groningen 1961.

[29] J.L. Krivine and B. Maurey, *Espaces de Banach stables.* Israel J. Math. **39** (1981), no. 4, 273–295.

[30] D.H. Leung, *Embedding $\ell^{p,\infty}$ into $L^{p,\infty}[0,1]$ complementably.* Bull. London Math. Soc. **23**, (1991), no. 6, 583–586.

[31] J. Lindenstrauss and L. Tzafriri, *On Orlicz sequence spaces. III.* Israel J. Math. **14** (1973), 368–389.

[32] J. Lindenstrauss and L. Tzafriri, *Classical Banach spaces. I.* Springer-Verlag, Berlin, 1977.

[33] J. Lindenstrauss and L. Tzafriri, *Classical Banach spaces. II.* Springer-Verlag, Berlin, 1979.

[34] P. Meyer-Nieberg, *Banach lattices.* Springer-Verlag, Berlin, 1991.

[35] V.D. Milman, *Operators of class C_0 and C_0^*.* (Russian) Teor. Funkcii Funkcional. Anal. i Prilozen. **10** (1970), 15–26.

[36] N.J. Nielsen, *On the Orlicz function spaces $L_M(0,\infty)$.* Israel J. Math. **20** (1975), no. 3–4, 237–259.

[37] S.Ja. Novikov, E.M. Semenov, and E.V. Tokarev, *The structure of subspaces of the spaces $\Lambda_p(\varphi)$.* (Russian) Dokl. Akad. Nauk SSSR **247** (1979), no. 3, 552–554.

[38] E.M. Semenov, *On the stability of the real interpolation method in the class of rearrangement invariant spaces.* Israel Mathem. Conference Proceedings **12** (Bar-Ilan University, Ramat Gan, 1999) 172–182.

[39] T. Schlumprecht, *Notes on class-1 and class-2 spaces.* (Unpublished.)

[40] P. Tradacete, *Spectral properties of disjointly strictly singular operators.* J. Math. Anal. Appl. **395** (2012), no. 1, 376–384.

[41] L. Weis, *Banach lattices with the subsequence splitting property,* Proc. Amer. Math. Soc. **105**, (1989) 87–96.

[42] W. Wnuk, *A note on the positive Schur property.* Glasgow Math. J. **31** (1989) 169–172.

[43] W. Wnuk, *Some characterizations of Banach lattices with the Schur property,* Rev. Mat. Univ. Complut. Madrid **2** (1989), suppl., 217–224.

[44] W. Wnuk, *Banach lattices with properties of the Schur type – a survey.* Confer. Sem. Mat. Univ. Bari **249** (1993).

Julio Flores
Departamento de Matemática Aplicada,
Ciencia e Ingeniería de Materiales
y Tecnología Electrónica, ESCET
Universidad Rey Juan Carlos
28933 Móstoles, Madrid, Spain
e-mail: julio.flores@urjc.es

Francisco L. Hernández
Departamento de Análisis Matemático
Universidad Complutense de Madrid
28040 Madrid, Spain
e-mail: pacoh@mat.ucm.es

Pedro Tradacete
Department of Mathematics
Universidad Carlos III de Madrid
28911 Leganés, Madrid, Spain.
e-mail: ptradace@math.uc3m.es

Ordered Structures and Applications: Positivity VII
Trends in Mathematics, 203–232

Monotonicity Properties of Banach Lattices and Their Applications – a Survey

Paweł Foralewski, Henryk Hudzik, Wojciech Kowalewski and Marek Wisła

Abstract. This is a survey of the most important results from the wide literature concerning monotonicity properties of Banach lattices and their applications.

Mathematics Subject Classification (2010). 46B42, 46B40, 46A40.

Keywords. Banach lattices, Köthe spaces, Calderón–Lozanovskiĭ spaces, generalized Calderón–Lozanovskiĭ spaces, Orlicz spaces, Musielak–Orlicz spaces, Köthe–Bochner spaces, Lorentz spaces, Orlicz–Lorentz spaces, Cesàro–Orlicz spaces, Orlicz–Sobolev spaces, spaces via sublinear operators, strict monotonicity, lower local uniform monotonicity, upper local uniform monotonicity, uniform monotonicity, order smoothness, uniform order smoothness, modulus of monotonicity, modulus of order smoothness, characteristic of monotonicity, rotundity, local uniform rotundity, uniform rotundity, complex rotundities, dominated best approximation, modified dominated best approximation, fixed point theory, ergodic theory.

1. Introduction

Applications of various good rotundity properties in various branches of mathematics, among others in Approximation Theory, Fixed Point Theory, Ergodic Theory, Control Theory, Probability Theory, Theory of Vector Analytic Functions and Theory of Generalized Inverses are well known. It is also known that monotonicity properties of Banach lattices X are restrictions of respective rotundity properties of X to the set of couples of comparable elements from the positive cone X_+ of X (see [41]). In consequence, if we restrict ourselves to Banach spaces being Banach lattices, then in many cases, good rotundity properties can be replaced successfully by respective monotonicity properties.

Uniform monotonicity was introduced and studied by Birkhoff in [7]. Since that time monotonicity properties and their applications to dominated best approximation, Fixed Point Theory, Ergodic Theory and complex rotundities were extensively investigated by many mathematicians.

The paper is organized as follows. First, in Section 2, the most important monotonicity properties of Banach lattices are defined and some theorems on the characterization of these properties in Banach lattices as well as in Köthe spaces are presented. Moreover, this section contains criteria for various monotonicity properties in particular classes of Köthe spaces as well as information about monotonicity properties of various particular classes of ordered function spaces are given.

In Section 3 some general results concerning the formulas for the modulus of monotonicity and for the characteristic of monotonicity in Banach lattices are presented. It is noted that the results of this section are strongly related to the Fixed Point Theory.

Section 4 contains results about relationships between some monotonicity and rotundity properties, as well as duality relationships between monotonicity properties and respective order smoothness properties and between the modulus of monotonicity of a Banach lattice X and the modulus of order smoothness of its dual X^*. Some applications of these results to special optimization problems are presented.

In the next section some interesting and natural relations between monotonicity properties of a real Köthe space E and complex rotundities of its complexification E^c are presented. These relations show the possibility of new applications of monotonicity properties because complex rotundity properties have some natural and important applications to the theory of vector-valued analytic functions.

Section 6 contains various results on applications of the monotonicity properties to the dominated and the modified dominated best approximation problems. In the last section some applications of the monotonicity properties to Ergodic Theory are presented.

2. Monotonicity properties in some Banach lattices, basic definitions and properties

First we will recall definitions of the most important monotonicity properties. Let $X = (X, \leq, \|\cdot\|)$ be a Banach lattice with a partial order $\leq$ (see [7, 65, 77]). By X_+, $B(X)$ and $S(X)$ we denote the positive cone, the unit ball and the unit sphere of X, respectively. Let us also denote $S_+(X) = S(X) \cap X_+$ and $B_+(X) = B(X) \cap X_+$.

A Banach lattice X is said to be strictly monotone (STM for short) if $x, y \in X_+$, $y \leq x$ and $y \neq x$ imply that $\|y\| < \|x\|$.

As usual, X is said to be lower (upper) locally uniformly monotone (LLUM or ULUM for short), whenever for any $x \in S_+(X)$ and $\varepsilon \in (0,1)$ (resp. $\varepsilon > 0$) there is $\delta = \delta(x, \varepsilon) \in (0,1)$ (resp. $\delta = \delta(x, \varepsilon) > 0$) such that the conditions $y \in X$, $0 \leq y \leq x$ (resp. $y \geq 0$) and $\|y\| \geq \varepsilon$ imply that $\|x - y\| \leq 1 - \delta$ (resp.

$\|x+y\| \geq 1+\delta$). In [42], an example of the lattice which is lower locally uniformly monotone, but not upper locally uniformly monotone was presented. However, it is still unknown if the first among the mentioned properties is weaker than the second one in general.

Finally, we say that X is uniformly monotone (UM for short) if for any $\varepsilon \in (0,1)$ there is $\delta(\varepsilon) \in (0,1)$ such that $\|x-y\| \leq 1-\delta(\varepsilon)$ whenever $x, y \in X_+$, $y \leq x$, $\|x\| = 1$ and $\|y\| \geq \varepsilon$. Recall that Birkhoff [7] defined a Banach lattice X to be uniformly monotone if for any $\varepsilon > 0$ there is $\eta(\varepsilon) > 0$ such that $\|x+y\| \geq 1+\eta(\varepsilon)$ whenever $x, y \in X_+$, $\|x\| = 1$ and $\|y\| \geq \varepsilon$. Kurc [61, Proposition 1.1] showed that the above two definitions are equivalent.

It is useful to formulate the monotonicity properties in terms of sequences. Namely, the lower (upper) local uniform monotonicity of a Banach lattice X is equivalent to requiring that for every $x \in X_+$ and any sequence $(x_n)_{n=1}^{\infty}$ in X such that $0 \leq x_n \leq x$ (resp. $x \leq x_n$) it is the case that $\|x_n - x\| \to 0$ as $n \to \infty$, whenever $\|x_n\| \to \|x\|$ as $n \to \infty$. Likewise, a Banach lattice X is uniformly monotone if and only if for any two sequences $(x_n)_{n=1}^{\infty}$ and $(y_n)_{n=1}^{\infty}$ in X_+ such that $y_n \leq x_n$ for any $n \in \mathbb{N}$, there holds $\|x_n - y_n\| \to 0$ as $n \to \infty$, whenever $\lim_{n\to\infty}(\|x_n\| - \|y_n\|) = 0$.

As we will indicate below, monotonicity properties of Banach lattices X are restrictions of respective rotundity properties of X to the set of the couples of comparable elements from the positive cone X_+. Recall, that a Banach space $(X, \|\cdot\|)$ is called rotund if for every $x, y \in S(X)$ with $x \neq y$, we have $\|(x+y)/2\| < 1$. X is said to be locally uniformly rotund if for any $x \in B(X)$ and any $\varepsilon > 0$ there is $\delta(x,\varepsilon) > 0$ such that for any $y \in B(X)$ the inequality $\|x-y\| \geq \varepsilon$ implies that $\|(x+y)/2\| \leq 1-\delta(x,\varepsilon)$. We say that X is uniformly rotund if for any $\varepsilon > 0$ there is $\delta_X(\varepsilon) > 0$ such that if $x, y \in B(X)$ and $\|x-y\| \geq \varepsilon$, then $\|(x+y)/2\| \leq 1-\delta_X(\varepsilon)$. We have the following.

Theorem 2.1 ([41, Theorem 1]). *Given a Banach lattice X the following hold true:*

(i) *If X_+ is rotund*[1], *then X is strictly monotone.*
(ii) *If X_+ is locally uniformly rotund, then X is upper and lower locally uniformly monotone.*
(iii) *If X_+ is uniformly rotund, then X is uniformly monotone.*
(iv) *In the set of the couples of comparable elements in the positive cone X_+ the converse of each of the above statements is also true.*

Let us note that statements (i) and (iii) follow also from the results of W. Kurc [61] presented at the beginning of Section 4.

[1]We say that a positive cone X_+ of a Banach lattice X is rotund if for every $x, y \in S_+(X)$ with $x \neq y$, we have $\|(x+y)/2\| < 1$. Analogously, we can define local uniform rotundity and uniform rotundity of X_+. It is well known that a Köthe space is rotund (locally uniformly rotund) [uniformly rotund] if and only if X_+ possesses the same property (see respectively [41, Theorems 2 and 3]) and [50, Lemma 1]).

A triple (T, Σ, μ) stands for a positive, nonatomic, complete and σ-finite measure space and $L^0 = L^0(T, \Sigma, \mu)$ denotes the space of all (equivalence classes of) Σ-measurable functions x from T to $\mathbb{R}$. For any $x, y \in L^0$ we write $y \leq x$ if and only if $y(t) \leq x(t)$ for μ-a.e. $t \in T$.

We will use also the counting measure space $(\mathbb{N}, 2^{\mathbb{N}}, m)$ and the space l^0 of all real sequences. Obviously, for any $x, y \in l^0$ we have $x \leq y$ if and only if $y(n) \leq x(n)$ for any $n \in \mathbb{N}$.

By $E = (E, \leq, \|\cdot\|_E)$ we denote a Köthe space over a nonatomic, σ-finite measure space (T, Σ, μ) or over the counting measure space $(\mathbb{N}, 2^{\mathbb{N}}, m)$, that is, E is a Banach subspace of L^0 or l^0, respectively, which satisfies the following conditions (see [51] and [65]):

(i) if $x \in E$, $y \in L^0$ and $|y(t)| \leq |x(t)|$ for μ-a.e. $t \in T$ or $y \in l^0$ and $|y(n)| \leq |x(n)|$ for any $n \in \mathbb{N}$, respectively, then $y \in E$ and $\|y\|_E \leq \|x\|_E$,

(ii) there exists a function x in E that is positive on the whole T or $\mathbb{N}$, respectively.

In the case of Köthe spaces one can give equivalent and simpler definitions of strict and uniform monotonicities. Namely, we have

Theorem 2.2 ([41, Theorem 8]). *For any Köthe space E the following conditions are equivalent:*

(i) *E is strictly monotone.*
(ii) *For any $x \in E \backslash \{0\}$ and $A \in \Sigma$ such that $\|x\chi_A\|_E > 0$, we have $\|x - x\chi_A\|_E < \|x\|_E$.*
(iii) *For every $x, y \in E \backslash \{0\}$ with $\mu(\{\operatorname{supp} x \cap \operatorname{supp} y\}) = 0$, we have $\|x + y\|_E > \max(\|x\|_E, \|y\|_E)$.*

Theorem 2.3 ([41, Theorem 6]). *Let E be a Köthe space. Then the following conditions are equivalent:*

(i) *E is uniformly monotone.*
(ii) *For any $\varepsilon \in (0, 1)$, there is $\sigma(\varepsilon) \in (0, 1)$ such that for any $x \in S_+(E)$ and for any $A \in \Sigma$, the inequality $\|x\chi_A\|_E \geq \varepsilon$ implies $\|x - x\chi_A\|_E \leq 1 - \sigma(\varepsilon)$.*
(iii) *For all $\varepsilon > 0$, there is $\tau(\varepsilon) > 0$ such that $x, y \in E_+$, $\|x\|_E = 1$, $\|y\|_E \geq \varepsilon$ and $\mu(\{\operatorname{supp} x \cap \operatorname{supp} y\}) = 0$, imply that $\|x + y\|_E \geq 1 + \tau(\varepsilon)$.*

It is known that lower locally uniformly monotone Köthe spaces are order continuous (see [23, Proposition 2.1]). Recall, that a Köthe space E is called order-continuous if for any element $x \in E$ and any sequence (x_n) in E_+ with $0 \leq x_n \leq |x|$ for all $n \in \mathbb{N}$ and $x_n \to 0$ μ-a.e. or coordinatewise, there holds $\|x_n\|_E \to 0$. Up to now it is not known if upper local uniform monotonicity of a Köthe space implies its order-continuity. However, we can characterize lower and upper local uniform monotonicity in terms of other properties.

For this purpose we define the Kadec–Klee type properties. Namely, we say that a Köthe function space E (E is a Köthe space over a nonatomic, σ-finite measure space (T, Σ, μ)) has the Kadec–Klee property with respect to the (local) convergence in measure, if for any $x \in E$ and any sequence (x_n) in E such that $\|x_n\|_E \to \|x\|_E$ and $x_n \to x$ (locally) in measure, we get $\|x_n - x\|_E \to 0$. Let us recall that local convergence in measure of (x_n) to x means that $x_n\chi_A \to x\chi_A$ in measure for any $A \in \Sigma$ with $\mu(A) < \infty$. Of course in the sequence case global convergence in measure coincides with the uniform convergence and local convergence in measure coincides with the pointwise convergence.

In the sequence case, we get the following

Theorem 2.4. *Let E be a Köthe sequence space, that is, a Köthe space over the counting measure space* $(\mathbb{N}, 2^{\mathbb{N}}, m)$. *Then:*

(i) [33, Theorem 2.7] *E is lower locally uniformly monotone if and only if E is strictly monotone and order continuous.*
(ii) [27, Theorem 4.1] *If E is strictly monotone and has the coordinatewise Kadec–Klee property, then E is upper locally uniformly monotone.*

As we will see below in the function case analogous results were obtained only for symmetric spaces. Let $L^0 = L^0([0,\gamma), \Sigma, m_L)$ be the space of all (equivalence classes of) Lebesgue measurable real-valued functions defined on the interval $[0,\gamma)$, where $0 < \gamma \leq \infty$. Given any $x \in L^0$ we define its distribution function $\mu_x : [0,+\infty) \to [0,\gamma]$ by

$$\mu_x(\lambda) = m_L(\{t \in [0,\gamma) : |x(t)| > \lambda\})$$

(see [5], [59] and [65]) and the non-increasing rearrangement $x^* : [0,\gamma) \to [0,\infty]$ of x as

$$x^*(t) = \inf\{\lambda \geq 0 : \mu_x(\lambda) \leq t\}$$

(with the convention that $\inf \emptyset = \infty$). We say that two functions $x, y \in L^0$ are equi-measurable if $\mu_x(\lambda) = \mu_y(\lambda)$ for all $\lambda \geq 0$. Then we obviously have $x^* = y^*$.

Recall that a Köthe function space E over the Lebesgue measure space $([0,\gamma), \Sigma, m_L)$ is called a symmetric space if E is rearrangement invariant which means that if $x \in E$, $y \in L^0$ and $x^* = y^*$, then $y \in E$ and $\|x\|_E = \|y\|_E$ (see [15]). For basic properties of symmetric spaces we refer to [5], [59] and [65].

Theorem 2.5. *The following statements are true:*

(i) [33, Theorem 2.6] *Suppose that E is a symmetric space over the Lebesgue measure space* $([0,\gamma), \Sigma, m_L)$. *Then E is lower locally uniformly monotone if and only if E is strictly monotone and order continuous (equivalently separable).*
(ii) [15, Theorem 3.2] *Let E be an order-continuous (equivalently separable) symmetric space over the Lebesgue measure space* $([0,\gamma), \Sigma, m_L)$. *Then the following statements are equivalent:*
   (a) *E is upper locally uniformly monotone.*
   (b) *E is strictly monotone and has the Kadec–Klee property with respect to the convergence in measure.*

If we assume additionally that $\gamma = \infty$, then both above conditions are equivalent to

(c) *E has the Kadec–Klee property with respect to the local convergence in measure.*

Now we will present criteria for monotonicity properties of Calderón–Lozanovskiĭ spaces in general and Orlicz and Orlicz–Lorentz spaces in particular.

In the remainder of the paper, φ denotes an Orlicz function (see [12, 68, 70]), that is, $\varphi : [-\infty, \infty] \to [0, \infty]$ (our definition is extended from $\mathbb{R}$ to $[-\infty, +\infty]$ by putting $\varphi(-\infty) = \varphi(\infty) = \infty$) and φ is convex, even, vanishing and continuous at zero, left continuous on $(0, \infty)$ and not identically equal to zero on $(-\infty, \infty)$. Denoting

$$a_\varphi = \sup\{u \geq 0 : \varphi(u) = 0\},$$
$$b_\varphi = \sup\{u \geq 0 : \varphi(u) < \infty\},$$

we get that the left continuity of φ on $(0, \infty)$ is equivalent to the fact that $\lim_{u \to (b_\varphi)^-} \varphi(u) = \varphi(b_\varphi)$.

Recall that an Orlicz function φ satisfies condition Δ_2 for all $u \in \mathbb{R}$ ($\varphi \in \Delta_2(\mathbb{R})$ for short) if there exists a constant $K > 0$ such that the inequality

$$\varphi(2u) \leq K\varphi(u) \tag{2.1}$$

holds for every $u \in \mathbb{R}$ (then we have $a_\varphi = 0$ and $b_\varphi = \infty$). In the same way the conditions denoted by $\varphi \in \Delta_2(\infty)$ and $\varphi \in \Delta_2(0)$ are obtained by requiring the existence of the constant $u_0 > 0$ such that $\varphi(u_0) < \infty$ (respectively $\varphi(u_0) > 0$) and such that the inequality (2.1) holds for all $u \geq u_0$ (respectively for all $u \in [0, u_0]$). In the first case we have $b_\varphi = \infty$ and in the second one $a_\varphi = 0$.

For a Köthe space E and an Orlicz function φ we say that φ satisfies condition Δ_2^E ($\varphi \in \Delta_2^E$ for short) if:

1) $\varphi \in \Delta_2(0)$ whenever $E \hookrightarrow L^\infty$,
2) $\varphi \in \Delta_2(\infty)$ whenever $L^\infty \hookrightarrow E$,
3) $\varphi \in \Delta_2(\mathbb{R})$ whenever neither $L^\infty \hookrightarrow E$ nor $E \hookrightarrow L^\infty$,

where the symbol $E \hookrightarrow F$ stands for the continuous embedding of E into F (see [10] and [40]).

Given any Orlicz function φ, we define on L^0 or l^0 a convex modular ϱ (see [70]) by

$$\varrho(x) = \begin{cases} \|\varphi \circ x\|_E & \text{if } \varphi \circ x \in E, \\ \infty & \text{otherwise,} \end{cases}$$

where $(\varphi \circ x)(t) = \varphi(x(t))$ for $t \in T$ or $(\varphi \circ x)(n) = \varphi(x(n))$ for $n \in \mathbb{N}$, respectively, and the Calderón–Lozanovskiĭ space

$$E_\varphi = \{x : \ \varphi \circ \lambda x \in E \text{ for some } \lambda > 0\}$$

(see [10], [40] and [68]) which becomes a normed space under the Luxemburg norm

$$\|x\|_\varphi = \inf\{\lambda > 0 : \varrho(x/\lambda) \leq 1\}.$$

Considering the space E_φ we shall assume that E has the Fatou property, that is, for any $x \in L^0$ or $x \in l^0$ and $(x_n)_{n=1}^\infty$ in E_+ such that $x_n \nearrow x$ μ-a.e. or coordinatewise and $\sup_n \|x_n\|_E < \infty$, we have $x \in E$ and $\|x\|_E = \lim_n \|x_n\|_E$ (see [51] and [65]). We also assume that there exists a function x in E_a that is positive on the whole T or $\mathbb{N}$, respectively, where E_a denotes the subspace of all order-continuous elements in E.

Now we will present criteria for monotonicity properties of Calderón–Lozanovskiĭ spaces.

Theorem 2.6. *Let E be a Köthe function spaces, that is, a Köthe space over a nonatomic, σ-finite measure space (T, Σ, μ). Then the following statements are true:*

(i) *The space E_φ is strictly monotone if and only if E is strictly monotone, $\varphi \in \Delta_2^E$ and φ vanishes only at zero.*
(ii) *The space E_φ is lower locally uniformly monotone if and only if E is lower locally uniformly monotone, $\varphi \in \Delta_2^E$ and φ vanishes only at zero.*
(iii) *If the space E_φ is order continuous, then E_φ is upper locally uniformly monotone if and only if E is upper locally uniformly monotone, $\varphi \in \Delta_2^E$ and φ vanishes only at zero.*
(iv) *The space E_φ is uniformly monotone if and only if E is uniformly monotone, $\varphi \in \Delta_2^E$ and φ vanishes only at zero.*

First we note that by the assumptions that E has the Fatou property and $\operatorname{supp} E_a = T$, we have $E \not\subset L^\infty$ and, in consequence, the condition $\varphi \in \Delta_2^E$ implies $b_\varphi = \infty$. Therefore, the statement (i) follows from [10, Theorem 1] and [40, Theorems 1 and 2]. In turn, the statements (ii) and (iii) have been shown in [33, Proposition 2.4(i) and Proposition 2.5(i), respectively]. Finally, the statement (iv) follows from [10, Theorem 2], [40, Theorems 1 and 2] and [41, Theorem 7].

In the case when E is a Köthe sequence space, that is, a Köthe space over the counting measure space $(\mathbb{N}, 2^{\mathbb{N}}, m)$, we can get analogous theorem assuming additionally that $\varphi(b_\varphi) \inf_{n\in\mathbb{N}} \|e_n\|_E \geq 1$, where e_n for $n \in \mathbb{N}$ are the basic unit vectors.

It is well known that if $E = L^1$, then $E_\varphi = (L^1)_\varphi = L^\varphi$, where L^φ is the Orlicz function space equipped with the Luxemburg norm. Since L^1 is uniformly monotone, by Theorem 2.6, we get

Corollary 2.7. *If $\mu(T) = \infty$, then the Orlicz function space L^φ is strictly monotone, equivalently uniformly monotone, if and only if $\varphi \in \Delta_2(\mathbb{R})$. Similarly, if $\mu(T) < \infty$, then the Orlicz function space L^φ is strictly monotone, equivalently uniformly monotone, if and only if $\varphi \in \Delta_2(\infty)$ and φ vanishes only at 0.*

Moreover, if $E = l^1$, then $E_\varphi = (l^1)_\varphi = l^\varphi$ is the Orlicz sequence space equipped with the Luxemburg norm. In this case we have

Corollary 2.8. *The Orlicz sequence space l^φ is strictly monotone, equivalently uniformly monotone, if and only if $\varphi \in \Delta_2(0)$ and $\varphi(b_\varphi) \geq 1$.*

If E is a Lorentz function space $\Lambda_{1,\omega}$, then E_φ is the corresponding Orlicz–Lorentz function space $\Lambda_{\varphi,\omega}$ (see [48, 49, 50]). Recall, that the Lorentz spaces $\Lambda_{1,\omega}$ are defined by

$$\Lambda_{1,\omega} = \left\{ x \in L^0([0,\gamma), \Sigma, m_L) : \|x\|_{\Lambda_{1,\omega}} := \int_0^\gamma x^*(t)\omega(t)dt < \infty \right\},$$

where ω denotes a nonnegative, non-increasing and locally integrable function on $[0,\gamma)$ (not identically 0), called a weight function. We say that a weight function ω is regular if there exists $\eta > 0$ such that $\int_0^{2t} \omega(t)dt \geq (1+\eta)\int_0^t \omega(t)dt$ for any $t \in [0, \gamma/2)$. Note that if the weight function ω is regular, then $\int_0^\infty \omega(t)dt = \infty$ in the case when $\gamma = \infty$ and $\omega(t) > 0$ for some $t > \gamma/2$ when $\gamma < \infty$.

It is well known that in the classes of Lorentz spaces and Orlicz–Lorentz spaces strict monotonicity and uniform monotonicity are not equivalent (see [33, Example 4.3]). More precisely, we have the following

Theorem 2.9 ([33, Proposition 4.1]). *The following conditions are equivalent:*
(i) $\int_0^\gamma \omega(t)dt = \infty$ *if* $\gamma = \infty$ *and* ω *is positive on* $[0,\gamma)$ *when* $\gamma < \infty$.
(ii) *The Lorentz space* $\Lambda_{1,\omega}$ *is strictly monotone.*
(iii) *The Lorentz space* $\Lambda_{1,\omega}$ *is lower locally uniformly monotone.*
(iv) *The Lorentz space* $\Lambda_{1,\omega}$ *is upper locally uniformly monotone.*

and

Theorem 2.10 ([39, Theorem 1]). *The Lorentz space* $\Lambda_{1,\omega}$ *is uniformly monotone if and only if the weight function* ω *is regular and* ω *is positive on* $[0,\gamma)$ *if* $\gamma < \infty$.

In consequence, we get

Corollary 2.11 ([33, Corollary 4.4]). *The following conditions are equivalent:*
(i) $\varphi \in \Delta_2(\mathbb{R})$ *and* $\int_0^\gamma \omega(t)dt = \infty$ *if* $\gamma = \infty$ *and* $\varphi \in \Delta_2(\infty)$, φ *vanishes only at* 0 *and* ω *is positive on* $[0,\gamma)$ *when* $\gamma < \infty$.
(ii) *The Orlicz–Lorentz space* $\Lambda_{\varphi,\omega}$ *is strictly monotone.*
(iii) *The Orlicz–Lorentz space* $\Lambda_{\varphi,\omega}$ *is lower locally uniformly monotone.*
(iv) *The Orlicz–Lorentz space* $\Lambda_{\varphi,\omega}$ *is upper locally uniformly monotone.*

and

Corollary 2.12 ([40, Theorem 10]). *If* $\gamma = \infty$, *then the Orlicz–Lorentz space* $\Lambda_{\varphi,\omega}$ *is uniformly monotone if and only if* $\varphi \in \Delta_2(\mathbb{R})$ *and the weight function* ω *is regular. Similarly, if* $\gamma < \infty$, *then the Orlicz–Lorentz space* $\Lambda_{\varphi,\omega}$ *is uniformly monotone if and only if* $\varphi \in \Delta_2(\infty)$, φ *vanishes only at* 0, *the weight function* ω *is regular and* ω *is positive on* $[0,\gamma)$.

In the case when $E = \lambda_{1,\omega}$ (the Lorentz sequence space), we get similar results, assuming additionally that $\varphi(b_\varphi)\omega(1) \geq 1$, where $\omega = (\omega(n))_{n=1}^\infty$ is the weight sequence (see [33, Proposition 4.2 and Corollary 4.5] and [9, Lemma 1]).

Finally, it is worth noticing that criteria for monotonicity properties were also given in other Banach lattices, among others in Musielak–Orlicz spaces [60, 61, 62,

42, 20] (a short survey of the results about monotonicity properties in Musielak–Orlicz spaces and of their subspaces of order-continuous elements was given in [36, pp. 31–34]), generalized Calderón–Lozanovskiĭ spaces E_φ (where the function φ is a Musielak–Orlicz function) [24, 28], Calderón–Lozanovskiĭ construction $\rho(X,Y)$ [53], Lorentz spaces $\Gamma_{p,\omega}$ [17], generalized Orlicz–Lorentz spaces [32, 25, 26, 27], Cesàro–Orlicz spaces [19], Banach spaces defined via sublinear operators [47] and Orlicz–Sobolev spaces [14].

Moreover, some other monotonicity properties were also defined, among others the properties CWLLUM, H^+STM, DUM and IUM which are important in the problems of the dominated and the modified dominated best approximation (see Section 6 where these notions will be defined and applied). Moreover, Kolwicz and Płuciennik [54] introduced a monotonicity property of normed lattices called uniform monotonicity in every order interval and they discovered that this property is useful in order to give a criterion for uniform rotundity in every direction of Calderón–Lozanovskiĭ spaces.

In the theory of Banach lattices the local monotonicity structure was also considered. Points of lower and upper strict monotonicity as well as points of lower and upper local uniform monotonicity were investigated.

A point $x \in S_+(X)$ is said to be a point of lower strict monotonicity if for any $y \in X_+$, $y \le x$, $y \ne x$ implies $\|y\| < \|x\|$. A point $x \in S_+(X)$ is said to be a point of upper strict monotonicity if $\|x+y\| > 1$ for any $y \in X_+ \setminus \{0\}$. In general, points of lower strict monotonicity differ from points of upper strict monotonicity. However, the fact that all points of $S_+(X)$ are points of lower strict monotonicity is equivalent to the fact that all points of $S_+(X)$ are points of upper strict monotonicity and both these facts are equivalent to strict monotonicity of X.

A point $x \in S_+(X)$ is said to be a point of lower local uniform monotonicity if for any $\varepsilon \in (0,1)$ there exists $\delta(x,\varepsilon) \in (0,1)$ such that $\|x-y\| \le 1-\delta(x,\varepsilon)$ whenever $y \in X_+$, $y \le x$ and $\|y\| \ge \varepsilon$. A point $x \in S_+(X)$ is said to be a point of upper local uniform monotonicity if for any $\varepsilon > 0$ there is $\eta(x,\varepsilon) > 0$ such that $\|x+y\| \ge 1+\eta(x,\varepsilon)$ whenever $y \in X_+$ and $\|y\| \ge \varepsilon$.

Monotonicity points defined as above were investigated in the papers [43, 44, 45, 21, 55, 56, 18].

3. Moduli and characteristics of monotonicity

In this section we will present the results concerning moduli and characteristics of monotonicity in Banach lattices. The problem of calculating the characteristic of monotonicity of a Banach lattice seems to be of great interest because of the result by Betiuk-Pilarska and Prus [6] stating that if a Banach lattice X has this characteristic strictly smaller than 1 and X is weakly orthogonal, then X has the weak normal structure, whence, consequently, X has the weak fixed point property (for the definition of these properties we refer to [52]). Therefore, it was quite natural that several papers were devoted to the problem of calculating the exact value or getting good estimates of the characteristic of monotonicity in various particular classes of Banach lattices.

For a given Banach lattice X, the function $\delta_{m,X} : [0,1] \to [0,1]$ defined by

$$\delta_{m,X}(\varepsilon) = \inf\{1 - \|x-y\| : 0 \le y \le x, \|x\| = 1, \|y\| \ge \varepsilon\} \tag{3.1}$$

is said to be the lower modulus of monotonicity of X. It is easy to show that (see [37])

$$\begin{aligned}\delta_{m,X}(\varepsilon) &= \inf\{1 - \|x-y\| :\ 0 \le y \le x, \|x\| = 1, \|y\| = \varepsilon\} \\ &= 1 - \sup\{\|x-y\| : 0 \le y \le x, \|x\| = 1, \|y\| = \varepsilon\}. \\ &= 1 - \sup\{\|x-y\| : 0 \le y \le x, \|x\| = 1, \|y\| \ge \varepsilon\}\end{aligned} \tag{3.2}$$

By definition, the lower modulus of monotonicity $\delta_{m,X}$ is a non-decreasing function. Moreover, the function $\delta_{m,X}$ is convex on the interval $[0,1]$ (see [63]), so $\delta_{m,X}$ is continuous on the interval $[0,1)$. It is also clear that $\delta_{m,X}(\varepsilon) \le \varepsilon$ for any $\varepsilon \in [0,1]$. Obviously, X is uniformly monotone if and only if $\delta_{m,X}(\varepsilon) > 0$ for every $\varepsilon \in (0,1]$. It is easy to see that a Banach lattice X is strictly monotone if and only if $\delta_{m,X}(1) = 1$.

The number $\varepsilon_{0,m}(X) \in [0,1]$ defined by

$$\varepsilon_{0,m}(X) = \sup\{\varepsilon \in [0,1] \colon \delta_{m,X}(\varepsilon) = 0\} = \inf\{\varepsilon \in (0,1] \colon \delta_{m,X}(\varepsilon) > 0\}$$

(where $\inf \emptyset := 1$) is said to be the characteristic of monotonicity of a Banach lattice X. Obviously, a Banach lattice X is uniformly monotone if and only if $\varepsilon_{0,m}(X) = 0$.

We can also define another characteristic of monotonicity of a Banach lattice X, namely

$$\widetilde{\varepsilon}_{0,m}(X) = \sup\{\varepsilon \ge 0 \colon \eta_{m,X}(\varepsilon) = 0\} = \inf\{\varepsilon > 0 \colon \eta_{m,X}(\varepsilon) > 0\},$$

where $\eta_{m,X}$ is the upper modulus of monotonicity defined, for all $\varepsilon \ge 0$, by the formula

$$\begin{aligned}\eta_{m,X}(\varepsilon) &= \inf\{\|x+y\| - 1 \colon x, y \in X_+, \|x\| = 1, \|y\| \ge \varepsilon\} \\ &= \inf\{\|x+y\| - 1 \colon x, y \in X_+, \|x\| = 1, \|y\| = \varepsilon\}\end{aligned}$$

(see [61] and [64]). It is clear by the norm triangle inequality that $\eta_{m,X}(\varepsilon) \le \varepsilon$ for all $\varepsilon \ge 0$. Obviously, a Banach lattice X is uniformly monotone if and only if $\eta_{m,X}(\varepsilon) > 0$ for all $\varepsilon > 0$ or, equivalently, if and only if $\widetilde{\varepsilon}_{0,m}(X) = 0$.

Let us also recall some relationships between two moduli of monotonicity $\delta_{m,X}$ and $\eta_{m,X}$ as well as relationships between the characteristics of monotonicity $\varepsilon_{0,m}(X)$ and $\widetilde{\varepsilon}_{0,m}(X)$. For arbitrary $\varepsilon \in (0,1)$ the following inequalities hold true (see [61]):

$$\frac{\delta_{m,X}(\varepsilon/(1+\varepsilon))}{1 - \delta_{m,X}(\varepsilon/(1+\varepsilon))} \le \eta_{m,X}(\varepsilon) \le \frac{\delta_{m,X}(\varepsilon)}{1 - \delta_{m,X}(\varepsilon)}. \tag{3.3}$$

Note that inequalities (3.3) are equivalent to the following ones

$$\frac{\eta_{m,X}(\varepsilon)}{1 + \eta_{m,X}(\varepsilon)} \le \delta_{m,X}(\varepsilon) \le \frac{\eta_{m,X}(\varepsilon/(1-\varepsilon))}{1 + \eta_{m,X}(\varepsilon/(1-\varepsilon))}$$

for any $\varepsilon \in (0,1)$. As a consequence, we get

$$\varepsilon_{0,m}(X) \leq \widetilde{\varepsilon}_{0,m}(X) \leq \min\{1, 2\varepsilon_{0,m}(X)\}.$$

Indeed, in [37], Theorem 1, it has been shown that $\varepsilon_{0,m}(X) \leq \widetilde{\varepsilon}_{0,m}(X) \leq 2\varepsilon_{0,m}(X)$. Simultaneously, since $\|x+y\| \geq \max(\|x\|, \|y\|)$ for any couple $x, y \geq 0$, by the triangle inequality, we have $\eta_{m,X}(\varepsilon) > 0$ for all $\varepsilon > 1$, whence we get $\widetilde{\varepsilon}_{0,m}(X) \leq 1$.

Moreover, for any Banach lattice X, the following formulas are true (see [37, Theorem 5] and [29, Proposition 2.10], respectively)

$$\begin{aligned}\varepsilon_{0,m}(X) = {} & \sup\{\limsup_{n\to\infty} \|x_n - y_n\| : 0 \leq y_n \leq x_n \text{ and } \|x_n\| = 1 \\ & \text{for any } n \in \mathbb{N} \text{ and } \|y_n\| \to 1\},\\ \widetilde{\varepsilon}_{0,m}(X) = {} & \sup\{\limsup_{n\to\infty} \|y_n - x_n\| : 0 \leq x_n \leq y_n \text{ and } \|x_n\| = 1 \\ & \text{for any } n \in \mathbb{N} \text{ and } \|y_n\| \to 1\}.\end{aligned}$$

Now we will present a few results, that are very useful while calculating the characteristic of monotonicity of particular Banach lattices. We start with the following theorem.

Theorem 3.1 ([29, Theorem 2.1]). *For any Banach lattice X the following equality holds true*

$$\varepsilon_{0,m}(X) = 1 - \delta_{m,X}(1^-), \tag{3.4}$$

where $\delta_{m,X}(1^-) = \lim_{\varepsilon\to 1^-} \delta_{m,X}(\varepsilon)$. Moreover,

$$\delta_{m,X}(1 - \delta_{m,X}(\varepsilon)) = 1 - \varepsilon$$

for arbitrary $\varepsilon \in (\varepsilon_{0,m}(X), 1]$ if $\varepsilon_{0,m}(X) < 1$ as well as also in the case when $\varepsilon = \varepsilon_{0,m}(X) = 1$.

Note that in equality (3.4), the value $\delta_{m,X}(1^-)$ cannot be replaced by $\delta_{m,X}(1)$. Namely, in [29, Examples 2.3 and 2.4], examples of Banach lattices X with $\delta_{m,X}(\varepsilon) = 0$ for any $\varepsilon \in [0,1)$ and $\delta_{m,X}(1) = 1$ were presented. Obviously, it is natural to ask if there exists a Banach lattice, for which $0 < \delta_{m,X}(1^-) < \delta_{m,X}(1)$. By equations (3.2) and (3.4) we get immediately that for an arbitrary Banach lattice X the following formulas hold:

$$\begin{aligned}\varepsilon_{0,m}(X) &= \lim_{\varepsilon\to 1^-} \left(\sup\{\|x-y\| : 0 \leq y \leq x, \|x\| = 1, \|y\| \geq \varepsilon\}\right)\\ &= \lim_{\varepsilon\to 1^-} \left(\sup\{\|x-y\| : 0 \leq y \leq x, \|x\| = 1, \|y\| = \varepsilon\}\right).\end{aligned}$$

In [29] another modulus of monotonicity and characteristic of monotonicity for Köthe spaces were introduced and it was proved that the characteristic of monotonicity corresponding to this new modulus coincide with the usual characteristic of monotonicity (see [29, Theorem 3.9] and [31, Theorem 2.3]). It is obvious from this result that, by using the new formula of the characteristic of monotonicity of Köthe spaces, it should be easier to calculate this coefficient in particular classes of Köthe spaces, e.g., in the case of Orlicz sequence spaces and Orlicz–Lorentz function spaces. In [29] and [31] just this new formula for the characteristic of monotonicity was applied successfully to find formulas for the characteristic of monotonicity of Orlicz spaces and Orlicz–Lorentz spaces.

Let us present now these new definitions of a modulus of monotonicity and the new formula for the characteristics of monotonicity in Köthe space. For any Köthe space E we define the modulus $\widehat{\delta}_{m,E}:[0,1]\to[0,1]$ by the formula

$$\widehat{\delta}_{m,E}(\varepsilon)=\inf\left\{1-\|x-x\chi_A\|_E : x\geq 0, \|x\|_E=1, A\in\Sigma, \|x\chi_A\|_E\geq\varepsilon\right\}.$$

Obviously, the modulus $\widehat{\delta}_{m,E}$ is a non-decreasing function with respect to $\varepsilon\in[0,1]$ and $\delta_{m,X}(\varepsilon)\leq\widehat{\delta}_{m,E}(\varepsilon)\leq\varepsilon$ for any $\varepsilon\in[0,1]$. Similarly as for the modulus $\delta_{m,X}$ in [37], it is also possible to prove that

$$\begin{aligned}\widehat{\delta}_{m,E}(\varepsilon)&=\inf\{1-\|x-x\chi_A\|_E:\ x\geq 0, \|x\|_E=1, A\in\Sigma, \|x\chi_A\|_E=\varepsilon\}\\ &=1-\sup\{\|x-x\chi_A\|_E : x\geq 0, \|x\|_E=1, A\in\Sigma, \|x\chi_A\|_E=\varepsilon\}.\\ &=1-\sup\{\|x-x\chi_A\|_E : x\geq 0, \|x\|_E=1, A\in\Sigma, \|x\chi_A\|_E\geq\varepsilon\}\end{aligned}$$

The characteristic of monotonicity $\widehat{\varepsilon}_{0,m}(E)$ corresponding to the modulus $\widehat{\delta}_{m,E}$ is defined by

$$\widehat{\varepsilon}_{0,m}(E)=\sup\left\{\varepsilon\in[0,1]\colon \widehat{\delta}_{m,E}(\varepsilon)=0\right\}=\inf\left\{\varepsilon\in(0,1]\colon \widehat{\delta}_{m,E}(\varepsilon)>0\right\},$$

(where $\inf\emptyset := 1$). By [29, Proposition 2.6], for arbitrary Köthe space E the following formula holds true

$$\begin{aligned}\widehat{\varepsilon}_{0,m}(E)=\ &\sup\{\limsup_{n\to\infty}\|x_n-x_n\chi_{A_n}\|_E : x_n\in S_+(E) \text{ and } A_n\in\Sigma\\ &\qquad\text{for any } n\in\mathbb{N} \text{ and } \|x_n\chi_{A_n}\|_E\to 1\}.\end{aligned}$$

Moreover, by Theorem 2.8 in [29], for any Köthe space E we have the equality $\varepsilon_{0,m}(E)=\widehat{\varepsilon}_{0,m}(E)$, whence we get

$$\begin{aligned}\varepsilon_{0,m}(E)&=\lim_{\varepsilon\to 1^-}\sup\left\{\|x-x\chi_A\|_E : x\in S_+(E), A\in\Sigma, \|x\chi_A\|_E\geq\varepsilon\right\}\\ &=\lim_{\varepsilon\to 1^-}\sup\left\{\|x-x\chi_A\|_E : x\in S_+(E), A\in\Sigma, \|x\chi_A\|_E=\varepsilon\right\}.\end{aligned}$$

(see Corollary 2.9 in [29]). Since $\varepsilon_{0,m}(E)=\widehat{\varepsilon}_{0,m}(E)$, we have $\delta_{m,X}(\varepsilon)=\widehat{\delta}_{m,E}(\varepsilon)=0$ for any $\varepsilon\in[0,\varepsilon_{0,m}(E))$ and $\lim_{\varepsilon\to 1^-}\delta_{m,X}(\varepsilon)=\lim_{\varepsilon\to 1^-}\widehat{\delta}_{m,E}(\varepsilon)$. In [29, Example 2.3] it has been shown that $\delta_{m,X}(\varepsilon)=\widehat{\delta}_{m,E}(\varepsilon)$ for any $\varepsilon\in[0,1]$ and $E=L^p([0,1],\Sigma,m)$, where $1\leq p<\infty$. So, it is natural to ask if these two moduli are equal in arbitrary Köthe spaces.

Now we will mention some results concerning the estimates and the calculations of the moduli and characteristics of monotonicity for particular Köthe spaces. First, the characteristic $\widetilde{\varepsilon}_{0,m}(L^\varphi)$ was calculated in [67], where L^φ denotes an Orlicz function space (over a nonatomic finite measure space) equipped with the Luxemburg norm as well as the Orlicz norm. In that paper the authors assumed for simplicity that the generating Orlicz function is an N-function. Next, in [37], lower and upper estimates of the lower modulus of monotonicity of Orlicz spaces equipped with the Luxemburg norm were given in terms of the Simonenko and

Lindberg indices. In turn, in [38] some estimates of the characteristics of monotonicity for Köthe–Bochner function spaces were presented. In [29, 31] the characteristics of monotonicity of Orlicz spaces and Orlicz–Lorentz spaces equipped with the Luxemburg norm were calculated. Finally, in [30] some estimates of the characteristics of monotonicity of Orlicz spaces equipped with the Orlicz norm were obtained.

4. Dual properties

In this section we will discuss relationships between some monotonicity and rotundity properties, as well as duality relationships between monotonicity properties and respective order smoothness properties and between the modulus of monotonicity of a Banach lattice X and the modulus of order smoothness of its dual X^* (see [61, 63]). Moreover, some applications of these results to particular optimization problems are presented.

It is easy to observe that a Banach space X is rotund if and only if, for every $x \in X$,

$$\max_{\pm} \|x \pm y\| > \|x\| \text{ for all } y \in X \setminus \{0\} \tag{4.1}$$

(see [61]). Indeed, if X is rotund and $x, y \in X$ with $y \neq 0$, then we have

$$\begin{aligned} \|x\| &= \max_{\pm} \|x \pm y\| \left\| \frac{x+y}{2\max_{\pm} \|x \pm y\|} + \frac{x-y}{2\max_{\pm} \|x \pm y\|} \right\| \\ &< \max_{\pm} \|x \pm y\|, \end{aligned}$$

so the condition (4.1) holds true.

On the other hand, if X is not rotund, then we can find $x, y \in X$ with $x \neq y$ and $\left\|\frac{1}{2}(x+y)\right\| = \|x\| = \|y\| > 0$. Then $\frac{x-y}{2} \neq 0$ and

$$\max_{\pm} \left\| \frac{x+y}{2} \pm \frac{x-y}{2} \right\| = \max\{\|x\|, \|y\|\} = \|x\| = \left\| \frac{x+y}{2} \right\|,$$

whence condition (4.1) is not satisfied. Thus, for any Banach lattice X, rotundity implies strict monotonicity.

Let us recall that a Banach space $(X, \|\cdot\|)$ is called uniformly rotund if for each $\varepsilon \in (0, 2]$ there exists $\delta_X(\varepsilon) \in (0, 1)$ such that $\left\|\frac{x+y}{2}\right\| \leq 1 - \delta_X(\varepsilon)$ whenever $\|x\|, \|y\| \leq 1$ and $\|x - y\| \geq \varepsilon$. This condition can be expressed equivalently in the following way (see [61]): for every $\varepsilon > 0$ we can find $\eta_X(\varepsilon) \in (0, 1)$ such that

$$\inf\left\{\max_{\pm} \|x \pm y\| : \|x\| = 1, \|y\| \geq \varepsilon\right\} \geq 1 + \eta_X(\varepsilon). \tag{4.2}$$

Indeed, let X be uniformly rotund, $\|x\| = 1$ and $\|y\| \geq \varepsilon$, where $\varepsilon \in (0, 1]$. We have

$$\left\| \frac{x+y}{\max_{\pm} \|x \pm y\|} - \frac{x-y}{\max_{\pm} \|x \pm y\|} \right\| = \frac{2\|y\|}{\max_{\pm} \|x \pm y\|} \geq \frac{2\|y\|}{1 + \|y\|} \geq \varepsilon$$

(the last inequality follows from the facts that function $f(a) = \frac{2a}{1+a}$ is increasing for $a \geq 0$ and $f(\varepsilon) \geq \varepsilon$ for $\varepsilon \in (0,1]$). Therefore, putting $\eta_X(\varepsilon) = \delta_X(\varepsilon)$, we obtain $\eta_X(\varepsilon) > 0$ and

$$\begin{aligned} 1 + \eta_X(\varepsilon) &\leq \|x\| + \max_{\pm} \|x \pm y\| \cdot \delta_X(\varepsilon) \\ &= \max_{\pm} \|x \pm y\| \left(\left\| \frac{1}{2} \left(\frac{x+y}{\max_{\pm} \|x \pm y\|} + \frac{x-y}{\max_{\pm} \|x \pm y\|} \right) \right\| + \delta_X(\varepsilon) \right) \\ &\leq \max_{\pm} \|x \pm y\| \left(1 - \delta_X(\varepsilon) + \delta_X(\varepsilon)\right) = \max_{\pm} \|x \pm y\| . \end{aligned}$$

Thus condition (4.2) holds true for $\varepsilon \in (0,1]$; for $\varepsilon > 1$ we can assume $\eta_X(\varepsilon) = \eta_X(1)$.

Conversely, if X is not uniformly rotund, then we can find $\varepsilon \in (0,2]$ and sequences (x_n) and (y_n) from $B(X)$ with $\|x_n - y_n\| \geq \varepsilon$ for any $n \in \mathbb{N}$ such that $\left\| \frac{x_n+y_n}{2} \right\| \to 1$. Hence $\left\| \frac{x_n-y_n}{\|x_n+y_n\|} \right\| \geq \frac{\varepsilon}{2} > 0$ for any $n \in \mathbb{N}$ and

$$\limsup_{n\to\infty} \max_{\pm} \left\| \frac{x_n+y_n}{\|x_n+y_n\|} \pm \frac{x_n-y_n}{\|x_n+y_n\|} \right\| = \limsup_{n\to\infty} \frac{2\max\{\|x_n\|, \|y_n\|\}}{\|x_n+y_n\|} \leq 1,$$

so condition (4.2) is not satisfied.

Therefore, for any Banach lattice X, uniform rotundity implies uniform monotonicity. Moreover, it is easy to show that for any Banach lattice X we have

$$0 \leq \delta_X(\varepsilon) \leq \delta_{m,X}(\varepsilon) \leq \varepsilon,$$

for any $\varepsilon \in (0,1]$ (see [63]), where $\delta_{m,X}(\cdot)$ denotes the modulus of monotonicity of the Banach lattice X that is considered in section 3 (see formula (3.1)). Note that for the real Lebesgue space $X = L^1(\mu)$ we have $\delta_{L^1(\mu)}(\varepsilon) = 0$ for any $\varepsilon \in (0,2]$, so $L^1(\mu)$ is not even rotund but, on the other hand, it is uniformly monotone and $\delta_{m,L^1(\mu)}(\varepsilon) = \varepsilon$ for every $\varepsilon \in (0.1]$, whence the Lebesgue space $L^1(\mu)$ is the most uniformly monotone Banach lattice among all Banach lattices.

A Banach lattice X is called order smooth, if for every $x \in S_+(X)$ and every order interval $[u^*, v^*] \subset \mathrm{Grad}(x)$ there holds $u^* = v^*$, where $\mathrm{Grad}(x) = \{x^* \in S_+(X^*) :< x, x^* >= \|x\|\}$. Analogously, a Banach lattice X is called order uniformly smooth, if $\rho_X(\tau)/\tau \to 0$, whenever $\tau \searrow 0$, where the modulus of smoothness $\rho_X(\tau)$, $\tau \in (0,1)$, is defined as follows:

$$\rho_X(\tau) = \sup\left\{\|x \vee \tau y\| - 1 : x, y \geq 0, \|x\| = \|y\| = 1\right\}.$$

Theorem 4.1 ([63, Theorem 1]). *Let X be a Banach lattice with the dual X^*. Then*

(a) *if X^* is strictly monotone, then X is order smooth,*

(b) *if X^* is order smooth, then X is strictly monotone.*

If X is reflexive then the converse implications hold true as well.

In order to give necessary and sufficient conditions for X^* to be UM the following relationships between modulus of monotonicity and modulus of smoothness are crucial.

Theorem 4.2 ([63, Theorem 3]). *For any Banach lattice X the following duality formulas hold true:*

(a) $\rho_X(\tau) = \rho_{X^{**}}(\tau)$,
(b) $\delta_{m,X}(\varepsilon) = \delta_{m,X^{**}}(\varepsilon)$,
(c) $\rho_{X^*}(\tau) = \sup_{0\le\varepsilon\le 1}\left(\varepsilon\tau - \delta_{m,X}(\varepsilon)\right)$,
(d) $\delta_{m,X}(\varepsilon) = \sup_{0\le\tau\le 1}\left(\varepsilon\tau - \rho_{X^*}(\tau)\right)$,

where $\varepsilon, \tau \in (0,1)$.

Using the above theorem, we can conclude that the order-uniform smoothness of a Banach lattice X is equivalent to the uniform monotonicity of X^*.

Theorem 4.3 ([63, Theorem 5]). *Let X be a Banach lattice. Then:*

(a) *X is uniformly monotone if and only if X^{**} is uniformly monotone,*
(b) *X is order uniformly smooth if and only if X^{**} is order uniformly smooth,*
(c) *X is uniformly monotone if and only if X^* is order uniformly smooth,*
(d) *X^* is uniformly monotone if and only if X is order uniformly smooth.*

Now we will present an application of the strict monotonicity property to some optimization problems. We will fix any $x \in X_+$ with $\|x\| = 1$ and consider the (convex) optimization problem

$$\begin{cases} \|x \vee y\| \to \min, \\ y \ge 0, \end{cases} \tag{4.3}$$

i.e., we will look for a $y \in X_+$ such that $\|x \vee y\| = \min_{y\in X_+} \|x \vee y\|$. Since X is a Banach lattice, $\min_{y\in X_+} \|x \vee y\| = \|x\|$, so the optimization problem reduces to describing the set $P_x \subset X_+$ of nonnegative solutions of the equation $\|x \vee y\| = \|x\|$. Evidently, the set P_x is nonempty because $[0,x] \subset P_x$, where $[0,x]$ denotes the order interval with the endpoints 0 and x. Define

$$\gamma(t) = \frac{\|x \vee ty\| - \|x\|}{t},$$

where $x, y \ge 0, t > 0$. The function $t \to \gamma(t)$ is convex, nonnegative and nondecreasing for $t > 0$. Moreover,

$$\inf_{t>0} \gamma(t) = \sup_{x^*,y^*\in \mathrm{Grad}(x), 0\le y^*\le x^*} (\langle y, x^* - y^*\rangle)$$

and the supremum on the right-hand side is attained at some $x^*, y^* \in \mathrm{Grad}(x)$. Therefore, we get the necessary condition for z to be a solution of the optimization problem (4.3).

Theorem 4.4 ([63, Theorem 7]). *A necessary condition for z to be a solution of the optimization problem* (4.3) *is*

$$\max_{x^*,y^*\in \mathrm{Grad}(x), 0\le y^*\le x^*} (\langle z, x^* - y^*\rangle) = 0.$$

The above condition is trivially satisfied if x is an order-smooth point, i.e., $\mathrm{Grad}(x) \cap X_+^*$ *contains no proper order interval.*

The next theorem gives the full characterizations of the solutions of the optimization problem (4.3).

Theorem 4.5 ([63, Theorem 8]). *Let X be a Banach lattice and let $x \in S_+(X)$. The element $y \geq 0$ is a solution of the optimization problem* (4.3) *if and only if there exists $x^* \in S_+(X^*)$ such that*

(a) *x^* attains its norm at x, i.e., $\langle x, x^* \rangle = 1$,*
(b) *x^* attains the value $\|x \vee y\|$ at $x \vee y$, i.e., $\langle x \vee y, x^* \rangle = \|x \vee y\|$,*
(c) *$< y - x, y^* > \leq 0$ for every $y^* \in B_+(X^*)$ with $0 \leq y^* \leq x^*$.*

Finally, it is worth noticing that in [76] the lattice structure of the gradient $\mathrm{Grad}(x)$ of $x \in X_+ \setminus \{0\}$ was studied. The sets $\mathrm{Grad}(x)$ and $\mathrm{Grad}(|x|)$ were compared up to isometry. It was proved that if the dual X^* is strictly monotone, then $\mathrm{Grad}(|x|)$ consists of positive elements only.

5. Relationships to complex rotundities

In this section we will give some results about relationships between monotonicity properties of real Banach lattices and complex rotundity properties of their complexifications. By the complexification of a real Köthe space $(E, \|\cdot\|_E)$ we mean the space

$$E^{\mathcal{C}} = \{z : T \to \mathcal{C} : z = x + iy, \quad x, y \in E\}$$

endowed with the norm

$$\|z\| = \left\| \sqrt{x^2 + y^2} \right\|_E = \| \, |z| \, \|_E \, .$$

Conversely, if $(F, \|\cdot\|)$ is the complexification of a real Köthe space E, then the space

$$F_r = \{z \in F : im(z) = 0\}$$

under the norm induced from F is a real Köthe space and it equals E. Therefore, $E = (E^{\mathcal{C}})_r$ and $F = (F_r)^{\mathcal{C}}$.

Let Z be a complex Banach space and $S(Z)$ be the unit sphere of Z. A point $z_0 \in S(Z)$ is called a complex extreme point (or $\mathcal{C}$-extreme point) if for any $z \in Z \setminus \{0\}$ we have $\sup_{|\lambda| \leq 1} \|z_0 + \lambda z\|_Z > 1$. A complex Banach space Z is said to be complex rotund ($\mathcal{C}$-rotund) if every $z \in S(Z)$ is a $\mathcal{C}$-extreme point. We say that a complex Banach space Z is complex uniformly rotund ($\mathcal{C}$-uniformly rotund) if for every $\varepsilon \in (0,1)$ there exists $\delta(\varepsilon) \in (0,1)$ such that

$$\|z\|_Z = 1 \text{ and } \|\tilde{z}\| \geq \varepsilon \implies \sup_{|\lambda| \leq 1} \|z + \lambda \tilde{z}\|_Z \geq 1 + \delta(\varepsilon) \tag{5.1}$$

for every $z, \tilde{z} \in Z$ or (see [22]) equivalently,

$$H_\infty^Z(\varepsilon) = \inf \left\{ \sup \left\{ \left\| z + e^{i\theta} \tilde{z} \right\| : 0 \leq \theta \leq 2\pi \right\} - 1 : \|z\| = 1, \|\tilde{z}\| = \varepsilon \right\} > 0.$$

A point $z \in S(Z)$ is called a point of complex local uniform rotundity ($\mathcal{C}$-LUR point) if for every $\varepsilon \in (0,1)$ there exists $\delta(z, \varepsilon) \in (0,1)$ such that the implication (5.1) holds true for $\delta(\varepsilon) = \delta(z, \varepsilon)$ and every $\tilde{z} \in Z$. If every point of the unit sphere $S(Z)$ is a $\mathcal{C}$-locally uniformly rotund point, then Z is called complex locally uniformly rotund ($\mathcal{C}$-LUR).

The notion of complex rotundity was introduced by Thorp and Whitley in [73], where they showed that the complex Lebesgue space $L^1(\Sigma,\mu)$ is $\mathcal{C}$-rotund (while the real Lebesgue space is not rotund). Globevnik [35] introduced the notion of uniform $\mathcal{C}$-rotundity and showed that the complex Lebesgue space $L^1(\Sigma,\mu)$ is $\mathcal{C}$-uniformly rotund. Next Wang and Teng [75] introduced the notion of local $\mathcal{C}$-uniform rotundity and investigated this property in Musielak–Orlicz spaces.

It turns out that there is a close relationship between monotonicity properties of a real Köthe space E and $\mathcal{C}$-rotundity properties of the complexification $E^{\mathcal{C}}$ of the space E (see [46]). The investigations of the relationships were continued in [64]. We start with the following theorem.

Theorem 5.1 ([46, Theorem 1]). *For any real Köthe space E, a point $z \in S(E^{\mathcal{C}})$ is a $\mathcal{C}$-extreme point of $B(E^{\mathcal{C}})$ if and only if $|z|$ is a point of upper strict monotonicity of E, i.e., $\| \, |z| \, \| < \| \, |z| + y\|$ for every $y \in E_+ \setminus \{0\}$.*

Applying the above result we get the following theorem on $\mathcal{C}$-rotundity of the complexification of real Köthe spaces.

Theorem 5.2 ([46, Corollary 1]). *Let E be a real Köthe function space. Then the space $E^{\mathcal{C}}$ is $\mathcal{C}$-rotund if and only if E is strictly monotone.*

The next theorem gives criteria for $\mathcal{C}$-uniform rotundity of the complexification $E^{\mathcal{C}}$ of E.

Theorem 5.3 ([46, Theorem 2]). *Let E be a real Köthe function space. The space $E^{\mathcal{C}}$ is $\mathcal{C}$-uniformly rotund if and only if E is uniformly monotone.*

However, for the $\mathcal{C}$-local uniform rotundity the equivalent monotonicity property is not known. But we do have the following:

Theorem 5.4 ([46, Lemma 1 and Corollary 2]). *For any real Köthe space E, if $z \in S(E^{\mathcal{C}})$ is a $\mathcal{C}$-locally uniformly rotund point then $|x|$ is a point of upper local uniform monotonicity in E. Thus, if $E^{\mathcal{C}}$ is a $\mathcal{C}$-locally uniformly rotund, then E is upper locally uniformly monotone.*

The above results have been applied to get criteria for complex rotundity properties of generalized Calderón–Lozanovskiĭ spaces over atomless and counting measure spaces. We state here the theorem for the atomless case only.

Theorem 5.5 ([46, Theorems 3 and 5]). *Let (T,Σ,μ) be a nonatomic σ-finite and complete measure space, E be an order-continuous real Köthe space with the Fatou property and φ be an Orlicz function. The complexification $E_\varphi^{\mathcal{C}}$ of the Calderón–Lozanovskiĭ space E_φ is $\mathcal{C}$-rotund (respectively, $\mathcal{C}$-uniformly rotund) if and only if $\varphi > 0$, $\varphi \in \Delta_2^E$ and E is a strictly monotone (respectively, uniformly monotone) space.*

In consequence, the following criteria for $\mathcal{C}$-rotundity and $\mathcal{C}$-uniform rotundity of Orlicz–Lorentz spaces have been obtained (see [46, Theorems 7 and 9] and also [16]).

Theorem 5.6. *Let (T, Σ, μ) be a nonatomic σ-finite and complete measure space and φ be an Orlicz function. The complexification $\Lambda^{\mathcal{C}}_{\varphi,\omega}$ of the Orlicz–Lorentz function space $\Lambda_{\varphi,\omega}$ is $\mathcal{C}$-rotund (respectively, $\mathcal{C}$-uniformly rotund) if and only if*

(i) *$\varphi > 0$ and $\varphi \in \Delta_2(\infty)$ if $\mu(T) < \infty$ and $\varphi \in \Delta_2(\mathbb{R})$ when $\mu(T) = \infty$,*

(ii) *$\omega > 0$ (respectively, and ω is regular) if $\mu(T) < \infty$ and $\int_0^{\mu(T)} \omega(t)dt = \infty$ (respectively, ω is regular) when $\mu(T) = \infty$.*

It is worth noticing that results from Theorems 5.2 and 5.3 were generalized to any Banach lattices in [64, Theorems 3.4 and 3.5]. Moreover, the following was also shown:

Theorem 5.7 ([64, Theorems 5.1 and 5.2]). *Let E be a real Köthe function space over a complete measure space (T, Σ, μ) and X be a non-trivial complex Banach space. The Köthe–Bochner space $E(X)$ is $\mathcal{C}$-rotund (respectively, $\mathcal{C}$-uniformly rotund) if and only if E is strictly monotone (resp., uniformly monotone) and X is $\mathcal{C}$-rotund (respectively, $\mathcal{C}$-uniformly rotund).*

6. Dominated best approximation in Banach lattices

Let us start with the definition of the best approximation problems in Banach spaces in order to see better the difference between the best approximation problems in Banach spaces and the dominated best approximation problems in Banach lattices.

Let $X = (X, \|\cdot\|)$ be a Banach space and A be its nonempty subset. Then for any $x \in X$ the number

$$d(x, A) = \inf\{\|x - y\| : y \in A\}$$

is called the distance of x from A. It is obvious that for any nonempty set A in X and any $x \in X$ the distance $d(x, A)$ is finite as well as that $d(x, A) = 0$ for any $x \in A$.

Given any $A \subset X$, $A \neq \emptyset$, the function $P_A(x) = X \to 2^X$ defined by

$$P_A(x) = \{z \in A : d(x, A) = \|x - z\|\}$$

is called the projection from X onto A, and for any $x \in X$ the set $P_A(x)$ is called the projection of x onto A.

It is important to know the following facts:

a) Under which additional assumptions about a fixed set A in X we know that $P_A(x) \neq \emptyset$ for any $x \in X$?

b) Under which assumptions on X, we know that $P_A(x) \neq \emptyset$, for any $x \in X$ and any nonempty, closed and convex subset A of X?

c) Under which assumptions about X we know that for any nonempty, closed and convex set A in X we have that $Card(P_A(x)) \leq 1$ for any $x \in X$?
d) Under which assumptions on X we know that $Card(P_A(x)) = 1$ for any $x \in X$ and any nonempty, closed and convex subset A of X?

It is known that the sufficient conditions for a) is that A is nonempty, convex and the intersections of A with all closed balls $B(0,r)$, $r > 0$ are weakly compact sets. Necessary and sufficient conditions for b), c) and d) are, respectively, that X is reflexive, X is rotund and X is rotund and reflexive.

It was natural to modify the above best approximation problems to the dominated best approximation problems when $X = (X, \leq, \|\cdot\|)$ is a Banach lattice. Recall following [61] that a subset A of X is called a sublattice of X if for every $x, y \in A$, $x \vee y \in A$ and $x \wedge y \in A$, and that for $z \in X$, we write $A \leq z$ (respectively $z \leq A$) if $x \leq z$ (respectively $z \leq x$) for all $x \in A$. The dominated best approximation problems are then obtained by requiring that the set A is a sublattice of X such that $z \geq A$ or $z \leq A$ for some $z \in X$ (see [61], [62], [42], [13] and [14]). In this more specific situation it was natural to expect that the conditions corresponding to the questions a), b), c), d) presented above would be weaker than for the general best approximation problems in Banach spaces without order structure. The results that were obtained later confirmed this prediction. Namely, when the class of the sets A in the dominated versions of the above questions b), c) and d) is restricted to closed sublattices of X, then the necessary and sufficient conditions for positive answers are, respectively, that X is order continuous, X is strictly monotone and X is strictly monotone and order continuous.

In [61] another two monotonicity properties denoted by CWLLUM and H$^+$STM, which play a crucial role in characterizations of the dominated best approximation problems, were defined. Property H$^+$STM can be viewed as a lattice version of the Kadec–Klee property together with rotundity, denoted in the literature as (HR). Namely, we say that X has the H$^+$ property if $\|x_n - x\| \to 0$ whenever $0 \leq x_n \leq x$ and $x_n \to x$ weakly. A Banach lattice X is said to have property H$^+$STM if it has both properties H$^+$ and STM. We say that a Banach lattice X has CWLLUM property if for any nonnegative $x^* \in X^*$ (the dual space of X) with $\|x^*\| = 1$, any $x \in S_+(X)$ and any sequence $(x_n)_{n=1}^{\infty}$ in X satisfying $0 \leq x_n \leq x$ for all $n \in \mathbb{N}$, the condition $x^*(x_n - x) \to \|x\|$ implies $\|x_n\| \to 0$. Both properties CWLLUM and H$^+$STM are weaker than lower local uniform monotonicity. In [42, Theorem 4.1] it has been proved that both properties H$^+$STM and CWLLLUM are equivalent to the combination of strict monotonicity and order-continuity and, in consequence, both properties H$^+$STM and CWLLLUM are equivalent to lower local uniform monotonicity in Köthe sequence spaces (see Theorem 2.4) and symmetric spaces (see Theorem 2.5).

We will present some of the results mentioned above together with their proofs. For Musielak–Orlicz spaces the proofs were given by W. Kurc (see [61]) and we adopt the proofs to general Banach lattices.

Theorem 6.1. *Let X be a Banach lattice. Then the following are equivalent:*

(a) *X is strictly monotone.*
(b) *For all $z \in X$ and any order interval $[x, y] \subset X$ satisfying $z \geq [x, y]$ or $z \leq [x, y]$, there holds $Card(P_{[x,y]}(z)) \leq 1$.*
(c) *For all $z \in X$ and any sublattice $A \subset X$ such that $z \geq A$ or $z \leq A$ there holds $Card(P_A(z)) \leq 1$, that is, the dominated best approximation problem with respect to A is unique whenever it is solvable.*

Proof. (a) $\Rightarrow$ (c) Let us assume that (a) is satisfied but (c) not, which means that, if $P_A(z) \neq \emptyset$, then there exist $u, w \in A$, $u \neq w$, such that $\|z - u\| = \|z - w\| = \inf_{v \in A} \|z - v\|$. Since A is a sublattice, we have $u \vee w \in A$. Assume without loss of generality that $z \geq A$ (the proof in the case $A \leq z$ is similar). Then, $0 \leq z - (u \vee w) \leq z - u$, so $u \vee w \in P_A(z)$. Since $u \neq w$, we have that either $u < u \vee w$ or $w < u \vee w$ (for example $u < u \vee w$ means that $u \leq u \vee w$ and $u \neq u \vee w$). In the first case we have the equalities $\|z - u\| = \|z - u \vee w\| = \|z - u - ((u \vee w) - u)\|$, a contradiction with strict monotonicity of X, because $(u \vee w) - u \geq 0$ and $(u \vee w) - u \neq 0$. In the second case the proof is similar.

The implication (c) $\Rightarrow$ (b) is obvious because order intervals are sublattices.

(b) $\Rightarrow$ (a) Let us assume that (b) is satisfied but (a) not, which means that there exists $z, w \in X$ such that $z \geq w \geq 0$, $w \neq 0$ and $\|z - w\| = \|z\|$. Let us define $A = [0, w]$. Then the previous equality means that $0, w \in P_A(z)$, which means that (b) does not hold, a contradiction. □

Theorem 6.2. *Assume that X is a σ-complete Banach lattice with an order-continuous norm, A is a closed sublattice of X, $z \in X$ and $z \geq A$ or $z \leq A$. Then $P_A(z) \neq \emptyset$.*

Proof. Let us consider first the case $z \geq A$ and let $(w_n)_{n=1}^{\infty}$ be a minimizing sequence in A with respect to z, that is, $d(z, A) := \inf_{w \in A} \|z - w\| = \lim_{n\to\infty} \|z - w_n\|$. Since A is a sublattice of X, there exist $u_n = \bigvee_{k=1}^{n} w_k \in A$ for any $n \in \mathbb{N}$. Moreover $0 \leq z - u_n \leq z - w_n$ for any $n \in \mathbb{N}$, which implies that $(u_n)_{n=1}^{\infty}$ is also a minimizing sequence in A for z. Since $u_n \leq z$ for any $n \in \mathbb{N}$ and X is σ-complete, there exists $u = \bigvee_{n=1}^{\infty} u_n \in X$ and $0 \leq u - u_n \downarrow 0$. By order-continuity of X, we conclude that $\|u - u_n\| \to 0$ as $n \to \infty$. Since A is closed, $u \in A$. We also have $d(z, A) \leq \|z - u\| \leq \|z - u_n\| + \|u_n - u\| \to d(z, A)$, whence $\|z - u\| = d(z, A)$, which means that $u \in P_A(z)$.

Consider now the case $z \leq A$. Let us define $B = 2z - A$. Taking any $x \in B$, one can find $u \in A$ such that $x = 2z - u$. Since $z \leq u$, we have $x = 2z - u \leq 2z - z = z$, which means that $B \leq z$. By the first case, which has been just proved here, there exists $u \in B$ such that $d(z, B) = \|z - u\|$. Let $v \in A$ be such that $u = 2z - v$. Then, we have $\|z - u\| = \|z - (2z - v)\| = \|v - z\| = \|z - v\|$. Let us note that

$$d(z, A) = \inf_{u \in A} \|z - (2z - u)\| = \inf_{u \in A} \| - z + u\| = \inf_{u \in A} \|z - u\| = d(z, A).$$

Therefore, the equality that has been proved already means that $d(z, A) = \|z - v\|$. Since $v \in A$, the proof of the second case is finished. □

The role of the upper and lower local uniform monotonicity in the dominated best approximation problems is illustrated by the next two theorems.

Let us recall (see [61]) that the dominated best approximation problem is called uniquely solvable if $Card(P_A(z)) = 1$, whenever $z \in X$ and $A \subset X$, where X is a Banach lattice.

Theorem 6.3. *Let X be a σ-complete Banach lattice. Then the lower local uniform monotonicity of X implies that both (lower and upper) dominated best approximation problems $A \leq z$ and $A \geq z$ are uniquely solvable.*

Proof. It is obvious that lower local uniform monotonicity of X implies its strict monotonicity. Moreover, local lower uniform monotonicity of X implies its order-continuity (see [23]). Therefore, the result follows by Theorems 6.1 and 6.2. □

In the next theorem we will use the notion of stability of the dominated best approximation problem. A dominated best approximation problem ($x \leq A$ or $x \geq A$) is said to be strongly solvable (see [42]) if it is uniquely solvable and stable. Stability of such problem means that for any minimizing sequence $(x_n)_{n=1}^{\infty}$ in A (that is a sequence such that $\|x - x_n\| \to d(x, A)$ as $n \to \infty$) we have that $d(x_n, P_A(x)) \to 0$ as $n \to \infty$.

Theorem 6.4. *Let X be a σ-complete Banach lattice and A be a closed sublattice of X. Then in both cases $A \leq z$ and $z \leq A$ the upper local uniform monotonicity and order-continuity of X yields that the lower and upper best dominated approximation problems are strongly solvable.*

Proof. Let the assumptions about X be satisfied and A be a closed sublattice of X. Let us take any $z \in X$ such that $A \leq z$. By the order-continuity of X and Theorem 6.2, the dominated best approximation problem is solvable. Since upper local uniform monotonicity of X implies its strict monotonicity, the problem is uniquely solvable, that is, there is a unique $u_0 \in A$ such that $\|z - u_0\| = \inf_{u \in A} \|z - u\|$, which means that $P_A(z) = \{u_0\}$.

Let $(u_n)_{n=1}^{\infty}$ be a minimizing sequence in A for z. Since A is a sublattice, $v_n := \bigvee_{k=1}^{n} u_k \in A$ exists for any $n \in \mathbb{N}$. Moreover, since X is a σ-complete Banach lattice, we have that $v_n \leq v := \bigvee_{n=1}^{\infty} v_n \leq z$. Hence $0 \leq v - v_n \downarrow 0$ and since X is order continuous, we get $\|v - v_n\| \to 0$ as $n \to \infty$, so that $v \in A$, since A is norm closed in X. Note that

$$d(z, A) = \inf_{u \in A} \|z - u\| \leftarrow \|z - u_n\| \geq \|z - v_n\| \geq \|z - v\| \geq \|z - u_0\| = d(z, A),$$

whence $v \in P_A(z)$. Now, the unique solvability of the best dominated problem implies that $v = u_0$. Since $0 \leq z - u_0 \leq z - u_n$ for any $n \in \mathbb{N}$ and $\|z - u_n\| \to \|z - u_0\|$, by the upper local uniform monotonicity of X, we get $\|u_0 - u_n\| \to 0$ as $n \to \infty$, which means that the dominated best approximation problem is stable. By the uniqueness of its solvability it is strongly solvable.

Let us consider now the case $z \leq A$. Since upper local uniform monotonicity implies strict monotonicity, so by order-continuity of X and by Theorems 6.1 and

6.2, our dominated best approximation problem is uniquely solvable, that is, there exists a unique $u_0 \in A$ such that $P_A(z) = \{u_0\}$. Let $(w_n)_{n=1}^{\infty}$ be a minimizing sequence in A with respect to z, that is, $d(x, A) = \lim_{n\to\infty} \|z - w_n\|$. Since A is a sublattice of X, there exist $u_n = \bigwedge_{k=1}^{n} w_k \in A$ for any $n \in \mathbb{N}$, whence $0 \le u_n - z \le w_n - z$ for any $n \in \mathbb{N}$ and, in consequence, $(u_n)_{n=1}^{\infty}$ is also a minimizing sequence for z. Since $z \le u_n$ for any $n \in \mathbb{N}$ and X is σ-complete, there exists $u = \bigwedge_{n=1}^{\infty} u_n \in X$. Since $u_{n+1} \le u_n$ for any $n \in \mathbb{N}$, we have $u \le u_n$ for any $n \in \mathbb{N}$ and $0 \le u_n - u \downarrow 0$.

Now, by the order-continuity of X, we get $\lim_{n\to\infty} \|u_n - u\| = 0$. Since A is closed in X, we get $u \in A$. We also have

$$d(z, A) \le \|z - u\| \le \|z - u_n\| + \|u_n - u\| \to d(z, A),$$

whence $\|z - u\| = d(z, A)$, and so $u \in P_A(z)$. Since our dominated best approximation problem is uniquely solvable, we get $u = u_0$. The conditions $0 \le u_0 - z \le u_n - z$ and $\|z - u_n\| \to \|z - u_0\|$ and upper local uniform monotonicity of X imply that $\|u_n - u_0\| \to 0$ as $n \to \infty$, which means that our dominated best approximation problem is strongly solvable. □

It is worth noticing that for Musielak–Orlicz function spaces $L^{\varphi}(\mu)$ and their subspaces $E^{\varphi}(\mu)$, equipped with the Luxemburg norm, the sets $P_A(f)$ were determined in terms of the subdifferential of the generating Musielak–Orlicz function (see [61] for atomless measure case). In sequence spaces the same was done in [62].

Two other monotonicity properties of Banach lattices, called decreasing (resp. increasing) uniform monotonicity, denoted by DUM (resp. IUM), were defined in [13] in connection with the modified dominated best approximation problems which were considered in that paper. Namely, a Banach lattice X is said to be decreasing (resp. increasing) uniformly monotone if for any sequences $(x_n)_{n=1}^{\infty}$ and $(y_n)_{n=1}^{\infty}$ in X_+ with $y_n \ge x_n \downarrow$ (resp. $x_n \ge y_n \uparrow$) and $\lim_{n\to\infty} \|x_n\| = \lim_{n\to\infty} \|y_n\|$ there holds $\|y_n - x_n\| \to 0$. Obviously

$$\begin{array}{ccccc} UM & \Rightarrow & IUM & \Rightarrow & LLUM \\ \Downarrow & & & & \Downarrow \\ DUM & \Rightarrow & ULUM & \Rightarrow & STM. \end{array}$$

Moreover, by Theorem 1.2 from [13] and Proposition 2.1 from [23], we know that in Köthe spaces the properties IUM and LLUM coincide. It has been proved in [13] that a Banach lattice X is DUM (resp. IUM) if and only if X is order continuous and ULUM (resp. LLUM). Properties DUM and IUM were applied in [13] to the modified best approximation problems defined below. Let us note that in such problems, in comparison with the usual dominated best approximation problems that were considered in [61, 62] and [42], in place of $f \ge A$ or $f \le A$, where A is sublattice of X, the authors used the conditions $f \ge A$ or $f \le A$, where A is any nonempty convex subset of X and $f \in D(A)$, where

$$D(A) = \{z \in X : \ A - z \text{ is an absolutely directed set}\}.$$

A subset B of X is said to be absolutely directed if for any $x, y \in B$ there exists $z \in B$ such that $|z| \leq |x| \wedge |y|$.

Now we will present seven theorems and a corollary concerning the modified dominated best approximation problems.

Theorem 6.5 ([13, Theorem 2.1]). *Let X be a Banach lattice. Then the following statements are equivalent:*

(1) *X is strictly monotone.*
(2) $\mathrm{Card}(P_A(z)) \leq 1$ *for any nonempty convex subset A of X and any $z \in D(A)$.*
(3) $\mathrm{Card}(P_A(z)) \leq 1$ *for any nonempty convex and closed subset A of X and any $z \in D(A)$.*

Remark 6.6 ([13, Remark 2.1]). If $P_A(z) = \{x\}$ and $z \in D(A)$, then for any $y \in A$, there exists $w \in A$ such that $|w - z| \leq |x - z| \wedge |y - z|$, which yield that $w \in P_A(z)$. Since $P_A(z) = \{x\}$, we have $w = x$. Therefore $|w - z| \leq |y - z|$ for all $y \in A$. This means that $|w - z|$ is the order infimum of $|A - z|$.

Let X be a Banach lattice and let us define for any $x \in X$,

$$x^{\perp} = \{z \in X : \ |z| \wedge |x| = 0\}.$$

Theorem 6.7 ([13, Theorem 2.2]). *Let X be an order-continuous Banach lattice and let for any $x \in X$, $x^{\perp}$ be the complemented sublattice of X defined as above. If X is LLUM (i.e., X is IUM according to the results mentioned before formulating Theorem 6.5), then:*

(1) *For all nonempty convex subsets A of X and all $z \in D(A)$, the set $A - z$ has a minimizing Cauchy sequence,*
(2) *For all nonempty convex and closed subsets A of X and all $z \in D(A)$,* $\mathrm{Card}(P_A(z)) = 1$.

Theorem 6.8 ([13, Theorem 2.3]). *Assume that X is a uniformly monotone Banach lattice. Then for any nonempty convex and closed subset A of X and any $z \in D(A)$,* $\mathrm{Card}(P_A(z)) = 1$.

Theorem 6.9 ([13, Theorem 3.1]). *Let X be a uniformly monotone Banach lattice and A be a nonempty convex and closed subset of X, $z_n \in D(A)$ and $P_A(z_n)) = \{x_n\}$ for any $n \in \mathbb{N}$. Then $\|z_n - z_0\| \to 0$ as $n \to \infty$ implies that $\|x_n - x_0\| \to 0$ as $n \to \infty$, i.e., the best approximation operator P_A is continuous.*

Theorem 6.10 ([13, Theorem 3.2]). *Let X be a Banach lattice. The following statements are equivalent:*

(1) *X is upper locally uniformly monotone.*
(2) *For any nonempty convex subset A of X and any $z \in D(A)$, there holds $\|x_n - x_0\| \to 0$ as $n \to \infty$, where $(x_n)_{n=1}^{\infty}$ is a minimizing sequence of $A - z$ and $x_0 + z \in P_A(z)$.*

Theorem 6.11 ([13, Theorem 3.3]). *Let X be a uniformly monotone Banach lattice and A be a nonempty convex and absolutely directed subset of X, that is, $0 \in D(A)$. Then any minimizing sequence of A is a Cauchy sequence.*

Theorem 6.12 ([13, Theorem 3.4]). *Let X be a Banach lattice. If for any nonempty convex and absolutely directed subset A of X all minimizing sequences of A are Cauchy sequences, then X is decreasing uniformly monotone, that is, order continuous and upper locally uniformly monotone.*

Corollary 6.13 ([13, Corollary 3.1]). *Let X be a uniformly monotone Banach lattice and A be a nonempty convex and closed subset of X and $z \in D(A)$. Then:*

(1) $\mathrm{Card}(P_A(z)) = 1$,
(2) *For any minimizing sequence $(x_n)_{n=1}^{\infty}$ of $A - z$ and $x_0 + z \in P_A(z)$, we have $\|x_n - x_0\| \to 0$ as $n \to \infty$,*
(3) *The best approximation operator $P_A : D(A) \to A$ is continuous.*

In [14] criteria for strict monotonicity, lower and upper local uniform monotonicity and uniform monotonicity of Orlicz–Sobolev spaces $W_{m,\varphi}$ equipped with the Luxemburg norm were given. Note that the Orlicz–Sobolev spaces $W_{m,\varphi}$ were endowed in [14] with the following natural semi-order: for any x and y from $W_{m,\varphi}$ we define that $y \le x$ if and only if for any $|\alpha| \le m$ we have $D^\alpha y \le D^\alpha x$ in L^φ, where $|\alpha|$ denotes degree of the derivative D^α, that is, $|\alpha| = \alpha_1 + \cdots + \alpha_n$ for $\alpha = (\alpha_1, \ldots, \alpha_n)$ with non-negative integers α_i, L^φ is an Orlicz function space and the derivatives $D^\alpha x$ are understood in the distributional sense. Next some results on the modified dominated best approximation problems in the spaces $W_{m,A}$ were presented. As an application, an interesting example about the approximation of a function $x \in W_{m,\varphi}$ by polynomials of degree $\le m$ were given.

Let us also note that in [18] some applications of the points of lower and upper monotonicity to the local dominated best approximation problems in Banach lattices were presented.

7. Some application to Ergodic Theory

Assume that X and Y are Banach lattices. An element $u \in X_+$ is said to be a *weak unit* if for any $f \in X_+$ the condition $|f| \wedge |u| = 0$ implies that $f = 0$.

Let (x_n) be a sequence in X_+. An element $\phi \in X_+$ is called a (*weak*) *truncated limit* of (x_n) (we denote such ϕ by $(W)TL\, x_n$), if for the weak unit u in X and every $k \in \mathbb{N}$, the sequence $(x_n \wedge ku)$ (weakly) converges to ϕ_k as $n \to \infty$ and $\phi_k \uparrow \phi$ as $k \to \infty$. Let us recall that a linear operator T from X into Y is called *positive* ($T \ge 0$ for short) if $TX_+ \subset X_+$, and T is called a *contraction* if $\|T\| \le 1$.

A sequence (x_n) is called TL *null* if $TL\,|x_n| = 0$. If X is the Lebesgue space L^1 over a measure space, then, as it has been shown in [2], TL *null* sequences are exactly the sequences of functions that converge to zero in measure on sets of finite measure.

Denote by A_n the Cesàro averages of the succsessive powers of an operator T, i.e.,

$$A_n = \frac{T + T^2 + \cdots + T^n}{n}.$$

Such operators have their origins in statistical mechanics and probability theory (see [1]). Questions about limits of the Cesàro averages are strongly related to the ergodic mean theorems. In general, we say that T is mean ergodic on X if for any $x \in X$, $A_n x$ converges strongly in X. In [3] Akcoglu and Sucheston studied the limiting behavior of these Cesáro averages in the case when T is a positive contraction. They were motivated by a few facts. Namely, if $X = L^1$ over a probability measure space, then for $x \in L^1$ $A_n x$, does not converge in L^1 or almost everywhere (see [11]), but converges in probability [58]. Moreover, a sequence (x_n) converges in probability to ϕ in L^1 if and only if $TL\, x_n = \phi$. In [3] the authors have proved two ergodic theorems for TL convergence in Banach lattices and in order to do so they introduced two monotonicity conditions for X. For a Banach lattice X we define the following two conditions:

(C_1) For $x, y \in X_+$, $y \neq 0$, we have $\|x + y\| > \|x\|$ (strict monotonicity of X).
(C) For every $\alpha > 0$ and $x' \in X_+$ there exists $\beta = \beta(x', \alpha)$ such that $\|x + y\| \geq \|y\| + \beta$ whenever $x, y \in X_+$, $x \leq x'$, $\|x\| \geq \alpha$, $\|y\| \leq 1$.

The condition (C) is stronger than the condition (C_1) and, in fact, it is a kind of uniform version of the condition (C_1) (see [4]). Both ergodic theorems proved in [3] use also two other assumptions concerning X, that are denoted for short by (A) and (B), respectively. The first one means that X has a weak unit, and the second one means that every norm-bounded increasing sequence in X converges in norm. Condition (B) is equivalent to the assumption that X is weakly sequentially complete or that X contains no isomorphic copy of c_0. Moreover, condition (B) implies that X is order-continuous.

Theorem 7.1. *Let a Banach lattice X satisfy conditions* (A), (B) *and* (C_1) *and let* $\|T\| \leq 1$, $\phi \in X_+$, $T\phi = \phi$, *and* $x \in X_+$. *Then* $A_n x \wedge \phi$ *converges strongly.*

Theorem 7.2. *Assume that a Banach lattice X satisfies conditions* (A), (B) *and* (C) *and let* $x \in X_+$. *Then the strong truncated limit* $TL\, A_n x = \phi$ *exists and* $T\phi = \phi$.

Subadditive ergodic theory in general function spaces was considered by Ghoussoub in [34]. A sequence (s_n) in X is said to be T-subadditive (resp. T-superadditive) if $s_{n+k} \leq s_n + T^n(s_k)$ (resp. $s_{n+k} \geq s_n + T^n(s_k)$) for all integers n, k. A sequence (s_n) is said to be T-additive if it is both superadditive and subadditive, i.e., $s_{n+k} = s_n + T^n(s_k)$ for all integers n, k. Moreover, (s_n) is T-additive if and only if $s_n = \sum_{i=0}^{n} T^i(s_i)$ for each $n \in \mathbb{N}$.

Theorem 7.3 ([34, Corollary II.2]). *Assume that T is a mean ergodic positive contraction on a Banach lattice X with order-continuous and strictly monotone norm. Then, for every positive T-subadditive process (s_n) in X, we have the strong convergence of the sequence* $\left(\frac{1}{n} s_n\right)$.

We say that a sequence (s_n) is of bounded T-variation if

$$\sup_m \frac{1}{m} \left\| \sum_{i=1}^{m} (s_i - T s_{i-1}) \right\| < \infty.$$

With this definition Ghossoub [34] obtained another theorem that links monotonicity of the norm and the ergodic theory.

Let us recall some definitions. A band in a vector lattice E is an ideal J with the property that if $A \subseteq J$ and x is the supremum of A in E then $x \in J$. A band B in E is a projection band if $E = B \oplus B_d$, where B_d is the disjoint complement of B in E. In this case if $x \in B$ and $y \in B_d$ then $x + y \geq 0$ if and only if $x \geq 0$ and $y \geq 0$. The map that takes $x + y$ to y in this case is the band projection onto B denoted by Q.

Suppose now E is a Banach lattice which is the range of a band projection Q in a Banach lattice G. We say that E is an L_p-ideal in G for some p $(1 \leq p < \infty)$ if $\|Qx\|^p + \|(I-Q)x\|^p \leq \|x\|^p$ for each $x \in G$. Typical examples of spaces which are L_p-ideals in their second duals are ([34]) reflexive Banach lattices (Q=Identity) and p-concave Banach lattices with p-concavity constant equal to one.

Theorem 7.4 ([34, Corollary II.4]). *Assume that T is a mean ergodic positive contraction on a Banach lattice X with strictly monotone norm such that X is an L^1-ideal in its second dual. Then, for every positive T-subadditive process (s_n) of bounded T-variation, we have the strong convergence of the sequence $\left(\frac{1}{n}s_n\right)$.*

References

[1] A.A. Albanese, J. Bonet and W.J. Ricker, *Mean ergodic semigroups of operators.* RACSAM **106**, 299–319.

[2] M.A. Akcoglu and L. Sucheston, *A stochastic ergodic theorem for superadditive processes.* Ergodic Theory and Dynamical Systems **3** (1983), 335–344.

[3] M.A. Akcoglu and L. Sucheston, *An ergodic theorem on Banach lattices.* Israel J. Math. **51(3)** (1985), 208–222.

[4] M.A. Akcoglu and L. Sucheston, *On uniform monotonicity of norms and ergodic theorems in function spaces.* Rend. Circ. Mat. **8(2)** (1985), 325–335.

[5] C. Bennett and R. Sharpley, *Interpolation of Operators.* Academic Press, Inc., New York, 1988.

[6] A. Betiuk-Pilarska and S. Prus, *Banach lattices which are order uniformly noncreasy.* J. Math. Anal. Appl. **342** (2008), 1271–1279.

[7] G. Birkhoff, *Lattice Theory.* Amer. Math. Soc., Providence, RI, 1979.

[8] B. Bru and H. Heinrich *Monotonies des espaces d'Orlicz.* C.R. Acad. Sci Paris **301** (1985), 893–894.

[9] J. Cerdà, H. Hudzik, A. Kamińska and M. Mastyło, *Geometric properties of symmetric spaces with applications to Orlicz–Lorentz spaces*, Positivity **2(4)** (1998), 311–337.

[10] J. Cerda, H. Hudzik and M. Mastyło, *On the geometry of some Calderón–Lozanovskiĭ interpolation spaces.* Indag. Math. (N.S.) **6(1)** (1995), 35–49.

[11] R.V. Chacon, *A class of linear transformations.* Proc. Am. Math. Soc. **15** (1964), 560–564.

[12] S. Chen, *Geometry of Orlicz Spaces.* Dissertationes Math. **356** (1996), 1–204.

[13] S. Chen, X. He and H. Hudzik, *Monotonicity and best approximation in Banach lattices.* Acta Mathematica Sinica, English Series, May, 2009, Vol. **25**, No. **5**, 785–794.

[14] S. Chen, X. He, H. Hudzik and A. Kamińska, *Monotonicity and best approximation in Orlicz–Sobolev spaces with the Luxemburg norm.* J. Math. Anal. Appl. **344** (2008), 687–698.

[15] V.I. Chilin, P.G. Dodds, A.A. Sedaev and F.A. Sukochev, *Characterizations of Kadec–Klee properties in symmetric spaces of measurable functions.* Trans. Amer. Math. Soc. **348(12)** (1996), 4895–4918.

[16] C. Choi, A. Kamińska, H.J. Lee *Complex convexities of Orlicz–Lorentz spaces and its applications.* Bull. Polish Acad. Sci. Math. **52** (2004), 19–38.

[17] M. Ciesielski, A. Kamińska, P. Kolwicz and R. Płuciennik, *Monotonicity and rotundity of Lorentz spaces* $\Gamma_{p,\omega}$. Nonlinear Anal. **75(5)** (2012), 2713–2723.

[18] M. Ciesielski, P. Kolwicz and A. Panfil, *Local monotonicity structure of symmetric spaces with applications*, J. Math. Anal. Appl. **409** (2014), 649–662.

[19] Y.A. Cui, H. Hudzik, N. Petrot, S. Suantai and A. Szymaszkiewicz, *Basic topological and geometric properties of Cesàro–Orlicz spaces*, Proc. Indian Acad. Sci. 115(4) (2005), 461–476.

[20] Y.A. Cui, H. Hudzik, L. Szymaszkiewicz and T. Wang, Criteria for monotonicity properties of Musielak–Orlicz spaces equipped with the Amemiya norm, J. Math. Anal. Appl. 303 (2005), 376–390.

[21] Y.A. Cui, H. Hudzik, M. Wisła, *Monotonicity properties and dominated best approximation problems in Orlicz spaces equipped with the p-Amemiya norm*, Journal of Mathematical Analysis and Applications **432** (2015), 1095–1105.

[22] W. Davis, D.J.H. Garling, N. Tomczak–Jagermann *The complex convexity of quasi-normed linear spaces.* J. Funct. Anal. **55** (1984), 110–150.

[23] T. Dominguez, H. Hudzik, G. López, M. Mastyło and B. Sims, *Complete characterization of Kadec–Klee properties in Orlicz spaces.* Houston J. Math. **29(4)** (2003), 1027–1044.

[24] P. Foralewski, *On some geometric properties of generalized Calderón–Lozanovskiĭ spaces*, Acta Math. Hungar. **80(1-2)** (1998), 55–66.

[25] P. Foralewski, *Some fundamental geometric and topological properties of generalized Orlicz–Lorentz function spaces.* Math. Nachr. **284(8-9)** (2011), 1003–1023.

[26] P. Foralewski, *On some geometric properties of generalized Orlicz–Lorentz function spaces.* Nonlinear Analysis **75** (2012), 6217–6236.

[27] P. Foralewski, *On some geometric properties of generalized Orlicz–Lorentz sequence spaces*, Indag. Math. (N.S.) **24(2)** (2013), 346–372.

[28] P. Foralewski and H. Hudzik, *On some geometrical and topological properties of generalized Calderón–Lozanovskiĭ sequence spaces*, Houston J. Math **25(3)** (1999), 523–542.

[29] P. Foralewski, H. Hudzik, R. Kaczmarek and M. Krbec, *Moduli and characteristics of monotonicity in some Banach lattices.* Fixed Point Theory and Applications, Volume **2010** (2010), Article ID 852346, doi:10.1155//2010//852346, 22 pages.

[30] P. Foralewski, H. Hudzik, R. Kaczmarek and M. Krbec, *Characteristic of monotonicity of Orlicz function spaces equipped with the Orlicz norm.* Commentationes Mathematicae **53(2)** (2013), 421–432. The Issue dedicated to Professor Julian Musielak Honorary Editor of this journal on the Occasion of his 85th Birthday.

[31] P. Foralewski, H. Hudzik, R. Kaczmarek, M. Krbec and M. Wójtowicz, *On the moduli and characteristic of monotonicity in Orlicz–Lorentz function spaces.* J. Convex Anal. **20(4)** (2013), 955–970.

[32] P. Foralewski, H. Hudzik and L. Szymaszkiewicz, *On some geometric and topological properties of generalized Orlicz–Lorentz sequence spaces*, Math. Nachr. 281(2) (2008), 181–198.

[33] P. Foralewski and P. Kolwicz, *Local uniform rotundity in Calderón–Lozanovskiĭ spaces.* J. Convex Anal. **14(2)** (2007), 395–412.

[34] N. Ghoussoub, *Subadditive ergodic theory in general function spaces.* Sémin. analyse fonctionnelle, Paris 1983–84, Publ. Math. Univ. Paris VII **20** (1984), 157–177.

[35] J. Globevnik, *On complex strict and uniform convexity.* Proc. Amer. Math. Soc. **47** (1975), 175–178.

[36] H. Hudzik, *Geometry of some classes of Banach function spaces*, Proceedings of the International Symposium on Banach and Function spaces (Kitakyushu, Japan, October 2–4, 2003), Yokohama Publishers, 17–57.

[37] H. Hudzik and R. Kaczmarek, *Moduli and characteristics of monotonicity in general Banach lattices and in Orlicz spaces in particular.* Nonlinear Analysis **70** (2009), 3407–3423.

[38] H. Hudzik and R. Kaczmarek, *Monotonicity characteristic of Köthe–Bochner spaces.* J. Math. Anal. Appl. **349** (2009), 459–468.

[39] H. Hudzik and A. Kamińska, *Monotonicity properties of Lorentz spaces.* Proc. Amer. Math. Soc. **123(9)** (1995), 2715–2721.

[40] H. Hudzik, A. Kamińska and M. Mastyło, *Geometric properties of some Calderón–Lozanovskiĭ spaces and Orlicz–Lorentz spaces.* Houston Journal of Mathematics **22(3)** (1996), 639–663.

[41] H. Hudzik, A. Kamińska and M. Mastyło, *Monotonicity and rotundity properties in Banach lattices.* Rocky Mountain Journal of Math. **30(3)** (2000), 933–950.

[42] H. Hudzik and W. Kurc, *Monotonicity properties of Musielak–Orlicz spaces and dominated best approximation in Banach lattices.* J. Approx. Theory **95(3)** (1998), 353–368.

[43] H. Hudzik, X.B. Liu, T.F. Wang, *Points of monotonicity in Musielak–Orlicz function spaces endowed with the Orlicz norm*, Publ. Math. Debrecen **60(3-4)** (2002), 385–403.

[44] H. Hudzik, X.B. Liu, T.F. Wang, *Points of monotonicity in Musielak–Orlicz function spaces endowed with the Luxemburg norm*, Arch. Math. (Basel) **82(6)** (2004), 534–545.

[45] H. Hudzik, A. Narloch, *Local monotonicity structure of Calderón–Lozanovskiĭ spaces.* Indag. Math. (N.S.) **15(2)** (2004), 245–255.

[46] H. Hudzik, A. Narloch, *Relationships between monotonicity and complex rotundity properties with some consequences.* Math. Scand. **96(2)** (2005), 289–306.

[47] H. Hudzik and K. Wlaźlak, *Monotonicity properties in Banach spaces via sublinear operators and corrigendum to "Rotundity properties in Banach spaces via sublinear operators"*, Nonlinear Anal. **64(6)** (2006), 1171–1188.

[48] A. Kamińska, *Some remarks on Orlicz–Lorentz spaces.* Math. Nachr. **147** (1990), 29–38.

[49] A. Kamińska, *Extremal points in Orlicz–Lorentz spaces.* Arch. Math. **55** (1990), 173–180.

[50] A. Kamińska, *Uniform convexity of generalized Lorentz spaces.* Arch. Math. **56** (1991), 181–188.

[51] L.V. Kantorovich and G.P. Akilov, *Functional Analysis.* Pergamon Press, Oxford, 1982.

[52] W.A. Kirk, B. Sims, *Handbook of metric fixed point theory.* Kluwer Academic Publishers, 2001.

[53] P. Kolwicz and K. Leśnik, *Topological and geometrical structure of Calderón–Lozanovskiĭ construction*, Math. Inequal. Appl. 13, 1 (2010), 175–196.

[54] P. Kolwicz and R. Płuciennik, *On uniform rotundity in every direction in Calderón–Lozanovskiĭ sequence spaces.* J. Convex Anal. **14(3)** (2007), 621–645.

[55] P. Kolwicz and R. Płuciennik, *Local $\Delta_2^E(x)$ condition as a crucial tool for local structure of Calderón–Lozanovskiĭ* , J. Math. Anal. Appl. **356** (2009), 605–614.

[56] P. Kolwicz and R. Płuciennik, *Points of upper local uniform monotonicity in Calderón–Lozanovskiĭ spaces*, J. Convex Anal. **17(1)** (2010), 111–130.

[57] M.A. Krasnoselski and Ya.B. Rutickiĭ, *Convex Functions and Orlicz Spaces.* Groningen, Nordhoff, 1961 (translation).

[58] U. Krengel, *On global limit behaviour of Markov chains and of general non-singular Markov processes.* Z. Wahrscheinlichkeitstheor. Verw. Geb. **6** (1966), 302–316.

[59] S.G. Krein, Ju.I. Petunin, and E.M. Semenov, *Interpolation of Linear Operators.* Transl. Math. Monogr. **5**, Amer. Math. Soc., Providence, 1982.

[60] W. Kurc, *Characterizations of some monotonicity properties of a lattice norm in Musielak–Orlicz spaces.* 17th Winter School on Abstract Analysis (Srní, 1989), Acta Univ. Carol. Math. Phys. **30(2)** (1989), 91–94.

[61] W. Kurc, *Strictly and uniformly monotone Musielak–Orlicz spaces and applications to best approximation.* J. Approx. Theory **69(2)** (1992), 173–187.

[62] W. Kurc, *Strictly and uniformly monotone sequential Musielak–Orlicz spaces.* Collect. Math. **50(1)** (1999), 1–17.

[63] W. Kurc, *A dual property to uniform monotonicity in Banach lattices*, Collect. Math. **44(1-3)** (1993), 155–165.

[64] H.J. Lee, *Monotonicity and complex convexity in Banach lattices.* J. Math. Anal. Appl. **307(1)** (2005), 86–101.

[65] J. Lindenstrauss and L. Tzafriri, *Classical Banach Spaces II, Function Spaces.* Springer-Verlag, Berlin-Heidelberg-New York, 1979.

[66] X.B. Liu and T.F. Wang, *Upper (lower) monotone coefficient of a point in Orlicz sequence space.* J. Heilongjiang Univ. Natur. Sci. **18(1)** (2001), 1–4 (in Chinese).

[67] Y.M. Lü, J.M. Wang and T.F. Wang, *Monotone coefficients and monotonicity of Orlicz spaces.* Rev. Mat. Complut. **12(1)** (1999), 105–114.

[68] L. Maligranda, *Orlicz Spaces and Interpolation.* Seminars in Math. 5, Universidade Estadual de Campinas, Campinas, SP, Brazil, 1989.

[69] A.M. Medzhitov and F.A. Sukochev, *The property (H) in Orlicz spaces.* Bull. Polish Acad. Sci. **40** (1992), 5–11.

[70] J. Musielak, *Orlicz Spaces and Modular Spaces.* Lecture Notes in Math. **1034**, Springer-Verlag, Berlin-Heidelberg-New York-Tokyo, 1983.

[71] M.M. Rao, Z.D. Ren, *Theory of Orlicz Spaces.* Marcel Dekker Inc., New York-Basel-Hong Kong, 1991.

[72] I. Singer, *The Theory of Best Approximation and Functional Analysis.* SIAM, 1974

[73] E. Thorp, R. Whitley, *The strong maximum modulus theorem for analytic functions into a Banach space.* Proc. Amer. Math. Soc. **18** (1967), 640–646.

[74] B. Turret, *Rotundity of Orlicz spaces.* Nederl. Akad. Wetensch. Proc. Ser. A **79**; Indag. Math. **38(5)** (1976), 462–469.

[75] T. Wang, Y. Teng, *Complex local uniform rotundity of Musielak–Orlicz spaces.* Science of China **30(1)** (2000), 24–31.

[76] M. Wójtowicz, *The positive aspects of smoothness in Banach lattices.* Positivity **17(2)** (2013), 257–263.

[77] A.C. Zaanen, *Introduction to Operator Theory in Riesz Spaces*, Springer-Verlag, Berlin-Heidelberg, 1997.

Paweł Foralewski, Henryk Hudzik, Wojciech Kowalewski and Marek Wisła
Faculty of Mathematics and Computer Science
Adam Mickiewicz University in Poznań
Umultowska 87
61–614 Poznań, Poland
e-mail: pawel.foralewski@amu.edu.pl
hudzik@amu.edu.pl
wojciech.kowalewski@amu.edu.pl
marek.wisla@amu.edu.pl

Ordered Structures and Applications: Positivity VII
Trends in Mathematics, 233–261

Spectrum of Weighted Composition Operators Part III: Essential Spectra of Some Disjointness Preserving Operators on Banach Lattices

A.K. Kitover

Abstract. We describe essential (in particular Fredholm and semi-Fredholm) spectra of operators on Banach lattices of the form $T = wU$, where w is a central operator and U is a disjointness preserving operator such that its spectrum $\sigma(U)$ is a subset of the unit circle.

Mathematics Subject Classification (2010). Primary 47B33; Secondary 47B48, 46B60.

Keywords. Disjointness preserving operators, spectrum, Fredholm spectrum, essential spectra.

1. Introduction

We recall the following well-known definition (see, e.g., [10, Ch. IV, Sec. 5, p. 230]).

Definition 1.1. A bounded linear operator T on a Banach space X is called *semi-Fredholm* if TX is closed in X and either $\mathrm{nul}(T) = \dim\ker(T) < \infty$ or $\mathrm{def}(T) = \dim(X/TX) < \infty$. The operator T is called *Fredholm* if $\mathrm{nul}(T) + \mathrm{def}(T) < \infty$.

In correspondence with Definition 1.1 the semi-Fredholm and Fredholm spectra of a bounded linear operator T are defined as follows. ($\sigma(T)$ as usual means the spectrum of a bounded linear operator T on a Banach space X.)

Definition 1.2. Let T be a bounded linear operator on a Banach space X. The semi-Fredholm spectrum of T is

$$\sigma_{sf}(T) = \{\lambda \in \sigma(T) : \text{ the operator } \lambda I - T \text{ is not semi-Fredholm}\}.$$

The Fredholm spectrum of T is

$$\sigma_f(T) = \{\lambda \in \sigma(T) : \text{ the operator } \lambda I - T \text{ is not Fredholm}\}.$$

It is well known (see [5]) that the Fredholm spectrum of a bounded linear operator T coincides with its spectrum in the Calkin algebra.

There are many different definitions of **essential spectrum** of a linear operator on a Banach space of which Fredholm and semi-Fredholm spectra are but two (though important) examples. We will follow the book [8] where five of them are discussed in details.

Definition 1.3 (See [8, Section I.4, p. 40]). Let T be a bounded linear operator on a Banach space X. We define the essential spectra of T as the following subsets of $\sigma(T)$.

- $\sigma_1(T) = \sigma(T) \setminus \{\zeta \in \mathbb{C} :$ the operator $\zeta I - T$ is semi-Fredholm $\}$.
- $\sigma_2(T) = \sigma(T) \setminus \{\zeta \in \mathbb{C} :$ the operator $\zeta I - T$ is semi-Fredholm and $\mathrm{nul}(\zeta I - T) < \infty\}$.
- $\sigma_3(T) = \sigma(T) \setminus \{\zeta \in \mathbb{C} :$ the operator $\zeta I - T$ is Fredholm $\}$.
- $\sigma_4(T) = \sigma(T) \setminus \{\zeta \in \mathbb{C} :$ the operator $\zeta I - T$ is Fredholm and $\mathrm{ind}(\zeta I - T) = 0\}$.
- $\sigma_5(T) = \sigma(T) \setminus \{\zeta \in \mathbb{C} :$ there is a component C of the set $\mathbb{C} \setminus \sigma_1(T)$ such that $\zeta \in C$ and the intersection of C with the resolvent set of T is not empty$\}$.

It is well known (see, e.g., [8] or [10]) that the sets $\sigma_i(T)$, $i \in [1, \dots, 5]$ are nonempty closed subsets of $\sigma(T)$ and that $\sigma_i(T) \subseteq \sigma_j(T)$, $1 \leq i < j \leq 5$, where all the inclusions can be proper. Nevertheless all the spectral radii $r_i(T)$, $i = 1, \dots, 5$ are equal to the same number (see [8, Theorem I.4.10]) which is called the essential spectral radius of T. It is also known (see [8]) that the spectra $\sigma_i(T)$, $i = 1, \dots, 4$ are invariant under compact perturbations, but $\sigma_5(T)$ in general is not.

The contents of the paper are as follows.

In Section 2 we describe the semi-Fredholm and Fredholm spectra of weighted composition operators

$$Tf = w(f \circ \varphi), f \in C(K), \tag{1}$$

where $C(K)$ is the space of all complex-valued continuous functions on a Hausdorff compact space K, $w \in C(K)$, and φ is a homeomorphism of K onto itself.

In Section 3 we, based on the results of Section 2, describe the spectra $\sigma_i(T), i = 1, \dots, 5$, where T is an operator of form (1).

In Section 4 we consider operators of the form $T = wU : X \to X$ where X is an arbitrary Banach lattice, w is a central operator on X, U is a d-isomorphism of X, and $\sigma(U)$ is a subset of the unit circle. We show that the study of essential spectra of such an operator can be reduced to the study of essential spectra of an appropriate operator of form (1) on $C(K)$, where K is the Stonean compact of the Dedekind completion of X.

In Section 5 we touch upon a much more difficult problem of describing essential spectra of weighted compositions induced by **non-invertible** maps and in a special case when φ is an open surjection on a compact K provide a criterion for $def(\lambda I - wT_\varphi) = 0$.

Finally, in the small appendix we provide some clarifications about the statement and the proof of Theorem 22 from [13] which is extensively used in the current paper.

2. When is the operator $\lambda I - T$ semi-Fredholm?

We start with recalling some results from [12].

Let T be as in (1). In [12, Theorem 3.29 and Theorem 3.31] we described two special cases when $\lambda \in \sigma(T)$ and either $def(\lambda I - T) = 0$ or $def(\lambda I - T') = 0$ where T' is the Banach conjugate of T. Because these results are crucial for our description of $\sigma_{sf}(T)$ we will reproduce them here, but first we need to recall a few notations from [12].

Let X be a Banach space and S be a bounded linear operator on X. We denote the spectrum of S by $\sigma(S)$ and consider the partition of $\sigma(S)$ into two subsets.

$$\sigma_{ap}(S) = \{\lambda \in \sigma(S) : \exists x_n \in X,\ \|x_n\| = 1,\ Sx_n - \lambda x_n \underset{n\to\infty}{\to} 0\},$$
$$\sigma_r(S) = \sigma(S) \setminus \sigma_{ap}(S).$$

Let φ be a homeomorphism of a compact Hausdorff space K onto itself. Then $\varphi^{(0)}$ will mean the identical map of K onto itself, $\varphi^{(m)} = \varphi \circ \varphi^{(m-1)}, m \in \mathbb{N}$, and $\varphi^{(-m)} = (\varphi^{(m)})^{-1}, m \in \mathbb{N}$.

Let $m \in \mathbb{N}$. We will denote by Π_m the subset of K that consists of all φ-periodic points of period less or equal to m.

Let U be a closed subset of K such that $\varphi(U) = U$. We will denote by T_U the operator induced by (1) on the space $C(U)$.

Theorem 2.1 ([12, Theorem 3.29]). *Let φ be a homeomorphism of the compact space K onto itself, $w \in C(K)$, and $(Tf)(k) = w(k)f(\varphi(k)), f \in C(K), k \in K$. Assume that the set of all φ-periodic points is of the first category in K. Let $\lambda \in \sigma(T)$. Consider the following statements.*

(*R*) *The operator $\lambda I - T$ has a right inverse, or equivalently $(\lambda I - T)C(K) = C(K)$, or equivalently $\lambda \in \sigma_r(T')$.*

(*L*) *The operator $\lambda I - T$ has a left inverse, or equivalently $\|(\lambda I - T)f\| \geq C\|f\|,\ f \in C(K), C > 0$, or equivalently $\lambda \in \sigma_r(T)$.*

(*A*) *$K = E \cup Q \cup F$ where the sets E, Q, and F are pairwise disjoint, the sets E and F are closed, $\varphi(E) = E$ and $\varphi(F) = F$, $\sigma(T_E) \subset \{\xi \in \mathbb{C} : |\xi| < |\lambda|\}$, and $\sigma(T_F) \subset \{\xi \in \mathbb{C} : |\xi| > |\lambda|\}$.*

(*B*) $\forall k \in Q \bigcap\limits_{n=0}^{\infty} cl\{\varphi^m(k) : m \geq n\} \subseteq F$ *and* $\forall k \in Q \bigcap\limits_{n=0}^{\infty} cl\{\varphi^{-m}(k) : m \geq n\} \subseteq E.$

(*C*) $\forall k \in Q \bigcap\limits_{n=0}^{\infty} cl\{\varphi^m(k) : m \geq n\} \subseteq E$ *and* $\forall k \in Q \bigcap\limits_{n=0}^{\infty} cl\{\varphi^{-m}(k) : m \geq n\} \subseteq F.$

Then the following equivalencies hold

(1) $R \Leftrightarrow A \wedge B$.

(2) $L \Leftrightarrow A \wedge C$.

Theorem 2.2 ([12, Theorem 3.31]). *Let φ be a homeomorphism of the compact space K onto itself, $M \in C(K)$, and $(Tf)(k) = M(k)f(\varphi(k), f \in C(K), k \in K$. Let $\lambda \in \sigma(T)$. The following conditions are equivalent.*

(1) $\lambda \in \sigma_r(T')$ (*respectively,* $\lambda \in \sigma_r(T)$).

(2) *There are* $m \in \mathbb{N}$ *and an open subset* P *of* K *such that* $P \subset \Pi_m$, $\varphi(P) = P$, $\lambda \notin \sigma(T_{clP})$, *and* K *can be partitioned as* $K = E \cup Q \cup F \cup P$ *where the sets* E, F, *and* Q *satisfy conditions A and B (respectively A and C) of Theorem* 2.1.

For any $m \in \mathbb{N}$ let us define $w_m \in C(K)$ as

$$w_m = w(w \circ \varphi) \cdots (w \circ \varphi^{(m-1)}).$$

We will need also the following lemma that follows from [12, Lemma 3.6 and Theorem 3.12.]

Lemma 2.3. *Let* T *be an operator of the form* (1) *on* $C(K)$ *and let* $\lambda \in \sigma_{ap}(T) \setminus \{0\}$. *Then there is a point* $k \in K$ *such that for every* $n \in \mathbb{N}$

$$|w_n(k)| \geq |\lambda|^n. \tag{2(a)}$$

and

$$|w_n(\varphi^{(-n)}(k))| \leq |\lambda|^n. \tag{2(b)}$$

Moreover, either

(I) k *is not a* φ*-periodic point,*

or at least one of the following conditions is satisfied.

(II) k *is a* φ*-periodic point and for any* $n \in \mathbb{N}$ *and any open neighborhood* V *of* k *there is a point* $v \in V$ *such that either* v *is not* φ*-periodic or its period is greater than* n.

(III) k *is a* φ*-periodic point of (the smallest) period* p *and* $w_p(k) = \lambda^p$.

Now we can start working on a complete description of Fredholm and semi-Fredholm spectra of operators of the form (1).

Lemma 2.4. *Let* T *be an operator of the form* (1). *Assume that* $\lambda \in \sigma_{ap}(T) \setminus \{0\}$, *that the operator* $\lambda I - T$ *is semi-Fredholm, and that* $\mathrm{nul}(\lambda I - T) < \infty$. *Let* $k \in K$ *be a point from the statement of Lemma* 2.3.

Then k *is an isolated point of* K.

Proof. Let us first assume that k satisfies condition (III) of Lemma 2.3. If k is not an isolated point in K then we can find a sequence of points $k_n \in K$ with the properties.

- If $m \neq n$ then $\varphi^{(i)}(k_m) \neq \varphi^{(j)}(k_n), 0 \leq i, j \leq p-1$.
- $w(\varphi^{(i)}(k_n)) \underset{n \to \infty}{\to} w(\varphi^{(i)}(k))$, $i = 0, 1, \ldots, p-1$.

Let u_n be the characteristic function of the singleton set $\{k_n\}$. Then u_n can be considered as an element of the second dual $C''(K)$ of norm one. Let $v_n = \sum\limits_{i=0}^{n-1} \lambda^{-i}(T'')^i u_n$. Then it is immediate to see that $\|v_n\| \geq \|u_n\| = 1$ and that $T''v_n - \lambda v_n \underset{n \to \infty}{\to} 0$. Moreover, v_n are pairwise disjoint elements of the Banach lattice $C''(K)$ and therefore the sequence v_n cannot contain a norm convergent subsequence. Therefore (see, e.g., [8, Corollary I.4.7, page 43]) the operator $\lambda I - T''$ cannot be semi-Fredholm and have a finite-dimensional null space. But then [8, Theorem I.3.7] the same is true for the operator $\lambda I - T$ and we come to a contradiction.

Next let us look at the case when k satisfies condition (II) of Lemma 2.3. In this case it is not difficult to see that we can find points $k_n \in K$ and positive integers $m(n)$ with the properties

(a) $m(n) \underset{n\to\infty}{\to} \infty$.
(b) For every $n \in \mathbb{N}$ all the points $\varphi^{(i)}(k_n), |i| \leq m(n)+1$, are distinct.
(c) The sets $E_n = \{\varphi^{(i)}(k_n) : |i| \leq m(n)+1\}, n \in \mathbb{N}$ are pairwise disjoint.
(d) For any $n \in \mathbb{N}$ the following inequalities hold

$$|w_i(k_n)| \geq \frac{1}{2}|\lambda|^i, \quad i = 1, 2, \dots, m(n)+1$$

and

$$|w_i(\varphi^{(-i)}(k_n))| \leq 2|\lambda|^i, \quad i = 1, 2, \dots, m(n)+1$$

Let u_n be the characteristic function of the set $\{\varphi^{(m(n))}(k_n)\}$ and let

$$v_n = \sum_{j=0}^{2m(n)+1} \Big(1 - \frac{1}{\sqrt{m(n)}}\Big)^{|j-m(n)|} \overline{\lambda}^{j} (T'')^j u_n.$$

Then simple estimates similar to the ones in [12] or [13] show that

$$\|T''v_n - \lambda v_n\| = o(\|v_n\|), n \to \infty.$$

Now we come to a contradiction exactly like on the previous step of the proof.

Finally let us look at the case when the point k is not φ-periodic. First let us notice that $k \notin cl\{\varphi^{(i)}(k), i \in \mathbb{N}\}$. Indeed, if k were a limit point of the sequence $\varphi^{(i)}(k), k \in \mathbb{N}$ then taking into consideration that all the points of this sequence are distinct we can easily produce points k_n from this sequence and positive integers $m(n)$ with properties (a)–(d) above, and come to a contradiction. Similarly we can prove that $k \notin cl\{\varphi^{(-i)}(k), i \in \mathbb{N}\}\}$. Thus k is an isolated point in the closure of its φ-trajectory. But we assumed that k is not isolated in K and then again simple topological reasons show that we can construct sequences $k_n \in K$ and $m(n) \in \mathbb{N}$ with properties (a)–(d). The resulting contradiction ends the proof of the lemma. □

Corollary 2.5. *Let T be an operator of form* (1) *on $C(K)$ and let $\lambda \in \mathbb{C} \setminus \{0\}$ be such that the operator $\lambda I - T$ is semi-Fredholm and* $\mathrm{nul}(\lambda I - T) < \infty$. *Then either*

$$\mathrm{nul}(\lambda I - T) = 0,$$

or there is a finite set $\{k_1, \dots, k_m\}$ of points isolated in K such that if $F = K \setminus \{\varphi^{(n)}(k_i), n \in \mathbb{Z}, i = 1, \dots, m\}$ then the operator $\lambda I - T_F$ is semi-Fredholm on $C(F)$ and

$$\mathrm{nul}(\lambda I - T_F) = 0.$$

Lemma 2.6. *Let T be an operator on $C(K)$ of form* (1), $\lambda \in \mathbb{C} \setminus \{0\}$, $\lambda I - T$ *be semi-Fredholm,* $\mathrm{nul}(\lambda I - T) < \infty$, *and let k be an isolated not φ-periodic point of K satisfying inequalities* 2(a) *and* 2(b) *from the statement of Lemma* 2.3. *Let*

$$R = \bigcap_{i=1}^{\infty} \{\varphi^{(n)}(k), n \geq i\} \quad \text{and} \quad L = \bigcap_{i=1}^{\infty} \{\varphi^{(-n)}(k), n \geq i\}.$$

Then

$$\sigma(T_R) \subset \{\xi \in \mathbb{C} : \ |\xi| > |\lambda|\} \tag{3(a)}$$

and

$$\sigma(T_L) \subset \{\xi \in \mathbb{C} : \ |\xi| < |\lambda|\}. \tag{3(b)}$$

Proof. First notice that $\lambda \notin \sigma_{ap}(T_R)$. Indeed, otherwise there would be a point $r \in R$ satisfying the inequalities 2(a) and 2(b). Because r is not an isolated point in K we come to a contradiction with Lemma 2.4. Therefore there are two possibilities: either $\lambda \in \sigma_r(T_R)$ or $\lambda \notin \sigma(T_R)$. Let us assume first that $\lambda \in \sigma_r(T_R)$. Then we can apply Theorem 2.2 to the operator T_R. Notice that if P is the subset of R from the statement of Theorem 2.2 the we have $\lambda\Gamma \cap \sigma(T_{clP}) = \emptyset$. Indeed, otherwise there would be a φ-periodic point $s \in clP$ such that $|w_{p-1}(s)| = |\lambda|^p$ where p is the period of s. But because in K the point s is a limit point of the sequence $\varphi^{(n)}(k), n \in \mathbb{N}$ we come to a contradiction with Lemma 2.4. Applying Theorem 3.10 from [12] we see that the set R can be partitioned as $R = E \cup Q \cup F$ where all the sets E, Q, F are nonempty and have the properties (A) and (C) from the statement of Theorem 2.1. To bring our assumption that $\lambda \in \sigma_r(T_R)$ to a contradiction let $m \in \mathbb{N}$ and $R_m = \bigcap_{i=1}^{\infty} \{\varphi^{(nm)}(k), n \geq i\}$. Obviously R_m is a closed subset of R and $\varphi^{(m)}(R_m) = R_m$. Notice that the set $R_m \cap Q$ is nonempty. Indeed, otherwise we would have $R_m \subset E \cup F$ and because $\varphi(E \cup F) = E \cup F$ it would follow that $R = E \cup F$ in contradiction with our assumption that $Q \neq \emptyset$. It follows from property (C) in the statement of Theorem 2.1 that the sets $R_m \cap E$ and $R_m \cap F$ are nonempty as well. If m is large enough then in some open neighbourhood of E in K we have the inequality $|w_m| \leq |\lambda|^m$ and therefore there is a $p \in \mathbb{N}$ such that $|w_m(\varphi^{(pm)}(k)) \leq |\lambda|^m$ and $|w_m(\varphi^{((p+1)m)}(k)) \geq |\lambda|^m$. Let u_m be the characteristic function of the singleton $\{\varphi^{((p+2)m)}(k)\}$ and

$$v_m = \sum_{j=0}^{2m} \left(1 - \frac{1}{\sqrt{m}}\right)^{|j-m|} \lambda^{-j} (T'')^j u_m.$$

Next notice that we can find the sequences $m(i), p(i)$, $i \in \mathbb{N}$ such that

$$\lim_{i\to\infty} m(i) = \infty, \ |w_{m(i)}(\varphi^{(pm(i))})(k)) \leq |\lambda|^{m(i)}, \ |w_{m(i)}(\varphi^{((p+1)m(i))})(k)) \geq |\lambda|^{m(i)},$$

and the elements $v_{m(i)}$ are pairwise disjoint in $C''(K)$. Finally notice that as in [12] and [13] we can show that $\|T''v_{m(i)} - \lambda v_{m(i)}\| = o(\|v_{m(i)}\|), i \to \infty$. in contradiction with our assumption that $\lambda I - T$ is semi-Fredholm and $\mathrm{nul}(\lambda I - T) < \infty$.

Thus now we have to consider the case when $\lambda \notin \sigma(T_R)$. In this case, by Theorem 3.10 from [12] we have $R = E \cup F$ where E and F are disjoint φ-invariant closed subsets of R, $\sigma(T_E) \subset \{\xi \in \mathbb{C} : |\xi| < |\lambda|\}$ and $\sigma(T_F) \subset \{\xi \in \mathbb{C} : |\xi| > |\lambda|\}$. The definition of R and elementary topological reasoning show that one of the sets E or F must be empty. If we assume that F is empty we immediately come to a contradiction with the inequality 2(a) whence $R = F$ and 3(a) is proved. Statement 3(b) can be proved in a similar way. □

Now we can state the first of our main results; before stating it let us notice that the case when the operator $\lambda I - T$ is semi-Fredholm and $\mathrm{nul}(\lambda I - T) = 0$ (in other words when $\lambda \in \sigma_r(T)$) is completely described by Theorems 2.1 and 2.2. We can therefore concentrate on the case when $\mathrm{nul}(\lambda I - T) > 0$.

Theorem 2.7. *Let T be an operator of form* (1) *on $C(K)$ and let $\lambda \in \sigma(T) \setminus \{0\}$. The following conditions are equivalent.*

(I) *The operator $\lambda I - T$ is semi-Fredholm and $0 < \mathrm{nul}(\lambda I - T) < \infty$.*

(II) *There is a finite subset $S = \{k_1, \dots, k_m, s_1, \dots, s_l\}$ of K with the properties*

1. *Every point of S is an isolated point in K.*
2. *The points $k_i, i = 1, \dots m$, are not φ-periodic and if the sets R_i and L_i are defined as in the statement of Lemma* 2.6 *then for each $i \in [1 : m]$ the conditions* 3(a) *and* 3(b) *are satisfied.*
3. *The points $s_1, \dots, s_l$ are φ-periodic and $\lambda^{p(i)} = w_{p(i)}(s_i)$ where $p(i)$ is the period of the point s_i.*
4. $\mathrm{nul}(\lambda I - T) = m + l$.
5. *Let $U = \bigcup_{j=-\infty}^{\infty} \varphi^{(j)}(S)$ and $V = K \setminus U$. Then either $\lambda \notin \sigma(T_V)$ or $\lambda \in \sigma_r(T_V)$.*

Proof. The implication (I) $\Rightarrow$ (II) follows from Lemma 2.4, Corollary 2.5, and Lemma 2.6. To prove the implication (II) $\Rightarrow$ (I) notice that by [8, Corollary I.4.7, p. 43] it is enough to prove that every sequence $u_n, n \in \mathbb{N}, u_n \in C(K)$ such that $\|u_n\| = 1$ and $\|Tu_n - \lambda u_n\| \underset{n\to\infty}{\to} 0$ contains a norm convergent subsequence. Condition (5) guarantees that $\|u_n\|_{C(V)} \underset{n\to\infty}{\to} 0$. Therefore we can and will assume that $u_n \equiv 0$ on V, $n \in \mathbb{N}$. Next, because the set $S_2 = \{s_1, \dots, s_l\}$ is a finite and clopen subset of K there is a subsequence of the sequence u_n that converges uniformly on S_2 and we can assume without loss of generality that u_n converges uniformly on S_2. Thus it remains to prove that if $k \in S$ is not φ-periodic then we can find a subsequence of u_n that converges uniformly on the set $A = cl\{\varphi^{(i)}(k), i \in \mathbb{Z}\}$. If $\|u_n\|_{C(A)} \underset{n\to\infty}{\to} 0$ then it is nothing to prove. Therefore we can assume without loss of generality that $\|u_n\|_{C(A)} = 1$. Let a_n be such a point in A that $|u_n(a_n)| = 1$ and let a be a limit point of the set $\{a_n\}$. The point a cannot belong to the set V. Indeed at that point we have inequalities 2(a) and 2(b). On the other hand if $a \in V$ then $a \in A \setminus \mathrm{Int}\, A$ in contradiction with condition (2). Thus $a \in \{\varphi^{(i)}(k), i \in \mathbb{Z}\}$

and we can assume without loss of generality that $a = k$ and that the sequence $u_n(a)$ converges to 1. We define the function u on K in the following way

$$u(x) = \begin{cases} 1, & \text{if } x = k, \\ \lambda^i / w_i(k), & \text{if } x = \varphi^{(i)}(k), i \in \mathbb{N}, \\ w_i(\varphi^{(-i)}(k)/\lambda^i, & \text{if } x = \varphi^{(-i)}(k), i \in \mathbb{N}, \\ 0, & \text{otherwise.} \end{cases}$$

Condition (2) guarantees that $u \in C(K)$. Clearly the sequence u_n converges to u pointwise on K and we will prove now that some subsequence of it converges to u uniformly. Indeed, otherwise we can find a positive constant c and a strictly increasing sequence of positive integers $i(n)$ such that either $|u_n(\varphi^{(i(n))})(k)| \geq c$ or $|u_n(\varphi^{(-i(n))})(k)| \geq c$. We will assume the first case because the second one can be considered absolutely similarly. Keeping in mind that $u_n \equiv 0$ on V and that u_n converge to u pointwise on K we can find another strictly increasing sequence of integers $j(n)$ such that $i(n) - j(n) \underset{n \to \infty}{\to} \infty$ and $|u_n(\varphi^{(i(n) \pm j(n))})| \leq c/n$. Let v_n be the restriction of u_n on the set $\{\varphi^{(d)}(k), d \in [i(n) - j(n) : i(n) + j(n)]\}$. Then $\|v_n\| \geq c$ and $\|Tv_n - \lambda v_n\| \underset{n \to \infty}{\to} 0$. Let r_n be a point in K where $|v_n|$ takes its maximum value and r be a limit point of the sequence r_n. Then $r \in clA \setminus A$ and at point r we have inequalities 2(a) and 2(b) in contradiction with condition (2) of the current theorem. □

Our next goal is to describe when the operator $\lambda I - T$, $\lambda \in \mathbb{C} \setminus \{0\}$, is semi-Fredholm and $\mathrm{def}(\lambda I - T) < \infty$. We start with the following remark.

Remark 2.8. The case when $\mathrm{def}(\lambda I - T) = 0$ was discussed in Theorems 2.1 and 2.2, and therefore we will assume that $0 < \mathrm{def}(\lambda I - T) < \infty$. In this case the operator $\lambda I - T'$ is semi-Fredholm and $\mathrm{nul}(\lambda I - T') = \mathrm{def}(\lambda I - T)$. (See, e.g., [8, Theorem I.3.7, p. 29].) Consider now an auxiliary operator

$$(\tilde{T}f)(x) = w(x) f(\varphi^{(-1)}(x)), \ f \in C(K), \ x \in K. \tag{$\bigstar$}$$

Then (see [12] and [13]) $\sigma(\tilde{T}) = \sigma(T)$ whence $\lambda \in \sigma(\tilde{T})$. Assume that $\lambda \in \sigma_{ap}(\tilde{T})$. Then (see [13]) there is a point $k \in K$ such that for every $n \in \mathbb{N}$

$$|w_n(k)| \leq |\lambda|^n. \tag{4(a)}$$

and

$$|w_n(\varphi^{(-n)}(k))| \geq |\lambda|^n. \tag{4(b)}$$

Notice now that k must be an isolated point in K. Indeed, otherwise similar to the proof of Lemma 2.3 we can construct a sequence of Borel regular measures μ_n on K (actually every μ_n is a finite linear combination of Dirac measures) such that the measures μ_n are pairwise disjoint in $C(K)'$ and $\|T'\mu_n - \lambda \mu_n\| = o(\|\mu_n\|), n \to \infty$ in contradiction with our assumption that $\lambda I - T'$ is semi-Fredholm and $\mathrm{nul}(\lambda I - T') < \infty$.

Corollary 2.9. *Let T be an operator of form* (1) *on $C(K)$ and let $\lambda \in \mathbb{C} \setminus \{0\}$ be such that the operator $\lambda I - T$ is semi-Fredholm and $0 < def(\lambda I - T) < \infty$. Then there is a finite set $S = \{k_1, \dots, k_m, s_1, \dots, s_l\}$ of points isolated in K such that the points $k_i, i = 1, \dots, m$ are not φ-periodic and satisfy conditions* 4(a) *and* 4(b). *Points $s_1, \dots, s_l$ are φ-periodic and $\lambda^{p(i)} = w_i(s_i), i = 1, \dots, s$, where $p(i)$ is the period of the point s_i. Moreover, if $F = K \setminus \bigcup_{i=-\infty}^{\infty} \varphi^{(i)}(S)$, then the operator $\lambda I - T_F$ is semi-Fredholm on $C(F)$ and $def(\lambda I - T_F) = 0$.*

Proof. The proof follows immediately from Remark 2.8 and Theorems 2.1 and 2.2. □

Lemma 2.10. *Assume conditions of Corollary* 2.9 *and let S be the set from the statement of that corollary. Assume also that $k \in S$ is not φ-periodic. Like in Lemma* 2.6 *let*

$$R = \bigcap_{i=1}^{\infty} \{\varphi^{(n)}(k), n \geq i\} \quad \text{and} \quad L = \bigcap_{i=1}^{\infty} \{\varphi^{(-n)}(k), n \geq i\}.$$

Then

$$\sigma(T_R) \subset \{\xi \in \mathbb{C} : |\xi| < |\lambda|\} \tag{5(a)}$$

and

$$\sigma(T_L) \subset \{\xi \in \mathbb{C} : |\xi| > |\lambda|\}. \tag{5(b)}$$

Proof. It follows from Remark 2.8 that $\lambda \notin \sigma_{ap}(\tilde{T}_{R \cup L})$ where $\tilde{T}$ is defined by the equation ($\bigstar$). The rest of the proof goes very similar to the proof of Lemma 2.6 and can be omitted. □

Theorem 2.11. *Let T be an operator of form* (1) *on $C(K)$ and let $\lambda \in \sigma(T) \setminus \{0\}$. The following conditions are equivalent.*

(I) *The operator $\lambda I - T$ is semi-Fredholm and* $0 < \operatorname{def}(\lambda I - T) < \infty$.

(II) *There is a finite subset $S = \{k_1, \dots, k_m, s_1, \dots, s_l\}$ of K with the properties*

1. *Every point of S is an isolated point in K.*
2. *The points $k_i, i = 1, \dots, m$, are not φ-periodic and if the sets R_i and L_i are defined as in the statement of Lemma* 2.6 *then for each $i \in [1 : m]$ the conditions* 5(a) *and* 5(b) *are satisfied.*
3. *The points $s_1, \dots, s_l$ are φ-periodic and $\lambda^{p(i)} = w_{p(i)}(s_i)$ where $p(i)$ is the period of the point s_i.*
4. $\operatorname{def}(\lambda I - T) = m + l$.
5. *Let $U = \bigcup_{j=-\infty}^{\infty} \varphi^{(j)}(S)$ and $V = K \setminus U$. Then either $\lambda \notin \sigma(T_V)$ or $\lambda \in \sigma_r(T'_V)$.*

Proof. The implication (I) $\Rightarrow$ (II) follows from Remark 2.8, Corollary 2.9, and Lemma 2.10.

To prove the implication (II) $\Rightarrow$ (I) we have to prove that if (II) is satisfied and μ_n is a sequence of regular Borel measures on K such that $\|\mu_n\| = 1$ and

$T'\mu_n - \lambda\mu_n \underset{n\to\infty}{\to} 0$ then the sequence μ_n contains a norm convergent subsequence. It follows from II(5) that without loss of generality we can assume that there is $k \in [k_1, \dots, k_m]$ such that supp $\mu_n \subset \{\varphi^i(k) : i \in \mathbb{Z}\}, n \in \mathbb{N}$. By the well-known criterion of compactness in l^1 (see, e.g., [7]) it is enough to prove that for any positive ε there is an $m = m(\epsilon) \in \mathbb{N}$ such that $|\mu_n|(\{\varphi^{(i)}(k), |i| > m\}) < \varepsilon, n \in \mathbb{N}$. Assume to the contrary that there is a positive ε and a subsequence $\nu_s = \mu_{n_s}$ of the sequence μ_n such that for any $l \in \mathbb{N}$ we have $|\nu_s|(\{\varphi^{(i)}(k) : |i| > l\}) \geq \varepsilon, s \in \mathbb{N}$. The operator T is the product of a central operator and a d-isomorphism and therefore the conjugate operator T' preserves disjointness. Thus $|T'||\nu_s| - |\lambda||\nu_s| \underset{s\to\infty}{\to} 0$. Let τ be a limit point of the sequence $|\nu_s|$ in the weak-$\star$ topology on $C(K)'$. Then τ is a probability measure on K and $|T'|\tau = |\lambda|\tau$. Let $Tr(k) = \{\varphi^{(i)}(k), i \in \mathbb{Z}\}$. Then clearly $\tau(clTr(k) \setminus Tr(k)) \geq \varepsilon$. Therefore $|\lambda| \in \sigma_{ap}(T', C(clTr(k) \setminus Tr(k))$. It follows from Theorems 2.1 and 2.2 that there is a point $l \in clTr(k) \setminus Tr(k)$ such that at point l we have the inequalities 5(a) and 5(b). Because l is a limit point of the set $Tr(k)$ we easily conclude that $\lambda \in \sigma_{ap}(T')$, a contradiction. □

Now we can answer the question when operator $\lambda I - T$ ($\lambda \neq 0$) is Fredholm. The next theorem follows directly from Theorems 2.7 and 2.11.

Theorem 2.12. *Let T be an operator of form* (1) *on $C(K)$ and $\lambda \in \sigma(T) \setminus \{0\}$. The following conditions are equivalent.*

(I) *The operator $\lambda I - T$ is Fredholm.*
(II) *There is a finite subset $S = \{k_1, \dots, k_m, l_1, \dots, l_n, s_1, \dots, s_q\}$ of K such that*
  (a) *Every point $k_i, i \in [1, \dots, m]$ is not φ-periodic and satisfies conditions* 2(a), 2(b), 3(a), *and* 3(b).
  (b) *Every point $l_i, i \in [1, \dots, n]$ is not φ-periodic and satisfies conditions* 4(a), 4(b), 5(a), *and* 5(b).
  (c) *Every point $s_i, i \in [1, \dots, q]$ is φ-periodic and $\lambda^{p(i)} = w_{p(i)}$ where $p(i)$ is the period of the point s_i.*
  (d) *Let $V = \bigcup_{i=-\infty}^{\infty} \varphi^{(i)}(S)$. Then $\lambda \notin \sigma(T_{K\setminus V})$.*

Moreover, if condition (II) *is satisfied then* $\mathrm{ind}(\lambda I - T) = n - m$.

Remark 2.13. It follows from Theorems 3.10 and 3.12 in [12] that condition $II(d)$ in the statement of Theorem 2.12 is equivalent to the following. The set $K \setminus V$ is the union of three disjoint subsets (of which two might be empty) E, F, and P with the properties.

- The set E is closed in K, $\varphi(E) = E$, and if $E \neq \emptyset$ then $\sigma(T_E) \subset \{\zeta \in \mathbb{C} : |\zeta| > |\lambda|\}$.
- The set F is closed in K, $\varphi(F) = F$, and if $F \neq \emptyset$ then $\sigma(T_F) \subset \{\zeta \in \mathbb{C} : |\zeta| < |\lambda|\}$.
- The set P is open in K, $\varphi(P) = P$, and if $P \neq \emptyset$ then P consists of φ-periodic points with periods bounded by some $N \in \mathbb{N}$ and $\lambda \notin \sigma(T_{clP})$.

To complete our description of Fredholm and semi-Fredholm operators of the form $\lambda I - T$ it remains to consider the case when $\lambda = 0$. Of course the only interesting case is $0 \in \sigma(T)$ and therefore we assume that the set $Z = \{k \in K : w(k) = 0\}$ is not empty. It follows immediately from Corollary I.4.7 in [8] that if T is semi-Fredholm then Z must be a clopen subset of K. Moreover, clearly Z must be a finite subset of K. Thus we obtain the following simple proposition which is most probably known.

Proposition 2.14. *Let T be an operator of form* (1) *on $C(K)$. The following conditions are equivalent.*

1. *T is semi-Fredholm.*
2. *T is Fredholm.*
3. *T is Fredholm and* $\operatorname{ind} T = 0$.
4. *The set $Z = \{k \in K : w(k) = 0\}$ is finite and consists of points isolated in K.*

3. Essential spectra of operators of form (1)

We are going now to describe the sets $\sigma_i(T), i = 1, \dots, 5$ for operators of form (1). To avoid unnecessary complicated and cumbersome statements we will start with the following remark.

Remark 3.1.

(1) In view of Proposition 2.14 it is enough to describe the sets $\sigma_i(T) \setminus \{0\}$.

(2) For any $p \in \mathbb{N}$ let Π^p be the set of all *non-isolated* φ-periodic points of period p. Then it follows from the results of the previous section that

$$cl\{\lambda \in \mathbb{C} : \exists p \in \mathbb{N}, \exists k \in \operatorname{Int}(\Pi^p)\ \lambda^p = w_p(k)\} \subseteq \sigma_1(T).$$

Therefore in our next theorem we will assume that $\operatorname{Int}(\Pi^p) = \emptyset, p \in \mathbb{N}$.

From the results in the previous section we obtain the following theorem.

Theorem 3.2. *Let T be an operator of form* (1) *on $C(K)$. Assume that* $\operatorname{Int}(\Pi^p) = \emptyset, p \in \mathbb{N}$ *where Π^p are the sets introduced in Remark* 3.1. *Then the sets $\sigma_i(T), i = 1, \dots, 5$ are rotation invariant. Moreover, if $\lambda \in \mathbb{C} \setminus 0$ then*

1. *The following conditions are equivalent.*
   (IA) $\lambda \in \sigma_1(T)$.
   (IB) *There are two (not necessarily distinct) points $k_1, k_2 \in K$ such that* none of them is an isolated point *in K, at k_1 we have inequalities* 2(a) *and* 2(b), *and at k_2 – inequalities* 4(a) *and* 4(b).
2. *The following conditions are equivalent.*
   (IIA) $\lambda \in \sigma_2(T)$.
   (IIB) *There is a* non-isolated point *$k \in K$ satisfying inequalities* 2(a) *and* 2(b).
3. *The following conditions are equivalent.*
   (IIIA) $\lambda \in \sigma_3(T)$.

(IIIB) *There is a* **non-isolated point** $k \in K$ *such that at this point either inequalities* 2(a) *and* 2(b) *or inequalities* 4(a) *and* 4(b) *are satisfied.*

4. *The following conditions are equivalent.*
   (IVA) $\lambda \in \sigma_4(T)$.
   (IVB) *Let* K_1 (*respectively* K_2) *be the set of all points in* K *that are not isolated* φ*-periodic points and satisfy* 2(a) *and* 2(b) (*respectively* 4(a) *and* 4(b)). *Then either at least one of these sets is infinite or they have distinct cardinalities.*

5. *The following conditions are equivalent.*
   (VA) $\lambda \in \sigma_5(T)$
   (VB) *There is a* $k \in K$ *such that* k *is not an isolated* φ*-periodic point and such that at this point either inequalities* 2(a) *and* 2(b) *or inequalities* 4(a) *and* 4(b) *are satisfied.*

The results of the previous section together with Theorems 2.1 and 2.2 provide the following corollary.

Corollary 3.3. *Let* T *be an operator of form* (1) *on* $C(K)$. *Then*

(1) *If* T *is band irreducible then* $\sigma_1(T) = \sigma(T)$.
(2) *If* K *has no* φ*-periodic isolated points then* $\sigma_5(T) \setminus \{0\} = \sigma(T) \setminus \{0\}$.
(3) *If* K *has no isolated points then* $\sigma_3(T) = \sigma(T)$.
(4) *Let* O *be the set of all isolated* φ*-periodic points in* K. *Then the essential spectral radius* $\rho_e(T)$ *of* T *can be computed in the following way* (*see, e.g.,* [12, Theorem 3.23])

$$\rho_e(T) = \rho(T_{K \setminus O}) = \max_{\mu \in \mathcal{M}_\varphi} \exp \int \ln |w| d\mu,$$

where $\mathcal{M}_\varphi$ *is the set of all* φ*-invariant regular Borel probability measures on* $K \setminus O$. *In particular, if* $O = \emptyset$ *then* $\rho_e(T) = \rho(T)$.

The diagrams on the next page illustrate our results from Theorem 3.2

Our next Theorem (that also follows from the results in [12] and in the previous section) provides an alternative description of the sets $\sigma_1(T)$, $\sigma_2(T)$, $\sigma_3(T)$, and $\sigma_5(T)$ that complements the one in Theorem 3.2. To state it let us introduce or recall the following notations.

- Γ is the unit circle. $\Gamma = \{\zeta \in \mathbb{C} : |\zeta| = 1\}$.
- O_1 is the set of all isolated φ-periodic points in K.
- $K_1 = K \setminus O_1$.
- $\Pi^p, p \in \mathbb{N}$ is the set of all φ-periodic points of period p in K_1.
- $\Sigma = cl\{\lambda \in \mathbb{C} : \exists p \in \mathbb{N}, \exists k \in \mathrm{Int}(\Pi^p)\ \lambda^p = w_p(k)\}$.
- $O_2 = \bigcup\limits_{p=1}^{\infty} \mathrm{Int}(\Pi^p)$.
- $K_2 = K_1 \setminus O_2$.

Diagram I: $\lambda \notin \sigma(T)$

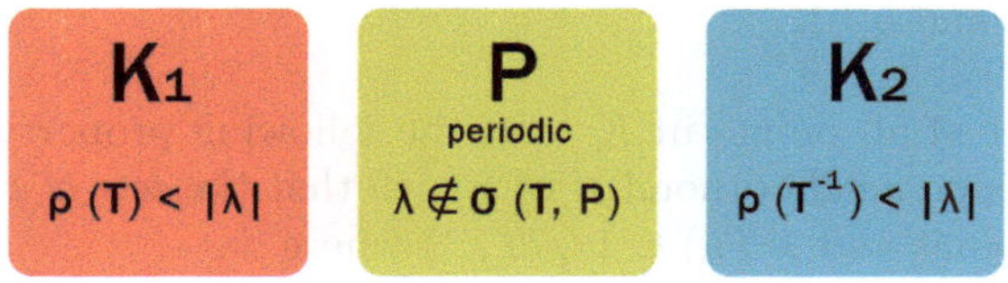

Diagram II: $\lambda \notin \sigma_5(T)$

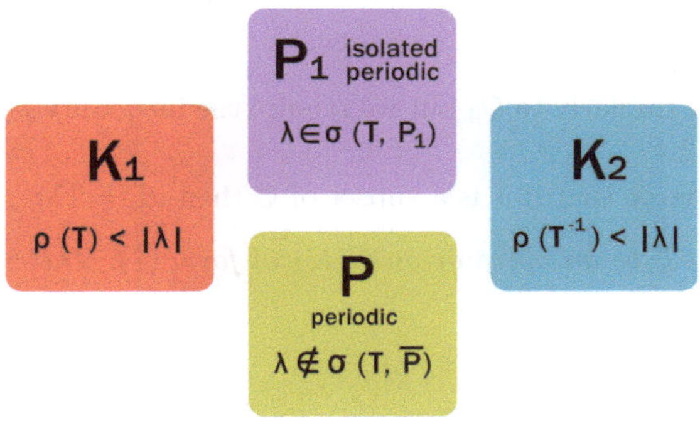

Diagram III: $\lambda \notin \sigma_4(T)$

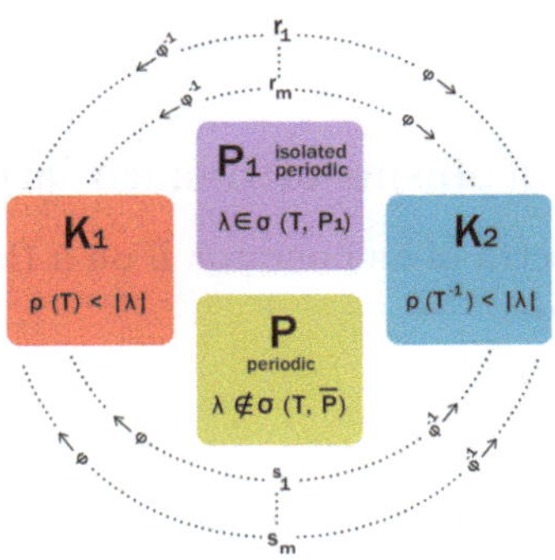

Diagram IV: $\lambda \notin \sigma_3(T)$

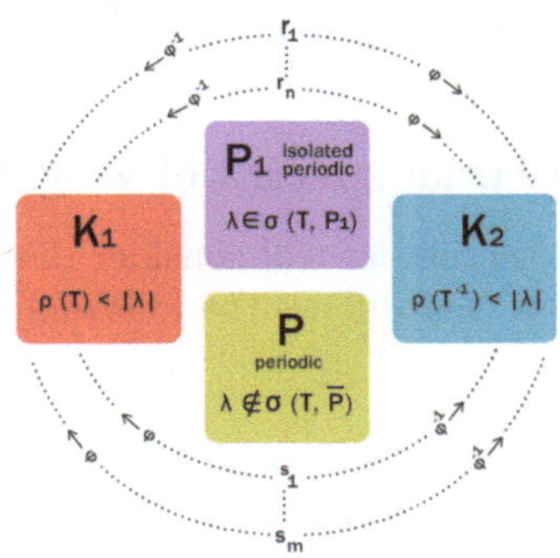

Diagram V: $\lambda \notin \sigma_2(T)$

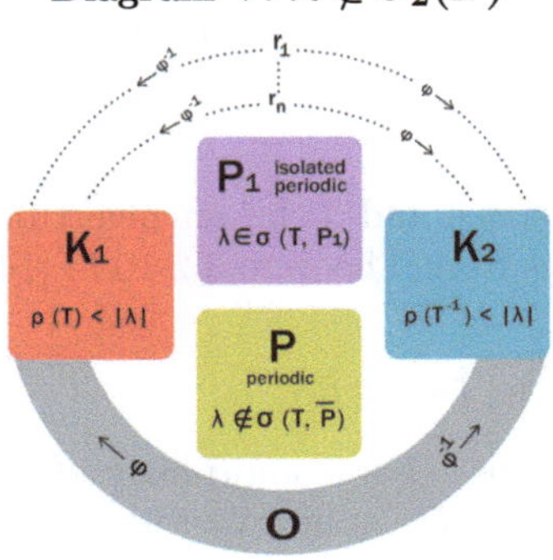

Diagram VI: $\lambda \notin \sigma_2(T')$

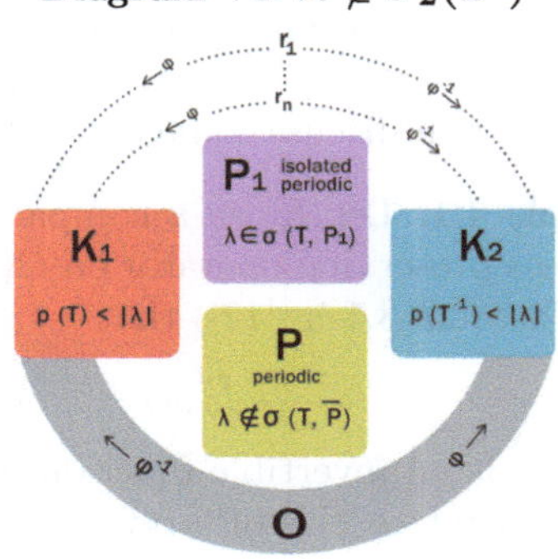

- O_3 is the set of all isolated points in K satisfying either conditions 3(a) and 3(b) or 5(a) and 5(b).
- $K_3 = K_2 \setminus O_3$.
- O_4 is the set of all points in K with the following property. If $k \in O_4$ then there is an open neighborhood V of k such that the sets $cl\, \varphi^{(i)}(V), i \in \mathbb{Z}$, are pairwise disjoint and $\rho(T_R) < 1/\rho(T_L^{-1})$ where

$$L = \bigcap_{n=1}^{\infty} cl \bigcup_{i=n}^{\infty} \varphi^{(-i)}(V) \quad \text{and} \quad R = \bigcap_{n=1}^{\infty} cl \bigcup_{i=n}^{\infty} \varphi^{(i)}(V).$$

- $K_4 = K_3 \setminus O_4$.
- O_5 is defined similarly to O_4 but we require the inequality $\rho(T_L) < 1/\rho(T_R^{-1})$.
- $K_5 = K_3 \setminus O_5$.

Let us also agree that if σ is a subset of $\mathbb{C}$ then $\sigma\Gamma = \{\lambda\gamma : \lambda \in \sigma, \gamma \in \Gamma\}$.

Theorem 3.4. *Let T be an operator on $C(K)$ of form* (1). *Then*

- $\sigma_5(T) = \sigma(T_{K_2})\Gamma \cup \Sigma$.
- $\sigma_3(T) = \sigma(T_{K_3})\Gamma \cup \Sigma$.
- $\sigma_2(T) = \sigma(T_{K_4})\Gamma \cup \Sigma$.
- $\sigma_1(T) = (\sigma(T_{K_4})\Gamma \cap \sigma(T_{K_5})\Gamma) \cup \Sigma$.

4. Essential spectra of weighted d-isomorphisms of Banach lattices

In this section we will consider the essential spectra of an operator T on a Banach lattice X that allows the following representation.

$$T = wU, \ w \in Z(X), \sigma(U) \subset \Gamma, \tag{6}$$

where U is a d-isomorphism of X.

The question when an invertible disjointness preserving operator on a Banach lattice can be represented in form (6) might be of independent interest. In particular, we can single out the class of Banach lattices on which **every** invertible disjointness preserving operator is of form (6). At the present no general criteria addressing either of the above questions are known (at least to the author) but below we will discuss some examples.

Example 4.1. Let K be a compact Hausdorff space and T be an invertible disjointness preserving operator on $C(K)$. It follows from Theorem 3.2.10 in [14] (see also Theorem 3.1 in [12]) that

$$(Tf)(k) = w(k)f(\varphi(k)), \ k \in K, f \in C(K),$$

where w is an invertible function in $C(K)$ and φ is a homeomorphism of K onto itself. Therefore T is of form (6) and moreover $U = T_\varphi$ is an invertible isometry.

Example 4.2. Let (E, Σ, μ) be a measure space, $1 \leq p < \infty$, and T an invertible disjointness preserving operator on $L^p(E, \Sigma, \mu)$. Then $T = wU$ where $w \in L^\infty(E, \Sigma, \mu)$ and U is an invertible isometry of $L^p(E, \Sigma, \mu)$.

Proof. We can assume without loss of generality (see, e.g., [11, p. 286]) that μ is a measure on the hyperstonian compact Q of the algebra $L^\infty(E, \Sigma, \mu)$ such that $\mu(F) = 0$ for any subset F of the first category in Q, $\mu(G) > 0$ for any G open in Q, and for any μ-measurable subset G of Q such that $\mu(G) = \infty$ there is an $H \subset G$ such that $0 < \mu(H) < \infty$. The map $f \to TfT^{-1}$ defines an isomorphism of $L^\infty(E, \Sigma, \mu) \approx C(Q)$ (see [4, Chapter 8] for more details). Let φ be the corresponding homeomorphism of Q onto itself. Notice that a subset F of Q is μ-measurable if and only if $\varphi(F)$ is μ-measurable and $\mu(F) = 0 \Leftrightarrow \mu(\varphi(F)) = 0$.

Assume for a moment that μ is a sigma-finite measure and define a new measure ν as $\nu(E) = \mu(\varphi^{-1}(E)$. Let h be the Radon–Nikodym derivative $\frac{d\nu}{d\mu}$. Then h can be identified with an invertible element of $C(Q)$ and the operator U, $Ug = (h)^{1/p}(g \circ \varphi)$, is an invertible isometry of L^p, where $1 \leq p < \infty$ (see, e.g., [9, Theorem 3.2.5]). It remains to notice that the operator TU^{-1} is a band preserving operator on L^p and therefore an operator of multiplication by a function from L^∞.

To reduce the general case to the one already considered notice that we can find a family of pairwise disjoint clopen subsets Q_α of Q with the properties

- $\varphi(Q_\alpha) = Q_\alpha$,
- $\mu|Q_\alpha$ is a sigma-finite measure,
- $\bigcup Q_\alpha$ is dense in Q. □

Example 4.3. Let $1 < p < \infty$ and let X be a rearrangement invariant Banach function space on $(0, 1)$ or $(0, \infty)$ such that the upper and lower Boyd indices (see [6]) of X are equal to p. Then it can be proved using Boyd's interpolation theorem (see, e.g., [6, Theorem 5.16, p. 153]) that every invertible disjointness preserving operator on X can be presented in form (6), but in general we cannot claim that U is an isometry of X.

On the other hand we have the following "negative" example.

Example 4.4. Let $1 < p < q < \infty$ and X be $L^p(0, \infty) \cap L^q(0, \infty)$ with the standard norm $\|x\| = \|x\|_p + \|x\|_q$. Let $a > 0$ and $a \neq 1$. The operator T defined as $(Tf)(t) = f(at), f \in X, t \in (0, \infty)$ is a bounded invertible disjointness preserving operator on X but it cannot be presented in form (6).

We return now to the study of essential spectra.
We will assume first that X is a Dedekind complete Banach lattice. Let K be the Stonean compact space of X; then the center $Z(X)$ can be identified with $C(K)$. The map $f \to UfU^{-1}$, $f \in C(K)$ defines an isomorphism of the algebra $C(K)$. Let φ be the corresponding homeomorphism of K onto itself. We consider the weighted composition operator S on $C(K)$ defined as

$$(Sf)(k) = w(k)f(\varphi(k)), \ f \in C(K), \ k \in K. \tag{7}$$

Our nearest goal is to prove the following result.

Theorem 4.5. *Let X be a Dedekind complete Banach lattice and $Z(X) \sim C(K)$ be the center of X. Let T be an operator on X defined by formula* (6) *and S be the corresponding operator on $C(K)$ defined by* (7). *Then the essential spectra of operators T and S coincide.*

$$\sigma_i(T) = \sigma_i(S),\ i = 1, \dots, 5.$$

Proof. (1) In this part we will prove that $0 \in \sigma_i(T) \Leftrightarrow 0 \in \sigma_i(S), i = 1, \dots, 5$. Assume that $0 \in \sigma(T)$ then obviously $0 \in \sigma(S)$. Assume that $0 \notin \sigma_1(S)$, then by Proposition 2.14 the set $Z(w)$ of zeros of w is finite and consists of points isolated in K. Let χ be the characteristic function of the set $Z(w)$, then the operator χU is finite dimensional while the operator $T + \chi U$ is invertible on X. Therefore T is a Fredholm operator. Notice that $\dim\ker(S) = \dim\ker(T) = \text{card}(Z(w))$. The operator w' is a central operator on the Banach dual X' of X and because the operator $T' + U'\chi'$ is invertible on X' we see that $\dim\ker(T') = \text{card}(Z(w')) \leq \text{card}(Z(w)) = \dim\ker(T)$. On the other hand a similar reasoning shows that $\dim\ker(T) = \dim\ker(T'') \leq \dim\ker(T')$ whence $\text{ind}(T) = 0$.

Now assume that $0 \in \sigma_1(S)$. By Proposition 2.14 the set $Z(w)$ contains a point that is not isolated in K. Then we can construct a sequence of pairwise disjoint elements $x_n \in X$ such that $\|x_n\| = 1$ and $\chi_n x_n = x_n$ where χ_n is the characteristic function of the set $\{k \in K : |w(k)| \leq 1/n\}$. Then it is immediate to see that $Tx_n \underset{n\to\infty}{\to} 0$ whence $0 \in \sigma_2(T)$. But the set $Z(w')$ must also contain a point that is not isolated in K' where K' is the Gelfand compact of the ideal center of X'. Indeed, otherwise the sum of T' and a finite-dimensional operator would be invertible and T (together with T') would be Fredholm in contradiction with our assumption. Thus $0 \in \sigma_2(T')$ whence $0 \in \sigma_1(T)$.

It follows from the two previous paragraphs that

$$0 \in \sigma_i(T) \Leftrightarrow 0 \in \sigma_i(S), i = 1, 2, 3, 4. \tag{8}$$

The equivalence $0 \in \sigma_5(T) \Leftrightarrow 0 \in \sigma_5(S)$ follows from Definition 1.3, from (8), and from the equality $\sigma(T) = \sigma(S)$ proved in [13, Theorem 22]

(2) Here we will prove that $\sigma_1(S) \subseteq \sigma_1(T)$. Let $\lambda \in \sigma_1(S) \setminus \{0\}$. Without loss of generality we can assume that $\lambda = 1$. Then it follows from Theorems 3.2 and 3.4 as well as from Frolik's theorem [16, Theorem 6.25, p. 150][1] that we have to consider two possibilities.

(2a) There are a point $k \in K$ and $p \in \mathbb{N}$ such that $k \in \text{Int}(\Pi^p)$ and $w_p(k) = 1$. We can find clopen nonempty subsets $E_n, n \in \mathbb{N}$ of $\text{Int}(\Pi^p)$ such that

- $\varphi^{(i)}(E_n) \cap E_n = \emptyset,\ n \in \mathbb{N},\ 1 \leq i \leq p - 1$.
- $\varphi^{(p)}(E_n) = E_n, n \in \mathbb{N}$.
- If $m \neq n$ then $\varphi^{(i)}(E_m) \cap \varphi^{(j)}(E_n) = \emptyset, 0 \leq i, j \leq p - 1$.
- $\max\limits_{k \in \bigcup\limits_{i=0}^{p-1} \varphi^{(i)}(E_n)} |w_p(k) - 1| < 1/n, n \in \mathbb{N}$.

[1]Frolik's theorem states that if K is an extremely disconnected compact space and φ is a homeomorphism of K into itself then the set of all fixed points of φ is clopen in K.

Let P_n be the band projection on X corresponding to the set E_n and $x_n \in X$ be such that $\|x_n\| = 1$ and $P_n x_n = x_n$. Let $y_n = \sum_{i=0}^{p-1} T^i x_n, n \in \mathbb{N}$. Then obviously the elements y_n are pairwise disjoint in X, whence $\|y_n\| \geq \|x_n\| = 1$, and a simple estimate (see also [13]) shows that $Ty_n - y_n \underset{n\to\infty}{\to} 0$. Thus $1 \in \sigma_2(T)$.

Next, consider the band C_n in X corresponding to the clopen set $\bigcup_{i=0}^{p-1} \varphi^{(i)}(E_n)$. The band C_n as well as its complementary band are T-invariant whence the band C'_n (the Banach dual of C_n) in X' is T'-invariant. The restriction of the operator T^p on C_n is a central operator on C_n whence $(T')^p$ is central on C'_n. Moreover, $\|Q'_n((T')^p - I)\| = \|Q_n(T^p - I)\| < 1/n, n \in \mathbb{N}$, where Q_n is the band projection on the band C_n. Like in the previous paragraph we can construct a sequence f_n of pairwise disjoint elements of X' such that $\|f_n\| = 1$ and $T'f_n - f_n \to 0$. Therefore $1 \in \sigma_2(T) \cap \sigma_2(T') = \sigma_1(T)$.

(2b). There are two non φ-periodic and non-isolated points $k_1, k_2 \in K$ (it can happen that $k_1 = k_2$) such that at k_1 we have inequalities 2(a) and 2(b) and at k_2 – inequalities 4(a) and 4(b) with $\lambda = 1$. Because k_1 is a non-isolated and non φ-periodic point we can find clopen subsets F_n of K such that

- $\varphi^{(i)}(F_n) \cap \varphi^{(j)}(F_n) = \emptyset, -(n+1) \leq i < j \leq n+1$.
- If $n \neq m$ then $\left(\bigcup_{i=-m-1}^{m+1} \varphi^{(i)}(E_m)\right) \cap \left(\bigcup_{i=-n-1}^{n+1} \varphi^{(i)}(E_n)\right) = \emptyset$.
- for any $k \in F_n$ we have $|w_i(k)| \geq 1/2, 0 \leq i \leq n+1$ and $|w_i(\varphi^{(-i)}(k)| \leq 2, 1 \leq i \leq n+1$.

Let x_n be an element of the band corresponding to the set $\varphi^{(n)}(F_n)$ such that $\|x_n\| = 1$ and let

$$y_n = \sum_{i=0}^{2n} \left(1 - \frac{1}{\sqrt{n}}\right)^{|n-i|} T^i x_n.$$

Then the estimates very similar to the ones employed in [13] show that $\|Ty_n - y_n\| = o(\|y_n\|), n \to \infty$ whence $1 \in \sigma_2(T)$.

Let us prove now that $1 \in \sigma_2(T')$. The map $h \to h', h \in Z(X)$ defines an isometric isomorphism of $Z(X)$ (which we identify with $C(K)$) onto a subalgebra of $Z(X')$ (which we identify with $C(K')$). To this isometric embedding corresponds a continuous surjection $\tau : K' \to K$, and it is not difficult to see that if $h \in C(K)$, $u, v \in K'$, and $\tau(u) = \tau(v)$ then $h'(u) = h'(v)$. The map $g \to (U^{-1})'gU'$ defines an isomorphism of $C(K')$. Let φ' be the homeomorphism of K' onto itself corresponding to this isomorphism. Then it is immediate to see that

$$\tau(\varphi'(u)) = \varphi^{(-1)}(\tau(u)). \tag{9}$$

Let $s \in \tau^{-1}(k_2)$. Let $w'_n(u) = w'(u)w'(\varphi'(u)) \cdots w'((\varphi')^{(n)}(u), n \in \mathbb{N}$. It follows from the fact that at k_2 we have inequalities (4a) and (4b) and from (9) that

$$|w'_n(s)| \geq 1, \quad \text{and} \quad |w'_n((\varphi')^{(-n)}(s))| \leq 1, \quad n \in \mathbb{N}.$$

It is obvious that s cannot be a φ'-periodic point in K'. Moreover, it is not difficult to see that because k_2 is not an isolated point in K the set $\tau^{-1}(k_2)$ must contain points that are not isolated in K'. Thus, by what we have already proved $1 \in \sigma_2(w'U')$. It remains to notice that $(T)' = U'w' = U'(w'U')(U')^{-1}$ whence $1 \in \sigma_2(T')$.

(3) In this part we prove that $\sigma_1(T) \subseteq \sigma_1(S)$. Let $1 \notin \sigma_1(S)$. Because by Theorem 22 from [13] $\sigma(T) = \sigma(S)$ we can assume without loss of generality that $1 \in \sigma(S)$. We have to consider several possibilities.

(3a) The operator $I - S$ is semi-Fredholm and $\dim\ker(I - T) = 0$. In other words we assume that $1 \in \sigma_r(S)$. But by Theorems 20 and 22 from [13] $\sigma_r(T) = \sigma_r(S)$ whence $1 \notin \sigma_1(T)$.

(3b) The operator $I-S$ is semi-Fredholm and $0 < \dim\ker(I-T) < \infty$. Then conditions II(1)–II(5) of Theorem 2.7 are satisfied. We will keep the notations from the statement of Theorem 2.7. Then clU is a clopen φ-invariant subset of K. Let X_1 and X_2 be the bands in X corresponding to the clopen sets clU and $K \setminus clU$, respectively. Clearly $TX_i \subseteq X_i, i = 1, 2$. Conditions II(1)–II(3) of Theorem 2.7 combined with Theorems 20 and 21 from [13] guarantee that the operator $(I-T)|X_1$ is Fredholm while condition II(5) together with the same theorems implies that $1 \in \sigma_r(T|X_2)$. Thus the operator $(I-T)$ is semi-Fredholm and $\mathrm{ind}(I-T) = \mathrm{ind}(I-S)$.

(3c) The operator $I - S$ is semi-Fredholm and $\mathrm{def}(I - S) = 0$. It follows from Theorem 2.2 and from Frolik's theorem that K can be partitioned as $K = E \cup Q \cup F \cup P$ where P is a clopen φ-invariant subset of K and there is an $m \in \mathbb{N}$ such that $P = \bigcup\limits_{i=1}^{m} \Pi^p$, while the sets E, F, and Q satisfy conditions A and B of Theorem 2.1. Let K', τ, and φ' be as in part (2b) of the proof. Let P', E', F', and Q' be the τ-preimages in K' of the corresponding sets in K. Then it is easy to see that P' is a clopen φ'-invariant subset of K' and the sets E', F', and Q' satisfy conditions A and C of Theorem 2.1 (Of course, we have to substitute K by K', w by w', and φ by φ', respectively.) Let $\tilde{S}$ be the operator on $C(K')$ defined as

$$(\tilde{S})f(s) = w'(s)f(\varphi'(s)),\ f \in C(K'),\ s \in K'.$$

By Theorem 2.2 $1 \in \sigma_r(\tilde{S})$, and then by part (3a) of the proof we have $1 \in \sigma_r(T')$ whence $(I - T)X = X$.

(3d) The remaining case when $I-S$ is semi-Fredholm and $0 < \mathrm{def}(I-T) < \infty$ can be considered similarly to part (3b) of the proof by using part (3c) and Theorem 2.11.

Thus we have proved that $\sigma_1(T) = \sigma_1(S)$.

(4) The arguments applied in parts (2) and (3) of the proof show immediately that $\sigma_i(T) = \sigma_i(S), i = 2, 3, 4$. Finally, the equality $\sigma_5(T) = \sigma_5(S)$ follows from $\sigma_4(T) = \sigma_4(S)$ and from $\sigma(T) = \sigma(S)$ ([13, Theorem 22]). □

Corollary 4.6. *Let T be an operator of form* (6) *on a Dedekind complete Banach lattice X. Then*

(1) *If the operator T is band irreducible then $\sigma_1(T) = \sigma(T)$.*
(2) *If the Banach lattice X has no atoms then $\sigma_3(T) = \sigma(T)$.*

As a special but important case of Corollary 4.6 we obtain

Corollary 4.7. *Let (E, Σ, μ) be a measure space and T be an invertible disjointness preserving operator on $L^p(E, \Sigma, \mu)$, $1 \leq p \leq \infty$. Assume that for any $F \in \Sigma$ such that $\chi_F T \chi_F = T\chi_F$ (where χ_F is the operator of multiplication by the characteristic function of F) either $\mu(F) = 0$ or $\mu(E \setminus F) = 0$. Then $\sigma_1(T) = \sigma(T)$.*

Our next goal is to extend the result of Theorem 4.5 on arbitrary Banach lattices.

Let X be a Banach lattice, $w \in Z(X)$, and U be a d-isomorphism on X such that $\sigma(U) \subseteq \Gamma$. Because the operators w and U are order continuous, by Veksler's theorem [15] (see also [3, Theorem 1.65, p. 55], or [17, Lemma 140.1, p. 651]) they have unique order-continuous extensions $\hat{w}$ and $\hat{U}$ to the Dedekind completion $\hat{X}$ of X. It is easy to see that $\hat{w} \in Z(\hat{X})$, that $\hat{U}$ is a d-isomorphism of $\hat{X}$, and that $\sigma(\hat{U}) \subseteq \Gamma$. Let $\hat{T} = \hat{w}\hat{U}$. Like in Theorem 4.5 we can consider operator S associated with $\hat{T}$ and defined by (7) where the compact space $\hat{K}$ is the Gelfand compact of $Z(\hat{X})$ and $\hat{\varphi}$ is the homeomorphism of $\hat{K}$ induced by the map $f \to \hat{U} f (\hat{U})^{-1}, f \in C(\hat{K})$. We will prove below in Theorem 4.11 that

$$\sigma_i(T) = \sigma_i(\hat{T}), i = 1, \ldots, 5. \tag{10}$$

We start with the following special case of (10).

Proposition 4.8. *Let K be a compact Hausdorff space and $\hat{K}$ be the absolute (or Stonean compact) of K. Let φ be a homeomorphism of K onto itself, $w \in C(K)$, and $(Tf)(k) = w(k)f(\varphi(k)), k \in K, f \in C(K)$. Let $\hat{T}$ be the unique order-continuous extension of T onto $\widehat{C(K)} = C(\hat{K})$. Then $\sigma_i(T) = \sigma_i(\hat{T}), i = 1, \ldots, 5$.*

Proof. $C(K)$ is isometrically and algebraically embedded into $C(\hat{K})$. Let τ be the surjection of $\hat{K}$ onto K induced by this embedding. The operator $\hat{T}$ is of the form $(\hat{T}f)(k) = \hat{w}(k)f(\hat{\varphi}(k)), k \in \hat{K}, f \in C(\hat{K})$ and we have that

$$\varphi(\tau(k)) = \tau(\hat{\varphi}(k)), k \in \hat{K}. \tag{11}$$

The equalities $\sigma_i(T) = \sigma_i(\hat{T}), i = 1, \ldots, 5$ follow easily from (11), Theorem 3.2, and Proposition 2.14. □

Our next step is to prove that (10) holds in the case of Banach lattices with a quasi-interior point.

Let us recall that a point u in a Banach lattice X is called **quasi-interior** if the principal ideal X_u is dense in X.

Lemma 4.9. *Let X be a Banach lattice with a quasi-interior point u. Let $Z(X) = C(K)$ and $Z(\hat{X}) = C(K_1)$ be the ideal centers of X and its Dedekind completion $\hat{X}$, respectively. Then $C(K_1)$ is isometrically and lattice isomorphic to $\widehat{C(K)} = C(\hat{K})$ where $\hat{K}$ is the absolute of K.*

Proof. The proof below was communicated to the author by A.W. Wickstead.

First observe that for a Dedekind complete Riesz space with weak order unit u the center is isomorphic to the ideal generated by u. If u is a quasi-interior point for a Banach lattice X then it is a weak order unit for the Dedekind completion $\hat{X}$. Observe that the Dedekind completion of the principal ideal X_u may be identified with $\hat{X}_u$. Then we can identify as follows:

$$\widehat{Z(X)} \equiv \widehat{X_u} \equiv \hat{X}_u \equiv Z(\hat{X}).$$

□

Theorem 4.10. *Let X be a Banach lattice with a quasi-interior point. Let $w \in Z(X)$, U be a d-isomorphism of X such that $\sigma(U) \subseteq \Gamma$, and $T = wU$. Let $\hat{T} = \hat{w}\hat{U}$ be the unique order-continuous extension of T onto $\hat{X}$. Then $\sigma_i(T) = \sigma_i(\hat{T})$, $i = 1, \dots, 5$.*

Proof. (1) $0 \in \sigma_i(T) \Leftrightarrow 0 \in \sigma_i(\hat{T}), i = 1, \dots, 5$.

If $0 \in \sigma(\hat{T}) \backslash \sigma_1(\hat{T})$ then there are $n \in \mathbb{N}$ and pairwise disjoint atoms $u_1, \dots, u_n$ in $\hat{X}$ such that $\ker \hat{T} = B$ where B is the band in $\hat{X}$ generated by $u_1, \dots, u_n$. But clearly $B \subseteq X$ and B is a projection band in X. Moreover, the operator $T + P_B$, where P_B is the band projection on B is invertible in X whence T is Fredholm and $\text{null}(T) = \text{null}(\hat{T})$. Then we can prove that $\text{ind}(T) = 0$ similar to part (1) of the proof of Theorem 4.5. Assume now that $0 \in \sigma_1(\hat{T}) = \sigma_1(S)$. Recall that for any nonzero $\hat{x} \in \hat{X}$ there is a nonzero $x \in X$ such that $|x| \leq |\hat{x}|$. Then we can see that $0 \in \sigma_1(T)$ in the same way as in the proof of Theorem 4.5. Moreover, the same reasoning as in the proof of Theorem 4.5 shows that $0 \in \sigma_i(T) \Leftrightarrow 0 \in \sigma_i(\hat{T})$, $i = 1, \dots, 5$.

(2) The inclusion $\sigma_2(T) \subseteq \sigma_2(\hat{T})$ is trivial.

(3) $\sigma_2(\hat{T}) \subseteq \sigma_2(T)$. Let $\lambda \in \sigma_2(\hat{T})$. By step (1) we can assume without loss of generality that $\lambda = 1$. By Theorem 3.2 there is a non isolated point $\hat{k}, \in \hat{K}$ such that at this point we have inequalities (2a) and (2b). It remains to repeat the reasoning from parts (2a) and (2b) of the proof of Theorem 4.5 keeping in mind that the elements we denoted there by x_n can be chosen from X.

(4) $\sigma_2(T') \subseteq \sigma_2(\hat{T}')$. This inclusion follows from Theorem 4.5, Theorem 3.2, the fact that there is one-to-one correspondence between bands (in particular, between atoms) in X and in $\hat{X}$, and finally from Theorem 22 in [13]. Notice also that it follows from Theorems 4.5 and 3.2 that if $\text{def}(I - \hat{T}) < \infty$ then $\text{def}(I - T) = \text{def}(I - \hat{T})$.

(5) $\sigma_2(\hat{T}') \subseteq \sigma_2(T')$. Assume that $1 \in \sigma_2(\hat{T}')$. Then by Theorems 4.5 and 3.2 there is a point $\hat{k} \in \hat{K}$ such that $\hat{k}$ is not an isolated point and at this point we have inequalities (4a) and (4b) (relatively to $\hat{\varphi}$ and $\hat{w}$). It follows from Lemma 4.9 that there is a point $k \in K$ such that it is not isolated in K and at this point we have inequalities (4a) and (4b). Next notice that $C(K) = Z(X)$ can be isometrically embedded into $Z(X')$ and we can repeat the reasoning from step (2b) of the proof of Theorem 4.5.

(6) The equalities $\sigma_i(T) = \sigma_i(\hat{T}), i = 1, 3, 4$ follow immediately from what we have already proved. Finally, the equality $\sigma_5(T) = \sigma_5(\hat{T})$ follows from $\sigma_1(T) = \sigma_1(\hat{T})$, $\sigma(T) = \sigma(\hat{T})$ (Theorem 22 in [13]), and from the definition of the set σ_5. □

Now we are ready to prove (10) in full generality.

Theorem 4.11. *Let X be a Banach lattice. Let $w \in Z(X)$, U be a d-isomorphism of X such that $\sigma(U) \subseteq \Gamma$, and $T = wU$. Let $\hat{T} = \hat{w}\hat{U}$ be the unique order-continuous extension of T onto $\hat{X}$. Then $\sigma_i(T) = \sigma_i(\hat{T}), i = 1, \dots, 5$.*

Proof. A look at the proof of Theorem 4.10 shows that we used the condition that X has a quasi interior point only at one place. $1 \in \sigma_2((\hat{T})') \Rightarrow 1 \in \sigma_2(T')$. To prove this implication in general case notice first that if $1 \in \sigma_2((\hat{T})')$ then by Theorems 3.2 and 4.5 there is a non isolated point u in the Gelfand compact $(\hat{K})'$ of $Z((\hat{X})')$ such that at this point we have inequalities (2a) and (2b) for $(\hat{w})'$ and the homeomorphism ψ' corresponding to $(\hat{U})'$. For any $x \in X$ let $J(x)$ be a closed $\hat{U}$ and$(\hat{U})^{-1}$-invariant ideal in $\hat{X}$ such that $\hat{x} \in J(x)$. Recall that the conjugate $J(x)'$ is a band in $(\hat{X})'$ which we will denote as B_x. There are two possibilities.

(a) There is an $x \in X$ such that $1 \in \sigma_2((\hat{T})'|B_x)$. Then we proceed as in the proof of Theorem 4.10.

(b) If we cannot find an x as in (a) then notice that the set $\bigcup\limits_{x \in X} \text{supp}(B_x)$ is dense in $(\hat{K})'$. Therefore we can find elements $x_n \in X$ and points $u_n \in \text{supp}(B_{x_n})$ such that the bands B_{x_n} are pairwise disjoint

$$|((\hat{w}_m))'(u_n)| \geq 1 - 1/n, \quad m = 1, \dots, n$$

$$|(\hat{w}_m)'((\psi')^{(-m)}(u_n))| \leq 1 + 1/n, \quad m = 1, \dots, n.$$

Applying Lemma 4.9 we can find points $v_n \in K'$ such that

$$|((w_m)'(u_n)| \geq 1 - 1/n, \quad m = 1, \dots, n$$

and

$$|((w_m)'((\varphi')^{(-m)}(u_n))| \leq 1 + 1/n, \quad m = 1, \dots, n.$$

Let v be any accumulation point of the set $\{v_n\}$. Then at v we have inequalities (2a) and (2b) whence $1 \in \sigma_2(T')$. □

Corollary 4.12. *The statement of Corollary 4.6 remains true without the assumption that X is a Dedekind complete Banach lattice.*

In connection with Theorem 4.11 the following question might be of interest.

Problem 4.13. Let X be a Banach lattice and T be an order-continuous linear bounded operator on X. Let $\hat{X}$ be the Dedekind completion of X and $\hat{T}$ be the unique order-continuous extension of T on $\hat{X}$.

Is it true that $\sigma_i(T) = \sigma_i(\hat{T}), i = 0, 1, \dots, 5$, where $\sigma_0(T) = \sigma(T)$.

5. $C(K)$ revisited. The case $(\lambda I - T)C(K) = C(K)$ for weighted compositions generated by non-invertible open surjections

The spectrum of arbitrary disjointness preserving operators on $C(K)$ was described in [12]. The problem of describing essential spectra of such operators in the case when the map $\varphi : K \to K$ is not invertible becomes considerably more complicated and its complete solution remains unknown to the author. In this section we will provide necessary and sufficient conditions for the equality $(\lambda I - wT_\varphi)C(K) = C(K)$ in the case when $|w| > 0$ on $C(K)$, the map $\varphi : K \to K$ is open, and $\varphi(K) = K$ (see Theorem 5.14 below).

Definition 5.1. Let K be a compact Hausdorff space and φ be a continuous map of K onto itself.

(1) We call a subset S of K a φ-string if $S = \{s_i : i \in \mathbb{Z}\}$ and $\varphi(s_i) = s_{i+1}, i \in \mathbb{Z}$

(2) Let S be a φ-string. We define the set $\overrightarrow{S}$ as

$$\overrightarrow{S} = \bigcap_{i=1}^{\infty} cl\{s_k : k \geq i\} \quad \text{and the set } \overleftarrow{S} \text{ as} \quad \overleftarrow{S} = \bigcap_{i=1}^{\infty} cl\{s_k : k \leq -i\}.$$

Lemma 5.2. *Let $T = wT_\varphi$ be a weighted composition operator on $C(K)$ and $\varphi(K) = K$. Let $\lambda \in \sigma_r(T^\star)$, i.e., $\lambda \in \sigma(T)$ and $(\lambda I - T)C(K) = C(K)$. Then $\lambda \neq 0$.*

Proof. If $TC(K) = C(K)$ then clearly $|w| > 0$ on K, whence the operator of multiplication by w is invertible in $C(K)$ and therefore $T_\varphi C(K) = C(K)$. It follows immediately that φ is injective and therefore a homeomorphism of K onto itself, whence T is invertible on $C(K)$. □

Lemma 5.3. *Let $T = wT_\varphi$ be a weighted composition operator on $C(K)$. Let $\varphi(K) = K$ and $|w| > 0$ on K. Let $\lambda \in \sigma_r(T^\star)$. Then there is an open subset U of K such that*

(i) $\varphi(U) = \varphi^{-1}(U) = U$

and for any φ-string $S = \{s_i, i \in \mathbb{Z}\}$ such that $S \subset U$ we have

(ii) $$\liminf_{n\to\infty} |w_n(s_0)|^{1/n} > |\lambda| \text{ and } \limsup_{n\to\infty} |w_n(s_{-n})|^{1/n} < |\lambda|. \tag{12}$$

Proof. Because $\lambda \in \sigma_r(T^\star)$ there is $f \in C(K)$ such that $f \neq 0$ and $Tf = \lambda f$. Let $U_1 = \{k \in K : f(k) \neq 0\}$. It follows easily from $|w| > 0$ on K that $\varphi(U_1) = \varphi^{-1}(U_1) = U_1$. Let $K_1 = cl(U_1)$ then $\varphi(K_1) = K_1$ and the formula $T = wT_\varphi$ shows that T induces a weighted composition operator $T_1 = w_1 T_{\varphi_1}$ on $C(K_1)$ where w_1 and φ_1 are restrictions of w and φ, respectively, on K_1. Clearly $f_1 = f|K_1 \in C(K_1)$ and $T_1 f_1 = \lambda f_1$. It follows from the fact that $(\lambda I - T)C(K) = C(K)$ and from the Tietze extension theorem that $(\lambda I - T_1)C(K_1) = C(K_1)$ whence $\lambda \in \sigma_r(T_1^\star)$. The set $\sigma_r(T_1^\star)$ is open in $\mathbb{C}$ and therefore there is $\gamma \in \sigma_r(T_1^\star)$ such that $|\gamma| > |\lambda|$. Let g be a nonzero function in $C(K_1)$ such that $T_1 g = \gamma g$ and let $V = \{k \in$

$K_1 : g(k) \neq 0\}$. Then V is an open subset of K_1 and $\varphi_1^{-1}(V) = \varphi_1(V) = V$. We cannot claim that either V is open in K or that $\varphi^{-1}(V) = V$ but it is true that $V \subset \varphi^{-1}(V)$. Consider $U_2 = U_1 \cap V$; then it is immediate that U_2 is an open nonempty subset of K and recalling that $\varphi^{-1}(U_1) = U_1$ and $\varphi_1^{-1}(V) = V$ we see that $\varphi^{-1}(U_2) = \varphi(U_2) = U_2$. Let $K_2 = clU_2$ and let T_2 be the operator on $C(K_2)$ defined similarly to the definition of T_1 above. Then $\lambda \in \sigma_r(T_2^\star)$ and we can find $\delta \in \sigma_r(T_2^\star)$ such that $|\delta| < |\lambda|$. Let $h \in C(K_2)$ be such that $h \neq 0$ and $T_2 h = \delta h$. Let $W = \{k \in K_2 : h(k) \neq 0\}$ and let $U = W \cap U_2$. As above we can show that U is an open subset of K and

$$\varphi^{-1}(U) = \varphi(U) = U. \tag{13}$$

Now let $S \subset U$ be a φ-string. For any $n \in \mathbb{N}$ we have $w_n(s_0)g(\varphi^n(s_0)) = \gamma^n g(s_0)$ whence

$$|w_n(s_0) \geq |g(s_0)||\gamma|^n/\|g\|_{C(K_1)}. \tag{14}$$

On the other hand for every $n \in \mathbb{N}$ we have $w_n(s_{-n})h(s_0) = \delta^n h(s_{-n})$ whence

$$|w_n(s_{-n} \leq |\delta|^n \|h\|_{C(K_2)}/h(s_0). \tag{15}$$

The statement of the lemma follows from (13)–(15). □

Definition 5.4. Assume conditions of Lemma 5.3. We will denote by $O(\lambda)$ the union of all open subsets of K with properties (i) and (ii) from the statement of Lemma 5.3. Clearly $O(\lambda)$ is the largest (by inclusion) open subset of K with these properties.

Lemma 5.5. *Assume conditions of Lemma* 5.3. *Let* $O(\lambda)$ *be from Definition* 5.4. *Then the set* $K_\lambda = K \setminus O(\lambda)$ *is not empty.*

Proof. If $K_\lambda = \emptyset$ then $O(\lambda) = K$ is a compact Hausdorff space. We can assume without loss of generality that $|\lambda| = 1$. It follows from the definition of $O(\lambda)$ that for every $k \in K$ there are an open neighborhood $V(k)$ of k and $m(k) \in \mathbb{N}$ such that $|w_{m(k)}(t)| > 2, t \in V(k)$. Let $\{V(k_1), \dots, V(k_s)\}$ be a finite subcover of the cover $\{V(k) : k \in K\}$ and let $m_i = m(k_i), i = 1, \dots, s$. Next let us fix $k_0 \in K$. Then for any $n \in \mathbb{N}$ we can find $p \in \mathbb{N}$ such that $0 \leq n - p \leq \max\{m_1, \dots, m_s\}$ and $|w_p(\varphi^{-n}(k_0)| > 2$ in obvious contradiction with the second inequality in (12). □

Lemma 5.6. *Assume conditions of Lemma* 5.3. *Let* $O(\lambda)$ *be from Definition* 5.4. *Let* T_λ *be the weighted composition operator induced by* T *on* $C(K_\lambda)$ *where* $K_\lambda = K \setminus O(\lambda)$.

Then $\lambda \notin \sigma(T_\lambda)$.

Proof. First of all notice that because $\varphi^{-1}(O(\lambda)) = O(\lambda)$ we have $\varphi(K_\lambda) = K_\lambda$ and the operator T_λ is correctly defined. Moreover, $(\lambda I - T_\lambda)C(K_\lambda) = C(K_\lambda)$. Assume contrary to the statement of the lemma that $\lambda \in \sigma(T_\lambda)$. Then $\lambda \in \sigma_r(T_\lambda^\star)$ and by Lemma 5.3 there is an open subset V of K_λ with properties (i) and (ii) from the statement of that lemma. The set $O(\lambda) \cup V$ is open in K and has properties (i) and (ii) in contradiction with maximality of $O(\lambda)$. □

Lemma 5.7. *Assume conditions of Lemma* 5.3. *Let* $F_\lambda = clO(\lambda) \setminus O(\lambda)$. *Let* T^λ *be the operator induced by* T *on* $C(F_\lambda)$.

Then $\sigma(T^\lambda, C(F_\lambda)) \cap \lambda\Gamma = \emptyset$.

Proof. We will prove first that $\lambda \notin \sigma(T^\lambda)$. Indeed, because $\varphi(F_\lambda) = F_\lambda$ we have $(\lambda I - T^\lambda)C(F_\lambda) = C(F_\lambda)$ and therefore, if $\lambda \in \sigma(T^\lambda)$, then there is a nonzero $g \in C(F_\lambda)$ such that $T^\lambda g = \lambda g$. But F_λ is a closed subset of K_λ and therefore g can be identified with an element $g^{\star\star}$ in $C(K_\lambda)^{\star\star}$ such that $T_\lambda^{\star\star} g^{\star\star} = \lambda g^{\star\star}$ in contradiction with Lemma 5.6.

Next assume that there is $\gamma \in \sigma(F_\lambda)$ such that $|\gamma| = |\lambda|$. Then by Theorem 3.12 in [12] there is a φ-periodic point $t \in F_\lambda$ such that $\gamma^p = w_p(t)$ where p is the smallest positive period of t. It follows from (1) and from $\varphi(O(\lambda)) = O(\lambda)$ that the set $O(\lambda)$ cannot contain eventually [2] φ-periodic points. Then it follows from the proof of Lemma 3.5 in [12] that $\lambda\Gamma \subset \sigma_{ap}(T^\star)$, a contradiction. □

Lemma 5.8. *There is a closed subset* G_λ *of* F_λ *with the following properties.*

1. $\varphi(G_\lambda) = G_\lambda$.
2. *The restriction of* φ *on* G_λ *is a homeomorphism of* G_λ *onto itself.*
3. *The operator* T_{G_λ} *defined on* $C(G_\lambda)$ *by the formula* $T(f|G_\lambda) = (Tf)|G_\lambda$ *is invertible and* $\rho(T_{G_\lambda}^{-1}) < 1/|\lambda|$.
4. $\exists m \in \mathbb{N}$ *such that* $G_\lambda \subset \mathrm{Int}_{F_\lambda} \varphi^{-m}(G_\lambda)$.
5. *Let* $H_\lambda = F_\lambda \setminus \bigcup_{n=1}^{\infty} \varphi^{-n}(G_\lambda)$. *Then* H_λ *is a closed subset of* F_λ, $\varphi(H_\lambda) = H_\lambda$, *and* $\rho(T_{H_\lambda}) < |\lambda|$.

Proof. The proof follows immediately from Lemma 5.7 and Theorem 3.10 in [12]. □

Lemma 5.9. *Let the map* φ *be open and let* G_λ *be the set from the statement of Lemma* 5.7. *Then there is an open neighborhood* V *of* G_λ *in* K *such that the map* $\varphi : V \to K$ *is one-to-one.*

Proof. We can assume without loss of generality that $\lambda = 1$. For any $n \in \mathbb{N}$ let $Q_n = \{k \in K : \ |w_{n+1}(k)| > 2\}$. It follows from the inequality $\rho(T_{G_\lambda}^{-1}) < 1$ that for any large enough $n \in \mathbb{N}$ the set Q_n is an open neighborhood of G_λ. Let $R_n = \varphi^n(Q_n)$, then R_n is an open neighborhood of G_λ because φ is open. We claim that for any large enough $n \in \mathbb{N}$ the map $\varphi : R_n \to K$ is one-to-one. Assume to the contrary that for any $N \in \mathbb{N}$ there are an $n > N$ and $p, q \in R_n$ such that $p \neq q$ but $\varphi(p) = \varphi(q)$. Let $s, t \in Q_n$ be such that $\varphi^n(s) = p$ and $\varphi^n(t) = q$. Let $\mu_1 = \frac{1}{w_{n+1}(s)}\delta_s$ and $\mu_2 = \frac{1}{w_{n+1}(t)}\delta_t$. Notice that $\|\mu_i\| \leq 1/2, i = 1, 2$, because $s, t \in Q_n$. Consider the discrete measure μ on K defined as

$$\mu = \sum_{i=0}^{n} \left(1 - \frac{1}{\sqrt{n}}\right)^{n-i} (T^\star)^i \mu_1 - \sum_{i=0}^{n} \left(1 - \frac{1}{\sqrt{n}}\right)^{n-i} (T^\star)^i \mu_2.$$

---

[2] A point $k \in K$ is called eventually φ-periodic if there is an $n \in \mathbb{N}$ such that the point $\varphi^{(n)}(k)$ is φ-periodic.

Notice that $(T^\star)^n\mu_1 = \frac{1}{w(p)}\delta_p$ and $(T^\star)^n\mu_2 = \frac{1}{w(q)}\delta_q$ whence $\|\mu\| \geq 2\|1/w\|_\infty$. Next notice that $(T^\star)^{n+1}\mu_1 = (T^\star)^{n+1}\mu_2 = \delta_{\varphi(p)}$ whence

$$T^\star\mu - \mu = \left(1 - \frac{1}{\sqrt{n}}\right)^n (\mu_1 - \mu_2) - \frac{1}{\sqrt{n}}\sum_{i=1}^{n}\left(1 - \frac{1}{\sqrt{n}}\right)^{n-i}(T^\star)^i\mu_1 + \frac{1}{\sqrt{n}}\sum_{i=1}^{n}\left(1 - \frac{1}{\sqrt{n}}\right)^{n-i}(T^\star)^i\mu_2.$$

Therefore $\|T^\star\mu - \mu\| \leq \left(1 - \frac{1}{\sqrt{n}}\right)^n + \frac{1}{\sqrt{n}}\|\mu\|$. Because n can be chosen arbitrary large we see that $1 \in \sigma_{a.p.}(T^\star)$, a contradiction. □

Lemma 5.10. *Assume conditions of Lemma 5.8. For any $n \in \mathbb{N}$ let Q_n be the set defined in the proof of Lemma 5.9. Then there is $N \in \mathbb{N}$ such that if $n \geq N$ then $\varphi^n(Q_n) \cap O(\lambda) \subset Q_n \cap O(\lambda)$.*

Proof. Assume contrary to our statement that for any $N \in \mathbb{N}$ there are $n > N$ and $k \in O(\lambda)$ such that

$$|w_n(k)| > 2 \text{ and } |w_n(\varphi^n(k))| \leq 2. \tag{16}$$

Let $\nu = \frac{1}{w_n(k)}\delta_k$ and

$$\mu = \sum_{i=0}^{2n-1}\left(1 - \frac{1}{\sqrt{n}}\right)^{|n-i|}(T^\star)^i\nu.$$

Notice that because k is not a φ-periodic or eventually φ-periodic point all the terms in the sum above represent pairwise disjoint point measures on K. Therefore $\|\mu\| \geq \|(T^\star)^n\nu\| = \|\delta_{\varphi^n(k)}\| = 1$. On the other hand

$$T^\star\mu - \mu = \left(1 - \frac{1}{\sqrt{n}}\right)^n \nu - \sum_{i=1}^{n}\frac{1}{\sqrt{n}}\left(1 - \frac{1}{\sqrt{n}}\right)^{n-i}(T^\star)^i\nu + \sum_{i=n+1}^{2n-1}\frac{1}{\sqrt{n}}\left(1 - \frac{1}{\sqrt{n}}\right)^{i-n}(T^\star)^i\nu + \left(1 - \frac{1}{\sqrt{n}}\right)^{n-1}(T^\star)^{2n}\nu$$

and in virtue of (16) $\|T^\star\mu - \mu\| \leq \left(1 - \frac{1}{\sqrt{n}}\right)^n + 2\left(1 - \frac{1}{\sqrt{n}}\right)^{n-1} + \frac{1}{\sqrt{n}}\|\mu\|$. Because n is arbitrary large we have $1 \in \sigma_{ap}(T^\star)$ in contradiction with our assumption. □

Corollary 5.11. *There are an open nonempty neighborhood V of G_λ in $cl(O_\lambda)$ and $n \in \mathbb{N}$ such that*

- *$\varphi(V) \subset V$.*
- *The map $\varphi : V \to \varphi(V)$ is a homeomorphism.*
- *$clV \cap H_\lambda = \emptyset$.*
- *$|w_n| > 1$ on clV.*

Proof. Fix a large enough $n \in \mathbb{N}$ and consider the set R_n introduced in the proof of Lemma 5.9. The set $V = \bigcap_{k=0}^{n-1} \varphi^k(R_n \cap O(\lambda))$ has the required properties. □

Lemma 5.12. *Assume conditions of Lemma* 5.8. *Let* V *be an open neighborhood in* K *of the set* G_λ *such that* $\varphi(V) \subset V$ *and* $V \cap H_\lambda = \emptyset$. *Then* $H = clO_\lambda \setminus \bigcup_{k=1}^{\infty} \varphi^{-k}(V) = H_\lambda$.

Proof. Clearly $\varphi(H) \subseteq H$ and $H_\lambda \subseteq H$. Assume that $H \setminus H_\lambda \neq \emptyset$. It follows from (5) in the statement of Lemma 5.8 that $H \setminus H_\lambda \subset O(\lambda)$. Let T_H be the weighted composition operator wT_φ considered on $C(H)$. We have to consider three possibilities.

(a) $\rho(T_H) < |\lambda|$. That contradicts the first inequality in (1) in the statement of Lemma 5.3.

(b) $\rho(T_H) > |\lambda|$ and $\lambda \notin \sigma(T_H)$. In this case it follows from [12] that there is a closed subset L of H such that $\varphi(L) = L$, the operator $T_L = wT_\varphi$ is invertible on $C(L)$, and $\rho(T_L^{-1} < 1/|\lambda|$. Clearly $L \cup H_\lambda = \emptyset$ whence $L \subset O(\lambda)$. But then we come to a contradiction with the second inequality in (1).

(c) $\lambda \in \sigma_r(T_H^\star)$. We bring this case to a contradiction similar to (b) by using statements (1)–(3) of Lemma 5.8. □

The previous statements provide necessary conditions for $\lambda \in \sigma_r(T^\star)$. As Theorem 5.14 below shows the combination of these conditions is also sufficient. Before we state and prove this theorem we need one simple result which is most probably known.

Lemma 5.13. *Let* K *be a compact Hausdorff space,* φ *be a map of* K *into itself, and* T_φ *be the corresponding composition operator on* $C(K)$. *The following statements are equivalent*

1. *The Banach dual operator* $T_\varphi^\star$ *preserves disjointness.*
2. *The map* φ *is one-to-one.*

Proof. The implication (1) ⇒ (2) is trivial. Assume that φ is a homeomorphism of K onto $\varphi(K)$. The operator T_φ induces a positive isometry of $C(\varphi(K))$ onto $C(K)$. Therefore the dual operator $T_\varphi^\star$ can be considered as a positive isometry of $C(K)^\star$ onto $C(\varphi(K))^\star$. To finish the proof we can use a theorem of Abramovich [1] that states that a positive surjective isometry between normed lattices preserves disjointness. Alternatively we can notice that $C(K)^\star$ and $C(\varphi(K))^\star$ are L^1-spaces and we can apply the theorem of Lamperti (see, e.g., [9, Chapter 3]) to see that $T_\varphi^\star$ preserves disjointness (because in the statement of Lamperti theorem the measure is assumed to be sigma-finite some simple additional reasoning is needed). □

Theorem 5.14. *Let* K *be a compact Hausdorff space and* φ *be an open continuous map of* K *onto itself. Let* w *be an invertible element of* $C(K)$. *Let* T *be the weighted* *composition operator*

$$(Tf)(k) = w(k)f(\varphi(k)), \ f \in C(K), \ k \in K.$$

Let $\lambda \in \sigma(T)$. *The following conditions are equivalent.* (I) $\lambda \in \sigma_r(T^\star)$ (*i.e.*, $\lambda \in \sigma(T)$ *and* $(\lambda I - T)C(K) = C(K)$). (II)

1. $\lambda \neq 0$.
2. *There is a nonempty open subset* O *of* K *such that* $\varphi(O) = O = \varphi^{-1}(O)$, *for every point* $k \in O$ *conditions* (1) *are satisfied,* $F = clO \setminus O \neq \emptyset$, *and* $\lambda \notin \sigma(T, C(K \setminus O)$.
3. $\lambda\Gamma \cap \sigma(T, C(F)) = \emptyset$.
4. *There are subsets* G *and* H *of* F *with properties* (1)–(5) *from the statement of Lemma* 5.8.
5. *There are an open neighborhood* V *of* G *in* clO *and* $m \in \mathbb{N}$ *such that* $V \cap H = \emptyset$, $\varphi(V) \subset V$, $clO \setminus \bigcup\limits_{n=1}^{\infty} \varphi^{-n}(V) = H$, *the map* $\varphi : V \to \varphi(V)$ *is a homeomorphism, and* $|w_m| > 1$ *on* clV.

Proof. The implication (I) $\Rightarrow$ (II) has been already proved.

Assume (II). It follows from II(1) and Theorem 3.10 in [12] that $\lambda \in \sigma(T)$. Assume contrary to our statement that $\lambda \in \sigma_{ap}(T^\star)$. We can assume without loss of generality that $\lambda = 1$. Then there is a sequence $\mu_n \in C(K)^\star, n \in \mathbb{N}$ such that $\|\mu_n\| = 1$ and $T^\star\mu_n - \mu_n \underset{n\to\infty}{\to} 0$.

It follows from II(2) that $|\mu_n|(K \setminus O) \underset{n\to\infty}{\to} 0$ and therefore we can assume that $|\mu_n|(K \setminus O) = 0, n \in \mathbb{N}$.

Consider the set V and the integer m from II(5). The ideal J of all functions from $C(K)$ that are equal 0 on V is T-invariant and it is easy to see from II(5) and the fact that $\rho(T, C(H)) < 1$ that $\rho(T|J) < 1$. Therefore $|\mu_n|(clO \setminus V) \underset{n\to\infty}{\to} 0$ and we can assume that $|\mu_n|(V) = 1, n \in \mathbb{N}$. Let T_V be the operator wT_φ considered on $C(clV)$. Then $T_V^\star\mu_n - \mu_n \underset{n\to\infty}{\to} 0$. By Lemma 5.13 the operator $T_V^\star$ preserves disjointness and therefore $|T_V|^\star|\mu_n| - |\mu_n| \underset{n\to\infty}{\to} 0$. Let μ be a probability measure on V which is an accumulation point of the sequence $|\mu_n|, n \in \mathbb{N}$ in the weak-$\star$ topology. Then $|T_V^\star|\mu = \mu$. Let $S = \operatorname{supp}\mu$. Then $\varphi(S) = S$ and the operator T_S induced by T on $C(S)$ is invertible. We have $(T_\varphi^m)^\star(|w_m|)^\star\mu = \mu$. But $(T_\varphi^m)^\star$ is an isometry on $C(S)^\star$ and $\|(|w_m|)^\star\mu\| > \|\mu\|$, a contradiction. □

6. Appendix

The purpose of this appendix is to clarify some details about the statement and the proof of Theorem 22 in [13] which was extensively used in the current paper.

(1) The aforementioned theorem states that if T is an operator on a Banach lattice X of the form (6) then

$$\sigma(T, X) = \sigma(\hat{T}, \hat{X}) = \sigma(S, C(K)) = \sigma(\hat{S}, C(\hat{K})).$$

While the equalities

$$\sigma(T, X) = \sigma(\hat{T}, \hat{X}) = \sigma(\hat{S}, C(\hat{K}))$$

and their proof in [13] are correct, the statement that any of these sets is equal to $\sigma(S, C(K))$ is in general false, as in particular follows from Example 32 in [13].

Fortunately, we did not use this equality in full generality, and in the case when X is a Banach lattice with a quasi-interior point it is true, as follows from Proposition 4.8 and Lemma 4.9.

(2) In the proofs of Theorems 22 and 15 in [13] the following fact was used. Let X be a Banach lattice and B be a band in X, then

$$\widehat{X/B} = \hat{X}/\hat{B}.$$

It was (and still is) my assumption that this fact must be known, but I was not able to find it in the literature. Therefore a short proof is provided below.

Proof. Consider the canonical map $T : X \to X/B$; $Tx = [x]$. Then [2, Problem 9.3.2, p. 29] the map T is order continuous. Therefore by Veksler's theorem [15] the operator T has the unique order-continuous extension $\hat{T} : \hat{X} \to \widehat{X/B}$. Notice that because $\hat{T}$ is order continuous and $T(X) = X/B$ we have $\hat{T}(\hat{X}) = \widehat{X/B}$. On the other hand $\ker \hat{T} = \hat{B}$. Indeed, obviously $B \subset \ker \hat{T}$ and because $\hat{T}$ is order continuous we have $\hat{B} \subset \ker \hat{T}$. On the other hand if $\hat{x} \in \hat{X} \setminus \hat{B}$ then there is $x \in B^d$ such that $x \neq 0$ and $|x| \leq |\hat{x}|$. The operators T and (therefore) $\hat{T}$ preserve disjointness whence $|\hat{T}\hat{x}| = \hat{T}|\hat{x}| \geq T|x| = |Tx| \neq 0$ and $\hat{x} \notin \ker \hat{T}$.

Thus we see that $\widehat{X/B}$ is isometrically and lattice isomorphic to $\hat{B}^d = \hat{X}/\hat{B}$. □

References

[1] Abramovich Y.A., *Isometries of normed lattices.* Optimizatsiya No. 43(60) (1988), 74–80. (Russian)

[2] Abramovich Y.A. and Aliprantis C.D., *Problems in Operator Theory*, Graduate Studies in Mathematics, v. 51, AMS, 2002.

[3] Aliprantis C.D. and Burkinshaw O., *Positive Operators*, Springer, 2006.

[4] Abramovich Y.A. and Kitover A.K. *Inverses of Disjointness Preserving Operators.* Memoirs of the AMS, 679, v. 143, 2000.

[5] Atkinson F.V., *The normal solubility of linear equations in normed spaces.* Mat. Sb. (N.S.) 28 (70) (1951), 3–14. (Russian)

[6] Bennett C. and Sharpley R., *Interpolation of Operators*, Academic Press, 1988.

[7] Dunford N. and Schwartz J.T., *Linear Operators, General Theory*, Interscience Publishers, NY, 1958.

[8] Edmunds D.E. and Evans W.D., *Spectral Theory and Differential Operators*, Clarendon Press, Oxford, 1987.

[9] Fleming R.J. and Jamison J.E., *Isometries of Banach spaces: function spaces*, CHAPMAN & HALL/CRC Monographs and Surveys in Pure and Applied Mathematics **129**, 2003.

[10] Kato T., *Perturbation theory for linear operators*, Springer, 1966.

[11] Kantorovich L.V. and Akilov G.P., *Fubctional Analysis.* Pergamon Press, 1982.

[12] Kitover A.K., *Spectrum of weighted composition operators: Part 1. Weighted composition operators on $C(K)$ and uniform algebras.* Positivity, v. 15, n. 4, (2011) 639–659.

[13] Kitover A.K., *Spectrum of weighted composition operators: Part 2. Spectrum of weighted composition operators on subspaces of Banach lattices.* Positivity, v. 17, n. 3 (2013) 655–676.

[14] Meyer-Niebrg P., *Banach Lattices.* Springer, 1991.

[15] Veksler A.I., *On the homomorphisms between the classes of regular operators in K-lineals and their completions.* Izv. Vyss. Ucebn. Zaved. Matematika, No. I (14) (1960), 48–57. (Russian)

[16] Walker R.C., *The Stone–Čech compactification.* Ergebnisse der Mathematik und ihrer Grenzgebiete, Band 83. Springer, Berlin (1974).

[17] Zaanen A.C., *Riesz Spaces II*, North-Holland, 1983.

A.K. Kitover
Community College of Philadelphia
1700 Spring Garden St.
Philadelphia, PA, USA
e-mail: akitover@ccp.edu

Ordered Structures and Applications: Positivity VII
Trends in Mathematics, 263–274

Bernoulli Processes in Riesz Spaces

Wen-Chi Kuo, Jessica J. Vardy and Bruce A. Watson

Abstract. The action and averaging properties of conditional expectation operators are studied in the, measure-free, Riesz space, setting of Kuo, Labuschagne and Watson [*Conditional expectations on Riesz spaces.* J. Math. Anal. Appl., **303** (2005), 509–521.] but on the abstract L^2 space, $\mathcal{L}^2(T)$ introduced by Labuschagne and Watson [*Discrete Stochastic Integration in Riesz Spaces.* Positivity, **14**, (2010), 859–575.]. In this setting the Bienaymé inequality is proved and from this foundation Bernoulli processes are considered. Bernoulli's strong law of large numbers and Poisson's theorem are given.

Mathematics Subject Classification (2010). Primary 60G20; Secondary 47B60.

Keywords. Riesz spaces, conditional expectation operators, f-algebra, averaging operators, Bernoulli processes, conditional independence, strong laws of large numbers.

1. Introduction

Various authors have considered generalizations of stochastic processes to vector lattices/Riesz spaces, with a variety of assumptions being made on the processes being considered. Most of this work has focused on martingale theory, see, for example, [4], [8], [11], [12], [18] and [19]. The abstract properties of conditional expectation operators have also been explored in various settings, see [5], [13], [16] and [17] and [21]. However, the more elementary processes such as Markov processes, see [20], Bernoulli processes and Poisson processes, which rely only on the concepts of a conditional expectation operator and independence, have received little attention. As these processes have less accessible structure, their study relies more heavily on properties of the underlying Riesz space, the representation of the conditional expectation operators and multiplication operations in Riesz spaces. If a Riesz space has a weak order unit, then the order ideal generated by a weak order unit is order dense in the space. However the order ideal generated by the weak order unit is an f-algebra, see [2], [3] and [22], giving a multiplicative structure on a dense subspace. Much of the work in this paper relies on a Riesz space vector analogue of L^2 and the action of conditional expectations in this space, see Theorem 3.2, and their averaging property, see Lemma 3.1. In particular the Bienaymé equality, Theorem 4.2, will be posed in this setting. The Bienaymé equality enables us to give a Riesz space analogue of Bernoulli's law of large numbers, Theorem 5.2. One important property of the ideal generated by the weak order unit is that it possesses a functional calculus which enables one to lift continuous real-valued functions on $[0,1]$ to the Riesz space, see [6] and [22]. This is critical for Poisson's theorem, Theorem 5.6. We refer the reader to [15] for the classical version of the Bienaymé equality, the Bernoulli law of large numbers and Poisson's theorem.

2. Riesz space preliminaries

We refer the reader to [1] and [22] for general Riesz space theory. The definitions and preliminaries presented here are specific to Riesz spaces with a weak order unit and a conditional expectation operator.

The notion of a conditional expectation operator in a Dedekind complete Riesz space, E, with weak order unit was introduced in [12] as a positive order-continuous projection $T : E \to E$, with range $\mathcal{R}(T)$ a Dedekind complete Riesz subspace of E, and having Te a weak order unit of E for each weak order unit e of E. Instead of requiring Te to be a weak order unit of E for each weak order unit e of E one can equivalently impose that there is a weak order unit in E which is invariant under T. Averaging properties of conditional expectation operators and various other structural aspects were considered in [13]. In particular if B is the band in E generated by $0 \leq g \in \mathcal{R}(T)$ and P is the band projection onto B, it was shown that $Tf \in B$, for each $f \in B$, $Pf, (I-P)f \in \mathcal{R}(T)$ for each $f \in \mathcal{R}(T)$, where I denotes the identity map, and $Tf \in B^d$, for each $f \in B^d$. A consequence of these relations and Freudenthal's theorem, [22], is that if B is the band in E generated by $0 \leq g \in \mathcal{R}(T)$, with associated band projection P, then $TP = PT$, see [13] for details.

To access the averaging properties of conditional expectation operators a multiplicative structure is needed. In the Riesz space setting the most natural multiplicative structure is that of an f-algebra. This gives a multiplicative structure that is compatible with the order and additive structures on the space. The ideal, E^e, of E generated by e, where e is a weak order unit of E and E is Dedekind complete, has a natural f-algebra structure. This is constructed by setting $(Pe)\cdot(Qe) = PQe = (Qe)\cdot(Pe)$ for band projections P and Q, and extending to E^e by use of Freudenthal's Theorem. In fact this process extends the multiplicative structure to the universal completion, E^u, of E. This multiplication is associative, distributive and is positive in the sense that if $x, y \in E^+$ then $xy \geq 0$. Here e is the multiplicative unit. For more information about f-algebras see [2, 3, 5, 7, 13, 22]. If T is a conditional expectation operator on the Dedekind complete Riesz space E with weak order unit $e = Te$, then restricting our attention to the f-algebra E^e, T is an averaging operator on E^e if $T(fg) = fTg$ for

$f, g \in E^e$ and $f \in R(T)$, see [5, 7, 13]. More will said about averaging operators in Section 3.

In a Dedekind complete Riesz space, E, with weak order unit and T a strictly positive conditional expectation on E. We say that the space is T-universally complete if for each increasing net (f_α) in E_+ with (Tf_α) order-bounded in the universal completion E^u, we have that (f_α) is order convergent. If this is not the case, then both the space and conditional expectation operator can be extended so that the extended space is T-universally complete with respect to the extended T, see [13]. The extended space is also know as the natural domain of T, denoted $\operatorname{dom}(T)$ of $\mathcal{L}^1(T)$, see [5, 7].

Let E be a Dedekind complete Riesz space with conditional expectation T and weak order unit $e = Te$. If P and Q are band projections on E, we say that P and Q are T-conditionally independent if

$$TPTQe = TPQe = TQTPe. \tag{2.1}$$

We say that two Riesz subspaces E_1 and E_2 of E are T-conditionally independent with respect to T if all band projections $P_i, i = 1, 2$, in E with $P_i e \in E_i, i = 1, 2$, are T-conditionally independent with respect to T. Equivalently (2.1) can be replaced with

$$TPTQw = TPQw = TQTPw \quad \text{for all} \quad w \in \mathcal{R}(T). \tag{2.2}$$

It should be noted that T-conditional independence of the band projections P and Q is equivalent to T-conditional independence of the closed Riesz subspaces $\langle Pe, \mathcal{R}(T)\rangle$ and $\langle Qe, \mathcal{R}(T)\rangle$ generated by Pe and $\mathcal{R}(T)$ and by Qe and $\mathcal{R}(T)$ respectively. From the Radon–Nikodým–Douglas–Andô type theorem was established in [21], if E is a T-universally complete, a subset F of E is a closed Riesz subspace of E with $\mathcal{R}(T) \subset F$ if and only if there is a unique conditional expectation T_F on E with $\mathcal{R}(T_F) = F$ and $TT_F = T = T_F T$. In this case $T_F f$ for $f \in E^+$ is uniquely determined by the property that

$$TPf = TPT_F f \tag{2.3}$$

for all band projections on E with $Pe \in F$. As a consequence of this, two closed Riesz subspaces E_1 and E_2 with $\mathcal{R}(T) \subset E_1 \cap E_2$ are T-conditionally independent, if and only if

$$T_1 T_2 = T = T_2 T_1, \tag{2.4}$$

where T_i is the conditional expectation commuting with T and having range $E_i, i = 1, 2$. Here (2.4) can be equivalently replaced by

$$T_i f = Tf, \quad \text{for all} \quad f \in E_{3-i}, \quad i = 1, 2, \tag{2.5}$$

see [20]. The concept of T-conditional independence can be extended to a family, say $(E_\lambda)_{\lambda\in\Lambda}$, of closed Dedekind complete Riesz subspaces of E with $\mathcal{R}(T) \subset E_\lambda$ for all $\lambda \in \Lambda$. We say that the family is T-conditionally independent if, for each pair of disjoint sets $\Lambda_1, \Lambda_2 \subset \Lambda$, we have that E_{Λ_1} and E_{Λ_2} are T-conditionally independent. Here $E_{\Lambda_j} := \left\langle \bigcup_{\lambda\in\Lambda_j} E_\lambda \right\rangle$. Finally, we say that a sequence (f_n) in E is T-conditionally independent if the family of closed Riesz subspaces $\langle \{f_n\} \cup \mathcal{R}(T)\rangle, n \in \mathbb{N}$, is T-conditionally independent.

3. Conditional expectation operators in $\mathcal{L}^2(T)$

In this work we assume that E is T-universally complete and in this case we have $\mathcal{L}^1(T) = E$, see [11]. As E^u, the universal completion of E, is an f-algebra, multiplication of elements of E is defined but does not necessarily result in an element of E. This leads us, as in [11], to define

$$\mathcal{L}^2(T) := \left\{x \in \mathcal{L}^1(T) | x^2 \in \mathcal{L}^1(T)\right\}.$$

If $f, g \in \mathcal{L}^2(T)$ then in the f-algebra E^u, $0 \le (f \pm g)^2 = f^2 \pm 2fg + g^2$. Thus $\pm 2fg \le f^2 + g^2$ and $2|fg| \le f^2 + g^2 \in \mathcal{L}^1(T) = E$. Hence $fg \in \mathcal{L}^1(T) = E$. As noted in [11], a consequence of this is that $\mathcal{L}^2(T)$ is a vector space.

The averaging property of conditional expectation operators only makes sense if it can be ensured that the all products involved remain in the space. Theorem 4.3 of [13] states that if E is a Dedekind complete Riesz space with weak order unit, T is a conditional expectation operator on E and E is also an f-algebra, then T is an averaging operator, i.e., $T(fg) = gTf$ for all $f \in E, g \in \mathcal{R}(T)$. The averaging property is revisited in [11, Theorem 2.1] without proof. The variant of [11, Theorem 2.1] drops the assumption that E is an f-algebra, but imposes the additional conditions that $fg \in E$ and that E is T-universally complete. A strengthened version of this is proved in Lemma 3.1. This however does not address whether $Sf \in \mathcal{L}^2(T)$ for $f \in \mathcal{L}^2(T)$ and S a conditional expectation operator on E with $TS = T = ST$. For this see Theorem 3.2 below. As a consequence of Lemma 3.1 and Theorem 3.2, we are able to conclude, see Theorem 3.3 below, that for such a conditional expectation operator, S, $S(fTg) = Tg \cdot Sf$ for all $f, g \in \mathcal{L}^2(T)$.

Lemma 3.1. *Let E be a Dedekind complete Riesz space with weak order unit, e, and T is a conditional expectation operator on E with $Te = e$. If $f, g, fg \in E$ with $g \in \mathcal{R}(T)$ then $g \cdot Tf \in E$ and $T(fg) = g \cdot Tf$.*

Proof. Case I: $f, g, fg \in E_+$ with $g \in \mathcal{R}(T)$. Let $f_n = f \wedge ne$ and $g_n = g \wedge ne$. Then $f_n \uparrow f$ and $g_n \uparrow g$. Here $f_n, g_n \in E_+^e$ with $g_n \in \mathcal{R}(T)$, so [13, Theorem 4.3] can be applied to give $T(f_n g_m) = g_m T(f_n), m, n \in \mathbb{N}$. Thus

$$g_m T(f_n) = T(f_n g_m) \le T(fg), \quad m, n \in \mathbb{N}. \tag{3.1}$$

Here $f_n g_m \uparrow fg$ in E, so from the order-continuity of T, $T(f_n g_m) \uparrow T(fg)$ in E. In the universal completion, E^u, of E, we have $g_m T(f_n) \uparrow gT(f)$, however, from (3.1), $g_m T(f_n)$ is bounded above by $T(fg) \in E$, so $g_m T(f_n) \uparrow gT(f)$ in E, giving $T(fg) = gT(f)$.

Case II: $f, g, fg \in E$ with $g \in \mathcal{R}(T)$. From Case I, $T(f^\pm g^\mp) = g^\mp T(f^\pm)$ and $T(f^\pm g^\pm) = g^\pm T(f^\pm)$, from which the result follows. □

The invariance of $\mathcal{L}^2(T)$ with respect to a conditional expectations S where $TS = S = ST$ is a consequence of Jensen's inequality, see [10, Theorem 4.4] for a general form of Jensen's inequality in Riesz spaces proved using functional calculus. However, for the special case of $\mathcal{L}^2(T)$ an elementary proof is available.

Theorem 3.2. *Let E be a T-universally complete Riesz space with weak order unit, e, where T is a strictly positive conditional expectation operator with $Te = e$ and let S be a conditional expectation operator on E with $TS = T = ST$. If $f \in \mathcal{L}^2(T)$ then $Sf \in \mathcal{L}^2(T)$.*

Proof. Let $f \in \mathcal{L}^2(T)$ and define $f_n = (ne \wedge |f|) \in E^e_+$, $n \in \mathbb{N}$. Working in the f-algebra E^e we have

$$0 \le S(f_n - Sf_n)^2 = Sf_n^2 - 2S(f_n \cdot Sf_n) + S(Sf_n)^2. \tag{3.2}$$

The averaging and projection properties of S applied in (3.2) give

$$Sf^2 \ge Sf_n^2 \ge (Sf_n)^2. \tag{3.3}$$

Taking the order limit as $n \to \infty$ completes the proof. □

Corollary 3.3. *Let E be a T-universally complete Riesz space with weak order unit, e, where T is a strictly positive conditional expectation operator with $Te = e$. Let S, J be conditional expectation operators on E with $TS = T = ST$, $TJ = T = JT$ and $JS = J = SJ$. If $f, g \in \mathcal{L}^2(T)$ then $S(f \cdot Jg) = Jg \cdot S(f)$.*

Proof. As $f, g \in \mathcal{L}^2(T)$ from Theorem 3.2 $Jg \in \mathcal{L}^2(T)$. Now $f, Jg \in \mathcal{L}^2(T)$ so $f, Jg, f \cdot Jg \in E$ so Lemma 3.1 gives $Jg \cdot Sf \in E$ and $S(f \cdot Jg) = Jg \cdot Sf$. □

For the readers convenience we state Čebyčev's inequality in $\mathcal{L}^2(T)$, see [9, Lemma 3.1] with $\alpha = 2$.

Theorem 3.4 (Čebyčev's Inequality). *Let E be a Dedekind complete Riesz space with conditional expectation T and weak order unit $e = Te$. Let $f \in \mathcal{L}^2(T), f \ge 0$, and $\epsilon \in \mathbb{R}, \epsilon > 0$, then*

$$TP_{(f-\epsilon e)^+}e \le \frac{1}{\epsilon^2}T(f^2).$$

4. Bienaymé equality

The Bienaymé equality of classical statistics gives that the variance of a finite sum of independent random variables coincides with sum of their variances. In this section we give a measure free conditional version of this result in $\mathcal{L}^2(T)$. Before we can proceed with this we require a result on T-conditionally independent random variables in $\mathcal{L}^2(T)$.

Lemma 4.1. *Let E be a T-universally complete Riesz space with weak order unit, $e = Te$, where T is a strictly positive conditional expectation operator on E. Let $f, g \in \mathcal{L}^2(T)$. If f and g are T-conditionally independent then*

$$Tfg = Tf \cdot Tg = Tg \cdot Tf.$$

Proof. Let T_f and T_g denote the conditional expectations with ranges $\langle \mathcal{R}(T), f\rangle = E_f$ and $\langle \mathcal{R}(T), g\rangle = E_g$ respectively. Here $\langle \mathcal{R}(T), g\rangle$ denotes the order-closed Riesz subspace of E generated by $\mathcal{R}(T)$ and g, and similarly for $\langle \mathcal{R}(T), f\rangle$. The existence and uniqueness of T_f and T_g are given by the Radon–Nikodým Theorem, see [21]. Here E_g and E_f are T-conditionally independent as the T-conditional independence of f and g is defined in terms of the independence of E_f and E_g, see Section 2. For each $h \in E$, $T_f h \in E_f$ and $T_g h \in E_g$. Now, as E_f and E_g are T-conditionally independent, from (2.5) with $E_1 = E_f$ and $E_2 = E_g$, we have

$$T_f(T_g h) = Th = T_g(T_f h). \tag{4.1}$$

Applying (4.1) with $h = fg$ gives

$$T(fg) = T_f T_g(fg)$$

As T_f and T_g are averaging operators in $\mathcal{L}^2(T)$, by Corollary 3.3, $T_g(fg) = gT_g f$. Thus

$$T(fg) = T_f T_g(fg) = T_f(gT_g f), \tag{4.2}$$

however taking (4.1) with $h = f$ yields $T_g f = Tf$, which along with (4.2) gives

$$T(fg) = T_f(gTf). \tag{4.3}$$

In (4.3), $Tf \in \mathcal{R}(T) \subset E_f$, so by Corollary 3.3, $T_f(gTf) = Tf \cdot T_f g$. Finally considering (4.1) with $h = g$ gives $T_f g = Tg$. Thus

$$T(fg) = T_f(gTf) = Tf \cdot T_f g = Tf \cdot Tg. \qquad \square$$

From Theorem 3.2, if $f \in \mathcal{L}^2(T)$ then $Tf \in \mathcal{L}^2(T)$ which gives $(f - Tf) \in \mathcal{L}^2(T)$. Hence, $(f - Tf)^2 \in \mathcal{L}^1(T)$ and so $T(f - Tf)^2$ exists for all $f \in \mathcal{L}^2(T)$. We now define the variance of f by

$$\mathrm{var}(f) = T(f - Tf)^2 = Tf^2 - (Tf)^2. \tag{4.4}$$

Theorem 4.2 (Bienaymé Equality). *Let E be a T-universally complete Riesz space with weak order unit, $e = Te$, where T is a strictly positive conditional expectation operator on E. If $(f_k)_{k\in\mathbb{N}}$, is a T-conditionally independent sequence in $\mathcal{L}^2(T)$, then*

$$\mathrm{var}\left(\sum_{k=1}^{n} f_k\right) = \sum_{k=1}^{n} \mathrm{var}(f_k),$$

for each $n \in \mathbb{N}$.

Proof. As

$$\left\langle f_{i_1}, \dots, f_{i_j}, \mathcal{R}(T)\right\rangle = \left\langle f_{i_1} - Tf_{i_1}, \dots, f_{i_j} - Tf_{i_j}, \mathcal{R}(T)\right\rangle,$$

for each subset $\{i_1, \dots, i_j\}$ of $\{1, \dots, n\}$, it follows that $f_k - Tf_k$, $k = 1, 2, \dots, n$ are T-conditionally independent. From Theorem 3.2, as $f_i \in \mathcal{L}^2(T)$ it follows that $Tf_i \in \mathcal{L}^2(T)$ and consequently from Lemma 4.1 that

$$T[(f_i - Tf_i)(f_j - Tf_j)] = [T(f_i - Tf_i)]\cdot[T(f_j - Tf_j)],$$

for $i \neq j$. However, as T is a projection, see Section 2,

$$T(f_k - Tf_k) = 0, \quad \text{for each} \quad k \in \mathbb{N},$$

giving

$$T[(f_i - Tf_i)(f_j - Tf_j)] = 0, \tag{4.5}$$

for $i \neq j$. From the definition of variance

$$\operatorname{var}\left(\sum_{k=1}^{n} f_k\right) = T\left(\sum_{k=1}^{n} f_k - T\sum_{k=1}^{n} f_k\right)^2 = T\left(\sum_{k=1}^{n}(f_k - Tf_k)\right)^2,$$

which can be expanded to give

$$\operatorname{var}\left(\sum_{k=1}^{n} f_k\right) = T\sum_{k=1}^{n}(f_k - Tf_k)^2 + T\sum_{j\neq k}(f_j - Tf_j)(f_k - Tf_k) \tag{4.6}$$

Now applying (4.5) to (4.6) gives

$$\operatorname{var}\left(\sum_{k=1}^{n} f_k\right) = \sum_{k=1}^{n} T(f_k - Tf_k)^2 = \sum_{k=1}^{n} \operatorname{var}(f_k).$$ □

5. Bernoulli and Poisson processes

In classical probability, a Bernoulli process is one in which the events at any given time are independent of the events at all other times. The payoff of an event occurring is 1 unit and 0 units for it not occurring. Thus in the Riesz space setting, the process can be described by the sequence of independent band projections P_k where k indexes time and the payoff at time k is $P_k e$. The probability of an event at time k occurring must be independent of k, in the measure theoretic terms, this can be expressed as the expectation of each event is independent of time. This can be generalized to the conditional expectation of the events being time invariant, which lead to the Riesz space setting requirement that $TP_k e = f$, for all $k \in \mathbb{N}$. Here T is some fixed conditional expectation operator. Thus we are led to the following formal definition of a Bernoulli process in Riesz spaces.

Definition 5.1. Let E be a Dedekind complete Riesz space with weak order unit, e, and conditional expectation operator T with $Te = e$. Let $(P_k)_{k\in\mathbb{N}}$ be a sequence of T-conditionally independent band projections. We say that $(P_k)_{k\in\mathbb{N}}$ is a Bernoulli process if

$$TP_k e = f \quad \text{for all } k \in \mathbb{N},$$

for some fixed $f \in E$.

The payoff at time n is thus

$$S_n = \sum_{j=1}^{n} P_j e.$$

We denote by $P_{S_n=je}$ the band projection on the band where $S_n = je$, in the notation used earlier $P_{S_n=je} = (I - P_{(S_n-je)^+})(I - P_{(S_n-je)^-})$.

Theorem 5.2. *Let E be a T-universally complete Riesz space with weak order unit, $e = Te$, where T is a strictly positive conditional expectation operator on E. Let $(P_j)_{j\in\mathbb{N}}$ be T-conditionally independent band projections with $TP_je = f$ for all $j \in \mathbb{N}$ and $S_n = \sum_{j=1}^{n} P_je$. Then*

$$TS_n = nf, \tag{5.1}$$

$$TP_{S_n=je}e = \frac{n!}{j!(n-j)!} f^j(e-f)^{n-j}, \tag{5.2}$$

$$\operatorname{var}(S_n) = nf(e-f). \tag{5.3}$$

Proof. As $TP_ie = f$, (5.1) follows directly from applying T to S_n.

Fix $n \in \mathbb{N}$ and let

$$Q_j = \frac{1}{j!(n-j)!} \sum_{\sigma\in\Lambda} P_{k_{\sigma(1)}} \cdots P_{k_{\sigma(j)}} (I - P_{k_{\sigma(j+1)}}) \cdots (I - P_{k_{\sigma(n)}}). \quad j = 0, \dots, n.$$

Here Λ denotes the set of all permutations of $\{1, \dots, n\}$ and the division by $j!(n-j)!$ is as there are $j!(n-j)!$ permutations which yield the same band projection $P_{k_{\sigma(1)}} \cdots P_{k_{\sigma(j)}} (I - P_{k_{\sigma(j+1)}}) \cdots (I - P_{k_{\sigma(n)}})$. Other permutations yield band projections disjoint from the above one. Thus Q_j is a band projection, $Q_0, \dots, Q_n$ partition the identity, I, in the sense that $Q_iQ_j = 0$ for all $i \neq j$, and $\sum_{i=0}^{n} Q_i = I$. Moreover, from the definition of Q_j, it follows that $Q_jS_n = jQ_je, j = 0, \dots, n$. Thus

$$S_n = \sum_{j=0}^{n} Q_jS_n = \sum_{j=0}^{n} jQ_je.$$

The T-conditional independence of $P_1, \dots, P_n$ and Lemma 4.1 applied iteratively give that

$$\begin{aligned}
&TP_{k_{\sigma(1)}} \cdots P_{k_{\sigma(j)}} (I - P_{k_{\sigma(j+1)}}) \cdots (I - P_{k_{\sigma(n)}})e \\
&= T((P_{k_{\sigma(1)}} \cdots P_{k_{\sigma(j)}} e) \cdot ((I - P_{k_{\sigma(j+1)}}) \cdots (I - P_{k_{\sigma(n)}})e)) \\
&= (T(P_{k_{\sigma(1)}} \cdots P_{k_{\sigma(j)}} e) \cdot (T(I - P_{k_{\sigma(j+1)}}) \cdots (I - P_{k_{\sigma(n)}})e)) \\
&= \prod_{i=1}^{j} TP_{k\sigma(i)}e \cdot \prod_{i=j+1}^{n} T(I - P_{k_{\sigma(i)}})e \\
&= f^j(e-f)^{n-j}.
\end{aligned}$$

Hence

$$TP_{S_n=je}e = TQ_je = \frac{1}{j!(n-j)!}\sum_{\sigma\in\Lambda} f^j(e-f)^{n-j},$$

from which (5.2) follows as the cardinality of Λ is $n!$.

As $P_1e,\dots,P_ne$ are T-conditionally independent and are in $\mathcal{L}^2(T)$, Bienaymé's equality applied to S_n gives

$$\operatorname{var}(S_n) = \sum_{k=1}^{n} \operatorname{var}(P_ke). \tag{5.4}$$

From (4.4) applied to P_ke we have

$$\operatorname{var}(P_ke) = TP_ke - (TP_ke)^2 = f - f^2. \tag{5.5}$$

As e is the multiplicative unit, combining (5.4) and (5.5) yields (5.3). □

Theorem 5.3 (Bernoulli Law of Large Numbers). *Let E be a T-universally complete Riesz space with weak order unit, $e = Te$, where T is a strictly positive conditional expectation operator on E. Let $(P_k)_{k\in\mathbb{N}}$ be a Bernoulli process with partial sums S_n and $TP_ke = f, k \in \mathbb{N}$. For each $\epsilon > 0$,*

$$TP_{(|\frac{S_n}{n}-f|-\epsilon e)^+}e \to 0,$$

relatively uniformly as $n \to \infty$.

Proof. By the Čebyčev inequality,

$$TP_{(|S-nf|-n\epsilon e)^+}e \le \frac{1}{n^2\epsilon^2}T|S_n - nf|^2. \tag{5.6}$$

However, from (4.4) and (5.1),

$$T|S_n - nf|^2 = \operatorname{var}(S_n). \tag{5.7}$$

Combining (5.6) with (5.7) and using (5.3) to simplify the result, gives

$$TP_{(|S-nf|-n\epsilon e)^+}e \le \frac{f(e-f)}{n\epsilon^2}, \tag{5.8}$$

from which the result follows upon observing that $P_{(|\frac{S_n}{n}-f|-\epsilon e)^+} = P_{(|S-nf|-n\epsilon e)^+}$. □

One of the interesting features of Bernoulli's law of large numbers is that it gives not just the convergence of $TP_{(|\frac{S_n}{n}-f|-\epsilon e)^+}e$ to zero. It also gives some indication of the size of the band on which $|\frac{S_n}{n} - f| > \epsilon e$ by bounding the conditional expectation of the band projection applied to e by $\frac{f(e-f)}{n\epsilon^2}$, hereby indicating an upper bound for the rate of convergence 'in probability'.

Using the results on martingale difference sequences developed for the study of mixingale in [14, Lemma 4.1] we obtain a weak law of large numbers for Bernoulli processes.

Theorem 5.4 (Weak law of large numbers). *Let E be a T-universally complete Riesz space with weak order unit, $e = Te$, where T is a strictly positive conditional expectation operator on E. Let $(P_j)_{j=1,\dots,n}$ be T-conditionally independent band projections with $TP_je = f$ for all $j = 1,\dots,n$ and $S_n = \sum_{j=1}^n P_je$, then, in order,*

$$\lim_{n\to\infty} T\left|f - \frac{S_n}{n}\right| = 0.$$

Proof. Setting $f_i = P_ie$ and T_i to be the conditional expectation with range $\langle \mathcal{R}(T), P_1e, \dots, P_ie\rangle$, it follows that (g_i, T_i), where $g_i := f_i - T_{i-1}f_i$, is a martingale difference sequence. Here $|f_i| \le e$. Thus from [14, Lemma 4.1],

$$\lim_{n\to\infty} T\left|\frac{1}{n}\sum_{i=1}^n g_i\right| = 0. \tag{5.9}$$

The independence of the band projections $P_i, i \in \mathbb{N}$, gives that $T_{i-1}f_i = Tf_i = f$. Hence (5.9) can be written as

$$\lim_{n\to\infty} T\left|f - \frac{1}{n}\sum_{i=1}^n P_ie\right| = 0,$$

from which the theorem follows. □

Before progressing further we need to define an exponential map on Riesz spaces.

Remark 5.5. Let $C([-1,1])$ denote the Riesz space of continuous real functions on $[-1,1]$. Set $f_n(t) := \left(1 - \frac{t}{n}\right)^n$, then $f_n \in C([-1,1])$ and $f_n(t) \to e^{-t} = f(t)$ in order and the supremum norm on $C([-1,1])$. Thus, by [6, Theorem 3.1], for each $g \in E^e$, $f_n(g) \to f(g)$ e-uniformly (and, thus, in order) as $n \to \infty$. In addition, by the functional calculus, $f(g)$ defines an element of E^e which we will denote by e^{-g}.

We now consider the sequences of Bernoulli processes known as Poisson sequences. Here the partial sums of each Bernoulli process form a Bernoulli process.

Theorem 5.6 (Poisson). *Let E be a T-universally complete Riesz space with weak order unit, $e = Te$, where T is a strictly positive conditional expectation operator on E. Let $P_{n,k}$, $k = 1,\dots,n$, $n \in \mathbb{N}$, be T-conditionally independent band projections with $TP_{n,k}e = g_n$ for all $k = 1,\dots,n$, $n \in \mathbb{N}$. If $S_n = \sum_{k=1}^n P_{n,k}e$ are T-conditionally independent with $TS_n = g$, $n \in \mathbb{N}$, then for each $j = 0, 1, \dots,$*

$$TP_{S_n=j}e \to \frac{g^j}{j!}e^{-g},$$

e-uniformly as $n \to \infty$.

Proof. From (5.2),

$$TP_{S_n=j}e = \frac{n!}{j!(n-j)!} g_n^j (e-g_n)^{n-j},$$

but (5.1) gives $g = TS_n = ng_n$. Hence

$$TP_{S_n=j}e = \frac{n!}{j!(n-j)!} \left(\frac{g}{n}\right)^j \left(e - \frac{g}{n}\right)^{n-j},$$

which can be expanded to give

$$\left(e - \frac{g}{n}\right)^j TP_{S_n=j}e = \frac{g^j}{j!} \left(1 - \frac{1}{n}\right) \left(1 - \frac{2}{n}\right) \cdots \left(1 - \frac{j-1}{n}\right) \left(e - \frac{g}{n}\right)^n. \quad (5.10)$$

Taking the limit as $n \to \infty$ in (5.10) gives

$$TP_{S_n=j}e = \frac{g^j}{j!} \lim_{n\to\infty} \left(e - \frac{g}{n}\right)^n,$$

which together with Remark 5.5 concludes the proof. □

Acknowledgment

Wen-Chi Kuo funded by an NRF post doctoral fellowship. Bruce A. Watson funded in part by NRF grant IFR2011032400120 and the Centre for Applicable Analysis and Number Theory.

We thank the referee for his/her valuable suggestions.

References

[1] C.D. Aliprantis, K. Border, *Infinite dimensional analysis, 3rd Editiion.* Springer, 2006.

[2] K. Boulabiar, G. Buskes, A. Triki, *Results in f-algebras.* Positivity, Trends in Mathematics (2007), 73–96.

[3] G. Buskes, A. van Rooij, *Almost f-algebras: Commutativity and the Cauchy–Schwartz inequality.* Positivity **4** (2000), 227–231.

[4] R. De Marr, *A martingale convergence theorem in vector lattices.* Canadian J. Math. **18** (1966), 424–432.

[5] P.G. Dodds, C.B. Huijsmans, B. de Pagter, *Characterizations of conditional expectation-type operators.* Pacific J. Math. **141** (1990), 55–77.

[6] J.J. Grobler, *On the functional calculus in Archimedean Riesz spaces with applications to approximation theorems.* Quaestiones Math. **11** (1988), 307–321.

[7] J.J. Grobler, B. de Pagter, *Operators representable as multiplication-conditional expectation operators.* J. Operator Theory **48** (2002), 15–40.

[8] J.J. Grobler, *Continuous stochastic processes in Riesz spaces: the Doob–Meyer decomposition.* Positivity **14** (2010), 731–751.

[9] J.J. Grobler, *The Kolmogorov-Čentsov theorem and Brownian motion in vector lattices.* J. Math. Anal. Appl. **410** (2014), 891–901.

[10] J.J. Grobler, *Jensen's and martingale in equalities in Riesz spaces.* Indag. Math. N.S. **25** (2014), 275–295.

[11] C.C.A. Labuschagne, B.A. Watson, *Discrete Stochastic Integration in Riesz Spaces.* Positivity **14**, (2010), 859–575.

[12] W.-C. Kuo, C.C.A. Labuschagne, B.A. Watson, *Discrete time stochastic processes on Riesz spaces.* Indag. Math. N.S. **15** (2004), 435–451.

[13] W.-C. Kuo, C.C.A. Labuschagne, B.A. Watson, *Conditional expectations on Riesz spaces.* J. Math. Anal. Appl. **303** (2005), 509–521.

[14] W.-C. Kuo, J.J. Vardy, B.A. Watson, *Mixingales on Riesz spaces.* J. Math. Anal. Appl. **402** (2013), 731–738.

[15] M. Loève, *Probability Theory I, 4th Edition.* Springer, 1977.

[16] W.A.J. Luxemburg, B. de Pagter, *Representations of positive projections II.* Positivity **9** (2004), 569–605.

[17] H.H. Schaefer, *Banach lattices and positive operators.* Springer, 1974.

[18] G. Stoica, *Martingales in vector lattices.* Bull. Math. Soc. Sci. Math. Roumanie. (N.S.) **34(82)** (1990), 357–362.

[19] V. Troitsky, *Martingales in Banach lattices.* Positivity **9** (2005), 437–456.

[20] J.J. Vardy, B.A. Watson, *Markov processes in Riesz spaces.* Positivity **16** (2012), 373–391 and erratum on page 393.

[21] B.A. Watson, *An Andô–Douglas type theorem in Riesz spaces with a conditional expectation.* Positivity **13** (2009), 543–558.

[22] A.C. Zaanen, *Introduction to Operator Theory in Riesz Space.* Springer, 1997.

Wen-Chi Kuo
School of Computational and Applied Mathematics
University of the Witwatersrand
Private Bag 3
P.O. WITS 2050, South Africa
e-mail: `wcakuo@gmail.com`

Jessica J. Vardy and Bruce A. Watson
School of Mathematics
University of the Witwatersrand
Private Bag 3
P.O. WITS 2050, South Africa
e-mail: `jessicajvardy@gmail.com`
`b.alastair.watson@gmail.com`

Ordered Structures and Applications: Positivity VII
Trends in Mathematics, 275–279

Involutions and Complex Structures on Real Vector Lattices

Zalina Kusraeva

Abstract. The purpose of this note is to show the existence of a band preserving complex structure and nontrivial band preserving involution in a universally complete vector lattice without locally one-dimensional bands.

Mathematics Subject Classification (2010). Primary 47B60; Secondary 47A99, 46A40.

Keywords. Involution, complex structure, universally complete vector lattice, d-basis, band projection, locally one-dimensional vector lattice.

A linear operator T on a vector lattice E is called *involutory* or an *involution* if $T \circ T = I_E$ (or, equivalently, $T^{-1} = T$) and is called a *complex structure* if $T \circ T = -I_E$ (or, equivalently, $T^{-1} = -T$). The operator $P - P^{\perp}$, where P is a projection operator on E and $P^{\perp} = I_E - P$, is an involution. The involution $P - P^{\perp}$, where P is a band projection is referred to as *trivial*. The main result of this note tells us that in a real universally complete vector lattice without locally one-dimensional bands there are band preserving complex structures and nontrivial band preserving involutions.

Theorem 1. *Let E be a universally complete real vector lattice without locally one-dimensional bands* (*see Def.* 2). *Then the following assertions hold:*

(1) *For every finite collection $\{x_1, \dots, x_n\} \subset E$ there exists a nontrivial band preserving involution T on E such that $T(x_i) = x_i$ for all $i = 1, \dots, n$.*
(2) *There exists a band preserving complex structure on E.*

Recall that a linear operator on E is said to be *band preserving* if it leaves every band invariant. If E is a vector lattice with a projection property, then a linear operator in E is band preserving if and only if it commutes with all band-projections. A *universally complete* vector lattice is a vector lattice, which is Dedekind complete and laterally complete. For the theory of vector lattices and positive operators we refer to the book [4].

Supported by a grant from Russian Foundation for Basic Research, project No. 09-01-00442.

Definition 2. Let E be a universally complete vector lattice. A subset $\mathcal{E} \subset E$ is called *d-independent*, if for each band projection ρ on E the set $\{\rho e : \rho e \neq 0,\ e \in \mathcal{E}\}$ is linearly independent, that is, the collection of all non-zero members of the set $\rho\mathcal{E}$ is linearly independent. Any maximal (by inclusion) set of d-independent vectors is called a *d-basis*. A universally complete vector lattice E is called *locally one-dimensional* if $\{\mathbf{1}\}$ is a d-basis in E.

Using the Hamel basis, it can be proved that there exists an additive involution on $\mathbb{R}$, see Kuczma [6]. For the proof of Theorem 1 we carry out similar constructions making use of d-basis instead of the Hamel basis. Before launching into details we state some needed properties of d-bases from Abramovich and Kitover [1].

Lemma 3. *Let $\mathcal{E}$ be a fixed d-basis in a universally complete vector lattice E. Then for each $x \in E$ there exists a full collection $(\rho_\xi)_{\xi\in\Xi}$ of pairwise disjoint band projections (depending on x) the following representation holds:*

$$x = \sum_{\xi\in\Xi} \sum_{e\in\mathcal{E}} \alpha_{\xi,e}\rho_\xi e, \tag{1}$$

where $\alpha_{\xi,e}$ are some scalars (depending on x), such that for each ξ only a finite number of coefficients $\alpha_{\xi,e}$ may be nonzero.

Proof. See [1, p. 33] and [7, Proposition 5.1.1.(3)]. □

The expression (1) is called a *d-expansion* of x with respect to d-basis $\mathcal{E}$. A d-expansion is not unique, as we always can subdivide any projection band E_i into the direct sum of two complementary projection bands.

Theorem 4. *If E is a universally complete vector lattice, then for each non-zero band B in E there is a non-zero band $B_0 \subseteq B$ such that there exists a d-basis in B_0 consisting of weak units in B_0.*

Proof. See [1, Theorem 6.4]. □

Theorem 5. *Let $\mathcal{E}$ be a d-basis in a universally complete vector lattice consisting of weak units. Then either $\mathcal{E}$ is a singleton, or $\mathcal{E}$ is of infinite cardinality.*

Proof. See [1, Theorem 6.8]. □

Lemma 6. *Let E be a universally complete vector lattice with a full collection of pairwise disjoint bands $(E_\xi)_{\xi\in\Xi}$. If $T_\xi : E_\xi \to E_\xi$ is a band preserving linear operator for all $\xi \in \Xi$, then there exists a unique band preserving linear operator $T : E \to E$ such that $T|_{E_\xi} = T_\xi$.*

Proof. Define the operator T on E as

$$Tx := \sum_\xi T_\xi x_\xi \quad (x \in E),$$

where $x_\xi \in E_\xi$, $\pi_\xi x = \pi_\xi x_\xi$ $(\xi \in \Xi)$, and π is a band projection corresponding to E_ξ. Obviously, T is a sought operator. □

The following result was obtained by Abramovich and Kitover in [1, Theorem 14.9] and by McPolin and Wickstead in [8, Theorem 3.2].

Theorem 7. *A universally complete vector lattice is locally one-dimensional if and only if every band preserving liner operator in it is order bounded.*

Lemma 8. *Let E be Dedekind complete vector lattice. Then there is no order-bounded band preserving complex structure in E and there is no nontrivial order-bounded band preserving involution in E.*

Proof. An order-bounded band preserving operator T on a universally complete vector lattice E with weak unit $\mathbf{1}$ is a multiplication operator: $Tx = ax$ $(x \in E)$ for some $a \in E$. It follows that T is an involution if and only if $a^2 = \mathbf{1}$ and hence there is a band projection P on E with $a = P\mathbf{1} - P^{\perp}\mathbf{1}$ or $T = P - P^{\perp}$. If T is a complex structure on E then the corresponding equation $a^2 = -\mathbf{1}$ has no solution. □

Now we are ready to give the proof of Theorem 1.

Proof. **Step 1.** According to Theorem 4 there exists a full family of pairwise disjoint non-zero band (B_ξ) in E each of which has a d-basis consisting of weak units. In view of Lemma 6 there is no loss of generality in assuming that there exists a d-basis in E, consisting of weak units.

Step 2. Consider a finite set $\{x_1, \ldots, x_n\} \in E$. According to Lemma 3 each of $x_i \in E$ has its own partition of unity that guarantee the d-expansion (1) for x_i $(i = 1, \ldots, n)$. Refining these partitions of unity we can choose a common partition of unity (ρ_ξ) guaranteeing d-expansion for every $x_1, \ldots, x_n$:

$$x_i = \sum_{e \in \mathcal{E}} \alpha^{(i)}_{\xi,e} \rho_\xi e, \tag{2}$$

where $\mathcal{E}_i := \{e \in \mathcal{E} : \alpha^{(i)}_{\xi,e} \neq 0\}$ is finite set. Put $\mathcal{E}_0 = \bigcup_{i=1}^{n} \mathcal{E}_i$. Since $\mathcal{E}_i$ is finite, so is also $\mathcal{E}_0$. Hence $\mathcal{E} \setminus \mathcal{E}_0$ is infinite. There exists a decomposition:

$$\mathcal{E} \setminus \mathcal{E}_0 = \mathcal{E}_1 \cup \mathcal{E}_2,$$

with $\mathcal{E}_1 \cap \mathcal{E}_2 = \varnothing$ and $\operatorname{card} \mathcal{E}_1 = \operatorname{card} \mathcal{E}_2$. So we have $\mathcal{E} = \mathcal{E}_0 \cup \mathcal{E}_1 \cup \mathcal{E}_2$. Hence there exists a one-to-one mapping g from $\mathcal{E}_1$ onto $\mathcal{E}_2$. Thus the function g^{-1} is defined on $\mathcal{E}_2$ and maps $\mathcal{E}_2$ onto $\mathcal{E}_1$.

Step 3. Now we define an operator $T : \mathcal{E} \to \mathcal{E}$ as follows:

$$T(e) = \begin{cases} g(e), & \text{for } e \in \mathcal{E}_1, \\ g^{-1}(e), & \text{for } e \in \mathcal{E}_2, \\ e, & \text{for } e \in \mathcal{E}_0. \end{cases}$$

Next, we define $T(\pi e) = \pi T e$ for all $\pi \in \mathbb{P}(E)$ and $e \in \mathcal{E}$. In particular, $g(\pi e) = \pi g(e)$ $(e \in \mathcal{E}_1)$ and $g^{-1}(\pi e) = \pi g^{-1}(e)$ $(e \in \mathcal{E}_2)$ by definition. Finally, if $x \in E$ has two distinct d-expansions (1), we put

$$Tx = \sum_{\xi \in \Xi} \Bigg(\sum_{e \in \mathcal{E}_0} \alpha_{\xi,e} \rho_\xi e + \sum_{e \in \mathcal{E}_1} \alpha_{\xi,e} \rho_\xi g(e) + \sum_{e \in \mathcal{E}_2} \alpha_{\xi,e} \rho_\xi g^{-1}(e) \Bigg). \tag{3}$$

The definition of T is found. Indeed, if x has two distinct d-expansions of the form (1) with two different partitions of unity (π_η) and (ρ_ξ), the two values of Tx defined by (3) using these partitions of unity, coincide with the value of Tx defined by (3) using the common refinement $(\pi_\eta\rho_\xi)$ of given partitions of unity.

To ensure that so-constructed operator T is band preserving we have to show that T commutes with every band projection:

$$\pi Tx = \sum_{\xi\in\Xi}\Big(\sum_{e\in\mathcal{E}_0}\alpha_{\xi,e}\rho_\xi\pi e + \sum_{e\in\mathcal{E}_1}\alpha_{\xi,e}\rho_\xi g(\pi e) + \sum_{e\in\mathcal{E}_2}\alpha_{\xi,e}\rho_\xi g^{-1}(\pi e)\Big) = T(\pi x).$$

In a view of (2) $x_i = \sum_{e\in\mathcal{E}_0}\alpha^{(i)}_{\xi,e}\rho_\xi e$ and therefore $Tx_i = \sum_{e\in\mathcal{E}}\alpha^{(i)}_{\xi,e}\rho_\xi Te = x_i$, since $Te = e$ for $e\in\mathcal{E}_0$. Check that T is involutory. Taking into account that $Tg(e) = e$ if $e\in\mathcal{E}_1$ and $Tg^{-1}(e) = e$ if $e\in\mathcal{E}_2$, we deduce:

$$\begin{aligned}\rho_\xi T^2x &= T\Big(\sum_{e\in\mathcal{E}_0}\alpha_{\xi,e}\rho_\xi e + \sum_{e\in\mathcal{E}_1}\alpha_{\xi,e}\rho_\xi g(e) + \sum_{e\in\mathcal{E}_2}\alpha_{\xi,e}\rho_\xi g^{-1}(e)\Big)\\ &= \sum_{e\in\mathcal{E}_0}\alpha_{\xi,e}\rho_\xi Te + \sum_{e\in\mathcal{E}_1}\alpha_{\xi,e}\rho_\xi Tg(e) + \sum_{e\in\mathcal{E}_2}\alpha_{\xi,e}\rho_\xi Tg^{-1}(e) = \rho_\xi x.\end{aligned}$$

By Lemma 6 we have $T^2x = x$ for all $x\in X$.

Step 4. The proof of the second part is similar. One have to repeat Step 1 and Step 3. The only difference is that $\mathcal{E}_0 = \varnothing$ and the operator T is defined by

$$T(e) = \begin{cases} -g(e), & \text{for } e\in\mathcal{E}_1,\\ g^{-1}(e), & \text{for } e\in\mathcal{E}_2.\end{cases}$$

Next, we put $T(\pi e) = \pi Te$ for all $\pi\in\mathbb{P}(E)$ and $e\in\mathcal{E}$. Given a d-expansion (1) of $x\in E$ we define Tx by (3) without the first sum in parentheses. By the same reasons T is a band preserving operator. Finally, $T^2 = -I_E$, since

$$\begin{aligned}\rho_\xi T^2x &= T\Big(-\sum_{e\in\mathcal{E}_1}\alpha_{\xi,e}\rho_\xi g(e) + \sum_{e\in\mathcal{E}_2}\alpha_{\xi,e}\rho_\xi g^{-1}(e)\Big)\\ &= -\sum_{e\in\mathcal{E}_1}\alpha_{\xi,e}\rho_\xi Tg(e) + \sum_{e\in\mathcal{E}_2}\alpha_{\xi,e}\rho_\xi Tg^{-1}(e) = -\rho_\xi x\end{aligned}$$

holds for all ξ. The proof is complete. □

Now we will state two immediate corollaries from Theorem 1. The first one is related to the *Wickstead problem* raised in [10]: Is any band preserving linear operator in a vector lattice automatically order bounded? An overview of the main ideas and results on the Wickstead problem and its variations, focusing primarily on the case of band preserving operators in a universally complete vector lattice see in [5].

Corollary 9. *Let E be a universally complete vector lattice. Then the following are equivalent:*

(1) *E contains a locally one-dimensional band.*
(2) *There is no nontrivial band preserving involution on E.*

(3) *There is no band preserving complex structure on E.*

Proof. The sufficiency of (1) is immediate from Theorem 1, while the necessity follows from Theorem 7 and Lemma 8. □

Corollary 10. *Let E be a universally complete real vector lattice without one-dimensional bands. Then E admits a structure of complex vector space with a band preserving complex multiplication.*

Proof. A complex structure T on E allows to define on E a structure of vector space over the field of complex numbers $\mathbb{C}$, by setting $(\alpha + i\beta)x = \alpha x + \beta T(x)$ for all $z = \alpha + i\beta \in \mathbb{C}$ and $x \in E$. If T is band preserving then the map $x \mapsto zx$ $(x \in E)$ is evidently band preserving for all $z \in \mathbb{C}$. □

Acknowledgment

My sincere acknowledgments to the referee for his insightful remarks.

References

[1] Abramovich Yu.A., Kitover A.K., *Inverses of disjointness preserving operators.* Memoirs of the American Mathematical Society, **143**(679) (2000).

[2] Abramovich Yu.A., Veksler A.I., Koldunov A.V., *Operators preserving disjointness.* Dokl. Academ. Nauk USSR **248**, 1033–1036, (1979).

[3] Abramovich Yu.A., Veksler A.I., and Koldunov A.V., *On disjointness preserving operators.* Dokl. Akad. Navuk SSSR, **289**(5), 1033–1036, (1979).

[4] Aliprantis C.D., Burkinshaw O., *Positive Operators.* New York: Academic Press, (1985).

[5] Gutman A.E., Kusraev A.G., Kutateladze S.S., *The Wickstead Problem.* Siberian Electronic Math. Reports, **5**, 293–333 (2008).

[6] Kuczma M., *An introduction to the theory of functional equations and inequalities.* Basel-Boston-Berlin: Birkhäuser (2009).

[7] Kusraev A.G., *Dominated operators*, Moscow. Nauka (2003).

[8] McPolin P.T.N. and Wickstead A.W., *The order boundedness of band preserving operators on uniformly complete vector lattices.* Math. Proc. Cambridge Philos. Soc., **97**(3), 481–487 (1985).

[9] Vulikh B.Z., *Introduction of the theory of partially ordered spaces*, Wolters-Noordhoff Sci. Publication, Groningen (1967).

[10] Wickstead A.W., *Representation and duality of multiplication operators on Archimedean Riesz spaces.* Compositio Math., **35**(3), 225–238 (1977).

Zalina Kusraeva
Southern Mathematical Institute of the
Vladikavkaz Scientific Center of the RAS
362027, Vladikavkaz, Russia
e-mail: `zali13@mail.ru`

Ordered Structures and Applications: Positivity VII
Trends in Mathematics, 281–306

Banach Space-valued Extensions of Linear Operators on L^∞

Nick Lindemulder

Abstract. Let E and G be two Banach function spaces, let $T \in \mathcal{L}(E, Y)$, and let $\langle X, Y \rangle$ be a Banach dual pair. In this paper we give conditions for which there exists a (necessarily unique) bounded linear operator $T_Y \in \mathcal{L}(E(Y), G(Y))$ with the property that

$$\langle x, T_Y e \rangle = T\langle x, e \rangle, \qquad e \in E(Y), x \in X.$$

The first main result states that, in case $\langle X, Y \rangle = \langle Y^*, Y \rangle$ with Y a reflexive Banach space, for the existence of T_Y it sufficient that T is dominated by a positive operator. We furthermore show that for Y within a wide class of Banach spaces (including the Banach lattices) the validity of this extension result for $E = \ell^\infty$ and $G = \mathbb{K}$ even characterizes the reflexivity of Y.

The second main result concerns the case that T is an adjoint operator on $L^\infty(A)$: we assume that $E = L^\infty(A)$ for a semi-finite measure space $(A, \mathscr{A}, \mu)$, that $\langle F, G \rangle$ is a Köthe dual pair, and that T is $\sigma(L^\infty(A), L^1(A))$-to-$\sigma(G, F)$ continuous. In this situation we show that T_Y also exists provided that T is dominated by a positive operator. As an application of this result we consider conditional expectation on Banach space-valued L^∞-spaces.

Mathematics Subject Classification (2010). Primary 46E40; Secondary 46E30, 46B10.

Keywords. Adjoint operator, Banach function space, Banach limit, conditional expectation, domination, dual pair, L^∞, positive operator, vector-valued extension, reflexivity, Schauder basis.

1. Introduction

Given two measure spaces $(A, \mathscr{A}, \mu)$ and $(B, \mathscr{B}, \nu)$, $p, q \in [1, \infty]$, a bounded linear operator $T \in \mathcal{L}(L^p(A), L^q(B))$, and a Banach space Y, one can ask the question whether T has a Y-valued extension $T_Y \in \mathcal{L}(L^p(A; Y), L^q(B; Y))$ in the sense that there exists a (necessarily unique) bounded linear operator $T_Y \in \mathcal{L}(L^p(A; Y), L^q(B; Y))$ satisfying

$$\langle T_Y f, y^* \rangle = T\langle f, y^* \rangle, \qquad f \in L^p(A; Y), y^* \in Y^*. \tag{1.1}$$

Note that T_Y (if it exists) extends the tensor extension $T \otimes I_Y$ of T, which is the linear operator from the algebraic tensor product $L^p(A) \otimes Y$ to the algebraic tensor product $L^q(B) \otimes Y$ determined by the formula

$$(T \otimes I_Y)(f \otimes y) = Tf \otimes y, \qquad f \in L^p(A), y \in Y.$$

For $p \in [1, \infty[$ it holds that $L^p(A) \otimes Y$ is dense in $L^p(A;Y)$, so that T_Y is just the unique extension of $T \otimes I_Y$ to a bounded linear operator from $L^p(A;Y)$ to $L^q(B;Y)$. It is well known that, in this case, the extension T_Y exists if T is dominated by a positive operator (i.e., there exists a positive operator $S \in \mathcal{L}(L^p(A), L^q(B))$ such that $|Tf| \leq S|f|$ for all $f \in L^p(A)$) or Y is (isomorphic to) a Hilbert space; this can, for instance, be found in [16, Subsection 4.5.c] (also see [15]). Another extension result says that, if $p = q \in [1, \infty[$, $A = B$, and Y is isomorphic to a closed linear subspace of a quotient of a space $L^p(C)$, then the extension T_Y exists for every $T \in \mathcal{L}(L^p(A))$; see [19]. There also exist examples in which T_Y does not exist. In fact, for some operators T the existence of the Y-valued extension T_Y characterizes Y as being isomorphic to a Hilbert space or characterizes different geometric properties of the Banach space Y; for example, the fact that the Fourier–Plancherel transform $\mathscr{F}$ on $L^2(\mathbb{R}^d)$ has a Y-valued extension $\mathscr{F}_Y$ on $L^2(\mathbb{R}^d; Y)$ if and only if Y is isomorphic to a Hilbert space is due to Kwapień [23], and the characterization of the UMD Banach spaces as those Banach spaces for which the Hilbert transform (on $L^p(\mathbb{R}^d)$) has an extension to a bounded linear operator on $L^p(\mathbb{R}^d; Y)$ for some/all $p \in]1, \infty[$ is due to Burkholder [8] (sufficiency of UMD) and Bourgain [5] (necessity of UMD) (see also the survey paper [9]). For Banach space-valued extension results for singular integral operators (in the UMD setting) we refer to [20] (and the references therein).

It seems that the extension problem (1.1) has not been considered in the literature for $p = \infty$. In this paper we will obtain analogues for $p = \infty$ of the just mentioned results for $p < \infty$ about Banach space-valued extensions of operators dominated by a positive operator and Hilbert space-valued extensions of arbitrary bounded linear operators; we will in fact consider the extension problem in more general settings then discussed in this introduction. In the Banach space setting we will mainly consider the extension problem in two directions.

The first direction is concerned with Y-valued extensions T for Y a reflexive Banach space, with as main result in this direction (Theorem 3.6) the existence of T_Y plus a norm estimate in case that T is dominated by a positive operator. Via a result of Zippin [32], which says that every separable reflexive Banach space embeds into a reflexive Banach space with a Schauder basis, we can reduce the situation to the case that Y is a reflexive Banach space with a Schauder basis. This basis can then be used to define T_Y. We show that for the special case $A = \mathbb{N}$, $B = \{0\}$, so that $L^\infty(A) = \ell^\infty$ and $L^q(B) = \mathbb{K}$ (the scalar field), and $Y \in \{c_0, \ell_1\}$, the extension T_Y fails to exist when $T \in \mathcal{L}(\ell^\infty, \mathbb{K}) = (\ell^\infty)^*$ is a Banach limit (so in particular $T \geq 0$). As a consequence of a generalization of a classical result due to Lozanovski on the reflexivity on Banach lattices we find that, given a Banach limit $T \in \mathcal{L}(\ell^\infty, \mathbb{K})$, for Y within a large class of Banach spaces (including the Banach lattices), Y is reflexive if and only if the Y-valued extension $T_Y \in \mathcal{L}(\ell^\infty(Y), Y)$ of T exists (Corollary 3.12).

In the second direction we consider arbitrary Y under the additional assumption that T is an adjoint operator. To be more precise, suppose that $(A, \mathscr{A}, \mu)$ and $(B, \mathscr{B}, \nu)$ are both σ-finite and that $q \in]1, \infty]$, so that we have canonical isometric isomorphisms $L^\infty(A) \cong (L^1(A))^*$ and $L^q(B) \cong (L^{q'}(B))^*$ (with $\frac{1}{q} + \frac{1}{q'} = 1$). Let $T = S^* \in \mathcal{L}(L^p(A), L^q(B))$ be the adjoint of $S \in \mathcal{L}(L^{q'}(B), L^1(A))$ and let Y be an arbitrary Banach space. As the main result (Theorem 3.13) in this direction we will show, in case that T is dominated by a positive operator, the existence of both T_Y and S_{Y^*} together with norm estimates plus the adjoint relation

$$\int_B \langle T_Y f, g \rangle \, d\nu = \int_A \langle f, S_{Y^*} g \rangle \, d\mu, \qquad f \in L^\infty(A; Y), g \in L^{q'}(B; Y^*).$$

The idea is to first obtain S_{Y^*} by bounded extension of $S \otimes I_{Y^*}$ and then show that the Banach space adjoint $(S_{Y^*})^*$ of this extension restricts to an operator $L^\infty(A; Y) \subset (L^1(A; Y^*))^* \longrightarrow L^q(B; Y) \subset (L^{q'}(B; Y^*))$, which is the desired extension T_Y. An example and motivation for this extension problem is the conditional expectation operator on Banach space-valued L^∞-spaces.

The paper is organized as follows. In Section 2 we will first treat some necessary preliminaries. In Section 3 we present the results of this paper, with a formulation of the general extension problem and some basics, in the second subsection the extension problem for reflexive Y, in the third subsection the extension problem of adjoint operators on L^∞ for general Banach dual pairs, and in the fourth (and last) subsection the extension problems in the Hilbert space setting. Next, the proof of Theorem 3.13 is given in Section 4. Finally, as an application and motivation, we consider the conditional expectation operator on Banach-valued L^∞-spaces in Section 5.

Conventions and notations. Throughout this paper we fix a field $\mathbb{K} \in \{\mathbb{R}, \mathbb{C}\}$ and assume that all spaces are over this field $\mathbb{K}$. For a normed space X we denote by B_X its closed unit ball. We furthermore write $\ell^p = \ell^p(\mathbb{N})$, $1 \le p \le \infty$. For two Banach lattices E and F we denote by $M(E, F)$ the set of all linear operators from E to F which are dominated by a positive operator.

2. Preliminaries

2.1. Banach dual pairs

For the general theory of dual systems we refer to [27].

A *Banach duality (pairing)* between two Banach spaces X and Y is a bounded bilinear form $\langle \cdot, \cdot \rangle : X \times Y \longrightarrow \mathbb{K}$ for which the induced linear maps $x \mapsto \langle x, \cdot \rangle, X \longrightarrow Y^*$ and $y \mapsto \langle \cdot, y \rangle, Y \longrightarrow X^*$ are injections. A *Banach dual pair* is a triple $(X, Y, \langle \cdot, \cdot \rangle)$ consisting of two Banach spaces X and Y together with Banach duality $\langle \cdot, \cdot \rangle$ between them. We write $\langle X, Y \rangle = (X, Y, \langle \cdot, \cdot \rangle)$. We call a

Banach dual pair $\langle X, Y\rangle$ *norming* if

$$\|x\| = \sup_{y\in B_Y}\langle x, y\rangle \quad \text{and} \quad \|y\| = \sup_{x\in B_X}\langle x, y\rangle.$$

Note that in this case X can be viewed as a closed subspace of Y^* and Y can be viewed as a closed subspace of X^*.

Let $\langle X, Y\rangle$ be a Banach dual pair. Then the locally convex Hausdorff topology on X generated by the family of seminorms $\{\,|\langle\,\cdot\,, y\rangle|\,\}_{y\in Y}$ is called the *weak topology* on X generated by the pairing $\langle X, Y\rangle$ and is denoted by $\sigma(X, Y)$. The *weak topology* on Y generated by $\langle X, Y\rangle$ is defined similarly and is denoted by $\sigma(Y, X)$. The topological dual of $(X, \sigma(X, Y))$ and $(Y, \sigma(Y, X))$ are Y and X, respectively; that is, $(X, \sigma(X, Y))' = \{\langle\,\cdot\,, y\rangle \mid y \in Y\}$ and $(Y, \sigma(Y, X))' = \{\langle x, \cdot\,\rangle \mid x \in X\}$. We shall always make the identifications $(X, \sigma(X, Y))' = Y$ and $(Y, \sigma(Y, X))' = X$.

A linear subspace Z of Y is $\sigma(Y, X)$ dense in Y if and only if Z separates the points of X, i.e., for every nonzero $x \in X$ there exists a $z \in Z$ with $\langle x, z\rangle \neq 0$.

Recall that $\sigma(X, X^*)$ is called the *weak topology* on X and that $\sigma(X^*, X)$ is called the *weak** topology on X^*.

Suppose that we are given two Banach dual pairs $\langle X_1, Y_1\rangle$ and $\langle X_2, Y_2\rangle$ and a linear operator S from X_1 to X_2. Viewing Y_i as vector subspace of the algebraic dual $X_i^\#$ of X_i $(i = 1, 2)$, S is continuous as an operator $S : (X_1, \sigma(X_1, Y_1)) \longrightarrow (X_2, \sigma(X_2, Y_2))$ if and only if its algebraic adjoint $S^\#$ maps Y_2 into Y_1; we say that S is $\sigma(X_1, Y_1)$-to-$\sigma(X_2, Y_2)$ continuous. In this situation, the restriction $S' : Y_2 \longrightarrow Y_1$ of $S^\#$ is called the *adjoint* of S with respect to the dualities $\langle X_1, Y_1\rangle$ and $\langle X_2, Y_2\rangle$, and it is a $\sigma(Y_2, X_2)$-to-$\sigma(Y_1, X_1)$ continuous linear operator whose adjoint is $S'' = (S')' = S$. The operators S and S' are automatically bounded operators as a consequence of the closed graph theorem. Finally, note that if we a priori know S to be bounded and view Y_i as vector subspace of the norm dual X_i^* of X_i $(i = 1, 2)$, then S is $\sigma(X_1, Y_1)$-to-$\sigma(X_2, Y_2)$ continuous if and only if its Banach space adjoint S^* maps Y_2 into Y_1.

2.2. Duality and Schauder bases for Banach spaces

In Subsection 3.1 we will use Schauder bases in order to define Banach space-valued extensions of linear operators; for the basis of the theory of Schauder bases we refer to [2]. The following well-known facts will be important for us in this direction.

Fact 2.1.

(I) *Let X be a Banach space. If X has Schauder basis $\{b_n\}_{n\in\mathbb{N}}$, then X is reflexive if and only if $\{b_n\}_{n\in\mathbb{N}}$ is both boundedly complete and shrinking.*

(II) *Every separable reflexive Banach space is isomorphic to a closed linear subspace of a reflexive Banach space with a Schauder basis.*

(III) *Let X be a closed linear subspace of a Banach lattice E. If X is complemented in E or E has an order-continuous norm, then the following statements are equivalent:*

(a) *X is reflexive.*

(b) *X does not have linear subspaces isomorphic to c_0 or ℓ^1.*

A reference for (I) is [2, Theorem 3.2.13]. (II) is due to Zippin [32]. (III) is a generalization due to Tzafriri and Meyer-Nieberg of a result of Lozanovski about the reflexivity of Banach lattices; see [30] (and the references therein). For a version of this reflexivity result for finitely generated Banach $C(K)$-modules we refer to [21].

2.3. Riesz spaces and Banach lattices

For to the theory of Riesz spaces and Banach lattices we refer to the books [3], [26]. Let us recall the following notation, definitions and facts.

Given a measure space $(A, \mathscr{A}, \mu)$, we denote by $L^0(A) = L^0(A, \mathscr{A}, \mu; \mathbb{K})$ the $\mathbb{K}$-Riesz space of all μ-a.e. equivalence classes of $\mathbb{K}$-valued $\mathscr{A}$-measurable functions on A with its natural lattice operations.

We say that a linear operator $T : E \longrightarrow F$ between two Banach lattices is *dominated* by a positive operator $S \in \mathcal{L}(E, F)$ if it holds that $|Te| \leq S|e|$ for all $e \in E$; we also say that S is a dominant for T and we write $T \lesssim S$. We denote by $\mathrm{maj}(T)$ the set of all dominants of T; then $\mathrm{maj}(T) \subset \mathcal{L}_b(E, F)^+$. If there is a least element in $\mathrm{maj}(T)$ with respect to the ordering of $\mathcal{L}_b(E, F)$ then it is called the *least dominant* of T and is denoted by $|T|$. We denote by $M(E, F)$ the space of all linear operators $T : E \longrightarrow F$ for which $\mathrm{maj}(T) \neq \emptyset$. Then $M(E, F) \subset \mathcal{L}(E, F)$. For $T \in M(E, F)$ we define

$$\|T\|_{M(E,F)} := \inf\{ \|S\| \ : \ S \in \mathrm{maj}(T)\}.$$

Then $\|T\| \leq \|T\|_{M(E,F)}$ for all $T \in M(X, Y)$ and $\|T\|_{M(E,F)} = \| \, |T| \, \|$ whenever $|T|$ exists; in particular $\|T\| = \|T\|_{M(E,F)}$ when $T \geq 0$.

A linear operator $T : E \longrightarrow F$ between two Banach lattices is called *regular* if it is a linear combination of positive operators. We denote by $\mathcal{L}_r(E, F)$ the space of all such operators. Then we have $\mathcal{L}_r(E, F) \subset M(E, F) \subset \mathcal{L}(E, F)$ and we write $\|T\|_r := \|T\|_{M(E,F)}$ for $T \in \mathcal{L}_r(E, F)$. If F is Dedekind complete, then we have $\mathcal{L}_r(E, F) = M(E, F)$.

A Banach lattice E is called a *KB-space* (*Kantorovich–Banach space*) if every increasing norm bounded sequence of E^+ is norm convergent. It is not difficult to see that a Banach lattice E is a KB-space if and only if every increasing norm bounded net of E^+ is norm convergent. Every reflexive Banach lattice is an example of a KB-space. Another example is the Lebesgue space $L^1(A)$.

A Banach lattice E is said to have a *Levi norm* if every increasing norm bounded net of E^+ has a supremum in E. When this property only holds for sequences, then we say that E has a *sequentially Levi norm.* KB-spaces are examples of Banach lattices having a Levi norm. An other example is $L^\infty(A)$ on a Maharam measure space $(A, \mathscr{A}, \mu)$; see the next subsection for the notion of Maharam measure space. Note that a Banach lattice with a Levi norm must be Dedekind complete.

A Banach lattice E is said to have a Fatou norm if $\sup_\alpha \|x_\alpha\| = \|x\|$ whenever $\{x_\alpha\}_\alpha \subset E$ is an increasing net with supremum x.

2.4. Measure theory

General measure theory. For the content of this paragraph we refer to [13].

A measure space $(A, \mathscr{A}, \mu)$ is called

- *semi-finite* if for every $B \in \mathscr{A}$ with $\mu(B) > 0$ there exists a $C \subset B, C \in \mathscr{A}$ with $0 < \mu(C) < \infty$;
- *decomposable* (or *strictly localizable*)[1] if there exists a family $\{A_i\}_{i \in I}$ of pairwise disjoint sets in $\mathscr{A}$ such that $\mu(A_i) \in]0, \infty[$ for all $i \in I$, and for each $B \in \mathscr{A}$ of finite measure there exists countable subset $I_0 \subset I$ of indices and a μ-null set $N \in \mathscr{A}$ such that $A = \bigcup_{i \in I_0}(B \cap A_i) \cup N$;
- *Maharam* (or *localizable*) if it is semi-finite and if for every $\mathscr{E} \subset \mathscr{A}$ there is a $H \in \mathscr{A}$ such that (i) $E \setminus H$ is negligible for every $E \in \mathscr{E}$ and (ii) if $G \in \mathscr{A}$ and $E \setminus G$ is negligible for every $E \in \mathscr{E}$, then $H \setminus G$ is negligible.

Regarding the relation between the different types of measure spaces, the following chain of implications holds true [13, Theorem 211L]:

$$\sigma\text{-finite} \implies \text{decomposable} \implies \text{Maharam} \implies \text{semi-finite}. \tag{2.1}$$

A more elegant equivalent definition of Maharam measure space is via the measure algebra of $(A, \mathscr{A}, \mu)$, which is obtained from $\mathscr{A}$ by identifying sets which are μ-a.e. equal: a measure space is Maharam if and only if its measure algebra is Maharam, i.e., is a semi-finite measure algebra which is Dedekind complete as a Boolean algebra (see [12] and [14]).

The canonical linear map $g \mapsto \Lambda_g, L^\infty(A) \longrightarrow (L^1(A))^*$ is an injection if and only if $(A, \mathscr{A}, \mu)$ is semi-finite, in which case it is an isometry, and this map is a bijection if and only if $(A, \mathscr{A}, \mu)$ is Maharam, in which case it is an isometric isomorphism; see [13, Theorem 243G]. The sufficiency of Maharamness in the latter statement is in fact a special case of Fact 2.2. Another important characterization of the Maharam measure spaces among the semi-finite measure spaces is [13, Theorem 241.G.(b)]: a semi-finite measure space $(A, \mathscr{A}, \mu)$ is Maharam if and only if $L^0(A)$ is Dedekind complete.

Banach space-valued measurability. Let $(A, \mathscr{A}, \mu)$ be a measure space and let X be a Banach space.

We denote by

$$\mathrm{St}(A; X) := \left\{ \sum_{j=1}^{n} 1_{A_j} \otimes x_j \; : \; A_j \in \mathscr{A} \text{ disjoint}, x_j \in X \right\}$$

the vector space of X-valued step functions; here we use the usual notational convention to view, given a function $f : A \longrightarrow \mathbb{K}$, $f \otimes x$ as the function $a \mapsto f(a)x$, $A \longrightarrow X$. A function $f : A \longrightarrow X$ is called strongly measurable if it is the pointwise limit of a sequence $(f_k)_{k \in \mathbb{N}} \subset \mathrm{St}(A; X)$; it can be shown that the sequence $(f_k)_k$ can be chosen such that $\|f_k\|_X \leq \|f\|_X$. The well-known Pettis measurability theorem says that a function $f : A \longrightarrow X$ is strongly measurable

[1] Such measure spaces are also said to satisfy the *direct sum property*.

if and only if f is separably valued and $\langle f, x^* \rangle$ is measurable for all x^* in some weak* dense subspace Z of X^*; consequently, if $f : A \longrightarrow X$ is strongly measurable and takes its values in a closed linear subspace Y of X, then f is also strongly measurable as a function $A \longrightarrow Y$. We denote by $L^0(A; X)$ the vector space of all μ-a.e. equivalence classes of strongly measurable functions $f : A \longrightarrow X$. We also view $L^0(A; X)$ as the vector space of all μ-a.e. equivalence classes of functions $g : A \longrightarrow X$ which are μ-a.e. equivalent to a strongly measurable function on $f : A \longrightarrow X$.

2.5. Banach function spaces

For the theory of Banach function spaces we refer to [31], [26] (σ-finite measure spaces) and [12], [14] (general measure spaces and, in particular, Maharam measure spaces).

A *Banach function space* on $(A, \mathscr{A}, \mu)$ is an ideal E of $L^0(A)$ which is equipped with a Banach lattice norm. Note that each Banach function space is σ-Dedekind complete, being an ideal in the σ-Dedekind complete $L^0(A)$. Examples of Banach function spaces are the L^p-spaces ($p \in [1, \infty]$), Orlicz spaces, Lorentz spaces, and Marcienkiewicz spaces.

A *Köthe dual pair (of Banach function spaces)* on $(A, \mathscr{A}, \mu)$ is a Banach dual pair $\langle E, F \rangle$ consisting of two Banach functions spaces E and F on $(A, \mathscr{A}, \mu)$ with $E \cdot F \subset L^1(A)$ for which the pairing $\langle \cdot\, , \cdot \rangle$ is given by

$$\langle e, f \rangle = \int_A fg \, d\mu, \qquad e \in E, f \in F.$$

Observe that the induced linear maps $e \mapsto \langle e, \cdot \rangle, E \longrightarrow F^*$ and $f \mapsto \langle \cdot , f \rangle, F \longrightarrow E^*$ are lattice isomorphisms onto their images. Examples of Köthe dual pairs are $\langle L^1(A), L^\infty(A) \rangle$ for $(A, \mathscr{A}, \mu)$ semi-finite or $\langle L^p(A), L^{p'}(A) \rangle$ for $p, p' \in]1, \infty[, \frac{1}{p} + \frac{1}{p'} = 1$; these two examples are even norming.

The *Köthe dual* of a Banach function space E on $(A, \mathscr{A}, \mu)$ is the ideal $E^\times$ of $L^0(A)$ defined by

$$E^\times := \{ f \in L^0(A) \, : \, fe \in L^1(A) \, \forall e \in E \},$$

and is equipped with the seminorm

$$\|f\|_{E^\times} := \sup \left\{ \left| \int_A fe \, d\mu \right| \, : \, e \in E, \|e\| \leq 1 \right\}.$$

Suppose that $(A, \mathscr{A}, \mu)$ is Maharam. Then every Banach function space E on $(A, \mathscr{A}, \mu)$ is Dedekind complete, being an ideal in the Dedekind complete $L^0(A)$, and has a well-defined *support* or *carrier* supp(E) in A, which is the smallest set supp(E) (with respect to μ-a.e. inclusion) such that every $e \in E$ vanishes μ-a.e. on $A \setminus \text{supp}(E)$. It holds that supp$(E) = A$ if and only if E is order dense in $L^0(A)$ if and only if for every $B \in \mathscr{A}$ there exists a $C \in \mathscr{A}$ such that $C \subset A$, $\mu(C) > 0$, and $1_C \in E$. In situation we have the following important duality result:

Fact 2.2. *Suppose that E is a Banach function space on the Maharam measure space $(A, \mathscr{A}, \mu)$ having full carrier (i.e.,* supp$(E) = A$*). Then $E^\times$ is a Banach* *function space on* $(A, \mathscr{A}, \mu)$ *with* $\mathrm{supp}(E^\times) = A$ *and* $\langle E, E^\times \rangle$ *is a Köthe dual pair on* $(A, \mathscr{A}, \mu)$ *for which the image of* $f \mapsto \langle \cdot, f \rangle, E^\times \longrightarrow E^*$ *is the band of order-continuous functionals in* E^*. *In particular,* $f \mapsto \langle \cdot, f \rangle, E^\times \longrightarrow E^*$ *is an isometric lattice isomorphism if and only if* E *has an order-continuous norm.*

Note that $E = L^\infty(A)$ does in general not have an order-continuous norm, in which case the norm dual $(L^\infty(A))^*$ has functionals which are not order continuous, or equivalently, functionals which do not belong to the Köthe dual $(L^\infty(A))^\times = L^1(A)$. In the special case of the counting measure space $(A, \mathscr{A}, \mu) = (\mathbb{N}, \mathcal{P}(\mathbb{N}), \#)$, so that $E = \ell^\infty$, examples of linear functionals belonging to $(\ell^\infty)^* \setminus \ell^1$ are the so-called Banach limits, whose existence can be established using Hahn–Banach (see [10, Section III.7]).

Definition 2.3. A bounded linear functional $\Lambda \in (\ell^\infty)^*$ is called a *Banach limit* if it has the following properties:

(a) If $\{x_n\}_{n\in\mathbb{N}} \in \ell^\infty$ is a convergent sequence with limit x (as $n \to \infty$), then $\Lambda(\{x_n\}_{n\in\mathbb{N}}) = x$.
(b) Λ is positive.
(c) $\Lambda(\{x_n\}_{n\in\mathbb{N}}) = \Lambda(\{x_{n+1}\}_{n\in\mathbb{N}})$ for all $\{x_n\}_{n\in\mathbb{N}} \in \ell^\infty$.

We will use Banach limits as a counterexample to the extension problem in Subsection 3.2; see Example 3.10.

2.6. Köthe–Bochner spaces

Given a Banach function space E on a measure space $(A, \mathscr{A}, \mu)$, we define the vector space

$$E(X) := \{ f \in L^0(A; X) \,:\, \|f\|_X \in E \}.$$

Endowed with the norm $\|f\| := \| \|f\|_X \|_E$, $E(X)$ becomes a Banach space which is called the *Köthe–Bochner space* associated with E and X. We denote by $E\tilde{\otimes}X$ the closure of $E \otimes X$ in $E(X)$; recall that we use the usual convention to view $e \otimes x$ as the function $a \mapsto e(a)x$. We have $E(X) = E\tilde{\otimes}X$ provided that E has an order-continuous norm; in fact, it is not difficult to show that the linear subspace

$$\begin{aligned} \mathrm{St}_E(A; X) &:= \left\{ \sum_{j=1}^n 1_{A_j} \otimes x_j \,:\, A_j \in \mathscr{A} \text{ disjoint}\,, 1_{A_j} \in E, x_j \in X \right\} \\ &= \mathrm{St}(A; X) \cap E(X) \end{aligned}$$

of step functions which are in $E(X)$ is already dense in $E(X)$ provided that E has an order-continuous norm (see [18]). We would like to mention that there are several cross-norms on $E \otimes X$ which coincide with the restricted norm coming from $E(X)$ (see [18] and the references therein).

Observe that for $E = L^p(A)$ $(p \in [1, \infty])$ we get the usual Lebesgue–Bochner space $E(X) = L^p(A; X)$.

If $\langle X,Y\rangle$ is a Banach dual pair and $\langle E,F\rangle$ a Köthe dual pair on $(A,\mathscr{A},\mu)$, then $\langle E(X),F(Y)\rangle$, $\langle E\tilde{\otimes}X,F(Y)\rangle$, $\langle E(X),F\tilde{\otimes}Y\rangle$ and $\langle E\tilde{\otimes}X,F\tilde{\otimes}Y\rangle$ are Banach dual pairs under the pairing

$$\langle e,f\rangle=\langle e,f\rangle_{\langle E(X),F(X)\rangle}:=\int_A\langle e(a),f(a)\rangle_{\langle X,Y\rangle}\,d\mu(a);$$

in fact $E\otimes X$ and $F\otimes Y$ are already separating for $F(Y)$ and $E(X)$, respectively.

Suppose that $\langle X,Y\rangle$ is norming. If $\langle E,F\rangle=\langle L^1(A),L^\infty(A)\rangle$ with $(A,\mathscr{A},\mu)$ semi-finite or $\langle E,F\rangle=\langle L^p(A),L^{p'}(A)\rangle$ with $1<p,p'<\infty,\frac{1}{p}+\frac{1}{p'}=1$, then the Banach dual pair $\langle E(X),F(Y)\rangle$ is norming; note that for the latter it suffices to consider the σ-finite case. In the case of a semi-finite measure space $(A,\mathscr{A},\mu)$ it can in fact be shown (with a slight modification of the proof of [6, Theorem 1.1]) that, if $\langle E,F\rangle$ is a norming Köthe dual pair on A, then the dual pair $\langle F(X),E(Y)\rangle$ is norming as well.

2.7. Banach space-valued extensions of linear operators between Banach function spaces

Given two Banach function spaces E and G, a bounded linear operator S from E to G and a Banach space X, we can define the tensor extension $S\otimes I_X$ from $E\otimes X$ to $G\otimes X$ as the linear operator determined by the formula

$$(S\otimes I_X)(f\otimes x):=Sf\otimes x,\qquad f\in E,x\in X.$$

It is a natural question whether $S\otimes I_X$ extends to a bounded linear operator from $E\tilde{\otimes}X$ to $G\tilde{\otimes}X$; recall that $F\tilde{\otimes}X$ denotes the closure of $F\otimes X$ in $F(X)$ when F is a Banach function space. If $S\lesssim R$ for a positive operator $R\in\mathcal{L}(E,G)$ ($R\geq 0$ dominates S), then it can be shown that

$$\|(S\otimes I_X)e\|_X\leq R\,\|e\|_X \tag{2.2}$$

for all $f\in E\otimes X$ (cf. Lemma 2.3 of [17]), from which it is immediate that:

Fact 2.4. *Let S be a bounded linear operator between two Banach function spaces E and G and let X be a Banach space. If $S\in M(E,G)$ (i.e., S is dominated by a positive operator), then $S\otimes I_X$ has a unique extension to a bounded linear operator S_X from $E\tilde{\otimes}X$ to $G\tilde{\otimes}X$ of norm $\|S_X\|\leq\|S\|_{M(E,G)}$.*

Note that if E has an order-continuous norm, so that $E\tilde{\otimes}X=E(X)$ (i.e., $E\otimes X$ is dense in $E(X)$), then the fact says that, for every $S\in M(E,G)$, the tensor extension $S\otimes I_X$ extends to a bounded linear operator $S_X\in\mathcal{L}(E(X),G(X))$, or equivalently, there exists a (necessarily unique) bounded linear operator $S_X\in\mathcal{L}(E(X),G(X))$ with the property that

$$\langle S_Xe,x^*\rangle=S\langle e,x\rangle,\qquad e\in E(X),x^*\in X^*.$$

The aim of this paper is to obtain analogues of this extension result (in the latter formulation) for E not (necessarily) having an order-continuous norm, with as main interest $E=L^\infty(A)$. Our two main results in this direction are Theorem 3.6 and Theorem 3.13.

In case G has a Levi norm (so that G must be Dedekind complete and thus $M(E,G) = \mathcal{L}_r(E,G)$) the converse of the above fact holds as well and is an easy consequence of the fact taken from [7] that, in this case, S is regular if and only if there exists a constant $C \geq 0$ such that, for all $e_1, \dots, e_N \in E$,

$$\left\| \sum_{n=1}^{N} |Se_n| \right\|_G \leq C \left\| \sum_{n=1}^{N} |e_n| \right\|_E .$$

Fact 2.5. *Let S be a bounded linear operator between two Banach function spaces F and G of which G has a Levi norm. Then the following assertions are equivalent.*

(a) *S is regular;*
(b) *$S \otimes I_{\ell^1}$ has an extension to (a necessarily unique) bounded linear operator $S_{\ell^1} \in \mathcal{L}(E\tilde{\otimes}\ell^1, G\tilde{\otimes}\ell^1)$;*
(c) *$S \otimes I_X$ has an extension to (a necessarily unique) bounded linear operator $S_X \in \mathcal{L}(E\tilde{\otimes}X, G\tilde{\otimes}X)$ for every Banach space X.*
In this situation we have $\|S_X\| \leq \|S\|_r \leq \|S_{\ell^1}\|$.

In case that $X = H$ is a Hilbert space, $E = L^{p_1}(A)$ and $G = L^{p_2}(B)$ with $1 \leq p_1, p_2 < \infty$, we do not need to impose any restrictions on the operator S for $S \otimes I_H$ to have a bounded extension. This result was proved in the 1930's by Marcinkiewicz and Zygmund using Gaussian techniques [25]: in fact, there exists a constant $0 < K \leq \max\{\frac{\|\gamma\|_{p_1}}{\|\gamma\|_{p_2}}, 1\}$, where γ denotes a standard Gaussian random variable, such that, for all operators $S \in \mathcal{L}(L^{p_1}(A), L^{p_2}(B))$, $S \otimes I_H$ has a bounded extension $S_H \in \mathcal{L}(L^{p_1}(A;H), L^{p_2}(B;H))$ of norm $\|S_H\| \leq K\|S\|$ for any Hilbert space H, or equivalently, we have the following square function estimate

$$\left\| \left(\sum_{k=1}^{n} |Se_n|^2 \right)^{1/2} \right\|_{L^{p_2}(B)} \leq K\|S\| \left\| \left(\sum_{k=1}^{n} |e_n|^2 \right)^{1/2} \right\|_{L^{p_1}(A)},$$

valid for all $e_1, \dots, e_n \in L^{p_1}(A)$ (see also [15]). Using the Grothendieck inequality, Krivine [22] showed that this inequality is in fact valid for general Banach lattices with as best possible constant K (working for all pairs of Banach lattices) the Grothendieck constant K_G (also see [24, p. 82]). As a consequence:

Fact 2.6. *Let S be a bounded linear operator between two Banach function spaces E and G and let H be a Hilbert space. Then $S \otimes I_H$ has a bounded extension $S_H \in \mathcal{L}(E\tilde{\otimes}H, G\tilde{\otimes}H)$ of norm $\|S_H\| \leq K_G\|S\|$.*

Again note (as after Fact 2.4) that if E has an order-continuous norm, then the result says that there exists a (necessarily unique) bounded linear operator $S_H \in \mathcal{L}(E(H), G(H))$ with the property that

$$(S_H e \,|\, h)_H = S\,(e \,|\, h)_H, \qquad e \in E(X), h \in H.$$

We will extend this result to general E not having an order-continuous norm under a mild assumption on G (Proposition 3.19); moreover, we will show that if S is an adjoint operator, then so is S_H (Corollary 3.20).

2.8. When are all bounded linear operators regular?

Regarding Banach space-valued extensions of operators between Banach function spaces, in view of Fact 2.4 (and Fact 2.5) it is interesting to know between which Banach function spaces every bounded linear operator is automatically regular. For the following Banach lattice theoretic result in this direction we refer to [1] and [29].

Fact 2.7. *Let E and F be two Banach lattices. In each of the following cases we have that every bounded linear operator from E to F is regular:*

(i) *F is Dedekind complete and has a strong order unit.*
(ii) *E is lattice isomorphic to an AL-space and F has a Levi norm.*
(iii) *E is lattice isomorphic to an atomic AL-space.*
(iv) *E is atomic with order-continuous norm and F is an AM-space.*
Moreover, in case (i) *and* (ii)*, if F has a Fatou norm, then we have $\|T\| = \|T\|_{\text{reg}}$ for all $T \in \mathcal{L}(E, F)$.*

Note that for example every bounded linear operator $T : L^p(A) \longrightarrow L^\infty(B)$, $p \in [1, \infty[$ and B Maharam, is regular by (i) and that every bounded linear operator $T : L^1(A) \longrightarrow L^q(B)$, $q \in [1, \infty[$, is regular by (ii), and thus have Y-valued extensions T_Y of norm $\|T_Y\| \leq \|T\|$ for every Banach space Y (by Fact 2.4).

3. Results

3.1. The extension problem

Let E and G be two Banach function spaces and let $T \in \mathcal{L}(E, G)$. Given a Banach dual pair $\langle X, Y\rangle$, we are interested in the question whether there exists a (necessarily unique) bounded linear operator $T_Y \in \mathcal{L}(E(Y), G(Y))$ with the property that

$$\langle x, T_Y e\rangle = T\langle x, e\rangle, \qquad e \in E(Y), x \in X. \tag{3.1}$$

We call the operator T_Y the Y-valued extension of T with respect to the pairing $\langle X, Y\rangle$.

In case E has an order-continuous norm, so that $E \otimes Y$ is dense in $E(Y)$ (i.e., $E\tilde{\otimes}Y = E(Y)$), T_Y is just the unique extension of $T \otimes I_Y$ to a bounded linear operator $T_Y \in \mathcal{L}(E(Y), G(Y))$. So, in this situation, we have existence of T_Y provided that T is dominated by a positive operator (Fact 2.4) or Y is a Hilbert space (Fact 2.6). In this paper we will consider the extension problem (3.1) for E not (necessarily) having an order-continuous norm, with as main interest $E = L^\infty(A)$, and obtain analogues of the two just mentioned extension results; see Theorem 3.6 and Theorem 3.13 for extensions of operators dominated by a positive operator and Proposition 3.19 and Corollary 3.20 for Hilbert space-valued extensions.

Remark 3.1. Note that if T_Y is a mapping $E(Y) \longrightarrow G(Y)$ satisfying (3.1), then T_Y is automatically a linear operator which is bounded by the closed graph theorem.

Moreover, if $T \in M(E,G)$ and $\langle X,Y\rangle$ is norming, then we have the norm estimate $\|T_Y\| \le \|T\|_{M(E,G)}$.

Proof of the norm estimate. Let $e \in E(Y)$ be given. Pick a positive operator $R \in \mathcal{L}(E,G)$ dominating T. Since $\langle X,Y\rangle$ is norming, we can pointwise estimate

$$\|T_Y e\|_Y = \sup_{x\in B_X} |\langle x, T_Y e\rangle| \overset{(3.1)}{=} \sup_{x\in B_X} |T\langle x,e\rangle| \le \sup_{x\in B_X} R|\langle x,e\rangle| \le R\,\|e\|_Y\,,$$

and thus $\|T_Y e\|_{G(Y)} \le \|R\|\,\|e\|_{E(Y)}$. Therefore, $\|T_Y\| \le \|T\|_{M(E,G)}$. □

The following simple lemma gives, in two situations, a suggestion how to obtain the Y-valued extension of T:

Lemma 3.2. *Let E and G be two Banach function spaces, $T \in \mathcal{L}(E,G)$, and $\langle X,Y\rangle$ a Banach dual pair. Assume that T has a Y-valued extension T_Y with respect to $\langle X,Y\rangle$.*

(i) *If Y has a Schauder basis $\{b_n\}_{n\in\mathbb{N}}$ with biorthogonal functionals $\{b_n^*\}_{n\in\mathbb{N}} \subset X$, then we must have*

$$T_Y e = \sum_{n=0}^{\infty} T\langle b_n^*, e\rangle \otimes b_n \tag{3.2}$$

pointwise in Y for every $e \in E(Y)$.

(ii) *Suppose that $\langle D,E\rangle$ and $\langle F,G\rangle$ are Köthe dual pairs and that T is $\sigma(E,D)$-to-$\sigma(G,F)$ continuous with adjoint $S \in \mathcal{L}(F,D)$. If $S\otimes I_X$ has an extension to a bounded linear operator $S_X \in \mathcal{L}(F\tilde{\otimes}X, D\tilde{\otimes}X)$, then T_Y is $\sigma(E(Y), D\tilde{\otimes}X)$-to-$\sigma(G(Y), F\tilde{\otimes}X)$ continuous with adjoint S_X.*

Proof. (i) is immediate from the definition of Schauder basis and (3.1). For (ii), let $e \in E(Y)$. For $f \in F$ and $x \in X$ we compute

$$\begin{aligned}\langle T_Y e, f\otimes x\rangle_{\langle G(Y),F\tilde{\otimes}X\rangle} &= \langle\langle T_Y e, x\rangle_{\langle Y,X\rangle}, f\rangle_{\langle G,F\rangle} \overset{(3.1)}{=} \langle T\langle e,x\rangle_{\langle Y,X\rangle}, f\rangle_{\langle G,F\rangle}\\ &= \langle\langle e,x\rangle_{\langle Y,X\rangle}, Sf\rangle_{\langle E,D\rangle} = \langle e, Sf\otimes x\rangle_{\langle E(Y),D\tilde{\otimes}X\rangle},\end{aligned}$$

so that, by linearity,

$$\langle T_Y e, \phi\rangle_{\langle G(Y),F\tilde{\otimes}X\rangle} = \langle e, (S\otimes I_X)\phi\rangle_{\langle E(Y),D\tilde{\otimes}X\rangle} = \langle e, S_X\phi\rangle_{\langle E(Y),D\tilde{\otimes}X\rangle}$$

for all $\phi \in F\otimes X$. By continuity and density this identity extends to all $\phi \in F\tilde{\otimes}X$, proving the desired result. □

In the setting of (i) in this lemma, if the basis $\{b_n\}_{n\in\mathbb{N}}$ is boundedly complete and if X is the closed linear span of $\{b_n^*\}_{n\in\mathbb{N}}$ in X^*, then we can use formula (3.2) to define T_Y:

Lemma 3.3. *Let E and G be two Banach function spaces and let $T \in \mathcal{L}(E,G)$. Suppose that Y is a Banach space having a boundedly-complete Schauder basis $\{b_n\}_{n\in\mathbb{N}}$ with biorthogonal functionals $\{b_n^*\}_{n\in\mathbb{N}}$. Define X as the closed linear span of $\{b_n^*\}_{n\in\mathbb{N}}$ in Y^*. If $T \in M(E,G)$, then it has a Y-valued extension*

$T_Y \in \mathcal{L}(E(Y), G(Y))$ *with respect to* $\langle X, Y\rangle$ *(in the sense of* (3.1)*) of norm* $\|T_Y\| \leq \|T\|_{M(E,G)}$.

Proof. Let $R \in \mathcal{L}(E,G)$ be a positive operator dominating T. For all $e \in E(Y)$ we can estimate

$$\left\|\sum_{n=0}^{N} T\langle b_n^*, e\rangle \otimes b_n\right\|_Y = \left\|(T\otimes I_Y)\left(\sum_{n=0}^{N}\langle b_n^*, e\rangle \otimes b_n\right)\right\|_Y \overset{(2.2)}{\leq} R\left\|\sum_{n=0}^{N}\langle b_n^*, e\rangle \otimes b_n\right\|_Y \leq KR\,\|e\|_Y\,,$$

where K is the basis constant of $\{b_n\}_{n\in\mathbb{N}}$.

Since the basis $\{b_n\}_{n\in\mathbb{N}}$ is boundedly complete, we can define $T_Y e \in L^0(A;Y)$ as the pointwise limit $\lim_{N\to\infty}\sum_{n=0}^{N} T\langle b_n^*, e\rangle \otimes b_n$ in Y to obtain an element $T_Y e \in G(Y)$ satisfying $\|T_Y e\|_Y \leq KR\,\|e\|_Y \in G$. It then clearly holds that $\langle b_n^*, T_Y e\rangle = T\langle b_n^*, e\rangle$ for all $e \in E(Y)$ and $n \in \mathbb{N}$, from which it follows that, in fact,

$$\langle x, T_Y e\rangle = T\langle x, e\rangle, \qquad e \in E(Y), x \in X.$$

Remark 3.1 now completes the proof. □

In the situation of the above lemma, the canonical map $j : Y \longrightarrow X^*$ given by $j(y)(x) = \langle y, x\rangle$, for all $y \in Y$ and $x \in X$, is an isomorphism, which is isometric in case $\{b_n\}_{n\in\mathbb{N}}$ is monotone; see [2, Theorem 3.2.10]. In particular, (possibly) up to an equivalence of norms, the above lemma is concerned with a special case of the situation $\langle X, Y\rangle = \langle X, X^*\rangle$. Regarding general Y-valued extensions with respect to $\langle X, Y\rangle = \langle X, X^*\rangle$, let us remark the following:

Remark 3.4. Let E and G be two Banach function spaces and let $T \in \mathcal{L}(E,G)$. Let X be a Banach space and put $Y := X^*$. In this situation we would like to simply define the Y-valued extension T_Y of T with respect to $\langle X, Y\rangle$ by (3.1). However, $\{\langle x, Te\rangle : x \in X\} \subset G$ is just a family of equivalence classes of measurable functions and it is not clear how to obtain an element $T_Y e \in G(Y, \sigma(Y,X))$. In case G is a Banach function space over $(B, \mathscr{B}, \nu) = (B, \mathcal{P}(B), \#)$ this problem does not occur. Moreover, if B is countable or Y is separable, then we obtain an element $T_Y e \in G(Y)$.

In view of Lemma 3.2.(ii) it is natural to consider the extension problem in the following lemma.

Lemma 3.5. *Let* $\langle D, E\rangle$ *and* $\langle F, G\rangle$ *be two Köthe dual pairs and let* $T \in \mathcal{L}(E,G)$ *be a* $\sigma(E,D)$*-to-*$\sigma(G,F)$ *continuous linear operator with adjoint* $S \in \mathcal{L}(F,D)$*. For any dual pair of Banach spaces* $\langle X, Y\rangle$*, the following are equivalent:*

(a) $T \otimes I_Y$ *extends to a (necessarily unique)* $\sigma(E(Y), D\tilde{\otimes}X)$*-to-*$\sigma(G(Y), F\tilde{\otimes}X)$ *continuous linear operator* $T_Y \in \mathcal{L}(E(Y), G(Y))$*.*

(b) $S \otimes I_X$ *extends to a (necessarily unique)* $\sigma(F\tilde{\otimes}X, G(Y))$*-to-*$\sigma(D\tilde{\otimes}X, E(Y))$ *continuous linear operator* $S_X \in \mathcal{L}(F\tilde{\otimes}X, D\tilde{\otimes}X)$*.*

In this situation, S_X *and* T_Y *are adjoints of each other and* T_Y *is the* Y*-valued extension of* T *with respect to* $\langle X, Y\rangle$ *(in the sense of* (3.1)*).*

Proof. Note that the uniqueness in (a) and (b) follows from the $\sigma(E(Y), D\tilde{\otimes}X)$-density of $E \otimes Y$ in $E(Y)$ and the $\sigma(F\tilde{\otimes}X, G(Y))$-density of $F \otimes X$ in $F\tilde{\otimes}X$. The adjoint part in the last statement is contained in the proof of the implications "(a)⇒(b)" and "(b)⇒(a)". That T_Y then is the Y-valued extension of T with respect to $\langle X, Y\rangle$ can be seen as follows: Given $e \in E(Y)$ and $x \in X$, we have

$$\begin{aligned}\langle f, \langle x, T_Y e\rangle_{\langle X,Y\rangle}\rangle_{\langle F,G\rangle} &= \langle f \otimes x, T_Y e\rangle_{\langle F\tilde{\otimes}X,G(Y)\rangle} = \langle Sf \otimes x, e\rangle_{\langle D\tilde{\otimes}X,E(Y)\rangle}\\ &= \langle Sf, \langle x, e\rangle_{\langle X,Y\rangle}\rangle_{\langle D,E\rangle} = \langle f, S\langle x, e\rangle_{\langle X,Y\rangle}\rangle_{\langle F,G\rangle}\end{aligned}$$

for every $f \in F$. As F is separating for G, this shows $\langle x, T_Y e\rangle = T\langle x, e\rangle$.

"(a) ⇒ (b)": Let $S_X := (T_Y)' \in \mathcal{L}(F\tilde{\otimes}X, D\tilde{\otimes}X)$ be the adjoint of T_Y. A computation similar to the above one yields that $\langle S_X f, e\rangle_{\langle D\tilde{\otimes}X,E(Y)\rangle} = \langle (S \otimes I_X)f, e\rangle_{\langle D\tilde{\otimes}X,E(Y)\rangle}$ for all $f \in F \otimes X$ and $e \in E \otimes Y$ while $E \otimes Y$ separates the points of $D\tilde{\otimes}X$, whence $S_X f = (S \otimes I_X)f$ for all $f \in F \otimes X$. This gives (b).

"(b) ⇒ (a)": Completely analogous to the implication "(a) ⇒ (b)". □

In the next two subsections we will use Lemma 3.3 and Lemma 3.5 to obtain our two main extension results, Theorem 3.6 and Theorem 3.13.

3.2. Extensions with respect to $\langle X, Y\rangle = \langle Y^*, Y\rangle$ with Y reflexive

Theorem 3.6. *Let E and G be two Banach function spaces and let $T \in M(E, G)$. If Y is a reflexive Banach space, then the Y-valued extension $T_Y \in \mathcal{L}(E(Y), G(Y))$ of T with respect to $\langle Y^*, Y\rangle$ (in the sense of* (3.1)*) exists and is of norm $\|T_Y\| \leq \|T\|_{M(E,G)}$.*

As an immediate consequence of this theorem and Fact 2.7 we have:

Corollary 3.7. *Let E and G be two Banach function spaces such that one of the following four conditions is satisfied:*

(i) *G has a strong order unit;*
(ii) *E is lattice isomorphic to an AL-space and G has a Levi norm;*
(iii) *E is lattice isomorphic to an atomic AL-space;*
(iv) *E is atomic with order-continuous norm and G is an AM-space.*

Then, for every $T \in \mathcal{L}(E, G)$ and every reflexive Banach space Y, the Y-valued extension $T_Y \in \mathcal{L}(E(Y), G(Y))$ of T with respect to $\langle Y^*, Y\rangle$ exists. Moreover, in case of (i) *and* (ii)*, if G has a Fatou norm, then we have the norm estimate $\|T_Y\| \leq \|T\|$.*

In combination with Fact 2.1.(II), the next lemma allows us to reduce the proof of the theorem to the case that Y is a reflexive Banach space with a Schauder basis.

Lemma 3.8. *Let E and G be two Banach function spaces, $T \in \mathcal{L}(E, G)$, and Y a Banach space.*

(i) *If T has a U-valued extension T_U with respect to $\langle U^*, U\rangle$ for every separable closed linear subspace U of Y, then T also has a Y-valued extension T_Y with respect to $\langle Y^*, Y\rangle$.*

(ii) *If Y is a closed linear subspace of a Banach space Z for which T has a Z-valued extension T_Z with respect to $\langle Z^*, Z\rangle$, then T also has Y-valued extension T_Y with respect to $\langle Y^*, Y\rangle$.*

Proof. (i) This follows easily from the fact that every $f \in E(Y)$ may be viewed as an element of $E(U) \subset E(Y)$ for some separable closed linear subspace U of Y in combination with Remark 3.1.

(ii) Viewing $E(Y)$ as closed linear subspace of $E(Z)$, T_Z restricts to an operator on $E(Y)$ as consequence of the fact that $Y = {}^\perp(Y^\perp)$. From Hahn–Banach it follows that $T_Y := T_Z\big|_{E(Y)}$ is a Y-valued extension of T with respect to $\langle Y^*, Y\rangle$. □

We are now ready to give a clean proof of the theorem.

Proof of Theorem 3.6. First, in view of (i) of the above lemma and the fact that a closed linear subspace of a reflexive Banach space is a reflexive Banach space on its own right, it suffices to consider the case that Y is a separable reflexive Banach space. Next, in view of (ii) of the above lemma and Fact 2.1.(II), it is in turn enough to treat the case that Y is a reflexive Banach space having a Schauder basis. By Fact 2.1.(I), this basis is both boundedly complete and shrinking. The existence of T_Y now follows from an application of Lemma 3.3; for the norm estimate we refer to Remark 3.1. □

Remark 3.9. The use of Fact 2.1.(II) in the above proof can be avoided in the special case that G is a Banach function space over $(B, \mathscr{B}, \nu) = (B, \mathcal{P}(B), \#)$; see Remark 3.4.

In the next example we show that for the non-reflexive Banach spaces $Y = c_0$ and $Y = \ell^1$ the statement of Theorem 3.6 does not hold. Note that in both cases Y has a Schauder basis (the standard basis), with in case $Y = c_0$ a basis which is shrinking but not boundedly complete and in case $Y = \ell^1$ a basis which is boundedly complete but not shrinking; also see Fact 2.1.(I).

Example 3.10. Let $E = \ell^\infty$ and $G = \mathbb{K}$. Take $T \in \mathcal{L}(\ell^\infty, \mathbb{K}) = (\ell^\infty)^*$ to be a Banach limit (see Definition 2.3). Then T is a positive operator, but for $Y \in \{c_0, \ell^1\}$ the Y-valued extension $T_Y \in \mathcal{L}(\ell^\infty(Y), Y)$ of T with respect to $\langle Y^*, Y\rangle$ does not exist.

Proof. Let us first treat the case $Y = c_0$. To the contrary we assume that T_{c_0} does exist. By Lemma 3.2.(i) (equation (3.2)) we must then have

$$(Tf_k)_{k\in\mathbb{N}} = T_{c_0} f \in c_0$$

for all $f = (f_k)_{k\in\mathbb{N}} \in \ell^\infty(c_0)$; here f_k is the kth coordinate in c_0 of f (with respect to the standard basis). But for

$$f = (f_k)_{k\in\mathbb{N}} \in \ell^\infty(c_0) \quad \text{given by} \quad f_k := 1_{\{k,k+1,\dots\}}$$

we have $(Tf_k)_{k\in\mathbb{N}} = \mathbf{1} \notin c_0$, a contradiction.

Next we treat the case $Y = \ell^1$. We again assume to the contrary that T_{ℓ^1} does exist. By Lemma 3.2.(i) (equation (3.2)) we must then have

$$(Tf_k)_{k\in\mathbb{N}} = T_{\ell^1} f \in \ell^1$$

for all $f = (f_k)_{k\in\mathbb{N}} \in \ell^\infty(\ell^1)$; here f_k is the kth coordinate in ℓ^1 of f (with respect to the standard basis). But for

$$f = (f_k)_{k\in\mathbb{N}} \in \ell^\infty(\ell^1) \quad \text{given by} \quad f_k := 1_{\{k\}}$$

and $\mathbf{1} \in \ell^\infty = (\ell^1)^*$ this yields

$$0 = \sum_{k=0}^{\infty} 0 = \sum_{k=0}^{\infty} Tf_k = \langle \mathbf{1}, T_{\ell^1} f\rangle = T\langle \mathbf{1}, f\rangle = T\mathbf{1} = 1,$$

a contradiction. □

Remark 3.11. In [4] it is shown that Banach spaces 1-complemented in their bidual admit vector-valued Banach limits, whereas c_0 does not. Since a Y-valued extension with respect to a norming dual pair $\langle X, Y\rangle$ of a Banach limit is a vector-valued Banach limit on Y, the latter also gives an explanation for the failure of the extension for $Y = c_0$ in the above example. The case $Y = \ell^1$ in this example shows that a vector-valued Banach limit on ℓ^1 cannot be obtained as an ℓ^1-valued extension with respect to $\langle \ell^\infty, \ell^1\rangle$ of a Banach limit; note, however, that ℓ^1-valued extensions with respect to $\langle c_0, \ell^1\rangle$ of Banach limits exist by Lemma 3.3 (also see Remark 3.4). Finally, observe that, by Theorem 3.6, every reflexive Banach space Y admits vector-valued Banach limits which are Y-valued extensions with respect to $\langle Y^*, Y\rangle$ of Banach limits.

Combining this example with Fact 2.1.(III), Lemma 3.8.(ii), and Theorem 3.6, we see that, for Y in a wide class of Banach spaces (including the Banach lattices), Theorem 3.6 even characterizes the reflexivity of Y:

Corollary 3.12. *Let Y be a closed linear subspace of a Banach lattice E such that: Y is complemented in E or E has an order-continuous norm. Given a Banach limit $T \in \mathcal{L}(\ell^\infty, \mathbb{K}) = (\ell^\infty)^*$, the following statements are equivalent.*

(a) *Y is reflexive;*
(b) *T has a Y-valued extension $T_Y \in \mathcal{L}(\ell^\infty(Y), Y)$ with respect to $\langle Y^*, Y\rangle$;*
(c) *Y does not have linear subspaces isomorphic to c_0 or ℓ_1.*

3.3. Extensions of adjoint operators on L^∞ with respect to arbitrary Banach dual pairs $\langle X, Y\rangle$

Theorem 3.13. *Let $(A, \mathscr{A}, \mu)$ be a semi-finite measure space, let $\langle F, G\rangle$ be a Köthe dual pair of Banach function spaces over a measure space $(B, \mathscr{B}, \nu)$, and let $T \in \mathcal{L}(L^\infty(A), G)$ be a $\sigma(L^\infty(A), L^1(A))$-to-$\sigma(G, F)$ continuous linear operator, say with adjoint $S \in \mathcal{L}(F, L^1(A))$. If $T \in M(L^\infty(A), G)$, then we have, for any dual pair of Banach spaces $\langle X, Y\rangle$, that $T \otimes I_Y$ has a unique extension to a $\sigma(L^\infty(A; Y), L^1(A; X))$-to-$\sigma(G(Y), F\tilde{\otimes}X)$ continuous linear operator $T_Y \in \mathcal{L}(L^\infty(A; Y), G(Y))$. In this situation, T_Y is the Y-valued extensions of T with respect to $\langle X, Y\rangle$*

and the adjoint $S_X \in \mathcal{L}(F\tilde{\otimes}X, L^1(A;X))$ *of* T_Y *is the unique bounded extension of* $S \otimes I_X$. *Moreover,* $S \in M(F, L^1(A)) = \mathcal{L}_r(F, L^1(A))$ *and these extensions are of norm* $\|S_X\| \leq \|S\|_r$ *and* $\|T_Y\| \leq \|T\|_{M(L^\infty(A),G)}$.

We will give the proof of this theorem in the next section. In Section 5 we will use this theorem to obtain the conditional expectation operator on Banach space-valued L^∞-spaces.

Remark 3.14. Note that for $T\otimes I_Y$ to have an extension to a $\sigma(L^\infty(A;Y), L^1(A;X))$-to-$\sigma(G(Y), F\tilde{\otimes}X)$ continuous linear operator $T_Y \in \mathcal{L}(L^\infty(A;Y), G(Y))$ it is necessary that S is regular. Indeed, from Lemma 3.5 it then follows that $S \otimes I_X$ extends to a bounded operator $S_X \in \mathcal{L}(F\tilde{\otimes}X, L^1(A;X))$ for any Banach space X (just take $\langle X, Y\rangle = \langle X, X^*\rangle$ as dual pair of Banach spaces), which by Fact 2.5 just means that S is regular.

We will in fact start the proof of this theorem by showing that S is regular, then extend $S \otimes I_X$ to a bounded linear operator $S_X \in \mathcal{L}(F\tilde{\otimes}X, L^1(A;X))$ and obtain T_Y by restriction of the Banach space adjoint

$$(S_X)^* \in \mathcal{L}((L^1(A;X))^*, (F\tilde{\otimes}X)^*).$$

Next, we consider situations in which the extension of $T \otimes I_X$ in Theorem 3.13 is for free. The idea is to impose conditions on $\langle F, G\rangle$ which guarantee that T is automatically regular, either via T being a bounded linear operator from $L^\infty(A)$ to G or via S and the following little lemma:

Lemma 3.15. *In addition to the assumptions of Theorem* 3.13, *suppose that the image of* $i : g \mapsto \langle\,\cdot\,, g\rangle, G \longrightarrow F^*$ *is a band in* F^*. *Then* T *is regular provided that* S *is regular.*

Proof. First note that $i(G)$ is a projection band in the Dedekind complete F^*. Let P be the associated band projection. Since i is a lattice isomorphism onto its image, this projection P induces a positive linear map $\pi : F^* \longrightarrow G$ such that $\pi \circ i = I_G$. Now note that $\pi \circ S^* : (L^1(A))^* \longrightarrow G$ extends T and is regular if S is so. □

Note that G must be Dedekind complete, being lattice isomorphic to a band in the Dedekind complete F^*.

Examples of Köthe dual pairs satisfying the hypotheses of this lemma are $\langle F, G\rangle = \langle L^p(B), L^q(B)\rangle$ with $1 < p, q < \infty$, $\frac{1}{p} + \frac{1}{q} = 1$ on an arbitrary measure space $(B, \mathcal{B}, \nu)$ or $\langle F, G\rangle = \langle F, F^\times\rangle$ with F a Banach function space on a Maharam measure space $(B, \mathscr{B}, \nu)$ having full carrier (see Fact 2.2).

Corollary 3.16. *Let* $(A, \mathscr{A}, \mu)$ *be a semi-finite measure space, let* $\langle F, G\rangle$ *be a Köthe dual pair of Banach function spaces over a measure space* $(B, \mathscr{B}, \nu)$, *and let* $T : L^\infty(A) \longrightarrow G$ *be a* $\sigma(L^\infty(A), L^1(A))$*-to-*$\sigma(G, F)$ *continuous linear operator. In each of the following cases* T *is automatically regular:*

(i) *G is Dedekind complete and has a strong order unit.*
(ii) *The image of* $g \mapsto \langle \cdot, g \rangle, G \longrightarrow F^*$ *is a band in* F^* *and* F *is lattice isomorphic to an AL-space.*

As a consequence, in each of these cases we have that, for any dual pair of Banach spaces $\langle X, Y \rangle$, $T \otimes I_Y$ *has a unique extension to a* $\sigma(L^\infty(A;Y), L^1(A;X))$*-to-*$\sigma(G(Y), F\tilde{\otimes}X)$ *continuous linear operator* $T_Y \in \mathcal{L}(L^\infty(A;Y), G(Y))$, *which is the* Y*-valued extension of* T *with respect to* $\langle X, Y \rangle$. *In this situation, denoting by* $S \in \mathcal{L}(F, L^1(A))$ *the adjoint of* T *with respect to the dualities* $\langle L^1(A), L^\infty(A) \rangle$ *and* $\langle F, G \rangle$ *and by* $S_X \in \mathcal{L}(F\tilde{\otimes}X, L^1(A;X))$ *the adjoint of* T_X *with respect to the dualities* $\langle L^1(A)(X), L^\infty(A;Y) \rangle$ *and* $\langle F\tilde{\otimes}X, G(Y) \rangle$, $S \in \mathcal{L}_r(F, L^1(A))$, *and* S_X *is the unique bounded extension of* $S \otimes I_X$. *Moreover, these extensions are of norm* $\|S_X\| \leq \|S\|_r$ *and* $\|T_Y\| \leq \|T\|_{M(L^\infty(A),G)}$.

Proof. Case (i) is an immediate consequence of Fact 2.7, whereas case (ii) follows from a combination Fact 2.7 and the above lemma. □

Examples of Köthe dual pairs $\langle F, G \rangle$ satisfying the hypothesis of this result are $\langle F, G \rangle = \langle L^1(B), L^\infty(B) \rangle$ on a Maharam measure space $(B, \mathscr{B}, \nu)$ and $\langle F, G \rangle = \langle L^p(B), L^{p'}(B) \rangle$, $p, p \in [1, \infty]$, $\frac{1}{p} + \frac{1}{p'}$, on a finite measure space $(B, \mathscr{B}, \nu)$.

Finally, we give two situations (involving some extra assumptions on $\langle G, F \rangle$) in which T being regular is not only a sufficient condition but a necessary condition as well. The idea is to impose conditions on $\langle G, F \rangle$ which allow us to obtain that T is regular, either via an application of Fact 2.5 to T or via an application of this theorem to S in combination with Lemma 3.15.

Proposition 3.17. *Suppose, in addition to the assumptions of Theorem* 3.13, *that either*

(i) *G has a Levi norm, or*
(ii) *the image of* $g \mapsto \langle \cdot, g \rangle, G \longrightarrow F^*$ *is a band in* F^*.

Then T *must be regular if, for some dual pair of Banach spaces* $\langle X, Y \rangle$ *with* $Y = \ell^1$ *in case (i) and* $X = \ell^1$ *in case (ii),* $T \otimes I_Y$ *has an extension to a* $\sigma(L^\infty(A;Y), L^1(A;X))$*-to-*$\sigma(G(Y), F\tilde{\otimes}X)$ *continuous linear operator*

$$T_Y \in \mathcal{L}(L^\infty(A;Y), G(Y)).$$

Proof. Let us first consider case (i). Note that it, in particular, $T \otimes I_{\ell^1}$ has a unique extension to a bounded linear operator from $L^\infty(A)\tilde{\otimes}\ell^1$ to $G\tilde{\otimes}\ell^1$. Fact 2.5 now yields that T is regular.

Next we consider case (ii). In view Lemma 3.15 it suffices to prove that S is regular. By Fact 2.5 and the fact that $L^1(A)$ has a Levi norm, for this it is in turn enough to show that $S \otimes I_{\ell^1}$ has an extension to a bounded linear operator $S_{\ell^1} \in \mathcal{L}(F\tilde{\otimes}\ell^1, L^1(A;\ell^1))$. But, by Lemma 3.5, the adjoint of the $\sigma(L^\infty(A;Y), L^1(A;\ell^1))$-to-$\sigma(G(Y), F\tilde{\otimes}\ell^1)$ continuous linear operator $T_Y \in \mathcal{L}(L^\infty(A;Y), G(Y))$ is an extension of $S \otimes I_{\ell^1}$. □

Corollary 3.18. *Suppose, in addition to the assumptions of Theorem* 3.13, *that the image of* $g \mapsto \langle \cdot, g\rangle, G \longrightarrow F^*$ *is a band in* F^*. *Then the following assertions are equivalent.*

(a) T *is regular.*
(b) S *is regular.*
(c) $T \otimes I_{\ell^\infty}$ *has an extension to a bounded linear operator from* $L^\infty(A)\tilde{\otimes}\ell^\infty$ *to* $G\tilde{\otimes}\ell^\infty$.
(d) $S \otimes I_{\ell^1}$ *has an extension to a bounded linear operator from* $F\tilde{\otimes}\ell^1$ *to* $L^1(S;\ell^1)$.
(e) *For any dual pair of Banach spaces* $\langle X, Y\rangle$, $T \otimes I_Y$ *has an extension to an operator* $T_Y \in \mathcal{L}(L^\infty(A;Y), G(Y))$ *which is* $\sigma(L^\infty(A;Y), L^1(A;X))$*-to-*$\sigma(G(Y), F\tilde{\otimes}X)$ *continuous.*
(f) *For any dual pair of Banach spaces* $\langle X, Y\rangle$, $S \otimes I_X$ *has an extension to an operator* $S_X \in \mathcal{L}(F\tilde{\otimes}X, L^1(A;X))$ *which is* $\sigma(F\tilde{\otimes}X, F^\times(Y))$*-to-*$\sigma(L^1(A;X),$ $L^\infty(A;Y))$ *continuous.*

In this situation, for which to occur it suffices that G has a strong order unit, S_X *and* T_Y *are adjoints of each other with respect to the dualities* $\langle F\tilde{\otimes}X, G(Y)\rangle$ *and* $\langle L^1(A;X), L^\infty(A;Y)\rangle$. *Moreover, these extensions are of norm* $\|S_X\| \le \|S\|_r$ *and* $\|T_Y\| \le \|T\|_r$.

Proof. Note that G must be Dedekind complete, being lattice isomorphic to a band in the Dedekind complete F^*.

"(a) ⇒ (c)": See Fact 2.4.

"(c) ⇒ (d)": Viewing $L^1(A;\ell^1)$ as a closed linear subspace of $(L^\infty\tilde{\otimes}\ell^\infty)^*$ via the isometric embedding

$$i : L^1(A;\ell^1) \longrightarrow (L^\infty(A)\tilde{\otimes}\ell^\infty)^*, h \mapsto \langle h, \cdot\rangle_{\langle L^1(A;\ell^1), L^\infty(A;\ell^\infty)\rangle}\big|_{L^\infty(A)\tilde{\otimes}\ell^\infty},$$

it is enough that $S \otimes I_{\ell^1}$ has an extension to a bounded linear operator from $F\tilde{\otimes}\ell^1$ to $(L^\infty\tilde{\otimes}\ell^\infty)^*$. For this let

$$j : F\tilde{\otimes}\ell^1 \longrightarrow (G\tilde{\otimes}\ell^\infty)^*, f \mapsto \langle f, \cdot\rangle_{\langle F(\ell^1), G(\ell^\infty)\rangle}\big|_{G\tilde{\otimes}\ell^\infty}$$

be the natural continuous inclusion and let $U \in \mathcal{L}((G\tilde{\otimes}\ell^\infty)^*, (L^\infty(A)\tilde{\otimes}\ell^\infty)^*)$ be the Banach spaced adjoint of the bounded extension of $T \otimes I_{\ell^\infty}$. Now observe that $U \circ j$ extends $S \otimes I_{\ell^1}$.

"(d) ⇒ (b)": This follows from Fact 2.5 and the fact that $L^1(A)$ has a Levi norm.

"(b) ⇒ (a)": This is precisely Lemma 3.15.

"(a) ⇔ (e)": Combine Theorem 3.13 with the above proposition.

"(e) ⇔ (f)": See Lemma 3.5.

The final assertion follows from Theorem 3.13 and Corollary 3.16. □

3.4. Extensions with respect to $\langle X, Y\rangle = \langle H^*, H\rangle$ for a Hilbert space H

Similar to Fact 2.6, for the existence of the extension in Theorem 3.6 we do not need to impose any conditions on T under the extra assumption that G has a sequentially Levi norm.

Proposition 3.19. *Let E and G be two Banach function spaces, $T \in \mathcal{L}(E,G)$, and H a Hilbert space. Suppose that G has a sequentially Levi norm. Then T has a H-valued extension $T_H \in \mathcal{L}(E(H), G(H))$ with respect to $\langle H^*, H\rangle$ (in the sense of (3.1)) which is of norm $\|T_Y\| \leq K_G \|T\|$, where K_G is the Grothendieck constant.*

Proof. We may without loss of generality assume that H is separable, see Lemma 3.8.(i). Now choose an orthonormal basis $\{h_n\}_{n\in\mathbb{N}}$ of H. Given an $e \in E(H)$, it suffices to show that $\sum_{n\in\mathbb{N}} T\langle h_n, e\rangle \otimes h_n$ converges pointwise a.e. in H to an element of norm $\leq K_G \|T\| \|e\|_{E(H)}$. But this follows from the hypothesis that G has a sequentially Levi norm in combination with the estimate

$$\left\| \left(\sum_{n=0}^{N} |T\langle h_n, e\rangle|^2 \right)^{1/2} \right\|_G \leq K_G \|T\| \left\| \left(\sum_{n=0}^{N} |\langle h_n, e\rangle|^2 \right)^{1/2} \right\|_E \leq K_G \|T\| \|e\|_{E(H)};$$

here we use the Grothendieck inequality for Banach lattices (see [24, p. 82]). □

As an immediate consequence of this proposition, Fact 2.6, and Lemma 3.2.(ii), we have something similar for Theorem 3.13:

Corollary 3.20. *Let $\langle D,E\rangle$ and $\langle F,G\rangle$ be two Köthe dual pairs, let $T \in \mathcal{L}(E,G)$ be a $\sigma(E,D)$-to-$\sigma(G,F)$ continuous linear operator with adjoint $S \in \mathcal{L}(F, L^1(A))$, and let H be a Hilbert space. Then it holds that $T \otimes I_H$ has a unique extension to a $\sigma(E(H), D\tilde{\otimes}H^*)$-to-$\sigma(G(H), F\tilde{\otimes}H^*)$ continuous linear operator $T_H \in \mathcal{L}(E(Y), G(H))$. In this situation, T_H is the H-valued extensions of T with respect to $\langle H^*, H\rangle$ and the adjoint $S_H \in \mathcal{L}(F\tilde{\otimes}H^*, F\tilde{\otimes}H^*)$ of T_H is the unique bounded extension of $S \otimes I_{H^*}$. Moreover, these extensions are of norm $\|S_{H^*}\| \leq K_G \|S\|$ and $\|T_H\| \leq K_G \|T\|$.*

4. Proof of Theorem 3.13

Let the notations and assumptions be as in Theorem 3.13. For the proof of this theorem we need three lemmas. Before we can state the first lemma, we have to define the notion of countable step function: a function $f : A \longrightarrow Y$ is called a *countable step function* if it is measurable and only assumes countably many values. Note that such a function is strongly measurable and can (in fact) be written as the pointwise limit $f = \sum_{k=0}^{\infty} 1_{A_k} y_k$ with $(A_k)_{k\in\mathbb{N}}$ a mutually disjoint sequence in $\mathscr{A}$ and $(y_n)_{n\in\mathbb{N}}$ a sequence in Y.

Lemma 4.1. *The subspace of countable step functions lying in $L^\infty(A;Y)$ is dense in $L^\infty(A;Y)$.*

Proof. See the proof of Proposition 1.9 in [28]. □

Lemma 4.2. *Let $\langle D,E\rangle$ be a Köthe dual pair of Banach function spaces on a measure space $(C, \mathscr{S}, \rho)$, suppose that $(e_k)_{k\in\mathbb{N}} \subset E$ is such that $\sum_{k=0}^{\infty} |e_k|$ is in E, and let $(y_k)_{k\in\mathbb{N}}$ be a bounded sequence in Y. Then $\sum_{k=0}^{\infty} e_k(c) y_k$ converges in Y for a.a. $c \in C$ and the resulting function $e : C \to Y$, defined by $e(c) := \sum_{k=0}^{\infty} e_k(c) y_k$,* *belongs to* $E(Y)$ *and is of norm* $\|e\| \le \|(y_k)\|_\infty \|\sum_{k=0}^\infty |e_k|\|$. *Moreover, we have* $e = \sum_{k=0}^\infty e_k \otimes y_k$ *with convergence in the* $\sigma(E(Y), D(X))$*-topology.*

Proof. Observing that, for a.a. $c \in C$,

$$\sum_{k=0}^\infty \|e_k(c)y_k\| \le \|(y_k)\|_\infty \sum_{k=0}^\infty |e_k(c)|,$$

we find that, for a.a. $c \in C$, $e(c) = \sum_{k=0}^\infty e_k(c)y_k$ converges in Y and $\|e(c)\| \le \|(y_k)\|_\infty \sum_{k=0}^\infty |e_k(c)|$. Therefore, $e \in E(Y)$ with $\|e\| \le \|(y_k)\|_\infty \|\sum_{k=0}^\infty |e_k|\|$.

To prove the final assertion, fix an $d \in D(X)$.

The sequence

$$\left(c \mapsto \left\langle d(c), \sum_{k=N+1}^\infty e_k(c)y_k \right\rangle\right)_{N\in\mathbb{N}} \subset L^1(C)$$

converges a.e. to 0 as $N \to \infty$ and is dominated by a scalar multiple of

$$\|d\|_X \left\|\sum_{k=0}^\infty |e_k|\right\| \|(y_k)\|_\infty \in L^1(C),$$

so that

$$\left\langle d, e - \sum_{k=0}^N e_k \otimes y_k \right\rangle = \int_C \left\langle d(c), \sum_{k=N+1}^\infty e_k(c)y_k \right\rangle d\rho(c) \overset{N\to\infty}{\longrightarrow} 0. \qquad \square$$

Lemma 4.3. *Viewing* $L^1(A)$ *as closed Riesz subspace of* $(L^\infty(A))^*$, $L^1(A)$ *is a band in* $(L^\infty(A))^*$.

Proof. Recalling that $\langle L^1(A), L^\infty(A)\rangle$ is a norming Köthe dual pair, we may view $L^1(A)$ and $L^\infty(A)$ as closed Riesz subspaces of $(L^\infty(A))^*$ and $(L^1(A))^*$, respectively. Accordingly, let $J : (L^\infty(A))^* \xrightarrow{\simeq} (L^1(A))^{**}/(L^\infty(A))^\perp$ be the canonical isometric lattice isomorphism and let $\pi : (L^1(A))^{**} \longrightarrow (L^1(A))^{**}/(L^\infty(A))^\perp$ be the natural map.

To see that $L^1(A)$ is an ideal in $(L^\infty(A))^*$, let $\Lambda \in (L^\infty(A))^*$ and $f \in L^1(A)$ be such that $0 \le \Lambda \le f$ in $(L^\infty(A))^*$. Then f viewed as a functional on $(L^1(A))^*$ is positive and its restriction to $L^\infty(A)$ dominates the positive $\Lambda \in (L^\infty(A))^*$. Hence, Λ has an extension to a functional $\tilde\Lambda$ on $(L^1(A))^*$ satisfying $0 \le \tilde\Lambda \le f$ in $(L^1(A))^{**}$. Since $L^1(A)$, having an order-continuous norm, is an ideal in $(L^1(A))^{**}$, it follows that $\tilde\Lambda \in L^1(A)$. Therefore, $\Lambda = J^{-1}(\pi(\tilde\Lambda)) \in J^{-1}(\pi(L^1(A))) = L^1(A)$.

It remains to be shown that the ideal $L^1(A)$ in $(L^\infty(A))^*$ is also order closed in $(L^\infty(A))^*$. To this end, let $\{f_\alpha\}_\alpha \subset L^1(A)$ be such that $0 \le f_\alpha \nearrow \Lambda \in (L^\infty(A))^*$. Then we in particular have that $\{f_\alpha\}$ is an increasing positive norm bounded net in $L^1(A)$. From the fact that $L^1(A)$ has a Levi norm it now follows that $f_\alpha \nearrow f$ for some $f \in L^1(A)$. But then we must have $\Lambda = f \in L^1(A)$, as desired. $\square$

We are now ready to prove Theorem 3.13

Proof of Theorem 3.13. We only need to establish existence of T_Y and the norm estimates.

First observe that S is regular. Indeed, letting $j : F \hookrightarrow G^*$ be the natural inclusion and letting $\pi : (L^\infty(A))^* \longrightarrow L^1(A)$ be the map induced by Lemma 4.3, we have that $S = \pi \circ T^* \circ j : F \longrightarrow L^1(A)$. Therefore, $S \in M(F, L^1(A)) = \mathcal{L}_r(F, L^1(A))$ as $\pi \geq 0$, $T^* \in M(G^*, (L^\infty(A))^*)$, and $j \geq 0$.

By Fact 2.4, as $S \in \mathcal{L}(F, L^1(A))$ is regular, $S \otimes I_X$ has an extension to an operator $S_X \in \mathcal{L}(F\tilde{\otimes}X, L^1(A;X))$ of norm $\|S\|_X \leq \|S\|_{reg}$. Letting $i : L^\infty(Y;X) \hookrightarrow L^1(A;X))^*$ and $j : G(Y) \hookrightarrow (F\tilde{\otimes}X)^*$ be the natural continuous inclusions, we claim that (i) $(S_X)^* \circ i$ extends $j \circ (T \otimes I_Y)$ and (ii) $(S_X)^*$ maps $i(L^\infty(Y;X))$ into $j(G(Y))$ and (iii) $j^{-1} \circ (S_X)^* \circ i \in \mathcal{L}(L^\infty(A;Y), G(Y))$ of norm $\leq \|T\|_{M(L^\infty(A),G)}$. Then (ii) tells us that S_X has an adjoint $(S_X)'$ w.r.t. the dualities $\langle F\tilde{\otimes}X, G(Y)\rangle$ and $\langle L^1(A;X), L^\infty(A;Y)\rangle$, which by (i) extends $T \otimes I_Y$. The norm inequality $\|T_Y\| \leq \|T\|_{M(L^\infty(A),G)}$ then follows from (iii).

For (i), let $h \in L^\infty(A) \otimes Y$ be arbitrary. Then an elementary computation shows that $\langle (T \otimes I_Y)h, f\rangle = \langle h, S_X f\rangle$ for all f in the dense subspace $F \otimes X$ of $F\tilde{\otimes}X$, which by continuity extends to all $f \in F(X)$. This gives (i).

For (ii) and (iii) we denote by V the linear space consisting of all countable step functions in $L^\infty(A;Y)$ equipped with the restricted norm of $L^\infty(A;Y)$. Let $R \in \mathcal{L}(L^\infty(A), G)$ be a positive operator dominating T and fix an arbitrary $h \in V$, say $h = \sum_{k=0}^\infty 1_{A_k} y_k$ with $(A_k)_{k\in\mathbb{N}}$ a mutually disjoint sequence in $\mathscr{A}$ and $(y_k)_{k\in\mathbb{N}}$ a bounded sequence in Y. Then note that, by Lemma 4.2, $h = \sum_{k=0}^\infty 1_{A_k} \otimes y_k$ with convergence in the $\sigma(L^\infty(A;Y), L^1(A;X))$-topology. From the weak* continuity of $(S_X)^*$ as a Banach space adjoint operator and (i) it follows that

$$(S_X)^* i(h) = \sum_{k=0}^\infty (S_X)^* i(1_{A_k} \otimes y_k) \overset{(i)}{=} \sum_{k=0}^\infty j(T1_{A_k} \otimes y_k) \tag{4.1}$$

with convergence in the weak*-topology.

For the sequence $(T1_{A_k})_{k\in\mathbb{N}} \subset G$, $\sum_{k=0}^\infty |T1_{A_k}| \in G$ follows from the ideal property of G and the estimate

$$\sum_{k=0}^\infty |T1_{A_k}| \leq \sum_{k=0}^\infty R1_{A_k} = \lim_{K\to\infty} \sum_{k=0}^K R1_{A_k} = \lim_{K\to\infty} R\left(\sum_{k=0}^K 1_{A_k}\right) \leq R1 \in G.$$

Via Lemma 4.2 we obtain convergence of the series $\sum_{k=0}^\infty T1_{A_k} \otimes y_k$ in $G(Y)$ w.r.t. the $\sigma(G(Y), F\tilde{\otimes}X)$-topology together with a norm estimate of the resulting element of $G(Y)$:

$$\left\|\sum_{k=0}^\infty T1_{A_k} \otimes y_k\right\| \leq \|R1\| \, \|(y_k)\|_\infty \leq \|R\| \, \|h\| .$$

In combination with (4.1) this gives $(S_X)^* i(h) \in j(G(Y))$ and $\left\|j^{-1}((S_X)^* ih)\right\| \leq \|R\| \, \|h\|$. As h and R were arbitrary, this shows that $(S_X)^* \circ i$ maps V into $G(Y)$ and that $j^{-1} \circ ((S_X)^* \circ i|_V) \in \mathcal{L}(V, G(Y))$ is of norm $\leq \|T\|_{M(L^\infty(A),G)}$. V being a dense subspace of $L^\infty(A;Y)$ (see Lemma 4.1), $j^{-1}\circ((S_X)^*\circ i|_V)$ has a unique extension to a bounded linear operator Q from $L^\infty(A;Y)$ to $G(Y)$ of norm $\le \|T\|_{M(L^\infty(A),G)}$. The observation that $j\circ Q$ and $(S_X)^*\circ i$ coincide on the dense subspace V of $L^\infty(A;Y)$ and consequently that $j\circ Q=(S_X)^*\circ i$ now yields (ii) and (iii). □

5. An application: Conditional expectation on Banach space-valued L^∞-spaces

Let $(A,\mathscr{A},\mu)$ be a measure space, $\mathscr{F}\subset\mathscr{A}$ a sub-σ-algebra, and X a Banach space. The conditional expectation operator on $L^1(A;X)$ with respect to $\mathscr{F}$ is the operator $\mathbb{E}^1_{\mathscr{F},X}\in\mathcal{L}(L^1(A))$ which assigns to an $f\in L^1(A;X)$ the unique $\mathbb{E}^1_{\mathscr{F},X}f\in L^1(A,\mathscr{F};X)$ satisfying

$$\int_F f\,d\mu=\int_F \mathbb{E}^1_{\mathscr{F},X}f\,d\mu,\qquad F\in\mathscr{F};\tag{5.1}$$

here we write $L^1(A,\mathscr{F};X)$ for the closed linear subspace of $L^1(A;X)$ consisting of all equivalence classes which have a strongly $\mathscr{F}$-measurable representative. This operator is a contractive projection with range $L^1(A,\mathscr{F};X)$ and it can be obtained via bounded tensor extension of the conditional expectation operator $\mathbb{E}^1_{\mathscr{F}}$ on $L^1(A)$, which is a positive operator. We refer to [17], where also pointwise convexity (Jensen-type) inequalities are proved for X-valued extensions of positive operators.

Now suppose that $(A,\mathscr{A},\mu)$ is semi-finite and that the restricted measure space $(A,\mathscr{F},\mu|_{\mathscr{F}})$ is Maharam; it can in fact be shown that $(A,\mathscr{A},\mu)$ is automatically semi-finite when $(A,\mathscr{F},\mu|_{\mathscr{F}})$ is Maharam. Given a Banach space Y, we would like to define the conditional expectation operator on $L^\infty(A;Y)$ with respect to $\mathscr{F}$ as the operator $\mathbb{E}^\infty_{\mathscr{F},Y}\in\mathcal{L}(L^\infty(A;Y))$ which assigns to an $f\in L^\infty(A;Y)$ the unique $\mathbb{E}^\infty_{\mathscr{F},Y}f\in L^\infty(A,\mathscr{F};Y)$ satisfying

$$\int_F f\,d\mu=\int_F \mathbb{E}^\infty_{\mathscr{F},Y}f\,d\mu,\qquad F\in\mathscr{F}(\mu);\tag{5.2}$$

here $\mathscr{F}(\mu)=\{F\in\mathscr{F}:\mu(F)<\infty\}$. In scalar case $Y=\mathbb{K}$ we can define $\mathbb{E}^\infty_{\mathscr{F}}=\mathbb{E}^\infty_{\mathscr{F},\mathbb{K}}$ by restriction to $L^\infty(A)$ of the Banach space adjoint $(\mathbb{E}^1_{\mathscr{F}})^*\in\mathcal{L}((L^1(A))^*)$:

Lemma 5.1. *The conditional expectation operator $\mathbb{E}^1_{\mathscr{F}}$ on $L^1(A)$ is a $\sigma(L^1(A),L^\infty(A))$-to-$\sigma(L^1(A),L^\infty(A))$ continuous linear operator whose adjoint is a positive contractive projection on $L^\infty(A)$ with range $L^\infty(A,\mathscr{F})$ satisfying the above definition of conditional expectation operator on $L^\infty(A)$.*

Proof. Recall that $\mathbb{E}^1_{\mathscr{G}}$ is a positive contractive projection on $L^1(A)$ with range $L^1(A,\mathscr{G})$; so $L^1(A)=L^1(A,\mathscr{G})\oplus U$ where $U:=(1-\mathbb{E}^1_{\mathscr{G}})L^1(A)$. Since $L^\infty(A,\mathscr{G})=(L^1(A,\mathscr{G}))^*$ (as $(A,\mathscr{G},\mu|_{\mathscr{G}})$ is Maharam), it follows that $(\mathbb{E}^1_{\mathscr{G}})^*$ is a positive contractive projection on $(L^1(A))^*=L^\infty(A,\mathscr{G})\oplus U^*$ with range $L^\infty(A,\mathscr{G})$. Identifying $L^\infty(A)$ with a closed subspace of $(L^1(A))^*$ ($(A,\mathscr{A},\mu)$ is semi-finite), $(\mathbb{E}^1_{\mathscr{G}})^*$ restricts to a contractive projection on $L^\infty(A)$ with range $L^\infty(A,\mathscr{G})$. □

We can now obtain $\mathbb{E}_{\mathscr{G}}^{\infty,Y}$ from $\mathbb{E}_{\mathscr{F}}^{\infty}$ via an application of Theorem 3.13:

Proposition 5.2. *Suppose that $(A, \mathscr{A}, \mu)$ is semi-finite and that the restricted measure space $(A, \mathscr{F}, \mu|_{\mathscr{F}})$ is Maharam. For every Banach space Y we have existence of the conditional expectation operator $\mathbb{E}_{\mathscr{F},Y}^{\infty}$ on $L^\infty(A;Y)$ (see (5.2)). $\mathbb{E}_{\mathscr{F},Y}^{\infty}$ is a contractive projection on $L^\infty(A;Y)$ with range $L^\infty(A,\mathscr{F};Y)$. Moreover, if $\langle X, Y\rangle$ is a Banach dual pair, then we have*

$$\int \langle f, \mathbb{E}_{\mathscr{F},Y}^{\infty} g\rangle \, d\mu = \int \langle \mathbb{E}_{\mathscr{F},X}^{1} f, g\rangle \, d\mu, \qquad f \in L^1(A;X), g \in L^\infty(A;Y).$$

Proof. Since there always exists a Banach space X for which there is Banach dual pairing $\langle X, Y\rangle$ (just take $X = Y^*$), we may prove the first and second assertion at the same time. Applying Theorem 3.13 to $S = \mathbb{E}_{\mathscr{F}}^{1} \in \mathcal{L}(L^1(A))$ and $T = \mathbb{E}_{\mathscr{F}}^{\infty} \in \mathcal{L}(L^\infty(A))$, we get contractions $S_X \in \mathcal{L}(L^1(A;X))$ and $T_Y \in \mathcal{L}(L^\infty(A;Y))$ with respect to the duality $\langle L^1(A;X), L^\infty(A;Y)\rangle$ such that S_X is the unique bounded extension of $S \otimes I_X$ and

$$\langle x, T_Y f\rangle = T\langle x, f\rangle, \qquad x \in X, f \in L^\infty(A;Y).$$

Recalling that $\mathbb{E}_{\mathscr{F},X}^{1} = S_X$, it is not difficult to see that we can take $\mathbb{E}_{\mathscr{F},Y}^{\infty} := T_Y$. □

For a different and more direct way to define the conditional expectation operator on Banach-valued L^∞-spaces we refer to [11] (also see the references therein).

Acknowledgment
The author would like to thank Mark Veraar for making him aware of Fact 2.1.(II).

References

[1] Y.A. Abramovich and C.D. Aliprantis. Positive operators. In *Handbook of the geometry of Banach spaces, Vol. I*, pp. 85–122. North-Holland, Amsterdam, 2001.

[2] F. Albiac and N.J. Kalton. *Topics in Banach space theory*, volume 233 of *Graduate Texts in Mathematics*. Springer, New York, 2006.

[3] C.D. Aliprantis and O. Burkinshaw. *Positive operators.* Springer, Dordrecht, 2006. Reprint of the 1985 original.

[4] R. Armario, F.J. Garsiya-Pacheko, and F.J. Peres-Fernandes. On vector-valued banach limits. *Functional Analysis and Its Applications*, 47(4):315–318, 2013.

[5] J. Bourgain. Vector-valued singular integrals and the H^1-BMO duality. In *Probability theory and harmonic analysis* (*Cleveland, Ohio*, 1983), volume 98 of *Monogr. Textbooks Pure Appl. Math.*, pp. 1–19. Dekker, New York, 1986.

[6] A.V. Buhvalov. The analytic representation of operators with an abstract norm. *Izv. Vysš. Učebn. Zaved. Matematika*, (11(162)):21–32, 1975.

[7] A.V. Buhvalov. The duality of functors that are generated by spaces of vector-valued functions. *Izv. Akad. Nauk SSSR Ser. Mat.*, 39(6):1284–1309, 1438, 1975.

[8] D.L. Burkholder. A geometric condition that implies the existence of certain singular integrals of Banach space-valued functions. In *Conference on harmonic analysis in honor of Antoni Zygmund, Vol. I, II (Chicago, Ill.,* 1981), Wadsworth Math. Ser., pp. 270–286. Wadsworth, Belmont, CA, 1983.

[9] D.L. Burkholder. Martingales and singular integrals in Banach spaces. In *Handbook of the geometry of Banach spaces, Vol. I*, pages 233–269. North-Holland, Amsterdam, 2001.

[10] J.B. Conway. *A course in functional analysis*, volume 96 of *Graduate Texts in Mathematics.* Springer-Verlag, New York, second edition, 1990.

[11] N. Dinculeanu. Conditional expectations for general measure spaces. *J. Multivariate Anal.*, 1:347–364, 1971.

[12] D.H. Fremlin. *Topological Riesz spaces and measure theory.* Cambridge University Press, London, 1974.

[13] D.H. Fremlin. *Measure theory. Vol.* 2. Torres Fremlin, Colchester, 2003. Broad foundations, corrected second printing of the 2001 original.

[14] D.H. Fremlin. *Measure theory. Vol.* 3. Torres Fremlin, Colchester, 2004. Measure algebras, corrected second printing of the 2002 original.

[15] J. García-Cuerva and J.L. Rubio de Francia. *Weighted norm inequalities and related topics*, volume 116 of *North-Holland Mathematics Studies.* North-Holland Publishing Co., Amsterdam, 1985. Notas de Matemática [Mathematical Notes], 104.

[16] L. Grafakos. *Classical and modern Fourier analysis.* Pearson Education, Inc., Upper Saddle River, NJ, 2004.

[17] M. Haase. Convexity inequalities for positive operators. *Positivity*, 11(1):57–68, 2007.

[18] W. Hensgen. Some properties of the vector-valued Banach ideal space $E(X)$ derived from those of E and X. *Collect. Math.*, 43(1):1–13, 1992.

[19] C. Herz. The theory of p-spaces with an application to convolution operators. *Trans. Amer. Math. Soc.*, 154:69–82, 1971.

[20] T.P. Hytönen. Vector-valued extension of linear operators, and Tb theorems. In *Vector measures, integration and related topics*, volume 201 of *Oper. Theory Adv. Appl.*, pp. 245–254. Birkhäuser Verlag, Basel, 2010.

[21] A. Kitover and M. Orhon. Reflexivity of Banach $C(K)$-modules via the reflexivity of Banach lattices. *Positivity*, 18(3):475–488, 2014.

[22] J.L. Krivine. Théorèmes de factorisation dans les espaces réticulés. In *Séminaire Maurey–Schwartz* 1973–1974*: Espaces L^p, applications radonifiantes et géométrie des espaces de Banach, Exp. Nos.* 22 *et* 23, p. 22. Centre de Math., École Polytech., Paris, 1974.

[23] S. Kwapień. Isomorphic characterizations of inner product spaces by orthogonal series with vector valued coefficients. *Studia Math.*, 44:583–595, 1972. Collection of articles honoring the completion by Antoni Zygmund of 50 years of scientific activity, VI.

[24] J. Lindenstrauss and L. Tzafriri. *Classical Banach spaces. II*, volume 97 of *Ergebnisse der Mathematik und ihrer Grenzgebiete* [*Results in Mathematics and Related Areas*]. Springer-Verlag, Berlin, 1979. Function spaces.

[25] J. Marcinkiewicz and A. Zygmund. Quelques inégalités pour les opérations linéaires. *Fund. Math*, pages 122–132, 1939.

[26] P. Meyer-Nieberg. *Banach lattices*. Universitext. Springer-Verlag, Berlin, 1991.

[27] H.H. Schaefer and M.P. Wolff. *Topological vector spaces*, volume 3 of *Graduate Texts in Mathematics*. Springer-Verlag, New York, second edition, 1999.

[28] N.N. Vakhania, V.I. Tarieladze, and S.A. Chobanyan. *Probability distributions on Banach spaces*, volume 14 of *Mathematics and its Applications (Soviet Series)*. D. Reidel Publishing Co., Dordrecht, 1987. Translated from the Russian and with a preface by Wojbor A. Woyczynski.

[29] A.W. Wickstead. Regular operators between Banach lattices. In *Positivity*, Trends Math., pages 255–279. Birkhäuser, Basel, 2007.

[30] M. Wójtowicz. Reflexivity and the separable quotient problem for a class of Banach spaces. *Bull. Polish Acad. Sci. Math.*, 50(4):383–394, 2002.

[31] A.C. Zaanen. *Integration*. North-Holland Publishing Co., Amsterdam, 1967. Completely revised edition of An introduction to the theory of integration.

[32] M. Zippin. Banach spaces with separable duals. *Trans. Amer. Math. Soc.*, 310(1): 371–379, 1988.

Nick Lindemulder
Delft Institute of Applied Mathematics
Delft University of Technology
P.O. Box 5031
NL-2600 GA Delft, The Netherlands
e-mail: `N.Lindemulder@tudelft.nl`

Ordered Structures and Applications: Positivity VII
Trends in Mathematics, 307–327

A Unified Approach to Topological Invariants of Lorentz–Orlicz–Marcinkiewicz Spaces

A.A. Mekler

Abstract. The paper provides a survey of results concerning topological invariants of the spaces in a Lorentz–Orlich–Marcinkiewicz triple such that their fundamental functions can be expressed explicitly one through another. In particular, the paper contains an interpretation of topological invariants through the distribution properties of a "base" sequence of natural numbers – the same for every space in the triple.

Mathematics Subject Classification (2010). Primary 46E30.

Keywords. Banach function spaces, isomorphism.

Introduction

Topological invariants of a Banach Function Space (BFS) are by definition the properties of such a space conserved by its linear isomorphisms; thus they are the properties not of a specific norm on that space but the properties of the whole class of equivalent norms on it.

Our approach to invariants is based on considering them as properties of *modulars*. A modular is a class of "norming" functions that induce equivalent norms on the spaces indicated in the title of the paper. It is well known, [1], that among the triples of Orlicz, Lorentz, and Marcinkiewicz spaces on $[0, \infty)$ there are such ones that the *fundamental function* of any of them defines the other two as sets (notice that in such a case the above-mentioned fundamental function defines the norm on the corresponding space equivalent to the original one). We call the spaces in such a triple *related*. Topological invariants of each of the related spaces are in a natural one to one correspondence but their interpretations might be quite different. Consider for example the well-known topological invariant of (always non separable) Marcinkiewicz space – the boundedness of the Hardy operator. For the related Orlicz space it corresponds to the property of being separable.

The goal of this paper is to present a survey of the author's work on topological invariants of related spaces and some algebraic structures of these invariants defined by the binary operation of *superposition* of norming functions. Some results were presented on 7th POSITIVITY Conference, Leiden, July, 2013, [30].

Definition ([23]).

1. A *norming* function is a real valued continuous function on $[0,\infty)$ such that $f(0)=0$, f is positive and strictly increasing on $(0,\infty)$, and $\lim\limits_{t\to\infty} f(t)=\infty$. The set of all norming functions will be denoted by $\mathcal{F}$.
2. We say that two functions $f_1, f_2 \in \mathcal{F}$ are $\overset{m}{\sim}$*equivalent* and write $f_1 \overset{m}{\sim} f_2$ if for some constant ν, $\nu \geq 1$, and for all $t \in [0,\infty)$ we have
$$\nu^{-1}\cdot f_2(\nu\cdot t) \leq f_1(t) \leq \nu \cdot f_2(\nu^{-1}\cdot t), \qquad (\overset{m}{\sim}).$$
If the above inequalities are satisfied only in a neighborhood of 0 or infinity then we will speak about $\overset{m}{\sim}$ equivalence in the corresponding neighborhood.
3. The class $\mathfrak{f}$ of all functions from $\mathcal{F}$, that are $\overset{m}{\sim}$equivalent to some $f \in \mathcal{F}$, we will call a $\overset{m}{\sim}$*modular*; The set of all $\overset{m}{\sim}$modulars will be denoted by $\mathfrak{F}$ ($\mathfrak{F}_0$, $\mathfrak{F}_\infty$, respectively).
4. An $\overset{m}{\sim}$*invariant* is by definition such a property of a norming function that holds for all the functions from its $\overset{m}{\sim}$modular.

Because in every of the related spaces the compression/dilation operator σ_s, $s \geq 0$
$$\sigma_s x(t) := x(t\cdot\nu),\ t\in[0,\infty),$$
is bounded, [1], the statements expressing topologically invariant properties of these spaces cannot depend on compressions/dilations of norming functions along the t-axis. In other words topologically invariant properties of a space must be formulated in terms of the modular of its norming function.

The properties of $\mathfrak{F}$ modulars (respectively, of $\mathfrak{F}_0$ or $\mathfrak{F}_\infty$ modulars) can be characterized by the limit behaviour of the corresponding norming functions at 0 and ∞ (only at 0, or only at ∞, respectively) under the action of operator σ_s. If the norming function is *symmetric* (see Definition 1.2) then it is enough to consider its limit behaviour at 0 (or at ∞ if it is more convenient). Thus instead of a space of functions on $[0,\infty)$ we can consider a pair of spaces on $[0,1]$ such that each of them is generated by its norming function restricted on $[0,1]$ (or, respectively, on $[1,\infty)$) and *symmetrically extended* on the whole interval $[0,\infty)$. These techniques (described in details in the main body of the paper) will allow us to restrict our considerations to the case of spaces on $[0,1]$.

It is worth noticing that properties of a single norming function correspond to the properties of the unit sphere of the space this function defines, i.e., such a property defines a *geometric* invariant, while a property of the modular of this function is a *topological* invariant. On the other hand if a geometric invariant P is fixed we can define the corresponding topological invariant $\mathcal{P}$ as follows.
$$\mathcal{F}\in\langle\mathcal{P}\rangle \iff \exists f\in\mathcal{F}, f\in\langle P\rangle.$$

Thus arises a problem of finding an analytic expression for a topological invariant corresponding to the given geometric one.

Let us briefly discuss the $\overset{m}{\sim}$invariants considered in the current paper.

A. In Section 1 we introduce modulars of Marcinkiewicz spaces (M-modulars) and Orlicz spaces (N-modulars) (see [20] and [23]). We introduce so-called *involutions* that establish one to one correspondences between the sets of all M-modulars and all N-modulars. These involutions provide one to one correspondences between topological invariants of Marcinkiewicz and Orlicz spaces. For example the property of *regularity* of an M-modular is equivalent to the Hardy–Littlewood property, [7], of the corresponding Marcinkiewicz spaces. This property is in correspondence with the Δ_2*-condition*, [4], for an N-modular, i.e., the condition of separability of the related Orlicz spaces.

B. The notion of regularly varying function with parameter α $(0 \leq \alpha \leq 1)$ came from Mathematical Physics, [5] and represents a geometric invariant. The usefulness of this notion for consideration of so-called *symmetric functionals* on Marcinkiewicz spaces becomes clear from the results in [8]–[10]. The problem of finding an analytic expression for the corresponding topological invariant was solved in [11], [16], and [17] for $\alpha = 0$ and $\alpha = 1$. For $0 < \alpha < 1$ this problem was solved in [18].

In this connection we want to notice that the $\overset{m}{\sim}$invariant of regular varying is stronger that the $\overset{m}{\sim}$invariant of regularity.

C. The property of *submultiplicativity* of M-modulars is stronger than the property of regular varying, [22], but for so-called *quickly varying at* 0 M-modulars (i.e., for $\alpha = 1$) these properties are equivalent. An important consequence is that if a Marcinkiewicz space has a nontrivial band of symmetric functionals then its norm is submultiplicative.

Another consequence is that if a Marcinkiewicz space equals as a set to the related Orlicz space then its modular is quickly varying at 0, [12]–[14].

It is also worth noticing that an M-modular is submultiplicative if and only if the corresponding N-modular satisfies the well-known Δ'-condition, [4].

D. One of the below-considered invariants is connected with the problem of p-*convexity*, [2]. Let ψ be a defined on $[0,1]$ concave symmetric norming function, and δ_ψ be its upper index of compression/dilation, [3]. It is known, [15], that for $1 < p \neq 1/\delta_\psi$ the Marcinkiewicz space $M_\psi([0,1])$ is p-convex (and therefore the Lorentz space $\Lambda_{\psi_*}([0,1])$ is q-concave, where $1/p + 1/q = 1$) if and only if $p < 1/\delta_\psi$. The last statement is equivalent to the following one: if a concave ψ is given and $1 < p \neq 1/\delta_\psi$ then the power $\psi^p \overset{m}{\sim}$ is equivalent to a concave function if and only if $p < 1/\delta_\psi$.

Then the following problem arises: find a characterization of the $\overset{m}{\sim}$invariant defined by the property that raising a concave ψ to the limit power $1/\delta_\psi$ also provides a function $\overset{m}{\sim}$equivalent to a concave one (below we call such a ψ and its

M-modular a *pseudopower*). It was proved in [27] that if a symmetric function ψ is a pseudopower then the corresponding to ψ *supremal* and *upper-limit* functions of compression/dilation are $\overset{m}{\sim}$equivalent. From this fact we obtain a criterion for any function on $[0,\infty)$, that is $\overset{m}{\sim}$equivalent to a concave function, to be pseudopower.

Thus we obtain that a pseudopower function is submultiplicative at 0 and therefore regularly varying at 0 (the inverse in general is false).

Because every norming functions maps $[0,\infty)$ onto itself the composition of two norming functions also is a norming function. Considering composition as a semigroup operation we can correctly define the *semigroup* of $\overset{m}{\sim}$modulars of norming functions, [29]. The sets of $\overset{m}{\sim}$modular corresponding to some $\overset{m}{\sim}$invariants can be considered as algebraic substructures of the above-defined semigroup. These substructures were studied in [20], [21], and [29].

In the papers [19], [21]–[24], and [26]–[29] the $\overset{m}{\sim}$isomorphisms of any M-modular are interpreted with the help of the limit behavior of some corresponding to it sequence of natural numbers (more precisely the equivalency class of such a sequence). The essence of such an approach consists in conversion to the logarithmic scale followed by discretization of values of norming functions; it allows for simultaneous interpretation of N and M-modulars through so-called *a-modulars*. To achieve it we introduce the notion of natural *base*. Such a base simultaneously describes both types of a-modulars. We would like to emphasize that such an interpretation is stable under the action of involutions connecting related N and M modulars. By using these techniques we can provide a simple description of the action at 0 of the operator of compression/dilation on $\overset{m}{\sim}$modulars.

The described above techniques can be applied to interpret any (in particular, discussed in this paper) topological invariants of related triples (Lorentz, Orlicz, and Marcinkiewicz spaces).

1. M-functions and M-modulars. Related triples/paires. Symmetric functions and modulars

Definition 1.1.

1. Let ψ be a continuous, **strictly concave** norming function on $[0,\infty)$ such that
   (a) $\psi(1) = 1$.
   (b) ψ is not $\overset{m}{\sim}$ equivalent to the function e, $e(t) \equiv t$ on $(0,\infty)$.
   (c) There is a norming function on $[0,\infty)$, $\psi_\star$ (called dual to ψ), satisfying conditions (a) and (b) such that

$$\psi_\star(t) \overset{m}{\sim} \frac{t}{\psi(t)}, t > 0.$$

Then we will denote the $\overset{m}{\sim}$modular of ψ by Ψ and call ψ a Marcinkiewicz function (briefly M-function), respectively, dual Marcinkiewicz modular (or M-modular) Ψ_*.

If two functions belong to the same M-modular we call them *equiconcave* functions.

2. Similarly, a continuous strictly convex norming function ϕ on $[0,\infty)$ which is not $\overset{m}{\sim}$equivalent to $e(t)$ is called an *Orlicz function* or *N-function*. The function complementary to ϕ (see [4]) is denoted $\phi^\star$. The $\overset{m}{\sim}$modular of an N-function ϕ is called Orlicz modular or N-modular and denoted by Φ. The elements of an N-modular are called *equiconvex* functions.

It is clear from the definition of $\overset{m}{\sim}$equivalence that all the information about $\overset{m}{\sim}$invariants can be expressed in terms of limit/supremal values of compression/dilation operators on equiconcave or equiconvex functions, e.g., upper and lower indices of compression/dilation (see below).

As we have already mentioned the set of all spaces of Lorentz, Marcinkiewicz, and Orlicz can be partitioned into *related triples* $(\Lambda_\varphi,\ \mathsf{M}_\psi,\ \mathsf{L}^*_\phi)$, such that the norming (i.e., fundamental) functions of these spaces are transformed into each other by natural involutions and thus induce involutions on their modulars. More precisely, we speak about involutions of *duality* $I_1 : I_1(\psi) := \psi_*$, *complementarity* $I_2 : I_2(\phi) := \phi^*$, *inverse* $I_3 : I_3(\phi) := \phi^{-1}$, and *inversion* $I_4 : I_4\xi(0) := 0$, $I_4\xi(t) := t\cdot\xi(\frac{1}{t})$, $0 < t < \infty$.

The fundamental functions of the Lorentz space Λ_φ and the dual to it Marcinkiewicz space M_ψ are mutually dual: $\varphi = \psi_*$, and therefore their $\overset{m}{\sim}$invariants can be expressed in a simple "mutually dual" form. The correspondence between $\overset{m}{\sim}$invariants of Marcinkiewicz and Orlicz spaces is a bit more complicated.

For any N-function ϕ we have the following two equivalencies of M-functions: $I_3I_2\phi \overset{m}{\sim} I_1I_3\phi$; $I_4I_3I_2\phi \overset{m}{\sim} I_4I_1I_3\phi$. For modulars these equivalencies can be expressed as two *involution formulas*

$$1)\qquad I_3I_2\Phi = I_1I_3\Phi,$$

$$2)\qquad I_4I_3I_2\Phi = I_4I_1I_3\Phi.$$

The involution identities provide a reason to call the M-modular $\Phi^\frown \ni \phi^\frown := I_4I_1I_3\phi$ and the N-modular Φ a *related pair of modulars*. Every M-modular Ψ is related to the N-modular $\Psi^\smile \ni \psi^\smile := (I_3)^{-1}I_1I_4\psi$, where the M-function ψ is in Ψ. The pairs of functions $(\phi^\frown,\phi)$, $(\psi,\psi^\smile)$, as well as the pairs of modulars $(\Phi^\frown,\Phi)$ and $(\Psi,\Psi^\smile)$ are called *mutually related.* $\overset{m}{\sim}$invariants of the modulars that are mutually related are also called *mutually related* and can be easily expressed one through the other. By adding to a related pair the Lorentz space dual to the corresponding Marcinkiewicz space we can speak about a *related* triple of spaces or, respectively, of topological invariants.

The following remark will play an important role in the sequel: $\overset{m}{\sim}$invariants of the Marcinkiewicz space $\mathsf{M}_\xi[0,\infty)$ can be identified with pairs of $\overset{m}{\sim}$invariants of the Marcinkiewicz space $\mathsf{M}_\xi(0,1)$. This can be done in the following way.

On the sets of M- and N-functions we define the *involution of symmetry* I_5: $I_5\xi(t) := \frac{1}{\xi(\frac{1}{t})}$, $t \in (0,\infty)$. This involution induces the corresponding involution on the sets of equiconcave (respectively, equiconvex) functions.

Definition 1.2. A function ξ and its modular $\Xi \ni \xi$ are called *symmetric* if the following $\overset{m}{\sim}$ equivalence holds

$$\xi(t) \overset{m}{\sim} I_5\xi(t),\ t \in [0,\infty).$$

It is immediate that the Marcinkiewicz space $\mathsf{M}_\xi[0,\infty)$ generated by an equiconcave symmetric function ξ can be identified with the Marcinkiewicz space $\mathsf{M}_\xi(0,1)$.

For an equiconcave function ξ on $[0,\infty)$ we define its symmetric extensions from $[0,1]$ onto $(0,\infty)$ and from $[1,\infty)$ onto $(0,\infty)$, respectively, as

$$\begin{cases} \xi^0(t) = \xi(t), \quad \text{if } t \in [0,1], \quad \xi^0(t) = I_5\xi(t), \text{ if } t \in [1,\infty); \\ \xi^\infty(t) = I_5\xi(t), \text{ if } t \in (0,1], \ \xi^\infty(t) = \xi(t), \quad \text{if } t \in [1,\infty). \end{cases}$$

Both ξ^0 and ξ^∞ are symmetric and equiconcave. They are called left and, respectively, right *symmetric brackets* for ξ. Thus we have a natural correspondence between the Mazrcinkiewicz space $\mathsf{M}_\xi(0,\infty)$ and the pair of Marcinkiewicz spaces $\mathsf{M}_{\xi^0}(0,1)$ and $\mathsf{M}_{\xi^\infty}(0,1)$, generated by the left and right symmetric bracket of ξ, respectively.

Similarly we define symmetric equiconvex functions and construct symmetric brackets for any equiconvex function.

It follows from the considered above equivalencies that to any $\overset{m}{\sim}$invariant of a Marcinkiewicz space $\mathsf{M}(0,\infty)$ corresponds a pair of $\overset{m}{\sim}$invariants of the space $\mathsf{M}(0,1)$, and vice versa. For a symmetric M-modular the $\overset{m}{\sim}$invariants in this pair are identical. Because related modulars are symmetric or nonsymmetric at the same time the involution formulas establish a one-to-one correspondence between $\overset{m}{\sim}$invariants of related spaces of Lorentz, $\Lambda_{(\phi^\frown)_*}(0,1)$, Marcinkiewicz, $\mathsf{M}_{\phi^\frown}(0,1)$, and Orlicz, $\mathsf{L}_\phi(0,1)$.

Definition 1.3. For a norming function ξ on $[0,\infty)$ we define three types of *supremal* compression/dilation functions depending on parameter $s \geq 0$, two types of *upperlimit* functions, and the lower (γ) and the upper (δ) *indices of dilation/compression*:

$$\begin{cases} \mathfrak{S}_\xi(s) := \sup\limits_{t\in[0,\infty)} \frac{\sigma_s\xi(t)}{\xi(t)}, & \gamma_\xi := \lim\limits_{s\to 0} \frac{\log_2 \mathfrak{S}_\psi(s)}{\log_2 s}; \quad \delta_\xi := \lim\limits_{s\to\infty} \frac{\log_2 \mathfrak{S}_\xi(s)}{\log_2 s}; \\ \mathfrak{S}^0_\xi(s) := \sup\limits_{t\in[0,1], s\cdot t\in[0,1]} \frac{\sigma_s\xi(t)}{\xi(t)}, & \gamma^0_\xi := \lim\limits_{s\to 0} \frac{\log_2 \mathfrak{S}^0_\xi(s)}{\log_2 s}; \quad \delta^0_\xi := \lim\limits_{s\to\infty} \frac{\log_2 \mathfrak{S}^0_\xi(s)}{\log_2 s}; \\ \mathfrak{S}^\infty_\xi(s) := \sup\limits_{t\geq 1,\ s\cdot t\geq 1} \frac{\sigma_s\xi(t)}{\xi(t)}\xi(t), & \gamma^\infty_\xi := \lim\limits_{s\to 0} \frac{\log_2 \mathfrak{S}^\infty_\xi(s)}{\log_2 s}; \ \delta^\infty_\xi := \lim\limits_{s\to\infty} \frac{\log_2 \mathfrak{S}^\infty_\xi(s)}{\log_2 s}; \\ \mathfrak{L}^0_\xi(s) := \limsup\limits_{t\to 0} \frac{\sigma_s\xi(t)}{\xi(t)}; & \mathfrak{L}^\infty_\xi(s) := \limsup\limits_{t\to\infty} \frac{\sigma_s\xi(t)}{\xi(t)}. \end{cases}$$

For a symmetric equiconcave function φ we have the following $\overset{m}{\sim}$equivalenvies

$$\mathfrak{S}_\varphi(s) \overset{m}{\sim} \mathfrak{S}^0_\varphi(s) \overset{m}{\sim} \mathfrak{S}^\infty_\varphi(s),\ \mathfrak{L}^0_\varphi(s) \overset{m}{\sim} \mathfrak{L}^\infty_\varphi(s),\ s \geq 0,$$

as well as equalities $\gamma_\varphi = \gamma^i_\varphi,\ \delta_\varphi = \delta^i_\varphi,\ i = 0, \infty$.

2. Examples of topological invariants of Marcinkiewicz and Orlicz spaces

Definition 2.1.

1. Recall that an N-function ϕ and its N-modular Φ satisfy the Δ_2 condition, [4], if ϕ is $\overset{m}{\sim}$equivalent at infinity to the function $\sigma_2\phi$.
2. An M-modular Ψ is called *$\mathcal{HLP}_M$-modular* (respectively, *$\mathcal{HLP}_\Lambda$-modular*), [7], if the Marcinkiewicz space $\mathsf{M}_\Psi[0,\infty)$ (respectively, the Lorentz space $\Lambda_{\Psi_*}[0,\infty)$) has the Hardy–Littlewood property, i.e., for any norming function $\xi \in \Psi$ (respectively, $\xi \in \Psi_*$) the function $\frac{\xi(t)}{t}$ belongs to the corresponding space. It is well known, [1], that for a symmetric M-function ψ its modular Ψ is a $\mathcal{HLP}_M$-modular (respectively, $\mathcal{HLP}_\Lambda$-modular) if and only if when $\gamma_\psi > 0$ (respectively, $\delta_\psi < 1$).

Theorem 2.1. *A symmetric M-modular Ψ is a $\mathcal{HLP}_M$-modular if and only if the related N-modular $\Psi^\smile$ satisfies the Δ_2-condition.*

Thus the separability of the Orlicz space $\mathsf{L}^*_\Phi(0,1)$ is equivalent to the fulfilment of the Hardy–Littlewood property in the related Marcinkiewicz space $\mathsf{M}_{\Phi^\frown}(0,1)$.

Definition 2.2. A norming function F is called *submultiplicative* (respectively, *supermultiplicative*) on some subset Q of $[0,\infty)$, [1], if there is a constant $c > 0$ such that

$$F(s \cdot t) \leq c \cdot F(s) \cdot F(t) \text{ (respectively, } F(s \cdot t) \geq c \cdot F(s) \cdot F(t)),\ s,t \in Q.$$

For equiconcave functions we have the following theorem.

Theorem 2.2.

1. *$\overset{m}{\sim}$invariant of submultiplicativity is stronger then $\mathcal{HLP}_M$ $\overset{m}{\sim}$invariant.*
2. *An equiconcave function ψ is submultiplicative on $[0,\infty)$ if and only if*
$$\psi(s) \overset{m}{\sim} \mathfrak{S}_\psi(s).$$
3 *A symmetric equiconcave function φ is submultiplicative on $[0,1]$ if and only if in some neighborhood of* 0
$$\varphi(s) \overset{m}{\sim} \mathfrak{S}^0_\varphi(s).$$
4. *A M-modular Ψ is submultiplicative if and only if the related N-modular $\Psi^\smile$ is submultiplicative as well.*

Thus the fulfilment of $\overset{m}{\sim}$invariant of submultiplicativity in $\mathsf{M}_\Psi(0,1)$ is equivalent to the fulfilment of Δ'-condition in the Orlicz space $\mathsf{L}^*_{\Psi^\smile}(0,1)$, [4].

Definition 2.3. If a norming function on $[0,\infty)$ is simultaneously submultipicative and supermultiplicative we call it and its modular *equimultiplicative.*

Theorem 2.3 ([28]).

1. *An equiconcave function ψ is equimultiplicative if and only there are constants $0<\alpha,\beta<1$ such that $\psi(t)\overset{m}{\sim}t^{\alpha}$ at 0, and $\psi(t)\overset{m}{\sim}t^{\beta}$ at ∞.*
2. *For a symmetric equimultiplicative function we have $\alpha=\beta$.*
3. *Similar statements take place for equiconvex equimultiplicative functions with $\alpha,\beta>1$.*

Definition 2.4. A norming function F is called a *regular variation function* with parameter α, $0<\alpha<\infty$, (shortly RV_α-function) at 0, respectively, at ∞, if

$$\lim_{t\to 0}\frac{\sigma_s F(t)}{F(t)}=s^{\alpha},\quad \text{respectively,}\quad \lim_{t\to\infty}\frac{\sigma_s F(t)}{F(t)}=s^{\alpha}.$$

In the case when we consider only one of singular points 0 (respectively, ∞) we apply the notation $F\in RV^0_\alpha$ (respectively, $F\in RV^\infty_\alpha$).

For an equiconcave function always $\alpha\in(0,1)$, while for an equiconvex one $\alpha>1$.

Equiconcave RV^0_1 as well as equiconvex RV^∞_∞-functions are called *rapidly growing*, while equiconcave RV^0_0 and equiconvex RV^∞_1-functions are called *slowly growing*, [5].

Let us emphasize that for an M-function ψ the RV-property is a property of the unit sphere of the Marcinkiewicz space $M_\psi(0,\infty)$. This property in general is not preserved for an equivalent norm. The topological invariant of $M_\psi(0,\infty)$ corresponding to the RV_α-property was first obtained for rapidly and slowly growing equiconcave functions, [11], [16], [17], and then generalized for an arbitrary $\alpha\in[0,1]$ in [18].

For equiconcave functions on $[0,\infty)$ we will denote the RV$\overset{m}{\sim}$invariants by mRV^0_α and mRV^∞_α, respectively. It is known, [8]–[10], that for $\alpha=0$ and $\alpha=1$ the corresponding RV$\overset{m}{\sim}$invariants can provide criterions of existence or not existence of some types of singular functionals (so-called *symmetric functionals*) on corresponding Marcinkiewicz spaces.

Theorem 2.4.

1. (a) *If for an equiconcave function ψ we have at 0*

$$\limsup_{t\to 0}\frac{\sigma_s\psi(t)}{\psi(t)}\overset{m}{\sim}s^{\alpha}$$

then there is an M-function $\varphi\overset{m}{\sim}\psi$ such that $\varphi\in RV^0_\alpha$.

(b) *If for an equiconcave function ψ we have at ∞*

$$\limsup_{t\to\infty}\frac{\sigma_s\psi(t)}{\psi(t)}\overset{m}{\sim}s^{\alpha}$$

then there is an M-function $\varphi\overset{m}{\sim}\psi$ such that $\varphi\in RV^\infty_\alpha$.

2. *If Ψ is a symmetric and submultiplicative on $[0,1]$ M-modular then there is an α such that $\Psi \in mRV_\alpha^0$.*
3. *If an M-modular $\Psi \in mRV_1^0$ then Ψ is submultiplicative on $[0,1]$.*
4. *If $\alpha > 1$ the involution formulas allow us to make similar conclusions for $\overset{m}{\sim}$invariants induced by RV_α^∞.*

Now we will consider the topological invariant of a related triple of Lorentz–Orlicz–Marcinkiewicz spaces connected with the topic of *p-convexity* of Banach spaces of measurable functions, [2]. Let ψ be a symmetric equiconcave function and δ_ψ be its upper index. It is known that for $1 < p \neq \frac{1}{\delta_\psi}$ the Marcinkiewicz space $M_\psi([0,1])$ is p-convex (or equivalently the Lorentz space $\Lambda_{\psi_*}([0,1])$ is q-concave, where $1/p + 1/q = 1$) if and only if $p < \frac{1}{\delta_\psi}$, [15]. The last statement is equivalent to the following one. For an equiconcave function ψ on $[0,\infty)$ and for $1 < p \neq \frac{1}{\delta_\psi}$ the power ψ^p is equiconcave if and only if $p < \frac{1}{\delta_\psi}$.

Definition 2.5. An equiconcave function ψ (as well as its modular Ψ) is called a *pseudopower* if the power $\psi^{\frac{1}{\delta_\psi}}$ is an equiconcave function.

By Theorem 2.3 an equimultiplicative function cannot be a pseudopower. We denote the union of disjoint classes of equimultiplicative and pseudopower $\overset{m}{\sim}$modulars by $(pPow)$.

As an example of a pseudopower function on $[0,\infty)$ we can consider an equiconcave function $(\varphi^0)^{\frac{1}{2}}$ where $\varphi^0(0) = 0$, $\varphi^0(t) = -t \cdot \log_2 \frac{t}{2}, 0 < t \leq 1$, and $\varphi^0(t) = \frac{1}{\varphi^0(\frac{1}{t})}$, $t > 1$.

On the other hand the following equiconcave function $(\varphi^\infty)^{\frac{1}{2}}$ is not a pseudopower, where $\varphi^\infty(t) := t \cdot \log_2 2t, 1 \leq t < \infty$, and $\varphi^\infty(t) = \frac{1}{\varphi^\infty(\frac{1}{t})}$, $t \in (0,1]$.

Theorem 2.5 ([26]).

1. *Let φ be an equiconcave function and φ^0, φ^1 be its left and right brackets, respectively. Each of the following conditions guarantees that φ is a pseudopower.*
   (a) *φ^0 is a pseudopower and $\delta_{\varphi^0} > \delta_{\varphi^1}$.*
   (b) *φ^0 and φ^1 are both pseudopowers and $\delta_{\varphi^0} = \delta_{\varphi^1}$.*
2. *Conversely, if φ is a pseudopower then its bracket that has the larger upper index is also a pseudopower. If the upper indices of the brackets are equal then both brackets are pseudopowers.*

Theorem 2.6 ([27]).

1. *Let ψ be an equiconcave function with the symmetric M-modular Ψ. The following conditions are equivalent (see Definition 1.3).*
   (a) *$\Psi \in (pPow)$.*
   (b) *$\mathfrak{S}_\psi^\infty(s) \overset{m}{\sim} \mathfrak{L}_\psi^\infty(s)$.*
2. *If Ψ is a symmetric pseudopower modular then Ψ is submultiplicative at 0.*

Definition 2.6. We say that the Marcinkiewicz space $M_\psi(0,1)$ has the property $\mathcal{P}^*$ if the non-increasing function $\frac{\psi(t)}{t}$ belongs to $L^1(0,1)$.

It is immediate that $\overset{m}{\sim}$invariant $\mathcal{P}^*$ is weaker than $(\mathcal{HLP})_M$-invariant. It is well known, [3], that for an M-function ψ the following is true.

Theorem 2.7. *The following inclusions are equivalent*

$$M_\psi(0,1) \in \mathcal{P}^* \quad \textit{and} \quad \frac{d\psi(t)}{dt} \in L\log^+ L(0,1).$$

It is known, [12]–[14], that the Orlicz space $\mathsf{L}_\phi(0,1)$ and the Marcinkiewicz space $\mathsf{M}_\psi(0,1)$ can under some conditions be equal as sets. The next theorem provides a criterion for it.

Theorem 2.8. *The following conditions are equivalent.*

1. $\mathsf{L}_\phi(0,1) = \mathsf{M}_\psi(0,1)$, *as sets.*
2. (a) $\phi(t) \overset{m}{\sim} I_5 I_3 I_4 \psi$, *and*
   (b) $\psi \in (\mathcal{HLP})_M$, *and*
   (c) $\exists\, \varepsilon > 0 : \sum_{n\geq 1} 2^n \cdot \psi_*^{-1}\big(\varepsilon \cdot \psi_*(2^{-n})\big) < \infty$.

3. The semigroup of M-modulars

In this section we consider the set of all M-modulars on $[0,1]$ which as was noticed before can be identified with the set of all symmetric M-modulars. To extend these considerations to the set of all M-modulars we can use the symmetric brackets of an arbitrary M-modular.

On the set $\mathfrak{M}$ of all M-modulars on $[0,1]$ we define the binary operation of composition as follows.

Definition 3.1. $\Psi_1 \circ \Psi_2$ is the modular of the composition of M-functions

$$\psi_1 \circ \psi_2(t) := \psi_1\Big(\psi_2(t)\Big),\ t \in [0,1],$$

where $\psi_i \in \Psi_i,\ i = 1,2$.

Theorem 3.1.

1. *Composition of M-modular is well defined.*
2. *The upper index of the composition of two equiconcave on $[0,1]$ functions is less or equal to the product of upper indices of the factors; moreover, if the factors are either $(pPow)$ or mRV^0-functions, we have the equality.*
3. *The set $\mathfrak{M}$ endowed with the operation $\circ$ is a non-abelian semigroup.*

Definition 3.2. Let $\mathcal{V}$ be a subset of the semigroup $(\mathfrak{M},\ \circ)$. We will call $\mathcal{V}$:

1. *subsemigroup*, if $v_1, v_2 \in \mathcal{V} \Rightarrow v_1 \circ v_2 \in \mathcal{V}$;
2. *closed subsemigroup*, if $v_1, v_2 \in \mathcal{V} \Leftrightarrow v_1 \circ v_2 \in \mathcal{V}$;
3. *right closed subsemigroup*, if $v_1, v_2 \in \mathcal{V} \Rightarrow v_1 \circ v_2 \in \mathcal{V} \Rightarrow v_2 \in \mathcal{V}$;

4. *two-sided ideal*, if $[v_1 \in \mathcal{V}$ or/and $v_2 \in \mathcal{V}] \Rightarrow v_1 \circ v_2 \in \mathcal{V}$;
5. *closed ideal*, if $\mathcal{V}$ is an ideal and $v_1 \circ v_2 \in \mathcal{V} \Leftrightarrow [v_1 \in \mathcal{V}$ or/and $v_2 \in \mathcal{V}]$;
6. *strongly closed ideal*, if $\mathcal{V}$ is an ideal and $v_1 \circ v_2 \in \mathcal{V} \Leftrightarrow [v_1 \in \mathcal{V},\ v_2 \in \mathcal{V}]$.

The M-modulars satisfying some $\overset{m}{\sim}$invariant $\mathcal{P}$ constitute a subset in $\mathfrak{M}$ which we will denote $\mathfrak{M}_{\mathcal{P}}$. Depending on $\mathcal{P}$ this subset can have different semigroup properties.

Theorem 3.2 ([29]). *Let $\mathcal{V}$ be a closed subsemigroup. Then its complement $\mathcal{V}^c := \mathfrak{M} \setminus \mathcal{V}$ is a closed ideal.*

Vice versa, the complement of a closed ideal is a closed subsemigroup.

The next theorem, [20], [29], characterizes the algebraic properties of the set $\mathfrak{M}_{\mathcal{P}}$ where $\mathcal{P}$ is one of the $\overset{m}{\sim}$invariants considered in Section 2.

Theorem 3.3.
1. *If $\mathcal{P}$ is $(\mathcal{HLP})_M$ or mRV_1^0 then $\mathfrak{M}_{\mathcal{P}}$ is a closed ideal.*
2. *If $\mathcal{P} = (\mathcal{HLP})_\Lambda$ then $\mathfrak{M}_{\mathcal{P}}$ is an ideal.*
3. *If $\mathcal{P} = mRV_0^0$ then $\mathfrak{M}_{\mathcal{P}}$ is a closed ideal.*
4. *If $\mathcal{P}$ is the $\overset{m}{\sim}$invariant of sub or supermultiplicativity then $\mathfrak{M}_{\mathcal{P}}$ is a subsemigroup.*
5. *If $\mathcal{P}$ is the $\overset{m}{\sim}$invariant of equimultiplicativity then $\mathfrak{M}_{\mathcal{P}}$ is a subsemigroup isomorphic to the commutative group of M-modulars generated by power functions.*
6. *If $\mathcal{P} = \bigcup_{0<\omega<1} mRV_\omega^0$ then $\mathfrak{M}_{\mathcal{P}}$ is a subsemigroup. Moreover, $(\psi \in mRV_\alpha^0$, $\varphi \in mRV_\beta^0$, $0 < \alpha, \beta < 1) \Rightarrow \psi \circ \varphi \in RV_{\alpha \cdot \beta}^0$.*
7. *If $\mathcal{P} = (pPow)$ then $\mathfrak{M}_{\mathcal{P}}$ is a right closed subsemigroup.*

4. Interpretation of $\overset{m}{\sim}$invariants via natural bases

We introduce the basic notions of the language of natural bases with the help of simple remarks and lemmas. Some of the statements are clarified by providing their proofs.

4.1. Natural bases

Definition 4.1.
1. $\mathfrak{P}(\mathbb{N})$ denotes the set of all subsets of the set of natural numbers $\mathbb{N}$. A set $K \in \mathfrak{P}(\mathbb{N})$ is called *biinfinite* if it as well as its complement, $\mathbb{N} \setminus K$, are infinite subsets of $\mathbb{N}$.
2. A strictly increasing sequence of natural numbers $\mathbf{b} = \{b_k\}_{1 \le k < \infty} := (b_k)$ is called a (*natural*) *base* if $b_0 = 0$ and $\{b_k\}_{1 \le k < \infty}$ is a biinfinite subset of $\mathbb{N}$. The symbol $\mathcal{B}$ denotes the set of all bases. Let $\mathbf{b} = (b_k)$ be a base. The subset $\mathbb{N} \setminus \{b_k\}_{k \ge 1} := \{b_{*i}\}_{i \ge 1}$ of $\mathbb{N}$ ordered as a strictly increasing sequence and complemented by $b_{*0} = 0$ is called the base *dual* to the base $\mathbf{b}$ and is denoted as $\mathbf{b}_*$, $b_* = (b_{*i})$.

3. Let $\mathbf{b}_1$ and $\mathbf{b}_2$ be two bases. We write $\mathbf{b}_1 \prec \mathbf{b}_2$ if $\mathbf{b}_1 \subseteq \mathbf{b}_2$ and the set $\mathbf{b}_2 \cap \mathbf{b}_{1*}$ is a base.
4. If a base $\mathbf{b}$ is given we can define the following two maps of $\mathbb{N}$ into itself.

   The *quantitative sequence*

$$\mathbf{q_b} = (q_{\mathbf{b}}(n)), \; q_{\mathbf{b}}(n) := (b_n - b_{n-1}) > 0, \; n \in \mathbb{N}, \tag{4.1}$$

and the *place-sequence*

$$\mathbf{p_b} = (p_{\mathbf{b}}(n)), \; p_{\mathbf{b}}(n) := \sum_{i=0}^{n-1} \chi_{\mathbf{b}}(i), \; n \in \mathbb{N}, \tag{4.2}$$

where $\chi_{\mathbf{b}}$ denotes the characteristic function of the set $\mathbf{b} \subset \mathbb{N} \cup \{0\}$.

It is immediate that

$$p_{\mathbf{b}}(1) = 1, p_{\mathbf{b}}(n) \leq p_{\mathbf{b}}(n+1) \leq p_{\mathbf{b}}(n) + 1, \; n \geq 1; \; \lim_{n\to\infty} p_{\mathbf{b}}(n) = \infty. \tag{4.3}$$

Remark 4.1. It is immediate that every positive sequence $\{\mathbf{q} = q(n)\}$ of naturals as well as every sequence $\{\mathbf{p} = p(n)\}$ satisfying (4.3) are quantitative sequence and place-sequence, respectively, for an appropriate base $\mathbf{b}$. Thus the correspondences between bases, their quantitative sequences, and their place-sequences are bijections.

Definition 4.2. The *superposition* of two bases $\mathbf{b}_1 \circ \mathbf{b}_2$ is the base that corresponds in the unique way (Remark 4.1) to the place-sequence

$$(\mathbf{p}_{\mathbf{b}_1} \circ \mathbf{p}_{\mathbf{b}_2})(j) := p_{\mathbf{b}_1}(p_{\mathbf{b}_2}(j)), \; j \geq 1. \tag{4.4}$$

(It is immediate that the right part of (4.4) is indeed a place-sequence, and therefore the superposition $\mathbf{b} := \mathbf{b}_1 \circ \mathbf{b}_2$. is well defined.)

Lemma 4.2. *Let $\mathbf{p}_1$ and $\mathbf{p}_2$ be arbitrary place-sequences and $\mathbf{b}_1$ and $\mathbf{b}_2$ be the corresponding bases. Then for the superposition $\mathbf{b} = \mathbf{b}_1 \circ \mathbf{b}_2$ of these bases we have*

$$\chi_{\mathbf{b}}(k) = \chi_{\mathbf{b}_2}(k) \cdot \chi_{\mathbf{b}_1}(p_2(k)), \; k \geq 1. \tag{4.5}$$

Conversely, if (4.5) *holds for the superposition $\mathbf{b}$ of the bases $\mathbf{b}_1$ and $\mathbf{b}_2$ then*

$$\mathbf{p_b} = \mathbf{p}_{\mathbf{b}_1} \circ \mathbf{p}_{\mathbf{b}_2}.$$

Corollary 4.3. *Let $\mathbf{b}_1 = (m_k)$, $\mathbf{b}_2 = (n_k)$ be two arbitrary bases. Then for their superposition $\mathbf{b} = \mathbf{b}_1 \circ \mathbf{b}_2$ the following statements are true:*

- $(\mathbf{p}_1 \circ \mathbf{p}_2) = \{p(n)\}$ *where* $p(n) \leq \min[p_1(n), p_2(n)], n \geq 1$;
- $\mathbf{b} = \{n_{m_k}\}_{k\geq 1}$;
- $\mathbf{b} \prec \mathbf{b}_2$.

Theorem 4.4. *Let $\mathbf{p}$ and $\mathbf{p}_2 = \{p_2(n)\}$ be two place-sequences defining the bases $\mathbf{b} = (b_n)$ and $\mathbf{b}_2$, respectively. Then the following statements are equivalent.*

- $\mathbf{b} \prec \mathbf{b}_2$;

- *The equation*

$$\mathbf{p} = x \circ \mathbf{p}_2$$

is solvable and its unique solution is the place-sequence $\mathbf{p}_1$ *corresponding to the base*

$$\mathbf{b}_1 := (p_2(b_n)).$$

Definition 4.3.

1. Let $q^{(1)} = \{q_n^{(1)}\}_{n\geq 1}$ and $q^{(2)} = \{q_n^{(2)}\}_{n\geq 1}$ be two sequences of real numbers. We will call these sequences $\overset{a}{\sim}$*equivalent* and write $q^{(1)} \overset{a}{\sim} q^{(2)}$ if there is a natural d such that

$$\sum_{i=1}^{n} q_i^{(1)} \leq \sum_{i=1}^{n+d} q_i^{(2)} \leq \sum_{i=1}^{n+2d} q_i^{(1)}, \ n \geq 1.$$

2. We will call two bases $\mathbf{b}^{(1)} = (b_k^{(1)})$ and $\mathbf{b}^{(2)} = (b_k^{(2)})$ $\overset{a}{\sim}$equivalent and write $\mathbf{b}^1 \overset{a}{\sim} \mathbf{b}^2$ if there is a natural d such that

$$b_k^{(1)} \leq b_{k+d}^{(2)} \leq b_{k+2d}^{(1)}, \ k \geq 1.$$

The set of all bases that are $\overset{a}{\sim}$equivalent to the base $\mathbf{b}$ is called the $\overset{a}{\sim}$modular of the base $\mathbf{b}$ and denoted by $\mathfrak{b}$. Finally, the set of all $\overset{a}{\sim}$modular is denoted by $\mathfrak{B}$.

3. Two place-sequences $\mathbf{p}_1 = \{p_1(n)\}_{n\geq 1}$ and $\mathbf{p}_2 = \{p_2(n)\}_{n\geq 1}$ are called $\overset{a}{\sim}$equivalent ($\mathbf{p}_1 \overset{a}{\sim} \mathbf{p}_2$) if there is a natural d such that

$$p_1(n) \leq p_2(n) + d \leq p_1(n) + 2d, \ n \geq 1.$$

Let $\mathbf{b}$ be a base. We will denote by $\{\mathbf{q}_{\mathbf{b}_*}\}$ and $\{\mathbf{p}_{\mathbf{b}_*}\}$ the qualitative and place-sequence corresponding to the dual base $\mathbf{b}_*$. It is immediate that $\mathbf{p}_{\mathbf{b}_*}(n) = n - p_{\mathbf{b}}(n) + 1, \ n \geq 1$. Equally immediate is the statement of the following lemma.

Lemma 4.5. *Let* $\mathbf{b}^{(1)}$ *and* $\mathbf{b}^{(2)}$ *be two arbitrary bases. The following statements are equivalent:*

$$\mathbf{b}^{(1)} \overset{a}{\sim} \mathbf{b}^{(2)}, \tag{4.6.1}$$

$$\mathbf{b}_*^{(1)} \overset{a}{\sim} \mathbf{b}_*^{(2)}, \tag{4.6.2}$$

$$\mathbf{q}_{\mathbf{b}^{(1)}} \overset{a}{\sim} \mathbf{q}_{\mathbf{b}^{(2)}}, \tag{4.6.3}$$

$$\mathbf{q}_{\mathbf{b}_*^{(1)}} \overset{a}{\sim} \mathbf{q}_{\mathbf{b}_*^{(2)}}, \tag{4.6.4}$$

$$\mathbf{p}_{\mathbf{b}^{(1)}} \overset{a}{\sim} \mathbf{p}_{\mathbf{b}^{(2)}}, \tag{4.6.5}$$

$$\mathbf{p}_{\mathbf{b}_*^{(1)}} \overset{a}{\sim} \mathbf{p}_{\mathbf{b}_*^{(2)}}. \tag{4.6.6}$$

Definition 4.4.

1. The map $\omega : \mathfrak{B} \to \mathfrak{P}(\mathbb{N})$ is called $\overset{a}{\sim}$*invariant* if on $\overset{a}{\sim}$equivalent bases ω takes $\overset{a}{\sim}$equivalent values.
2. Let $\mathbf{b} = (b_k)$ be a base. We will consider the corresponding *supremal sequence*

$$S_{\mathbf{b}}(m) := \sup_{0 \le n < \infty} \sum_{j=n+1}^{n+m} \chi_{\mathbf{b}}(j), \ m \ge 0$$

and the *upperlimit sequence*

$$L_{\mathbf{b}}(m) := \limsup_{n \to \infty} \sum_{j=n+1}^{n+m} \chi_{\mathbf{b}}(j), \ m \ge 0.$$

It is easy to see that $L_{\mathbf{b}}(m) \le S_{\mathbf{b}}(m) \le m, \ m \ge 1$.

Remark 4.6. It is immediate that for $\{S_{\mathbf{b}}(m)\}$ are satisfied all of the relations (4.3) and for $\{L_{\mathbf{b}}(m)\}$ – the first two of those relations.

By Remark 4.1 if we know the sequence $\{S_{\mathbf{b}}(m)\}$ then the base $\mathbf{b}_{S_b}$ is uniquely defined.

By Lemma 4.5 the map $\mathbf{b} \to \mathbf{b}_{S_b}$ is $\overset{a}{\sim}$invariant.

Finally, if the monotonic sequence $\{L_{\mathbf{b}}(m)\}$ is unbounded then its values uniquely define the base $\mathbf{b}_{L_b}$ and the map $\mathbf{b} \to \mathbf{b}_{L_{\mathbf{b}}}$ is $\overset{a}{\sim}$invariant.

Theorem 4.7. *The upperlimit sequence* $L_{\mathbf{b}}(m)$ *is bounded by* m *if and only if there is a natural base* $\tilde{\mathbf{b}} \overset{a}{\sim} \mathbf{b}$ *such that*

$$\lim_{n \to \infty} \mathbf{q}_{\tilde{\mathbf{b}}}(n) = \infty. \tag{4.7}$$

Definition 4.5. Let $\mathbf{b} = \{b_k\}$ be a base. We define the lower, $\gamma_{\mathbf{b}}$, and the upper, $\delta_{\mathbf{b}}$, indices of this base in the following way

$$\begin{cases} \gamma_{\mathbf{b}} := \lim\limits_{m \to \infty} \inf\limits_{0 \le n < \infty} \dfrac{\sum\limits_{j=n+1}^{n+m} \chi_{\mathbf{b}}(j)}{m}; \\ \delta_{\mathbf{b}} := \lim\limits_{m \to \infty} \sup\limits_{0 \le n < \infty} \dfrac{\sum\limits_{j=n+1}^{n+m} \chi_{\mathbf{b}}(j)}{m}. \end{cases}$$

It is immediate that these indices exist, that they coincide for two $\overset{a}{\sim}$equivalent bases, and that

$$0 \le \gamma_{\mathbf{b}} \le \delta_{\mathbf{b}} \le 1; \ \delta_{\mathbf{b}} = 1 - \gamma_{\mathbf{b}_*}; \ \gamma_{\mathbf{b}} = 1 - \delta_{\mathbf{b}_*}.$$

Definition 4.6.

1. Let $\mathbf{b} = \{b_k\}_{k \ge 0}$ be a base. The *n-shift* of this base is by definition the base $\mathbf{b}^{[n]} = (b_k^{[n]})$, where $b_0^{[n]} := 0$; $b_k^{[n]} := b_{k+n}, \ k \ge 1$. Notice that $\overset{a}{\sim}$invariant properties of $\overset{a}{\sim}$modular are exactly the properties invariant under shifts.

2. A base $\mathbf{b} = (b_k)$ is called *uniform* if it is $\overset{a}{\sim}$equivalent to its 1-shift. Clearly two $\overset{a}{\sim}$equivalent bases are simultaneously uniform or not.

Lemma 4.8. *A base* $\mathbf{b} = (b_n)$ *is uniform if and only if*

$$\sup_{n>0}[b_n - b_{n-1}] < \infty. \tag{4.8}$$

Definition 4.7. Let R be a natural number, $R > 1$. A sequence of natural numbers such that all its terms are equal either to 1 or to R is called *R-reduced.*

Lemma 4.9. *Let* $\mathbf{b}$ *be a uniform base. Then there are a natural number* R *and a base* $\widetilde{\mathbf{b}}$ *such that* $\widetilde{\mathbf{b}} \overset{a}{\sim} \mathbf{b}$ *and the quantitative sequence* $\mathbf{q}_{\widetilde{\mathbf{b}}}$ *is* R*-reduced.*

Let us fix a base $\mathbf{b}$ and let $\mathbf{p_b}$ be the corresponding place-sequence.

Definition 4.8. 1. A base $\mathbf{b} = (b_k)$ is called *subadditive*, respectively, *superadditive* if there is a natural d such that for all natural m and n we have the following inequalities:

$$b_{n+m-d} \le b_n + b_m, \quad n + m > d,$$

respectively,

$$b_{n+m+d} \ge b_n + b_m, \quad n + m > d. \tag{4.9}$$

2. We call the place sequence $\mathbf{p_b}$ *subadditive*, respectively, *superadditive* if there is a natural d such that for all natural m and n we have the following inequalities.

$$p_{\mathbf{b}}(n + m) \ge p_{\mathbf{b}}(n) + p_{\mathbf{b}}(m) - d,$$

respectively,

$$p_{\mathbf{b}}(n + m) \le p_{\mathbf{b}}(n) + p_{\mathbf{b}}(m) + d. \tag{4.10}$$

Because $p_{\mathbf{b}}^{(m)} \uparrow_{m\uparrow\infty} \infty$ the conditions of super and subadditivity can be expressed in the following way. There is a natural d such that

$$\begin{cases} p_{\mathbf{b}}(n + m) \ge p_{\mathbf{b}}(n - d) + p_{\mathbf{b}}(m - d), \ n, m \ge d; \\ \qquad\qquad \text{respectively,} \\ p_{\mathbf{b}}(n + m) \le p_{\mathbf{b}}(n + d) + p_{\mathbf{b}}(m + d), \ n, m \ge 1. \end{cases} \tag{4.11}$$

Remark 4.10.

1. A base $\mathbf{b}$ is superadditive (respectively, subadditive) if and only if its place-sequence is subadditive (respectively, superadditive).
2. Every subadditive base is uniform.

Lemma 4.11. *If* $\mathbf{b} = (b_n)$ *is a subadditive base then*

$$\beta := \inf_{n\ge1} \frac{b_n}{n} = \lim_{n\to\infty} \frac{b_n}{n}.$$

Proof. We have $1 \le \beta \le \infty$ and without loss of generality we can assume that $\beta < \infty$. Let us fix an arbitrary small $\varepsilon > 0$ and chose a natural n_ε such that

$$b_{n_\varepsilon}/n_\varepsilon < \beta + \varepsilon. \tag{4.12}$$

Let m be a natural number such that $m > n_\varepsilon + d$. we can chose a natural $k = k_m$ such that

$$(k_m + 1) \cdot n_\varepsilon \le m - d \le (k_m + 2) \cdot n_\varepsilon. \tag{4.13}$$

Because $m - d = k_m \cdot n_\varepsilon + m - d - k_m \cdot n_\varepsilon$ in virtue of subadditivity we have

$$\beta \le \frac{b_{m-d}}{m-d} \le \frac{b_{k_m \cdot n_\varepsilon}}{m-d} + \frac{b_{m-k_m \cdot n_\varepsilon - d}}{m-d} \le \cdots \tag{$*$}$$

Because

$$k_m \cdot n_\varepsilon = \underbrace{n_\varepsilon + \cdots + n_\varepsilon}_{k_m}$$

we can apply subadditivity, (4.12), (4.13), as well as the monotonicity of the base (that implies that $b_{m-k_m \cdot n_\varepsilon} \le b_{2n_\varepsilon + d}$) to extend the inequalities ($*$) and to obtain

$$\cdots \le \frac{k_m \cdot n_\varepsilon \cdot (\beta + \varepsilon)}{(k_m + 1) \cdot n_\varepsilon} + \frac{b_{m-k_m \cdot n_\varepsilon}}{m-d} \le \frac{\beta + \varepsilon}{1 + \frac{1}{k_m}} + \frac{2n_\varepsilon + d}{m-d}. \tag{4.14}$$

By taking the limit in (4.14) when $m \to \infty$ we obtain the statement of the lemma. □

Theorem 4.12. *Let* $\mathbf{b} = (b_n)$ β *be a subadditive base and* $\beta = \lim_{n\to\infty} \frac{b_n}{n}$. *Then for the repeated limit is true*

$$\lim_{n\to\infty} \lim_{m\to\infty} \frac{b_{n+m} - b_n}{m} = \beta. \tag{4.15}$$

Proof. By changing the notations of variables we can rewrite (4.9) as

$$b_{n+m} \le b_n + b_{m-d}, \text{ where } m > d,\ n \ge 1. \tag{4.16}$$

Let us fix $\varepsilon > 0$. By the previous lemma and by (4.16) for any n and for m, such that $m > d$ and $\frac{b_{m-d}}{m} = \frac{b_{m-d}}{m-d} \cdot \frac{m-d}{m} \le \beta + \varepsilon$, we have

$$\frac{b_{n+m} - b_n}{m} \le \frac{b_{m-d}}{m} \le \beta + \varepsilon.$$

On the other hand for large n, such that $n' \ge n$ and $|\frac{b_{n'}}{n'} - \beta| < \varepsilon$, and for arbitrary natural m we have $\frac{b_{n+m}}{n+m} > \beta - \varepsilon$. Therefore for such n and arbitrary m we have

$$\begin{aligned}\frac{b_{n+m} - b_n}{m} &= \frac{b_{n+m}}{n+m}(1 + \frac{n}{m}) - \frac{b_n}{n}\frac{n}{m} \\ &> (\beta - \varepsilon)(1 + \frac{n}{m}) - (\beta + \varepsilon)\frac{n}{m} > \beta - (1 + 2\frac{n}{m})\varepsilon,\end{aligned}$$

whence $|\frac{b_{n+m} - b_n}{m} - \beta| \le (1 + 2\frac{n}{m})\varepsilon$. Because for the repeated limit is true

$$\lim_{n\to\infty} \lim_{m\to\infty} \frac{n}{m} = 0$$

and ε is arbitrary small we obtain (4.15). □

Definition 4.9. A base $\mathbf{b} = (b_n)$ is called *condensifying* if there is a natural d such that for any natural m and n there is a natural n', $n' = n'(m) > n$, such that

$$\sum_{i=n+1}^{n+m} \chi_{\mathbf{b}} \leq \sum_{i=n'+1}^{n'+m+d} \chi_{\mathbf{b}}.$$

4.2. Bases and equiconcave/equiconvex functions

The constructions considered in the current subsection are contained in [19] and [24], see also [6].

Definition 4.10. Let $\mathbf{b}^0 = \{n_k^0\}_{k=0,1,\ldots}$ and $\mathbf{b}^\infty = \{n_k^\infty\}_{k=0,1,\ldots}$ be two arbitrary natural bases. Let us fix a nonnegative integer j. Then we can find two numbers $k_0(j)$ and $k_\infty(j)$, such that

$$n^0_{k_0(j)} \leq j < n^0_{k_0(j)} + 1, \; n^\infty_{k_\infty(j)} \leq j < n^\infty_{k_\infty(j)} + 1.$$

We will now construct two binary-measurable functions: φ^0 on $[0,1)$ and φ^∞ on $[1,\infty)$. By binary-measurable we mean that the function φ^0 is measurable relatively to the partition of $[0,1]$ by the points $2^{-\nu}$ and φ^∞ is measurable relatively to the partition of $[1,\infty)$ by the points 2^ν where $\nu = 0,1,2,\ldots$. For $2^{-j} \leq t < 2^{-j+1}$, $j \geq 1$, we define $\varphi^0(t)$ as $\varphi^0(t) = 2^{-n^0_{k_0(j)}}$. On the other hand, for $2^j \leq t < 2^{j+1}$, $j \geq 0$, we define $\varphi^\infty(t)$ as $\varphi^\infty(t) = 2^{n^\infty_{k_\infty(j)}}$.

Next, let $\varphi(t) = \varphi^0(t)$ if $t \in [0,1)$ and $\varphi(t) = \varphi^\infty(t)$ if $t \in [1,\infty)$. It is easy to see that the function φ defined in such a way on $[0,\infty)$ is equiconcave. We call it the function *generated by the pair of natural bases* $(\mathbf{b}^0, \mathbf{b}^\infty)$. Moreover, $\mathbf{b}^0$ (respectively, $\mathbf{b}^\infty$) is called *the left* (respectively, *the right*) base for φ.

We will now show that the converse is true as well: an arbitrary M-function ψ (and therefore an arbitrary equiconcave function) can be generated uniquely (up to $\overset{m}{\sim}$equivalency) by a pair of natural bases via the construction described in Definition 4.10.

Definition 4.11. Let ψ be an M-function defined on $[0,\infty)$. For any natural n let D_n^- be the diadic half-segment $[2^{-n}, 2^{-n+1})$, while D_n^+ be the diadic half-segment $[2^n, 2^{n+1})$. The points $\psi(2^j)$, where j is an arbitrary integer, will be called ψ-points. Let us introduce two functions $\mathbf{p}^0_\psi$ and $\mathbf{p}^\infty_\psi$ by corresponding each ψ-point with the number of the diadic segment that contains it.

$$\mathbf{p}^0_\psi(j) = [-\log_2 \psi(2^{-j})], \; \mathbf{p}^\infty_\psi(j) = [\log_2 \psi(2^j)], \; j \geq 0, \tag{4.17}$$

where $[r]$ means the integer part of a real number r. It is easy to see that for an M-function ψ both functions $\mathbf{p}^0_\psi : \mathbb{N} \mapsto \mathbb{N}$ and $\mathbf{p}^\infty_\psi : \mathbb{N} \mapsto \mathbb{N}$ are surjective maps and each of them satisfies relations (4.3). Thus these functions are place-sequences defining uniquely the bases $\mathbf{b}^0_\psi$, $\mathbf{b}^\infty_\psi$, respectively. This bases, as can be readily seen, correspond to the left and the right base, respectively, of some equiconcave function that is $\overset{m}{\sim}$equivalent to ψ. The quantitative sequences of these bases we denote by $\mathbf{q}^0_\psi(n)$ $\mathbf{q}^\infty_\psi(n)$, respectively.

Remark 4.13. It is obvious that the corresponding bases of two M-functions are pairwise $\overset{a}{\sim}$equivalent if and only if the functions themselves are $\overset{m}{\sim}$equivalent. Therefore we can define the left and the right base of an equiconcave function uniquely up to $\overset{a}{\sim}$equivalency as the corresponding base of an arbitrary M-function from its M-modular. In other words there is a bijective correspondence between $\overset{m}{\sim}$modulars of equiconcave functions and pairs of $\overset{a}{\sim}$modulars of bases. This correspondence can be expressed in the following way.

$$\mathbf{b}_{\varphi_{\mathbf{b}}} \overset{a}{\sim} \mathbf{b}\varphi_{\mathbf{b}_{\varphi}} \overset{a}{\sim} \mathbf{b},$$

where by $\mathbf{b}$ we understand a pair of bases. Clearly this correspondence conserves duality in the classes of $\overset{m}{\sim}$modulars and pairs of $\overset{a}{\sim}$modulars.

Lemma 4.14. *An equiconcave function ψ is symmetric if and only if its left and right bases are $\overset{a}{\sim}$equivalent:*

$$\mathbf{b}_{\psi}^{\infty} \overset{a}{\sim} \mathbf{b}_{\psi}^{0}.$$

Remark 4.15. It follows from (4.6) that for a symmetric M-function ψ we have

$$\gamma_{\psi} = \gamma_{\mathbf{b}_{\psi}^{0}} = \gamma_{\mathbf{b}_{\psi}^{\infty}}, \ \delta_{\psi} = \delta_{\mathbf{b}_{\psi}^{0}} = \delta_{\mathbf{b}_{\psi}^{\infty}}. \tag{4.18}$$

Let ϕ be an N-function. We introduce the following notation

$$b_n^{\varphi} := [\log_2 \varphi(2^n)], \ n \geq 0.$$

Then $0 = b_0^{\varphi} < b_1^{\varphi} < \cdots$, and it follows from the convexity of ϕ that the strictly increasing sequence $\{b_n^{\varphi}\}_{n>0}$ is a biinfinte subset in $\mathbb{N}$.

Definition 4.12. The subset $\mathbf{b}_{\phi}^{\infty} := \{b_n^{\varphi}\}_{n>0}$ is called the *right base* of the N-function ϕ as well as of its modular $\Phi \ni \phi$.

The *left base* of an N-function ϕ and its modular Φ is defined as the right base of the N-function $I_5\phi$.

Remark 4.16. For an arbitrary N-modular Φ, $\Phi \ni \phi$, its right base $\mathbf{b}_{\phi}^{\infty}$ is $\overset{a}{\sim}$equivalent to the left base $\mathbf{b}_{\phi^{\frown}}^{0}$ of the related M-modular. Thus mutually related M and N-functions (modulars) have $\overset{a}{\sim}$equivalent bases.

4.3. Tables

Definition 4.13.

1. Let $\mathbf{b} = (b_n)$ be a natural base. The following table of natural numbers (with infinite number of both rows and columns) is called the *lower table* or *table of base compression* of base $\mathbf{b}$.

$$\mathfrak{T}_{\mathbf{b}}(n,m) := \sum_{i=b_n+1}^{\mathbf{b}_{n+m}} \chi_{\mathbf{b}}(i), \ n,m \geq 0.$$

2. The *upper table*, or *table of base dilation* of the base $\mathbf{b}$ is the table

$$\mathfrak{T}^{\mathbf{b}}(n,m) := b_{n+m} - b_n, \ n,m \geq 0.$$

Notice that the lower table characterizes the distribution of the base as a subset of the naturals, while the upper one – the distribution of the dual base.

In what follows we will restrict our considerations to the case of one base $\mathbf{b} = (b_n)$ corresponding to a symmetric equiconcave function, in other words when M-functions are defined on $[0,1]$. Regarding the general case of an equiconcave function ψ on $[0,\infty)$ notice that it can be reduced to the consideration of the pair of bases corresponding to the left and right symmetric brackets of ψ.

In terms of limit and extremal relations for the upper (as well as for the lower) table of the base of a equicincave function ψ (or an equiconvex function ϕ) we can interpret all the $\overset{m}{\sim}$invariants of the triple of related spaces Λ_{ψ_*}, M_ψ, L^*_ϕ, where $\psi = \phi^\frown$. As examples we will consider $\overset{m}{\sim}$invariants of M-modulars on $[0,1]$ considered in Section 2.

Theorem 4.17. *The* $\overset{m}{\sim}$*invariants of* M*-modulars allow the following interpretation via the tables.*

I. 1. $\limsup_{n\geq 0} \mathfrak{T}^{\mathbf{b}_\psi}(n,1) < \infty \Leftrightarrow \gamma_{\mathbf{b}_\psi} > 0 \Leftrightarrow \psi^\smile \in (\Delta_2)$.

2. $\limsup_{n\geq 0} \mathfrak{T}^{\mathbf{b}_{\psi_*}}(n,1) < \infty \Leftrightarrow \delta_{\mathbf{b}_\psi} < 1$.

II. Submultiplicativity ψ : $\sup_{m,n\geq 1} \left(\mathfrak{T}^{\mathbf{b}_\psi}(0,m) - \mathfrak{T}^{\mathbf{b}_\psi}(n,m)\right) < \infty$.

Supermultiplicativity $\psi : \sup_{m,n\geq 1} \left(\mathfrak{T}^{\mathbf{b}_\psi}(n,m) - \mathfrak{T}^{\mathbf{b}_\psi}(0,m)\right) < \infty$.

III. Equimultiplicativity $\psi \Leftrightarrow \sup_{n;\ k,\ m} |\mathfrak{T}^{\mathbf{b}_\psi}(k,m) - \mathfrak{T}^{\mathbf{b}_\psi}(n,m)| < \infty \Leftrightarrow \psi(t) \overset{m}{\sim} t^{\gamma_{\mathbf{b}_\psi}}$.

IV. Repeated limit $\lim_{n\to\infty} \lim_{m\to\infty} \dfrac{m}{\mathfrak{T}^{\mathbf{b}_\psi}(n,m)} = \alpha \Leftrightarrow \psi \in RV^0_\alpha,\ \alpha \in [0,1]$.

V. $\psi \in \Psi \in (pPow) \Leftrightarrow \sup_{n\geq 0} \mathfrak{T}^{\mathbf{b}_{\psi_*}}(n,m) \overset{a}{\sim} \limsup_{n\to\infty} \mathfrak{T}^{\mathbf{b}_{\psi_*}}(n,m)$.

For a symmetric M-modular Ψ a criterion for $\Psi \in pPow$ can be formulated directly in terms of bases $\mathbf{b}_\psi$, $\psi \in \Psi$:

$$\Psi \in (pPow) \Leftrightarrow \text{ the base } \mathbf{b}_\psi \textbf{ is condensifying}.$$

VI. $\mathsf{M}_\psi(0,1) \in \mathcal{P}^* \Leftrightarrow \sum_{n\geq 1} \mathfrak{T}^{b_\psi}(n,1) \cdot 2^{-n} < \infty$.

VII. $\mathsf{M}_\psi(0,1) = \mathsf{L}^*_\phi(0,1) \Leftrightarrow \mathbf{b}_{\psi_*} \overset{a}{\sim} \mathbf{b}_{\phi^{-1}}\quad \exists \varphi \overset{m}{\sim} \psi : \sum_{n=0}^\infty 2^{-\mathfrak{T}^{b_{\varphi_*}}(n,1)} < \infty$.

Thus for the segment $[0,1]$ and every of the spaces from the related triple (Lorentz, Marcinkiewicz, Orlicz) all the information about their topological invariants is contained in limit and extremal values of the variable representing the distribution of some sequence of natural numbers – the common base of these spaces (and the dual base).

For the triple of such spaces on $[0,\infty)$ the information is contained in the corresponding distributions of both the right and the left bases. The converse is also true: by choosing an increasing biinfinite sequence of natural numbers we can define any of those three related spaces on $[0, 1]$ (and thus the other two as well). Moreover, by changing the distribution of the defining sequence in $\mathbb{N}$ we can obtain arbitrary related topological invariants of the constructed spaces.

Similarly, by choosing two such sequences we can construct related triples of spaces on $[0, \infty)$ such that one of these spaces has the given $\overset{m}{\sim}$invariant and two others can be reconstructed from the first one.

References

[1] Krein S.G., Petunin Y.I., Semenov E.M. Interpolation of Linear Operators. Translations of Mathematical Monographs, vol. 54, American Mathematical Sociaty, Providence. R.I., 1982.

[2] Lindenstrauss J., Tzafriri L. Classical Banach Spaces II, Springer, Berlin, 1979.

[3] Bennett C., Sharpley R. Interpolation of Operators, Academic Press, Pure and applied mathematics, v. 129, New York – London, 1988.

[4] M.A. Krasnosel'skiĭ, Ya.B. Rutickiĭ. Convex functions and Orlicz spaces, Groningen, 1961.

[5] Bingham N.H., Goldie C.M., Teugels J.L. Regular Variations. Cambridge University Press, Cambridge, 1987.

[6] Odinets V.P., Shlenzak V.A. An introduction to convex analysis (Russian), Moscow ◇ Izhevsk, RXD, 2011 (Russian).

[7] Lorentz G.G. Majorants in Spaces of Integrable Functions. Amer. J. Math. 77(1955), 484–492.

[8] Dodds P.G., de Pagter B., Semenov E.M., Sukochev F.A. Symmetric Functionals and Singular Traces. Positivity, 2, 1998, 47–75.

[9] Dodds, P.G., de Pagter, B., Sedaev, A.A., Semenov, E.M., Sukochev, F.A. Singular symmetric functionals. Zap. Nauchn. Sem. St.-Petersburg. Otdel. Mat. Inst. Steklov. (POMI) 290 (2002), 42–71. English translation in J. Math. Sci. (N.Y.) 124 (2) 2004, 4867–4885.

[10] Kalton N., Sukochev F. Rearrangement-Invariant Functionals with Applications to Traces on Symmetrically Normed Ideals. Canad. Math. Bull., 51 (1), 2008, 67–80.

[11] Drasin D., Seneta E. A generalization of slowly varying function, Proc. Amer. Math. Soc. 96, 1986, 470–472.

[12] Lorentz G.G. Relations between Function Spaces, Proc. Amer. Math. Soc. 12 (1), 1961, 127–132.

[13] Rutickiĭ Ya.B., On some classes of measurable functions, Uspekhi Mat. Nauk, 20 (4), 1965, 205–208 (Russian).

[14] Sedaev A.A., Smurov V.A. On finding one numerical characteristic of Marcinkiewicz spaces. In "Methods of solution of operator equations", VGU, Voronezh, 1978, 135–142 (Russian).

[15] Novikov S.Y. Cotype and type of Lorentz function spaces. Mat. Zametki., 32 (2), 1982, 213–221 (Russian). English translation: Novikov, S.Ya. "Cotype and type of Lorentz function spaces." Mathematical Notes 32 (2) 1982, 586–590.

[16] Mekler A.A. On existence of $\lim_{t\to 0} \frac{\psi(2t)}{\psi(t)}$ for a concave function ψ (Russian), in "Theses of X pansoviet school on theory of operators in functional spaces", v. 2, Chelyabinsk, 1986, 117–118 (Russian).

[17] Mekler A.A. Duality in the semigroup of concave functions (Russian), in "IX simposium on the problem of redundance in informatic systems", Pansoviet Organization on kibernetics of the soviet Academy of Sciences, v. 2, Leningrad, 1986, 117–118 (Russian).

[18] Abakumov E.V., Mekler A.A. A Concave Regularly Varying Leader for Equi-concave Functions, J. Math. Anal. Appl. 187 (3), 1994, 943–951.

[19] Mekler A.A. On Regularity and Weak Regularity of Functions Generating Marcinkiewicz Spaces, in Proc. Intern. Conf. FUNCTION SPACES V, Poznan, Poland, July 2000, H. Hudzik and L. Skrzypczak eds., Marcel Dekker, Lect. Not. Pur. Appl. Math. Ser., **213**, July 2000, 379–387.

[20] Mekler A.A. On a semigroup of modular functions with involution. Zap. Nauchn. Sem. St.-Petersburg. Otdel. Mat. Inst. Steklov. (POMI) 315 (2004), Issled. po Linein. Oper. i Teor. Funkts. 32, 121–131 (Russian). English translation in J. Math. Sci. (N.Y.) 134 (2006), no. 4, 2305–2310.

[21] Mekler A.A. Representation of the superposition of concave modulars on a double logarithmic scale. Some current problems in modern mathematics and education in mathematics 121–128, Ross. Gos. Ped. Univ., St. Petersburg, 2007 (Russian).

[22] Mekler A.A., On natural characteristics of regularly varying quasi-concave modulars. Vestn. Syktyvkar. Univ. Ser. 1. 8, 2008, 27–38 (Russian).

[23] Mekler A.A. Notes on some invariant properties of Marcinkiewicz and Orlicz spaces, I. Vestn. Syktyvkar. Univ. Ser. 1, 14, 2011, 33–48 (Russian).

[24] Mekler A.A. Notes on some invariant properties of Marcinkiewicz and Orlicz spaces, II. Vestn. Syktyvkar. Univ. Ser. 1, 14, 2011, 49–66 (Russian).

[25] Mekler A.A. A scale of invariants for Orlich and Marcinkiewicz spaces, in "Dynamical Systems, Differential Equations and Applications". Proceedings of 8th AIMS Conference in Dresden, Germany, May, 2010.

[26] Mekler A.A. On modulars of Marcinkiewicz spaces on $[0,1]$ and $[0,\infty)$. Vestn. Syktyvkar. Univ. Ser. 1, 15, 2012, 95–112 (Russian).

[27] Mekler A.A. On modulars of Marcinkiewicz spaces on $[0,1]$ and $[0,\infty)$, II. Vestn. Syktyvkar. Univ. Ser. 1 Mat. Mekh. Inform., 16, 2012, 101–111 (Russian).

[28] Mekler A.A. On multiplicativity of Marcinkiewicz' modulars. Vestn. Syktyvkar. Univ. Ser. 1, 17, 2013, 66–79 (Russian).

[29] Mekler A.A. On the semigroup of Marcinkiewicz' modulars. Vestn. Syktyvkar. Univ. Ser. 1, 17, 2013, 80–94 (Russian).

[30] Mekler A.A. http://websites.math.leidenuniv.nl/positivity2013/presentations/

A.A. Mekler
Wilsdruffer Str. 7
D-01067 Dresden, Germany
e-mail: `alexandre.mekler@o2online.de`

Ordered Structures and Applications: Positivity VII
Trends in Mathematics, 329–337

On Poles of the Resolvents of Dominated Elements in Ordered Banach Algebras

Kelvin Muzundu

Abstract. Let x and y be elements of an ordered Banach algebra (OBA) such that $0 \leq x \leq y$. In this paper we will discuss results that give conditions under which the spectral radius of x is a pole of the resolvent of x, given that the spectral radius of y is a pole of the resolvent of y. The results obtained will be used to establish some spectral and asymptotic properties of dominated elements in OBAs.

Mathematics Subject Classification (2010). Primary 46H05; Secondary 47A10, 47B65, 06F25.

Keywords. Ordered Banach algebra, positive element, spectrum.

1. Introduction

If x and y are elements of an ordered Banach algebra (OBA) such that $0 \leq x \leq y$, the general problem of finding conditions under which certain properties of y will be inherited by x has been studied by various authors (cf. [1], [2], [3], [5], [6], [7], [8], [9]). This problem is referred to in the literature as the domination problem. For positive operators on a Banach lattice, the domination problem is classical. It has been extensively studied and various authors have made contributions (see [10] for a survey of some of the results).

In the context of the domination problem, a question that naturally arises is the following: if x and y are elements of an OBA such that $0 \leq x \leq y$ and if the spectral radius of y is a pole of the resolvent of y, when do we get the spectral radius of x to be a pole of the resolvent of x? In this paper we will provide results that answer this question. These results will then be applied to establish some spectral and asymptotic properties of dominated elements in OBAs, which are complementary to the results in [6] and [8].

Throughout A will be a complex Banach algebra with unit 1. The spectrum and spectral radius of an element x in A will be denoted by $\sigma(x)$ and $r(x)$ respectively. A point $\alpha \in \sigma(x)$ is called an *eigenvalue* of x if there exists a $0 \neq u \in A$ such that $xu = \alpha u$ or $ux = \alpha u$. Then u is said to be an *eigenvector* corresponding to α. The set of all isolated points of $\sigma(x)$ is denoted by **iso** $\sigma(x)$. If $\alpha \in$ **iso** $\sigma(x)$, the spectral projection corresponding to x and α will be denoted by $p(x, \alpha)$. A point $\alpha \in$ **iso** $\sigma(x)$ is said to be a pole of order k of the resolvent function $\lambda \mapsto R(\lambda, x) = (\lambda - x)^{-1}$, $\lambda \in \mathbb{C} \setminus \sigma(x)$ if k is the smallest positive integer such that $(\alpha - x)^k p(x, \alpha) = 0$. An element x in A is said to be *ergodic* if the sequence of sums $\sum_{k=0}^{n-1} \frac{x^k}{n}$ converges.

All ideals considered will be two-sided. A point $\alpha \in$ **iso** $\sigma(x)$ is said to be a Riesz point of $\sigma(x)$ relative to an ideal F if the corresponding spectral projection $p(x, \alpha)$ belongs to F. An ideal I in A is called *inessential* if the spectrum of every element in I is either finite or a sequence converging to zero. Let F be a closed ideal of A. An element x in A is said to be Riesz relative to F if $\sigma(x + F)$ in the quotient algebra A/F consists of zero.

We will denote by $\mathcal{L}(X)$ the algebra of all bounded linear operators on a Banach space X and by $\mathcal{K}(X)$ the ideal of compact operators on X. Let E be a Banach lattice. A linear operator $T : E \to E$ is regular if it can be written as a linear combination over $\mathbb{C}$ of positive operators. The space $\mathcal{L}^r(E)$ of all regular operators on E is a linear subspace of $\mathcal{L}(E)$. When $\mathcal{L}^r(E)$ is equipped with the norm

$$||T||_r = \inf\{||S|| : S \in \mathcal{L}(E), |Tx| \leq S|x| \quad \text{for all } x \in E\},$$

it becomes a Banach algebra which contains the unit of $\mathcal{L}(E)$ (see [9] and the references given there). We will denote by $\mathcal{K}^r(E)$ the closure in $\mathcal{L}^r(E)$ of the ideal of finite rank operators on E. It is well known that this is a closed inessential ideal of A.

This paper consists of five sections. Section 2 recalls the definition and properties of an ordered Banach algebra (OBA). In Section 3 we consider the problem of determining when the spectral radius of a dominated element in an OBA is a pole of the resolvent of the element, given that the spectral radius of the dominating element is a pole of the resolvent of this element. The main result here is Theorem 3.3, which shows that the conditions that are required come naturally with the OBA structure.

In Section 4, the results of Section 3 will be applied to establish properties of ergodic elements in OBAs, which are are complementary to ([6], Theorems 5.1, 5.2 and 5.5). Finally, in Section 5, we apply the results of Section 3 in relation to eigenvalues and eigenvectors to obtain results complementary to ([8], Theorems 4.3 and 4.4).

2. Ordered Banach algebras

In ([9], Section 3) an algebra cone C of a Banach algebra A was defined and it was shown that C induces an ordering on A which is compatible with the algebraic structure of A. Such a Banach algebra is called an ordered Banach algebra (OBA).

We now recall those definitions and the additional properties that algebra cones may have.

A nonempty subset C of a Banach algebra A is called a *cone* if C satisfies the following:

(i) $C + C \subseteq C$,
(ii) $\lambda C \subseteq C$ for all scalars $\lambda \geq 0$.

If C also satisfies the property $C \cap -C = \{0\}$, then it is called a *proper* cone. We say that C is *closed* if it a closed (in the topological sense) subset of A.

Every cone C in a Banach algebra A induces an ordering $\leq$ defined by $x \leq y$ if and only if $y - x \in C$, for $x, y \in A$. This ordering is reflexive and transitive. In addition, C is proper if and only if the ordering is antisymmetric. In view of the fact that C induces an ordering on A, we find that $C = \{x \in A : x \geq 0\}$. Therefore the elements of C are called *positive*.

A cone C in a Banach algebra A is called an *algebra cone* if it satisfies the following:

(i) $C.C \subseteq C$,
(ii) $1 \in C$, where 1 is the unit of A.

A Banach algebra ordered by an algebra cone is called an *ordered Banach algebra* (OBA). We will denote by (A, C) a Banach algebra A ordered by an algebra cone C.

Let (A, C) be an OBA. If there exists a real number $\alpha > 0$ such that $||x|| \leq \alpha||y||$ whenever $0 \leq x \leq y$ w.r.t. C, then we say that C is a *normal* algebra cone of A. It is easy to show that every normal algebra cone is proper.

Let (A, C) be an OBA. If $0 \leq x \leq y$ w.r.t. C implies that $r(x) \leq r(y)$, then we say that the spectral radius in (A, C) is *monotone*. It is well known that if C is normal, then the spectral radius in (A, C) is monotone (see [9], Theorem 4.1).

If (A, C) is an OBA, F a closed ideal in A and $\pi : A \to A/F$ the canonical homomorphism, then $(A/F, \pi C)$ is an OBA. Normality of the algebra cone πC in A/F is defined in the usual way. We will say that the spectral radius is monotone in $(A/F, \pi C)$ if $0 \leq x \leq y$ w.r.t. C implies that $r(x + F, A/F) \leq r(y + F, A/F)$.

We now give some examples of ordered Banach algebras.

Example ([4, Example 3.5]). Let $A = M_n(\mathbb{C})$ be the Banach algebra of $n \times n$ complex matrices. Let C be the subset of A consisting of matrices with only non-negative entries and C' the subset of A consisting of diagonal matrices with only non-negative entries. Then C and C' are closed, normal algebra cones of A, and so (A, C), (A, C') are OBAs.

Example ([4, Example 3.8]). Let $A = \ell^\infty$ and $C = \{(c_1, c_2, \dots) \in A : c_i \geq 0$ for all $i \in \mathbb{N}\}$. Then C is a closed, normal algebra cone of A. Therefore (A, C) is an OBA.

Example ([3, Example 3.3]). Let A be a commutative C^*-algebra, $C = \{x \in A : x = x^*$ and $\sigma(x) \subseteq [0, \infty)\}$ and F a closed ideal of A. Then C is a closed, normal algebra cone of A and πC is a normal algebra cone of A/F. Therefore (A, C) and $(A/F, \pi C)$ are OBAs.

Example ([4, Example 3.4]). Let E be a complex Banach lattice and let $C = \{x \in E : x = |x|\}$. If $K = \{T \in \mathcal{L}(E) : TC \subseteq C\}$, then K is a closed, normal algebra cone of $\mathcal{L}(E)$. Therefore $(\mathcal{L}(E), K)$ is an OBA.

Example ([3, Example 3.2]). Let E be a Dedekind complete Banach lattice, $C = \{x \in E : x \geq 0\}$ and $K = \{T \in \mathcal{L}(E) : TC \subseteq C\}$. Then $(\mathcal{L}^r(E), K)$ is an OBA with a closed, normal algebra cone and $(\mathcal{L}^r(E)/\mathcal{K}^r(E), \pi K)$ is an OBA such that the spectral radius in $(\mathcal{L}^r(E)/\mathcal{K}^r(E), \pi K)$ is monotone.

3. Poles of the resolvent

This section considers the primary question under consideration: if x and y are elements of an OBA such that $0 \leq x \leq y$ and $r(y)$ is a pole of the resolvent of y, when is $r(x)$ a pole of the resolvent of x? The main result is Theorem 3.3.

We begin with the following proposition, which is related to the problem at hand. It tells us how to conclude that a pole of the resolvent of an element in a Banach algebra is a simple pole.

Proposition 3.1. *Let A be a Banach algebra, $x \in A$ and suppose that λ_0 is a pole of the resolvent of x. If (λ_n) is a sequence in $\mathbb{C}$ such that $\lambda_n \to \lambda_0$ as $n \to \infty$ and $\overline{\lim\limits_{n\to\infty}} \|(\lambda_n - \lambda_0)R(\lambda_n, x)\|$ exists, then λ_0 is a simple pole of the resolvent of x.*

Proof. Suppose that λ_0 is a pole of order k of the resolvent of x, with $k > 1$. Then for $n \in \mathbb{N}$, the resolvent of x has a Laurent series expansion

$$R(\lambda_n, x) = \frac{x_{-k}}{(\lambda_n - \lambda_0)^k} + \frac{x_{-k+1}}{(\lambda_n - \lambda_0)^{k-1}} + \cdots + \frac{x_{-1}}{\lambda_n - \lambda_0} + x_0 + x_1(\lambda_n - \lambda_0) + \cdots,$$

on a deleted neighbourhood of λ_0, and $x_{-k} \neq 0$. Multiplying both sides of the expansion by $(\lambda_n - \lambda_0)^k$ and then taking limits, it follows from continuity of the norm that $\lim\limits_{n\to\infty} \|(\lambda_n - \lambda_0)^k R(\lambda_n, x)\| = \|x_{-k}\|$. Thus $\overline{\lim\limits_{n\to\infty}} \|(\lambda_n - \lambda_0)^k R(\lambda_n, x)\|$ exists and equals $\|x_{-k}\|$. From the hypothesis, it follows that

$$\|x_{-k}\| = \left(\overline{\lim_{n\to\infty}} |\lambda_n - \lambda_0|^{k-1}\right)\left(\overline{\lim_{n\to\infty}} \|(\lambda_n - \lambda_0)R(\lambda_n, x)\|\right) = 0,$$

which contradicts $x_{-k} \neq 0$. Hence $k = 1$. □

To prove the main result of this section, we will make use of the following basic property of resolvents of elements in OBAs.

Proposition 3.2. *Let A be an OBA with a closed algebra cone C and suppose that the spectral radius in (A, C) is monotone. If $x, y \in A$ are such that $0 \leq x \leq y$, then $R(\lambda, x) \leq R(\lambda, y)$ for $\lambda > r(y)$.*

Proof. By monotonicity, $r(x) \leq r(y)$ and so for any $\lambda > r(y)$, the resolvent of x has a Neumann series representation $R(\lambda, x) = \frac{1}{\lambda}\sum_{k=0}^{\infty}\left(\frac{x}{\lambda}\right)^k$ and the resolvent of y has a Neumann series representation $R(\lambda, y) = \frac{1}{\lambda}\sum_{k=0}^{\infty}\left(\frac{y}{\lambda}\right)^k$. Since C is closed, it follows that $\frac{1}{\lambda}\sum_{k=0}^{\infty}\frac{1}{\lambda^k}(y^k - x^k) \in C$, which implies that $R(\lambda, x) \leq R(\lambda, y)$. □

We now present the main result, which shows that under natural conditions that come with the OBA structure, the spectral radius $r(x)$ of a dominated element x is a pole of the resolvent of x, if the spectral radius $r(y)$ of the dominating element y is a pole of the resolvent of y. In addition, the order of the pole $r(x)$ is dominated by the order of the pole $r(y)$.

Theorem 3.3. *Let A be an OBA with a closed, normal algebra cone C and let $x, y \in A$ such that $0 \leq x \leq y$. Suppose that $r(x) = r(y)$ and $r(x) \in$ **iso** $\sigma(x)$. If $r(y)$ is a pole the of resolvent of y, then $r(x)$ is a pole of the resolvent of x. Moreover, the order of $r(x)$ is at most equal to the order of $r(y)$.*

Proof. Suppose that $r(y)$ is a pole of order m of the resolvent of y. Since C is closed, from $0 \leq x \leq y$ and the Neumann series representations of $R(\lambda, x)$ and $R(\lambda, y)$ for $\lambda > r(y)$, we have that

$$0 \leq \lim_{\lambda \downarrow r(x)}[(\lambda - r(x))^m R(\lambda, x)] \leq \lim_{\lambda \downarrow r(y)}[(\lambda - r(y))^m R(\lambda, y)]. \qquad (*)$$

Now suppose that $r(x)$ is not a pole of the resolvent of x or $r(x)$ is a pole of order $k > m$ of the resolvent of x. If we consider the Laurent series expansions of the resolvents of x and y about $r(x)$ and $r(y)$ respectively and use the fact the norm is a continuous function, we obtain that $||\lim_{\lambda \downarrow r(x)}(\lambda - r(x))^k R(\lambda, x)|| = \infty$ whereas $||\lim_{\lambda \to r(y)}(\lambda - r(y))^k R(\lambda, y)|| = ||b_{-m}||$, where b_{-m} is the coefficient in the Laurent series expansion of the resolvent of y corresponding to the power $(\lambda - r(y))^{-m}$. But considering $(*)$ and the fact that C is a normal algebra cone, this is not possible. □

From Example 2 and Theorem 3.3, we obtain the following result for positive operators on a Banach lattice.

Corollary 3.4. *Let E be a complex Banach lattice and let $S, T \in K$ such that $0 \leq S \leq T$ and $r(S) = r(T)$, where $K = \{U \in \mathcal{L}(E) : UC \subseteq C\}$ and $C = \{x \in E : x = |x|\}$. If $r(S)$ is an isolated point of $\sigma(S)$ and if $r(T)$ is a pole of order m of the resolvent of T, then $r(S)$ is a pole of order at most m of the resolvent of S.*

The next corollary also follows immediately from Theorem 3.3.

Corollary 3.5. *Let A be an OBA with a closed, normal algebra cone C and let $x, y \in A$ such that $0 \leq x \leq y$. Suppose that $r(x) = r(y)$ and $r(x) \in$ **iso** $\sigma(x)$. If $r(y)$ is a simple pole of the resolvent of y, then $r(x)$ is a simple pole of the resolvent of x.*

Example 2 and Corollary 3.5 yield the following result regarding positive operators on a Banach lattice.

Corollary 3.6. *Let E be a complex Banach lattice and let $S, T \in K$ such that $0 \leq S \leq T$ and $r(S) = r(T)$, where $K = \{U \in \mathcal{L}(E) : UC \subseteq C\}$ and $C = \{x \in E : x = |x|\}$. If $r(S)$ is an isolated point of $\sigma(S)$ and if $r(T)$ is a simple pole of the resolvent of T, then $r(S)$ is a simple pole of the resolvent of S.*

4. Ergodic elements

In this section, $f_n(x)$ will denote the element $\sum_{k=0}^{n-1} \frac{x^k}{n}$ in the Banach algebra A. In [6] the problem of determining when a dominated element in an OBA is ergodic, given that the dominating element is ergodic, was investigated. The following three theorems are the main results.

Theorem 4.1 ([6], Theorem 5.1). *Let (A, C) be an OBA with C closed and proper, and let $x, y \in A$ such that $0 \leq 1 \leq x \leq y$. Suppose that $r(y) = 1 \in \textbf{iso}\,\sigma(x)$ and $r(y)$ is a simple pole of the resolvent of y. If y is ergodic with $f_n(y) \to p(y, r(y))$ and if $p(x, r(x)) = p(y, r(y))$, then x is ergodic with $f_n(x) \to p(x, r(x))$.*

Theorem 4.2 ([6], Theorem 5.2). *Let A be an OBA with a closed, normal algebra cone C and let $x, y \in A$ such that $0 \leq x \leq y$. Suppose that $1 \in \textbf{iso}\,\sigma(x)$ is a pole of the resolvent of x. If y is ergodic, then x is ergodic.*

Theorem 4.3 ([6, Theorem 5.5]). *Let A be a semisimple OBA with a closed, normal algebra cone C and let $x, y \in A$ such that $0 \leq x \leq y$. Let I be a closed inessential ideal of A such that the spectral radius in $(A/I, \pi C)$ is monotone. If y is ergodic and if $r(y)$ is a Riesz point of $\sigma(y)$, then x is ergodic.*

The next two theorems are complementary to Theorems 4.1, 4.2, and 4.3. We start with the following, which strengthens Theorems 4.2 and 4.3.

Theorem 4.4. *Let A be an OBA with a closed, normal algebra cone C and $x, y \in A$ such that $0 \leq x \leq y$. Suppose that $r(x)$ is a pole of the resolvent of x. If y is ergodic, then x is ergodic.*

Proof. Normality of C implies that the spectral radius in (A, C) is monotone by ([9], Theorem 4.1). Since y is ergodic, it follows from ([6] Lemma 5.3) that $r(x) \leq r(y) \leq 1$. This leads to four possibilities: $r(x) < r(y) < 1$, $r(x) < r(y) = 1$, $r(x) = r(y) < 1$, $r(x) = r(y) = 1$. Now, from $0 \leq x \leq y$, we obtain that $0 \leq \frac{x^n}{n} \leq \frac{y^n}{n}$ (for all $n \in \mathbb{N}$). Since y is ergodic, $\frac{y^n}{n} \to 0$ as $n \to \infty$ by ([6], Proposition 4.9 and Lemma 4.8). From normality of C, it follows that $\frac{x^n}{n} \to 0$ as $n \to \infty$. In the first three cases, we get that $1 \notin \sigma(a)$. Thus $\sum_{k=0}^{n-1} \frac{x^k}{n} \to 0$ as $n \to \infty$ by ([6], Lemma 5.4), which means that x is ergodic. For the case $r(x) = r(y) = 1$, the result follows immediately from Theorem 4.2. $\square$

Theorem 4.2 is a corollary of ([6], Lemma 5.3) and Theorem 4.4, while Theorem 4.3 is a corollary of ([6], Lemma 5.3), ([3], Lemma 2.1, Theorem 4.3), and Theorem 4.4.

The next theorem, which strengthens Theorem 4.1 when the algebra cone is normal, is obtained using a result from Section 3.

Theorem 4.5. *Let A be an OBA with a closed, normal algebra cone C and $x, y \in A$ such that $0 \leq x \leq y$. Suppose that $r(y) = 1 \in \textbf{iso}\ \sigma(y)$ and $r(x) \in \textbf{iso}\ \sigma(x)$. If y is ergodic with $f_n(y) \to p(y, r(y))$, then x is ergodic with $f_n(x) \to p(x, r(x))$.*

Proof. If $r(x) < r(y)$, the fact that y is ergodic, together with ([6], Lemma 5.4) imply that x ergodic with its sequence of ergodic sums converging to 0. If $r(x) = r(y)$, the result follows by applying ([6], Theorem 4.10), then Corollary 3.5 and ([6], Theorem 4.10) again. □

5. Eigenvalues and eigenvectors

The next two theorems about eigenvalues and eigenvectors for dominated elements in OBAs are some of the main results in [8].

Theorem 5.1 ([8], Theorem 4.3). *Let A be an OBA with a proper, closed algebra cone C and let $x, y \in A$. Suppose that $0 < r(x) = r(y)$ and that $0 < r(y) \leq x \leq y$. If $r(y)$ is a pole of the resolvent of y, then $r(y)$ is an eigenvalue of y with a corresponding eigenvector u, and $r(x)$ is an eigenvalue of x with corresponding eigenvector u.*

Theorem 5.2 ([8], Theorem 4.4). *Let (A, C) be a semisimple OBA with C closed and such that the spectral radius in (A, C) is monotone. Let I be a closed inessential of A such that the spectral radius in $(A/I, \pi C)$ is monotone. Suppose that $x, y \in A$ such that $0 < x \leq y$ and $0 < r(x) = r(y)$. If $r(y)$ is an eigenvalue of y with a positive corresponding eigenvector and with $r(y)$ a Riesz point of $\sigma(y)$, then $r(x)$ is an eigenvalue of x with a positive corresponding eigenvector.*

The next corollary is complementary to Theorem 5.1. It shows how to conclude that the spectral radius of the dominated element is an eigenvalue of the element, given the less restrictive inequality $0 < x \leq y$ instead of $0 < r(y) \leq x \leq y$.

Corollary 5.3. *Let A be an OBA with a closed, normal algebra cone C and let $x, y \in A$ such that $0 < x \leq y$, $r(x) = r(y)$ and $0 \neq r(x) \in \textbf{iso}\ \sigma(x)$. If $r(y)$ is a pole of the resolvent of y, then $r(y)$ is an eigenvalue of y with a positive corresponding eigenvector, and $r(x)$ is an eigenvalue of x with a positive corresponding eigenvector.*

Proof. The result follows from ([7], Theorem 3.2), then Theorem 3.3 and ([7], Theorem 3.2) again. □

The next result shows when $r(y)$ in Corollary 5.3 is a simple pole of the resolvent of y, the spectral projections associated with $r(x)$ and $r(y)$ are the corresponding eigenvectors.

Corollary 5.4. *Let A be an OBA with a closed, normal algebra cone C and let $x, y \in A$ such that $0 < x \leq y$. Suppose that $r(x) = r(y)$, $r(x) \in \textbf{iso}\ \sigma(x)$, and $r(y)$ a simple pole of the resolvent of y. Then $r(y)$ is an eigenvalue of y, with $p(y, r(y))$ a corresponding eigenvector, and $r(x)$ is an eigenvalue of x with $p(x, r(x))$ a corresponding eigenvector.*

Proof. By ([7], Theorem 3.2), $r(y)$ is an eigenvalue of y, with corresponding eigenvector $p(y, r(y))$. By Corollary 3.5 and ([7], Theorem 3.2), $r(x)$ is an eigenvalue of x, with a corresponding eigenvector $p(x, r(x))$. $\square$

The following corollary complements Corollary 5.4.

Corollary 5.5. *Let A be an OBA with a closed, normal algebra cone C and let $x, y \in A$ such that $0 < x \leq y$, with $r(x) = r(y) \in \textbf{iso}\ \sigma(x)$. If $r(y)$ is an eigenvalue of y with corresponding eigenvector $p(y, r(y))$, then $r(x)$ is an eigenvalue of x with corresponding eigenvector $p(x, r(x))$.*

Proof. If $r(y)$ is an eigenvalue of y with corresponding eigenvector $p(y, r(y))$, then $r(y)$ is a simple pole of the resolvent of y. It follows from Corollary 3.5 and ([7], Theorem 4.2) that $r(x)$ is an eigenvalue of x with corresponding eigenvector $p(x, r(x))$. $\square$

Another complementary result to Theorems 5.1 and 5.2 is the following.

Corollary 5.6. *Let A be a semisimple OBA with a closed, normal algebra cone C and let $x, y \in A$ such that $0 < x \leq y$, with $0 < r(x)$. Suppose that I is a closed inessential ideal of A such that the spectral radius in $(A/I, \pi C)$ is monotone. If y is Riesz relative to I, then $r(y)$ is an eigenvalue of y with a positive corresponding eigenvector, and $r(x)$ is an eigenvalue of x with a positive corresponding eigenvector.*

Proof. That $r(y)$ is an eigenvalue of y with a positive corresponding eigenvector follows directly from ([7], Theorem 3.7). From ([9], Theorem 6.2) and ([7], Theorem 3.7), we obtain that $r(x)$ is an eigenvalue of x with a positive corresponding eigenvector. $\square$

The next result proves that if $r(x) = r(y)$ is assumed, we are still able to make the conclusion of Corollary 5.6 without any assumptions about the quotient algebra.

Corollary 5.7. *Let A be a semisimple OBA with a closed, normal algebra cone C and let $x, y \in A$ such that $0 < x \leq y$, with $0 \neq r(x) = r(y) \in \textbf{iso}\ \sigma(a)$. Suppose that I is a closed inessential ideal of A such that y is Riesz relative to I. Then $r(y)$ is an eigenvalue of y with a positive corresponding eigenvector, and $r(x)$ is an eigenvalue of x with a positive corresponding eigenvector.*

Proof. Clearly, $r(y)$ is an eigenvalue of y with a positive corresponding eigenvector. Now, since A is semisimple, ([7], Theorem 3.11) yields that $r(y)$ is a pole of the resolvent of y. To obtain that $r(x)$ is an eigenvalue of x with a positive corresponding eigenvector, we apply Theorem 3.3 and ([7], Theorem 3.2). □

In [5] we defined a commutatively ordered Banach algebra (COBA) as a Banach algebra A ordered by a cone C which is closed under positive scalar multiplication and under multiplication of commuting elements. A COBA is a more general structure than an OBA and obviously, every OBA is a COBA. We then showed that much of the known spectral theory in OBAs is obtainable in COBAs when appropriate modifications compatible with the COBA structure are adopted. In the same spirit, we can establish COBA counterparts of the OBA results in this paper when suitable adjustments are made.

Acknowledgment

We would like to sincerely thank the anonymous reviewer for the valuable suggestions that improved the paper.

References

[1] D. Behrendt and H. Raubenheimer, *On domination of inessential elements in ordered Banach algebras.* Illinois J. Math. **51(3)** (2007), 927–936.

[2] H. du T. Mouton and S. Mouton, *Domination properties in ordered Banach algebras.* Studia Math. **149 (1)** (2002), 63–73.

[3] S. Mouton, *Convergence properties of positive elements in Banach algebras.* Math. Proc. R. Ir. Acad. **102 A (2)** (2002), 149–162.

[4] S. Mouton, *A spectral problem in ordered Banach algebras.* Bull. Austral. math. Soc. **67** (2003), 131–144.

[5] S. Mouton and K. Muzundu, *Commutatively ordered Banach algebras.* Quaest. Math. **36** (2013), 1–29.

[6] S. Mouton and K. Muzundu, *Domination by ergodic elements in ordered Banach algebras.* Positivity **18 (1)** (2014), 119–130.

[7] S. Mouton and H. Raubenheimer, *More spectral theory in ordered Banach algebras.* Positivity **1** (1997), 305–317.

[8] K. Muzundu, *Generalized domination in ordered Banach algebras.* Journal of Mathematics and System Science **3** (2013), 335–341.

[9] H. Raubenheimer and S. Rode, *Cones in Banach algebras.* Indag. Math. **7** (1996), 489–502.

[10] A.C. Zaanen, *Riesz Spaces II.* North-Holland, Amsterdam, 1983.

Kelvin Muzundu
Department of Mathematics and Statistics
University of Zambia, P.O. Box 32379
Lusaka, Zambia
e-mail: kmzundu@gmail.com

Ordered Structures and Applications: Positivity VII
Trends in Mathematics, 339–347

A Note on Universal Operators

Timur Oikhberg

Abstract. An operator T is called universal for the complement of the ideal $\mathfrak{A}$ if T does not belong to $\mathfrak{A}$, and factors through every element of the complement of $\mathfrak{A}$. We show that the complements of many ideals (such as the ideal of strictly (co)singular operators, or any maximal normed ideal) have no universal operators. On the other hand, the complement of the ideal of finitely strictly singular operators has a universal operator. Moreover, we show that, for many ideals $\mathfrak{A}$, any positive operator which factors positively through any positive member of the complement of $\mathfrak{A}$ must be compact.

Mathematics Subject Classification (2010). Primary: 47L20; Secondary: 46B42, 47B10, 47B65.

Keywords. Operator ideal, universal operator, positive operator.

1. Introduction

Suppose $\mathfrak{A}$ is an operator ideal (see, e.g., [2] for a concise introduction). We denote by $C\mathfrak{A}$ the complement of $\mathfrak{A}$. We say that $T \in B(X,Y)\backslash\mathfrak{A}(X,Y)$ is *universal* for $C\mathfrak{A}$ if it *factors* through every operator $S \in C\mathfrak{A}$ – that is, for any Banach spaces X_0 and Y_0, and any $S \in B(X_0,Y_0)\backslash\mathfrak{A}$, there exist $A \in B(X,X_0)$ and $B \in B(Y_0,Y)$ so that $T = BSA$. In the cases mentioned below, the existence of universal operators is known:

1. $\mathfrak{A}$ is the ideal of compact operators: the formal identity from ℓ_1 to ℓ_∞ is a universal operator for $C\mathfrak{A}$ [8].
2. $\mathfrak{A}$ is the ideal of weakly compact operators: the summing operator from ℓ_1 to ℓ_∞ is a universal operator for $C\mathfrak{A}$ [9].

The author was partially supported by Simons Foundation (travel grant 210060). He wishes to thank the staff of Carle Hospital in Urbana, IL, and of Heartland Rehabilitation Center in Champaign, IL, where part of this work was carried out. Last but not least, the author is grateful to the anonymous referee for many helpful comments, and in particular, for bringing [4], [6], and [7] to the author's attention.

3. $\mathfrak{A}$ is the ideal of ℓ_p-strictly singular or c_0-strictly singular operators: any isomorphic embedding from ℓ_p (or c_0) to ℓ_∞ is a universal operator for $\mathrm{C}\mathfrak{A}$, cf. [2, Section 1].
4. $\mathfrak{A}$ is the ideal of (c_0, p, q)-summing operators: the formal identity from ℓ_{p^*} to ℓ_q is a universal operator for $\mathrm{C}\mathfrak{A}$ [7].

The situation is more complicated for Dunford–Pettis operators [5].

In this note, we further study the existence of universal operators. In Section 2, we examine some ideals closed in the operator norm – namely, the ideals of strictly singular, strictly cosingular, and finitely strictly singular operators. We show that the complements of the first two ideals have no universal operators, while the complement of the third one does. In Section 3 we prove that the complements of quasi-normed ideals, verifying certain (very general) conditions, have no universal elements. These results apply, for instance, to all maximal Banach ideals. In Section 4, we switch to the Banach lattice setting. Suppose X, Y, X_0, Y_0 are Banach lattices. We say that $T \in B(X,Y)_+$ *positively factors* through $S \in B(X_0, Y_0)_+$ if there exist $A \in B(X, X_0)_+$ and $B \in B(Y_0, Y)_+$ so that $T = BSA$. $T \in \mathrm{C}\mathfrak{A}(X,Y)_+$ is *positively universal* for $\mathrm{C}\mathfrak{A}$ if it positively factors through every member of $\mathrm{C}\mathfrak{A}_+$. We establish the non-existence of positively universal operators, for a wide class of operator ideals.

2. Ideals closed in the operator norm

First recall some definitions: an operator $T \in B(X,Y)$ is called

1. *strictly singular* if, for any infinite-dimensional $E \subset X$, the restriction $T|_E$ is not an isomorphism.
2. *strictly cosingular* if, for any infinite-dimensional quotient $q : Y \to F$, qT is not surjective.
3. *finitely strictly singular* if, for every $\varepsilon > 0$, there exists $N \in \mathbb{N}$ so that, for any N-dimensional subspace $E \subset X$, we can find a norm one $e \in E$ so that $\|Te\| < \varepsilon$.

We refer the reader to [1] for the first two ideals, and to [12] for the third one. Note that recently, the class of finitely strictly singular operators has been studied intensively, see, e.g., [4] and [6].

First we show that the complements of strictly (co)singular operators possess no universal objects.

Proposition 2.1. *If an operator T factors through every non-strictly (co)singular operator, then T itself is strictly (co)singular.*

Proof. Suppose, for the sake of contradiction, that $T \in B(X,Y)$ is not strictly singular, and factors through any non-strictly operator. By restricting the domain of X, and extending its range, we may assume that $T : X \to \ell_\infty(\Gamma)$ is an isomorphism into. Clearly T factors through I_{ℓ_p} for $1 \leqslant p \leqslant \infty$, and through I_{c_0}. Thus,

X embeds isomorphically into ℓ_p for any p. Thus, X is ℓ_p-saturated for any p, which is clearly impossible.

The case of strictly cosingular operators is handled similarly. Suppose $T \in B(X,Y)$ is not strictly cosingular, and factors through any non-strictly cosingular operator. By composing X with quotient maps $\ell_1(\Gamma) \to X$ and $q : Y \to Y/Y_0$, we can assume that $T \in B(\ell_1(\Gamma), Y)$ is such that $T(\mathbf{B}(\ell_1(\Gamma))) \supset c\mathbf{B}(Y)$ (here and below, $\mathbf{B}(Z)$ stands for the closed unit ball of the normed space Z). T factors through ℓ_p for $1 < p < \infty$, hence the same is true for $T^* \in B(Y^*, \ell_\infty(\Gamma))$. As T^* is an isomorphism, be obtain a contradiction, as in the strictly singular case. □

The situation is different for finitely strictly singular operators.

Proposition 2.2. *There exists an operator $T \in B((\sum_n \ell_2^n)_{\ell_1}, \ell_\infty)$ which is not finitely strictly singular, and factors through every non-finitely strictly singular operator.*

The folklore result below (needed to prove Proposition 2.2) can be established by emulating the proof of [10, Lemma 1.a.6].

Lemma 2.3. *Suppose E is a finite-dimensional subspace of an infinite-dimensional space X. For any $c > 1$, X contains a finite codimensional subspace Y so that, for any $e \in E$ and $y \in Y$, $c\|e + y\| \geqslant \|e\|$.*

Proof of Proposition 2.2. Let $id : (\sum_n \ell_2^n)_{\ell_1} \to (\sum_n \ell_2^n)_{c_0}$ be the formal identity, let $j : (\sum_n \ell_2^n)_{c_0} \to \ell_\infty$ be an isometric embedding, and set $T = j \circ id$. Clearly, T is not finitely strictly singular. We establish its universality.

Suppose a contraction $S \in B(X,Y)$ is not FSS. By Dvoretzky Theorem, X contains 2-Hilbertian subspaces E of arbitrarily large dimension so that $\|Sx\| \geqslant c\|x\|$ for any $x \in E$ (here $c > 0$ is a constant depending only on S). We now construct a sequence of 2-Hilbertian subspaces E_n, of dimension n, so that $\|Sx\| \geqslant c\|x\|$ for any $x \in E_n$, and let $F_n = S(E_n)$. The sequence (E_n) must have a special property: for any finite sequences $e_1 \in E_1, \ldots, e_n \in E_n, f_1 \in F_1, \ldots, f_n \in F_n$, we have $\|e_1 + \cdots + e_k\| \leqslant 2\|e_1 + \cdots + e_n\|$, $\|f_1 + \cdots + f_k\| \leqslant 2\|f_1 + \cdots + f_n\|$ whenever $k \leqslant n$. Once this is done, we obtain a factorization of T through S: for each n, pick a contraction $u_n \in B(\ell_2^n, E_n)$ so that $\|u_n^{-1}\| \leqslant 2$. Then $U : (\sum_n \ell_2^n)_1 \to \mathrm{span}[E_1, E_2, \ldots] \subset X : (\xi_1, \xi_2, \ldots) \mapsto \sum_k u_k \xi_k$ is a contraction. Let $v_n = u_n^{-1}(S|_{E_n})^{-1} : F_n \to \ell_2^n$, and observe that $\|v_n\| \leqslant 2c^{-1}$. Define $V : \mathrm{span}[F_1, F_2, \ldots] \to (\sum_n \ell_2^n)_{c_0}$ via $V(f_1 + \cdots + f_n) = (v_1 f_1, \ldots, v_n f_n)$. We claim that V is a bounded well-defined operator. Indeed, by construction, for $k \leqslant n$ we have

$$\|f_k\| \leqslant \|f_1 + \cdots + f_k\| + \|f_1 + \cdots + f_{k-1}\| \leqslant 4\|f_1 + \cdots + f_n\|.$$

Consequently,

$$\|V(f_1 + \cdots + f_n)\| = \max_{1 \leqslant k \leqslant n} \|v_k f_k\| \leqslant 8c^{-1}\|f_1 + \cdots + f_n\|.$$

Now extend V to $W : Y \to \ell_\infty$ using injectivity. Then $T = WSU$.

The sequences (E_n) and (F_n) are constructed recursively. E_1 can be selected arbitrarily. Set $F_1 = S(E_1)$. Use Lemma 2.3 to find finite codimensional $X_1 \subset X$ and $Y_1 \subset Y$ so that $2\|e_1 + x_1\| \geqslant \|e_1\|$ and $2\|f_1 + y_1\| \geqslant \|f_1\|$ whenever $e_1 \in E_1$, $f_1 \in F_1$, $x_1 \in X_1$, and $y_1 \in Y_1$.

Suppose $E_1, \dots, E_n$, $F_1, \dots, F_n$, $X_1, \dots, X_n$, $Y_1, \dots, Y_n$ have already been constructed in such a way that, for $1 \leqslant k \leqslant n$:

1. E_k and $F_k = S(E_k)$ are finite dimensional, while X_k and Y_k are finite codimensional.
2. For $k > 1$, $E_k \cup X_k \subset X_{k-1}$, and $F_k \cup Y_k \subset Y_{k-1}$.
3. For any $x \in X_k$, $e \in E_1 + \dots + E_k$, $y \in Y_k$, and $f \in F_1 + \dots + F_k$, we have $\|e\| \leqslant 2\|e + x\|$ and $\|f\| \leqslant 2\|f + y\|$.

Pick a 2-Hilbertian $E \subset X$, of dimension $n + 1 + \operatorname{codim} X_n + \operatorname{codim} Y_n$, so that $\|Se\| \geqslant c\|e\|$ for any $e \in E$. Then the dimension of $E' = E \cap X_n \cap S^{-1}(Y_n)$ is at least $n + 1$. Clearly, any $(n + 1)$-dimensional subspace of E' can serve as E_{n+1}. Complete the procedure by selecting $X_{n+1} \subset X_n$ and $Y_{n+1} \subset Y_n$ using Lemma 2.3. □

3. Normed ideals

In this section, $(\mathfrak{A}, \boldsymbol{\alpha})$ is a quasi-Banach ideal of operators on Banach spaces. It is well known that such ideals have an equivalent ideal p-norm, for some $p \in (0, 1]$ (see, e.g., [2, Section 2]). Throughout this section, we assume that $\mathfrak{A}$ is p-normed, and satisfies two conditions:

$$\sup\left\{\boldsymbol{\alpha}(I_E) : \dim E < \infty\right\} = \infty. \tag{3.1}$$

and

$$\text{If } T \notin \mathfrak{A}(X, Y), \text{ then } \sup_{E \subset X, \dim E < \infty} \boldsymbol{\alpha}(T|_E) = \infty. \tag{3.2}$$

These conditions hold, for instance, for all maximal Banach operator ideals, different from the ideal of all bounded operators (see, e.g., [3, Chapter 6]).

Theorem 3.1. *If* $(\mathfrak{A}, \boldsymbol{\alpha})$ *is a quasi-Banach operator ideal satisfying* (3.1) *and* (3.2), *then* $\mathrm{C}\mathfrak{A}$ *has no universal operator.*

Begin by establishing a simple lemma.

Lemma 3.2. *Suppose* $(\mathfrak{A}, \boldsymbol{\alpha})$ *is a p-normed quasi-Banach operator ideal. Then there exists a constant* C *so that, for any finite-dimensional normed space* E, $\boldsymbol{\alpha}(I_E) \leqslant C(\dim E)^{1/p}$.

Proof. Note first that, for any rank one operator $x^* \otimes y \in B(X, Y)$ (X and Y are Banach spaces), $\boldsymbol{\alpha}(x^* \otimes y) = C\|x^*\|\|y\|$, where the constant C depends solely on $\mathfrak{A}$. Indeed, let $C = \boldsymbol{\alpha}(I_{\mathbb{F}})$, where $\mathbb{F}$ is the underlying field of scalars ($\mathbb{R}$ or $\mathbb{C}$).

Now consider non-zero $x^* \in X^*$ and $y \in Y$. To show that $\boldsymbol{\alpha}(x^* \otimes y) \geqslant C\|x^*\|\|y\|$, consider the contractions $j : \mathbb{F} \to X : 1 \mapsto x$ (with $\|x\| = 1$) and $q : Y \to \mathbb{F} : z \mapsto \langle y^*, z\rangle$ (with norm one y^* selected in such a way that $\|y\| = \langle y^*, y\rangle$).

Note that $q \circ (x^* \otimes y) \circ j = \langle x^*, x \rangle \|y\| I_{\mathbb{F}}$, hence $\boldsymbol{\alpha}(x^* \otimes y) \geqslant \boldsymbol{\alpha}\big(q \circ (x^* \otimes y) \circ j\big) \geqslant |\langle x^*, x \rangle| \|y\| \boldsymbol{\alpha}(I_{\mathbb{F}}) = C |\langle x^*, x \rangle| \|y\|$. Taking the supremum over all $x \in X$ of norm one, we obtain $\boldsymbol{\alpha}(x^* \otimes y) \geqslant C \|x^*\| \|y\|$. The inequality $\boldsymbol{\alpha}(x^* \otimes y) \leqslant C \|x^*\| \|y\|$ is proved in a similar fashion.

Now suppose E is an n-dimensional space. Using the Auerbach basis of E, write $I_E = P_1 + \cdots + P_n$, where $P_1, \ldots, P_n$ are contractive projections of rank one. Then $\boldsymbol{\alpha}(I_E) \leqslant \big(\sum_i \boldsymbol{\alpha}(P_i)^p\big)^{1/p} = C n^{1/p}$. □

Next we quantify the "non-belonging" to $\mathfrak{A}$. For $T \in B(X, Y)$ and $n \in \mathbb{N}$, define

$$\beta_n(T) = \sup \big\{ \boldsymbol{\alpha}(T|_E) : E \subset X, \dim E = n \big\}.$$

Note that, by the ideal property, $\boldsymbol{\alpha}(T|_E) \geqslant \boldsymbol{\alpha}(T|_F)$ if $F \subset E$, hence, in the definition of β_n, we could have taken the supremum over all $E \subset X$ of dimension not exceeding n. We have:

Lemma 3.3. *If $S \in B(X_0, Y_0)$ factors through $T \in B(X, Y)$, then there exists a constant c so that $\beta_n(S) \leqslant c \beta_n(T)$ for every n.*

Proof. Write $S = VTU$. If E is a finite-dimensional subspace of X_0, then, by the ideal property, $\boldsymbol{\alpha}(S|_E) \leqslant \|V\| \boldsymbol{\alpha}(T|_{U(E)}) \|U\|$, and therefore,

$$\begin{aligned} \beta_n(S) = \sup_{\dim E = n} \boldsymbol{\alpha}(S|_E) &\leqslant \|V\| \sup_{\dim E = n} \boldsymbol{\alpha}(T|_{U(E)}) \|U\| \\ &\leqslant \|U\| \|V\| \sup_{\dim F \leqslant n} \boldsymbol{\alpha}(T|_F) = \|U\| \|V\| \beta_n(T), \end{aligned}$$

as claimed. □

The following technical lemma is crucial.

Lemma 3.4. *Suppose the quasi-normed ideal $(\mathfrak{A}, \boldsymbol{\alpha})$ satisfies (3.1). Then, for any sequence $0 < \alpha_1 < \alpha_2 < \cdots$, increasing without a bound, there exists an operator T, so that $\lim_n \beta_n(T) = \infty$, and $\beta_n(T) \leqslant \alpha_n$ for infinitely many values of n.*

Proof. By the discussion in the beginning of the section, we can assume that $\boldsymbol{\alpha}$ is a p-norm. Furthermore, by Lemma 3.2 and its proof, we can assume that $\boldsymbol{\alpha}(x^* \otimes y) = \|x^*\| \|y\|$ for any $x^* \in X$ and $y \in Y$. Consequently, $\boldsymbol{\alpha}(I_E) \leqslant (\dim E)^{1/p}$ for any finite-dimensional normed space E. We shall find finite-dimensional spaces E_i and an operator $T = \oplus \gamma_i I_{E_i}$, acting on $X = \big(\sum_i E_i\big)_2$, with the desired properties. We use the notation $k_i = \dim E_i$, $c_i = \boldsymbol{\alpha}(I_{E_i})$, and $\sigma_i = \gamma_i c_i$. The parameters will be selected in such a way that:

$$\begin{aligned} &\sigma_i < \frac{\sigma_{i+1}}{2^{1/p}} \quad \text{for any } i. \\ &\gamma_n < \frac{\sigma_i}{2^{(n-i)/p} k_i^{1/p}} \quad \text{whenever } n > i. \\ &\sigma_i < \frac{\alpha_{k_i}}{3} \quad \text{for any } i. \end{aligned} \tag{3.3}$$

Suppose for a moment the selection has already been made. As $E_n = T(E_n)$ is contractively complemented in X,

$$\beta_{k_n}(T) \geqslant \boldsymbol{\alpha}(T|_{E_n}) = \sigma_n,$$

and we conclude that $\lim_n \beta_{k_n}(T) = \infty$. On the other hand, suppose F is a subspace of X, with $\dim F = k_n$. Let P_i be the canonical projection onto E_i, and set $F_i = P_i(F)$. Then $\boldsymbol{\alpha}(T|_F)^p \leqslant \sum_i \boldsymbol{\alpha}(T|_{F_i})^p$. Note that, for $i \leqslant n$, $\boldsymbol{\alpha}(T|_{F_i}) \leqslant \boldsymbol{\alpha}(T|_{E_i}) = \sigma_i$, while for $i > n$, by the conditions imposed on $\mathfrak{A}$,

$$\boldsymbol{\alpha}(T|_{F_i}) \leqslant \gamma_i (\dim F_i)^{1/p} \leqslant \gamma_i k_n^{1/p} \leqslant \frac{\sigma_n}{2^{(i-n)/p}}.$$

Therefore,

$$\boldsymbol{\alpha}(T|_F)^p \leqslant \sum_{i \leqslant n} \sigma_i^p + \sum_{i>n} \frac{\sigma_n^p}{2^{i-n}} \leqslant \sum_{i \leqslant n} \frac{\sigma_n^p}{2^{n-i}} + \sum_{i>n} \frac{\sigma_n^p}{2^{i-n}} \leqslant 3\sigma_n^p < \alpha_{k_n}^p.$$

It remains to construct the sequences satisfying (3.3). To "prime the pump", select $k_1 = 1$. Then $c_1 = 1$. Let $\gamma_1 = \alpha_1 / c_1$.

Now suppose $E_1, \gamma_1, \ldots, E_s, \gamma_s$ have already been selected to satisfy (3.3). Let

$$\gamma = \min_{1 \leqslant i \leqslant s} 2^{i-1-s} \sigma_i k_i^{-1/p}, \quad \sigma = 2^{1/p} \sigma_s, \text{ and } c = \frac{\sigma}{\gamma}.$$

Find $k > 3k_s$ so that $\boldsymbol{\alpha}(I_E) > c$ for some k-dimensional E, and $\alpha_k > 8^{1/p}\sigma$. Set $E_{s+1} = E$. Then $k_{s+1} = k$, and $c_{s+1} = \boldsymbol{\alpha}(I_E)$. Finally, set

$$\gamma_{s+1} = \min\left\{\gamma, \frac{\alpha_{k_{s+1}}}{4c_{s+1}}\right\}.$$

Then

$$\sigma_{s+1} = \gamma_{s+1} c_{s+1} = \min\left\{\gamma c_{s+1}, \frac{\alpha_{k_{s+1}}}{4}\right\} \geqslant \sigma \geqslant 2^{1/p} \sigma_s.$$

Thus, the first part of (3.3) holds for $s+1$. It is even easier to check the second and third parts. □

Proof of Theorem 3.1. Suppose, for the sake of contradiction, that $\mathbf{C}\mathfrak{A}$ contains a universal operator S. By Lemma 3.3, for any $T \in \mathbf{C}\mathfrak{A}$ there exists $c > 0$ so that $\beta_n(T) \geqslant c\beta_n(S)$ for any n. Furthermore, by the conditions imposed of $\mathfrak{A}$, $\lim_n \beta_n(S) = \infty$. However, by Lemma 3.4, there exists $T \in \mathbf{C}\mathfrak{A}$ for which $\beta_n(T) \leqslant \sqrt{\beta_n(S)}$ infinitely often. This yields a contradiction. □

4. Positively universal operators

In this section we narrow our attention to positive operators on Banach lattices.

Theorem 4.1. *Suppose $\mathfrak{A}$ is an operator ideal, not containing an isomorphic embedding of ℓ_1 into ℓ_∞. If T positively factors through every $S \in \mathbf{C}\mathfrak{A}$, then T is compact.*

Throughout this section, we work with real lattices. The complex case can be obtained with minor modifications.

Remark 4.2. (1) If an ideal $\mathfrak{A}$ contains an into isomorphism $S : \ell_1 \to \ell_\infty$, then $B(\ell_1, \ell_\infty) \subset \mathfrak{A}$. Indeed, consider $T \in B(\ell_1, \ell_\infty)$. Find $U : S(\ell_1) \to \ell_1$, so that $US = I_{\ell_1}$. By the injectivity of ℓ_∞, the operator $V = TU : S(\ell_1) \to \ell_\infty$ has an extension $W : \ell_\infty \to \ell_\infty$. It is easy to see that $T = WS$, hence T belongs to $\mathfrak{A}$.

(2) Any into isomorphism $S : \ell_1 \to \ell_\infty$ is universal for the complement of the ideal of ℓ_1-strictly singular operators, see [2, 1.18].

The following lemma (needed to establish Theorem 4.1) may be folklore, but we have not seen it in the literature. As before, we denote by $\mathbf{B}(Z)$ the unit ball of Z.

Lemma 4.3. *Suppose Z is a Banach lattice.*

1. *There exists a set I and a contractive positive map $q : \ell_1(I) \to Z$ so that $\mathbf{B}(Z) \subset 2q(\mathbf{B}(\ell_1(I)))$.*
2. *There exists a set J and a contractive positive map $j : Z \to \ell_\infty(J)$ so that $\|jz\| \geqslant \|z\|/2$ for any $z \in Z$.*

Proof. (1) Denote by I the positive part of $\mathbf{B}(Z)$. Then the map $q : \ell_1(I) \to Z : \delta_z \mapsto z$ is positive and contractive. Moreover, $\mathbf{B}(Z) \subset I - I \subset 2q(\mathbf{B}(\ell_1(I)))$.

(2) Let J be the set of all contractive positive functionals on Z, and define $j : Z \to \ell_\infty(J) : z \mapsto \big(z^*(z)\big)_{z^* \in J}$. Clearly, j is positive and contractive. Moreover, for any $z \in Z$ there exists $z^* \in \mathbf{B}(Z^*)_+$ so that $|z^*(z)| \geqslant \|z\|/2$. Indeed, consider the disjoint decomposition $z = z_+ - z_-$. Without loss of generality, assume that $\|z_+\| \geqslant 1/2$. Consider the subspace $W \subset Z$, spanned by z_+ and all elements w disjoint with z_+. By [11, Theorem 1.1.1], $\|z_+ + w\| \geqslant \|z_+\|$, for any w like this. By the Hahn–Banach Theorem, there exists a norm one functional $w^* \in W^*$ so that $w^*(z_+) = \|z_+\|$, and $w^*(w) = 0$ whenever $w \perp z_+$. Then $z^* = w^*_+$ has norm not exceeding 1. Recall that (cf. [11, Section 1.3]), for any $u \in Z_+$, $z^*(u) = \sup\big\{w^*(v) : 0 \leqslant v \leqslant u\big\}$. Consequently, $z^*(z_+) = \|z_+\|$, and $z^*(z_-) = 0$. We conclude that $|z^*(z)| \geqslant \|z\|/2$. Therefore, $\|jz\| \geqslant \|z\|/2$ for any $z \in Z$. □

Proof of Theorem 4.1. Suppose, for the sake of contradiction, that a non-compact $T \in B(X, Y)_+$ positively factors through every $S \in B(\ell_1, \ell_\infty)_+ \backslash \mathfrak{A}$. By Lemma 4.3, there exists a set Γ, and positive contractive maps $q : \ell_1(\Gamma) \to X$ and $j : Y \to \ell_\infty(\Gamma)$, so that $\mathbf{B}(X) \subset 2q(\mathbf{B}(\ell_1(\Gamma)))$, and $\|jy\| \geqslant \|y\|/2$ for any $y \in Y$. If T positively factors through S, then so does jTq. Furthermore, jTq is not compact. Thus, we can assume that $X = \ell_1(\Gamma)$, and $Y = \ell_\infty(\Gamma)$.

Next we show that we can take Γ to be countable. To this end, denote the canonical transfinite basis of $\ell_1(\Gamma)$ by $(\delta_i)_{i \in \Gamma}$. The convex hull of a relatively compact set is relatively compact, hence there exists a countably infinite $I \subset \Gamma$ so that the family $(T\delta_i)_{i \in I}$ is c-separated, for some $c > 0$ (that is, $\|T\delta_i - T\delta_j\| > c$ for any distinct $i, j \in I$). Moreover, there exists a countable $G \subset \Gamma$ so that, for any distinct $i, j \in I$, there exists $g \in G$ with $|(T\delta_i - T\delta_j)_g| > c$.

For $\Lambda \subset \Gamma$, let $P_\Lambda \in B(\ell_1(\Gamma))$ be the corresponding coordinate projection. By the above, the family $(P_G^* T P_I \delta_i)_{i \in I}$ is c-separated, and in particular, $P_G^* T P_I$ is not compact. We establish our claim by identifying $G \cup I$ with $\mathbb{N}$.

Find a family of infinite subsets $A_i \subset \mathbb{N}$ $(i \in \mathbb{N})$ so that, for any $N \in \mathbb{N}$, and any sequence $(\varepsilon_i)_{i=1}^N \subset \{-1, 1\}^N$, $\cap_{i=1}^N A_i^{(\varepsilon_i)} \neq \emptyset$ (here, $A^{(1)} = A$, and $A^{(-1)} = \mathbb{N} \backslash A$). One can, or instance, let (q_i) be an enumeration of prime numbers, and define A_i as the set of multiples of q_i.

Then $U : \ell_1 \to \ell_\infty : \delta_i \mapsto u_i = \chi_{A_i} - \chi_{\mathbb{N} \backslash A_i}$ is an isometric embedding. Now for each $n \in \mathbb{N}$ define the operator $U_n : \ell_1 \to \ell_\infty : \delta_i \mapsto (1 - 2^{-n})\mathbf{1} + 2^{-n} u_i$ ($(\delta_i)_{i \in \mathbb{N}}$ is the canonical basis of ℓ_1). In other words, $U_n = (1 - 2^{-n}) U_0 + 2^{-n} U$, where, for $x = (x_1, x_2, \ldots) \in \ell_1$, $U_0 x = (\sum_i x_i)\mathbf{1}$. Note that U_n is a contraction. Moreover, $U_n \notin \mathfrak{A}$, for any natural n. Indeed, U_0 has rank 1, and operator ideals are stable under finite rank perturbations.

Suppose, for the sake of contradiction, that there exists a non-compact $T \in B(\ell_1, \ell_\infty)_+$ so that, for every $n \in \mathbb{N}$, $T = B_n U_n A_n$, for some positive A_n and B_n. Let $v_i = T\delta_i$. By using TP_I instead of T if necessary, we can assume $\inf \|v_i\| > 0$. By considering TD, where D is a diagonal operator, we can further assume that $\|v_i\| = 1$ for any i. Finally, in light of the previous reasoning, we can assume that the sequence (v_i) is c-separated, for some $c \in (0, 1/2)$. Then, for $i \neq j$, either $v_i - (1 - c/2) v_j$ or $v_j - (1 - c/2) v_i$ is not positive. To establish this, write $v_i = (v_{is})_{s \in \mathbb{N}}$. For $i \neq j$, there exists $s \in \mathbb{N}$ so that $|v_{is} - v_{js}| > c$. If $v_{is} - v_{js} > c$, then

$$v_{js} - \left(1 - \frac{c}{2}\right) v_{is} < v_{js} - \left(1 - \frac{c}{2}\right)(v_{js} + c)$$
$$= c\left[\frac{v_{js}}{2} - \left(1 - \frac{c}{2}\right)\right] \leqslant c\left[\frac{1}{2} - \left(1 - \frac{c}{2}\right)\right] < 0.$$

The case of $v_{js} - v_{is} > c$ is handled similarly.

Now fix $n > \log_2(4/c)$. Let $w_i = U_n A_n \delta_i$. We can write $A_n \delta_i = (a_{ik})_{k=1}^\infty$. Then $a_{ik} \geqslant 0$ for any i and k, and

$$\|A_n\| \geqslant \|A_n \delta_i\| = \alpha_i := \sum_k a_{ik} \geqslant \frac{\|B_n U_n A_n \delta_i\|}{\|B_n U_n\|} \geqslant \frac{1}{\|B_n\|}.$$

Pick $\lambda > 1$ so that $(1 - 2^{1-n})/\lambda > 1 - c/2$. By compactness, we can find $i \neq j$ so that $\lambda^{-1} < \alpha_i / \alpha_j < \lambda$. The desired contradiction will be achieved once we prove that both $w_i - (1 - c/2) w_j$ and $w_j - (1 - c/2) w_i$ are positive (indeed, then the positive operator B_n cannot take w_i and w_j to v_i and v_j, respectively).

Recall that $-\mathbf{1} \leqslant u_i \leqslant \mathbf{1}$, hence

$$w_i = \sum_k U_n a_{ik} \delta_k = (1 - 2^{-n})\alpha_i + 2^{-n} \sum_k a_{ik} u_k \in [(1 - 2^{1-n})\alpha_i \mathbf{1}, \alpha_i \mathbf{1}],$$

and similarly, $(1 - 2^{1-n})\alpha_j \mathbf{1} \leqslant w_j \leqslant \alpha_j \mathbf{1}$. Then

$$w_i - \left(1 - \frac{c}{2}\right) w_j \geqslant (1 - 2^{1-n})\alpha_i \mathbf{1} - \left(1 - \frac{c}{2}\right)\alpha_j \mathbf{1}$$
$$\geqslant \alpha_j \left[(1 - 2^{1-n})\lambda^{-1} - \left(1 - \frac{c}{2}\right)\right]\mathbf{1} \geqslant 0,$$

and $w_j - (1 - c/2)w_i$ is tackled similarly. □

References

[1] P. Aiena, *Fredholm and local spectral theory, with applications to multipliers.* Kluwer Academic Publishers, Dordrecht, 2004.

[2] J. Diestel, H. Jarchow, and A. Pietsch, *Operator ideals.* Handbook of the geometry of Banach spaces, Vol. I, 437–496, North-Holland, Amsterdam, 2001.

[3] J. Diestel, H. Jarchow, and A. Tonge, *Absolutely summing operators.* Cambridge University Press, Cambridge, 1995.

[4] J. Flores, F. Hernández, and Y. Raynaud. *Super strictly singular and cosingular operators and related classes.* J. Operator Theory 67 (2012), 121–152.

[5] M. Girardi and W. Johnson, *Universal non-completely-continuous operators.* Israel J. Math. 99 (1997), 207–219.

[6] F. Hernández, Y. Raynaud, and E. Semenov, *Bernstein widths and super strictly singular inclusions.* A panorama of modern operator theory and related topics, 359–376, Oper. Theory Adv. Appl., 218, Birkhäuser/Springer, Basel, 2012.

[7] A. Hinrichs and A. Pietsch, *The closed ideal of* (c_0, p, q)*-summing operators.* Integral Equations Operator Theory 38 (2000), 302–316.

[8] W. Johnson, *A universal non-compact operator.* Colloq. Math. 23 (1971), 267–268.

[9] J. Lindenstrauss and A. Pełczynski, *Absolutely summing operators in* L_p*-spaces and their applications.* Studia Math. 29 (1968), 275–326.

[10] J. Lindenstrauss and L. Tzafriri, *Classical Banach spaces I.* Springer, Berlin, 1977.

[11] P. Meyer-Nieberg, *Banach lattices.* Springer, Berlin, 1991.

[12] A. Plichko. *Superstrictly singular and superstrictly cosingular operators.* Functional analysis and its applications, 239–255, Elsevier, Amsterdam, 2004.

Timur Oikhberg
Department of Mathematics
University of Illinois
Urbana IL 61801, USA
e-mail: oikhberg@illinois.edu

Ordered Structures and Applications: Positivity VII
Trends in Mathematics, 349–361

Matrix Monotone Functions and the Generalized Powers–Størmer Inequality

Hiroyuki Osaka

Abstract. In this note a generalization of Powers–Størmer inequality for operator monotone functions on $[0, +\infty)$ and for positive linear functional on general C^*-algebras will be introduced and be shown that the generalized Powers–Størmer inequality characterizes the tracial functionals on C^*-algebras and the monotonicity for a given function.

Mathematics Subject Classification (2010). Primary 46L30; Secondary 15A45.

Keywords. Matrix monotone functions, matrix convex functions, Powers–Størmer inequality, trace.

1. Introduction

Let $n \in \mathbf{N}$ and M_n be the algebra of $n \times n$ matrices. We call a function f matrix convex of order n or n-convex in short whenever the inequality

$$f(\lambda A + (1-\lambda)B) \leq \lambda f(A) + (1-\lambda)f(B), \ \lambda \in [0,1]$$

holds for every pair of selfadjoint matrices $A, B \in M_n$ such that all eigenvalues of A and B are contained in an interval I ($\subset \mathbf{R}$). Matrix monotone functions on I are similarly defined as the inequality

$$A \leq B \Longrightarrow f(A) \leq f(B)$$

for an arbitrary selfadjoint matrices $A, B \in M_n$ such that $A \leq B$ and all eigenvalues of A and B are contained in I.

We denote the spaces of operator monotone functions and of operator convex functions by $P_\infty(I)$ and $K_\infty(I)$ respectively. The spaces for n-monotone functions

This work was completed with the support by the JSPS grant for Scientific Research No. 20540220.

and n-convex functions are written as $P_n(I)$ and $K_n(I)$. We have then

$$P_1(I) \supseteq \cdots \supseteq P_{n-1}(I) \supseteq P_n(I) \supseteq P_{n+1}(I) \supseteq \cdots \supseteq P_\infty(I)$$
$$K_1(I) \supseteq \cdots \supseteq K_{n-1}(I) \supseteq K_n(I) \supseteq K_{n+1}(I) \supseteq \cdots \supseteq K_\infty(I).$$

Here we meet the facts that $\cap_{n=1}^\infty P_n(I) = P_\infty(I)$ and $\cap_{n=1}^\infty K_n(I) = K_\infty(I)$. We regard these two decreasing sequences as noncommutative counterpart of the classical piling sequence $\{C^n(I), C^\infty(I), \mathrm{Anal}(I)\}$, where the class $\mathrm{Anal}(I)$ denotes the set of all analytic functions over I. We could understand that the class of operator monotone functions $P_\infty(I)$ corresponds to the class $\{C^\infty(I), \mathrm{Anal}(I)\}$ by the famous characterization of those functions by Loewner as the restriction of Pick functions.

In these circumstances, it will be well recognized that we should not stick our discussions only to those classes $P_\infty(I)$ and $K_\infty(I)$, that is, the class of operator monotone functions and that of operator convex functions. Those classes $\{P_n(I)\}$ and $\{K_n(I)\}$ are not merely optional ones to $P_\infty(I)$ and $K_\infty(I)$. They should play important roles in the aspect of noncommutative calculus as the ones $\{C^n(I)\}$ play in usual (commutative) calculus.

The first basic question is whether $P_{n+1}(I)$ (resp. $K_{n+1}(I)$) is strictly contained in $P_n(I)$ (resp. $K_n(I)$) for every n. In [31] the gap for $n = 2$, that is, $P_3([0,\infty)) \subsetneq P_2([0,\infty))$, was pointed out. This gap problem for arbitrary n, however, has been solved only recently ([9], [25], [12]). (See Section 2.)

On the other hand, there are basic equivalent assertions known only at the level of operator monotone functions and operator convex functions by [10], [11]. We shall discuss those (equivalent) assertions as the correlation problem between two kinds of piling structures $\{P_n(I)\}$ and $\{K_n(I)\}$, that is, we are planning at first to discuss relations between those assertions at each level n.

In [26] (resp. [17]) we discussed about the following 3 assertions at each level n among them in order to see clear insight of the aspect of the problems:

(i) $f(0) \le 0$ and f is n-convex (resp. n-concave) in $[0,\alpha)$,

(ii) For each matrix A with its spectrum in $[0,\alpha)$ and a contraction C in the matrix algebra M_n,

$$f(C^*AC) \le C^* f(A) C,$$

(resp. $f(C^*AC) \le C^* f(A) C$)

(iii) The function $\frac{f(t)}{t}$ (resp. $\frac{t}{f(t)}$) $(= g(t))$ is n-monotone in $(0,\alpha)$.

Then we showed that for each n the condition (ii) is equivalent to the condition (iii) and the assertion that f is n-convex with $f(0) \le 0$ implies that $g(t)$ is $(n-1)$-monotone (resp. f is n-concave with $f(0) \ge 0$ implies that $g(t)$ is $(n-1)$-monotone). (See Section 3.)

Powers–Størmer inequality (see, for example, [29, Lemma 2.4], [28, Theorem 11.19]) asserts that for $s \in [0,1]$ the following inequality

$$2\,\mathrm{Tr}(A^s B^{1-s}) \ge \mathrm{Tr}(A + B - |A - B|) \tag{1.1}$$

holds for any pair of positive matrices A, B. This is a key inequality to prove the upper bound of Chernoff bound, in quantum hypothesis testing theory [1]. This inequality was first proven in [1], using an integral representation of the function t^s. After that, N. Ozawa gave a much simpler proof for the same inequality, using fact that for $s \in [0,1]$ function $f(t) = t^s$ $(t \in [0,+\infty))$ is an operator monotone ([18, Proposition 1.1]). Recently, Y. Ogata in [24] extended this inequality to standard von Neumann algebras. The motivation for the present paper is to investigate whether replacing the function $f(t) = t^s$ by another operator monotone function (this class is intensively studied, see [9][25]) can yield a smaller upper bound for $\mathrm{Tr}(A + B - |A - B|)$ than what is used in quantum hypothesis testing. Based on N. Ozawa's proof we formulate Powers–Størmer's inequality for an arbitrary operator monotone function on $[0, +\infty)$ in the context of general C^*-algebras. (See Section 4.)

As applications, the generalized Powers–Størmer inequality characterizes the trace property for a normal linear positive functional on a von Neumann algebras and for a linear positive functional on a C^*-algebra. (See Section 5.) It also characterizes the monotonicity of a given function in this inequality. (See Section 6.)

2. Preliminary

We shall sometimes use the standard regularization procedure, cf. for example Donoghu [6, p11]. Let ϕ be a positive and even C^∞-function defined on the real axis, vanishing outside the closed interval $[-1, 1]$ and normalized such that

$$\int_{-1}^{1} \phi(x) = 1.$$

For any locally integrable function f defined in an open interval (a, b) we form its regularization

$$f_\varepsilon(t) = \frac{1}{\varepsilon}\int_a^b \phi(\frac{t-s}{\varepsilon})f(s)ds, \quad t \in \mathbf{R}$$

for small $\varepsilon > 0$, and realize that it is infinitely many times differentiable. For $t \in (a+\varepsilon, b-\varepsilon)$ we may also write

$$f_\varepsilon(t) = \int_{-1}^{1} \phi(s)f(t-\varepsilon s)ds.$$

If f is continuous, then f_ε converges uniformly to f on any compact subinterval of (a, b). If in addition f is n-convex (or n-monotone) in (a, b), then f_ε is n-convex (or n-monotone) in the slightly smaller interval $(a+\varepsilon, b-\varepsilon)$. Since the pointwise limit of a sequence of n-convex (or n-monotone) functions is again n-convex (or n-monotone), we may therefore in many applications assume that an n-convex or n-monotone function is sufficiently many times differentiable.

For a sufficiently smooth function $f(t)$ we denote its nth divided difference for n-tuple of points $\{t_1, t_2, \ldots, t_n\}$ defined as, when they are all different,

$$[t_1, t_2]_f = \frac{f(t_1) - f(t_2)}{t_1 - t_2}, \quad \text{and inductively}$$
$$[t_1, t_2, \ldots, t_n]_f = \frac{[t_1, t_2, \ldots, t_{n-1}]_f - [t_2, t_3, \ldots, t_n]_f}{t_1 - t_n}.$$

And when some of them coincides such as $t_1 = t_2$ and so on, we put as

$$[t_1, t_1]_f = f'(t_1) \quad \text{and inductively} \quad [t_1, t_1, \ldots, t_1]_f = \frac{f^{(n-1)}(t_1)}{(n-1)!}.$$

When there appears no confusion we often skip the referring function f. We notice here the most important property of divided differences is that it is free from permutations of $\{t_1, t_2, \ldots, t_n\}$ in an open interval I.

Proposition 2.1.

(1) (Ia) *Monotonicity* (*Loewner* 1934 [21])

$$f \in P_n(I) \Longleftrightarrow ([t_i, t_j]) \geq 0 \textit{ for any } \{t_1, t_2, \ldots, t_n\}$$

(IIa) *Convexity* (*Kraus* 1936 [20])

$$f \in K_n(I) \Longleftrightarrow ([t_1, t_i, t_j]) \geq 0 \textit{ for any } \{t_1, t_2, \ldots, t_n\},$$

where t_1 can be replaced by any (fixed) t_k.

(2) (Ib) *Monotonicity* (*Loewner* 1934 [21], *Dobsch* 1937 [5]-*Donoghue* 1974 [6]) *For $f \in C^{2n-1}(I)$*

$$f \in P_n(I) \Longleftrightarrow M_n(f; t) = \left(\frac{f^{(i+j-1)}(t)}{(i+j-1)!}\right) \geq 0 \ \forall t \in I.$$

(IIb) *Convexity* (*Hansen–Tomiyama* 2007 [12]) *For $f \in C^{2n}(I)$*

$$f \in K_n(I) \Longrightarrow K_n(f; t) = \left(\frac{f^{(i+j)}(t)}{(i+j)!}\right) \geq 0 \ \forall t \in I.$$

In particular, for $n = 2$ the converse is also true.

We remind that to prove the implication $M_n(f; t) \geq 0 \Rightarrow f \in P_n(I)$ in (Ib) the local property for the monotonicity plays an essential role. Similarly to prove the converse implication in the criterion of convexity in (IIb) in the above proposition we need **the local property conjecture for the convexity**, that is, if f is n-convex in the intervals (a, b) and (c, d) $(a < c < b < d)$, then f is n-convex on (a, d).

Now we have only a partial sufficiency, that is, if $K_n(f; t_0)$ is positive, then there exists a neighborhood of t_0 on which f is n-convex. (See [12, Theorem 1.2] for example.)

The method for the implication (II_b) $\Rightarrow$ (IIa) under the assumption of the local property theorem for the convexity may be familiar for some specialist.

Proposition 2.2. *Let $f \in C^{2n}(I)$ such that $K_n(f;t) = \left(\frac{f^{(i+j)}(t)}{(i+j)!}\right) \geq 0 \ \forall t \in I$. Suppose that n-convexity has the local property. Then $f \in K_n(I)$.*

3. Double piling structure

As we have mentioned in the introduction, there are basic equivalent assertions known for operator monotone functions and operator convex functions (cf. [10]). Namely we have

Theorem A. *For $0 < \alpha \leq \infty$, the following assertions for a real-valued continuous function f in $[0, \alpha)$ are equivalent:*

(1) *f is operator convex and $f(0) \leq 0$.*

(2) *For an operator A with its spectrum in $[0, \alpha)$ and a contraction C,*

$$f(C^*AC) \leq C^*f(A)C.$$

(3) *For two operators A, B with their spectra in $[0, \alpha)$ and two contractions C, D such that $C^*C + D^*D \leq 1$ we have the inequality*

$$f(C^*AC + D^*BD) \leq C^*f(A)C + D^*f(B)D.$$

(4) *For an operator A with its spectrum in $[0, \alpha)$ and a projection P we have the inequality,*

$$f(PAP) \leq Pf(A)P$$

(5) *The function $g(t) = \frac{f(t)}{t}$ is operator monotone in the open interval $(0, \alpha)$.*

In this section, we shall discuss mutual relationships of the above assertions when we restrict the property of the function f at each fixed level n, that is, when f and g are assumed to be only n-matrix convex and n- matrix monotone. We regard the problem as the problem of double piling structure of those decreasing sequences $\{P_n(I)\}$ and $\{K_n(I)\}$ down to $P_\infty(I)$ and $K_\infty(I)$ respectively. In this sense, standard double piling structure known for these assertions before is the following. We describe these implications using the following convention: if the fact that the statement (A) holds for the matrix algebra M_m implies that statement (B) holds for the matrix algebra M_n, then we write $(A)_m \to (B)_n$.

Theorem A is proved in the following way.

$$(1)_{2n} \to (2)_n \to (5)_n \to (4)_n, (2)_{2n} \to (3)_n \to (4)_n, \text{and } (4)_{2n} \to (1)_n.$$

Therefore, those assertions become equivalent when f is operator convex and g is operator monotone by the piling structure.

Thus, the basic problem for double piling structure is to find the minimum difference of degrees between those gaped assertions. Since, however, even single piling problems are clarified only recently, as we have mentioned above, in spite of a long history of monotone matrix functions and convex matrix functions, little is known for the double piling structure except the result by Mathias ([22]), which asserts that a $2n$-monotone function in the positive half-line $[0, \infty)$ becomes n-concave.

Now in order to make our investigations more transparently we mainly concentrate our discussions to the relationships between (1), (2) and (5). In fact, we need not say anything about (4) when $n = 1$, and for the reason choosing (2) instead of (3) we just borrow the witty expression in [10], "correctness must bow to applicability". Before going into our discussions, we state each assertion in a precise way but skipping the condition of the spectrum of a matrix A. Namely, in the interval $[0, \alpha)$ we consider the following assertions.

(i) $f(0) \leq 0$, and f is n-convex.

(ii) For each positive semidefinite element A with its spectrum in $[0, \alpha)$ and a contraction C in M_n, we have

$$f(C^*AC) \leq C^*f(A)C.$$

(iii) The function $g(t) = \frac{f(t)}{t}$ is n-monotone in the interval $(0, \alpha)$.

We shall show then the equivalency of the assertions (ii) and (iii). Hence the problem is reduced to the relationship between (i) and (iii) (or (ii)). Namely, we have the following

Theorem 3.1 ([26]). *Let $n \in \mathbf{N}$.*

1. *The assertions* $(\mathrm{ii})_n$ *and* $(\mathrm{iii})_n$ *are equivalent,*
2. *The assertion* $(\mathrm{i})_n$ *implies the assertion* $(\mathrm{iii})_{n-1}$.

When f is a convex function, $-f$ is a concave function. Hence we have the following.

Theorem 3.2 ([17]). *Let $f\colon [0, \alpha) \to \mathbf{R}$ $(0 < \alpha \leq \infty)$ be a continuous function. Consider the following three assertions:*

(i) *$f(0) \geq 0$, and f is n-concave,*

(ii) *For each positive semidefinite element A with its spectrum in $[0, \alpha)$ and a contraction C in M_n, we have*

$$f(C^*AC) \geq C^*f(A)C.$$

(iii) *The function $g(t) = \frac{t}{f(t)}$ is n-monotone in the interval $(0, \alpha)$.*

Then we have for each $n \in \mathbf{N}$

1. *The assertions* $(\mathrm{ii})_n$ *and* $(\mathrm{iii})_n$ *are equivalent,*
2. *The assertion* $(\mathrm{i})_n$ *implies the assertion* $(\mathrm{iii})_{n-1}$.

4. Generalized Powers–Størmer inequality

One of the most basic tasks in quantum statistics is the discrimination of two different quantum states. In the quantum hypothesis testing problem, one has to decide between two states of a system. The state ρ_0 is the null hypothesis and ρ_1 is alternative hypothesis.

The problem is to decide which hypothesis is true. The decision is performed by a two-valued measurement $\{T, I-T\}$, where $0 \leq T \leq I$ is an observable. T corresponds to the acceptance of ρ_0 and $I-T$ corresponds to the acceptance of ρ_1. T is called a test.

The total error $\mathrm{Err}(T)$ of T is

$$\begin{aligned}\mathrm{Err}(T) &= \frac{1}{2}\operatorname{Tr}(\rho_0(I-T)) + \frac{1}{2}\operatorname{Tr}(\rho_1 T)\\ &= \frac{1}{2}\{1 - \operatorname{Tr}(T(\rho_0 - \rho_1))\}.\end{aligned}$$

Then asymptotic error exponent for ρ_0 and ρ_1 is

$$\lim_{n\to\infty} \frac{1}{n}\log \mathrm{Err}_n(T_{(n)}),$$

where for all $n \in \mathbf{N}$ $T_{(n)}$ is a $d^n \times d^n$ quantum multiple test, and

$$\mathrm{Err}_n(T_{(n)}) := \frac{1}{2}\left\{1 - \operatorname{Tr}(T_{(n)}(\rho_0^{\otimes n} - \rho_1^{\otimes n}))\right\}.$$

If the limit $\lim_{n\to\infty} \dfrac{1}{n}\log \mathrm{Err}_n(T_{(n)})$ exists, we refer to it as *the asymptotic error exponent*.

The lower band and upper bounds for the asymptotic error exponent are given by he following.

Theorem 4.1 ([1], [23]). *Let $\{\rho_0, \rho_1\}$ be hypothetic states on $\mathbf{C}^d$, $T_{(n)}$ be quantum multiple test, and $Q_{(n)}$ be a support projections on $(\rho_0^{\otimes n} - \rho_1^{\otimes n})$. Then one has*

(i) *(M. Nussbaum and A. Szkola)*

$$\liminf_{n\to\infty} \frac{1}{n}\log \mathrm{Err}_n(T_{(n)}) \geq \inf\{\log \operatorname{Tr}(\rho_0^{1-s}\rho_1^s) \mid 0 \leq s \leq 1\}.$$

(ii) *(K.M.R. Audenaert, et al.)*

$$\limsup_{n\to\infty} \frac{1}{n}\log \mathrm{Err}_n(Q_{(n)}) \leq \inf\{\log \operatorname{Tr}(\rho_0^{1-s}\rho_1^s) \mid 0 \leq s \leq 1\}.$$

In the proof of the previous Theorem 4.1(ii) the following inequality played a key role.

Theorem 4.2 ([1]). *For any positive matrices A and B on $\mathbf{C}^d$ we have*

$$\frac{1}{2}(\operatorname{Tr} A + \operatorname{Tr} B - \operatorname{Tr}|A-B|) \leq \operatorname{Tr}(A^{1-s}B^s)\ (s \in [0,1]).$$

If we consider a function $f(t) = t^{1-s}$ and $g(t) = t^s = \dfrac{t}{f(t)}$, then both functions f and g are operator monotone. The inequality, then, can be reformed by

$$\operatorname{Tr} A + \operatorname{Tr} B - \operatorname{Tr}|A-B| \leq 2\operatorname{Tr}(f(A)^{\frac{1}{2}} g(B) f(A)^{\frac{1}{2}}).$$

Theorem 4.3 ([16], [17]). *Let f be a $2n$-monotone function (or $(n+1)$-concave function) on $[0,\infty)$ such that $f((0,\infty)) \subset (0,\infty)$. Then for any pair of positive matrices $A, B \in M_n(\mathbf{C})$*

$$\mathrm{Tr}(A) + \mathrm{Tr}(B) - \mathrm{Tr}(|A-B|) \leq 2\,\mathrm{Tr}(f(A)^{\frac{1}{2}} g(B) f(A)^{\frac{1}{2}}),$$

where $g(t) = \frac{t}{f(t)}$.

We give a sketch of the proof.

Let A, B be positive matrices and, let

$$A - B = (A-B)_+ - (A-B)_- = P - Q$$

and $|A-B| = P + Q$. We may, then, show that

$$\mathrm{Tr}(A) - \mathrm{Tr}(f(A)^{\frac{1}{2}} g(B) f(A)^{\frac{1}{2}}) \leq \mathrm{Tr}(P)$$

holds as follows:

$$\begin{aligned}
&\mathrm{Tr}(A) - \mathrm{Tr}(f(A)^{\frac{1}{2}} g(B) f(A)^{\frac{1}{2}}) \\
&= \mathrm{Tr}(f(A)^{\frac{1}{2}} g(A) f(A)^{\frac{1}{2}}) - \mathrm{Tr}(f(A)^{\frac{1}{2}} g(B) f(A)^{\frac{1}{2}}) \\
&\leq \mathrm{Tr}(f(A)^{\frac{1}{2}} g(B+P) f(A)^{\frac{1}{2}}) - \mathrm{Tr}(f(A)^{\frac{1}{2}} g(B) f(A)^{\frac{1}{2}}) \\
&\leq \mathrm{Tr}(f(B+P)^{\frac{1}{2}} (g(B+P) - g(B)) f(B+P)^{\frac{1}{2}}) \\
&\leq \mathrm{Tr}(f(B+P)^{\frac{1}{2}} g(B+P) f(B+P)^{\frac{1}{2}}) - \mathrm{Tr}(f(B)^{\frac{1}{2}} g(B) f(B)^{\frac{1}{2}}) \\
&= \mathrm{Tr}(P).
\end{aligned}$$

In particular we have

Corollary 4.4. *Let f be an operator monotone function on $[0,\infty)$ such that $f((0,\infty)) \subset (0,\infty)$. Then for any pair of positive matrices $A, B \in M_n(\mathbf{C})$*

$$\mathrm{Tr}(A) + \mathrm{Tr}(B) - \mathrm{Tr}(|A-B|) \leq 2\,\mathrm{Tr}(f(A)^{\frac{1}{2}} g(B) f(A)^{\frac{1}{2}}),$$

where $g(t) = \frac{t}{f(t)}$.

Since any C^*-algebra can be realized as a closed selfadjoint $*$-algebra of $B(H)$ for some Hilbert space H. We can generalize Corollary 4.4 in the framework of C^*-algebras.

Theorem 4.5. *Let τ be a tracial functional on a C^*-algebra $\mathcal{A}$, f be a strictly positive, operator monotone function on $[0,\infty)$. Then for any pair of positive elements $A, B \in \mathcal{A}$*

$$\tau(A) + \tau(B) - \tau(|A-B|) \leq 2\tau(f(A)^{\frac{1}{2}} g(B) f(A)^{\frac{1}{2}}), \tag{4.1}$$

where $g(t) = tf(t)^{-1}$.

Proof. Let π be the universal representation of $\mathcal{A}$ and $\hat{\tau}$ be a positive linear functional on $\pi(\mathcal{A})$ by $\hat{\tau}(\pi(A)) = \tau(A)$ for $A \in \mathcal{A}$. Then $\hat{\tau}$ has the trace property. Since

g is operator monotone on $(0, \infty)$ by [10, Corollary 6], through the same steps in the proof of Theorem 4.3 we have that for any positive operators A and B in $\mathcal{A}$

$$\hat{\tau}(\pi(A)) + \hat{\tau}(\pi(B)) - \hat{\tau}(\pi(|A-B|)) \leq 2\hat{\tau}(f(\pi(A))^{\frac{1}{2}} g(\pi(B)) f(\pi(A))^{\frac{1}{2}}),$$

that is,

$$\tau(A) + \tau(B) - \tau(|A-B|) \leq 2\tau(f(A)^{\frac{1}{2}} g(B) f(A)^{\frac{1}{2}}).$$

□

5. Characterization of the trace property

In this section we shall show that the generalized Powers–Størmer inequality in the previous section guarantees the trace property for a positive linear functional on operator algebras.

Lemma 5.1 ([16]). *Let φ be a positive linear functional on M_n and f be a continuous function on $[0, \infty)$ such that $f(0) = 0$ and $f((0, \infty)) \subset (0, \infty)$. If the following inequality*

$$\varphi(A+B) - \varphi(|A-B|) \leq 2\varphi(f(A)^{\frac{1}{2}} g(B) f(A)^{\frac{1}{2}}) \tag{5.1}$$

holds true for all $A, B \in M_n^+$, then φ should be a positive scalar multiple of the canonical trace Tr *on M_n, where*

$$g(t) = \begin{cases} \frac{t}{f(t)} & (t \in (0, \infty)) \\ 0 & (t = 0) \end{cases}.$$

By analogy with a number of other similar cases (see [7] or [32]), the proof for the trace property of a positive normal functional satisfying the inequality (5.1) on a von Neumann algebra can be reduced to the case of the algebra M_2 of all matrices of order 2×2.

Theorem 5.2 ([16]). *Let φ be a positive normal linear functional on a von Neumann algebra $\mathcal{M}$ and f be a continuous function on $[0, \infty)$ such that $f(0) = 0$ and $f((0, \infty)) \subset (0, \infty)$. If the following inequality*

$$\varphi(A) + \varphi(B) - \varphi(|A-B|) \leq 2\varphi(f(A)^{\frac{1}{2}} g(B) f(A)^{\frac{1}{2}}) \tag{5.2}$$

holds true for any pair $A, B \in \mathcal{M}^+$, then φ is a trace, where

$$g(t) = \begin{cases} \frac{t}{f(t)} & (t \in (0, \infty)) \\ 0 & (t = 0) \end{cases}.$$

Proof. By [19, Proposition 8.1.1] we have only to show that $\varphi(P_1) = \varphi(P_2)$ for any pair of nonzero equivalent projections P_1 and P_2. Moreover, we may assume that P_1 and P_2 are mutually orthogonal. Indeed, considering mutually orthogonal equivalent projections $P_1' = P_1 \vee P_2 - P_1$ and $P_2' = P_1 - P_1 \wedge P_2$ we can show that

$$\varphi(P_1) = \frac{\varphi(P_1 \wedge P_2) + \varphi(P_1 \vee P_2)}{2}.$$

By symmetry we have $\varphi(P_1) = \varphi(P_2)$.

Hence we assume that P_1 and P_2 are nonzero mutually orthogonal equivalent projections in $\mathcal{M}$. Note that $(P_1 + P_2)\mathcal{M}(P_1 + P_2)$ is isomorphic to M_2. Then the inequality (5.2) still holds true for the operators in $\mathcal{N}$ and for the restriction of the functional φ to $\mathcal{N}$. According to Lemma 5.1, this restriction is a tracial functional on $\mathcal{N}$, and hence $\varphi(P_1) = \varphi(P_2)$. □

Corollary 5.3. *Let φ be a positive linear functional on a C^*-algebra $\mathcal{A}$ and f be a continuous function on $[0, \infty)$ such that $f(0) = 0$ and $f((0, \infty)) \subset (0, \infty)$. If the following inequality*

$$\varphi(A) + \varphi(B) - \varphi(|A - B|) \leq 2\varphi(f(A)^{\frac{1}{2}} g(B) f(A)^{\frac{1}{2}}) \tag{5.3}$$

holds true for any pair $A, B \in \mathcal{A}^+$, then φ is a tracial functional, where

$$g(t) = \begin{cases} \frac{t}{f(t)} & (t \in (0, \infty)) \\ 0 & (t = 0) \end{cases} .$$

The following is inspired by [30, Theorem 2.2].

Proposition 5.4 ([17]). *Let $n \in \mathbb{N}$ $(n \geq 2)$, and φ a positive linear functional on M_n. Let f be a strictly positive, continuous function on $(0, \infty)$. Assume that the function g on $(0, \infty)$ defined by $g(t) = \frac{t}{f(t)}$, is differentiable and strictly increasing on $(0, \infty)$. Suppose that*

$$\varphi(A) \leq \varphi(f(A)^{\frac{1}{2}} g(B) f(A)^{\frac{1}{2}}) \tag{5.4}$$

for any positive invertible $A, B \in M_n$ such that $0 < A \leq B$.

Then φ has the trace property if g satisfies the condition:

$$\inf_{\lambda > \mu} \frac{\sqrt{g'(\lambda) g'(\mu)}}{\frac{g(\lambda) - g(\mu)}{\lambda - \mu}} = 0. \tag{5.5}$$

6. Characterization of operator monotonicity

In this section, following the idea from [4] we give a characterization of operator monotonicity of matrix functions by the generalized Powers–Størmer type inequality. The following lemma is obvious.

Lemma 6.1. *Let $A = (a_{ij}), B = (b_{ij})$ be positive invertible in M_n and S a non-finite rank density operator on an infinite-dimensional, separable Hilbert space H. Suppose that $a_{11} > b_{11}$. Then there exist an orthogonal system $\{\xi_i\}_{i=1}^{\infty} \subset H$ and $\{\lambda_i\}_{i=1}^{\infty} \subset [0, 1)$ such that $\sum_{i=1}^{\infty} \lambda_i = 1$, $S\xi_i = \lambda_i \xi_i$, and $\sum_{i=1}^{n} a_{ii}\lambda_i > \sum_{i=1}^{n} b_{ii}\lambda_i$.*

Theorem 6.2 ([17]). *Let H be an infinite-dimensional, separable Hilbert space and φ a normal state on $B(H)$ such that its corresponding density operator S_φ is not finite rank. Let f be a strictly positive, continuous function on $(0, \infty)$, and g be a function on $(0, \infty)$ defined by $g(t) = \frac{t}{f(t)}$. Suppose that*

$$\varphi(A + B) - \varphi(|A - B|) \leq 2\varphi(f(A)^{\frac{1}{2}} g(B) f(A)^{\frac{1}{2}}) \tag{6.1}$$

for any positive invertible $A, B \in B(H)$. Then both of functions f and g on $(0,\infty)$ are operator monotone.

If $f(t) = \lambda t$ for some $\lambda > 0$, then g is constant on $(0,\infty)$. In this case, the inequality (6.1) automatically holds. When the range of the density operator S_φ, however, is proper subspace in a Hilbert space H, the inequality (6.1) does not hold for non-invertible positive operators.

Proposition 6.3 ([17]). *Let H be a separable Hilbert space and φ be a normal state on $B(H)$. Let f be a strictly positive, continuous function on $[0,\infty)$ with $f(0) = 0$, g a function on $(0,\infty)$ defined by $g(t) = \frac{t}{f(t)}$ on $(0,\infty)$ and $g(0) = 0$. Suppose that the range of the density operator S_φ of φ is a proper subspace of H. Then there exist positive non-invertible operators A and B which do not satisfy the inequality*

$$\varphi(A+B) - \varphi(|A-B|) \leq 2\varphi(f(A)^{\frac{1}{2}} g(B) f(A)^{\frac{1}{2}}). \tag{6.2}$$

Proof. We shall give a sketch of the proof.

Let $\{\xi_i\}_{i\in\mathbf{N}}$ be an orthogonal system and $\{\lambda_i\} \subset [0,1)$ such that $\lambda_1 \geq \lambda_2 \geq \cdots \geq 0$, $S_\varphi \xi_i = \lambda_i \xi$ and $\sum_{i=1}^{\infty} \lambda_i = 1$.

Since $S_\varphi(H) \subsetneq H$, we take ξ_{i_0} such that $S_\varphi(\xi_{i_0}) = 0$. For $\delta, \varepsilon > 0$ such that $\delta > \varepsilon$ we set

$$A = \varepsilon|\xi_1\rangle\langle\xi_1| + \sqrt{\varepsilon(\delta-\varepsilon)}(|\xi_1\rangle\langle\xi_{i_0}| + |\xi_{i_0}\rangle\langle\xi_1|) + (\delta-\varepsilon)|\xi_{i_0}\rangle\langle\xi_{i_0}|$$

and

$$B = \varepsilon|\xi_1\rangle\langle\xi_1| - \sqrt{\varepsilon(\delta-\varepsilon)}(|\xi_1\rangle\langle\xi_{i_0}| + |\xi_{i_0}\rangle\langle\xi_1|) + (\delta-\varepsilon)|\xi_{i_0}\rangle\langle\xi_{i_0}|.$$

We have then

$$\begin{aligned}
\varphi(A+B) &= 2\,\mathrm{Tr}(S_\varphi(\varepsilon|\xi_1\rangle\langle\xi_1| + (\delta-\varepsilon)|\xi_{i_0}\rangle\langle\xi_{i_0}|)) \\
&= 2\,\mathrm{Tr}(\lambda_1\varepsilon|\xi_1\rangle\langle\xi_1|) = 2\lambda_1\varepsilon \\
\varphi(|A-B|) &= 2\,\mathrm{Tr}(S_\varphi\sqrt{\varepsilon(\delta-\varepsilon)}(|\xi_1\rangle\langle\xi_1| + |\xi_{i_0}\rangle\langle\xi_{i_0}|)) \\
&= 2\lambda_1\sqrt{\varepsilon(\delta-\varepsilon)} \\
\varphi(f(A)^{1/2}g(B)f(A)^{1/2}) &= \varepsilon\lambda_1\frac{(\delta-2\varepsilon)^2}{\delta^2}.
\end{aligned}$$

Therefore, if positive operators A and B satisfy the inequality (6.2), we have

$$\varepsilon - \sqrt{\varepsilon(\delta-\varepsilon)} \leq \varepsilon\frac{(\delta-2\varepsilon)^2}{\delta^2}.$$

But we have a contradiction if we take $\delta = \dfrac{4\varepsilon}{3}$. □

The following problem is plausible.

Problem 6.4. Let f and g be the functions. Suppose that for any n and any positive matrices $A, B \in M_n$

$$\mathrm{Tr}(A+B) - \mathrm{Tr}(|A-B|) \leq 2\,\mathrm{Tr}(f(A)^{\frac{1}{2}} g(B) f(A)^{\frac{1}{2}}).$$

Is the function f operator monotone?

References

[1] K.M.R. Audenaert, J. Calsamiglia, L.I. Masanes, R. Munoz-Tapia, A. Acin, E. Bagan, F. Verstraete, *Discriminating states: The Quantum Chernoff Bound*, Phys. Rev. Lett. **98**(2007) 160501.

[2] R. Bhatia, *Matrix analysis*, Graduate texts in mathematics, Springer New York, 1997.

[3] R. Bhatia and T. Sano, *Loewner matrices and operator convexity*, Math. Ann. 344(2007), 36–54.

[4] D.T.Hoa, O.E. Tikhonov, *To the theory of operator monotone and operator convex functions*, Izv. Vyssh. Uchebn. Zaved. Mat. **3**(2010) 9–14. [Translation: Russian Mathematics. **54**(3)(2010) 7–11.]

[5] O. Dobsch, *Matrixfunktionen beschränkter Schwankung*, Math. Z. **43**(1937) 353–388.

[6] W.F. Donoghue, *Monotone matrix function and analytic continuation*, Springer 1874.

[7] L.T. Gardner, *An inequality characterizes the trace*, Canad. J. Math. **31**(1979), 1322–1328.

[8] F. Hansen, *An operator inequality*, Math. Ann. **246**(1980), 249–250.

[9] F. Hansen, G. Ji, and J. Tomiyama, *Gaps between classes of matrix monotone functions*, Bull. London Math. Soc. **36**(2004), 53–58.

[10] F. Hansen and G.K. Pedersen, *Jensen's inequality for operators and Loewner's theorem*, Math. Ann. **258**(1982), 229–241.

[11] F. Hansen and G.K. Pedersen, *Jensen's operator inequality*, Bull. London Math. Soc. **35**(2003), 553–564.

[12] F. Hansen and J. Tomiyama, *Differential analysis of matrix convex functions*, Linear Alg. and its Appl. **420**(2007), 102–116,

[13] F. Hansen and J. Tomiyama, *Differential analysis of matrix convex functions II*, Inequal. Pure Appl. Math. 10(2009), no. 2, 5 pp.

[14] F. Hiai and T. Sano, *Loewner matrices of matrix convex and monotone functions*, J. Math. Soc. Japan **64**(2012), no. 2, 343–364.

[15] R.A. Horn and C.R. Johnson, *Topics in matrix analysis*, Cambridge Univ. Press 1991.

[16] D.T. Hoa, H. Osaka, H.M. Toan, *On generalized Powers–Størmer's inequality*, Linear Algebra and its Applications. **438**(1)(2013) 242–249.

[17] D.T. Hoa, H. Osaka, and J.Tomiyama, *Characterization of operator monotone functions by Powers–Størmer type inequalities*, Linear and Multilinear Algebra **63**(2015), no. 8, 1577–1589.

[18] V. Jaksic, Y. Ogata, C. -A. Pillet, R. Seiringer, *Quantum hypothesis testing and non-equilibrium statistical mechanics*, Rev. Math. Phys. **24**(2012), no. 6, 1230002, 67 pp. arXiv:1109.3804v1 [math-ph].

[19] R.V. Kadison, J.R. Ringrose, *Fundamentals of the Theory of Operator Algebras, Vol. II*, Academic Press., 1986.

[20] F. Kraus, *Über konvexe Matrixfunktionen*, Math. Z. **41**(1936), 18–42.

[21] K. Loewner, *Über monotone Matrixfunktionen*, Math. Z. **38**(1934), 177–216.

[22] R. Mathias, *Concavity of monotone matrix functions of finite order*, Linear and Multilinear Algebra **27**(1990), 129–138,

[23] M. Nussbaum and A. Szkola, *The Chernoff lower bound for symmetric quantum hypothesis testing*, Ann. Statist. **37**(2009), no. 2, 1040–1057, quant-ph/0607216.

[24] Y. Ogata, *A Generalization of Powers–Størmer Inequality*, Letters in Mathematical Physics, 97:3 (2011) 339–346.

[25] H. Osaka, S. Silvestrov, and J. Tomiyama, *Monotone operator functions, gaps and power moment problem*, Math. Scand. **100** (2007), no. 1, 161–183.

[26] H. Osaka and J. Tomiyama, *Double piling structure of matrix monotone functions and of matrix convex functions*, Linear and its Applications **431**(2009), 1825–1832.

[27] H. Osaka and J. Tomiyama, *Double piling structure of matrix monotone functions and of matrix convex functions II*, Linear and its Applications **437**(2012), 735–748.

[28] D. Petz, *Quantum information theory and quantum statistics*, Theoretical and Mathematical Physics. Springer-Verlag, Berlin, 2008.

[29] R.T. Powers, E. Størmer, *Free States of the Canonical Anticommutation Relations*, Commun. math. Phys., **16**(1970) 1–33.

[30] T. Sano and T. Yatsu, *Characterizations of tracial property via inequalities*, Journal of Ineq. in Pure and Appl. Math. **7**(2006), no. 1, Article 36, 11 pp.

[31] G. Sparr, *A new proof of Löwner's theorem on monotone matrix functions*, Math. Scand. **47**(1980), 2, 266–274.

[32] O.E. Tikhonov, *Subadditivity inequalities in von Neumann algebras and characterization of tracial functional*, Positivity. **9**(2005) 259–264.

Hiroyuki Osaka
Noji-Higashi 1-1-1
Kusatsu 525-8577, Japan
e-mail: osaka@se.ritsumjei.ac.jp

Ordered Structures and Applications: Positivity VII
Trends in Mathematics, 363–370

Inverses of Operator Convex Functions

Henrik L. Pedersen and Mitsuru Uchiyama

Abstract. The inverse of an operator convex function is shown to be operator monotone and due to the wide applications of the class of Nevanlinna–Pick functions we formulate a function theoretic version of the results.

Mathematics Subject Classification (2010). Primary 30D15; Secondary 33B15; 30E20.

Keywords. Operator monotone function, Pick function, inverse function, logarithmic derivative.

1. Introduction

The class of Nevanlinna–Pick, or simply Pick, functions enters in, e.g., operator theory, orthogonal polynomials and special functions. In operator theory the class of functions that preserve the partial ordering induced by the cone of positive definite operators have attracted much attention. These functions turn out to be a subclass of the Pick functions. Jacobi matrices are studied through the so-called m-function, which is a Stieltjes transform related to the resolvent of self-adjoint operators in Hilbert space, see, e.g., [8], and Stieltjes transforms are closely related to Pick functions. Furthermore, Pick functions are used as parameter space in Nevanlinna's parametrization of all solutions to the classical Hamburger moment problem on the real line. See [1]. They have also been a fruitful tool when investigating special functions, such as Euler's Gamma function, see, e.g., [6], and they enter in areas of geometric function theory, in connection with universally convex and starlike functions. See, e.g., [15].

For a real-valued continuous function g defined on an interval I of the real line, $g(\mathbf{A})$ is defined by the functional calculus for any bounded linear self-adjoint operator $\mathbf{A}$ having its spectrum in I. A function g is called operator monotone

The research of the first author was supported by grant 10-083122 from The Danish Council for Independent Research | Natural Sciences.
The research of the second auhtor was supported by (JSPS) KAKENHI25400116.

on I if $g(\mathbf{A}) \leq g(\mathbf{B})$ for all bounded linear self-adjoint operators $\mathbf{A}$ and $\mathbf{B}$ having their spectra in I and such that $\mathbf{A} \leq \mathbf{B}$. This class of functions is denoted $\mathbf{P}(I)$.

A continuous function $g : I \to \mathbb{R}$ is called operator convex on I if

$$g(s\mathbf{A} + (1-s)\mathbf{B}) \leq sg(\mathbf{A}) + (1-s)g(\mathbf{B})$$

for every $s \in (0,1)$ and for every pair of bounded self-adjoint operators $\mathbf{A}$ and $\mathbf{B}$ whose spectra are both in I. Operator concave functions are defined likewise. For an introduction to operator monotone functions (and Pick functions) see [9] or [17].

For the reader's convenience we state some fundamental facts concerning Pick functions. A Pick function is a holomorphic function f defined in the upper half-plane $\mathbb{C}_+$ for which $\Im f(z) \geq 0$ for all $z \in \mathbb{C}_+$. It is well known that there is a one-to-one correspondence between Pick functions f and triples (a, b, μ), where $a \geq 0$, $b \in \mathbb{R}$, and μ is a positive Borel measure on $\mathbb{R}$ such that

$$\int_{-\infty}^{\infty} \frac{d\mu(t)}{t^2+1} < \infty.$$

The correspondence is given by the formula

$$f(z) = az + b + \int_{-\infty}^{\infty} \left(\frac{1}{t-z} - \frac{t}{t^2+1} \right) d\mu(t).$$

Any Pick function f can be extended to $\mathbb{C} \setminus \mathbb{R}$ by reflection. For an open interval I of $\mathbb{R}$ it can be proved that f has a holomorphic extension across I, that is from $\mathbb{C} \setminus \mathbb{R}$ to $\mathbb{C} \setminus (\mathbb{R} \setminus I)$ if and only if the corresponding measure μ has no support in I.

The results in this paper are to a large extent based on properties of (conditionally) positive definite kernels. A real-valued function K defined on $I \times I$ is said to be a positive definite kernel if for any n, any points $t_1, \ldots, t_n \in I$ and any complex numbers $z_1, \ldots, z_n$,

$$\sum_{k=1}^{n} K(t_i, t_j) z_i \overline{z_j} \geq 0.$$

A real-valued function K defined on $I \times I$ is said to be a conditionally positive definite kernel if it is symmetric and for any n, any $t_1, \ldots, t_n \in I$ and any complex numbers $z_1, \ldots, z_n$ such that $\sum_{j=1}^{n} z_j = 0$,

$$\sum_{k=1}^{n} K(t_i, t_j) z_i \overline{z_j} \geq 0.$$

The negative of a conditionally positive definite kernel is called a conditionally negative definite kernel, or sometimes just negative definite. See, e.g., [5]. For a C^1-function $f : I \to \mathbb{R}$ the *Löwner kernel* K_f is defined as

$$K_f(s,t) = \frac{f(s) - f(t)}{s - t}, \quad s \neq t, \qquad K_f(t,t) = f'(t).$$

The following theorem, first proved by Löwner (see [13]), is a fundamental result linking operator monotone functions, positive definite kernels and Pick functions together. We refer the reader to [17], [9] or [16] for more recent proofs of this result.

Theorem 1.1. *Let I be an open interval of $\mathbb{R}$. The following are equivalent for a function $f : I \to \mathbb{R}$.*

(i) *f belongs to $\mathbf{P}(I)$;*
(ii) *The kernel K_f is positive definite on $I \times I$;*
(iii) *f has a holomorphic extension to $\mathbb{C} \setminus (\mathbb{R} \setminus I)$ such that f is a Pick function.*

Another important result is Theorem 1.2 below, due to Krauss (see [11]), and furnishing a connection between operator convex and operator monotone functions. For more recent proofs, see [3] or [18].

Theorem 1.2. *Let I be an open interval of $\mathbb{R}$. If g is operator convex on I, then g is of class $C^2(I)$ and for each $c \in I$ the function $t \mapsto K_g(t, c)$ is operator monotone on I. Conversely, if $g \in C^2(I)$ and there exists $c \in I$ such that $K_g(t, c)$ is operator monotone on I, then g is operator convex.*

It is known that a function h is operator monotone on a half-line (a, ∞) if and only if h is operator concave and $\lim_{t\to\infty} h(t) > -\infty$. See [18].

In the next section we give our main results concerning operator convex functions and in the following section the results are discussed in the framework of Pick functions. Apart from the main new result in Proposition 2.2 it is our aim to exploit some of the connections between operator theory and complex analysis and in particular the application of Pick functions in the theory of special functions, by stressing the power of Löwner's theorem.

2. Results for operator convex functions

A source of motivation for the results in this section is a recent result stating that the so-called principal inverse of Euler's Gamma function is operator monotone. See [19]. The methods relied on certain conditionally negative definite kernels. The connection is briefly explained in Remark 3.5. First, we state a well-known result.

Lemma 2.1. *If K is a conditionally negative definite kernel on $I \times I$ and if $K(x, y) > 0$ for all $x, y \in I$, then $1/K$ is infinitely divisible.*

We invoke this to prove our main result.

Proposition 2.2. *Let g be a C^1-function on I with $g'(t) > 0$ for all $t \in I$. If $t \mapsto K_g(t, c)$ is operator concave for some $c \in I$, then the inverse function g^{-1} belongs to $\mathbf{P}(g(I))$.*

Proof. Since $g'(t) > 0$, K_g is a strictly positive function on $I \times I$. Put $f(t) = K_g(t, c)$ and then $h(t) = K_f(t, c)$. Since $-f(t)$ is operator convex on I, by Theorem 1.2, $-h(t)$ is operator monotone. By Theorem 1.1, $-K_h(t, s)$ is positive definite. Since $g(t) = f(t)(t - c) + g(c)$,

$$K_g(t, s) = f(t) + f(s) - f(c) + (t - c)K_h(t, s)(s - c) \quad (\text{for } t \neq s),$$

and this equality also holds for $t = s$. The kernel $K_g(t, s)$ is therefore conditionally negative definite, and thus by Lemma 2.1, $1/K_g(t, s)$ is infinitely divisible, and hence positive definite. Thus

$$K_{g^{-1}}(x, y) = \frac{1}{K_g(g^{-1}(x), g^{-1}(y))}$$

is also infinitely divisible, especially positive definite on $g(I) \times g(I)$. Theorem 1.1 yields that $g^{-1} \in \mathbf{P}(g(I))$. □

Proposition 2.3. *Let g be an increasing operator convex function on an infinite interval $I = (a, \infty)$, with $-\infty \le a$. Then $g^{-1} \in \mathbf{P}(g(I))$.*

Proof. Since g is operator convex then by Theorem 1.2, $t \mapsto K_g(t, c)$ is operator monotone for all $c \in I$, and by the remark following Theorem 1.2, $K_g(t, c)$ is operator concave. Since g is increasing, Proposition 2.2 yields that $g^{-1} \in \mathbf{P}(g(I))$. □

Remark 2.4. In the case where $I = [0, \infty)$ and $g(I) = [0, \infty)$, namely for $g(t) = tf(t)$ with $f \in \mathbf{P}((0, \infty))$, Proposition 2.3 was first shown in [2] in a different way.

Corollary 2.5. *Let g be an increasing function on $I = (a, \infty)$, with $a \ge -\infty$. Then*
(i) $g^{-1} \in \mathbf{P}(g(I))$ *if* $g > 0$ *and* $\log g$ *is operator convex,*
(ii) $g^{-1} \in \mathbf{P}(g(I))$ *if* $f \circ g$ *is operator convex for some* $f \in \mathbf{P}(g(I))$.

Proof. Suppose $\mathbf{A} \le \mathbf{B}$ for $\mathbf{A}$ and $\mathbf{B}$ with spectra in $g(I)$. Put $\mathbf{A}' = g^{-1}(\mathbf{A})$, $\mathbf{B}' = g^{-1}(\mathbf{B})$. From $g(\mathbf{A}') = \mathbf{A} \le \mathbf{B} = g(\mathbf{B}')$ it follows that $\log g(\mathbf{A}') \le \log g(\mathbf{B}')$. By Proposition 2.3, $\mathbf{A}' \le \mathbf{B}'$. We consequently get (i); (ii) similarly follows. □

Corollary 2.6. *Let g be a positive and increasing function on $I = (a, \infty)$ with $a \ge -\infty$. Suppose $g'/g \in \mathbf{P}(I)$. Then $g^{-1} \in \mathbf{P}(g(I))$.*

Proof. It is known that for any $c > a$, $\log g(t) = \int_c^t (g'(s)/g(s))ds + \log g(c)$ is operator convex. By Corollary 2.5 we get the required result. □

3. Pick function counterparts

In this section we discuss briefly what these results say in the framework of Pick functions. The equivalence of (ii) and (iii) in Löwner's theorem describes the connection between Pick functions and positive definite kernels. We feel that it is worthwhile to include proofs of some of the results based only on this equivalence.

Proposition 3.1. *A function $g : (0, \infty) \to [0, \infty)$ has an extension to a Pick function in $\mathbb{C} \setminus (-\infty, 0]$ if and only if it admits the representation*

$$g(z) = az + \int_0^\infty \frac{z}{z + t}\, d\sigma(t),$$

where $a \ge 0$, and σ is a positive measure on $[0, \infty)$ such that $\int_0^\infty d\sigma(t)/(t+1) < \infty$.

Remark 3.2. The proposition describes the class of positive operator monotone functions in $(0,\infty)$. For a proof see [4] or [1, Addenda and Problems to Chapter 3].

The proposition can be used to give a direct proof of the following Lemma 3.3 about Pick functions. We remark that $f = \int f'(t)dt$ is operator convex, so the lemma also follows from the previous section.

Lemma 3.3. *Suppose that f is holomorphic in the cut plane $\mathbb{C} \setminus (-\infty, 0]$, and that f is real valued and increasing on $(0,\infty)$. If f' is a Pick function then f^{-1} can be extended from $f((0,\infty))$ to a Pick function defined in $\mathbb{C} \setminus (\mathbb{R} \setminus f((0,\infty)))$.*

Proof. The kernel K_f is conditionally negative definite on $(0,\infty) \times (0,\infty)$. Indeed, using the representation of f' in Proposition 3.1,

$$\begin{aligned} f(x) - f(c) &= \int_c^x \left(\alpha s + \int_0^\infty \frac{s}{s+t} d\sigma(t) \right) ds \\ &= \frac{\alpha(x^2 - c^2)}{2} + \int_0^\infty (x - c - t(\log(x+t) - \log(c+t)))\, d\sigma(t). \end{aligned}$$

This gives

$$K_f(x,y) = \frac{\alpha}{2}(x+y) + \int_0^\infty \left(1 - t\frac{\log(x+t) - \log(y+t)}{x-y}\right) d\sigma(t).$$

Since $(x,y) \mapsto x+y$ is conditionally negative definite and $K_{\log(\cdot+t)}$ is positive definite, K_f is conditionally negative definite. By Lemma 2.1, $1/K_f$ is positive definite. If $x, y \in f((0,\infty))$, we choose $t, s \in (0,\infty)$ such that $f(t) = x$ and $f(s) = y$, and then we have

$$K_{f^{-1}}(x,y) = \frac{f^{-1}(x) - f^{-1}(y)}{x-y} = \frac{t-s}{f(t) - f(s)}.$$

We have just seen that this is a positive definite kernel and thus f^{-1} has an extension across $f((0,\infty))$ to a Pick function. □

Remark 3.4. Corollary 2.6 can also be proved using the composition of two Pick functions: Indeed, if g satisfies the conditions of Corollary 2.6 then, by Lemma 3.3, the inverse h of $\log g$ is extended to a Pick function in the cut plane $\mathbb{C} \setminus (\mathbb{R} \setminus \log g((0,\infty)))$. Since log is also a Pick function then $h(\log z)$ is a Pick function in $\mathbb{C} \setminus (\mathbb{R} \setminus g((0,\infty)))$, and

$$g(h(\log z)) = e^{(\log g)(h(\log z))} = e^{\log z} = z.$$

Hence g has an inverse on $g((0,\infty))$ that can be extended to a Pick function.

Remark 3.5. It is a classical fact that Euler's Gamma function Γ increases on the interval (α,∞), where α denotes the only positive zero of the ψ-function (the logarithmic derivative of Γ). In [19] it was shown that the inverse of $\Gamma|_{(\alpha,\infty)}$ has an extension to a Pick function in the complex plane cut along the half-line $(-\infty, \Gamma(\alpha)]$.

A central ingredient in the proof consisted in showing that the kernel $K_{\log\Gamma}$ is conditionally negative definite. This result follows from Corollary 2.5 since

$$\log\Gamma(t) = -\log t - \gamma t + \sum_{n=1}^{\infty}\left(\frac{t}{n} - \log\left(1+\frac{t}{n}\right)\right)$$

(where γ is Euler's constant) is operator convex, and Γ is increasing on (α,∞). The result about the Gamma function has been generalized to a certain class of entire functions of genus 1, see [14], and this is also a consequence of Corollary 2.5 (or 2.6).

We end the paper with some remarks concerning the class of Laguerre–Pólya functions. An entire function is said to be of Laguerre–Pólya class if it can be approximated uniformly over compact subsets of the complex plane by polynomials with real zeros only. This class consists of exactly those entire functions f admitting the Weierstraß factorization

$$f(z) = z^m e^{az^2+bz+c}\prod_{k=1}^{\infty}\left(1-\frac{z}{z_n}\right)e^{z/z_n},$$

where $z_n \in \mathbb{R}\setminus\{0\}$ such that $\sum_{n=1}^{\infty}|z_n|^{-2} < \infty$, $a \le 0$, $b, c \in \mathbb{R}$ and $m \ge 0$. For such a function f, $\log f$ is univalent in $\mathbb{C}_+$ and the conformal image of $\mathbb{C}_+$ is a so-called comb-domain, obtained by removing certain horizontal half-lines from the complex plane (or from a horizontal strip of the complex plane). Conversely, any such comb-domain is the conformal image of the logarithm of a suitable function from the Laguerre–Pólya class. See [10].

The class of Pólya functions is usually defined as the class of holomorphic functions f defined in $\mathbb{C}_+$ for which $\partial_y|f(x+iy)| \ge 0$ for $y > 0$. (This class goes back to de Branges, see [7].) By the Cauchy–Riemann equations it is not difficult to see that the condition is equivalent to $-f'/f$ being a Pick function. It is also easy to see that the class of Pólya functions contains the Laguerre–Pólya functions.

A domain $\mathcal{D}$ is a $\mathcal{P}$-domain if for any $w \in \mathcal{D}$ and any $t \ge 0$ we have $w - t \in \mathcal{D}$. It is known that the conformal image of the upper half-plane under $\log p$ is a so-called $\mathcal{P}$-domain for any function p in the Pólya class. See [12].

In order to stress the connection to complex analysis we end the paper by formulating a result for the Pólya class.

Corollary 3.6. *Suppose that a function p from the Pólya class can be holomorphically extended through $(0,\infty)$ and is positive and decreasing there. Then the inverses of $1/p$ and of $-\log p$ can be extended to Pick functions. Furthermore, the conformal image $\log p(\mathbb{C}_+)$ covers the whole of the lower half-plane.*

Proof. This follows immediately, noting the general fact that if $f(g(z)) = z$ for all $z \in \mathbb{C}_+$ then the image $f(\mathbb{C}_+)$ must contain $\mathbb{C}_+$. □

References

[1] N.I. Akhiezer, The classical moment problem and some related questions in analysis, Hafner Publishing Co., New York, 1965.

[2] T. Ando, *Comparison of norms* $|||f(A) - f(B)|||$ *and* $|||f(|A - B|)|||$, Math. Z., **197** (1988), 403–409.

[3] J. Bendat and S. Shermann, *Monotone and convex operator functions*, Trans. Amer. Math. Soc., **79** (1955), 58–71.

[4] C. Berg, Stieltjes–Pick–Bernstein–Schoenberg and their connection to complete monotonicity. In: Positive definite functions. From Schoenberg to Space-Time Challenges. Ed. J. Mateu and E. Porcu. Dept. of Mathematics, University Jaume I, Castellon, Spain, 2008.

[5] C. Berg, J.P.R. Christensen and P. Ressel, Harmonic analysis on semigroups. Theory of positive definite and related functions, Graduate Texts in Mathematics vol. 100. Springer-Verlag 1984.

[6] C. Berg and H.L. Pedersen, *A one-parameter family of Pick functions defined by the gamma function and related to the volume of the unit ball in n-space*, Proc. Amer. Math. Soc., **139** (2011), 2121–2132.

[7] L. de Branges, Hilbert Spaces of Entire Functions, Prentice Hall, Englewood Cliffs, NJ, 1968.

[8] J.S. Christiansen, B. Simon and M. Zinchenko, *Finite gap Jacobi matrices, I. The isospectral torus.* Constr. Approx. **32** (2010), 1–65.

[9] W.F. Donoghue, Monotone Matrix Functions and Analytic Continuation, Springer-Verlag, 1974.

[10] A. Eremenko and P. Yuditskii, *Comb Functions*, Contemp. Math., **578** (2012), 99–118.

[11] F. Krauss, *Über konvexe Matrixfunktionen*, Math. Zeit., **41** (1936), 18–42.

[12] H. Grunsky, *Zur konformen Abbildung von Gebieten, die in einer Richtung konvex sind*, J. Math. Anal. Appl., **34** (1971), 685–701.

[13] K. Löwner, *Über monotone Matrixfunktionen*, Math. Z., **38** (1934), 177–216.

[14] H.L. Pedersen, *Extension of inverses of entire functions of genus* 1 *and* 2 *to the upper half-plane*, Comput. Methods Funct. Theory **13** (2013), 263–275.

[15] S. Ruscheweyh, L. Salinas and T. Sugawa, *Completely monotone sequences and universally prestarlike functions*, Isr. J. Math., **171** (2009), 285–304.

[16] R.L. Schilling, R. Song and Z. Vondraček, Bernstein functions. Theory and applications, de Gruyter Studies in Mathematics **37**, Walter de Gruyter & Co., 2010.

[17] B. Simon, Convexity. An analytic viewpoint, Cambridge Tracts in Mathematics **187**, Cambridge University Press, Cambridge, 2011.

[18] M. Uchiyama, *Operator monotone functions, positive definite kernel and majorization*, Proc. Amer. Math. Soc. **138** (2010), 3985–3996.

[19] M. Uchiyama, *The principal inverse of the gamma function*, Proc. Amer. Math. Soc., **140** (2012), 1343–1348.

Henrik L. Pedersen
Department of Mathematical Sciences
University of Copenhagen
Universitetsparken 5
DK-2100 Copenhagen, Denmark
e-mail: henrikp@math.ku.dk

Mitsuru Uchiyama
Department of Mathematics
Shimane University
Matsue City
Shimane, Japan

Current address:

Department of Mathematical Sciences
Ritsumeikan University
Kusatsu, Shiga 525–8577, Japan
e-mail: uchiyama@riko.shimane-u.ac.jp
uchimitsuyama@gmail.com
uchiyama@fc.ritsumei.ac.jp

Ordered Structures and Applications: Positivity VII
Trends in Mathematics, 371–394

Some Remarks on Approximation Properties with Applications

Oleg Reinov

Abstract. We study some known approximation properties and introduce and investigate several new approximation properties, closely connected with different quasi-normed tensor products. These are the properties like the AP_s or $AP_{(s,w)}$ for $s \in (0, 1]$, which give us the possibility to identify the spaces of s-nuclear and (s, w)-nuclear operators with the corresponding tensor products (e.g., related to Lorentz sequence spaces). Some applications are given (in particular, we present not difficult proofs of the trace-formulas of Grothendieck–Lidskiĭ type for several ideals of nuclear operators).

Mathematics Subject Classification (2010). 46B28.

Keywords. Nuclear operator, tensor product, approximation property, eigenvalue.

Introduction

In 1955 A. Grothendieck [7] has introduced the notion of the projective tensor product of the locally convex vector spaces and developed the corresponding theory. It was a very deep generalization of Schatten–von Neumann theory of S_p-spaces and the corresponding theory of tensor products of Hilbert spaces [29]. One of the nice properties of Grothendieck's projective tensor product of type $E\widehat{\otimes}F$ is that its topological dual can be described as the space $B(E, F)$ of all continuous bilinear forms on $E \times F$. In the particular case of the projective product $Y^*\widehat{\otimes}X$ of Banach spaces Y^* (dual to Y) and X, one can identify the Banach dual to $Y^*\widehat{\otimes}X$ with the space $L(X, Y^{**})$ of all linear continuous operators from X to Y^{**} in a natural way (by using a linear continuous functional "trace"; see below). It turned out that the topological (locally convex) dual to the subspace $L(X, Y)$ of $L(X, Y^{**})$, equipped with the topology of compact convergence ($L_c(X, Y)$ in notations of [7]), can be identified with a quotient of $Y^*\widehat{\otimes}X$ ([7], Chap. I, Prop. 22). As was shown by A. Grothendieck, it follows from the last statement that the injectivity of the canonical map $Y^*\widehat{\otimes}X \to L(Y,X)$ is equivalent to the density of the set $X^* \otimes Y$ of all finite rank operators from X to Y in $L_c(X,Y)$. If $X = Y$, then the last is equivalent to the fact that the identity map is in the closure of the set of finite rank operators in the topology of compact convergence. This leads to the famous Grothendieck's definition of the notion of the AP (approximation property) for a Banach space: Following A. Grothendieck, we say that a Banach space X has the AP, if for every compact subset K of X and for any $\varepsilon > 0$ one can find a finite rank operator R in X such that $\sup_{x\in K} ||Rx - x|| \le \varepsilon$. The property is so important that we can find its applications in a great number of papers devoted to the theory of operators in Banach spaces.

Let us reformulate (following [7]) the definition of the AP in terms of tensor products: X has the AP iff the natural mapping $j : X^*\widehat{\otimes}X \to L(X,X)$ is one-to-one (here by "natural map" we mean the unique extension of the natural linear inclusion $X^* \otimes X \to L(X,X)$ from the normed space $(X^* \otimes X, ||\cdot||_\wedge)$ (i.e., with the projective norm) to the completion $X^*\widehat{\otimes}X$). This fact becomes evident if we note that the map j is one-to-one iff the closure of $X^* \otimes X$ in $L_c(X,X)$ is the whole space $L(X,X)$ (or, what is the same, the identity map id_X is in this closure). The image of the tensor product $X^*\widehat{\otimes}X$ in $L(X,X)$ is, by definition, the space $N(X,X)$ of nuclear operators in X (with the norm induced from $X^*\widehat{\otimes}X$). Thus, X has the AP iff (as we can write) $N(X,X) = X^*\widehat{\otimes}X$.

On the space $X^* \otimes X$, the usual linear functional "trace" is defined which is continuous on the normed space $(X^* \otimes X, ||\cdot||_\wedge)$. After extension to the projective tensor product $X^*\widehat{\otimes}X$, this linear functional is still bounded and it can be seen that a Banach space Y has the AP iff every tensor element $z \in Y^*\widehat{\otimes}Y$ which generates a 0-operator in $L(Y,Y)$ has the property that trace $U \circ z = 0$ for every $U \in L(Y, Y^{**})$.

In Chapter II of [7], A. Grothendieck has generalized the notion of nuclear operators and also considered the more general tensor products: In the terminology of [7], an element of $X^*\widehat{\otimes}Y$ is said to be a "noyau de Fredholm de puissance p.ème sommable" ($p \in (0,1]$), if it is of the form $\sum_i \lambda_i\, x'_i \otimes y_i$, where $(\lambda_i) \in l_p$ and (x_i) (resp. y_i) is a bounded sequence in X^* (resp. Y). We will use the notation $X^*\widehat{\otimes}_pY$ for the corresponding tensor product (in [7], it is denoted by $X^* \overset{(p)}{\otimes} Y$). This is a linear subspace of the projective tensor product $X^*\widehat{\otimes}Y$, and with the natural metric (see [7], Chapter II, §1) it is a complete metric space.

If $z \in X^*\widehat{\otimes}_pY$, then the associated operator $\widetilde{z}$ from X to Y is called (by A. Grothendieck) as "une application de puissance p.ème sommable". The natural inclusion $X^*\widehat{\otimes}_pY \hookrightarrow X^*\widehat{\otimes}Y$ is one-to-one for any pair of Banach spaces X and Y (a priori, it is not evident; see [7], Chapter II, §1 for an explanation).

One of the interesting questions considered in [7], is the connection between the "order" of a tensor element $z \in X^*\widehat{\otimes}_pY$ and the "order" of the sequence of all eigenvalues of the corresponding operator $\widetilde{z}$ (evidently, $\widetilde{z}$ is compact). Among the results in this direction, let us mention only the following facts (which were obtained in [7], Chapter II, §1, Section 4): Let X and Y be Banach spaces and $0 <$

$p \leq 2/3$. Then the canonical map $j_p : X^*\widehat{\otimes}_p Y \to L(X,Y)$ is injective. Moreover, for every $z \in X^*\widehat{\otimes}_p X$ the sequence (z_i) of all eigenvalues of the associated operator (counted according to their multiplicities) is absolutely summable and trace $z = \sum_i z_i$ (the assertions are true for any locally convex vector space X).

Remembering one of the equivalent definitions of the *AP*, we see that it was, in fact, shown by A. Grothendieck that every Banach space has some approximation properties "of type p" for all $p \in (0, 2/3]$. We can use, e.g., a notation "AP_p". Thus, A. Grothendieck considered the notion of "p-approximation property" already in 1955, though implicitly.

Let us mention that A. Grothendieck (applying deep results of complex analysis and H. Weyl's [31] theorem on the Schatten–von Neumann classes S_p of compact operators in Hilbert spaces) has proved firstly the "eigenvalue theorem" for the case where $0 < p \leq 2/3$ and then, as a consequence, obtained the injectivity of the above maps j_p (surely, the main case is $p = 2/3$). In the paper [20] of the author, the reader can find a more simple proof of these theorems, where it was shown firstly that the map $j_{2/3}$ is one-to-one and then the eigenvalue result was obtained (by applying the Lidskiĭ theorem for the trace-class operators in Hilbert spaces [12]).

The question about the injectivity of the maps j_p for $p \in (2/3, 1)$ was not considered by A. Grothendieck in [7] explicitly. He posed the corresponding question only for the case $p = 1$. This famous approximation problem was solved in negative in 1972 by Per Enflo [6] (for the further information see [3], [14], [16], [30].

It seems that the notion of the approximation property "of type p" (for $0 < p < 1$) was (explicitly) considered firstly by the author in the paper [21], where it appeared as the "approximation property of order p". In [21], we used the tensor product definition (i.e., the injectivity of the map j_p). Some simple facts and different (counter)examples were presented in [21]. Instead of the term "une application de puissance p.ème sommable", we used there the name "a p-nuclear operator". Later p-nuclear operators (for $p \in (0,1)$) were studied, e.g., in [22], [23], [27], [8]. We refer the reader to these papers for the further information.

Our aim in these notes is to discuss several old and new definitions of different approximation properties and to formulate (and, partially, to present the proofs of) some results in this direction. Also, we give applications (in particular, to eigenvalues problems).

Preliminaries

All the spaces under considerations $(X, Y, \dots)$ are Banach, all linear mappings (operators) are continuous; as usual, $X^*, X^{**}, \dots$ are Banach duals (to X), and $x', x'', \dots$ (or $y', \dots$) are the functionals on $X, X^*, \dots$ (or on $Y, \dots$). If $x \in X, x' \in X^*$ then $\langle x, x' \rangle = \langle x', x \rangle = x'(x)$. $L(X,Y)$ stands for the Banach space of all linear bounded operators from X to Y. Every Banach space is considered as a Banach subspace of its second dual. If needed, by π_Y we denote the natural isometric injection of Y into Y^{**}.

We consider the algebraic tensor product $X^* \otimes Y$ as the linear space of all continuous finite rank operators from X to Y. The projective tensor product $X^*\widehat{\otimes}Y$ of the spaces X^* and Y is the completion of $X^* \otimes Y$ with respect to the norm $||z||_\wedge := \inf\{\sum |\lambda_k|\}$, where the infimum is taken over all finite representations of $z \in X^* \otimes Y$ in the form $z = \sum \lambda_k\, x'_k \otimes y_k$ with $||x'_k|| = ||y_k|| = 1$. Every element $z \in X^*\widehat{\otimes}Y$ admits a representation $z = \sum_{k=1}^{\infty} \lambda_k x'_k \otimes y_k$ such that $\sum |\lambda_k| < \infty$ and $||x'_k|| = ||y_k|| = 1$. If $X = Y$, then the functional "trace" on the tensor product $X^*\widehat{\otimes}X$ is well defined by the formula trace $z := \sum \lambda_k\, \langle x'_k, y_k\rangle$. The Banach dual to $X^*\widehat{\otimes}Y$ can be identify with the space $L(Y, X^{**})$ with duality given by "trace": for $z \in X^*\widehat{\otimes}Y$ and $U \in L(Y, X^{**})$ we put $\langle U, z\rangle :=$ trace $U \circ z = \sum \lambda_k\, \langle x_k, Uy_k\rangle$.

We use standard notations for the classical Banach spaces such as $L_p(\mu)$, $C(K)$, l_p, c_0 etc. For the theory of (sequence) Lorentz spaces, we refer to [1], [16], [17, Section 2.1]; see also [8, Section 5]. For the definitions of the notions of type and cotype, see any of these references: [4], [16], [18], [19] (Rademacher type p = Gauss type p and Rademacher cotype q = Gauss cotype q; so, we can apply results from G. Pisier's lecture [19], assuming that we are working with Rademacher notions).

Let us collect some facts we need. Recall that a subspace E of a Banach space X is b-complemented ($b > 0$) in X, if there exists a linear continuous projection P from X onto E such that $||P|| \le b$. As usual, if $p \in [1, \infty]$, then p' is the conjugate exponent: $1/p + 1/p' = 1$.

Let X be a Banach space and $1 < p \le 2$, $2 \le q < \infty$. 1) If X is of type p (cotype q) then every subspace is of type p (cotype q); 2) [4, Proposition 11.11] If X is of type p then any quotient of X is of type p; 3) [4, Proposition 11.10] If X is of type p then X^* is of cotype p'; 4) If X^* is of type p then X is of cotype p'; 5) If X is of type p then any subspace of any quotient (and any quotient of any subspace) of X is of type p; 6) [4, Corollary 11.9] A Banach space has the same type or cotype as its bidual; 7) [4, Corollary 11.7] Each L_r-space ($1 \le r < \infty$) has type $\min\{r, 2\}$ and cotype $\max\{r, 2\}$; 8) [19, see Theorem 4.1 and its Corollaries] If X is of type p and of cotype q then there is a constant $D_{p,q} > 0$ such that every finite-dimensional subspace E of X is $D_{p.q}\,(\dim E)^{1/p-1/q}$-complemented in X.

Recall also the well-known general fact (due to M.J. Kadec and M.G. Snobar [9]; see also [16, 28.2.6. Lemma]): in any Banach space every n-dimensional subspace is $n^{1/2}$-complemented.

Our main reference is [16]. All the notions, notations and facts, given here without any explanation, can be found in [1, 5, 8, 14, 16, 17, 18].

Contents

In Section 1 we reformulate the definition of Grothendieck approximation property in terms of 0-sequences that leads us to the consideration of more general approximation properties $\widetilde{AP_s}$ for $s \in (0, 1]$. We define them in terms of approximation of the identity maps in Banach spaces by finite rank operators on l_p-sequences (p depends on s). It seems that such properties, for the first time, were considered by the author in [22] (cf. Lemma 2.1 there). The simplest examples of Banach spaces possessing the properties of such a kind are subspaces of quotients of L_p-spaces.

In Section 2 we reformulate the $\widetilde{AP_s}$ in terms of tensor products showing that $\widetilde{AP_s} = AP_s$, where AP_s is the approximation property of order s introduced in the author paper [21] and investigated also in [2], [23], [27].

In Section 3 we introduce and investigate new notions of the approximation properties $AP_{t;p,r}$ and $AP_{(r,w)}$ defined by some "Lorentz tensor products" (tensor products generated by Lorentz sequence spaces). These notions are new and considered here for the first time. In particular, we obtain a characterization of the $AP_{(r,1)}$ in terms of approximation of the identity maps by finite rank operators on some (Lorentz) 0-sequences (Theorem 3.3).

In Section 4, with the help of a theorem of M.C. White [32], we prove Theorem 4.1 which gives us sufficient conditions for the famous Grothendieck–Lidskiĭ trace formula to be valid for certain quasi-Banach operator ideals. We present an application of the previous results about approximation properties to some eigenvalue problems. Theorem 4.1 is applied then to the case of $(2/3, 1)$-nuclear operators (related to the Lorentz space $l_{2/3\,1}$).

In Section 5 the results of Section 1 about subspaces of quotients of L_p-spaces together with White's theorem are applied for proving some more theorems concerning the distribution of eigenvalues of the nuclear operators. We give new relatively simple proofs of some recent results from the papers [25] and [28].

In Section 6 we prove two statements about the approximation properties considered in Sections 1–V. For instance, it is well known that if X^* has the AP of Grothendieck, then X has the AP too. We show that the same is true for all the natural approximation properties considered here, such that $AP_s, AP_{(r,w)}$ etc.

In Section 7 we introduce, following [24] and [26], two more notions of the approximation properties by using the spaces of so-called (r,p)-nuclear operators (a partial case of a class of (s,q,t)-nuclear operators from [16, 18.1]). Two theorems about eigenvalues of the (r,p)-nuclear operators are proved. In these theorems, trace formulas of Grothendieck–Lidskiĭtype are established for the cases where $1/r - 1/p = 1/2$. The first one was proved in [24] and [26] with the help of Fredholm Theory; the second theorem (Theorem 7.3) was obtained before by the same authors, again by using Fredholm Theory (but its proof was unpublished). Here, the different (more simple) method is used. Firstly, we show that every Banach space has the corresponding approximation properties $AP_{[r,p]}$ and $AP^{[r,p]}$. After this, we obtain the eigenvalue results by using simple arguments.

Finally, in Section 8, examples are given (they are taken from the paper [26]). The examples give a possibility to conclude that all the positive results of Sections 1–7 concerning the approximation properties and trace formulas are sharp.

1

It is well known that every compact subset of a Banach space is contained in the closed convex hull of a sequence converging to 0 (see, e.g., [7], p. 112 in

Ch. I, Lemme 12, or [14], p. 30, Proposition 1.e.2). Therefore, the Grothendieck approximation property for a Banach space X can be defined as follows: X has the AP iff for every sequence $(x_n)_{n=1}^{\infty} \subset X$ tending to zero, for any $\varepsilon > 0$ there exists a finite rank (continuous) operator R in X such that for each $n \in \mathbb{N}$ one has $||Rx_n - x_n|| \leq \varepsilon$. Consider a natural question: for which sequences $(x_n) \in c_0(X)$, under some additional assumptions, the identity map id_X can be approximated by finite rank operators, as above, and which of those conditions are sharp (or, if one wishes, optimal)?

One of the simplest fact (we think, known for more than 30 years) is that

$$\text{if } (x_n) \in l_2(X), \quad X \text{ is any, then the answer is positive.} \tag{$*$}$$

Here is a reason of this: Assuming $||x_n|| \searrow 0$, take any $N \in \mathbb{N}$ and consider the linear span $E_N := \mathrm{span}[x_n]_1^N$ as a subspace of X. Define, fixing an $\varepsilon > 0$, a finite rank operator R to be a projection from X onto E_N whose norm $\leq \sqrt{N}$.

Now if N is such that, for every $n \geq N$, we have $||x_n|| \leq \frac{\varepsilon}{\sqrt{N}+1}$, then

$$||Rx_n - x_n|| = 0 \quad \text{if } n \leq N,$$

and

$$||Rx_n - x_n|| \leq (||R|| + 1)\, ||x_n|| \leq \varepsilon \quad \text{if } n \geq N.$$

Of course, instead of $(*)$ we can consider the statement

$$\begin{array}{l}\text{if } (x_n) \in l^0_{2,\infty}(X) \ \ \text{(Lorentz space with “o”} \\ \text{small} - l^{min}_{(2,\infty)}(X) \text{ in notations of [16, 13.9.3 Remark]),} \\ X \text{ is any, then the answer is affirmative.}\end{array} \tag{$**$}$$

The idea of the above proof is very simple and can be applied in some more general situations. For instance, every subspace of finite dimension n of an L_p-space is $n^{|1/2-1/p|}$-complemented in that L_p-space. So, if $p \in [1, \infty]$, $\alpha = |\frac{1}{2} - \frac{1}{p}|$ and X is a subspace of an L_p-space, then

$$\begin{array}{l}\text{for every sequence } (x_n) \in l^0_{q,\infty}(X), \text{ where} \\ 1/q = \alpha, \text{ the answer is affirmative.}\end{array} \tag{$*\!*\!*$}$$

Remark 1.1. About sharpness: it will be discussed below.

Remark 1.2. The statement $(***)$ has, as a matter of fact, the following quantitative aspect: Given $\alpha \in [0, 1/2]$ and a Banach space X with the property that every finite-dimensional subspace F of X is contained in a finite-dimensional subspace $E \subset X$, which in turn is $C\,(\dim F)^{\alpha}$-complemented in X, we have

$$\begin{array}{l}\text{for every sequence } (x_n) \in l^0_{q,\infty}(X), \text{ where} \\ 1/q = \alpha, \text{ for any } \varepsilon > 0 \text{ there is a finite rank} \\ \text{operator } R \text{ in } X \text{ so that } \sup_n ||Rx_n - x_n|| \leq \varepsilon.\end{array} \tag{$*\!*\!*$}'$$

Particular cases:

(i) $q = 2$ and $\alpha = 1/2$ or $q = \infty$ and $\alpha = 0$;
(ii) $(x_n) \in l_q(X), q \in [2, \infty)$.

For a while let us introduce the notions of the corresponding approximation properties for a Banach space X (taking into account that the possibility of approximations on c_0-sequences by finite rank operators gives us the Grothendieck's approximation property AP) : Let $0 < q \leq \infty$ and $1/s = 1/q + 1$. We say that X has the $\widetilde{AP}_s$ [resp., the $\widetilde{AP}_{s,\infty}$] if for every $(x_n) \in l_q(X)$ [resp., $l^0_{q,\infty}(X)$] (where $l_q(X)$ means $c_0(X)$ for $q = \infty$) and for every $\varepsilon > 0$ there exists a finite rank operator $R \in X^* \otimes X$ such that $\sup_n ||Rx_n - x_n|| \leq \varepsilon$. Trivially, e.g., $\widetilde{AP}_{s_2} \implies \widetilde{AP}_{s_1}$ if $s_1 \leq s_2$. Thus, $\widetilde{AP}_1 (= AP)$ implies any $\widetilde{AP}_s$.

The statement $(*)$ (and $(**)$) says that every Banach space has the above property $\widetilde{AP}_{2/3}$ (and even the $\widetilde{AP}_{2/3,\infty}$). The statement $(***)$ gives the corresponding result for L_p-subspaces. Moreover, the assertion mentioned in Remark 1.2, shows that, for instance, any subspace of any quotient (= any quotient of any subspace) of a Banach space of type 2 (resp., of cotype 2) and of cotype p, $p \in [2, \infty)$ (resp., of type p'), possesses the $\widetilde{AP}_s$ (even the $\widetilde{AP}_{s,\infty}$) with $1/s = 1 + |1/2 - 1/p|$.

2 Let us recall that the notion of the AP of Grothendieck can be reformulated in terms of the projective tensor products "$\widehat{\otimes}$". Namely, a Banach space X has the AP iff for every Banach space Y the canonical (natural) mapping $Y^*\widehat{\otimes}X \to L(Y, X)$ is one-to-one (or, what is the same, the natural mapping $X^*\widehat{\otimes}X \to L(X) := L(X, X)$ is injective). In [7], A. Grothendieck has considered also some other tensor products (linear subspaces of "$\widehat{\otimes}$'s"), which we will denote by "$\widehat{\otimes}_s$" for $0 < s \leq 1$ (so that $\widehat{\otimes} = \widehat{\otimes}_1$) : For Banach spaces X and Y, let $Y^*\widehat{\otimes}_s X$ be a subspace of the projective tensor product $Y^*\widehat{\otimes}X$ consisting of the tensors $z \in Y^*\widehat{\otimes}X$, which admit representations of the form

$$z = \sum_{n=1}^{\infty} \lambda_n y'_n \otimes x_n,$$

where $(\lambda_n) \in l_s$, (y'_n) and (x_n) are bounded sequences from Y^* and X respectively. With a natural "quasi-norm" (see [16]) the linear subspace $Y^*\widehat{\otimes}_s X$ of the space $Y^*\widehat{\otimes}X$ can be considered as a "quasi-normed tensor product" (it is then a complete metric space [7]).

One of the nice (with a nontrivial proof in [7]) theorem of Grothendieck is the fact that the natural map from $Y^*\widehat{\otimes}_{2/3}X$ into $L(Y, X)$ is injective for any Banach spaces X, Y. Let us compare this Grothendieck's result with a simple assumption in Section 1, where "$s = 2/3$" appeared. Clearly, it is not a chance coincidence, and we really have

Theorem 2.1. *For $s \in (0, 1]$ and for a Banach space X, the following statements are equivalent:*

1) *X has the $\widetilde{AP}_s$ in the sense of the definition in Section 1;*
2) *X has the AP_s in the sense of the definition in [23], i.e., for every Banach space Y the natural mapping $Y^*\widehat{\otimes}_s X \to L(Y, X)$ is one-to-one.*

Moreover, the following statement (AP_s) *takes place:*

(AP_s) *A Banach space* X *has the* AP_s, $0 < s \leq 1$, *iff the canonical map* $X^*\widehat{\otimes}_s X \to L(X)$ *is one-to-one (or, what is the same, there exists no tensor element* $z \in X^*\widehat{\otimes}_s X$ *with* trace $z = 1$ *and* $\widetilde{z} = 0$, *where* $\widetilde{z}$ *is the associated (with* z*) operator from* X *to* X*).*

Maybe analogous theorems and facts are valid for the $\widetilde{AP}_{s,\infty}$ and the $AP_{s,\infty}$ from [23] (see Section 3 for a discussion).

Proof of the assertion (AP_s). Suppose that the AP_s-condition holds for X, but there exists a Banach space Y such that the natural map $Y^*\widehat{\otimes}_s X \to L(Y,X)$ is not one-to-one. Take an element $z \in Y^*\widehat{\otimes}_s X$ which is not zero, but generates a zero operator $\widetilde{z} : Y \to X$. Then we can find an operator $U \in L(X, Y^{**})$ so that trace $U \circ z = 1$. If $z = \sum_{k=1}^{\infty} \lambda_k\, y'_k \otimes x_k$ is a representation of z in $Y^*\widehat{\otimes}_s X$ $((\lambda_k) \in l_s, (x_k)$ and (y'_k) are bounded), then

$$1 = \text{trace}\, U \circ z = \sum_{k=1}^{\infty} \lambda_k \langle U x_k, y'_k\rangle = \sum_{k=1}^{\infty} \lambda_k\, \langle x_k, U^* y'_k\rangle$$

and $\sum_{k=1}^{\infty} \lambda_k\, U^* y'_k(x) x_k = 0$ for every $x \in X$. Put $x'_k := \lambda_k U^* y'_k$, $z_0 := \sum_{k=1}^{\infty} x'_k \otimes x_k \in X^*\widehat{\otimes}_s X$. We have

$$\text{trace}\, z_0 = 1,\ \ \widetilde{z}_0 \neq 0$$

(by the assumption about X). Consider a one-dimensional operator $R = x' \otimes x$ in X with the property that trace $R \circ z_0 > 0$. Then

$$0 < \text{trace}\, R \circ z_0 = \sum_{k=1}^{\infty} \langle x'_k, x\rangle \langle x', x_k\rangle = \sum_{k=1}^{\infty} \lambda_k\, \langle U^* y'_k, x\rangle \langle x', x_k\rangle$$
$$= \left\langle \sum_{k=1}^{\infty} \lambda_k\, \langle Ux, y'_k\rangle x_k, x' \right\rangle = \left\langle x', \sum_{k=1}^{\infty} \lambda_k\, U^* y'_k(x) x_k \right\rangle = 0. \qquad \square$$

Proof of Theorem 2.1. We will use the assertion (AP_s).

1) $\Longrightarrow$ 2). Let $z \in X^*\widehat{\otimes}_s X$ and trace $z = 1$. Write $z = \sum \lambda_k\, x'_k \otimes x_k$, where the sequences (x'_k) and (x_k) are bounded and $(\lambda_k) \in l_s$, $\lambda_k \geq 0$, (λ_k) is non-increasing. Then

$$z = \sum_{k=1}^{\infty} (\lambda_k^s\, x'_k) \otimes (\lambda_k^{1-s}\, x_k)$$

(recall that $1/s = 1 + 1/q$; so $1 - s = s/q$). The sequence $(\lambda_k^{1-s} x_k)$ is in $l_q(X)$. By 1), for every $\varepsilon > 0$ there exists a finite rank operator $R \in X^* \otimes X$ such that $||R(\lambda_k^{1-s} x_k) - \lambda_k^{1-s} x_k|| \leq \varepsilon$ for each $k \in \mathbb{N}$. It follows that, for this operator R,

$$|\,\text{trace}\,(z - R \circ z)| = \left| \sum_{k=1}^{\infty} \langle \lambda_k^s x'_k, \lambda_k^{1-s} x_k - R(\lambda_k^{1-s} x_k)\rangle \right| \leq \sum_{k=1}^{\infty} \lambda_k^s ||x'_k|| \cdot \varepsilon \leq c \cdot \varepsilon.$$

Hence, for small $\varepsilon > 0$ we have that, for the operator $R \in X^* \otimes X$,

$$|\,\text{trace}\, R \circ z| \geq 1/2$$

and therefore z generates a non-zero operator $\widetilde{z}$.

Before consider a proof of the implication 2) $\Longrightarrow$ 1) we will make some additional remarks. We collect the remarks in

Lemma 2.1. *Let $s \in (0,1]$, $q \in (0,\infty]$, $1/s = 1 + 1/q$. For $a := (a_k) \in l_1$ and $b := (b_k) \in l_q$ we have*

$$\left(\sum_{k=1}^{\infty} |a_k b_k|^s\right)^{1/s} \leq \sum_{k=1}^{\infty} |a_k| \cdot \left(\sum_{k=1}^{\infty} |b_k|^q\right)^{1/q}. \tag{0.1}$$

Moreover,

$$||a||_{l_1} = \sup_{||b||_{l_q}=1} \left(\sum_{k=1}^{\infty} |a_k b_k|^s\right)^{1/s}$$

(if $q = \infty$, evident changes must be made in (2.1)*).*

Proof of Lemma 2.1. We may consider the case where $q \in (0,\infty)$. Putting $p := 1/s$ (then $1/p' = 1 - s = s/q$ and $sp' = q$), we obtain

$$\sum_{k=1}^{\infty} |a_k b_k|^s \leq \left(\sum_{k=1}^{\infty} |a_k|^{sp}\right)^{1/p} \cdot \left(\sum_{k=1}^{\infty} |b_k|^{sp'}\right)^{1/p'} = \left(\sum_{k=1}^{\infty} |a_k|\right)^{s} \cdot \left(\sum_{k=1}^{\infty} |b_k|^q\right)^{s/q}.$$

For the second part: Let $a = (a_k) \in l_1$. Take $b_k := \frac{|a_k|^{1/q}}{||a||_{l_1}^{1/q}}$. Then

$$\sum_{k=1}^{\infty} |b_k|^q = \sum_{k=1}^{\infty} \frac{|a_k|}{||a||_{l_1}} = 1$$

and

$$\begin{aligned}
\left(\sum_{k=1}^{\infty} |a_k b_k|^s\right)^{1/s} &= \left(\sum_{k=1}^{\infty} \frac{|a_k|^{s/q}}{||a||_{l_1}^{s/q}} |a_k|^s\right)^{1/s} = \left(\sum_{k=1}^{\infty} \frac{|a_k|^{s/q+s}}{||a||_{l_1}^{s/q}}\right)^{1/s} \\
&= \left(\sum_{k=1}^{\infty} \frac{|a_k|^{s(1+1/q)}}{||a||_{l_1}^{s/q}}\right)^{1/s} = \left(\sum_{k=1}^{\infty} \frac{|a_k|}{||a||_{l_1}^{s/q}}\right)^{1/s} \\
&= \frac{(\sum_{k=1}^{\infty} |a_k|)^{1/s}}{||a||_{l_1}^{1/q}} = \left(\sum_{k=1}^{\infty} |a_k|\right)^{1/s-1/q} = ||a||_{l_1}. \qquad \square
\end{aligned}$$

Proof of Theorem 2.1 (continuation). 2) $\Longrightarrow$ 1). Suppose that X does not have the $\widetilde{AP}_s$, $1/s = 1 + 1/q$. Then there is a sequence $(x_n) \in l_q(X)$ (if $q = \infty$, we consider a sequence from $c_0(X) = l_\infty^0(X)$) such that there exists an $\varepsilon > 0$ with the property that for any finite rank operator $R \in X^* \otimes X$ the inequality $\sup_n ||Rx_n - x_n|| > \varepsilon$ is valid. Consider the space $C_0(K; X)$ for $K := \{x_n\}_{n=1}^{\infty} \cup \{0\}$.

Every operator U in X can be considered as a continuous function on K with values in X by setting $f_U(k) := U(k)$ for $k \in K$. In particular, for the identity map id in X and for any $R \in X^* \otimes X$ we have

$$||f_{\mathrm{id}} - f_R||_{C_0(K;X)} \geq \varepsilon.$$

The subset $\mathcal{R} := \overline{\{f_R : R \in X^* \otimes X\}}^{C_0(K;X)}$ of $C_0(K;X)$ is a closed linear subspace in $C_0(K;X)$. So, there exists an X^*-valued measure

$$\mu = (x'_k)_{k=1}^{\infty} \in C_0^*(K;X) = l_1(\{x_n\}_{n=1}^{\infty}) \cup \{0\}; X)$$

such that $\mu|_{\mathcal{R}} = 0$ and $\mu(f_{id}) = 1$. In other words, we can find a sequence (x'_k) with $\sum_{k=1}^{\infty} ||x'_k|| < \infty$ such that $\sum_{k=1}^{\infty} \langle x'_k, x_k \rangle = 1$ and $\sum_{k=1}^{\infty} \langle x'_k, Rx_k \rangle = 0$ for any $R \in X^* \otimes X$.

Define a tensor element $z \in X^* \widehat{\otimes} X$ by $z := \sum_{k=1}^{\infty} x'_k \otimes x_k$. Since $(x_k) \in l_q(X)$ and $(x'_k) \in l_1(X^*)$, we get from Lemma 2.1 that

$$\left(\sum_{k=1}^{\infty} ||x'_k||^s \, ||x_k||^s \right)^{1/s} \leq \sum_{k=1}^{\infty} ||x'_k|| \cdot \left(\sum_{k=1}^{\infty} ||x_k||^q \right)^{1/q}.$$

Therefore, $z \in X^* \widehat{\otimes}_s X$, trace $z = \sum_{k=1}^{\infty} \langle x'_k, x_k \rangle = 1$ and trace $R \circ z = 0$ for every $R \in X^* \otimes X$. This means that X does not have the AP_s. □

After Theorem 2.1 has been proved, we can make a conclusion: $AP_s = \widetilde{AP}_s$ for any $s \in (0, 1]$. Let us mention that this equality appeared firstly (without proofs) in [22, Lemma 2.1].

3

Now we are going to discuss some questions around the properties $\widetilde{AP}_{s,\infty}$ and $AP_{s,\infty}$. The $\widetilde{AP}_{s,\infty}$ was defined above. Recall the definition of the $AP_{s,\infty}$ from, e.g., [23]: We say that a Banach space X has the $AP_{s,\infty}$, $0 < s < 1$, if for every Banach space Y the natural mapping $Y^* \widehat{\otimes}_{s\infty} X \to L(Y, X)$ is one-to-one, where

$$Y^* \widehat{\otimes}_{s\infty} X$$
$$= \left\{ z \in Y^* \widehat{\otimes} X : \ z = \sum_{k=1}^{\infty} \lambda_k y'_k \otimes x_k, \ (x_k) \text{ and } (y'_k) \text{ are bounded, } (\lambda_k) \in l^0_{s\infty} \right\}.$$

Let us consider the connections between the $AP_{s,\infty}$ and the $\widetilde{AP}_{s,\infty}$. For a partial discussion of this we need a lemma, which follows from Lemma 2.1 by interpolation in Lorentz spaces.

Lemma 3.1. *Let $s \in (0, 1), q \in (0, \infty), 1/s = 1 + 1/q, r \in (0, \infty]$. If $a = (a_k) \in l_1$, $b = (b_k) \in l_{qr}$, then $ab := (a_k b_k)_{k=1}^{\infty} \in l_{sr}$. In particular, for $a \in l_1$ and $b \in l_{q\infty}$ the sequence ab is in $l_{s\infty}$ (thus, evidently, in $l^0_{s\infty}$).*

The *proof of Lemma* 3.1 consists of the application of Lemma 2.1 and the general interpolation theorem for the multiplication operator $\widetilde{a}$, induced by a fixed sequence $a = (a_k) \in l_1 : \widetilde{a}$ maps (b_k) to $(a_k b_k)$.

Namely, fix $s \in (0,1), q \in (0,\infty)$ with $1/s = 1 + 1/q$. Take $s_1, s_2 \in (0,1)$ and $q_1, q_2 \in (0,\infty)$ so that for some $\theta \in (0,1)$ we have

$$\frac{1}{q} = (1-\theta)\frac{1}{q_1} + \frac{1}{q_2},\ 0 < \frac{1}{s_2} < \frac{1}{s} < \frac{1}{s_1} < \infty,\ 0 < \frac{1}{q_2} < \frac{1}{q} < \frac{1}{q_1} < \infty,$$

and

$$\frac{1}{s_1} = 1 + \frac{1}{q_1},\ \frac{1}{s_2} = 1 + \frac{1}{q_2}.$$

By Lemma 2.1, $\widetilde{a}$ maps $l_{q_1q_1}$ into $l_{s_1s_1}$ and $\widetilde{a}$ maps $l_{q_2q_2}$ into $l_{s_2s_2}$. Applying, e.g., Theorem 5.3.1 from [1] or other results from the pages 113-114 in [1], we get that $\widetilde{a}$ maps l_{qr} into l_{sr}, $0 < r \le \infty$ (note that $1/s = 1 + 1/q = 1 + (1-\theta)/q_1 + \theta/q_2 = (1-\theta)+\theta+(1-\theta)/q_1+\theta/q_2 = (1-\theta)(1+1/q_1)+\theta(1+1/q_2) = (1-\theta)/s_1+\theta/s_2)$. □

Remark 3.1. As a matter of fact, $l_1 \cdot l_{q\infty} = l_{s1}$ in Lemma 3.1. We need now only the above inclusion.

Now let $t \in (0,1]$, $p \in (0,\infty]$, $r \in (0,\infty]$ and consider a tensor product $\widehat{\otimes}_{t;p,r}$, defined in the following way: For a couple of Banach spaces X, Y the tensor product $Y^*\widehat{\otimes}_{t;p,r}X$ consists of those elements z of the projective tensor product $Y^*\widehat{\otimes}X$ which admit representations of the type

$$z = \sum_{k=1}^{\infty} a_k b_k\, y'_k \otimes x_k;\ (y'_k) \text{ and } (x_k) \text{ are bounded, } (a_k) \in l_t,\ (b_k) \in l_{pr}$$

(recall that everywhere here we consider $l^0_{p\infty}$ in the case $r = \infty$).

Remark 3.2. As was noted in Remark 3.1, $l_1 \cdot l_{q\infty} = l_{s1} (\subset l^0_{s\infty} \subset l_{s\infty})$, where $0 < s < 1, 1/s = 1 + 1/q$. We have also

$$l_{s1} = l_1 \cdot l^0_{q\infty} \text{ and } l_1 \cdot l_{q\infty} = l_1 \cdot l^0_{q\infty}$$

(so, for example, in the definition of $\widehat{\otimes}_{1;q,\infty}$ one can assume that $(a_k) \in l_1$ and $(b_k) \in l^0_{q\infty}$). Indeed, if we use the equality $l_1 \cdot l_{q\infty} = l_{s1}$, take $d \in l_{s1}$ (assuming $d = d^* = (d^*_k)$). Then $\sum_{k=1}^\infty k^{1/s}\, d^*_k/k < \infty$, i.e., $\sum_{k=1}^\infty k^{1/q}\, d^*_k < \infty$. Let $\varepsilon = (\varepsilon_k)$ be a scalar sequence such that $\varepsilon_k \searrow 0$ and $\sum_{k=1}^\infty \varepsilon_k^{-1} d^*_k k^{-1/q} < \infty$. Put

$$\alpha_k := \frac{d^*_k k^{1/q}}{\varepsilon_k},\ \beta_k := \frac{\varepsilon_k}{k^{1/q}}.$$

Then $\alpha := (\alpha_k) \in l_1$ and $\beta := (\beta_k) \in l^0_{q\infty}$. So, $d = \alpha\beta \in l_1 \cdot l^0_{q\infty}$. Another way (not using "l_{s1}"): Let $0 < q < \infty$, $\alpha \in l_1$, $\beta \in l_{q\infty}$ (assuming, without loss of generality, that $\beta = \beta^*$). Consider a sequence $\varepsilon := (\varepsilon_k)$ such that $\varepsilon_k \searrow 0$ and $(\alpha_k/\varepsilon_k) \in l_1$. Put $\widetilde{\alpha} := \alpha/\varepsilon = (\alpha_k/\varepsilon_k)$ and $\widetilde{\beta} := \varepsilon\beta = (\varepsilon_k\beta_k)$. Then $\widetilde{\alpha} \in l_1$, $\widetilde{\beta} \in l^0_{q\infty}$ and $\alpha\beta = \widetilde{\alpha}\widetilde{\beta} \in l_1 \cdot l^0_{q\infty}$.

Let us say that X has the $AP_{t;p,r}$, if for every Banach space Y and for t, p, r as above the canonical mapping $Y^*\widehat{\otimes}_{t;p,r}X \to L(Y,X)$ is one-to-one.

By Lemma 3.1, if $s \in (0,1)$ and $1/s = 1+1/q$, then $\widehat{\otimes}_{1;q,\infty} \subset \widehat{\otimes}_{s,\infty}$. Therefore, we get

Corollary 3.1. *If $s \in (0,1)$ and $1/s = 1+1/q$, then $AP_{s,\infty} \implies AP_{1;q,\infty}$.* □

Evidently, also $AP_{s,\infty} \implies AP_s$ (for $s \in (0,1)$).

Theorem 3.2. *Let $s \in (0,1), q \in (0,\infty)$ and $1/s = 1+1/q$. If X has the $AP_{1;q,\infty}$, then X has the $\widetilde{AP}_{s,\infty}$. In particular, $AP_{s,\infty} \implies \widetilde{AP}_{s,\infty}$.*

Proof. It is enough to repeat word for word the proof of the implication 2) $\implies$ 1) of Theorem 2.1 ("continuation"), just changing "$\mathrm{m}l_q(X)$" by "$l^0_{q,\infty}$" (no necessity to apply Lemma 2.1 or Lemma 3.1). □

Remark 3.3. In this moment (when I am writing the text) I do not know whether the implication "$\widetilde{AP}_{s,\infty} \implies AP_{s,\infty}$" is true, for Banach spaces. Of course, no questions about the cases where $0 < s \leq 2/3$ (but the reason is only that every Banach space has the $\widetilde{AP}_{2/3,\infty}$ and the $AP_{2/3,\infty}$).

Let $0 < r < 1$ and $0 < w \leq \infty$, or $r = 1$ and $0 < w \leq 1$. For Banach spaces X, Y denote by $Y^*\widehat{\otimes}_{(r,w)}X$ the subset of $Y^*\widehat{\otimes}X$ consisting of tensors z such that

$$z = \sum_{k=1}^{\infty} \lambda_k \, y'_k \otimes x_k, \text{ where } (y'_k) \text{ and } (x_k) \text{ are bounded and } (\lambda_k) \in l_{rw}.$$

As was noted in Remark 3.1, if $s \in (0,1), q \in (0,\infty), 1/s = 1+1/q$, then $l_1 \cdot l_{q\infty} = l_{s1}$ (in the sense of the product in Lemma 3.1). In general case, where $0 < q_1, q_2, t_1, t_2 \leq \infty$, one has

$$l_{q_1 t_1} \cdot l_{q_2 t_2} = l_{s,t} \text{ provided that: } \frac{1}{q_1} + \frac{1}{q_2} = \frac{1}{s} \text{ and } \frac{1}{t_1} + \frac{1}{t_2} = \frac{1}{t} \tag{0.2}$$

(cf. [17], 2.1.13 Proposition). We can introduce a new definition of approximation properties, which are connected with Lorentz sequence spaces, namely: Let $0 < r < 1$ and $0 < w \leq \infty$. or $r = 1$ and $0 < w \leq 1$. A Banach space X has the $AP_{(r,w)}$, if for every Banach space Y the natural map $Y^*\widehat{\otimes}_{(r,w)}X \to L(Y,X)$ is one-to-one.

It follows (from Remark 3.1 or from (3.1)) that $AP_{1;q,\infty} = AP_{(s,1)}$ (for $s \in (0,1)$ and $1/s = 1+1/q$) and, more generally, $AP_{t;p,r} = AP_{(s,u)}$ for $1/t+1/p = 1/s$ and $1/t+1/r = 1/u$ $(t \in (0,1])$.

Therefore, we have (for $s \in (0,1)$)

$$AP_{s,\infty} \implies AP_{(s,1)} \implies \widetilde{AP}_{s,\infty}.$$

Moreover, taking into account the equality $\widehat{\otimes}_{1;q,\infty} = \widehat{\otimes}_{(s,1)}$ and applying the arguments from the proof of the implication "$\widetilde{AP}_s \implies AP_s$" of Theorem 2.1, we easily get

Theorem 3.3. $AP_{(s,1)} = \widetilde{AP}_{s,\infty}$.

Proof. As was mentioned above, $AP_{(s,1)} \implies \widetilde{AP}_{s,\infty}$. Let X have the $\widetilde{AP}_{s,\infty}$, i.e., for every sequence $(x_n) \in l^0_{q,\infty}$ (where $1/s = 1 + 1/q$) and every $\varepsilon > 0$ there exists a finite rank operator $R \in X^* \otimes X$ such that $\sup_n ||Rx_n - x_n|| < \varepsilon$. Since $AP_{(s,1)} = AP_{1;q,\infty}$, it is enough to show that if Y is a Banach space, $z \in Y^*\widehat{\otimes}_{1;q,\infty}X$ and $z \neq 0$, then the corresponding operator $\widetilde{z} : Y \to X$ is not zero too.

Let $z = \sum_{k=1}^{\infty} a_k b_k\, y'_k \otimes x_k$ be a representation of z with $(x_k), (y'_k)$ bounded, $(a_k) \in l_1$, $(b_k) \in l^0_{q\infty}$ and $b_k \searrow 0$. Then $(\widetilde{x}_k := b_k x_k) \in l^0_{q\infty}$ and, for an $\varepsilon > 0$ small enough (to be chosen), we can find an operator $R \in X^* \otimes X$ with the property that $\sup_n ||R\widetilde{x}_n - \widetilde{x}_n|| \leq \varepsilon$. Since $z \neq 0$, we can find an operator $V \in L(Y^*, X^*)$ such that $\sum_{k=1}^{\infty} a_k \langle V y'_k, \widetilde{x}_k \rangle = 1$. Now, when V is chosen, we have

$$1 = \sum_{k=1}^{\infty} a_k \langle V y'_k, \widetilde{x}_k - R\widetilde{x}_k \rangle + \sum_{k=1}^{\infty} a_k \langle V y'_k, R\widetilde{x}_k \rangle$$
$$\leq \varepsilon\, ||(a_k)||_{l_1}\, ||V|| \cdot \mathrm{const} + \left| \sum_{k=1}^{\infty} a_k b_k \langle R^* V y'_k, x_k \rangle \right|,$$

and, if ε is small enough, we get for the finite rank operator $R^*V : Y^* \to X^*$ that

$$|\,\mathrm{trace}\, z^t \circ (R^*V)| = |\,\mathrm{trace}\, (R^*V) \circ z^t| = \left| \sum_{k=1}^{\infty} a_k b_k \langle R^* V y'_k, x_k \rangle \right| > 0.$$

The last sum is the nuclear trace of the tensor element $\sum_{k=1}^{\infty} a_k b_k\, R^*V y'_k \otimes x_k$, which is a composition $R \circ z_0$ of the finite rank operator R and the tensor element $\sum_{k=1}^{\infty} a_k b_k\, V y'_k \otimes x_k$, that belongs to the tensor product $X^*\widehat{\otimes}_{1;q,\infty}X$. It follows that both z_0 and z generate the non-zero operators $\widetilde{z}_0$ and $\widetilde{z}$. □

Remark 3.4. Because of the equality $\widehat{\otimes}_{1;q,\infty} = \widehat{\otimes}_{(s,1)}$, it follows from the proof of Theorem 3.3 that X has the $AP_{(s,1)}$ iff the canonical mapping $X^*\widehat{\otimes}_{(s,1)}X \to L(X)$ is one-to-one (just like in the case of the classical Grothendieck approximation property).

Remark 3.5. Of course, it follows from Theorem 3.3 that every Banach space has the $AP_{(2/3,1)}$, but it is trivial because of the implication

$$AP^0_{(2/3,\infty)} \equiv AP_{2/3,\infty} \implies AP_{(2/3,w)} \text{ for any } w < \infty$$

(and, again, since every X has the $AP_{2/3,\infty}$!).

Our question in Remark 3.3 can be reformulated now as:

$$\text{Is it true that the } AP_{(s,1)} \text{ implies the } AP_{s,\infty}? \qquad (*)$$

4 Let us consider an application of the previous considerations. Now we know, in particular, that every Banach space has the $AP_{(2/3,1)}$. On the other hand, the corresponding operator ideal $N_{(2/3,1)}$ (related to the Lorentz space $l_{2/3\,1}$) has the eigenvalue type l_1 (see, e.g., [8, p. 243]). Since the continuous trace is unique on $\widehat{\otimes}_{(2/3,1)}$ and $\widehat{\otimes}_{(2/3,1)} = N_{(2/3,1)}$, it follows from White's results [32] that for each

Banach space X and for every operator $T \in N_{(2/3,1)}(X,X)$ the (nuclear) trace of T is well defined and equals the sum of all eigenvalues of T :

$$\text{trace } T = \sum_{k=1}^{\infty} \mu_k(T) \text{ (eigenvalues)} \,\forall\, X, \ \forall T \in N_{(2/3,1)}(X)$$

(on the right is the so-called "spectral sum" of T). More precisely, the last statement follows from Theorem 4.1 below.

Let us explain in more details how we apply a result of M.C. White. To do this, we formulate and prove a theorem which is almost an immediate consequence of White's theorem.

Theorem 4.1. *Let A be a quasi-Banach operator ideal, X be a Banach space, for which the set of all finite rank operators is dense in the space $A(X)$. Suppose that the natural functional "trace" is bounded on the subspace of all finite rank operators of $A(X)$ (and, therefore, can be extended to a continuous functional "trace$_A$" on the whole space $A(X)$). If the quasi-Banach operator ideal A is of eigenvalue type l_1, then the spectral trace (= "spectral sum") is continuous on the space $A(X)$ and for any operator $T \in A(X)$ we have*

$$\text{trace}_A(T) = \sum_{n=1}^{\infty} \mu_n(T).$$

where $(\mu_n(T))_{n=1}^{\infty}$ is the sequence of all eigenvalues of T counted according to their multiplicities.

Proof. Let $T \in A(X)$. By the assumption, the sequence $\{\mu_n(T)\}_{n=1}^{\infty}$ of all eigenvalues of T (counted according to their multiplicities) is in l_1. Since the quasi-normed ideal A is of spectral (= eigenvalue) type l_1, we can apply the main result from the paper [32] of M.C. White, which asserts:

If J is a quasi-Banach operator ideal with eigenvalue type l_1, then the spectral sum is a trace on the ideal J. $({}^{*}{}_{*}{}^{*})$

Recall (see [18], 6.5.1.1, or Definition 2.1 in [32]) that a *trace* on an operator ideal J is a class of complex-valued functions, all of which they write as τ, one for each component $J(E,E)$ (where E is a Banach space) so that

(i) $\tau(e' \otimes e) = \langle e', e \rangle$ for all $e' \in E^*, e \in E$;
(ii) $\tau(AU) = \tau(UA)$ for all Banach spaces F and operators $U \in J(E,F)$ and $A \in L(F,E)$;
(iii) $\tau(S+U) = \tau(S) + \tau(U)$ for all $S, U \in J(E,E)$;
(iv) $\tau(\lambda U) = \lambda \tau(U)$ for all $\lambda \in \mathbb{C}$ and $U \in J(E,E)$.

Our operator T belongs to the space $A(X,X) = A(X)$ and A is of eigenvalue type l_1. Thus, the assertion $({}^{*}{}_{*}{}^{*})$ implies that the spectral sum μ defined by $\mu(U) := \sum_{n=1}^{\infty} \mu_n(U)$ for $U \in A(E,E)$ is a trace on A.

By the principle of uniform boundedness (see [17], 3.4.6 (page 152), or [15]), there exists a constant $C > 0$ such that

$$|\mu(U)| \leq ||\{\mu_n(U)\}||_{l_1} \leq C\, a(U)$$

for all Banach spaces E and operators $U \in A(E, E)$.

Now, remembering that all operators in $A(X)$ can be approximated by finite rank operators and taking in account the conditions (iii)–(iv) for $\tau = \mu$, we obtain that the A-trace, i.e., $\mathrm{trace}_A\, T$, of our operator T coincides with $\mu(T)$ (recall that the continuous trace is uniquely defined in such a situation, that is on the space $A(X)$; cf. [18], 6.5.1.2). □

Since $\widehat{\otimes}_{1;2,\infty} = \widehat{\otimes}_{(2/3,1)}$ (see Theorem 3.3), we can reformulate the result, which we considered in the very beginning of this section, as

Corollary 4.1. *For each Banach space X and for any operator $T \in N_{1;2,\infty}(X)$ the nuclear trace of T is well defined and equals the sum of all eigenvalues of T :*

$$\mathrm{trace}\, T = \sum_{k=1}^{\infty} \mu_k(T)\ (\mathit{eigenvalues})\, \forall\, X,\ \ \forall T \in N_{(1;2,\infty)}(X). \qquad \square$$

Remark 4.1. Recall that A. Grothendieck [7] has obtained the last assertion for the case of 2/3-nuclear operators, i.e., for the ideal $N_{2/3} = N_{(2/3,\, 2/3)}$ (note that $l_{2/3} \subset l_{2/3\, 1}$).

5

The discussion in Section 1 shows that, for $p \in [1, \infty]$, any subspace of any quotient (= any quotient of any subspace) of an L_p-space possesses the $\widetilde{AP}_s$ (even the $\widetilde{AP}_{s,\infty}$) with $1/s = 1 + |1/2 - 1/p|$. We apply now these facts together with White's theorem for proving some more theorems concerning the distributions of eigenvalues of the nuclear operators. Below we will use Theorem 2.1 and, therefore, the fact that any subspace of any quotient of an L_p-space possesses the AP_s (where p, s as above). Thus, for such Banach spaces X, we can identify the tensor product $X^*\widehat{\otimes}_s X$ with its canonical image in the space $L(X) = L(X, X)$, i.e., with the space $N_s(X)$ of all s-nuclear operators in X, equipped with the quasi-norm induced from $X^*\widehat{\otimes}_s X$.

We give below relatively simple proofs of some recent results from the papers [25] and [28].

Theorem 5.1. *Let X be a subspace of an L_p-space, $1 \leq p \leq \infty$. If $T \in N_s(X, X)$, where $1/s = 1 + |1/2 - 1/p|$, then*

1. *the (nuclear) trace of T is well defined,*
2. *$\sum_{n=1}^{\infty} |\mu_n(T)| < \infty$, where $\{\mu_n(T)\}$ is the system of all eigenvalues of the operator T (written in according to their algebraic multiplicities) and*

$$\mathrm{trace}\, T = \sum_{n=1}^{\infty} \mu_n(T).$$

Proof. Let X be a subspace of an L_p-space $L_p(\mu)$ and $T \in N_s(X,X)$ with an s-nuclear representation

$$T = \sum_{k=1}^{\infty} \lambda_k x'_k \otimes x_k,$$

where $||x'_k||, ||x_k|| = 1$ and $\lambda_k \geq 0$, $\sum_{k=1}^{\infty} \lambda_k^s < \infty$. By Hahn–Banach, we can find the functionals $\widetilde{x}'_k \in L_p^*(\mu)$ $(k = 1, 2, \dots)$ with the same norms as for the corresponding functionals x'_k and so that $\widetilde{x}'_k|_X = x'_k$ for every k. Denote by $\widetilde{T}$ the operator

$$\widetilde{T} : L_p(\mu) \to X,\ \widetilde{T} := \sum_{k=1}^{\infty} \lambda_k \widetilde{x}'_k \otimes x_k,$$

and by $j : X \to L_p(\mu)$ the natural injection. Since the space X has the property AP_s, we have $N_s(L_p(\mu), X) = L_p^*(\mu)\widehat{\otimes}_s X$ and, therefore, the nuclear traces of the operators $j\widetilde{T}$ and $\widetilde{T}j$ are well defined. We have a diagram

$$X \xrightarrow{j} L_p(\mu) \xrightarrow{\widetilde{T}} X \xrightarrow{j} L_p(\mu),$$

in which $\widetilde{T}j = T \in N_s(X)$. Hence, the complete systems of eigenvalues of the operators $T = \widetilde{T}j$ and $j\widetilde{T} \in N_s(L_p(\mu))$ coincide. Applying Theorem 2.b.13 from [10] (see also [25]), we obtain that the sequence $(\mu_k(j\widetilde{T}))$ is summable. Therefore, we have $\mu_k(T) \in l_1$ and we can apply Theorem 4.1. But we apply the theorem firstly for the simplest case (later on we will continue the proof of our Theorem 5.1).

The first assertion of the next theorem is due to A. Grothendieck [7], the second one was proved by H. König in [11]. Surprisingly, but we could not find anywhere the main statement of the theorem about coincidence of the nuclear and spectral traces, neither in the monographs, nor in the mathematical journals. So we have no reference for this statement and have to formulate and to prove the next theorem here. Let us remark that, in any case, this theorem was proved (as a partial case of the proved there our Theorem 5.1) in [25].

Theorem 5.1′. *Let L be an L_p-space, $1 \leq p \leq \infty$. If $T \in N_s(L, L)$, where $1/s = 1 + |1/2 - 1/p|$, then*

1. *the (nuclear) trace of T is well defined,*
2. *$\sum_{n=1}^{\infty} |\mu_n(T)| < \infty$, where $\{\mu_n(T)\}$ is the system of all eigenvalues of the operator T (written in according to their algebraic multiplicities) and*

$$\text{trace}\ T = \sum_{n=1}^{\infty} \mu_n(T).$$

Proof. As we have said above, the assertions 1 and 2 are well known. To prove the last equality, consider the Banach operator ideal $\mathcal{L}_p$ of all operators which can be factored through L_p-spaces. Then the product $\mathcal{L}_p \circ N_s$ is a quasi-Banach operator ideal of spectral (=eigenvalue) type l_1 (e.g., by the assertion 2, proved earlier by H. König [11]). Now it is enough to apply Theorem 4.1 to finish the proof. □

Proof of Theorem 5.1 (continuation). As we have said, the complete systems of eigenvalues of the operators $T = \widetilde{T}j$ and $j\widetilde{T} \in N_s(L_p(\mu))$ coincide. By Theorem 5.1′,

$$\operatorname{trace} j\widetilde{T} = \sum_{k=1}^{\infty} \lambda_k \langle \widetilde{x}'_k, jx_k \rangle = \sum_{n=1}^{\infty} \mu_n(j\widetilde{T}),$$

the last sum is equal to

$$\sum_{n=1}^{\infty} \mu_n(T)$$

and the sum in the middle is

$$\sum_{k=1}^{\infty} \lambda_k \langle \widetilde{x}'_k, jx_k \rangle = \sum_{k=1}^{\infty} \lambda_k \langle x'_k, x_k \rangle = \operatorname{trace} T.$$

The (nuclear) trace of the operator T is well defined, because the space X has the AP_s. Therefore,

$$\operatorname{trace} T = \sum_{n=1}^{\infty} \mu_n(T),$$

and we are done. □

If Y is a quotient of an L_p-space, then, for a compact operator $U \in L(E,E)$, the adjoint U^* is also a compact operator and these two operators have the same eigenvalues $\mu \neq 0$ with the same multiplicities (see, e.g., [17], Theorem 3.2.26, or [5], Exercise VII.5.35). Also, any quotient of an L_p-space has the AP_s (where p, s are as above). So, it follows immediately from the just proved Theorem 5.1

Corollary 5.1. *Let Y be a quotient of an L_p-space, $1 \le p \le \infty$. If $T \in N_s(Y,Y)$, where $1/s = 1 + |1/2 - 1/p|$, then*

1. *the (nuclear) trace of T is well defined,*
2. *$\sum_{n=1}^{\infty} |\mu_n(T)| < \infty$, where $\{\mu_n(T)\}$ is the system of all eigenvalues of the operator T (written in according to their algebraic multiplicities) and*

$$\operatorname{trace} T = \sum_{n=1}^{\infty} \mu_n(T).$$
□

We used above some facts from Section 1. After Theorem 5.1 and its consequence have been proved, we are ready to present a simple prove of the corresponding result on the subspaces of quotients of the L_p-spaces (recall that, again, all such spaces have the AP_s with s and p satisfying the same conditions).

Theorem 5.2. *Let W be a quotient of a subspace (= a subspace of a quotient) of an L_p-space, $1 \le p \le \infty$. If $T \in N_s(W,W)$, where $1/s = 1 + |1/2 - 1/p|$, then*

1. *the (nuclear) trace of T is well defined,*

2. $\sum_{n=1}^{\infty} |\mu_n(T)| < \infty$, *where* $\{\mu_n(T)\}$ *is the system of all eigenvalues of the operator* T *(written in according to their algebraic multiplicities) and*

$$\text{trace } T = \sum_{n=1}^{\infty} \mu_n(T).$$

Proof. Let $L_p(\mu)$ be an L_p-space. Take Banach subspaces $X_0 \subset X \subset L_p(\mu)$ and consider the quotient X/X_0.

If $T \in N_s(X/X_0, X/X_0)$ $(=(X/X_0)^* \widehat{\otimes}_s X/X_0)$, then T admits a factorization of the type

$$X/X_0 \overset{A}{\to} c_0 \overset{D}{\to} l_1 \overset{B}{\to} X/X_0,$$

where A, B are continuous and D is a diagonal operator with a diagonal from l_s. Denoting by $\varphi : X \to X/X_0$ the factor map from X onto X/X_0 and taking a lifting $\Phi : l_1 \to X$ for B with $B = \varphi\Phi$, we obtain that the maps $\varphi\Phi DA : X/X_0 \to X/X_0$ and $\Phi DA\varphi : X \to X$ have the same eigenvalues $\mu \neq 0$ with the same multiplicities:

$$X \overset{\varphi}{\to} X/X_0 \overset{A}{\to} c_0 \overset{D}{\to} l_1 \overset{\Phi}{\to} X \overset{\varphi}{\to} X/X_0.$$

The spaces X and X/X_0 have the AP_s. Therefore, we have (cf. the proof of Theorem 5.1)

$$\text{trace } \varphi\Phi DA = \text{trace } \Phi DA\varphi.$$

Since X is a subspace of $L_p(\mu)$, we have, by Theorem 5.1,

$$\text{trace } \Phi DA\varphi = \sum_{n=1}^{\infty} \mu_n(\Phi DA\varphi).$$

Therefore,

$$\text{trace } T = \text{trace } BDA = \text{trace } \varphi\Phi DA = \sum_{n=1}^{\infty} \mu_n(\Phi DA\varphi)$$

$$= \sum_{n=1}^{\infty} \mu_n(\varphi\Phi DA) = \sum_{n=1}^{\infty} \mu_n(BDA) = \text{trace } T. \qquad \square$$

6

As is well known, in the classical case of the Grothendieck approximation property *AP*, if X^* has the *AP* then the space X also has this property. We will show now that the same is true for all approximation properties which are under consideration in this paper.

Denote by $\widehat{\otimes}_\alpha$ any of the tensor product $\widehat{\otimes}_s$, $\widehat{\otimes}_{s,\infty}$, $\widehat{\otimes}_{t;p,r}$, $\widehat{\otimes}_{(r,w)}$ with the parameters (see above), for which all these tensor products are the linear subspaces of the projective tensor product $\widehat{\otimes}$. Also, let us say that a Banach space X has the AP_α, if it is possesses the corresponding approximation property (i.e., AP_s, $AP_{s,\infty}$ etc.).

We need the following auxiliary result which may be of its own interest (compare with Remark 3.4).

Proposition 6.1. *A Banach space X has the AP_α if and only if the canonical map $X^*\widehat{\otimes}_\alpha X \to L(X)$ is one-to-one.*

Proof. Suppose that the canonical map $X^*\widehat{\otimes}_\alpha X \to L(X)$ is one-to-one, but there exists a Banach space Y such that the natural map $Y^*\widehat{\otimes}_\alpha X \to L(Y,X)$ is not injective. Let $z \in Y^*\widehat{\otimes}_\alpha X$ be such that $z \neq 0$ and the associated operator $\widetilde{z}$ is a 0-operator. Then we can find an operator V from $L(Y^*, X^*)$ (the dual space to the projective tensor product $Y^*\widehat{\otimes}X$) so that trace $V \circ z^t = 1$, where, as usual, z^t is the transposed tensor element, $z * t \in X\widehat{\otimes}Y^*$. Since $V \circ z^t \in X\widehat{\otimes}X^*$ and trace $V\circ z^t = 1$, the tensor element $(V\circ z^t)^t$ (which, evidently, belongs to $X^*\widehat{\otimes}_\alpha X$) is not zero. On the other hand, the operator induced by this element must be a 0-operator. Contradiction. □

Proposition 6.2. *With the above understanding, if the dual space Y^* has the AP_α, then Y has the AP_α too.*

Proof. We use Proposition 6.1. As it is known [7], the projective tensor product $Y^*\widehat{\otimes}Y$ is a Banach subspace of the tensor product $Y^*\widehat{\otimes}Y^{**}$. The tensor product $Y^*\widehat{\otimes}_\alpha Y$ is a linear subspace of $Y^*\widehat{\otimes}Y$, as well as $Y^*\widehat{\otimes}_\alpha Y^{**}$ is a linear subspace of $Y^*\widehat{\otimes}Y^{**}$. Therefore, the natural map $Y^*\widehat{\otimes}_\alpha Y \to Y^*\widehat{\otimes}_\alpha Y^{**}$ is one-to-one. Now if Y^* has the AP_α, then the canonical map $Y^{**}\widehat{\otimes}_\alpha Y^* \to L(Y^*, Y^*)$ is one-to-one. Since we can identify the tensor product $Y^{**}\widehat{\otimes}_\alpha Y^*$ with the tensor product $Y^*\widehat{\otimes}_\alpha Y^{**}$ (because of the "symmetries" in the definitions of the corresponding tensor products), it follows that the natural map $Y^*\widehat{\otimes}_\alpha Y \to L(Y,Y)$ is one-to-one. Thus, if Y^* has the AP_α, then Y has the AP_α too. □

Remark 6.1. The inverse statement is not true. For example, if $s \in (2/3, 1]$, then there exists a Banach space, possessing the Grothendieck approximation property, whose dual does not have the AP_s (it is well known for the case where $s = 1$). Moreover, if $s \in (2/3, 1]$, then we can find a Banach space W such that W has a Schauder basis and W^* does not have the AP_s. Indeed, let E be a separable reflexive Banach space without the AP_s (see [21] or [23]). Let Z be a separable space such that Z^{**} has a basis and there exists a linear homomorphism φ from Z^{**} onto E^* so that the subspace $\varphi^*(E)$ is complemented in Z^{***} and, moreover, $Z^{***} \cong \varphi^*(E) \oplus Z^*$ (see [13, Proof of Corollary 1]). Put $W := Z^{**}$. This (second dual) space W has a Schauder basis and its dual W^* does not have the AP_s.

7

Let us consider some more notions of the approximation properties associated with some other tensor products. For Banach spaces X and Y and $r \in (0,1], p \in [1,2]$, define a quasi-norm $||\cdot||_{N_{[r,p]}}$ on the tensor product $X^* \otimes Y$ by

$$\|u\|_{N_{[r,p]}} := \inf \left\{ \|(x_i')_{i=1}^n\|_{\ell_r(X^*)} \cdot \|(y_i)_{i=1}^n\|_{\ell^w_{p'}(Y)} \, : \, u = \sum_{i=1}^n x_i' \otimes y_i \right\}.$$

Here we denote, as usual, by $l_r(X^*)$ and $l^w_q(Y)$ the spaces of r-absolutely summable and weakly q-summable sequences, respectively.

Denote by $X^*\widehat{\otimes}_{[r,p]}Y$ the completion of the space $\left(X^*\otimes Y,\ \|\cdot\|_{N_{[r,p]}}\right)$. We have a natural continuous injection

$$j_{[r,p]}: X^*\widehat{\otimes}_{[r,p]}Y \to X^*\widehat{\otimes}Y$$

with $||j_{[r,p]}|| \leq 1$.

Every element $u \in X^*\widehat{\otimes}_{[r,p]}Y$ has a representation of the type $u = \sum_{i=1}^{\infty} x_i' \otimes y_i$, where $(x_i')_{i=1}^{\infty} \in \ell_r(X^*)$ and $(y_i)_{i=1}^{\infty} \in \ell_{p'}^w(Y)$. Consider the natural mappings

$$X^*\widehat{\otimes}_{[r,p]}Y \stackrel{j_{[r,p]}}{\to} X^*\widehat{\otimes}Y \stackrel{j}{\to} L(X,Y).$$

The image of the tensor product $X^*\widehat{\otimes}_{[r,p]}Y$ under the composition $\widetilde{j}_{[r,p]} := j \circ j_{[r,p]}$ is denoted by $N_{[r,p]}(X,Y)$. This is a quasi-Banach space of the (r,p)-nuclear operators (the quasi-norm is induced from the tensor product $X^*\widehat{\otimes}_{[r,p]}Y$). It is not difficult to see that every operator $T \in N_{[r,p]}(X,Y)$ admits a factorization of the kind

$$X \stackrel{A}{\to} c_0 \stackrel{D_r}{\to} l_1 \stackrel{i}{\to} l_p \stackrel{B}{\to} Y, \tag{0.3}$$

where A, B are compact, i is the injection, D_r is a diagonal operator with a diagonal from l_r. Also, every operator, which can be factored in such a way, is in $N_{[r,p]}(X,Y)$.

By the analogous way, we define the tensor product $X^*\widehat{\otimes}^{[r,p]}Y$ and the quasi-normed operator ideals $N^{[r,p]}(X,Y)$. Namely, $X^*\widehat{\otimes}^{[r,p]}Y$ is a linear subspace of the projective tensor product $X^*\widehat{\otimes}Y$, consisting of tensor elements z which admit a representation

$$u = \sum_{i=1}^{\infty} x_i' \otimes y_i,$$

where $(x_i')_{i=1}^{\infty} \in \ell_{p'}^w(X^*)$ and $(x_i)_{i=1}^{\infty} \in \ell_r(Y)$. Its canonical image in $L(X,Y)$ is the quasi-normed space $N^{[r,p]}(X,Y)$. It is not difficult to see that every operator $T \in N^{[r,p]}(X,Y)$ admits a factorization of the kind

$$X \stackrel{A}{\to} l_{p'} \stackrel{D_r}{\to} c_0 \stackrel{i}{\to} l_1 \stackrel{B}{\to} Y,$$

where A, B are compact, i is the injection, D_r is a diagonal operator with a diagonal from l_r. Also, every operator, which can be factored in such a way, is in $N^{[r,p]}(X,Y)$.

It is clear that $T \in N_{[r,p]}(X,Y)$ implies $T^* \in N^{[r,p]}(Y^*,X^*)$ and $T \in N^{[r,p]}(X,Y)$ implies $T^* \in N_{[r,p]}(Y^*,X^*)$. Inverse is not true (see, e.g., Example 8.3 below).

Now we can define the notions of the corresponding approximation properties by the usual way. We say that the space X has the $AP_{[r,p]}$ (respectively, the $AP^{[r,p]}$) if for every Banach space Y the natural mapping $Y^*\widehat{\otimes}_{[r,p]}X \to L(Y,X)$ (respectively, $Y^*\widehat{\otimes}^{[r,p]}X \to L(Y,X)$) is one-to-one. It can be seen that a Banach space X has the $AP_{[r,p]}$ (or $AP^{[r,p]}$) iff the canonical map $X^*\widehat{\otimes}_{[r,p]}X \to L(X)$ (or $X^*\widehat{\otimes}^{[r,p]}X \to L(X)$) is one-to-one (the proof is essentially the same as the proof of Theorem 6.1). Also, if X^* has the $AP_{[r,p]}$ (or $AP^{[r,p]}$) then X has the $AP^{[r,p]}$ (or $AP_{[r,p]}$) (the proof is the same as in Theorem 6.2).

Theorem 7.1. *Let $1/r - 1/p = 1/2$. Every Banach space has the properties $AP_{[r,p]}$ and $AP^{[r,p]}$.*

Proof. Suppose that $X \notin AP_{[r,p]}$ where $1/r - 1/p = 1/2$. Let $z \in X^* \widehat{\otimes}_{[r,p]} X$ be an element such that trace $z = 1, \tilde{z} = 0$. Since $z = \sum x'_k \otimes x_k$, where $(x'_k) \in l_r(X^*)$ and (x_k) is weakly p'-summable, the operator $\tilde{z}$ can be factored as

$$\tilde{z} : X \xrightarrow{A} l_\infty \xrightarrow{\Delta} l_1 \xrightarrow{j} l_p \xrightarrow{V} X,$$

where all the operators are continuous, j is an injection, Δ is a diagonal operator with a diagonal from l_r. Since $\tilde{z} = 0$, we have $V|_{j\Delta A(X)} = 0$. Consider $S := j\Delta AV : l_p \to l_p$. Evidently, $S^2 = 0$ and trace $S =$ trace $z = 1$. Since $S \in N_r(l_p, l_p)$, its nuclear trace equals the sum of all its eigenvalues (see Theorem 5.1′ above). This contradicts the fact that $S^2 = 0$. Now, let Y be another Banach space and put $X := Y^*$. We have shown that X has the $AP_{[r,p]}$. Therefore (see remarks before the formulation of Theorem 7.1), Y has the $AP^{[r,p]}$. □

We are ready to apply the above results to the investigation of eigenvalues problems for $N_{[r,p]}$- and $N^{[r,p]}$-operators. The first theorem below was proved in [26] by using Fredholm Theory. The same proof can be applied for the second theorem. Below we present very different simple proofs of them.

Theorem 7.2. *Let $1/r - 1/p = 1/2$. For every Banach space X and every operator $T \in N_{[r,p]}(X)$, trace T is well defined and if $(\mu_i)_{i=1}^\infty$ is a system of all eigenvalues of T, then $(\mu_i)_{i=1}^\infty \in l_1$ and*

$$\text{trace } T = \sum_{i=1}^\infty \mu_i.$$

Theorem 7.3. *Let $1/r - 1/p = 1/2$. For every Banach space X and every operator $T \in N^{[r,p]}(X)$, trace T is well defined and if $(\mu_i)_{i=1}^\infty$ is a system of all eigenvalues of T, then $(\mu_i)_{i=1}^\infty \in l_1$ and*

$$\text{trace } T = \sum_{i=1}^\infty \mu_i.$$

Both theorems can be proved by the analogues methods and the proofs are almost the same as the proof of Theorem 5.2 (by using Theorem 7.1). On the other hand, Theorem 7.3 is a consequence of Theorem 7.2 and vice versa. Let us give, firstly, a simple proof of Theorem 7.2 and then deduce Theorem 7.3.

Proof. Suppose that $T \in N_{[r,p]}(X)$ and consider a factorization (7.1) (in which $Y = X$) of the operator T. The sequence (μ_k) of all eigenvalues of T is the same as the sequence of all eigenvalues of the operator iD_rAB, which maps l_p into l_p. The last operator is r-nuclear, where $p \in [1,2]$ and $1/r = 1/p + 1/2$. By Theorem 5.1′,

$(\mu_k) \in l_1$ and trace $iD_r AB = \sum \mu_k$. Since the trace of T is well defined (Theorem 7.1), it is clear that trace $T =$ trace $iD_r AB = \sum \mu_k$.

Now, let $T \in N^{[r,p]}(X)$. Then $T^* \in N_{[r,p]}(X^*)$ and trace $T^* =$ trace T (apply Theorem 7.1). Since the operators T and T^* have the same systems of eigenvalues, Theorem 7.3 follows from just proved statement of Theorem 7.2. □

8

The next examples are taken from [26], where one can find the corresponding proofs. They show that all the above affirmative results concerning approximation properties and trace-formulas are sharp.

Example 8.1. Let $r \in (2/3, 1], p \in (1, 2], 1/r - 1/p = 1/2$. There exist Banach spaces E and V, $z_0 \in E^* \widehat{\otimes} V, S \in L(V, E)$ so that for every $p_0 \in [1, p)$

1) $z_0 \in E^* \widehat{\otimes}_{[r,1]} V$;
2) V has a basis;
3) V is the space of type p_0 and of cotype 2;
4) $S \circ z_0 \in E^* \widehat{\otimes}_{[r,p_0]} E$;
5) trace $S \circ z_0 = 1$;
6) the corresponding operator $\widetilde{S \circ z_0}$ is a 0-operator and, therefore, has no nonzero eigenvalues.

Example 8.2. Let $r \in [2/3, 1), p \in [1, 2), 1/r - 1/p = 1/2$. There exist Banach spaces E and V, $z_0 \in E^* \widehat{\otimes} V, S \in L(V, E)$ so that for every $\epsilon > 0$

1) $z_0 \in E^* \widehat{\otimes}_{[r+\epsilon,1]} V$;
2) V has a basis;
3) $S \circ z_0 \in E^* \widehat{\otimes}_{[r+\epsilon,p]} E$;
4) trace $S \circ z_0 = 1$;
5) the corresponding operator $\widetilde{S \circ z_0}$ is a 0-operator and therefore, has no nonzero eigenvalues.

Example 8.3. Let $r \in (2/3, 1]$, $p \in (1, 2], 1/r - 1/p = 1/2$. There exist two separable Banach spaces X and Z so that

(i) Z^{**} has a basis;
(ii) $\exists\, V \in X^* \widehat{\otimes} Z^{**} : V = \sum_{k=1}^{\infty} x'_k \otimes z''_k$; (x'_k) weakly p'_0-summable for each $p_0 \in [1, p)$; $(z''_k) \in l_r(Z^{**})$;
(iii) $V(X) \subset Z$; the operator V is not nuclear as a map from X into Z. Moreover, there exists an operator $U : Z^{**} \to Z$ such that
   (α) $\pi_Z U \in N^{[r,p_0]}(Z^{**}, Z^{**}) = Z^{***} \widehat{\otimes}^{[r,p_0]} Z^{**}$, $\forall\, p_0 \in [1, p)$;
   (β) U is not nuclear as a map from Z^{**} into Z;
   (γ) trace $\pi_Z U = 1$;
   (δ) $\pi_Z U : Z^{**} \to Z^{**}$ has no nonzero eigenvalues.

Acknowledgment

The author would like to thank the referee for helpful remarks and Alexander Alenitsyn for his technical help.

References

[1] J. Bergh, J. Löfström, *Interpolation spaces.* Berlin, Heidelberg, New York: Springer-Verlag, 1976.

[2] J. Bourgain, O.I. Reinov, *On the approximation properties for the space* H^∞. Math. Nachr. **122**, No. 2 (1985), 19–27.

[3] A.M. Davie, *The approximation problem for Banach spaces.* Bull. London Math. Soc. **5** (1973), 261–266.

[4] J. Diestel, H. Jarchow, A. Tonge, *Absolutely summing operators.* Cambridge Univ. Press, 1995.

[5] N. Dunford, J. Schwartz, *Linear operators, Part I.* Interscience, New York, 1958.

[6] P. Enflo, *A counterexample to the approximation property in Banach spaces.* Acta Math. **130** (1973), 309–317.

[7] A. Grothendieck, *Produits tensoriels topologiques et éspaces nucléaires.* Mem. Amer. Math. Soc. **16** (1955).

[8] A. Hinrichs, A. Pietsch, *p-nuclear operators in the sense of Grothendieck.* Math. Nachr. **283**, No. 2 (2010), 232–261.

[9] M.J. Kadec, M.G. Snobar, *Certain functionals on the Minkowski compactum* (Russian). Mat. Zametki **10** (1971), 453–458.

[10] H. Konig, *Eigenvalue distribution of compact operators.* Birkhäuser, 1986.

[11] H. Konig, *On the eigenvalue spectrum of certain operator ideals.* Coll. Math. **44** (1981), 15–28.

[12] V.B. Lidskiĭ, *Nonselfadjoint operators having a trace* (Russian). Dokl. Akad. Nauk SSSR **125** (1959), 485–487.

[13] J. Lindenstrauss, *On James' paper "Separable Conjugate Spaces".* Israel J. Math. **9** (1971), 279–284.

[14] J. Lindenstrauss, L. Tzafriri, *Classical Banach spaces, vol.* 1*: Sequence spaces.* Berlin-Heidelberg-New York, 1977.

[15] A. Pietsch, *Eigenwertverteilungen von Operatoren in Banachräumen.* Hausdorff-Festband: Theory of sets and topology, Berlin: Akademie-Verlag (1972), 391–402.

[16] A. Pietsch, *Operator Ideals.* North-Holland, 1980.

[17] A. Pietsch, *Eigenvalues and s-numbers.* Cambridge Univ. Press, 1987.

[18] A. Pietsch, *History of Banach spaces and linear operators.* Birkhäuser, 2007.

[19] G. Pisier, *Estimations des distances à un espace euclidien et des constantes de projéction des espaces de Banach de dimensions finie.* Séminaire d'Analyse Fonctionelle 1978–1979, Centre de Math., Ecole Polytech., Paris (1979), exposé 10, 1–21.

[20] O.I. Reinov, *A simple proof of two theorems of A. Grothendieck* (Russian). Vestn. Leningr. Univ. **7** (1983), 115–116.

[21] O.I. Reinov, *Disappearance of tensor elements in the scale of p-nuclear operators* (Russian). Theory of Operators and Theory of Functions, Leningrad, LGU, vol. 1 (1983), 145–165.

[22] O.I. Reinov, *A survey of some results in connection with Grothendieck approximation property.* Math. Nachr. **119** (1984), 257–264.

[23] O.I. Reinov, *Approximation properties AP_s and p-nuclear operators* (*the case* $0 < s \leq 1$). Journal of Mathematical Sciences **115**, No. 3 (2003), 2243–2250.

[24] O.I. Reinov, *Some more remarks on Grothendieck–Lidskiĭ trace formulas.* Journal of Prime Research in Mathematics **8** (2012), 148–154.

[25] O. Reinov, Q. Latif, *Grothendieck–Lidskiĭ theorem for subspaces of L_p-spaces.* Math. Nachr. **286**, No. 2-3 (2013), 279–282.

[26] O. Reinov, Q. Latif, *Distribution of eigenvalues of nuclear operators and Grothendieck–Lidskiĭ type formulas.* Journal of Mathematical Sciences **193**, No. 2 (2013), 312–329.

[27] O.I. Reinov, *On linear operators with s-nuclear adjoints,* $0 < s \leq 1$. J. Math. Anal. Appl. **415** (2014) 816–824.

[28] O.I. Reinov, Q. Latif, *Grothendieck–Lidskiĭ theorem for subspaces of quotients of L_p-spaces.* Banach Center Publications (to appear).

[29] R. Schatten, *A theory of cross-spaces.* Princeton University Press, 1950.

[30] A. Szankowski, *Subspaces without approximation property.* Isr. J. Math. **30** (1978), 123–129

[31] H. Weyl, *Inequalities between the two kinds of eigenvalues of a linear transformation.* Proc. Nat. Acad. Sci. U.S.A. **35** (1949), 408–411.

[32] M.C. White, *Analytic multivalued functions and spectral trace.* Math. Ann. **304** (1996), 665–683.

Oleg Reinov
St. Petersburg State University
Universitetskii pr. 28
Petrodvorets
198504 St. Petersburg, Russia
e-mail: orein51@mail.ru

Ordered Structures and Applications: Positivity VII
Trends in Mathematics, 395–414

An Inequality Type Condition for Quasinearly Subharmonic Functions and Applications

Juhani Riihentaus

Abstract. Generalizing older works of Domar and Armitage and Gardiner, we give an inequality for quasinearly subharmonic functions. As an application of this inequality, we improve Domar's, Rippon's and our previous results concerning the existence of the largest subharmonic minorant of a given function. Moreover, and as an another application, we give a sufficient condition for a separately quasinearly subharmonic function to be quasinearly subharmonic. Our result contains the previous results of Lelong, of Avanissian, of Arsove, of Armitage and Gardiner, and of ours.

Mathematics Subject Classification (2010). 31C05, 31B25,31B05.

Keywords. Subharmonic, quasinearly subharmonic, families of quasinearly subharmonic functions, domination conditions, separately quasinearly subharmonic functions.

1. Introduction

1.1. Let D be a domain in $\mathbb{R}^N$, $N \geq 2$, and let $u : D \mapsto [-\infty, +\infty)$ be subharmonic. (We consider the function $u \equiv -\infty$ subharmonic.) Then u is upper semicontinuous and satisfies the mean value inequality

$$u(x) \leq \frac{1}{\nu_N\, r^N} \int\limits_{B^N(x,r)} u(y)\, dm_N(y)$$

for all balls $\overline{B^N(x,r)} \subset D$. It is an important fact that if u is also nonnegative and $p > 0$, then there exists a constant $C = C(N,p)$ such that

$$u(x)^p \leq \frac{C}{\nu_N\, r^N} \int\limits_{B^N(x,r)} u(y)^p\, dm_N(y). \tag{1.1}$$

As a matter of fact, Fefferman and Stein [11], Lemma 2, p. 172, proved this inequality for absolute values of harmonic functions. See also [12], Lemma 3.7,

p. 116, [17], Theorem 1, p. 529, and [1], (1.5), p. 210 (also all these authors considered only absolute values of harmonic functions). However, the proof of Fefferman and Stein apply verbatim also in the more general case of nonnegative subharmonic functions. See [23], Lemma, p. 69, and also [25], [26], [28] and the references therein. A possibility for an essentially different proof was pointed out already in [33], pp. 188–190. Later other different proofs were given in [20], p. 18, and Theorem 1, p. 19 (see also [21], Theorem A, p. 15), [24], Lemma 2.1, p. 233, and [25], Theorem, p. 188. Observe that the results in [20], [24] and [25] hold, in addition to nonnegative subharmonic functions, also for more general quasinearly subharmonic functions.

The inequality (1.1) has many applications. Among others, it has been applied to the weighted boundary behavior of subharmonic functions, to the nonintegrability of subharmonic and superharmonic functions, and to the subharmonicity of separately subharmonic functions, see [23], [6], [26], [28], [10], [30], and the references therein.

1.2. In order to improve the above-referred results on the subharmonicity of separately subharmonic functions, we give below in Section 2 a rather general inequality type result which is related to the inequality (1.1), at least partly, and which applies more generally also to quasinearly subharmonic functions. This result has its origin in the previous considerations of Armitage and Gardiner [2], proof of Proposition 2, pp. 257–259, and in [27], Lemma 3.2, p. 5, [29], Lemma 2.2, p. 6. Observe, however and as already Armitage and Gardiner have pointed out, this inequality is based on an old argument of Domar [7], proof of Proposition 2, pp. 257–259, and proof of Theorem 1, pp. 258–259. In Section 3 we will then apply the obtained inequality type result to domination conditions for families of quasinearly subharmonic functions, improving Domar's, Rippon's and our previous results, see [7], Theorem 1 and Theorem 2, pp. 430–431, [31], Theorem 5, p. 128, and [29], Theorem 2.1, pp. 4–5. In addition, in Section 4 we apply this inequality to the quasinearly subharmonicity of separately quasinearly subharmonic functions, slightly improving our previous results [27], Theorem 4.1, pp. 8–9, [28], Theorem 3.3.1, pp. e2621–e2622.

Though we indeed give improvements to our previous results, our presentation is, nevertheless and at least in some sense, of a survey type. Our notation is rather standard, see, e.g., [14], [26], [27], [28] and [29]. However and for the convenience of the reader, we begin by recalling here the definitions of nearly subharmonic functions and quasinearly subharmonic functions.

1.3. Nearly subharmonic functions and quasinearly subharmonic functions

We say that a function $u : D \to [-\infty, +\infty)$ is *nearly subharmonic*, if u is Lebesgue measurable, $u^+ \in \mathcal{L}^1_{\mathrm{loc}}(D)$, and for all $\overline{B^N(x,r)} \subset D$,

$$u(x) \le \frac{1}{\nu_N\, r^N} \int\limits_{B^N(x,r)} u(y)\, dm_N(y).$$

Observe that in the standard definition of nearly subharmonic functions one uses the slightly stronger assumption that $u \in \mathcal{L}^1_{\mathrm{loc}}(D)$, see, e.g., [14], p. 14. However, our above, slightly more general definition seems to be more practical, see, e.g., [26], Proposition 2.1 (iii) and Proposition 2.2 (vi), (vii), pp. 54–55, and [28], Proposition 1.5.1 (iii) and Proposition 1.5.2 (vi), (vii), p. e2615. The following lemma emphasizes this fact still more:

Lemma 1.1 ([26], Lemma, p. 52). *Let $u : D \to [-\infty, +\infty)$ be Lebesgue measurable. Then u is nearly subharmonic (in the sense defined above) if and only if there exists a function u^*, subharmonic in D such that $u^* \geq u$ and $u^* = u$ almost everywhere in D. Here u^* is the upper semicontinuous regularization of u:*

$$u^*(x) = \limsup_{x' \to x} u(x').$$

Proof. The proof follows at once from [14], proof of Theorem 1, pp. 14–15 (and referring also to [26], Proposition 2.1 (iii) and Proposition 2.2 (vii), pp. 54–55). □

We say that a Lebesgue measurable function $u : D \to [-\infty, +\infty)$ is K-*quasinearly subharmonic*, if $u^+ \in \mathcal{L}^1_{\mathrm{loc}}(D)$ and if there is a constant

$$K = K(N, u, D) \geq 1$$

such that for all $\overline{B^N(x,r)} \subset D$,

$$u_M(x) \leq \frac{K}{\nu_N\, r^N} \int\limits_{B^N(x,r)} u_M(y)\, dm_N(y) \tag{1.2}$$

for all $M \geq 0$, where $u_M := \max\{u, -M\} + M$. A function $u : D \to [-\infty, +\infty)$ is *quasinearly subharmonic*, if u is K-quasinearly subharmonic for some $K \geq 1$.

For basic properties of quasinearly subharmonic functions, see [26], [28], [22], and the references therein. We recall here only that this function class includes, among others, subharmonic functions, and, more generally, quasisubharmonic and nearly subharmonic functions (see, e.g., [14], pp. 14, 26), also functions satisfying certain natural growth conditions, especially certain eigenfunctions, and polyharmonic functions. Also, the class of Harnack functions is included, thus, among others, nonnegative harmonic functions as well as nonnegative solutions of some elliptic equations. In particular, the partial differential equations associated with quasiregular mappings belong to this family of elliptic equations. See, e.g., [34].

Quasinearly subharmonic functions (perhaps with a different terminology, and sometimes in certain special cases, or just the corresponding generalized mean value inequalities (1.1) or (1.2)) have been considered in many papers, see, e.g., [23], [20], [26], [27], [28], [29], [22], [15], [8], [9], [16], [10], and the references therein. However and as a matter of fact, already Domar [7] considered (essentially) nonnegative quasinearly subharmonic functions.

1.4. Two additional notational remarks

The below presented proofs for our results, that is, the proofs of Theorem 2.1, Theorem 3.3 and Theorem 4.1, are quite much based on our previous arguments in [27], proofs of Lemma 3.2, pp. 5–7, and of Theorem 4.1, pp. 8–12, and [29], proof of Theorem 2.1, pp. 4–8. Therefore, and in order to make the possible comparison and checking etc. easier for the reader, we will use, as previously, certain constants s_0, s_1, s_2, s_3, s_4 and s_5.

Below in Examples 1, 2, 3, 4 and 5, we consider increasing functions $\phi : [0, +\infty] \to [0, +\infty]$, say, and which are of a certain form far away, that is, for big values of the argument. In such a case, we take the liberty to use the convention that the function is then automatically defined for small values of the argument in an appropriate way. As an example, if we write, for $p > 0$, $q \in \mathbb{R}$,

$$\phi(t) = \frac{t^p}{(\log t)^q},$$

we mean the following function:

$$\varphi(t) := \begin{cases} \frac{t^p}{(\log t)^q}, & \text{when } t \geq t_1, \\ \frac{t}{t_1}\phi(t_1), & \text{when } 0 \leq t < t_1, \end{cases}$$

where $t_1 \geq 2$ is some suitable integer in $\mathbb{N}$, say.

2. An inequality for quasinearly subharmonic functions

As pointed out already above, our previous result [27], Lemma 3.2, p. 5, was a generalized version of Armitage's and Gardiner's argument [2], proof of Proposition 2, pp. 257–258, and as such, it was based on an old argument of Domar [7], Lemma 1, pp. 431–432 and 430. The following is another variant of this inequality type result:

Theorem 2.1. *Let $K \geq 1$. Let $\varphi : [0, +\infty] \to [0, +\infty]$ and $\psi : [0, +\infty] \to [0, +\infty]$ be increasing functions such that there are s_0, $s_1 \in \mathbb{N}$, $s_0 < s_1$, such that*

(i) *the inverse functions φ^{-1} and ψ^{-1} are defined on $[\min\{\varphi(s_1 - s_0), \psi(s_1 - s_0)\}, +\infty]$,*
(ii) *$2K(\psi^{-1} \circ \varphi)(s - s_0) \leq (\psi^{-1} \circ \varphi)(s)$ for all $s \geq s_1$,*
(iii) *$\int\limits_{s_1}^{+\infty} \left[\frac{(\psi^{-1} \circ \varphi)(s+2)}{(\psi^{-1} \circ \varphi)(s)} \frac{1}{\varphi(s - s_0)}\right]^{\frac{1}{N-1}} ds < +\infty$.*

Let $u : D \to [0, +\infty)$ be a K-quasinearly subharmonic function. Let $\tilde{s}_1 \in \mathbb{N}$, $\tilde{s}_1 \geq s_1$, be arbitrary. Then for each $x \in D$ and $r > 0$ such that $\overline{B^N(x, r)} \subset D$ either

$$u(x) \leq (\psi^{-1} \circ \varphi)(\tilde{s}_1 + 1)$$

or

$$\Phi(u(x)) \leq \frac{C}{r^N} \int\limits_{B^N(x,r)} \psi(u(y))\, dm_N(y)$$

where $C = C(N, K, s_0)$ *and* $\Phi : [0, +\infty) \to [0, +\infty)$,

$$\Phi(t) := \begin{cases} \left(\int\limits_{(\varphi^{-1}\circ\psi)(t)-2}^{+\infty} \left[\frac{(\psi^{-1}\circ\varphi)(s+2)}{(\psi^{-1}\circ\varphi)(s)} \cdot \frac{1}{\varphi(s-s_0)} \right]^{\frac{1}{N-1}} ds \right)^{1-N}, & \text{when } t \geq s_3, \\ \frac{t}{s_3}\Phi(s_3), & \text{when } 0 \leq t < s_3, \end{cases}$$

where $s_3 := \max\{ s_1+3, s_2, (\psi^{-1}\circ\varphi)(s_1+3) \}$ *and* $s_2 := \max\{ s_1, (\psi^{-1}\circ\varphi)(s_1+1) \}$.

Proof. The proof follows at once from [27], proof of Lemma 3.2, pp. 5–7. As a matter of fact, it is sufficient to observe the following:

- Instead of functions $\varphi : [0, +\infty) \to [0, +\infty)$ and $\psi : [0, +\infty) \to [0, +\infty)$ one can equally well consider functions $\varphi : [0, +\infty] \to [0, +\infty]$ and $\psi : [0, +\infty] \to [0, +\infty]$. See [29], p. 5.
- If $(\psi^{-1} \circ \varphi)(j_0) \leq t < (\psi^{-1} \circ \varphi)(j_0 + 1)$, then

$$\sum_{k=j_0}^{+\infty} \left[\frac{(\psi^{-1} \circ \varphi)(k+1)}{(\psi^{-1} \circ \varphi)(k)} \cdot \frac{1}{\varphi(k - s_0)} \right]^{\frac{1}{N-1}}$$

$$\leq \int\limits_{(\varphi^{-1}\circ\psi)(t)-2}^{+\infty} \left[\frac{(\psi^{-1} \circ \varphi)(s+2)}{(\psi^{-1} \circ \varphi)(s)} \cdot \frac{1}{\varphi(s - s_0)} \right]^{\frac{1}{N-1}} ds. \qquad \square$$

3. Domination conditions for families of quasinearly subharmonic functions

We begin by recalling the results of Domar and Rippon. Let $F : D \to [0, +\infty]$ be an upper semicontinuous function. Let $\mathcal{F}$ be a family of subharmonic functions $u : D \to [0, +\infty)$, which satisfy the condition

$$u(x) \leq F(x) \text{ for all } x \in D.$$

Write

$$w(x) := \sup_{u\in\mathcal{F}} u(x),$$

and let $w^* : D \to [0, +\infty]$ be the upper semicontinuous regularization of w, that is,

$$w^*(x) := \limsup_{y\to x} w(y).$$

Improving the original results of Sjöberg [32] and Brelot [5], cf. also Green [13], Domar [7], Theorem 1 and Theorem 2, pp. 430–431, gave the following result:

Theorem 3.1. *If for some* $\epsilon > 0$,

$$\int\limits_D [\log^+ F(x)]^{N-1+\epsilon} \, dm_N(x) < +\infty,$$

then w *is locally bounded above in* D, *and thus* w^* *is subharmonic in* D.

As Domar points out, his method of proof applies also to more general functions, that is, to the above-defined nonnegative quasinearly subharmonic functions. Much later Rippon [31], Theorem 1, p. 128, generalized Domar's result in the following form:

Theorem 3.2. *Let* $\varphi : [0, +\infty] \to [0, +\infty]$ *be an increasing function such that*

$$\int_1^{+\infty} \frac{dt}{\varphi(t)^{\frac{1}{N-1}}} < +\infty.$$

If

$$\int_D \varphi(\log^+ F(x))\, dm_N(x) < +\infty,$$

then w *is locally bounded above in* D*, and thus* w^* *is subharmonic in* D.

As pointed out by Domar [7], pp. 436–440, and by Rippon [31], p. 129, the above results are for many particular cases sharp.

In [29], Theorem 2.1, pp. 4–5, we gave a general and at the same time flexible result which includes both Domar's and Rippon's results. Now we improve our result still further:

Theorem 3.3. *Let* $K \geq 1$. *Let* $\varphi : [0, +\infty] \to [0, +\infty]$ *and* $\psi : [0, +\infty] \to [0, +\infty]$ *be increasing functions for which there are* s_0, $s_1 \in \mathbb{N}$, $s_0 < s_1$, *such that*

(i) *the inverse functions* φ^{-1} *and* ψ^{-1} *are defined on* $[\min\{\varphi(s_1 - s_0), \psi(s_1 - s_0)\}, +\infty]$,
(ii) $2K(\psi^{-1} \circ \varphi)(s - s_0) \leq (\psi^{-1} \circ \varphi)(s)$ *for all* $s \geq s_1$,
(iii) *the following integral is convergent:*

$$\int_{s_1}^{+\infty} \left[\frac{(\psi^{-1} \circ \varphi)(s+2)}{(\psi^{-1} \circ \varphi)(s)} \cdot \frac{1}{\varphi(s - s_0)} \right]^{\frac{1}{N-1}} ds < +\infty.$$

Let $\mathfrak{F}_K$ *be a family of* K*-quasinearly subharmonic functions* $u : D \to [-\infty, +\infty)$ *such that*

$$u(x) \leq F_K(x) \text{ for all } x \in D,$$

where $F_K : D \to [0, +\infty]$ *is a Lebesgue measurable function. If for each compact set* $E \subset D$,

$$\int_E \psi(F_K(x))\, dm_N(x) < +\infty,$$

then the family $\mathfrak{F}_K$ *is locally (uniformly) bounded in* D. *Moreover, the function* $w^* : D \to [0, +\infty)$ *is* K*-quasinearly subharmonic. Here*

$$w^*(x) := \limsup_{y \to x} w(y),$$

where

$$w(x) := \sup_{u \in \mathfrak{F}_K} u^+(x).$$

Proof. Let E be an arbitrary compact subset of D. Write $\rho_0 :=\text{dist}(E, \partial D)$. Clearly $\rho_0 > 0$. Write

$$E_1 := \bigcup_{x \in E} \overline{B^N\Big(x, \frac{\rho_0}{2}\Big)}.$$

Then E_1 is compact, and $E \subset E_1 \subset D$. Take $u \in \mathfrak{F}_K^+$ arbitrarily, where

$$\mathfrak{F}_K^+ := \{\, u^+ \, : \, u \in \mathfrak{F}_K \,\}.$$

Let $\tilde{s}_1 = s_1 + 2$, say. Take $x \in E$ arbitrarily and suppose that $u(x) > \tilde{s}_3$, where $\tilde{s}_3 := \max\{\, s_1 + 5, (\psi^{-1} \circ \varphi)(s_1 + 5)\,\}$, say. By Theorem 2.1 we have,

$$\Phi(u(x)) \leq \frac{C}{\left(\frac{\rho_0}{2}\right)^N} \int\limits_{B^N(x, \frac{\rho_0}{2})} \psi(u(y))\, dm_N(y) \leq \frac{C}{\left(\frac{\rho_0}{2}\right)^N} \int\limits_{E_1} \psi(F_K(y))\, dm_N(y) < +\infty,$$

where

$$\Phi(t) := \left(\int\limits_{(\varphi^{-1} \circ \psi)(t) - 2}^{+\infty} \left[\frac{(\psi^{-1} \circ \varphi)(s+2)}{(\psi^{-1} \circ \varphi)(s)} \cdot \frac{1}{\varphi(s - s_0)} \right]^{\frac{1}{N-1}} ds \right)^{1-N}.$$

Now we know that

$$\int\limits_{s_1}^{+\infty} \left[\frac{(\psi^{-1} \circ \varphi)(s+2)}{(\psi^{-1} \circ \varphi)(s)} \cdot \frac{1}{\varphi(s - s_0)} \right]^{\frac{1}{N-1}} ds < +\infty.$$

Therefore, the set of function values

$$u(x), \quad x \in E, \quad u \in \mathfrak{F}_K^+,$$

is bounded above.

The rest of the proof goes then as in [29], proof of Theorem 2.1, pp. 7–8. □

Remark 3.4. In Theorem 3.3 one can, instead of the assumption (iii), use also the following:

(iii′) *the following series is convergent:*

$$\sum_{j=s_1+1}^{+\infty} \left[\frac{(\psi^{-1} \circ \varphi)(j+1)}{(\psi^{-1} \circ \varphi)(j)} \cdot \frac{1}{\varphi(j - s_0)} \right]^{\frac{1}{N-1}} < +\infty.$$

Instead of the above-used function Φ one uses then the function

$$\Phi_1 : [s_2, +\infty) \to [0, +\infty),$$

$$\Phi_1(t) := \left(\sum_{k=j_0}^{+\infty} \left[\frac{(\psi^{-1} \circ \varphi)(k+1)}{(\psi^{-1} \circ \varphi)(k)} \cdot \frac{1}{\varphi(k - s_0)} \right]^{\frac{1}{N-1}} \right)^{1-N},$$

where $j_0 \in \{\, s_1, s_1 + 2, \dots \}$ is such that

$$(\psi^{-1} \circ \varphi)(j_0) \leq t < (\psi^{-1} \circ \varphi)(j_0 + 1).$$

The function Φ_1 may of course be extended to the whole interval $[0,+\infty)$, for example as follows:

$$\Phi_1(t) := \begin{cases} \Phi_1(t), & \text{when } t \geq s_3, \\ \frac{t}{s_3}\Phi_1(s_3), & \text{when } 0 \leq t < s_3. \end{cases}$$

Before giving examples, we write Theorem 3.3 in the following, perhaps more concrete form, see also [29], Remark 2.5, pp. 8–9.

Corollary 3.5. *Let $K \geq 1$. Let $\varphi : [0,+\infty] \to [0,+\infty]$ and $\phi : [0,+\infty] \to [0,+\infty]$ be strictly increasing surjections for which there are s_0, $s_1 \in \mathbb{N}$, $s_0 < s_1$, such that*

(i) $2K\phi^{-1}(e^{s-s_0}) \leq \phi^{-1}(e^s)$ *for all* $s \geq s_1$,

(ii) *the following integral is convergent:*

$$\int_{s_1}^{+\infty} \left[\frac{\phi^{-1}(e^{s+2})}{\phi^{-1}(e^s)} \frac{1}{\varphi(s-s_0)}\right]^{\frac{1}{N-1}} ds < +\infty.$$

Let $\mathcal{F}_K$ be a family of K-quasinearly subharmonic functions $u : D \to [-\infty,+\infty)$ such that

$$u(x) \leq F_K(x) \ \ \textit{for all} \ \ x \in D,$$

where $F_K : D \to [0,+\infty]$ is a Lebesgue measurable function. If for each compact set $E \subset D$,

$$\int_E \varphi(\log^+ \phi(F_K(x)))\, dm_N(x) < +\infty,$$

then the family $\mathcal{F}_K$ is locally (uniformly) bounded in D. Moreover, the function $w^* : D \to [0,+\infty)$ is K-quasinearly subharmonic. Here

$$w^*(x) := \limsup_{y\to x} w(y),$$

where

$$w(x) := \sup_{u\in\mathcal{F}_K} u^+(x).$$

Proof. For the proof, just choose $\psi(t) = \varphi(\log^+ \phi(t))$. Since only big values count, we may simply use the formula $\psi(t) = \varphi(\log \phi(t))$. One sees easily that for some $\tilde{s}_2 \geq s_1$,

$$(\psi^{-1} \circ \varphi)(s) = \phi^{-1}(e^s) \ \ \text{for all} \ \ s \geq \tilde{s}_2.$$

It is then easy to see that the assumptions of Theorem 3.3 are satisfied. □

Example 1. Let $\varphi : [0,+\infty] \to [0,+\infty]$ be a strictly increasing surjection such that (for some $s_0 \in \mathbb{N}$),

$$\int_{s_1}^{+\infty} \frac{ds}{\varphi(s-s_0)^{\frac{1}{N-1}}} < +\infty.$$

Choosing then various functions ϕ which, together with φ, satisfy the conditions (i) and (ii) of Corollary 3.5, one gets more concrete results. If ϕ and ϕ^{-1} satisfy (at least far away) the Δ_2-condition, then the conditions (i) and (ii) are surely satisfied (see also [29], Remark 2.5, p. 8). Typical choices for ϕ might be, say, the following:

$$\phi(t) := \frac{t^p}{(\log t)^q}, \; p > 0,\, q \in \mathbb{R}.$$

The choice $p = 1$, $q = 0$ gives then the results of Domar and Rippon, Theorem 3.1 and Theorem 3.2 above. Choosing $0 < p < 1$ and $q \geq 0$, one gets (at least formal) improvements.

Example 2. Let $\phi : [0, +\infty] \to [0, +\infty]$ be a strictly increasing surjection for which there is $t_1 > 1$ such that

$$s = \phi(t) = \log t \quad \text{for} \quad t \geq t_1$$

and

$$t = \phi^{-1}(s) = e^s \quad \text{for} \quad s \geq \log t_1.$$

One sees easily that the condition (i) of Corollary 3.5 holds. In this case ϕ^{-1} does not satisfy a Δ_2-condition. Therefore, in order to apply Corollary 3.5 to a family $\mathcal{F}_K$ of K-quasinearly subharmonic functions, we must choose an appropriate strictly increasing surjection $\varphi : [0, +\infty] \to [0, +\infty]$ such that for some $\tilde{s}_1 \in \mathbb{N}$,

$$\int_{\tilde{s}_1}^{+\infty} \left[\frac{\phi^{-1}(e^{s+2})}{\phi^{-1}(e^s)} \cdot \frac{1}{\varphi(s - s_0)} \right]^{\frac{1}{N-1}} ds = \int_{\tilde{s}_1}^{+\infty} \frac{e^{\frac{e^2-1}{N-1} e^s}}{\varphi(s - s_0)^{\frac{1}{N-1}}} ds < +\infty.$$

Therefore, we have two restrictions for φ. As a first condition the above quite strong restriction and, as a second one, the following, at least seemingly mild condition: For each compact set $E \subset D$,

$$\int_E \varphi(\log^+(\log F_K(x)))\, dm_N(x) < +\infty.$$

4. On the subharmonicity of separately subharmonic functions and generalizations

Wiegerinck [35], Theorem, p. 770, see also [36], Theorem 1, p. 246, has shown that separately subharmonic functions need not be subharmonic. On the other hand, Armitage and Gardiner [2], Theorem 1, p. 256, showed that a separately subharmonic function u in a domain Ω of $\mathbb{R}^{m+n}$, $m \geq n \geq 2$, is subharmonic, provided $\phi \circ \log^+ u^+ \in \mathcal{L}^1_{\text{loc}}(\Omega)$, where $\phi : [0, +\infty) \to [0, +\infty)$ is an increasing function such that

$$\int_1^{+\infty} \frac{s^{\frac{n-1}{m-1}}}{\phi(s)^{\frac{1}{m-1}}} ds < +\infty.$$

Armitage's and Gardiner's result included all the previous existing results, that is, the results of Lelong [18], Théorème 1 bis, p. 315, of Avanissian [4], Théorème 9,

p. 140, see also [19], Proposition 3, p. 24, and [14], Theorem, p. 31, of Arsove [3], Theorem 1, p. 622, and ours [23], Theorem 1, p. 69. Though Armitage's and Gardiner's result was close to being sharp, see [2], p. 255, it was, however, possible to improve their result slightly further. This was done in [27], Theorem 4.1, pp. 8–9, with the aid of quasinearly subharmonic functions. See also [28], Theorem 3.3.1 and Corollary 3.3.3, pp. e2621-e2622. Now we improve our result still further:

Theorem 4.1. *Let $K \geq 1$. Let Ω be a domain in $\mathbb{R}^{m+n}$, $m \geq n \geq 2$. Let $u : \Omega \to [-\infty, +\infty)$ be a Lebesgue measurable function. Suppose that the following conditions are satisfied:*

(a) *For each $y \in \mathbb{R}^n$ the function*

$$\Omega(y) \ni x \mapsto u(x, y) \in [-\infty, +\infty)$$

is K-quasinearly subharmonic.

(b) *For each $x \in \mathbb{R}^m$ the function*

$$\Omega(x) \ni y \mapsto u(x, y) \in [-\infty, +\infty)$$

is K-quasinearly subharmonic.

(c) *There are increasing functions $\varphi : [0, +\infty) \to [0, +\infty)$ and $\psi : [0, +\infty) \to [0, +\infty)$ and s_0, $s_1 \in \mathbb{N}$, $s_0 < s_1$, such that*

(c1) *the inverse functions φ^{-1} and ψ^{-1} are defined on $[\min\{\,\varphi(s_1 - s_0), \psi(s_1 - s_0)\,\}, +\infty)$,*

(c2) $2K(\psi^{-1} \circ \varphi)(s - s_0) \leq (\psi^{-1} \circ \varphi)(s)$ *for all $s \geq s_1$,*

(c3) *the following integral is convergent:*

$$\int_{s_5}^{+\infty} \left[\frac{(\psi^{-1} \circ \varphi)(s+2)}{(\psi^{-1} \circ \varphi)(s)} \cdot \frac{1}{\varphi(s - s_0)}\right]^{\frac{1}{m-1}} \cdot \left(\int_{s_5+s_0+2}^{s+s_0+2} \left[\frac{(\psi^{-1} \circ \varphi)(t+2)}{(\psi^{-1} \circ \varphi)(t)}\right]^{\frac{1}{n-1}} dt\right)^{\frac{n-1}{m-1}} ds < +\infty,$$

where $s_5 := \max\{\, s_0 + s_1 + 3, s_0 + (\psi^{-1} \circ \varphi)(s_1 + 3), s_0 + (\varphi^{-1} \circ \psi)(s_1 + 3)\,\}$,

(c4) $\psi \circ u^+ \in \mathcal{L}^1_{\mathrm{loc}}(\Omega)$.

Then u is quasinearly subharmonic in Ω.

Proof. Recall that $s_2 = \max\{\, s_1, (\psi^{-1} \circ \varphi)(s_1 + 1)\,\}$ and $s_3 = \max\{\, s_1 + 3, s_2, (\psi^{-1} \circ \varphi)(s_1 + 3)\,\}$. Write $s_4 := \max\{\, s_0 + s_3, (\varphi^{-1} \circ \psi)(s_1 + 3)\,\}$, say. Clearly, $s_0 < s_1 < s_3 < s_4 < s_5$. (We may of course suppose that s_3, s_4 and s_5 are integers.) One may replace u by $\max\{\, u^+, M\,\}$, where $M = \max\{\, s_5 + 3, (\psi^{-1} \circ \varphi)(s_4 + 3), (\varphi^{-1} \circ \psi)(s_4 + 3)\,\}$, say. We continue to denote u_M by u.

Step 1 *Use of Theorem* 2.1.

Take $(x_0, y_0) \in \Omega$ and $r > 0$ arbitrarily such that $\overline{B^m(x_0, 2r) \times B^n(y_0, 2r)} \subset \Omega$. Take $(\xi, \eta) \in B^m(x_0, r) \times B^n(y_0, r)$ arbitrarily. We know that $u(\cdot, y)$ is K-quasinearly subharmonic for each $y \in B^n(y_0, 2r)$. In order to apply Theorem 2.1, it is clearly sufficient to show that

$$\int_{s_5+s_0+2}^{+\infty} \left[\frac{(\psi^{-1}\circ\varphi)(s+2)}{(\psi^{-1}\circ\varphi)(s)} \cdot \frac{1}{\varphi(s-s_0)}\right]^{\frac{1}{m-1}} ds < +\infty.$$

But this follows at once from the assumption (c3), since for all $s \geq s_5 + s_0 + 2$,

$$\left(\int_{s_5+s_0+2}^{s+s_0+2} \left[\frac{(\psi^{-1}\circ\varphi)(t+2)}{(\psi^{-1}\circ\varphi)(t)}\right]^{\frac{1}{n-1}} dt\right)^{\frac{n-1}{m-1}} \geq \left(\int_{s_5+s_0+2}^{s+s_0+2} 1\, dt\right)^{\frac{n-1}{m-1}}$$
$$= (s-s_5)^{\frac{n-1}{m-1}} \geq (s_0+2)^{\frac{n-1}{m-1}}.$$

From Theorem 2.1 it then follows that for all $y \in B^n(y_0, 2r)$

$$\tilde{\Phi}(u(\xi,y)) \leq \frac{C}{r^m} \int_{B^m(\xi,r)} \psi(u(x,y))dm_m(x), \tag{4.1}$$

where

$$\tilde{\Phi}(t) := \begin{cases} \left(\int_{(\varphi^{-1}\circ\psi)(t)-2}^{+\infty} \left[\frac{(\psi^{-1}\circ\varphi)(s+2)}{(\psi^{-1}\circ\varphi)(s)} \frac{1}{\varphi(s-s_0)}\right]^{\frac{1}{m-1}} ds\right)^{1-m}, & \text{when } t \geq s_3, \\ \frac{t}{s_3}\tilde{\Phi}(s_3), & \text{when } 0 \leq t < s_3. \end{cases}$$

Step 2 *Take mean values on both sides of* (4.1).

Taking (generalized) mean values with respect to the variable y over the ball $B^n(\eta, r)$ on both sides of (4.1), we get:

$$\begin{aligned} \frac{C}{r^n} \int_{B^n(\eta,r)} \tilde{\Phi}(u(\xi,y))dm_n(y) &\leq \frac{C}{r^n} \int_{B^n(\eta,r)} \left[\frac{C}{r^m} \int_{B^m(\xi,r)} \psi(u(x,y))dm_m(x)\right] dm_n(y) \\ &\leq \frac{C}{r^{m+n}} \int_{B^m(\xi,r)\times B^n(\eta,r)} \psi(u(x,y))dm_{m+n}(x,y) \\ &\leq \frac{C}{r^{m+n}} \int_{B^m(x_0,2r)\times B^n(y_0,2r)} \psi(u(x,y))dm_{m+n}(x,y). \end{aligned}$$

Here one must of course check that both $\psi \circ u(\cdot,\cdot)$ and $\tilde{\Phi}(u(\xi,\cdot))$ are Lebesgue measurable!

Step 3 *In order to apply Theorem* 2.1 *once more, define new functions* φ_1 *and* ψ_1.

Write $\psi_1 : [0,+\infty) \to [0,+\infty)$,

$$\psi_1(t) := \tilde{\Phi}(t) = \begin{cases} \left(\int_{(\varphi^{-1}\circ\psi)(t)-2}^{+\infty} \left[\frac{(\psi^{-1}\circ\varphi)(s+2)}{(\psi^{-1}\circ\varphi)(s)} \frac{1}{\varphi(s-s_0)}\right]^{\frac{1}{m-1}} ds\right)^{1-m}, & \text{when } t \geq s_3, \\ \frac{t}{s_3}\tilde{\Phi}(s_3), & \text{when } 0 \leq t < s_3. \end{cases}$$

It is easy to see that ψ_1 is defined, strictly increasing and continuous. Write then $\varphi_1 : [0,+\infty) \to [0,+\infty)$,

$$\varphi_1(t) := \begin{cases} \psi_1((\psi^{-1}\circ\varphi)(t)) = \tilde{\Phi}(\psi^{-1}(\varphi(t))), & \text{when } t \geq s_3, \\ \frac{t}{s_3}\psi_1((\psi^{-1}\circ\varphi)(s_3)) = \frac{t}{s_3}\tilde{\Phi}(\psi^{-1}(\varphi(s_3))), & \text{when } 0 \leq t < s_3. \end{cases}$$

Also φ_1 is defined, strictly increasing and continuous. This follows from the facts that ψ_1 is defined, strictly increasing and continuous (similarly as the functions $\varphi|[s_1 - s_0, +\infty)$ and $\psi|[s_1 - s_0, +\infty)$). Observe here that for $t \geq s_4$, say,

$$\varphi_1(t) = \left(\int\limits_{(\varphi^{-1}\circ\psi)((\psi^{-1}\circ\varphi)(t))-2}^{+\infty} \left[\frac{(\psi^{-1}\circ\varphi)(s+2)}{(\psi^{-1}\circ\varphi)(s)} \frac{1}{\varphi(s-s_0)} \right]^{\frac{1}{m-1}} ds \right)^{1-m}$$
$$= \left(\int\limits_{t-2}^{+\infty} \left[\frac{(\psi^{-1}\circ\varphi)(s+2)}{(\psi^{-1}\circ\varphi)(s)} \frac{1}{\varphi(s-s_0)} \right]^{\frac{1}{m-1}} ds \right)^{1-m}.$$

One sees easily that $(\psi_1^{-1}\circ\varphi_1)(t) = (\psi^{-1}\circ\varphi)(t)$ for all $t \geq s_3$, thus $2K(\psi_1^{-1}\circ\varphi_1)(s-s_0) \leq (\psi_1^{-1}\circ\varphi_1)(s)$ for all $s \geq \overline{s}_1 \geq s_4$.

To show that

$$\int\limits_{s_5+s_0+2}^{+\infty} \left[\frac{(\psi_1^{-1}\circ\varphi_1)(s+2)}{(\psi_1^{-1}\circ\varphi_1)(s)} \frac{1}{\varphi_1(s-s_0)} \right]^{\frac{1}{n-1}} ds < +\infty,$$

we proceed as follows.

Write $F : [s_5,+\infty) \times [s_5+s_0+2,+\infty) \to [0,+\infty)$,

$$F(s,t) := \begin{cases} \left[\frac{(\psi_1^{-1}\circ\varphi_1)(t+2)}{(\psi_1^{-1}\circ\varphi_1)(t)} \frac{(\psi^{-1}\circ\varphi)(s+2)}{(\psi^{-1}\circ\varphi)(s)} \frac{1}{\varphi(s-s_0)} \right]^{\frac{1}{m-1}}, \\ \quad \text{when } s_5+s_0+2 \leq t-s_0-2 \leq s, \\ 0, \quad \text{when } s_5 \leq s < t-s_0-2. \end{cases}$$

Suppose that $m > n \geq 2$. Then just calculate, use Minkowski's inequality and assumption (c3):

$$\left(\int\limits_{s_5+s_0+2}^{+\infty} \left[\frac{(\psi_1^{-1}\circ\varphi_1)(t+2)}{(\psi_1^{-1}\circ\varphi_1)(t)} \frac{1}{\varphi_1(t-s_0)} \right]^{\frac{1}{n-1}} dt \right)^{\frac{n-1}{m-1}}$$
$$= \left(\int\limits_{s_5+s_0+2}^{+\infty} \left[\frac{(\psi_1^{-1}\circ\varphi_1)(t+2)}{(\psi_1^{-1}\circ\varphi_1)(t)} \right]^{\frac{1}{n-1}} \right.$$
$$\left. \times \left(\int\limits_{t-s_0-2}^{+\infty} \left[\frac{(\psi^{-1}\circ\varphi)(s+2)}{(\psi^{-1}\circ\varphi)(s)} \frac{1}{\varphi(s-s_0)} \right]^{\frac{1}{m-1}} ds \right)^{-\frac{1-m}{n-1}} dt \right)^{\frac{n-1}{m-1}}$$

$$= \left(\int_{s_5+s_0+2}^{+\infty} \left[\frac{(\psi_1^{-1} \circ \varphi_1)(t+2)}{(\psi_1^{-1} \circ \varphi_1)(t)} \right]^{\frac{1}{n-1}} \right.$$

$$\left. \times \left(\int_{t-s_0-2}^{+\infty} \left[\frac{(\psi^{-1} \circ \varphi)(s+2)}{(\psi^{-1} \circ \varphi)(s)} \frac{1}{\varphi(s-s_0)} \right]^{\frac{1}{m-1}} ds \right)^{\frac{m-1}{n-1}} dt \right)^{\frac{n-1}{m-1}}$$

$$= \left(\int_{s_5+s_0+2}^{+\infty} \left(\int_{t-s_0-2}^{+\infty} \left(\left[\frac{(\psi_1^{-1} \circ \varphi_1)(t+2)}{(\psi_1^{-1} \circ \varphi_1)(t)} \right. \right. \right. \right.$$

$$\left. \left. \left. \left. \cdot \frac{(\psi^{-1} \circ \varphi)(s+2)}{(\psi^{-1} \circ \varphi)(s)} \frac{1}{\varphi(s-s_0)} \right]^{\frac{1}{m-1}} \right) ds \right)^{\frac{m-1}{n-1}} dt \right)^{\frac{n-1}{m-1}}$$

$$= \left(\int_{s_5+s_0+2}^{+\infty} \left[\int_{s_5}^{+\infty} F(s,t)\, ds \right]^{\frac{m-1}{n-1}} dt \right)^{\frac{n-1}{m-1}} \leq \left(\int_{s_5}^{+\infty} \left[\int_{s_5+s_0+2}^{+\infty} F(s,t)^{\frac{m-1}{n-1}} dt \right]^{\frac{n-1}{m-1}} ds \right)$$

$$= \int_{s_5}^{+\infty} \left(\int_{s_5+s_0+2}^{s+s_0+2} \left(\left[\frac{(\psi_1^{-1} \circ \varphi_1)(t+2)}{(\psi_1^{-1} \circ \varphi_1)(t)} \right]^{\frac{1}{m-1}} \right. \right.$$

$$\left. \left. \cdot \left[\frac{(\psi^{-1} \circ \varphi)(s+2)}{(\psi^{-1} \circ \varphi)(s)} \frac{1}{\varphi(s-s_0)} \right]^{\frac{1}{m-1}} \right)^{\frac{m-1}{n-1}} dt \right)^{\frac{n-1}{m-1}} ds$$

$$= \int_{s_5}^{+\infty} \left[\frac{(\psi^{-1} \circ \varphi)(s+2)}{(\psi^{-1} \circ \varphi)(s)} \frac{1}{\varphi(s-s_0)} \right]^{\frac{1}{m-1}}$$

$$\times \left(\int_{s_5+s_0+2}^{s+s_0+2} \left[\frac{(\psi_1^{-1} \circ \varphi_1)(t+2)}{(\psi_1^{-1} \circ \varphi_1)(t)} \right]^{\frac{1}{n-1}} dt \right)^{\frac{n-1}{m-1}} ds$$

$$= \int_{s_5}^{+\infty} \left[\frac{(\psi^{-1} \circ \varphi)(s+2)}{(\psi^{-1} \circ \varphi)(s)} \frac{1}{\varphi(s-s_0)} \right]^{\frac{1}{m-1}}$$

$$\times \left(\int_{s_5+s_0+2}^{s+s_0+2} \left[\frac{(\psi^{-1} \circ \varphi)(t+2)}{(\psi^{-1} \circ \varphi)(t)} \right]^{\frac{1}{n-1}} dt \right)^{\frac{n-1}{m-1}} ds < +\infty.$$

The case $m = n$ is considered similarly, just replacing Minkowski's inequality with Fubini's theorem.

Step 4 *Apply Theorem 2.1 to conclude that* $u(\cdot, \cdot)$ *is bounded in* $B^m(x_0, r) \times B^n(y_0, r)$.

With the aid of Theorem 2.1 we get

$$\Psi(u(\xi,\eta)) \leq \frac{C}{r^n} \int\limits_{B^n(\eta,r)} \tilde{\Phi}(u(\xi,y))dm_n(y)$$
$$\leq \frac{C}{r^{m+n}} \int\limits_{B^m(x_0,2r)\times B^n(y_0,2r)} \psi(u(x,y))dm_{m+n}(x,y),$$

where now

$$\Psi(t) := \begin{cases} \left(\int\limits_{(\varphi_1^{-1}\circ\psi_1)(t)-2}^{+\infty} \left[\frac{(\psi_1^{-1}\circ\varphi_1)(s+2)}{(\psi_1^{-1}\circ\varphi_1)(s)} \frac{1}{\varphi_1(s-s_0)} \right]^{\frac{1}{n-1}} ds \right)^{1-n}, & \text{when } t \geq s_3, \\ \frac{t}{s_3}\Psi(s_3), & \text{when } 0 \leq t < s_3, \end{cases}$$

or equivalently

$$\Psi(t) := \begin{cases} \left(\int\limits_{(\varphi^{-1}\circ\psi)(t)-2}^{+\infty} \left[\frac{(\psi_1^{-1}\circ\varphi_1)(s+2)}{(\psi_1^{-1}\circ\varphi_1)(s)} \frac{1}{\varphi_1(s-s_0)} \right]^{\frac{1}{n-1}} ds \right)^{1-n}, & \text{when } t \geq s_3, \\ \frac{t}{s_3}\Psi(s_3), & \text{when } 0 \leq t < s_3. \end{cases}$$

Observe that we know that

$$\int\limits_{s_5+s_0+2}^{+\infty} \left[\frac{(\psi_1^{-1}\circ\varphi_1)(s+2)}{(\psi_1^{-1}\circ\varphi_1)(s)} \frac{1}{\varphi_1(s-s_0)} \right]^{\frac{1}{n-1}} ds$$
$$\leq \int\limits_{s_5}^{+\infty} \left[\frac{(\psi^{-1}\circ\varphi)(s+2)}{(\psi^{-1}\circ\varphi)(s)} \frac{1}{\varphi(s-s_0)} \right]^{\frac{1}{m-1}}$$
$$\times \left(\int\limits_{s_5+s_0+2}^{s+s_0+2} \left[\frac{(\psi^{-1}\circ\varphi)(t+2)}{(\psi^{-1}\circ\varphi)(t)} \right]^{\frac{1}{n-1}} dt \right)^{\frac{n-1}{m-1}} ds,$$

and that by assumption (c3),

$$\int\limits_{s_5}^{+\infty} \left[\frac{(\psi^{-1}\circ\varphi)(s+2)}{(\psi^{-1}\circ\varphi)(s)} \frac{1}{\varphi(s-s_0)} \right]^{\frac{1}{m-1}}$$
$$\left(\int\limits_{s_5+s_0+2}^{s+s_0+2} \left[\frac{(\psi^{-1}\circ\varphi)(t+2)}{(\psi^{-1}\circ\varphi)(t)} \right]^{\frac{1}{n-1}} dt \right)^{\frac{n-1}{m-1}} ds < \infty.$$

Hence the set of function values

$$(\varphi_1^{-1}\circ\psi_1)(u(\xi,\eta)) - 2 = (\varphi^{-1}\circ\psi)(u(\xi,\eta)) - 2, \ (\xi,\eta) \in B^m(x_0,r)\times B^n(y_0,r),$$

must be bounded. Thus the function $u(\cdot,\cdot)$ is bounded above in $B^m(x_0,r) \times B^n(y_0,r)$. By [26], Proposition 3.1, p. 57, (or by [28], Proposition 3.2.1, p. e2620) we see that $u(\cdot,\cdot)$ is quasinearly subharmonic. □

Corollary 4.2. *Let $K \geq 1$. Let Ω be a domain in $\mathbb{R}^{m+n}$, $m \geq n \geq 2$. Let $u : \Omega \to [-\infty,+\infty)$ be a Lebesgue measurable function. Suppose that the following conditions are satisfied:*

(a) *For each $y \in \mathbb{R}^n$ the function*
$$\Omega(y) \ni x \mapsto u(x,y) \in [-\infty,+\infty)$$
is K-quasinearly subharmonic.

(b) *For each $x \in \mathbb{R}^m$ the function*
$$\Omega(x) \ni y \mapsto u(x,y) \in [-\infty,+\infty)$$
is K-quasinearly subharmonic.

(c) *There are strictly increasing surjections $\varphi : [0,+\infty) \to [0,+\infty)$ and $\phi : [0,+\infty) \to [0,+\infty)$ and s_0, $s_1 \in \mathbb{N}$, $s_0 < s_1$, such that*

(c1) $2K\phi^{-1}(e^{s-s_0}) \leq \phi^{-1}(e^s)$ *for all $s \geq s_1$,*

(c2) *the following integral is convergent:*
$$\int_{s_5}^{+\infty} \left[\frac{\phi^{-1}(e^{s+2})}{\phi^{-1}(e^s)} \frac{1}{\varphi(s-s_0)}\right]^{\frac{1}{m-1}} \left(\int_{s_5+s_0+2}^{s+s_0+2} \left[\frac{\phi^{-1}(e^{t+2})}{\phi^{-1}(e^t)}\right]^{\frac{1}{n-1}} dt \right)^{\frac{n-1}{m-1}} ds < +\infty,$$

(c3) $\varphi \circ \log^+ \phi(u^+) \in \mathcal{L}^1_{\mathrm{loc}}(\Omega)$.

Then u is quasinearly subharmonic in Ω.

Corollary 4.3. *Let $K \geq 1$. Let Ω be a domain in $\mathbb{R}^{m+n}$, $m \geq n \geq 2$. Let $u : \Omega \to [-\infty,+\infty)$ be a Lebesgue measurable function. Suppose that the following conditions are satisfied:*

(a) *For each $y \in \mathbb{R}^n$ the function*
$$\Omega(y) \ni x \mapsto u(x,y) \in [-\infty,+\infty)$$
is K-quasinearly subharmonic.

(b) *For each $x \in \mathbb{R}^m$ the function*
$$\Omega(x) \ni y \mapsto u(x,y) \in [-\infty,+\infty)$$
is K-quasinearly subharmonic.

(c) *There are strictly increasing surjections $\varphi : [0,+\infty) \to [0,+\infty)$ and $\phi : [0,+\infty) \to [0,+\infty)$ and s_0, $s_1 \in \mathbb{N}$, $s_0 < s_1$, such that*

(c1) $2K\phi^{-1}(e^{s-s_0}) \leq \phi^{-1}(e^s)$ *for all $s \geq s_1$,*

(c2) ϕ^{-1} *satisfies a Δ_2-condition,*

(c3) *the following integral is convergent:*

$$\int_{s_1}^{+\infty} \frac{s^{\frac{n-1}{m-1}}}{\varphi(s-s_0)^{\frac{1}{m-1}}}\, ds < +\infty,$$

(c4) $\varphi \circ \log^+ \phi(u^+) \in \mathcal{L}^1_{\mathrm{loc}}(\Omega)$.

Then u is quasinearly subharmonic in Ω.

Example 3. Let u be separately subharmonic in Ω. Let $\varphi : [0,+\infty) \to [0,+\infty)$ be a strictly increasing surjection such that

$$\int_{s_1}^{+\infty} \frac{s^{\frac{n-1}{m-1}}}{\varphi(s-s_0)^{\frac{1}{m-1}}}\, ds < +\infty.$$

Choosing then various functions ϕ, which, together with φ and u, satisfy the conditions (c1), (c2) and (c4) of Corollary 4.3, one gets more concrete results. Possible choices are, e.g.,

$$\phi(t) = \frac{t^p}{(\log t)^q},\; p > 0,\; q \in \mathbb{R}.$$

The case $p = 1$ and $q = 0$ gives the result of Armitage and Gardiner.

Example 4. Let u be separately subharmonic in Ω and $\varphi : [0,+\infty) \to [0,+\infty)$ be a strictly increasing surjection. Let $p > 0$, $q \geq 0$. Let $\phi : [0,+\infty) \to [0,+\infty)$ be a strictly increasing surjection for which there is $t_1 > 1$ such that

$$s = \phi(t) = e^{\left(\frac{\log t}{p}\right)^{\frac{1}{q+1}}} = e^{\sqrt[q+1]{\frac{\log t}{p}}} \quad \text{for} \quad t \geq t_1,$$

thus $t = \phi^{-1}(s) = e^{p(\log s)^{q+1}}$. One sees easily that the condition (c1) of Corollary 4.3 is satisfied, but the condition (c2) not. As a matter of fact, and as one easily sees,

$$\frac{\phi^{-1}(e^{s+2})}{\phi^{-1}(e^s)} \to +\infty \quad \text{as} \quad s \to +\infty.$$

Therefore, in this case one cannot use Corollary 4.3 to conclude that u is subharmonic. However, using Corollary 4.2 we see that u is subharmonic, provided that

$$\int_{s_5}^{+\infty} \frac{e^{\frac{p}{m-1}[(s+2)^{q+1}-s^{q+1}]}}{\varphi(s-s_0)^{\frac{1}{m-1}}} \left(\int_{s_5+s_0+2}^{s+s_0+2} e^{\frac{p}{n-1}[(t+2)^{q+1}-t^{q+1}]} dt \right)^{\frac{n-1}{m-1}} ds < +\infty, \tag{4.2}$$

and (this is just (c3))

$$\varphi \circ \left(\left[\frac{\log^+ u}{p} \right]^{\frac{1}{q+1}} \right) \in \mathcal{L}^1_{\mathrm{loc}}(\Omega).$$

The condition (4.2) is of course complicated, but it is easy to get simpler (but) stronger conditions, e.g., just estimating the inner integral.

Example 5. Let u be separately subharmonic in Ω and $\varphi : [0,+\infty) \to [0,+\infty)$ be a strictly increasing surjection. Let $p > 0$ and $\phi : [0,+\infty) \to [0,+\infty)$ be a strictly increasing surjection for which there is $t_1 > 1$ such that

$$s = \phi(t) = (\log t)^p \quad \text{for} \quad t \geq t_1, \qquad \text{and thus} \quad t = \phi^{-1}(s) = e^{s^{\frac{1}{p}}}.$$

Corollary 4.3 cannot now be applied, but from Corollary 4.2 it follows that u is subharmonic, provided that, in addition to the integrability condition (c3),

$$\varphi \circ \log^{+} \circ((\log^{+} u^{+})^{p}) \in \mathcal{L}^{1}_{\mathrm{loc}}(\Omega),$$

also the condition (c2) holds. One possibility to replace the, again rather complicated, condition (c2) by a simpler, but stronger one, is the following (we leave the details to the reader):

$$\int\limits_{s_5}^{+\infty} e^{\frac{2}{m-1}[e^{\frac{1}{p}(s+s_0+4)} - e^{\frac{1}{p}(s+s_0+2)}]} \frac{s^{\frac{n-1}{m-1}}}{\varphi(s-s_0)^{\frac{1}{m-1}}} ds < +\infty.$$

Remark 4.4. As is seen above, the proof of Theorem 4.1 is essentially based on a previous simple result of separately quasinearly subharmonic functions, namely on [26], Proposition 3.1, p. 57, see also [28], Proposition 3.2.1, p. e2620. The situation is of course similar in the special case of separately subharmonic functions: Armitage and Gardiner [2], proof of Theorem 1, pp. 257–259, base their result on the classical result of Avanissian [4], Théorème 9, p. 140. Equally well one might of course base the result on any of the following later results: [19], Proposition 3, p. 24, [14], Theorem, p. 31, Arsove [3], Theorem 1, p. 622, or [23], Theorem 1, p. 69. See also Lelong [18], Théorème 1 bis, p. 315, and Cegrell and Sadullaev [6], Theorem 3.1, p. 82. Therefore, there are indeed good reasons to improve also these old basic results. In this connection, we point out the following recent improvement:

Theorem 4.5 ([30], Theorem 2, pp. 367–368). *Let $K_1, K_2 \geq 1$. Let Ω be a domain in $\mathbb{R}^{m+n}$, $m, n \geq 2$. Let $u : \Omega \to [-\infty,+\infty)$ be such that*

(a) *for each $y \in \mathbb{R}^n$ the function*

$$\Omega(y) \ni x \mapsto u(x,y) \in [-\infty,+\infty)$$

is K_1-quasinearly subharmonic, and, for almost every $y \in \mathbb{R}^n$, subharmonic,

(b) *for each $x \in \mathbb{R}^m$ the function*

$$\Omega(x) \ni y \mapsto u(x,y) \in [-\infty,+\infty)$$

is upper semicontinuous, and, for almost every $x \in \mathbb{R}^m$, K_2-quasinearly subharmonic,

(c) *for some $p > 0$ there is a function $v \in \mathcal{L}^{p}_{\mathrm{loc}}(\Omega)$ such that $u^{+} \leq v$.*

Then for each $(a,b) \in \Omega$,

$$\limsup_{(x,y)\to(a,b)} u(x,y) \leq K_1 K_2\, u^{+}(a,b).$$

Observe that the proof of the above (quasinearly subharmonicity) result, and thus also the proof of the following special case result, is simpler than the proofs of the older subharmonicity results. (See also the previous versions [26], Corollary 3.2 and Corollary 3.3, p. 61, and [28], Corollary 3.2.4 and Corollary 3.2.5, p. e2621).

Corollary 4.6 ([30], Corollary 2, p. 369). *Let Ω be a domain in $\mathbb{R}^{m+n}$, $m, n \geq 2$. Let $u : \Omega \to [-\infty, +\infty)$ be such that*

(a) *for each $y \in \mathbb{R}^n$ the function*

$$\Omega(y) \ni x \mapsto u(x, y) \in [-\infty, +\infty)$$

is nearly subharmonic, and, for almost every $y \in \mathbb{R}^n$, subharmonic,

(b) *for each $x \in \mathbb{R}^m$ the function*

$$\Omega(x) \ni y \mapsto u(x, y) \in [-\infty, +\infty)$$

is upper semicontinuous, and, for almost every $x \in \mathbb{R}^m$, (nearly) subharmonic,

(c) *for some $p > 0$ there is a function $v \in \mathcal{L}^p_{\mathrm{loc}}(\Omega)$ such that $u^+ \leq v$.*

Then u is upper semicontinuous and thus subharmonic in Ω.

Proof. It is easy to see that for each $M \geq 0$, the function $u_M := \max\{u, -M\} + M$ satisfies the assumptions of Theorem 4.5. Thus u_M is upper semicontinuous. Since by [26], Corollary 3.1, p. 59 (see also [28], Corollary 3.2.3, pp. e2620–e2621, and [30], Corollary 1, p. 367), u_M is anyway nearly subharmonic, it is in fact subharmonic. Using then a standard result, see, e.g., [14], a), p. 8, one sees that u is subharmonic and thus also upper semicontinuous. □

Remark 4.7. Observe that Corollary 4.6 is partially related to the result [14], Proposition 2, pp. 34–35. Though our assumptions are partly slightly stronger, our proof (see [30], pp. 367–370) is, on the other hand, different and shorter.

References

[1] P. Ahern, W. Rudin, *Zero sets of functions in harmonic Hardy spaces*, Math. Scand. **73** (1993), 209–214.

[2] D.H. Armitage, S.J. Gardiner, *Conditions for separately subharmonic functions to be subharmonic*, Potential Anal. **2** (1993), 255–261.

[3] M.G. Arsove, *On subharmonicity of doubly subharmonic functions*, Proc. Amer. Math. Soc. **17** (1966), 622–626.

[4] V. Avanissian, *Fonctions plurisousharmoniques et fonctions doublement sousharmoniques*, Ann. Sci. École Norm. Sup. **78** (1961), 101–161.

[5] M. Brelot, *Minorantes sous-harmoniques, extrémales et capacités*, J. Math. Pures Appl. **24** (1945), 1–32.

[6] U. Cegrell, A. Sadullaev, *Separately subharmonic functions*, Uzbek. Math. J. **1** (1993), 78–83.

[7] Y. Domar, *On the existence of a largest subharmonic minorant of a given function*, Arkiv mat. **3**, nr. **39** (1957), 429–440.

[8] O. Dovgoshey, J. Riihentaus, *Bi-Lipschitz mappings and quasinearly subharmonic functions*, Int. J. Math. Math. Sci. **2010** (2010), 1–8.

[9] O. Dovgoshey, J. Riihentaus, *A remark concerning generalized mean value inequalities for subharmonic functions*, International Conference Analytic Methods of Mechanics and Complex Analysis, Dedicated to N.A. Kilchevskii and V.A. Zmorovich on the Occasion of their Birthday Centenary, Kiev, Ukraine, June 29–July 5, 2009, in: Transactions of the Institute of Mathematics of the National Academy of Ukraine **7**, no. **2** (2010), 26–33.

[10] O. Dovgoshey, J. Riihentaus, *Mean value type inequalities for quasinearly subharmonic functions*, Glasgow Math. J. **55** (2013), 349–368.

[11] C. Fefferman, E.M. Stein, *H^p spaces of several variables*, Acta Math. **129** (1972), 137–192.

[12] J.B. Garnett, *Bounded Analytic Functions*, Springer, New York, 2007 (Revised First Edition).

[13] J.W. Green, *Approximately subharmonic functions*, Proc. Amer. Math. Soc. **3** (1952), 829–833.

[14] M. Hervé, *Analytic and Plurisubharmonic Functions in Finite and Infinite Dimensional Spaces*, Lecture Notes in Mathematics 198, Springer-Verlag, Berlin, 1971.

[15] V. Kojić, *Quasi-nearly subharmonic functions and conformal mappings*, Filomat. **21**, no. **2** (2007), 243–249.

[16] P. Koskela, V. Manojlović, *Quasi-nearly subharmonic functions and quasiconformal mappings*, Potential Anal. **37** (2012), 187–196.

[17] Ü. Kuran, *Subharmonic behavior of $\mid h \mid^p$, ($p > 0$, h harmonic)*, J. London Math. Soc. (2) **8** (1974), 529–538.

[18] P. Lelong, *Les fonctions plurisousharmoniques*, Ann. Sci. École Norm. Sup. **62** (1945), 301–338.

[19] P. Lelong, *Plurisubharmonic Functions and Positive Differential Forms*, Gordon and Breach, London, 1969.

[20] M. Pavlović, *On subharmonic behavior and oscillation of functions on balls in $\mathbb{R}^n$*, Publ. Inst. Math. (Beograd) **55 (69)** (1994), 18–22.

[21] M. Pavlović, *Subharmonic behavior of smooth functions*, Math. Vesnik **48** (1996), 15–21.

[22] M. Pavlović, J. Riihentaus, *Classes of quasi-nearly subharmonic functions*, Potential Anal. **29** (2008), 89–104.

[23] J. Riihentaus, *On a theorem of Avanissian–Arsove*, Expo. Math. **7** (1989), 69–72.

[24] J. Riihentaus, *Subharmonic functions: non-tangential and tangential boundary behavior*, in: Function Spaces, Differential Operators and Nonlinear Analysis (FSDONA'99), Proceedings of the Syöte Conference 1999, V. Mustonen, J. Rákosnik (eds.), Math. Inst., Czech Acad. Science, Praha, 2000, pp. 229–238. (ISBN 80-85823-42-X)

[25] J. Riihentaus, *A generalized mean value inequality for subharmonic functions*, Expo. Math. **19** (2001), 187–190.

[26] J. Riihentaus, *Subharmonic functions, generalizations and separately subharmonic functions*, The XIVth Conference on Analytic Functions, July 22–28, 2007, Chełm, Poland, in: Scientific Bulletin of Chełm, Section of Mathematics and Computer Science **2** (2007), 49–76. (ISBN 978-83-61149-24-8) (arXiv:math/0610259v5 [math.AP] 8 Oct 2008)

[27] J. Riihentaus, *Quasi-nearly subharmonicity and separately quasi-nearly subharmonic functions*, J. Inequal. Appl. **2008**, Article ID 149712, 15 pages, 2008. (doi: 10.1155/2008/149712)

[28] J. Riihentaus, *Subharmonic functions, generalizations, weighted boundary behavior, and separately subharmonic functions: A survey*, Fifth World Congress of Nonlinear Analysts (WCNA 2008), Orlando, Florida, USA, July 2–9, 2008, in: Nonlinear Analysis, Series A: Theory, Methods & Applications **71**, no. **12** (2009), pp. e2613–e2627. (doi: 10.1016/j.na.2009.05.077)

[29] J. Riihentaus, *Domination conditions for families of quasinearly subharmonic functions*, Int. J. Math. Math. J. **2011**, Article ID 729849, 9 pages, 2011. (doi: 10.1155/2011/729849)

[30] J. Riihentaus, *Remarks on separately subharmonic functions*, International Conference Complex Analysis, Potential Theory and Applications, Dedicated to the Memory of Professor P.M. Tamrazov, Kiev, Ukraine, August 19–23, 2013, in: Transactions of the Mathematics of the National Academy of Sciences of Ukraine, Special Issue to Commemorate Professor P.M. Tamrazov, **10**, no. **1** (2013), 362–371.

[31] P.J. Rippon, *Some remarks on largest subharmonic minorants*, Math. Scand. **49** (1981), 128–132.

[32] N. Sjöberg, *Sur les minorantes sousharmoniques d'une fonction donnée*, Neuvième Congrès des mathématiciens Scandinaves, Helsingfors, pp. 309–319, 1938.

[33] A. Torchinsky, *Real-Variable Methods in Harmonic Analysis*, Academic Press, London, 1986.

[34] M. Vuorinen, *On the Harnack constant and the boundary behavior of Harnack functions*, Ann. Acad. Sci. Fenn. Ser. AI **7**, no. **2** (1982), 259–277.

[35] J. Wiegerinck, *Separately subharmonic functions need not be subharmonic*, Proc. Amer. Math. Soc. **104** (1988), 770–771.

[36] J. Wiegerinck, R. Zeinstra, *Separately subharmonic functions: when are they subharmonic*, in: Proceedings of Symposia in Pure Mathematics Vol. **52**, part 1, Eric Bedford, John P. D'Angelo, Robert E. Greene, Steven G. Krantz (eds.), Amer. Math. Soc., Providence, Rhode Island, 1991, pp. 245–249.

Juhani Riihentaus
University of Oulu, Department of Mathematical Sciences
P.O. Box 3000
FI-90014 Oulun yliopisto, Finland

and

University of Eastern Finland, Department of Physics and Mathematics
P.O. Box 111
FI-80101 Joensuu, Finland
e-mail: `riihentaus@member.ams.org`

Ordered Structures and Applications: Positivity VII
Trends in Mathematics, 415–467

Spectral Radii and Collatz–Wielandt Numbers for Homogeneous Order-preserving Maps and the Monotone Companion Norm

Horst R. Thieme

Abstract. It is well known that an ordered normed vector space X with normal cone X_+ has an order-preserving norm that is equivalent to the original norm. Such an equivalent order-preserving norm is given by

$$\sharp x \sharp = \max\{d(x, X_+), d(x, -X_+)\}, \qquad x \in X.$$

This paper explores the properties of this norm and of the half-norm $\psi(x) = d(x, -X_+)$ independently of whether or not the cone is normal. We use ψ to derive comparison principles for the solutions of abstract integral equations and compare Collatz–Wielandt numbers, bounds, and order-spectral radii for order-preserving homogeneous maps and give conditions for a local upper Collatz–Wielandt radius to have a lower positive eigenvector. For illustration, we consider a rank-structured population with mating.

Mathematics Subject Classification (2010). Primary 47H07, 47J10; Secondary 47N60.

Keywords. Ordered normed vector space, homogeneous map, normal cone, cone spectral radius, Collatz–Wielandt numbers, Krein–Rutman theorem, nonlinear eigenvectors, comparison theorems, basic reproduction number, population models with mating, rank structure.

1. Introduction and exposé of concepts and results

For models in the biological, social, or economic sciences, there is a natural interest in solutions that are positive in an appropriate sense, i.e., they take their values in the cone of an ordered normed vector space.

1.1. Cones and their properties

A closed subset X_+ of a normed real vector space X is called a *wedge* if

(i) X_+ is convex,
(ii) $\alpha x \in X_+$ whenever $x \in X_+$ and $\alpha \in \mathbb{R}_+$.

A wedge is called a *cone* if $X_+ \cap (-X_+) = \{0\}$. A wedge is called *solid* if it contains interior points. A wedge X_+ is called *generating* if $X = X_+ - X_+$, and *total* if X is the closure of $X_+ - X_+$. A cone X_+ is called *normal*, if there exists some $\delta > 0$ such that

$$\|x + z\| \geq \delta \text{ whenever } x \in X_+, z \in X_+, \|x\| = 1 = \|z\|. \tag{1.1}$$

Equivalent conditions for a cone to be normal are given in Theorem 2.1.

Nonzero points in a cone or wedge are called *positive*. In function spaces, typical cones are formed by the nonnegative functions.

X_+ is called an inf-*semilattice* [1] (or *minihedral* [24]) if $x \wedge y = \inf\{x, y\}$ exist for all $x, y \in X_+$. X_+ is called a sup-*semilattice* if $x \vee y = \sup\{x, y\}$ exist for all $x, y \in X_+$. X_+ is called a *lattice* if $x \wedge y$ and $x \vee y$ exist for all $x, y \in X_+$.

X is called a *lattice* if $x \vee y$ exist for all $x, y \in X$. Since $x \wedge y = -((-x) \vee (-y))$, also $x \wedge y$ exist for all x, y in a lattice X.

If X_+ is a cone in X, we introduce a *partial order* on X by $x \leq y$ if $y - x \in X_+$ for $x, y \in X$ and call X an *ordered normed vector space.*

Every ordered normed vector space carries the monotone companion half-norm $\psi(x) = d(x, -X_+)$,

$$\psi(x) \leq \psi(y), \qquad x, y \in X, x \leq y,$$

and the monotone companion norm $\sharp x \sharp = \max\{\psi(x), \psi(-x)\}$. See [2], [25, (4.2)], [9, L.4.1]. The cone X_+ is normal if and only if the monotone companion norm is equivalent to the original norm. By the open mapping theorem, X cannot be complete with respect to both norms unless X_+ is normal. The properties of the companion (half-) norm are studied in Section 3. In the following, we describe its applications to homogeneous order-preserving maps. Another application, positivity of solutions to abstract integral inequalities is given in Section 6.

1.2. Spectral radii for homogeneous order-preserving maps

For a linear bounded map B on a complex Banach space, the spectral radius of B is defined as

$$\mathbf{r}(B) = \sup\{|\lambda|; \lambda \in \sigma(B)\}, \tag{1.2}$$

where $\sigma(B)$ is the spectrum of B, $\sigma(B) = \mathbb{C} \setminus \rho(B)$, and $\rho(B)$ the resolvent set of B, i.e., the set of those $\lambda \in \mathbb{C}$ for which $\lambda - B$ has a bounded everywhere defined inverse. The following alternative formula holds,

$$\mathbf{r}(B) = \inf_{n \in \mathbb{N}} \|B^n\|^{1/n} = \lim_{n \to \infty} \|B^n\|^{1/n}, \tag{1.3}$$

which is also meaningful in a real Banach space. If B is a compact linear map on a complex Banach space and $\mathbf{r}(B) > 0$, then there exists some $\lambda \in \sigma(B)$ and $v \in X$ such that $|\lambda| = \mathbf{r}(B)$ and $Bv = \lambda v \neq 0$. Such a λ is called an eigenvalue of B. This raises the question whether $\mathbf{r}(B)$ could be an eigenvalue itself. There is a positive answer, if B is a positive operator and satisfies some generalized compactness assumption.

Throughout the rest of this paper, let X be an ordered normed vector space with cone X_+. We use the notation

$$\dot{X} = X \setminus \{0\} \quad \text{and} \quad \dot{X}_+ = X_+ \setminus \{0\}.$$

Definition 1.1. Let X and Z be ordered vector space with cones X_+ and Z_+ and $U \subseteq X$. $B : U \to Z$ is called *positive* if $B(U \cap X_+) \subseteq Z_+$.

B is called *order-preserving* (or monotone or increasing) if $Bx \leq By$ whenever $x, y \in U$ and $x \leq y$.

Positive linear maps from X to Y are order-preserving. Positive linear maps on X have the remarkable property that their spectral radius is a spectral value [8] [38, App. 2.2] if X is a Banach space and X_+ a normal generating cone.

The celebrated Krein–Rutman theorem [27], which generalizes parts of the Perron–Frobenius theorem to infinite dimensions, establishes that a compact positive linear map B with $\mathbf{r}(B) > 0$ on an ordered Banach space X with total cone X_+ has an eigenvector $v \in \dot{X}_+$ such that $Bv = \mathbf{r}(B)v$ and a positive bounded linear eigenfunctional $v^* : X \to \mathbb{R}$, $v^* \neq 0$, such that $v^* \circ B = \mathbf{r}(B)v^*$. This theorem has been generalized into various directions by Bonsall [7] and Birkhoff [5], Nussbaum [34, 35], and Eveson and Nussbaum [15] (see these papers for additional references).

1.2.1. Homogenous maps. In the following, X, Y and Z are ordered normed vector spaces with cones X_+, Y_+ and Z_+ respectively.

Definition 1.2. $B : X_+ \to Y$ is called (*positively*) *homogeneous* (*of degree one*), if $B(\alpha x) = \alpha B(x)$ for all $\alpha \in \mathbb{R}_+$, $x \in X_+$.

Since we do not consider maps that are homogeneous in other ways, we will simply call them homogeneous maps. If follows from the definition that $B(0) = 0$. Homogeneous maps are not Fréchet differentiable at 0 unless $B(x + y) = B(x) + B(y)$ for all $x, y \in X_+$. For the following holds.

Proposition 1.3. *Let $B : X_+ \to Y$ be homogeneous. Then the directional derivatives of B exist at 0 in all directions of the cone and*

$$\partial B(0, x) = \lim_{t \to 0+} \frac{B(tx) - B(0)}{t} = B(x), \qquad x \in X_+.$$

There are good reasons to consider homogeneous maps. Here is a mathematical one.

Theorem 1.4. *Let $F : X_+ \to Y$ and $u \in X$. Assume that the directional derivatives of F at u exist in all directions of the cone. Then the map $B : X_+ \to Y$, $B = \partial F(u, \cdot)$, is homogeneous,*

$$B(x) = \partial F(u, x) = \lim_{t \to 0+} \frac{F(u + tx) - F(u)}{t}, \qquad x \in X_+.$$

Proof. Let $\alpha \in \mathbb{R}_+$. Obviously, if $\alpha = 0$, $B(\alpha x) = 0 = \alpha B(x)$. So we assume $\alpha \in (0, \infty)$. Then

$$\frac{F(u + t[\alpha x]) - F(u)}{t} = \alpha \frac{F(u + [t\alpha]x) - F(u)}{t\alpha}.$$

As $t \to 0$, also $\alpha t \to 0$ and so the directional derivative in direction αx exists and $\partial F(u, \alpha x) = \alpha F(u, x)$. □

Another good reason are mathematical population models that take into account that, for many species, reproduction involves a mating process between two sexes. The maps involved therein are not only homogeneous but also order-preserving.

The eigenvector problem for homogeneous maps is quite different from the one for linear maps. Consider the following simple two sex population model $x_n = Bx_{n-1}$ where $x_n = (f_n, m_n)$ and $Bx = (B_1 x, B_2 x)$,

$$B_1(f, m) = p_f f + \beta_f \frac{fm}{f + m}, \qquad B_2(f, m) = p_m m + \beta_m \frac{fm}{f + m}. \tag{1.4}$$

This system models a population of females and males which reproduce once a year with f_n and m_n representing the number of females and males at the beginning of year n. The numbers p_f, p_m are the respective probabilities of surviving one year. The harmonic mean describes the mating process and the parameters β_f and β_m scale with the resulting amount of offspring per mated pair.

B always has the eigenvectors $(1, 0)$ and $(0, 1)$ associated with p_f and p_m respectively. Looking for a different eigenvalue, λ, of B, we can assume that $f + m = 1$ and obtain

$$\frac{\lambda - p_f}{\beta_f} = m, \qquad \frac{\lambda - p_m}{\beta_m} = f, \qquad m + f = 1.$$

We add and solve for λ,

$$\lambda = \frac{\beta_f \beta_m + \beta_m p_f + \beta_f p_m}{\beta_f + \beta_m},$$

and then for m and f,

$$m = \frac{\beta_m + p_m - p_f}{\beta_f + \beta_m}, \qquad f = \frac{\beta_f + p_f - p_m}{\beta_f + \beta_m}.$$

The eigenvector (m, f) lies in the biological relevant positive quadrant if and only if the eigenvalue λ is larger than the two other eigenvalues p_f and p_m.

We notice that we generically have three linearly independent eigenvectors (rather than at most two as for a 2×2 matrix) with the third being biological relevant if and only if it is associated with the largest eigenvalue.

The spectral radius of a positive linear map has gained considerable notoriety because of its relation to the basic reproduction number of population models which have a highly dimensional structure but implicitly assume a one to one sex ratio [3, 12, 13, 41, 48]. A spectral radius for homogeneous order-preserving maps should play a similar role as an extinction versus persistence threshold parameter for structured populations with two sexes [20, 21, 22, 23].

1.2.2. Cone norms for homogeneous bounded maps. For a homogeneous map $B : X_+ \to Y$, we define

$$\|B\|_+ = \sup\{\|B(x)\|; x \in X_+, \|x\| \leq 1\} \tag{1.5}$$

and call B *bounded* if this supremum is a real number. Since B is homogeneous,

$$\|B(x)\| \leq \|B\|_+ \, \|x\|, \qquad x \in X_+. \tag{1.6}$$

Let $H(X_+, Y)$ denote the set of bounded homogeneous maps $B : X_+ \to Y$, $H(X_+, Y_+)$ denote the set of bounded homogeneous maps $B : X_+ \to Y_+$, and $\mathrm{HM}(X_+, Y_+)$ the set of those maps in $H(X_+, Y_+)$ that are order-preserving.

$H(X_+, Y)$ is a real vector space and $\| \cdot \|_+$ is a norm on $H(X_+, Y)$ called the cone-norm. $H(X_+, Y_+)$ and $\mathrm{HM}(X_+, Y_+)$ are cones in $H(X_+, Y)$. We write $H(X_+) = H(X_+, X_+)$ and $\mathrm{HM}(X_+) = \mathrm{HM}(X_+, X_+)$.

If $B \in H(X_+, Y_+)$ and $C \in H(Y_+, Z_+)$, then $CB \in H(X_+, Z_+)$ and

$$\|CB\|_+ \leq \|C\|_+ \, \|B\|_+. \tag{1.7}$$

1.2.3. Cone and orbital spectral radius. Let $B \in H(X_+)$ and define $\phi : \mathbb{Z}_+ \to \mathbb{R}$ by $\phi(n) = \ln \|B^n\|_+$. Then $\phi(m + n) \leq \phi(m) + \phi(n)$ for all $m, n \in \mathbb{Z}_+$, and a well-known result implies the following formula for the *cone spectral radius*

$$\mathbf{r}_+(B) := \inf_{n \in \mathbb{N}} \|B^n\|_+^{1/n} = \lim_{n \to \infty} \|B^n\|_+^{1/n}, \tag{1.8}$$

which is analogous to (1.3). Mallet-Paret and Nussbaum [31, 32] suggest an alternative definition of a spectral radius for homogeneous (not necessarily bounded) maps $B : X_+ \to X_+$. First, define asymptotic least upper bounds for the geometric growth factors of B-orbits,

$$\gamma(x, B) := \gamma_B(x) := \limsup_{n \to \infty} \|B^n(x)\|^{1/n}, \qquad x \in X_+, \tag{1.9}$$

and then

$$\mathbf{r}_o(B) = \sup_{x \in X_+} \gamma_B(x). \tag{1.10}$$

Here $\gamma_B(x) := \infty$ if the sequence $(\|B^n(x)\|^{1/n})$ is unbounded and $\mathbf{r}_o(B) = \infty$ if $\gamma_B(x) = \infty$ for some $x \in X_+$ or the set $\{\gamma_B(x); x \in X_+\}$ is unbounded.

The number $\mathbf{r}_+(B)$ has been called *partial spectral radius* by Bonsall [8], X_+ spectral radius by Schaefer [37, 38], and *cone spectral radius* by Nussbaum [35]. Mallet-Paret and Nussbaum [31, 32] call $\mathbf{r}_+(B)$ the *Bonsall cone spectral radius* and $\mathbf{r}_o(B)$ the cone spectral radius. For $x \in X_+$, the number $\gamma_B(x)$ has been called *local spectral radius* of B at x by Förster and Nagy [16].

We will follow Nussbaum's older terminology [35] which shares the spirit with Schaefer's [37] term X_+ *spectral radius* and stick with *cone spectral radius* for $\mathbf{r}_+(B)$ and call $\mathbf{r}_o(B)$ the *orbital spectral radius* of B. Later, we will also introduce a *Collatz–Wielandt* radius. We refer to all of them as *order-spectral radii*. One readily checks that

$$\mathbf{r}_+(\alpha B) = \alpha \mathbf{r}_+(B), \quad \alpha \in \mathbb{R}_+, \qquad \mathbf{r}_+(B^m) = (\mathbf{r}_+(B))^m, \qquad m \in \mathbb{N}. \tag{1.11}$$

The same properties hold for $\mathbf{r}_o(B)$ though proving the second property takes some more effort [31, Prop. 2.1]. Actually, as we show in Section 3, if B is bounded,

$$\gamma(x, B^m) = (\gamma(x, B))^m, \qquad m \in \mathbb{N}, x \in X_+, \tag{1.12}$$

which implies

$$\mathbf{r}_o(B^m) = (\mathbf{r}_o(B))^m, \qquad m \in \mathbb{N}. \tag{1.13}$$

The cone spectral radius and the orbital spectral radius are meaningful if B is just positively homogeneous and bounded, but as in [31, 32] we will be mainly interested in the case that B is also order-preserving and continuous.

The two concepts coincide for many practical purposes. Gripenberg [17], gives an example for $\mathbf{r}_o(B) < \mathbf{r}_+(B)$.

Theorem 1.5. *Let X be an ordered normed vector with cone X_+ and $B : X_+ \to X_+$ be bounded, homogeneous and order-preserving.*

Then $\mathbf{r}_+(B) \ge \mathbf{r}_o(B) \ge \gamma_B(x)$, $x \in X_+$.

Further $\mathbf{r}_o(B) = \mathbf{r}_+(B)$ if one of the following hold:

(i) *X_+ is complete and normal.*
(ii) *B is continuous and power compact.*
(iii) *X_+ is complete and B is continuous and additive ($B(x + y) = B(x) + B(y)$ for all $x \in X_+$).*
(iv) *X_+ is normal and a power of B is uniformly order-bounded.*

The inequality is a straightforward consequence of the respective definitions. For the concepts and the proof of (iv) see Section 10 and Theorem 12.10. The conditions (ii) and (iii) have been verified in [31, Sec. 2] as has the condition (i) under the assumption that B is continuous (the overall assumption of [31] that X is a Banach space is not used in the proofs). Statement (i), without continuity of B, has been proved in [17] where also the normality assumption is shifted from the cone X_+ to the map B. See Theorem 3.1 for a further step in that direction.

Under assumptions (i), (ii) and (iv), there exists some $x \in X_+$ such that $\mathbf{r}_+(B) = \gamma_B(x)$, but only under (iv) the element x is known. See Theorems 3.1, 3.2 and 12.10.

For a bounded positive linear operator on an ordered Banach space, the spectral radius and cone spectral radius coincide provided that the cone is generating [32, Thm. 2.14]. This is not true if the cone is only total [8, Sec. 2.8].

1.2.4. Lower Collatz–Wielandt numbers. For homogeneous order-preserving $B : X_+ \to X_+$, the *lower Collatz–Wielandt number* of B at $x \in \dot{X}_+$ is defined as [16]

$$[B]_x = \sup\{\lambda \ge 0; B(x) \ge \lambda x\}. \tag{1.14}$$

Since X_+ is closed, $[B]_x \in \mathbb{R}_+$ and it is a lower eigenvalue of B,

$$B(x) \geq [B]_x x, \qquad x \in \dot{X}_+. \tag{1.15}$$

Lemma 1.6. *Let $B, C : X_+ \to X_+$ be homogeneous and order-preserving. Let $x \in \dot{X}_+$. Then $[CB]_x \geq [C]_x[B]_x$.*

This obvious result implies

$$b_{n+m} \geq b_n b_m, \quad b_n = [B^n]_x, \qquad n, m \in \mathbb{N}. \tag{1.16}$$

The *lower local Collatz–Wielandt radius* of B at $x \in \dot{X}_+$ is defined as

$$\eta_x(B) =: \sup_{n\in\mathbb{N}} [B^n]_x^{1/n}. \tag{1.17}$$

This implies

$$\eta_x(B^n) \leq (\eta_x(B))^n, \qquad n \in \mathbb{N}. \tag{1.18}$$

The *lower Collatz–Wielandt bound* is defined as

$$cw(B) = \sup_{x\in\dot{X}_+} [B]_x = \sup\{\lambda \geq 0;\ \exists x \in \dot{X}_+ : B(x) \geq \lambda x\}, \tag{1.19}$$

and the *Collatz–Wielandt radius* of B is defined as

$$\mathbf{r}_{cw}(B) = \sup_{x\in\dot{X}_+} \eta_x(B). \tag{1.20}$$

From the definitions, $cw(B) \leq \mathbf{r}_{cw}(B)$ and, by (1.18),

$$\mathbf{r}_{cw}(B^n) \leq (\mathbf{r}_{cw}(B))^n, \qquad n \in \mathbb{N}. \tag{1.21}$$

A homogeneous, bounded, order-preserving map $B : X_+ \to X_+$ is also bounded with respect to the monotone companion norm and $\sharp B \sharp_+ \leq \|B\|_+$. See Section 5. So we can define the companion cone spectral radius, $\mathbf{r}_+^\sharp(B)$, and the companion orbital spectral radius, $\mathbf{r}_o^\sharp(B)$, in full analogy to (1.8) and (1.10). If X_+ is normal (and the original and the companion norm are equivalent), the companion radii coincide with the original ones.

The monotonicity of the companion norm makes it possible to connect the Collatz–Wielandt radius and the other spectral radii by inequalities (Section 8).

Theorem 1.7. *Let $B : X_+ \to X_+$ be homogeneous, bounded, and order-preserving. Then*

$$cw(B) \leq \mathbf{r}_{cw}(B) \leq \mathbf{r}_o^\sharp(B) \leq \left\{ \begin{array}{c} \mathbf{r}_o(B) \\ \mathbf{r}_+^\sharp(B) \end{array} \right\} \leq \mathbf{r}_+(B).$$

If X_+ is complete (with respect to the original norm), then $\mathbf{r}_+^\sharp(B) \leq \mathbf{r}_o(B)$.

If B is also compact, the following progressively more general results have been proved over the years which finally lead to an extension of the Krein–Rutman theorem from linear to homogeneous maps.

Theorem 1.8. *Let B be compact, continuous, homogeneous and order-preserving and let r be any of the numbers $cw(B), \mathbf{r}_{cw}(B), \mathbf{r}_+(B)$. Then, if $r > 0$, there exists some $v \in \dot{X}_+$ with $B(v) = rv$. More specifically,*

$r = cw(B)$	:	Krein–Rutman 1948 [27],
$r = \mathbf{r}_{cw}(B)$	:	Krasnosel'skii 1964 [24, Thm. 2.5],
$r = \mathbf{r}_+(B)$	:	Nussbaum 1981 [34], Lemmens Nussbaum [30].

Actually, if $r = \mathbf{r}_+(B)$, then $cw(B) = \mathbf{r}_{cw}(B) = \mathbf{r}_+(B)$.

A proof for $r = cw(B)$ can also been found in [9]. [24, Thm. 2.5] is only formulated for the case that B is defined and linear on X, but the proof also works under the assumptions made above. The case $r = \mathbf{r}_{cw}(B)$ can also been found in [34, Cor. 2.1]. The case $r = \mathbf{r}_+(B)$ is basically proved in [34, Thm. 2.1], but some finishing touches are contained in the introduction of [30]. If $B(v) = rv$ with $v \in \dot{X}_+$ and $r = \mathbf{r}_+(B)$, then $r \le [B]_v \le cw(B)$ and equality of all three numbers follows.

It is well known that the compactness of B can be substantially relaxed though not completely dropped if B is linear [34]. This is also possible (to a lesser degree) if B is just homogeneous using homogeneous measures of noncompactness [31, 32]. We only mention two special cases of [31, Thm. 3.1] and [32, Thm. 4.9], respectively.

Theorem 1.9. *Let X_+ be complete. Let $B : X_+ \to X_+$ and $B = K + H$ where $K : X_+ \to X_+$ is homogeneous, continuous, order-preserving and compact and $H : X_+ \to X_+$ is homogeneous, continuous, and order-preserving.*

Then there exists some $v \in \dot{X}_+$ with $B(v) = \mathbf{r}_+(B)v$ if $\mathbf{r}_+(B) > 0$ and one of the two following conditions is satisfied in addition:

(a) *H is Lipschitz continuous on X_+, $\|H(x) - H(y)\| \le \Lambda\|x - y\|$ for all $x, y \in X_+$, with $\Lambda < \mathbf{r}_+(B)$.*
(b) *X_+ is normal, H is cone-additive ($H(x+y) = H(x)+H(y)$ for all $x, y \in X_+$), and $\mathbf{r}_+(H) < \mathbf{r}_+(B)$.*

The following observation is worth mentioning [25, Thm. 9.3] [35, Thm. 2.2].

Proposition 1.10. *Let $B : X_+ \to X_+$ be* positively linear, *i.e., $B(\alpha x+y) = \alpha B(x)+B(y)$ for all $x, y \in X_+$ and $\alpha \ge 0$. If $r > 0$, $w \in \dot{X}_+$, $m \in \mathbb{N}$ and $B^m(w) = r^m w$, then $B(v) = rv$ for some $v \in \dot{X}_+$.*

Simply set

$$v = \sum_{j=10}^{m-1} r^{-j} B^j(w). \tag{1.22}$$

It seems to be an open problem (with a beer barrel scent [33, 47]) whether any of the results in Theorem 1.8 holds if B^2 or some higher power of B is compact rather than B itself. For additional conditions to make such a result hold, see [1, Thm. 7.3], conditions (i) and (iii), and [42, Sec. 7].

The construction in (1.22) can be modified to yield the following result.

Proposition 1.11. *Let X_+ be a sup-semilattice and $B : X_+ \to X_+$ be homogeneous and order-preserving. If $r > 0$, $w \in \dot{X}_+$, $m \in \mathbb{N}$ and $B^m(w) \geq r^m w$, then $B(v) \geq rv$ for some $v \in \dot{X}_+$.*

This time, choose v as in the proof of [1, Thm. 5.1],

$$v = \bigvee_{j=0}^{m-1} r^{-j} B^j(w). \tag{1.23}$$

Notice that the number r in Proposition 1.11 satisfies $r \leq [B]_v \leq cw(B)$.

This observation provides conditions for equality to hold in Theorem 1.7. In order not to burden our representation with technical language or a list of various special cases and to include possible further developments, we make the following definition.

Definition 1.12. A homogeneous bounded map $B : X_+ \to X_+$ has the *KR property* (the Krein–Rutman property) if there is some $v \in \dot{X}_+$ with $B(v) = \mathbf{r}_+(B)v$ whenever $\mathbf{r}_+(B) > 0$.

B has the *lower KR property* if there is some $v \in \dot{X}_+$ with $B(v) \geq \mathbf{r}_+(B)v$ whenever $\mathbf{r}_+(B) > 0$.

The maps in Theorem 1.9 are examples that satisfy the KR property while power-compact, homogeneous, order-preserving, continuous maps on sup-semilattices are examples that satisfy the lower KR property (Proposition 1.11). $B : X_+ \to X_+$ is called *power-compact* if B^m is compact for some $m \in \mathbb{N}$.

We emphasize the lower KR property in addition to the KR property because of our interest in population dynamics. If B is the homogeneous first order approximation of a nonlinear map F at 0, the condition $\mathbf{r}_+(B) < 1$ is often enough to guarantee the local (and some sometime the global) stability of 0 for the dynamical system induced by F whether or not $\mathbf{r}_+(B)$ is an eigenvalue associated with a positive eigenvector. To prove persistence of the dynamical system, it is very helpful to have a positive lower eigenvector $B(v) \geq rv$ with $r > 1$ [21, 22, 23]. So the lower KR property of B (more generally the equality $cw(B) = \mathbf{r}_+(B)$) guarantees $\mathbf{r}_+(B)$ to be a sharp threshold parameter that separates local stability of the extinction equilibrium 0 from persistence of the population. It should be mentioned that the KR property only refers to one part of the Krein–Rutman theorem, the existence of an eigenvector associated with the cone spectral radius. As for the existence of a homogeneous order-preserving eigenfunctional, see [42, 44].

The lower KR property turns some of the inequalities in Theorem 1.7 into equalities.

Theorem 1.13. *Let B be bounded, homogeneous, order-preserving and B^m have the lower KR property for some $m \in \mathbb{N}$. Then*

$$\mathbf{r}_{cw}(B) = \mathbf{r}_o^\sharp(B) = \mathbf{r}_+^\sharp(B) = \mathbf{r}_o(B) = \mathbf{r}_+(B).$$

If X_+ is a sub-semilattice or $m = 1$, then also $cw(B) = \mathbf{r}_+(B)$ and B has the lower KR property.

The significance of $cw(B) = \mathbf{r}_+(B)$ is that the lower Collatz–Wielandt numbers $[B]_x$ provide lower estimates for $\mathbf{r}_+(B)$ that get arbitrarily sharp by choosing $x \in \dot{X}_+$ in the right way. See Section 14 for attempts in that direction for a rank-structured population model with mating.

Proof. The equalities follow from Theorem 1.7 if $\mathbf{r}_+(B) = 0$. So we can assume that $\mathbf{r}_+(B) > 0$. If B has the lower KR property, it follows immediately from the definitions that $\mathbf{r}_+(B) \le cw(B)$ and equality follows from Theorem 1.7. If B^m has the lower KR property, by (1.11) and (1.21),

$$(\mathbf{r}_+(B))^m = \mathbf{r}_+(B^m) \le cw(B^m) \le \mathbf{r}_{cw}(B^m) \le (\mathbf{r}_{cw}(B))^m,$$

and the equalities follow again from Theorem 1.7. If $r = \mathbf{r}_+(B) > 0$, there exists some $w \in \dot{X}_+$ with $B^m(w) \ge r^m w$. By Proposition 1.11, $B(v) \ge rv$ for some $v \in \dot{X}_+$ and $r \ge cw(B)$. □

The following monotone dependence of the various spectral radii on the map can be shown without assuming normality of the cone.

Theorem 1.14. *Let $A, B : X_+ \to X_+$ be bounded and homogeneous. Assume that $A(x) \le B(x)$ for all $x \in X_+$.*

If A is order-preserving and has the lower KR property, then $\mathbf{r}_+(A) = cw(A) \le cw(B)$.

1.2.5. Upper Collatz–Wielandt numbers. For $x \in \dot{X}_+$, the upper *Collatz–Wielandt numbers* of B at x [16] is defined as

$$\|B\|_x = \inf\{\lambda \ge 0; B(x) \le \lambda x\},$$

with the convention that $\inf \emptyset = \infty$, and the *local upper Collatz–Wielandt radius* $\eta^u(B)$ at $u \in \dot{X}_+$ as

$$\eta^u(B) = \inf_{n \in \mathbb{N}} \|B^n\|_u^{1/n}.$$

Collatz–Wielandt numbers, without this name, became more widely known when Wielandt used them for a new proof of the Perron–Frobenius theorem [49]. We will use them closer to Collatz' original purpose proving inclusion theorems (Einschließungssätze) for $\mathbf{r}_+(B)$ which generalize those in [10, 11].

If the cone X_+ is solid, with nonempty interior $\breve{X}_+$, one can define an *upper Collatz–Wielandt bound* (cf. [1, Sec. 7]) by

$$CW(B) = \inf_{x \in \breve{X}_+} \|B\|_x.$$

These new numbers, bound, and radius relate to the former bounds and radii in the following way and show that upper Collatz–Wielandt numbers taken at interior points provide upper estimates of the cone spectral radius. For this exposé, we illustrate our results for the special case of a solid cone.

Theorem 1.15. *Let X_+ be solid with nonempty interior $\breve{X}_+$ and $B : X_+ \to X_+$ be homogeneous, bounded and order-preserving. Then, for all $u \in \breve{X}_+$,*

$$\gamma_B(u) = \lim_{n\to\infty} \|B^n(u)\|^{1/n}, \qquad \eta^u(B) = \lim_{n\to\infty} \|B^n\|_u^{1/n},$$

and

$$cw(B) \le \mathbf{r}_{cw} \le \mathbf{r}_+^\sharp(B) \le \eta^u(B) \le \begin{cases} \gamma_u(B), \\ CW(B). \end{cases}$$

We obtain some equalities in the following cases.

- *If X_+ is normal, $\mathbf{r}_+(B) = \gamma_u(B) = \eta^u(B) \le CW(B)$ for all $u \in \breve{X}_+$.*
- *If B^m has the lower KR property for some $m \in \mathbb{N}$, $\mathbf{r}_{cw}(B) = \mathbf{r}_+(B) = \gamma_u(B) = \eta^u(B) \le CW(B)$ for all $u \in \breve{X}_+$.*
- *If B has the lower KR property, $cw(B) = \mathbf{r}_{cw}(B) = \mathbf{r}_+(B) = \gamma_u(B) = \eta^u(B) \le CW(B)$ for all $u \in \breve{X}_+$.*

Notice that the equality $\mathbf{r}_+(B) = \gamma_u(B) = \lim_{n\to\infty} \|B^n(u)\|^{1/n}$ means that the cone spectral radius can be determined by following the growth factors of an arbitrarily chosen $u \in \breve{X}_+$.

The estimates from above by upper Collatz–Wielandt numbers $\|B\|_u$ with $u \in \breve{X}_+$ can be arbitrarily sharp by appropriate choice of u if the map is compact. The next result should be compared to [1, Thm. 7.3]. Notice that $CW(B)$ here is $cw(B)$ in [1, Thm. 7.3]. Use of the companion half-norm makes it possible to drop the normality of the cone.

Theorem 1.16. *Let X_+ be solid and complete. Let $B : X_+ \to X_+$ and $B = K + A$ where $K : X_+ \to X_+$ is homogeneous, continuous, order preserving and compact and $A : X \to X$ is linear, positive, and bounded and $\mathbf{r}(A) < \mathbf{r}_+(B)$.*

Then $cw(B) = \mathbf{r}_{cw}(B) = \mathbf{r}_+(B) = \gamma_B(u) = \eta^u(B) = CW(B)$ for all $u \in \breve{X}_+$. If $r = CW(B) > 0$, then there exists some $v \in \dot{X}_+$ such that $B(v) = rv$.

The equality $\mathbf{r}_+(B) = CW(B)$ has also been established in [28, Thm. 4.6] for $B : \breve{X}_+ \to \breve{X}_+$ without any compactness assumption for B and without normality of the cone.

In Section 12, we prove more general versions of these theorems replacing solidity of the cone by uniform order-boundedness of the map.

Using the monotone companion half-norm, we show for certain classes of homogeneous maps B for which it is not clear whether they have the lower KR property that there is some $v \in \dot{X}_+$ such that $Bv \ge rv$ for $r = \eta^u(B)$ provided that $r > 0$ (Section 13).

In Section 14, we illustrate some of our results in a discrete model for a rank-structured population with mating.

A preliminary version of this paper has been posted on arXiv under a slightly different title [43].

2. More about cones

The cone X_+ of an ordered normed vector space is called *regular* if any decreasing sequence in X_+ converges. X_+ is called *fully regular* if any increasing bounded sequence in X_+ converges.

The norm of X is called *additive* on X_+ if $\|x + z\| = \|x\| + \|z\|$ for all $x, z \in X_+$. If the norm is additive, then the cone X_+ is normal.

2.1. Normal cones

The following result is well known [24, Sec. 1.2].

Theorem 2.1. *The following three properties are equivalent:*

(i) *X_+ is normal: There exists some $\delta > 0$ such that $\|x + z\| \geq \delta$ whenever $x \in X_+$, $z \in X_+$ and $\|x\| = 1 = \|z\|$.*

(ii) *The norm is semi-monotonic: There exists some $M \geq 0$ such that $\|x\| \leq M\|x + z\|$ for all $x, z \in X_+$.*

(iii) *There exists some $\tilde{M} \geq 0$ such that $\|x\| \leq \tilde{M}\|y\|$ whenever $x \in X$, $y \in X_+$, and $-y \leq x \leq y$.*

Remark 2.2. If X_+ were just a wedge, property (iii) would be rewritten as

> There exists some $\tilde{M} \geq 0$ such that $\|x\| \leq \tilde{M}\|y\|$ whenever $x \in X$, $y \in X_+$, and $y + x \in X_+$, $y - x \in X_+$.

Notice that this property implies that X_+ is cone: If $x \in X_+$ and $-x \in X_+$, then $0 + x \in X_+$ and $0 - x \in X_+$ and (iii) implies $\|x\| \leq \tilde{M}\|0\| = 0$.

Definition 2.3. An element $u \in X_+$ is called a *normal point* of X_+ if the set $\{\|x\|; x \in X_+, x \leq u\}$ is a bounded subset of $\mathbb{R}$; $u \in X_+$ is called a *regular point* of X_+ if every monotone sequence (x_n) in X_+ with $x_n \leq u$ for all $n \in \mathbb{N}$ converges.

Each regular point in X_+ is a normal point [44]. The following is proved in [25, Thm. 4.1].

Theorem 2.4. *Let X be an ordered normed vector space with cone X_+. If X_+ is a normal cone, then all elements of X_+ are normal points and all sets $\{\|x\| \in X; u \leq x \leq v\}$ with $u, v \in X$, $u \leq v$, are bounded subsets of $\mathbb{R}$.*

If all elements of X_+ are normal points and X_+ is complete or fully regular, then X_+ is a normal cone.

Proof. The first statement is obvious. Assume that X_+ is not a normal cone but complete or fully regular. Then there exist sequences (x_n) and (y_n) in X_+ such that $x_n \leq y_n$ and $\|x_n\| \geq 4^n\|y_n\|$ for all $n \in \mathbb{N}$. Set $v_n = \frac{y_n}{2^n\|y_n\|}$ and $u_n = \frac{x_n}{2^n\|y_n\|}$. Then $u_n \leq v_n$ and $\|v_n\| \leq 2^{-n}$ and $\|u_n\| \geq 2^n$. Since X_+ is complete or fully regular, the series $v = \sum_{n=1}^{\infty} v_n$ converges in X_+ and $u_n \leq v$ for all $n \in \mathbb{N}$. So v is not a normal point. □

Completeness or full regularity of X_+ are necessary for the normality of X_+ as shown by the forthcoming Example 2.11. Connections between completeness, normality, regularity and full regularity of cones are spelt out in the next result. The proofs in [24, 1.5.2] and [24, 1.5.3] only need completeness of X_+.

Theorem 2.5. *Let X be an ordered normed vector space with cone X_+.*

(a) *If X_+ is complete and regular, then X_+ is normal.*
(b) *If X_+ is complete and fully regular, then X_+ is normal.*
(c) *If X_+ is normal and fully regular, then X_+ is regular.*
(d) *If X_+ is complete and fully regular, then X_+ is regular.*

Theorem 2.6. *Let X be an ordered normed vector space with cone X_+. If X_+ is complete with additive norm, then X_+ is fully regular.*

Proof. Let (x_n) be an increasing sequence in X_+ such that there is some $c > 0$ such that $\|x_n\| \le c$ for all $n \in N$. Define $y_n = x_{n+1} - x_n$. Then $y_n \in X_+$ and, for $m \ge j$, $\sum_{k=j}^{m} y_n = x_{m+1} - x_j$. Since the norm is additive on X_+,

$$\sum_{k=1}^{m} \|y_n\| = \left\|\sum_{k=1}^{m} y_n\right\| = \|x_{m+1} - x_1\| \le 2c, \qquad m \in \mathbb{N}.$$

So (x_n) is a Cauchy sequence in the complete cone X_+ and converges. □

The cones of nonnegative functions of the Banach spaces $L^p(\Omega)$, $1 \le p < \infty$, are regular and completely regular, while the cones of $\mathrm{BC}(\Omega)$, the Banach space of bounded continuous functions, and of $L^\infty(\Omega)$ are neither regular nor completely regular. The cone of nonnegative measures in the Banach spaces of signed measures is fully regular because the norm is additive on the cone. All these cones are normal and generating. Forthcoming examples will present a cone that is regular, but neither completely regular, normal, nor complete.

2.2. An example where the cone is not normal: The space of sequences of bounded variation

Recall the Banach sequence spaces ℓ^∞, c, c_0 of bounded sequences, converging sequences and sequences converging to 0 with the supremum norm and the space ℓ^1 of summable sequences with the sum-norm. The cones of c_0 and ℓ^1 are regular (and thus normal), and the cones of ℓ^∞ and c are solid and normal.

The subsequent example for an ordered Banach space whose cone is not normal follows a suggestion by Wolfgang Arendt. A sequence (x_n) in $\mathbb{R}^\mathbb{N}$ is called of *bounded variation* if the following series converges

$$|x_1| + \sum_{n=1}^{\infty} |x_{n+1} - x_n| =: \|(x_n)\|_{bv}. \tag{2.1}$$

The sequences of bounded variation form a vector space, bv, over $\mathbb{R}$ with $\|\cdot\|_{bv}$ being a norm called the *variation-norm* [14, IV.2.8]. Notice

$$\ell^1 \subseteq bv \subseteq c, \qquad \begin{cases} \|x\|_\infty \le \|x\|_{bv}, & x \in bv, \\ \|x\|_{bv} \le 2\|x\|_1, & x \in \ell_1. \end{cases} \tag{2.2}$$

Lemma 2.7. *bv with the variation-norm is a Banach space.*

This is easily seen from (2.2) and the fact that ℓ^∞ is complete under the sup-norm. Notice that bv contains all constant sequences and $\|(x_n)\|_{bv} = |x_1| = \|(x_n)\|_\infty$ if (x_n) is a constant sequence. Actually, all monotone bounded nonnegative sequences are of bounded variation.

Lemma 2.8. *If (x_n) is a nonnegative bounded increasing sequence, then (x_n) is of bounded variation and $\|(x_n)\|_{bv} = \|(x_n)\|_\infty = \lim_{n\to\infty} x_n$.*

If (x_n) is a nonnegative decreasing sequence, then (x_n) is of bounded variation and $\|(x_n)\|_{bv} = 2x_1 - \lim_{n\to\infty} x_n$.

There are several cones we can consider in bv. The one we are going to consider here is the cone of nonnegative sequences of bounded variation, bv_+. Others are the cone of nonnegative bounded increasing sequences and the cone of nonnegative decreasing sequences.

Proposition 2.9. *bv_+ is generating, solid, but not normal (and not regular and not fully regular). X is a lattice and the lattice operations are continuous. If $x = (x_j)$, then $|x| = (|x_j|) \in bv$ and $\big\||x|\big\|_{bv} \le \|x\|_{bv}$ with strict inequality being possible.*

Every monotone nonnegative sequence that is bounded away from zero is in the interior of bv_+ as can be seen from (2.2). The space of Lipschitz continuous functions with Lipschitz norm is an example of an ordered normed vector space and lattice where the cone of nonnegative functions is not normal and the lattice operations are not continuous [31, p. 535].

That bv_+ is not normal follows from the following result that characterizes the normal points. Since bv_+ is complete, bv_+ is then also not regular and not fully regular (Theorem 2.5).

Theorem 2.10. *The following are equivalent for $x = (x_n) \in bv_+$:* (i) *x is a normal point,* (ii) $x \in \ell_1$, (iii) *x is a regular point. If $x \in \ell_1$, then*

$$\|x\|_1 - (1/2)x_1 \le \sup\{\|v\|_{bv}; v \in bv_+, x - v \in bv_+\} \le 2\|x\|_1.$$

Proof. Assume that $x = (x_n) \in bv_+$ is a normal element. Let y be the sequence $(x_1, 0, x_3, 0, x_5, \ldots)$. Then $0 \le y \le x$ and $y \in bv$ and $\|y\|_{bv} = |x_1| + 2\sum_{j=1}^\infty |x_{2j+1}|$. Further let z be the sequence $(0, x_2, 0, x_4, 0, \ldots)$. Then $0 \le z \le x$ and $z \in bv$ and $\|z\|_{bv} = 2\sum_{j=1}^\infty \|x_{2j}\|$. Hence $x \in \ell^1$ and

$$2\|x\|_1 - x_1 = \|y\|_{bv} + \|z\|_{bv} \le 2\sup\{\|v\|_{bv}; 0 \le v \le x\} \le 4\|x\|_1.$$

Now let $x = (x_j) \in \ell^1$ and (x^n) be a monotone sequence in ℓ^1_+ with $x^n \le x$, $x^n = (x^n_j)_{j\in\mathbb{N}}$. Then $x^n_j \le x_j$ for all $j, n \in \mathbb{N}$ and all sequences $(x^n_j)_{n\in\mathbb{N}}$ are monotone. So there exists some $y = (y_j)$ in ℓ^1_+ such that, for each $j \in \mathbb{N}$, $x^n_j \to y_j$ as $n \to \infty$, $y \le x$. By the dominated convergence theorem, $x^n \to y$ in ℓ^1 and so also in bv. Recall (2.2).

It is a general fact that each regular point is a normal point [44]. □

Example 2.11 (Wolfgang Arendt). The cone ℓ^1_+ in ℓ^1 with the variation norm is not normal though all points in ℓ^1_+ are normal elements. It is regular, but not fully regular.

Proof. Let x^m be the sequence where $x^m_j = 0$ for $j > 2m$, $x^m_j = 0$ for all odd indices and $x^m_j = 1$ otherwise. Let u^m be the sequence where $u^m_j = 0$ for $j > 2m$ and $u^m_j = 1$ otherwise. Then $\|u^m\|_{bv} = 2$ and $\|x^m\|_{bv} = 2m$. For all $m \in \mathbb{N}$, $x^m \le u^m$ but $\|x^m\|_{bv} = m\|u^m\|_{bv}$. So ℓ^1_+ is not normal under the variation norm. Theorem 2.4 implies that ℓ^1_+ is not fully regular under the variation norm. Let (x_n) be a decreasing sequence in ℓ^1_+. Since ℓ^1_+ is regular under the sum-norm, there exists some $x \in \ell^1_+$ with $\|x_n - x\|_1 \to 0$. By (2.2), $\|x_n - x\|_{bv} \to 0$. □

2.3. A convergence result

The following convergence principle, which has been distilled from the proof of [1, Thm. 5.2], will be applied several times.

Proposition 2.12. *Let X be an ordered normed vector space. Let $S \subseteq X^{\mathbb{N}}$ be a set of sequences with terms in X with the property that with $(x_n) \in S$ also all subsequences $(x_{n_j}) \in S$. Assume that every increasing (decreasing) sequence $(x_n) \in S$ has a convergent subsequence. Then every increasing (decreasing) sequence $(x_n) \in S$ converges.*

Proof. Let $(x_n) \in S$ be increasing (the decreasing case is similar). Then there exist a subsequence (x_{n_j}) and some $x \in X$ such that $x_{n_j} \to x$. Suppose that (x_n) does not converge to x. Then there exist some $\epsilon > 0$ and a strictly increasing sequence (k_i) in $\mathbb{N}$ such that $\|x_{k_i} - x\| \ge \epsilon$ for all $i \in \mathbb{N}$. By assumption, $(x_{k_i}) \in S$; further it is an increasing sequence. So, after choosing a subsequence, $x_{k_i} \to y$ for some $y \ne x$. Fix n_j. If i is sufficiently large, $k_i \ge n_j$ and $x_{n_j} \le x_{k_i}$. Taking the limit $i \to \infty$ yields $x_{n_j} \le y$ because the cone is closed. Now we let $j \to \infty$ and obtain $x \le y$. By symmetry, $y \le x$, a contradiction. □

3. More on cone and local spectral radii

Let $B : X_+ \to X_+$ be homogeneous and bounded, and $x \in X_+$. We prove (1.12), $\gamma(x, B^m) = (\gamma(x, B))^m$ for all $m \in \mathbb{N}$.

From the properties of the limit superior,

$$\begin{aligned}\gamma(x, B^m) &= \limsup_{k\to\infty} \|(B^m)^k(x)\|^{1/k} = \lim_{j\to\infty} \sup_{k\ge j} \|B^{mk}(x)\|^{m/(mk)} \\ &\le \lim_{j\to\infty} \sup_{n\ge mj} (\|(B^n(x)\|^{1/n})^m = \Big(\limsup_{n\to\infty} \|(B^n(x)\|^{1/n}\Big)^m = (\gamma(x, B))^m.\end{aligned}$$

To prove the opposite inequality, suppose that $\gamma(x, B^m) < (\gamma(x, B))^m$. Then there exists some $s \in (0, 1)$ such that $\gamma(x, B^m) < s^m(\gamma(x, B))^m$. By definition of $\gamma(x, B^m)$, there exists some $N \in \mathbb{N}$ such that

$$\|B^{mn}(x)\|^{1/(mn)} < s\gamma(x, B), \qquad n \ge N. \tag{3.1}$$

Choose a sequence (k_j) in $\mathbb{N}$ such that $k_j \to \infty$ and $\|B^{k_j}(x)\|^{1/k_j} \to \gamma(x, B)$. Then there exist sequences n_j and p_j such that $n_j \to \infty$ and $0 \le p_j < m$ and $k_j = mn_j + p_j$. By (1.7),

$$\|B^{k_j}(x)\|^{1/k_j} \le \|B^{p_j}\|_+^{1/k_j} \, \|B^{mn_j}(x)\|^{1/k_j}.$$

By the properties of the limit superior,

$$\gamma(x, B) \le \limsup_{j\to\infty}(\|B^{p_j}\|_+^{1/k_j}) \limsup_{j\to\infty}(\|B^{mn_j}(x)\|^{1/(mn_j)})^{(k_j-p_j)/k_j}.$$

By (3.1) and $p_j/k_j \to 0$,

$$\begin{aligned}\gamma(x, B) &\le \limsup_{j\to\infty}(s\gamma(x, B))^{(k_j-p_j)/k_j} \\ &\le \limsup_{j\to\infty}(s\gamma(x, B))^{1-(p_j/k_j)} = s\gamma(x, B),\end{aligned}$$

a contradiction.

The next results revisit conditions that imply the equality of cone and orbital spectral radius [17, 31] stressing the point that the cone spectral radius equals one of the local spectral radii (geometric growth factors).

Theorem 3.1. *Let X be an ordered normed vector space with complete cone X_+. Let $B : X_+ \to X_+$ be homogeneous, bounded, and order-preserving and satisfy the following normality condition: There exist some $m \in \mathbb{N}$ and $c > 0$ such that $\|B^m(x)\| \le c\|B^m(y)\|$ for all $x, y \in X_+$ with $x \le y$. Then there exists some $x \in X_+$ such that $\mathbf{r}_+(B) = \gamma_B(x) = \mathbf{r}_o(B) = \lim_{k\to\infty} \|B^k(x)\|^{1/k}$.*

Since B is order-preserving, the normality condition for B is weaker than condition (iv) in [17, Thm. 2.1] (there are $m \in \mathbb{N}$ and $c > 0$ such that $\|x\| \le c\|y\|$ for all $x, y \in B^m(X_+)$ with $x \le y$). The proof remains almost the same, but requires a modification. Since it is not too long, we give it here for the ease of the reader.

Proof. We can assume that $\mathbf{r}_+(B) > 0$. For each $k \in \mathbb{N}$, there exists some $x_k \in X_+$ with $\|x_k\| = 1$ and $\|B^k\|_+ \le \frac{k+1}{k}\|B^k(x_k)\|$. Since X_+ is complete, the series $x = \sum_{k=1}^\infty k^{-2}x_k$ converges and $x \in X_+$. Since B is order-preserving and $x_k \le k^2 x$,

$$B^j(x_k) \le B^j(k^2 x), \qquad j \in \mathbb{Z}_+.$$

For $k \ge m$,

$$\begin{aligned} B^k(x_k) = B^m(y_k), \qquad & B^k(k^2x) = B^m(z_k), \\ y_k := B^{k-m}(x_k) \le\; & B^{k-m}(k^2x) =: z_k. \end{aligned}$$

By the normality condition for B, $\|B^k(x_k)\| \le c\|B^k(k^2x)\|$ for $k \ge m$. Since B is homogeneous,

$$\|B^k\|_+ \le \frac{k+1}{k}\|B^k(x_k)\| \le c\frac{k+1}{k}k^2\|B^k(x)\|, \qquad k > m.$$

So

$$\mathbf{r}_+(B) \le \liminf_{k\to\infty} \left([c(k+1)k]^{1/k} \|B^k(x)\|^{1/k}\right) = \liminf_{k\to\infty} \|B^k(x)\|^{1/k}.$$

Since $\limsup_{k\to\infty} \|B^k(x)\|^{1/k} \le \mathbf{r}_+(B)$, $\mathbf{r}_+(B) = \lim_{k\to\infty} \|B^k(x)\|^{1/k}$. □

We combine ideas from [31, Thm. 2.3] and [17, Thm. 2.1] to obtain the same result for compact continuous B.

Theorem 3.2. *Let X be an ordered normed vector space with cone X_+. Let $B : X_+ \to X_+$ be homogeneous, continuous, order-preserving, and power compact. Then there exists some $x \in X_+$ such that*

$$\mathbf{r}_+(B) = \mathbf{r}_o(B) = \gamma_B(x) = \lim_{k\to\infty} \|B^k(x)\|^{1/k}.$$

Proof. Suppose $\liminf_{k\to\infty} \|B^k(x)\|^{1/k} < \mathbf{r}_+(B)$ for all $x \in X_+$ Since B is homogeneous, we can scale B such that $\mathbf{r}_+(B) = 1 > \liminf_{k\to\infty} \|B^k(x)\|^{1/k}$ for all $x \in X_+$. By (1.12) and (1.11), we can assume that B is compact.

For each $k \in \mathbb{N}$, there exists some $x_k \in X_+$ with $\|x_k\| = 1$ and $\|B^k\|_+ = 1 \le \frac{k+1}{k} \|B^k(x_k)\|$.

We define $y_k = \sum_{j=1}^k j^{-2} x_j$, $k \in \mathbb{N}$. Then (y_k) is an increasing bounded sequence in X_+. Since B is compact and order-preserving, $B(y_k) \to x \in X_+$ as $k \to \infty$ and $B(y_k) \le x$ for all $k \in \mathbb{N}$. Since B is order-preserving and homogeneous and $x_k \le k^2 y_k$,

$$B^k(x_k) \le k^2 B^k(y_k) \le k^2 B^{k-1}(x), \qquad k \in \mathbb{N}.$$

Since $\liminf_{k\to\infty} \|B^k(x)\|^{1/k} < 1$, there exists a strictly increasing sequence (k_i) in $\mathbb{N}$ such that $k_i^2 B^{k_i - 1}(x) \to 0$ as $i \to \infty$. By the proof of [31, Thm. 2.3], $\{B^k(x_k); k \in \mathbb{N}\}$ has compact closure and, after choosing another subsequence, $B^{k_i}(x_{k_i}) \to z$ for some $z \in X_+$. On the one hand, $\|z\| \ge 1$, and on the other hand, since the cone is closed, $z = 0$, a contradiction. So $\mathbf{r}_+(B) \le \liminf_{k\to\infty} \|B^k(x)\|^{1/k}$. Since $\limsup_{k\to\infty} \|B^k(x)\|^{1/k} \le \mathbf{r}_+(B)$, the assertion follows. □

It is worth mentioning that a normality condition as in Theorem 3.1 makes order-preserving homogeneous maps bounded. In fact, the normality condition can be weakened.

Definition 3.3. Let $u \in \dot{X}_+$ and $B : X_+ \to X_+$ be homogeneous. Then u is called a *normal point for B* if there exists some $c = c_u > 0$ such that $\|B(x)\| \le c$ for all $x \in X_+$ with $x \le u$.

Notice that u is a normal point for B if B is order-preserving and $B(u)$ is a normal point of X_+.

Theorem 3.4. *Let X be an ordered normed vector space with complete cone X_+. Let $B : X_+ \to X_+$ be homogeneous and assume that every point in X_+ is a normal point for B. Then B is bounded.*

Proof. Suppose that B is not bounded. There there exists a sequence (x_n) in X_+ such that $\|x_n\| = 1$ and $\|B(x_n)\| \geq n^4$ for all $n \in \mathbb{N}$. Since X_+ is complete, the series $u = \sum_{n=1}^{\infty} n^{-2} x_n$ converges. For all $n \in \mathbb{N}$, $n^{-2}x_n \leq u$; by assumption, u is a normal point for B. So there exists some $c > 0$ such that $c \geq \|B(n^{-2}x_n)\| = n^{-2}\|B(x_n)\| \geq n^2$, a contradiction. □

Normal points for B (or rather powers of B) will return in Section 12.

4. Monotone companion norm and half-norm

Every ordered normed vector space carries an order-preserving half-norm which we call the (*monotone*) *companion half-norm* (called the *canonical half-norm* in [2]).

Proposition 4.1 (cf. [2], [25, (4.2)], [9, L.4.1]). *We define the* (*monotone*) *companion half-norm* $\psi : X \to \mathbb{R}_+$ *by*

$$\psi(x) = \inf\{\|x+z\|; z \in X_+\} = d(x, -X_+) \tag{4.1}$$

$$= \inf\{\|y\|; x \leq y \in X\}, \qquad x \in X. \tag{4.2}$$

Then the following hold:

(a) ψ *is positively homogenous and order-preserving on* X.

(b) ψ *is subadditive on* X $(\psi(x+y) \leq \psi(x) + \psi(y),\ x, y \in X)$,

$$|\psi(x) - \psi(y)| \leq \|x - y\|, \qquad x, y \in X.$$

(c) *For* $x \in X$, $\psi(x) = 0$ *if and only if* $x \in -X_+$. *In particular* ψ *is strictly positive:* $\psi(x) > 0$ *for all* $x \in \dot{X}_+$.

(d) X_+ *is normal if and only if there exists some* $\delta > 0$ *such that* $\delta\|x\| \leq \psi(x)$ *for all* $x \in X_+$.

(e) *If the original norm* $\|\cdot\|$ *is order-preserving on* X_+, *then* $\|x\| = \psi(x)$ *for all* $x \in X_+$.

Here $d(x, -X_+)$ denotes the distance of x from $-X_+$.

Proof. The functional ψ inherits positive homogeneity from the norm. That ψ is order-preserving is immediate from (4.2). For all $x \in X$, $x \leq x$ and so $\|x\| \geq \psi(x)$.

Most of the other properties follow from (4.1) and the assumption that X_+ is a cone. Since ψ is subadditive,

$$|\psi(x) - \psi(y)| \leq \psi(x - y) \leq \|x - y\|.$$

If $-x \in X_+$, then $x \leq 0$ and $\psi(x) \leq \|0\| = 0$.

Assume that $x \in X$ and $\psi(x) = 0$. By definition, there exists a sequence (y_n) in X with $\|y_n\| \to 0$ and $y_n \geq x$ for all $n \in \mathbb{N}$. Then $y_n - x \in X_+$. Since X_+ is closed, $-x = \lim_{n\to\infty}(y_n - x) \in X_+$.

The strict positivity of ψ follows from $X_+ \cap (-X_+) = \{0\}$.

(d) Assume that X_+ is normal. By Theorem 2.1 (ii), there exists some $c > 0$ such that $\|x\| \le c\|y\|$ whenever $x, y \in X_+$ and $x \le y$. Hence $\|x\| \le c\psi(x)$ for all $x \in X_+$. Set $\delta = 1/c$.

The converse follows from ψ being order-preserving and $\psi(x) \le \|x\|$ for all $x \in X_+$. □

The functional ψ induces a monotone norm on X which is equivalent to the original norm if and only if X_+ is a normal cone.

Theorem 4.2 (cf. [25, Thm. 4.4]). *With ψ from Proposition 4.1, define*

$$\sharp x \sharp = \max\{\psi(x), \psi(-x)\}, \qquad x \in X.$$

Then $\sharp \cdot \sharp$ is a norm on X with the following properties.

- *$\sharp x \sharp \le \|x\|$ for all $x \in X$, $\sharp x \sharp = \psi(x)$ if $x \in X_+$, and $\sharp x \sharp = \psi(-x)$ if $-x \in X_+$.*
- *$\sharp \cdot \sharp$ is order-preserving on X_+: $\sharp x \sharp \le \sharp y \sharp$ for all $x, y \in X_+$ with $x \le y$. Moreover, for all $x, y, z \in X$ with $x \le y \le z$,*

$$\sharp y \sharp \le \max\{\sharp x \sharp, \sharp z \sharp\}. \tag{4.3}$$

- *X_+ is a normal cone if and only if the norm $\sharp \cdot \sharp$ is equivalent to the original norm.*
- *If the original norm $\| \cdot \|$ is order-preserving on X_+, then $\|x\| = \sharp x \sharp$ for all $x \in X_+ \cup (-X_+)$.*

Proof. It is easy to see from the properties of ψ that $\sharp \alpha x \sharp = |\alpha| \sharp x \sharp$ for all $\alpha \in \mathbb{R}$, $x \in X$, and that $\sharp \cdot \sharp$ is subadditive. Now $\psi(x) \le \|x\|$ and $\psi(-x) \le \|-x\| = \|x\|$ and so $\sharp x \sharp \le \|x\|$.

To prove (4.3), let $x, y, z \in X$ and $x \le y \le z$. Then $y \le z$ and $-y \le -x$. Since ψ is order-preserving on X,

$$\psi(y) \le \psi(z) \le \sharp z \sharp, \qquad \psi(-y) \le \psi(-x) \le \sharp x \sharp.$$

(4.3) now follows from the definition of ψ as do most of the remaining assertions. That $\| \cdot \|$ and $\sharp \cdot \sharp$ are equivalent norms if the cone is normal is shown in [25, Thm. 4.4]. The converse follows from Theorem 2.1 (ii). □

Definition 4.3. The norm $\sharp \cdot \sharp$ is called the (*monotone*) *companion norm* on the ordered normed vector space X.

Corollary 4.4. *Let X_+ be a normal cone. Then there exists some $c \ge 0$ such that $\|y\| \le c \max\{\|x\|, \|z\|\}$ for all $x, y, z \in X$ with $x \le y \le z$.*

Proof. Let $\sharp \cdot \sharp$ be the monotone companion norm from Theorem 4.2, which is equivalent to the original norm because X_+ is normal. Choose $c \ge 0$ such that $\sharp x \sharp \le \|x\| \le c \sharp x \sharp$ for all $x \in X$. Let $x \le y \le z$. By Theorem 4.2,

$$\|y\| \le c \sharp y \sharp \le c \max\{\sharp x \sharp, \sharp z \sharp\} \le c \max\{\|x\|, \|z\|\}.$$ □

Corollary 4.5 (Squeezing theorem [25, Thm. 4.3]). *Let X_+ be a normal cone. Let $y \in X$ and (x_n), (y_n), (z_n) be sequences in X with $x_n \le y_n \le z_n$ for all $n \in \mathbb{N}$ and $x_n \to y$ and $z_n \to y$. Then $y_n \to y$.*

Proof. Notice that $x_n - y \le y_n - y \le z_n - y$. By Corollary 4.4, with some $c \ge 0$ that does not depend on n,

$$\|y_n - y\| \le c \max\{\|z_n - y\|, \|x_n - y\|\} \to 0. \qquad \square$$

Example 4.6. (a) Let $X = \mathbb{R}$ with the absolute value. Then the monotone companion norm is also the absolute value and the companion half-norm is the positive part, $\psi(x) = \max\{x, 0\}$ for all $x \in \mathbb{R}$.

(b) Let $X = \mathbb{R}^2$ with the maximum norm and $X_+ = \mathbb{R}^2_+$. Then the monotone companion norm is also the maximum norm.

(c) Let $X = \mathbb{R}^2$ with $\|\cdot\|$ being either the Euclidean norm or the sum norm and $X_+ = \mathbb{R}^2_+$. Then $\sharp x \sharp = \|x\|$ if $x \in X_+ \cup (-X_+)$ while $\sharp x \sharp$ is the maximum norm of x otherwise.

Example 4.7. Let X be a normed vector lattice [39, II.5]. Since $x \le x^+$, $\psi(x) \le \|x^+\|$. Let $x \le y$. Then $x^+ \le y^+$, and $\|x^+\| \le \|y^+\| \le \|y\|$. Hence $\|x^+\| \le \psi(x)$. In combination,

$$\psi(x) = \|x^+\|, \qquad x \in X.$$

Further $\psi(-x) = \|(-x)^+\| = \|x^-\|$. So

$$\sharp x \sharp = \max\{\|x^+\|, \|x^-\|\}, \qquad x \in X,$$

and $\sharp x \sharp = \|x\|$ for all $x \in X_+ \cup (-X_+)$. If X is an abstract M-space [39, II.7],

$$\sharp x \sharp = \|x^+ \vee x^-\| = \||x|\| = \|x\|.$$

We turn to ordered Banach space the cone of which may be not normal.

Example 4.8. Let $X \subseteq \tilde{X}$ where $X, \tilde{X}$ are ordered normed vector spaces with norms $\|\cdot\|$ and $\|\cdot\|^\sim$ and $\|x\| \ge \|x\|^\sim$ for all $x \in X$. Let $\psi, \tilde{\psi}$ be the respective monotone companion half-norms. Then $\psi(x) \ge \tilde{\psi}(x)$ for all $x \in X$.

Assume that $(\tilde{X}, \|\cdot\|^\sim)$ is an abstract M-space. By the previous example,

$$\sharp x \sharp \ge \|x\|^\sim, \qquad x \in X.$$

Now assume that there exists some $u \in X$ with $\|u\| \le 1$ and $|x| \le \|x\|^\sim u$ for all $x \in X$. Then $\psi(\pm x) \le \|x\|^\sim$ and $\sharp x \sharp \le \|x\|^\sim$. In combination,

$$\sharp x \sharp = \|x\|^\sim, \qquad x \in X.$$

Example 4.9. To determine the monotone companion (half-) norm in a concrete case where the cone is not normal we revisit the space bv of sequences of bounded variation with the variation-norm.

Recall that $bv \subseteq \ell^\infty$ and $\|x\|_{bv} \ge \|x\|_\infty$ for all $x \in bv$. ℓ^∞ is an abstract M-space under the sup-norm. We apply the previous example with u being the sequence all the terms of which are 1. Then $\|u\|_{bv} = 1$ and $|x| \le \|x\|_\infty u$. We obtain that $\sharp x \sharp = \|x\|_\infty$ for all $x \in bv$.

5. Order-preserving maps and the companion norm

In the following, let X and Y be ordered normed vector spaces. We use the same symbols $\|\cdot\|$, ψ and $\sharp\cdot\sharp$ for the norms and the monotone companion (half-) norms on X and Y.

Theorem 5.1. *Let $B : X_+ \to Y_+$ be bounded, homogeneous, and order-preserving. Then $\psi(B(x)) \le \|B\|_+ \psi(x)$ for all $x \in X_+$. In particular, B is bounded with respect to the monotone companion norms and $\sharp B \sharp_+ \le \|B\|_+$.*

Proof. Since B is order-preserving, for $x \in X_+$,

$$\{y \in Y; B(x) \le y\} \supseteq \{B(z); x \le z \in X_+\}.$$

By definition of ψ,

$$\begin{aligned}\psi(B(x)) &\le \inf\{\|B(z)\|; z \in X_+, x \le z\} \le \inf\{\|B\|_+\|z\|; x \le z \in X\} \\ &= \|B\|_+ \psi(x).\end{aligned}$$

□

Theorem 5.2. *Let $B : X \to Y$ be bounded, linear and positive. Then B is bounded with respect to the monotone companion norm, $\psi(B(x)) \le \|B\|\psi(x)$ for all $x \in X$, and $\sharp B \sharp \le \|B\|$.*

Proof. By a similar proof as for Theorem 5.1, since B is order-preserving, $\psi(B(x)) \le \|B\|\psi(x)$ for all $x \in X$ and $\psi(-B(x)) = \psi(B(-x)) \le \|B\|\psi(-x)$. The assertion now follows from $\sharp x \sharp = \max\{\psi(x), \psi(-x)\}$. □

Various concepts of continuity are preserved if one switches from the original norm to the monotone companion norm. We only look at the most usual concept here.

Proposition 5.3. *Let $B : X \to Y$ be order-preserving and continuous at x with respect to the original norms. Then B is continuous at x with respect to the monotone companion norms.*

Proof. Let $x \in X$ and B be continuous at x. Let $y \in X$ as well. For any $z \ge y - x$, $z \in X$, we have

$$B(y) - B(x) = B(x + y - x) - B(x) \le B(x + z) - B(x).$$

By definition of the monotone companion half-norm,

$$\psi(B(y) - B(x)) \le \|B(x + z) - B(x)\|, \qquad y - x \le z \in X.$$

Let $\epsilon > 0$. Then there exists some $\delta_+ > 0$ such that $\|B(x + z) - B(x)\| < \epsilon$ for all $z \in X$, $\|z\| < \delta_+$. Now let $y \in X$ and $\psi(y - x) < \delta_+$. Then there exists some $z \in X$ such that $y - x \le z$ and $\|z\| < \delta_+$. Then $\|B(x + z) - B(x)\| < \epsilon$ and $\psi(B(y) - B(x)) < \epsilon$.

Also, for any $z \in X$ with $x - y \le z$, we have

$$B(x) - B(y) = B(x) - B(x - (x - y)) \le B(x) - B(x - z).$$

By definition of the monotone companion half-norm

$$\psi(B(x) - B(y)) \le \|B(x) - B(x - z)\|, \qquad z \in X, x - y \le z.$$

Let $\epsilon > 0$. Then there exists some $\delta_- > 0$ such that $\|B(x) - B(x - z)\| < \epsilon$ for all $z \in X$, $\|z\| < \delta_-$. Now let $y \in Y$ and $\psi(x - y) < \delta_-$. Then there exists some $z \in X$ such that $x - y \le z$ and $\|z\| = \| - z\| < \delta_-$. Thus $\psi(B(x) - B(y)) < \epsilon$.

Set $\delta = \min\{\delta_+, \delta_-\}$ and $\sharp y - x \sharp < \delta$. Then $\psi(\pm(y - x)) < \delta_\pm$ and $\psi(\pm(B(y) - B(x)) < \epsilon$. This implies $\sharp B(y) - B(x) \sharp < \epsilon$. □

Proposition 5.4. *Let $\phi : X_+ \to \mathbb{R}_+$ be homogeneous.*

(a) *If ϕ is bounded with respect to the monotone companion norm, then it is bounded with respect to the original norm and $\|\phi\|_+ \le \sharp\phi\sharp_+$.*
(b) *If ϕ is order-preserving and bounded with respect to the original norm, then it is bounded with respect to the monotone companion norm and $\|\phi\|_+ = \sharp\phi\sharp_+$. Further $\phi(x) \le \psi(x)\|\phi\|_+$ for all $x \in X_+$.*

Proof. (a) Let $x \in X_+$. Since ϕ is bounded with respect to the monotone companion norm,

$$\phi(x) \le \sharp\phi\sharp_+ \, \sharp x \sharp \le \sharp\phi\sharp_+ \, \|x\|.$$

(b) This follows from part (a) and Theorem 5.1. □

Proposition 5.5. *Let $\phi : X \to \mathbb{R}$ be linear.*

(a) *If ϕ is bounded with respect to the monotone companion norm, then it is bounded with respect to the original norm and $\|\phi\| \le \sharp\phi\sharp$.*
(b) *If ϕ is positive and bounded with respect to the original norm, then it is bounded with respect to the monotone companion norm and $\|\phi\| = \sharp\phi\sharp$ and $\phi(x) \le \|\phi\|\psi(x)$ for all $x \in X$.*

Proof. (a) The proof is similar to the one for Proposition 5.4.

(b) follows from part (a) and Theorem 5.2 and the fact that the monotone half-norm on $\mathbb{R}$ is given by the positive part. □

Let X_+^* be the dual wedge of positive linear functionals on X that are bounded with respect to the original norm $\|\cdot\|$. By Proposition 5.5, X_+^* is also the dual wedge of linear functions that are bounded with respect to the monotone companion norm $\sharp \cdot \sharp$. The respective norms induced on X^* are the same on X_+^*.

Proposition 5.6. *Let $x \in X$. Then there exists $x^* \in X_+^*$ such that $x^*x = \psi(x)$ and $\|x^*\| = \sharp x^* \sharp \le 1$. If $x \in X_+$, $\|x^*\| = 1$ can be achieved. Actually,*

$$\begin{aligned} \psi(x) &= \max\{x^*x; x^* \in X_+^*, \|x^*\| \le 1\}, \qquad x \in X, \\ \psi(x) &= \max\{x^*x; x^* \in X_+^*, \|x^*\| = 1\}, \qquad x \in X_+. \end{aligned} \tag{5.1}$$

Further [25, (4.3)], *for all $x \in X$,*

$$\sharp x \sharp = \max\{|x^*x|; x \in X_+^*, \|x^*\| = 1\} = \max\{|x^*x|; x \in X_+^*, \|x^*\| \le 1\}. \tag{5.2}$$

Proof. Recall that X_+^* and its norm do not depend on whether we consider X with the original norm or its monotone companion norm (Proposition 5.5).

Let $u \in X$. By [50, IV.6], we find $x^* \in X^*$ with $x^*u = \psi(u)$ and

$$-\psi(-x) \leq x^*x \leq \psi(x), \qquad x \in X,$$

and so

$$|x^*x| \leq \sharp x \sharp \leq \|x\|, \qquad x \in X.$$

Since $\psi(-x) = 0$ for all $x \in X_+$, $x^* \in X_+^*$. By Proposition 5.5, $\sharp x^* \sharp = \|x^*\| \leq 1$.

Now let $u \in X_+$. Then $x^*u = \psi(u) = \sharp u \sharp$ and so $\sharp x^* \sharp = 1$. Notice that we have proved $\leq$ in (5.1).

Suppose that $\psi(u) > \psi(-u)$. Then there exists $x^* \in X_+^*$ with $\sharp x^* \sharp \leq 1$ such that $x^*u = \psi(u) = \sharp u \sharp$. Hence $\sharp x^* \sharp = 1$.

If $\psi(u) < \psi(-u)$, there exists $x^* \in X_+^*$ with $\sharp x^* \sharp \leq 1$ such that $x^*(-u) = \sharp u \sharp = |x^*u|$. Again $\sharp x^* \sharp = 1$.

If $\psi(u) = \psi(-u) > 0$, then $u, -u \notin X_+$ and we can make the same conclusion.

It remains the case $\psi(u) = \psi(-u) = 0$. But then $u = 0$, and all equalities hold trivially. So the $\leq$ inequalities hold in (5.2). The $\geq$ inequalities follow from Proposition 5.5 (b). □

Corollary 5.7. *X_+ is closed with respect to the monotone companion norm.*

Proof. Let (x_n) be a sequence in X_+, $x \in X$ and $\sharp x_n - x \sharp \to 0$. Suppose that $x \notin X_+$. By a theorem of Mazur [50, IV.6., Thm. 3′], there exists a bounded linear $\phi : X \to \mathbb{R}$ such that $\phi(x) < -1$ and $\phi(y) \geq -1$ for all $y \in X_+$. Let $z \in X_+$, $n \in \mathbb{N}$. Then $nz \in X_+$ and $\phi(nz) \geq -1$. So $\phi(z) \geq -1/n$. We let $n \to \infty$ and obtain $\phi(z) \geq 0$. By Proposition 5.5, ϕ is continuous with respect to the monotone companion norm. So $0 \leq \phi(x_n) \to \phi(x)$ and $\phi(x) \geq 0$, a contradiction. □

6. Positivity of solutions to abstract integral inequalities

We consider integral inequalities of the following kind on an interval $[0, b]$, $0 < b < \infty$,

$$u(t) \geq \int_0^t K(t,s)u(s)ds, \qquad t \in [0,b]. \tag{6.1}$$

Here $u : [0, b) \to X$ is a continuous function, $K(t, s)$, $0 \leq s \leq t \leq b$, are bounded linear positive operators such that, for each $x \in X$, $K(t, s)x$ is a continuous function of (t, s), $0 \leq s \leq t \leq b$.

Theorem 6.1. *Let X be an ordered Banach space. Let $u : [0, b] \to X$ be a continuous solution of the inequality* (6.1). *Then $u(t) \in X_+$ for all $t \in [0, b]$.*

Proof. We define $v : [0, b) \to X$ by $v(t) = -u(t)$. Then

$$v(t) \leq \int_0^t K(t,s)v(s)ds, \qquad t \in [0,b]. \tag{6.2}$$

Since the monotone companion half-norm ψ is order-preserving,

$$\psi(v(t)) \le \psi\Big(\int_0^t K(t,s)v(s)ds\Big), \qquad t \in [0,b].$$

Since ψ is convex and homogeneous,

$$\psi(v(t)) \le \int_0^t \psi(K(t,s)v(s))ds, \qquad t \in [0,b]. \tag{6.3}$$

By Theorem 5.2,

$$\psi(K(t,s)v(s)) \le \|K(t,s)\|\psi(s), \qquad 0 \le s \le t \le b.$$

By the uniform boundedness theorem, there exists some $c \ge 0$ such that $\|K(t,s)\| \le c$ whenever $0 \le s \le t \le b$. So

$$\psi(K(t,s)v(s)) \le c\psi(v(s)), \qquad 0 \le s \le t \le b. \tag{6.4}$$

We substitute the last inequality into (6.3),

$$\psi(v(t)) \le c\int_0^t \psi(v(s))ds, \qquad t \in [0,b].$$

Let $\lambda > 0$. Then

$$e^{-\lambda t}\psi(v(t)) \le c\int_0^t e^{-\lambda(t-s)}e^{-\lambda s}\psi(v(s))ds, \qquad t \in [0,b].$$

Define $\alpha(\lambda) = \sup_{0\le t\le b} e^{-\lambda t}\psi(v(t))$. Then

$$e^{-\lambda t}\psi(v(t)) \le c\int_0^t e^{-\lambda(t-s)}\alpha(\lambda)ds \le c\frac{\alpha(\lambda)}{\lambda}, \quad 0 \le t \le b, \quad \text{and} \quad \alpha(\lambda) \le c\frac{\alpha(\lambda)}{\lambda}.$$

Choosing $\lambda > 0$ large enough, $\alpha(\lambda) \le 0$ and, since it is nonnegative, $\alpha(\lambda) = 0$. This implies $\psi(v(t)) = 0$ for all $t \in [0,b]$. By Proposition 4.1, $v(t) \in -X_+$ and so $u(t) = -v(t) \in X_+$ for all $t \in [0,b]$. □

7. The space of certain order-bounded elements and some functionals

Definition 7.1. Let $x \in X$ and $u \in X_+$. Then x is called *u-bounded* if there exists some $c > 0$ such that $-cu \le x \le cu$. If x is u-bounded, we define

$$\|x\|_u = \inf\{c > 0; -cu \le x \le cu\}. \tag{7.1}$$

The set of u-bounded elements in X is denoted by X_u. If $x, u \in X_+$ and x is not u-bounded, we define

$$\|x\|_u = \infty.$$

Two elements v and u in X_+ are called *comparable* if v is u-bounded and u is v-bounded, i.e., if there exist $\epsilon, c > 0$ such that $\epsilon u \le v \le cu$. Comparability is an equivalence relation for elements of X_+, and we write $u \sim v$ if u and v are comparable. Notice that $X_u = X_v$ if and only if $u \sim v$.

If X is a space of real-valued functions on a set Ω,

$$\|x\|_u = \sup\Big\{\frac{|x(\xi)|}{u(\xi)}; \xi \in \Omega, u(\xi) > 0\Big\}.$$

Since the cone X_+ is closed,

$$-\|x\|_u u \le x \le \|x\|_u u, \qquad x \in X_u. \tag{7.2}$$

X_u is a linear subspace of X, $\|\cdot\|_u$ is a norm on X_u, and X_u, under this norm, is an ordered normed vector space with cone $X_+ \cap X_u$ which is normal, generating, and has nonempty interior.

Lemma 7.2. *Let $u \in \dot{X}_+$. Then the following hold:*

(a) $\sharp x \sharp \le \|x\|_u \sharp u \sharp = \|x\|_u \max\{d(u, -X_+), d(u, X_+)\}, \qquad x \in X_u.$

(b) *u is a normal point of X_+ if and only if there exists some $c \ge 0$ such that $\|x\| \le c\|x\|_u$ for all $x \in X_u \cap X_+$.*

(c) *If X_+ is solid and u in the interior of X_+, then $X = X_u$, $d(u, X \setminus X_+) > 0$, and*

$$\|x\| \ge \|x\|_u \, d(u, X \setminus X_+), \qquad x \in X.$$

(d) *In turn, if $X_u = X$ and there exists some $\epsilon > 0$ such that $\|x\| \ge \epsilon\|x\|_u$ for all $x \in X$, then u is an interior point of X_+.*

(e) *If X_+ is solid and u is both an interior and a normal point of X_+, then $\|\cdot\|$ and $\|\cdot\|_u$ are equivalent.*

Proof. (a) By (7.2), if $x \in X_u$,

$$x, -x \le \|x\|_u u.$$

Since the companion functional is order-preserving on X,

$$\psi(x) \le \|x\|_u \psi(u), \qquad \psi(-x) \le \|x\|_u \psi(u)$$

and so $\sharp x \sharp \le \|x\|_u \sharp u \sharp$.

(b) Let u be normal point of X_+. Then there exists some $c > 0$ such that $\|y\| \le c$ for all $y \in X_+$ with $y \le u$. For $x \in X_+ \cap X_u$, $x \le \|x\|_u u$. If $x \ne 0$ in addition, $\|x\|_u^{-1} x \le u$. Hence $\big\| \|x\|_u^{-1} x \big\| \le c$.

The other direction is obvious.

(c) Let u be an interior element of X_+. Then $d(u, X \setminus X_+) > 0$. For any $\delta \in (0, d(u, X \setminus X_+))$, we have $u \pm \frac{\delta}{\|x\|} x \in X_+$ for all $x \in \dot{X}$. So $\pm x \le \frac{\|x\|}{\delta} u$ and so $x \in X_u$ and $\|x\|_u \le \frac{\|x\|}{\delta}$. Since this holds for any $\delta \in (0, d(u, X \setminus X_+)$, it also holds for $\delta = d(u, X \setminus X_+)$.

(d) Assume that $X_u = X$ and there exists some $\epsilon > 0$ such that $\|x\| \ge \epsilon\|x\|_u$ for all $x \in X$. This means that

$$\pm x \le \|x\|_u u \le (1/\epsilon)\|x\| u.$$

Hence $u \pm \frac{\epsilon}{\|x\|} x \in X_+$ for all $x \in \dot{X}$. This implies that u is an interior point of X_+.

(e) follows from combining (b) and (c). □

If X_+ is normal, by Theorem 2.1, there exists some $M \geq 0$ such that

$$\|x\| \leq M\|x\|_u\|u\|, \qquad x \in X_u. \tag{7.3}$$

If X_+ is a normal and complete cone of X, then $X_+ \cap X_u$ is a complete subset of X_u with the metric induced by the norm $\|\cdot\|_u$. For more information see [24, 1.3] [6, I.4], [25, 1.4].

For $u \in X_+$, one can also consider the functionals

$$\left.\begin{aligned}(x/u)^\diamond &= \inf\{\alpha \in \mathbb{R}; x \leq \alpha u\}\\ (x/u)_\diamond &= \sup\{\beta \in \mathbb{R}; \beta u \leq x\}\end{aligned}\right\} x \in X,$$

with the convention that $\inf(\emptyset) = \infty$ and $\sup(\emptyset) = -\infty$. If X is a space of real-valued functions on a set Ω,

$$\left.\begin{aligned}(x/u)^\diamond &= \sup\left\{\tfrac{x(\xi)}{u(\xi)}; \xi \in \Omega, u(\xi) > 0\right\}\\ (x/u)_\diamond &= \inf\left\{\tfrac{x(\xi)}{u(\xi)}; \xi \in \Omega, u(\xi) > 0\right\}\end{aligned}\right\} x \in X.$$

Many other symbols have been used for these two functionals in the literature; see Thompson [45] and Bauer [4] for some early occurrences. For $x \in X_+$, $\|x\|_u = (x/u)^\diamond$. Since we will use this functional for $x \in X_+$ only, we will stick with the notation $\|x\|_u$. Again for $x \in X_+$, $(x/u)_\diamond$ is a nonnegative real number, and we will use the leaner notation

$$[x]_u = \sup\{\beta \geq 0; \beta u \leq x\}, \qquad x, u \in X_+. \tag{7.4}$$

Since the cone X_+ is closed,

$$x \geq [x]_u u, \qquad x, u \in X_+. \tag{7.5}$$

Further $[x]_u$ is the largest number for which this inequality holds.

Lemma 7.3. *Let $u \in \dot{X}_+$. Then the functional $\phi = [\cdot]_u : X_+ \to \mathbb{R}_+$ is homogeneous, order-preserving and concave. It is bounded with respect to the original norm on X and also to the monotone companion norm,*

$$[x]_u \leq \frac{\sharp x \sharp}{\sharp u \sharp} \leq \frac{\|x\|}{\sharp u \sharp}, \quad x \in X_+, \quad \textit{and} \quad \|\phi\|_+ = \sharp \phi \sharp_+ \leq \frac{1}{\sharp u \sharp}.$$

ϕ is upper semicontinuous with respect to the original norm and

$$\big|[y]_u - [x]_u\big| \leq \|y - x\|_u, \qquad y, x \in X_u \cap X_+. \tag{7.6}$$

Recall that $\sharp u \sharp = d(u, -X_+)$.

Proof. We apply the monotone companion norm to (7.5),

$$\sharp x \sharp \geq [x]_u \sharp u \sharp, \qquad x \in X_+.$$

The equality $\|\phi\|_+ = \sharp \phi \sharp_+$ follows from Proposition 5.4. (7.6) has been proved in [44]. The other properties are readily derived from the definitions. □

See [26] for an in-depth treatment of this functional.

8. Companion spectral radii

If $B : X_+ \to X_+$ is homogeneous and bounded with respect to the original norm, then, by Proposition 5.1, B is also bounded with respect to the monotone companion norm $\sharp\cdot\sharp$ and $\sharp B\sharp_+ \le \|B\|_+$. So, we can define *the companion cone spectral radius*, the *companion growth bounds*, and the *companion orbital spectral radius* by

$$\mathbf{r}_+^\sharp(B) = \inf_{n\in\mathbb{N}} \sharp B^n \sharp_+^{1/n} = \lim_{n\to\infty} \sharp B^n \sharp_+^{1/n} \tag{8.1}$$

and

$$\mathbf{r}_o^\sharp(B) = \sup_{x\in X_+} \gamma_B^\sharp(x), \qquad \gamma_B^\sharp(x) = \limsup_{n\to\infty} \sharp B^n(x)\sharp^{1/n}. \tag{8.2}$$

Since $\sharp x\sharp \le \|x\|$ for all $x \in X$, we have the estimates

$$\begin{aligned} \mathbf{r}_o^\sharp(B) &\le \mathbf{r}_+^\sharp(B) \le \mathbf{r}_+(B), \\ \mathbf{r}_o^\sharp(B) &\le \mathbf{r}_o(B) \ \le \mathbf{r}_+(B). \end{aligned} \tag{8.3}$$

If the cone X_+ is normal, the companion norm is equivalent to the original norm and the respective spectral radii equal their companion counterparts. The following proposition implies Theorem 1.7. Recall Section 1.2.4, in particular

$$[B]_x = \sup\{\lambda \ge 0; B(x) \ge \lambda x\} = [B(x)]_x. \tag{8.4}$$

Proposition 8.1. *Let B be bounded, homogeneous and order-preserving. Then, for all $x \in \dot{X}_+$,*

$$[B]_x \le \eta_x(B) \le \gamma_B^\sharp(x) \le \gamma_B(x).$$

Further $cw(B) \le \sharp B\sharp \le \|B\|$ and

$$cw(B) \le \mathbf{r}_{cw}(B) \le \mathbf{r}_o^\sharp(B) \le \left\{ \begin{matrix} \mathbf{r}_+^\sharp(B) \\ \mathbf{r}_o(B) \end{matrix} \right\} \le \mathbf{r}_+(B).$$

Proof. Let $x \in \dot{X}_B$. The first inequality follows from (1.17). By (8.4), $B(x) \ge [B]_x x$. By induction $B^n(x) \ge [B]_x^n x$.

We apply the monotone companion half-norm ψ, which is homogeneous, and obtain $\psi(B^n(x)) \ge [B]_x^n \psi(x)$. Since $\psi(x) > 0$, $\gamma_x(B) \ge \gamma_x^\sharp(B) \ge [B]_x$. Since, for all $n \in \mathbb{N}$, B^n is homogeneous and order-preserving,

$$[B^n]_x \le \gamma_x^\sharp(B^n) \le \gamma_x^\sharp(B)^n.$$

The last equality follows from (1.12). Since this holds for all $n \in \mathbb{N}$, by (1.17) $\eta_x(B) \le \gamma_x^\sharp(B)$.

The remaining assertions follow directly from the definitions. □

The following criteria for the positivity of the lower Collatz–Wielandt bound and the Collatz–Wielandt radius are obvious from their definitions.

Lemma 8.2. *Let $B : X_+ \to X_+$ be homogeneous and order-preserving.*

Then $cw(B) > 0$ if and only if there exist $\epsilon > 0$ and $x \in \dot{X}_+$ such that $B(x) \ge \epsilon x$.

Further $\mathbf{r}_{cw}(B) > 0$ *if and only if there exists* $\epsilon > 0$, $m \in \mathbb{N}$, *and* $x \in \dot{X}_+$ *such that* $B^m(x) \geq \epsilon x$.

The proof of the following results uses ideas from [17].

Theorem 8.3. *Let* X *be an ordered normed vector space with complete cone* X_+ *and* $B : X_+ \to X_+$ *be homogeneous, bounded and order-preserving. Then there exists some* $x \in X_+$ *such that*

$$\mathbf{r}_+^\sharp(B) = \mathbf{r}_o^\sharp(B) = \gamma_B^\sharp(x) \leq \gamma_B(x) \leq \mathbf{r}_o(B).$$

Proof. We can assume that $\mathbf{r}_+^\sharp(B) > 0$. For each $k \in \mathbb{N}$, there exists some $x_k \in X_+$ with $\psi(x_k) = 1$ and $\sharp B^k \sharp_+ \leq \frac{k+1}{k}\psi(B^k(x_k))$. By (4.2), there exists some $y_k \in X_+$ with $x_k \leq y_k$ and $\|y_k\| \leq 2$. Since X_+ is complete, the series $x = \sum_{k=1}^\infty k^{-2} y_k$ converges, $x \in X_+$. Since ψ and B are order-preserving and homogeneous and $y_k \leq k^2 x$,

$$\sharp B^k \sharp_+ \leq \frac{k+1}{k}\psi(B^k(y_k)) \leq \frac{k+1}{k} k^2 \psi(B^k(x)).$$

Thus

$$\mathbf{r}_+^\sharp(B) \leq \limsup_{k\to\infty} ((k+1)k)^{1/k} \gamma^\sharp(x, B) = \gamma^\sharp(x, B). \qquad \square$$

9. Strictly increasing maps

We introduce a class of order-preserving homogeneous maps for which the lower KR property implies the KR property (see Definition 1.12). The following definition is similar to the one in [6, III.2.1].

Definition 9.1. Let $\theta : X_+ \to \mathbb{R}_+$ be order-preserving and homogeneous. An order-preserving map $B : X_+ \to X_+$ is called strictly θ-increasing if for any $x, y \in X_+$ with $x \leq y$ and $\theta(x) < \theta(y)$ there exists some $\epsilon > 0$ and some $m \in \mathbb{N}$ such that $B^m(y) \geq (1+\epsilon)B^m(x)$.

B is called strictly increasing if B is strictly θ-increasing where θ is the restriction of the norm to X_+.

Theorem 9.2. *Let* $\theta : X_+ \to X_+$ *be order-preserving and homogeneous and* $B : X_+ \to X_+$ *be continuous, homogeneous, and strictly* θ*-increasing. Assume that there is some* $p \in \mathbb{N}$ *such that* $(B^p(x_n))_{n\in\mathbb{N}}$ *has a convergent subsequence for any increasing sequence* (x_n) *in* X_+ *where* $\{\theta(x_n); n \in \mathbb{N}\}$ *is bounded. Then* B *has the KR property whenever it has the lower KR property.*

Proof. Since B is homogeneous, we can assume that $\mathbf{r}_+(B) = 1$ and that there exists some $x \in \dot{X}_+$ such that $B(x) \geq x$. Then the sequence (x_n) defined by $x_n = B^n(x)$ is increasing. We claim that $(\theta(x_n))$ is bounded. If not, then there exists some $n \in \mathbb{N}$ with $\theta(x_{n-1}) < \theta(x_n)$ where $x_0 = x$. Since B is strictly θ-increasing, there exists some $\epsilon > 0$ and some $m \in \mathbb{N}$ such that $B^m(x_n) \geq (1+\epsilon)B^m(x_{n-1})$.

By definition of (x_n), $B(y) \geq (1+\epsilon)y$ for $y = B^m(x_{n-1}) \geq x$. Since $y \in \dot{X}_+$, $\mathbf{r}_+(B) \geq cw(B) \geq 1+\epsilon$, a contradiction.

Choose $p \in \mathbb{N}$ according to the assumption of the theorem. We apply the convergence principle in Proposition 2.12. Let S be the set of sequences $(B^p(y_n))$ where (y_n) is a increasing sequence in X_+ such that $(\theta(y_n))$ is bounded. Then S has the property required in Proposition 2.12 and so every increasing sequence in S converges.

Since (x_n) is increasing and bounded, $(B^p(x_n)) \in S$ converges with limit v. Since $B^p(x_n) = x_{n+p}$, $x_n \to v$. Since $x_{n+1} = B(x_n)$ and B is continuous, $B(v) = v$. □

We mention some interesting properties of strictly θ-increasing maps.

Proposition 9.3. *Let $\theta : X_+ \to \mathbb{R}_+$ be order-preserving and homogeneous and $B : X_+ \to X_+$ be homogeneous and strictly θ-increasing. Let $r, s > 0$ and $v, w \in X_+$ with $\theta(v) > 0$ and $\theta(w) > 0$.*

(a) *If $B(v) \geq rv$ and $B(w) \leq sw$ and v is w-bounded, then $r \leq s$ and $r = s$ implies $w \geq \frac{\theta(w)}{\theta(v)} v$.*

(b) *If $B(v) = rv$ and $B(w) = sw$ and v and w are comparable, then $r = s$ and $w = \frac{\theta(w)}{\theta(v)} v$.*

Proof. We first assume that $\theta(v) = 1 = \theta(w)$.
(a) Since v is w-bounded, $w \geq [w]_v v$ with $[w]_v > 0$.

Case 1: $\theta(w) = \theta([w]_v v)$.

Then $1 = [w]_v$ and $w \geq v$. So $rv \leq B(v) \leq B(w) \leq sw$. We apply θ and obtain $r \leq s$.

Case 2: $\theta(w) > \theta([w]_v v)$

Since B is strictly θ-increasing, there exists some $\delta > 0$ such that

$$sw \geq B(w) \geq (1+\epsilon)B([w]_v v) = (1+\epsilon)[w]_v B(v) \geq (1+\epsilon)[w]_v rv,$$

which implies that $[w]_v \geq (1+\epsilon)(r/s)[w]_v$ and so $r < s$.

In either case $r \leq s$. If $r = s$, the second case cannot occur and the first case holds where $w \geq v$.

(b) From part (a), by symmetry, $r = s$ and $w \geq v$ and then $w = v$.

If just $\theta(v) > 0$ and $\theta(w) > 0$, we set $\tilde{v} = \frac{1}{\theta(v)} v$ and $\tilde{w} = \frac{1}{\theta(w)} w$. Then $\theta(\tilde{v}) = 1 = \theta(\tilde{w})$ and $B\tilde{v} \geq r\tilde{v}$ and $B\tilde{w} \leq s\tilde{w}$. We apply the previous considerations to $\tilde{v}$ and $\tilde{w}$ and obtain (a) and (b) also in the general case. □

10. Order-bounded maps

The following terminology has been adapted from various works by Krasnosel'skii [24, Sec. 2.1.1] and coworkers [25, Sec. 9.4] though it has been modified.

Definition 10.1. Let $B : X_+ \to X_+$, $u \in X_+$. B is called *pointwise u-bounded* if, for any $x \in X_+$, there exist some $n \in \mathbb{N}$ and $\gamma > 0$ such that $B^n(x) \le \gamma u$. The point u is called a *pointwise order bound* of B.

B is called *uniformly u-bounded* if there exists some $c > 0$ such that $B(x) \le c\|x\|u$ for all $x \in X_+$. The element u is called a *uniform order bound* of B.

B is called *uniformly order-bounded* if it is uniformly u-bounded for some $u \in X_+$. B is called *pointwise order-bounded* if it is pointwise u-bounded for some $u \in X_+$.

If $B : X_+ \to X_+$ is bounded and X_+ is solid, then B is uniformly u-bounded for every interior point u of X_+.

Uniform order boundedness is preserved if the original norm $\|\cdot\|$ is replaced by its monotone companion norm $\sharp \cdot \sharp$.

Proposition 10.2. *Let $B : X_+ \to X_+$ be order-preserving. Let $u \in X_+$ and B be uniformly u-bounded. Then B is also uniformly u-bounded with respect to the monotone companion norm.*

Proof. Let $x \in X_+$. By (4.2), for each $n \in \mathbb{N}$ there exists some $y_n \in X_+$ such that $x \le y_n$ and $\psi(x) \le \|y_n\| \le \psi(x) + (1/n)$. Since B is order-preserving and uniformly u-bounded,

$$B(x) \le B(y_n) \le c\|y_n\|u, \qquad n \in \mathbb{N}.$$

We take the limit as $n \to \infty$ and obtain $B(x) \le c\psi(x)u$. □

Proposition 10.3. *Let X_+ be a complete cone, $u \in X_+$, and $B : X_+ \to X_+$ be continuous, order-preserving and homogeneous. Then the following hold.*

(a) *B is uniformly u-bounded if for any $x \in X_+$ there exists some $c = c_x \ge 0$ such that $B(x) \le cu$.*
(b) *If B be pointwise u-bounded, then some power of B is uniformly u-bounded.*

Proof. We prove (b); the proof of (a) is similar. Define

$$M_{n,k} = \{x \in X_+; B^n(x) \le ku\}, \qquad n, k \in \mathbb{N}.$$

Since B is continuous and X_+ is closed, each set $M_{n,k}$ is closed. Since B is assumed to be pointwise u-bounded, $X_+ = \bigcup_{k,n\in\mathbb{N}} M_{n,k}$. Since X_+ is a complete metric space, by the Baire category theorem, there exists some $n, k \in \mathbb{N}$ such that $M_{n,k}$ contains a relatively open subset of X_+: There exists some $y \in X_+$ and $\epsilon > 0$ such that $y+\epsilon z \in M_{n,k}$ whenever $z \in X$, $\|z\| \le 1$, and $y+\epsilon z \in X_+$. Now let $z \in X_+$ and $\|z\| \le 1$. Since B is order-preserving and $y + \epsilon z \in X_+$, $B^n(\epsilon z) \le B^n(y + \epsilon z) \le ku$. Since B is homogeneous, for all $x \in \dot{X}_+$,

$$B^n(x) = \frac{\|x\|}{\epsilon} B^n\Big(\frac{\epsilon}{\|x\|}x\Big) \le \frac{k}{\epsilon}\|x\|u.$$ □

11. Upper semicontinuity of the companion spectral radius

In order to be able to compare the companion spectral radius to the upper Collatz-Wielandt bound which will be defined later, some results on the upper semicontinuity of the cone spectral radius are useful. For more results of that kind see [30].

In the following, let X be an ordered normed vector space with cone X_+.

Lemma 11.1. *Let B be bounded and homogeneous, $x \in X_+$, and B be continuous at $B^n(x)$ for all $n \in \mathbb{N}$. Let (B_k) be a sequence of bounded homogeneous maps such that $\|B_k - B\|_+ \to 0$ as $k \to \infty$ and (x_k) be a sequence in X_+ such that $x_k \to x$. Then, for all $n \in \mathbb{N}$, $B_k^n(x_k) \to B^n(x)$ as $k \to \infty$.*

Proof. For $k \in \mathbb{N}$,

$$\begin{aligned}\|B_k(x_k) - B(x)\| &\leq \|B_k(x_k) - B(x_k)\| + \|B(x_k) - B(x)\| \\ &\leq \|B_k - B\|_+ \, \|x_k\| + \|B(x_k) - B(x)\| \overset{k\to\infty}{\longrightarrow} 0. \qquad \square\end{aligned}$$

This provides the basis step for an induction proof. The induction step follows in the same way.

Theorem 11.2. *Let $u \in X_+$, $u \neq 0$. Let $B : X_+ \to X_+$ be homogeneous and bounded and B be continuous at $B^n(u)$ for all $n \in \mathbb{N}$. Let (B_k) be a sequence of bounded, homogeneous, order-preserving maps such that $\|B_k - B\|_+ \to 0$ as $k \to \infty$. Assume that there exist $m \in \mathbb{N}$ and $c \geq 0$ such that $B_k^m(x) \leq c\|x\|u$ for all $k \in \mathbb{N}$ and all $x \in X_+$. Then*

$$\limsup_{k\to\infty} \mathbf{r}_+^\sharp(B_k) \leq \gamma_B^\sharp(u) \leq \mathbf{r}_o^\sharp(B) \leq \mathbf{r}_+^\sharp(B).$$

Proof. Choose $m \in \mathbb{N}$ and $c \geq 0$ as in the statement of the theorem. By Proposition 10.2 and its proof,

$$B_k^m(x) \leq c\sharp x\sharp u, \qquad k \in \mathbb{N}, x \in X_+.$$

Since B_k is order-preserving and homogeneous,

$$B_k^{n+m}(x) \leq c\sharp x\sharp B_k^n(u), \qquad k, n \in \mathbb{N}, x \in X_+.$$

We apply the monotone companion norm,

$$\sharp B_k^{n+m}(x)\sharp \leq c\sharp x\sharp \, \sharp B_k^n(u)\sharp, \qquad k, n \in \mathbb{N}, x \in X_+.$$

So

$$\sharp B_k^{n+m}\sharp_+ \leq c\sharp B_k^n(u)\sharp, \qquad n \in \mathbb{N}.$$

Let $r > \gamma_B^\sharp(u)$. Then there exists some $N \in \mathbb{N}$ such that

$$\sharp B^n(u)\sharp < r^n, \qquad n > N.$$

Let $n \in \mathbb{N}$, $n > N$. Since $B_k^n(u) \overset{k\to\infty}{\longrightarrow} B^n(u)$ by Lemma 11.1, there exists some $k_n \in \mathbb{N}$ such that

$$\sharp B_k^n(u)\sharp < r^n, \qquad k \geq k_n.$$

We combine the inequalities.

$$\sharp B_k^{n+m}\sharp_+ \le c r^n, \qquad k \ge k_n.$$

Then

$$\mathbf{r}_+^\sharp(B_k) \le \sharp B_k^{n+m}\sharp_+^{1/(n+m)} \le c^{-(n+m)} r^{n/(n+m)}, \qquad k \ge k_n.$$

So, for any $n > N$,

$$\limsup_{k\to\infty} \mathbf{r}_+^\sharp(B_k) \le c^{-(n+m)} r^{n/(n+m)}.$$

We take the limit as $n \to \infty$ and obtain

$$\limsup_{k\to\infty} \mathbf{r}_+^\sharp(B_k) \le r.$$

Since this holds for any $r > \gamma_B^\sharp(u)$, the assertion follows. □

12. Upper Collatz–Wielandt numbers

From Section 1.2.5, recall the *upper Collatz–Wielandt number* of B at $x \in \dot{X}_+$,

$$\|B\|_x = \|B(x)\|_x = \inf\{r \ge 0; B(x) \le rx\}. \tag{12.1}$$

where $\|B\|_x = \infty$ if $B(x)$ is not x-bounded. Also recall the *upper local Collatz–Wielandt radius* of B at $x \in X_+$,

$$\eta^x(B) = \inf_{n\in\mathbb{N}} \|B^n\|_x^{1/n}. \tag{12.2}$$

Lemma 12.1. *Let $B, C : X_+ \to X_+$ be homogeneous and order-preserving. Let $x \in \dot{X}_+$, and $B(x)$ and $C(x)$ be x-bounded. Then $C(B(x))$ is x-bounded and $\|CB\|_x \le \|C\|_x\|B\|_x$.*

Proof. By (7.2), $C(x) \le \|C\|_x x$ and $B(x) \le \|B\|_x x$ and $CB(x) \le \|B\|_x C(x) \le \|B\|_x\|C\|_x x$; so $\|CB\|_x \le \|B\|_x\|C\|_x$. □

Lemma 12.2. *Let $u \in \dot{X}_+$.*

(a) *Then $\eta^u(B^m) \ge (\eta^u(B))^m$ for all $m \in \mathbb{N}$.*

(b) *If there exists some $k \in \mathbb{N}$ such that $B^m(u)$ is u-bounded for all $m \ge k$, then $\eta^u(B) = \lim_{n\to\infty} \|B^n\|_u^{1/n} < \infty$ and $\eta^u(B^m) = (\eta^u(B))^m$ for all $m \in \mathbb{N}$.*

Proof. (a) Let $u \in \dot{X}_+$, $m \in \mathbb{N}$. Then

$$\eta^u(B^m) = \inf_{n\in\mathbb{N}} \|B^{mn}\|_u^{1/n} = (\inf_{n\in\mathbb{N}} \|B^{mn}\|_u^{1/(mn)})^m \ge (\inf_{k\in\mathbb{N}} \|B^k\|_u^{1/k})^m.$$

(b) Now let $k \in \mathbb{N}$ such that $B^m(u)$ is u-bounded for all $m \ge k$. By Lemma 12.1,

$$c_{n+m} \le c_n c_m, \qquad n, m \ge k, \qquad c_n = \|B^n\|_u.$$

Let r be an arbitrary number such that $\eta^u(B) = \inf_{n\in\mathbb{N}} \|B^n\|_u^{1/n} < r$. Then there exists some $m \in \mathbb{N}$ such that $c_m^{1/m} = \|B^m\|_u^{1/m} \le r$. So $B^m(u) \le r^m u$. By applying B^m as often as necessary, we can assume that $m \ge k$.

Any number $n \in \mathbb{N}$ with $n \geq 2m$ has a unique representation $n = pm + q$ with $p \in \mathbb{N}$ and $m \leq q < 2m$. Then, for $n \geq 2m$,

$$c_n \leq c_m^p c_q \leq r^{mp} c_q.$$

If $c_q = 0$, then both the limit inferior and the limit are zero and equal. So we can assume that $c_q \neq 0$. We have

$$c_n^{1/n} \leq r^{pm/n} c_q^{1/n}.$$

As $n \to \infty$, $pm/n \to 1$ and $\limsup_{n\to\infty} c_n^{1/n} \leq r$. Since r was any number larger than the infimum, the limes superior and inferior coincide and the limit exists and equals the infimum.

The second equality in (b) follows from the fact that every subsequence of a convergent sequence converges to the same limit. □

Remark 12.3. If $B(u)$ is u-bounded, $\|B\|_u$ is the cone norm of B in X_u with u-norm, and $\eta^u(B)$ is the cone spectral radius of B taken in X_u.

Recall the concepts of a normal point $u \in \dot{X}_+$ for B in Definition 3.3 (inspired by [17]) and of a normal point of X_+ in Definition 2.3.

Remark 12.4.

(a) If $u \in \dot{X}_+$ is a normal point of X_+, then u is a normal point for all bounded homogeneous $B : X_+ \to X_+$.

(b) If $u \in \dot{X}_+$ and $B(u)$ is a normal point of X_+ and B is order-preserving, then u is a normal point for B.

Proof. (a) There exists some $c > 0$ such that $\|x\| \leq c$ for all $x \in X_+$ with $x \leq u$. Then $\|B(x)\| \leq \|B\|_+ c$ for all $x \in X_+$ with $x \leq u$.

(b) There exists some $c > 0$ such that $\|y\| \leq c$ for all $y \in X_+$ with $y \leq B(u)$. Let $x \in X_+$ and $x \leq u$. Since B is order-preserving, $B(x) \leq B(u)$ and $\|B(x)\| \leq c$. □

Theorem 12.5. *Let $B : X_+ \to X_+$ be homogeneous, bounded and order-preserving. Let $u \in X_+$, $\alpha \in \mathbb{R}_+$ and $k \in \mathbb{N}$ such that $B^k(u) \leq \alpha^k u$.*

Let u be a normal point for some power of B or B be power-compact. Then $\gamma_B(u) \leq \alpha$.

Proof. Since $\gamma(u, B^k) = (\gamma(u, B))^k$ by (1.12), we can assume that $k = 1$. Since B is homogenous, it is enough to show that $B(u) \leq u$ implies that $\gamma_B(u) = \gamma(u, B) \leq 1$.

Let $B(u) \leq u$. Then $B^n(u) \leq u$ for all $n \in \mathbb{N}$.

We first assume that u is a normal point for some power of B. By Definition 3.3, there exist some $\tilde{c} > 0$ and $m \in \mathbb{N}$ such that $\|B^m(x)\| \leq \tilde{c}$ for all $x \in X_+$ with $x \leq u$. Then $\|B^{m+n}(u)\| \leq \tilde{c}$ for all $n \in \mathbb{N}$. This implies $\gamma_B(u) \leq 1$.

Now assume that B^ℓ is compact for some $\ell \in \mathbb{N}$ and that $\gamma_B(u) > 1$. Then the sequence (a_n) with

$$a_n = \|B^n(u)\| \tag{12.3}$$

is unbounded. By a lemma by Bonsall [8], there exists a subsequence a_{n_j} such that

$$a_{n_j} \to \infty, j \to \infty, \qquad a_k \le a_{n_j}, \qquad k = 1, \dots, n_j, \quad j \in \mathbb{N}.$$

Set $v_j = \frac{1}{a_{n_j}} B^{n_j}(u)$. Then

$$v_j = B^\ell(w_j), \qquad w_j = \frac{1}{\|B^{n_j}(u)\|} B^{n_j - \ell}(u).$$

Now

$$\|w_j\| \le \frac{a_{n_j - \ell}}{a_{n_j}} \le 1.$$

So, after choosing a subsequence, (v_j) converges to some $v \in X_+$, $\|v\| = 1$. Since $B^n(u) \le u$ for all $n \in \mathbb{N}$,

$$v_j \le \frac{1}{a_{n_j}} u.$$

Since $a_{n_j} \to \infty$ and X_+ is closed, we have $v \le 0$, a contradiction. □

Corollary 12.6. *Let $B : X_+ \to X_+$ be homogeneous, bounded and order-preserving. Let $u \in \dot{X}_+$ and $B^k(u)$ be u-bounded for all but finitely many $k \in \mathbb{N}$. Assume that u is a normal point for some power of B or that B is power-compact. Then $\gamma_B(u) \le \eta^u(B)$.*

Proof. By definition, $B^k(u) \le \|B^k\|_u u$ for all $k \in \mathbb{N}$ with $\|B^k\|_u < \infty$. By Theorem 12.5, $\gamma_B(u) \le \|B^k\|_u^{1/k}$ for all $k \in \mathbb{N}$. So $\gamma_B(u) \le \eta^u(B)$. □

Theorem 12.7. *Let $B : X_+ \to X_+$ be homogeneous, bounded and order-preserving. Let $u \in X_+$, $\alpha \in \mathbb{R}_+$ and $k \in \mathbb{N}$ such that $B^k(u) \le \alpha^k u$.*

(a) *Let u be a normal point for some power of B. Then $\mathbf{r}_o(B) \le \alpha$ if B is pointwise u-bounded, and $\mathbf{r}_+(B) \le \alpha$ if some power of B is uniformly u-bounded,*
(b) *Let B be pointwise u-bounded and power-compact. Then $\mathbf{r}_o(B) \le \alpha$.*

Proof. As in the proof of Theorem 12.5, we can reduce the proof to the implication

$$B(u) \le u \implies \mathbf{r}_o(B) \le 1.$$

Assume that $B(u) \le u$. Let $x \in X_+$. Since B is pointwise u-bounded, there exists some $m \in \mathbb{N}$ and $c > 0$ such that $B^m(x) \le cu$. For all $n \in \mathbb{N}$, $B^{m+n}(x) \le cB^n(u) \le cu$. Since u is a normal point for some power of B, by Definition 3.3, there exist some $k \in \mathbb{N}$ and some $\tilde{c} > 0$ such that $\|B^{k+m+n}(c^{-1}x)\| \le \tilde{c}$ for all $n \in \mathbb{N}$. This implies $\gamma_B(x) \le 1$. Since $x \in X_+$ has been arbitrary, $\mathbf{r}_o(B) \le 1$.

If B is bounded and B^m is uniformly u-bounded, we can replace c by $c\|x\|$ and we obtain $\mathbf{r}_+(B) \le 1$.

Now let B be power-compact and assume that there is some $x \in X_+$ with $\gamma_B(x) > 1$. The same proof as for Theorem 12.5 provides a sequence (n_j) in $\mathbb{N}$ with $\|B^{n_j}(x)\| \to \infty$ and $v_j = \|B^{n_j}(x)\|^{-1} B^{n_j}(x) \to v$ with some $v \in X_+$, $\|v\| = 1$. But, for large enough j, $B^{n_j}(x) \le cu$ and $v_j \le c\|B^{n_j}(x)\|^{-1} u$ and so $v \le 0$, a contradiction □

Similarly as for Corollary 12.6, this yields the following result.

Corollary 12.8. *Let $B : X_+ \to X_+$ be homogeneous, bounded and order-preserving.*

(a) *Let u be a normal point for some power of B. Then $\mathbf{r}_o(B) \le \eta^u(B)$ if B is pointwise u-bounded, and $\mathbf{r}_+(B) \le \eta^u(B)$ if B is bounded and some power of B is uniformly u-bounded.*

(b) *If B is power-compact and pointwise u-bounded, then $\mathbf{r}_o(B) \le \eta^u(B)$.*

For those $x \in X_+$ for which the sequence $\sharp B^n(x)\sharp^{1/n}$ is bounded, we extend the definition of the *companion growth bound* of the B-orbit of x by

$$\gamma_B^\sharp(x) := \limsup_{n\to\infty} \sharp B^n(x)\sharp^{1/n} \tag{12.4}$$

and set it equal to infinity otherwise. We extend the definition of the *orbital companion spectral radius* of B by

$$\mathbf{r}_o^\sharp(B) := \sup_{x\in X_+} \gamma_B^\sharp(x). \tag{12.5}$$

Theorem 12.9. *Let $B : X_+ \to X_+$ be homogeneous and order-preserving, $u \in X_+$. Let B be pointwise u-bounded and assume that there is some $\ell \in \mathbb{N}$ such that $B^n(u)$ is u-bounded for all $n \ge \ell$. Then*

$$\eta_x(B) \le \gamma_B^\sharp(x) \le \gamma_B^\sharp(u) \le \eta^u(B)$$

for all $x \in X_+$ and

$$cw(B) \le \mathbf{r}_{cw}(B) \le \mathbf{r}_o^\sharp(B) = \gamma_B^\sharp(u) \le \eta^u(B).$$

If B is bounded and B^m has the lower KR property for some $m \in \mathbb{N}$, then $\gamma_B(u) \le \mathbf{r}_+(B) \le \eta^u(B)$.

Proof. Let $x \in X_+$. We can assume $x \ne 0$. Since B is pointwise u-bounded, there exists some $k = k(x) \in \mathbb{N}$ and some $c = c(x) > 0$ such that $B^k(x) \le cu$. Since B is order-preserving and homogeneous, $B^n(x) \le cB^{n-k}(u)$ for all $n > k$. This implies

$$B^n(x) \le cB^{n-k}(u) \le c\|B^{n-k}\|_u u.$$

We apply the monotone companion norm,

$$\sharp B^n(x)\sharp \le c\sharp B^{n-k}(u)\sharp \le c\|B^{n-k}\|_u \sharp u\sharp.$$

So

$$\sharp B^n(x)\sharp^{1/n} \le c^{1/n}\sharp B^{n-k}(u)\sharp^{1/n} \le \|B^{n-k}\|_u^{1/n} \big(c\sharp u\sharp\big)^{1/n}.$$

We take the limit superior as $n \to \infty$, use Lemma 12.2 (b), recall Proposition 8.1 and obtain the first inequality. The second then follows by taking the supremum over $x \in X_+$ and recalling $cw(B) \le \mathbf{r}_{cw}(B)$ from Proposition 8.1.

Assume that B is bounded and B^m has the lower KR property for some $m \in \mathbb{N}$. We can assume that $r = \mathbf{r}_+(B) > 0$. Then $B^m(v) \ge r^m v$ with $r = \mathbf{r}_+(B)$ and some $v \in \dot{X}_+$. Since B is pointwise u-bounded, there exists some $c > 0$ and $k \in \mathbb{N}$

(which depend on v) such that $B^k(v) \le cu$. So $r^{m+k}v = B^{m+k}(v) \le cB^m(u)$. For all $n \in \mathbb{N}$, $r^{m+k+n}(v) \le cB^{m+n}(u)$. Then

$$r^{k+j}\|v\|_u \le c\|B^j\|_u, \qquad j \in \mathbb{N},\ j \ge m.$$

So

$$r \le (c/r^k\|v\|_u)^{1/j}\|B^j\|_u^{1/j}, \quad j \in \mathbb{N},\ j \ge m.$$

Taking the limit as $j \to \infty$ yields the desired result. □

Theorem 12.10. *Let $B : X_+ \to X_+$ be homogeneous, bounded and order-preserving, $u \in X_+$. Let some power of B be uniformly u-bounded. Then*

$$cw(B) \le \mathbf{r}_{cw}(B) \le \mathbf{r}_o^\sharp(B) = \mathbf{r}_+^\sharp(B) = \gamma_B^\sharp(u) = \eta^u(B) \le \gamma_B(u) \le \mathbf{r}_+(B).$$

Under additional assumptions, the following hold:

- *If u is a normal point for some power of B, then $\eta^u(B) = \mathbf{r}_o(B) = \mathbf{r}_+(B) = \gamma_B(u) = \lim_{n\to\infty} \|B^n(u)\|^{1/n}$.*
- *If some power of B has the lower KR property, then $\mathbf{r}_{cw}(B) = \eta^u(B) = \gamma_B(u) = \mathbf{r}_o(B) = \mathbf{r}_+(B)$.*
- *If B has the lower KR property, then $cw(B) = \mathbf{r}_{cw}(B) = \eta^u(B) = \gamma_B(u) = \mathbf{r}_o(B) = \mathbf{r}_+(B)$.*

Proof. Let $k \in \mathbb{N}$ such that B^k is uniformly u-bounded. By Proposition 10.2, B^k is also uniformly u-bounded with respect to the monotone companion norm. Then there exists some $c > 0$ such that $B^k(x) \le c\sharp x\sharp u$ for all $x \in X_+$. For all $n \in \mathbb{N}$, $B^{k+n}(x) = B^k(B^n(x)) \le c\sharp B^n(x)\sharp u$. By definition of upper Collatz–Wielandt numbers, with $x = u$,

$$\|B^{k+n}\|_u \le \sharp B^n(u)\sharp c, \qquad n \in \mathbb{N}.$$

By (12.2),

$$(\eta^u(B))^{(k+n)/n} \le \sharp B^n(u)\sharp^{1/n} c^{1/n}, \qquad n \in \mathbb{N}.$$

We take the limit as $n \to \infty$,

$$\eta^u(B) \le \liminf_{n\to\infty} \sharp B^n(u)\sharp^{1/n} \le \liminf_{n\to\infty} \|B^n(u)\|^{1/n}. \tag{12.6}$$

This implies $\eta^u(B) \le \gamma_B^\sharp(u)$. The other inequalities and equalities follow from Theorem 12.9.

Let u be a normal point. By Corollary 12.8,

$$\eta^u(B) \ge \mathbf{r}_+(B) \ge \mathbf{r}_o(B) \ge \gamma_B(u) = \limsup_{n\to\infty} \|B^n(u)\|^{1/n}.$$

Together with (12.6), this implies equalities.

Since the companion norm is order-preserving,

$$\sharp B^{k+n}(x)\sharp \le c\sharp x\sharp\ \sharp B^n(u)\sharp \quad \text{and} \quad \sharp B^{k+n}\sharp \le c\sharp B^n(u)\sharp, \quad n \in \mathbb{N}.$$

Since B is bounded, $\mathbf{r}_+^\sharp(B) \le \gamma_B^\sharp(u)$.

The other statements now follow from the previous theorems. □

In view of estimating the cone spectral radius from above the following observation may be of interest.

Corollary 12.11. *Let $B : X_+ \to X_+$ be a homogeneous, bounded, order-preserving map. Assume that X_+ is normal and complete or some power of B has the lower KR property.*

Then $\mathbf{r}_+(B)$ is a lower bound for all upper Collatz–Wielandt numbers $\|B\|_u$ where $u \in \dot{X}_+$, $B(u)$ is u-bounded and B is pointwise u-bounded.

Proof. Combine the previous theorems with (12.2) and recall that $\mathbf{r}_o^\sharp(B) = \mathbf{r}_+(B)$ if X_+ is normal and complete. □

12.1. The upper Collatz–Wielandt bound

Let $u \in \dot{B}_+$ and $B(u)$ be u-bounded. Then $B(x)$ is x-bounded for any u-comparable $x \in X_+$. Recall Definition 7.1.

So we define the upper Collatz–Wielandt bound with respect to u by

$$CW_u(B) = \inf\{\|B\|_x; x \in X_+, x \sim u\}. \tag{12.7}$$

If $x \in X_+$ and $x \sim u$, $\eta^x(B) = \eta^u(B)$. Since $\eta^x(B) \leq \|B\|_x$,

$$\eta^u(B) \leq CW_u(B). \tag{12.8}$$

We have the following inequalities from Theorem 12.9 and Theorem 12.10.

Theorem 12.12. *Let B be homogeneous and order-preserving. Let $u \in \dot{X}_+$, $B(u)$ be u-bounded and B be pointwise u-bounded. Then*

$$cw(B) \leq \mathbf{r}_{cw}(B) \leq \eta^u(B) \leq CW_u(B).$$

Lower KR property of the map turns some of the inequalities in equalities (Theorem 12.10).

Theorem 12.13. *Let B be homogeneous, bounded, and order-preserving and some power of B have the lower KR property. Let $u \in \dot{X}_+$ and some power of B be uniformly u-bounded. Then*

$$\mathbf{r}_{cw}(B) = \mathbf{r}_o(B) = \mathbf{r}_+(B) = \eta^u(B) \leq CW_u(B).$$

12.2. Monotonicity of order-spectral radii and the Collatz–Wielandt radius

If the cone X_+ is normal, the cone and orbital spectral radius are increasing functions of the homogeneous bounded order-preserving maps (cf. [1, L.6.5]). Collatz–Wielandt numbers, bounds, and radii and the companion radii are increasing functions of the map even if the cone is not normal.

Theorem 12.14. *Let $A, B : X_+ \to X_+$ be bounded and homogeneous. Assume that $A(x) \leq B(x)$ for all $x \in X_+$ and that A or B are order-preserving.*

Then $cw(A) \leq cw(B)$, $\mathbf{r}_{cw}(A) \leq \mathbf{r}_{cw}(B)$, $\mathbf{r}_+^\sharp(A) \leq \mathbf{r}_+^\sharp(B)$, $\mathbf{r}_o^\sharp(A) \leq \mathbf{r}_o^\sharp(B)$. Further, for all $x \in X_+$, $\|A\|_x \leq \|B\|_x$ and $\eta^x(A) \leq \eta^x(B)$.

If $u \in \dot{X}_+$ and $A(u)$ and $B(u)$ are u-bounded, then $CW_u(A) \leq CW_u(B)$.

If X_+ is a normal cone, then also $\mathbf{r}_+(A) \leq \mathbf{r}_+(B)$ and $\mathbf{r}_o(A) \leq \mathbf{r}_o(B)$.

Proof. We claim that $A^n(x) \le B^n(x)$ for all $x \in X_+$ and all $n \in \mathbb{N}$. For $n = 1$, this holds by assumption. Now let $n \in \mathbb{N}$ and assume the statement holds for n. If A is order-preserving, then, for all $x \in X_+$, since $B^n(x) \in X_+$,

$$A^{n+1}(x) = A(A^n(x)) \le A(B^n(x)) \le B(B^n(x)) = B^{n+1}(x).$$

If B is order-preserving, then, for all $x \in X_+$, since $A^n(x) \in X_+$,

$$A^{n+1}(x) = A(A^n(x)) \le B(A^n(x)) \le B(B^n(x)) = B^{n+1}(x).$$

Since the companion norm is order-preserving, $\sharp A^n(x)\sharp \le \sharp B^n(x)\sharp$ for all $x \in X_+$, $n \in \mathbb{N}$. Further $\sharp A^n\sharp_+ \le \sharp B^n\sharp_+$ for all $n \in \mathbb{N}$ and $\mathbf{r}_+^\sharp(A) \le \mathbf{r}_+^\sharp(B)$. Further $\gamma_A^\sharp(x) \le \gamma_B^\sharp(x)$ and so $\mathbf{r}_o^\sharp(A) \le \mathbf{r}_o^\sharp(B)$.

If X_+ is normal, the respective order radii taken with the original norm coincide with those taken with the companion norm.

As for the Collatz–Wielandt radius,

$$B^n(x) \ge A^n(x) \ge [A^n]_x x.$$

By (1.14), $[B^n]_x \ge [A^n]_x$, and the claim follows from (1.17), (1.19), and (1.20). The proofs for $\|\cdot\|_x$, η^x, and CW_u are similar. □

Proof of Theorem 1.14. By Theorem 12.14, $cw(A) \le cw(B)$. The assertion now follows from Theorem 1.13. □

12.3. The upper Collatz–Wielandt bound as eigenvalue

Conditions which make $CW_u(B)$ an eigenvalue of B with positive eigenvector and imply equality between all these numbers including $CW_u(B)$ can be found in [1, Thm. 7.3]. Using the companion half-norm ψ, one can drop that the cone is normal and complete provided that the map is compact. Solidity of the cone can be replaced by the weaker assumption that the map is uniformly u-bounded.

Theorem 12.15. *Let $B : X_+ \to X_+$ be continuous, compact, homogeneous, and order-preserving. Let $u \in \dot{X}_+$ and B be uniformly u-bounded.*

Then $cw(B) = \mathbf{r}_{cw}(B) = \mathbf{r}_+(B) = \eta^u(B) = CW_u(B)$.

If $r = CW_u(B) > 0$, then there exists some $v \in \dot{X}_u$ such that $B(v) = rv$.

Remark 12.16. If X_+ is complete, we also obtain this result if we replace compactness of B by the assumptions in Theorem 1.9 with part (a) or by assumption (ii) in [1, Thm. 7.3].

More generally, the following holds.

Theorem 12.17. *Let $B : X_+ \to X_+$ be homogeneous and order-preserving. Let $u \in \dot{X}_+$ and B be uniformly u-bounded and continuous at $B^n(u)$ for all $n \in \mathbb{N}$. Assume there is some $\epsilon_0 > 0$ such that, for all $\epsilon \in (0, \epsilon_0)$, the maps B_ϵ, $B_\epsilon(x) = B(x) + \epsilon\psi(x)u$, have eigenvectors $B_\epsilon(v_\epsilon) = \lambda_\epsilon v_\epsilon$ with $v_\epsilon \in \dot{X}_+$ and $\lambda_\epsilon > 0$.*

Then $\mathbf{r}_+(B) \ge \gamma_u(B) \ge CW_u(B) = \eta^u(B)$ with equality holding everywhere if u is a normal point for some power of B or some power of B has the lower KR property.

Further, if B has the KR property and $CW_u(B) > 0$, there exists some $v \in \dot{X}_+$ such that $B(v) = CW_u(B)v$.

Proof. Choose a sequence (ϵ_n) in $(0, \epsilon_0)$ with $\epsilon_n \to 0$. Set $B_n = B_{\epsilon_n}$. The maps B_n inherit uniform u-boundedness from B.

By assumption, there exist $v_n \in \dot{X}_+$ and $r_n > 0$ such that $B(v_n)+\epsilon_n\psi(v_n)u = r_n v_n$. Since B is uniformly u-bounded and $\psi(v_n) > 0$, v_n is u-comparable. By (12.7), $r_n \geq CW_u(B_n)$. Also $r_n \leq cw(B_n)$ by (1.19). By Theorem 12.12 and Theorem 12.10, $\eta^u(B_n) = \mathbf{r}_+^\sharp(B_n) = CW_u(B_n)$ for all $n \in \mathbb{N}$. Further, $CW_u(B_n) \geq CW_u(B)$.

Suppose that $\mathbf{r}_+^\sharp(B) < CW_u(B)$. Since $\epsilon_n \to 0$, $\|B_n - B\|_+ \to 0$. By Theorem 11.2, $\mathbf{r}_+^\sharp(B_n) < CW_u(B)$ for large n, a contradiction.

So $\mathbf{r}_+^\sharp(B) \geq CW_u(B)$. By Theorem 12.10, also $\eta^u(B) = \mathbf{r}_+^\sharp(B)$. Since $\eta^u(B) \leq CW_u(B)$, we have $\eta^u(B) = CW_u(B)$. The other inequalities follow from Theorem 12.10.

If some power of B has the lower KR property, equality holds by Theorem 12.13. If u is a normal point of X_+, equality follows from Theorem 12.10. Assume that B has the KR property and $CW_u(B) > 0$. Then $\mathbf{r}_+(B) = CW_u(B) > 0$ and there exists some $v \in \dot{X}_+$ such that $B(v) = \mathbf{r}_+(B)v$. □

The equality $\mathbf{r}_+(B) = CW_u(B)$ guarantees that, at least in theory, one can get arbitrarily sharp estimates of $\mathbf{r}_+(B)$ from above in terms of upper Collatz–Wielandt numbers $\|B\|_x$ by choosing an appropriate $x \in \dot{X}_+$ for which $B(x)$ is x-bounded and B pointwise x-bounded (Corollary 12.11). Crude attempts in this direction are made for the rank-structured discrete population model with mating in Section 14.

The idea of perturbing the map B as above or in a similar way is quite old; see [36, Satz 3.1] and [46, Thm. 3.6].

Theorem 12.18. *Let X_+ be complete. Assume that $B = K + A$ where $K : X_+ \to X_+$ is compact, homogeneous, continuous and order preserving and $A : X \to X$ is linear, positive and bounded and $\mathbf{r}(A) < \mathbf{r}_+(B)$. Let $u \in \dot{X}_+$ and B be uniformly u-bounded.*

Then $cw(B) = \mathbf{r}_{cw}(B) = \mathbf{r}_+(B) = \eta^u(B) = CW_u(B)$.

If $r = CW_u(B) > 0$, then there exists some $v \in \dot{X}_u$ such that $B(v) = rv$.

Proof. For $\epsilon \in [0, 1]$, we define $B_\epsilon : X_+ \to X_+$ by $B_\epsilon(x) = B(x) + \epsilon\psi(x)u$ where ψ is the companion half-norm. Then $B_\epsilon = K_\epsilon + A$ with $K_\epsilon(x) = K(x) + \epsilon\psi(x)u$, and K_ϵ is compact, continuous, order-preserving and homogeneous.

Since A is linear and bounded, $B_\epsilon^n = K_{n,\epsilon} + A^n$ with compact, continuous, homogeneous, order-preserving maps $K_{n,\epsilon}$.

If n is chosen large enough, $\|A^n\| < \mathbf{r}_+(B^n)$. By Theorem 1.9 (a), some power of B has the KR property, and $\mathbf{r}_+(B) = \eta^u(B) = \gamma_B(u) = \mathbf{r}_{cw}(B) \leq CW_u(B)$ by Theorem 12.13. Since $B(x) \leq B_\epsilon(x)$ for all $x \in X_+$, $\mathbf{r}_+(B) = \mathbf{r}_{cw}(B) \leq \mathbf{r}_{cw}(B_\epsilon) \leq \mathbf{r}_+(B_\epsilon)$ by Theorem 12.14.

So $\mathbf{r}(A) < \mathbf{r}_+(B) \leq \mathbf{r}_+(B_\epsilon)$. By Theorem 1.9 (a), for large enough n, there exist eigenvectors $v_\epsilon \in \dot{X}_+$ such that $B_\epsilon^n(v_\epsilon) = r_\epsilon^n v_\epsilon$ with $r_\epsilon = \mathbf{r}_+(B_\epsilon)$.

For $\epsilon \in (0,1]$, B_ϵ is strictly ψ-increasing (Definition 9.1), and B_ϵ^n inherits this property. Set $w_\epsilon = B_\epsilon(v_\epsilon)$. Then $B_\epsilon^n(w_\epsilon) = r_\epsilon^n w_\epsilon$. Since B_ϵ is uniformly u-bounded and $\psi(v_\epsilon) > 0$, w_ϵ is u-comparable. Since $B_\epsilon^k(v_\epsilon) \in \dot{X}_+$ for all $k \in \mathbb{N}$, $B_\epsilon^n(v_\epsilon)$ is u-comparable and so v_ϵ and w_ϵ are comparable.

By Proposition 9.3, $B_\epsilon(v_\epsilon) = \alpha_\epsilon v_\epsilon$ for some $\alpha_\epsilon > 0$ which must then equal r_ϵ. Since v_ϵ is u-comparable $r_\epsilon = cw(B_\epsilon) = \mathbf{r}_{cw}(B_\epsilon) = \mathbf{r}_+(B_\epsilon) = CW_u(B_\epsilon)$ for all $\epsilon \in (0,1]$. By Theorem 12.17, $CW_u(B) \leq \mathbf{r}_+(B) = \eta^u(B) = \gamma^u(B)$ and equality holds everywhere by our earlier inequality.

Now choose a decreasing sequence (ϵ_k) in $(0,1]$ with $\epsilon_k \to 0$. Then $r_k = \mathbf{r}_+(B_{\epsilon_k})$ form a decreasing sequence with $r_k \geq CW_u(B)$. Suppose $CW_u(B) > 0$. Set $v_k = v_{\epsilon_k}$. We can assume that $\|v_k\| = 1$ for all $k \in \mathbb{N}$. Then

$$r_k v_k = K(v_k) + A v_k + \epsilon_k \psi(v_k) u.$$

Let $r = \lim_{k\to\infty} r_k$. Since K is compact,

$$(r - A) v_k = (r - r_k) v_k + K(v_k) + \epsilon_k \psi(v_k) u$$

converges as $k \to \infty$ after choosing a subsequence. Since $r \geq CW_u(B) = \mathbf{r}_+(B) > \mathbf{r}_+(A)$, $(r - A)^{-1} = \sum_{j=0}^{\infty} (1/r)^{j+1} A^k$ exists as a continuous additive homogeneous map and acts as the inverse of $r - A$. This implies that $v_k \to v$ for some $v \in X_+$, $\|v\| = 1$. Since B is continuous, $rv = B(v)$ which implies that $r = \mathbf{r}_+(B)$. Then $r \leq cw(B)$ and equality follows. If $CW_u(B) = 0$, equality holds anyway. □

13. Monotonically compact maps on semilattices

As before, let X be an ordered normed vector space with cone X_+.

Definition 13.1. Let $u \in \dot{X}_+$. $B : X_+ \to X_+ \cap X_u$ is called *antitonically u-compact* if $(B(x_n))$ has a convergent subsequence for each decreasing sequence (x_n) in X_+ for which there is some $c > 0$ such that $x_n \leq cu$ for all $n \in \mathbb{N}$.

B is called *monotonically u-compact* if $(B(x_n))$ has a convergent subsequence for each monotone sequence (x_n) in X_+ for which there is some $c > 0$ such that $x_n \leq cu$ for all $n \in \mathbb{N}$.

If $u \in X_+$ and $B(u)$ is a regular point and B is order-preserving, then B is monotonically u-compact. If X_+ is regular (Section 2), then every order-preserving homogeneous continuous $B : X_+ \to X_+$ is monotonically u-compact for any $u \in X_+$.

Definition 13.2. Let $u \in \dot{X}_+$, $B : X_+ \to X_+$ and $B(u)$ be u-bounded. B is called *antitonically continuous* at $x \in X_+$ if, for for any decreasing sequence (x_k) in X_+ with $x \leq x_k \leq cu$ for all $k \in \mathbb{N}$ (with some $c > 0$ independent of k) and $\|x_k - x\| \to 0$, we have $\psi(B(x_k) - B(x)) \to 0$.

B is called *monotonically continuous* at $x \in X_+$ with for every monotone sequence (x_k) in X_u with $x_k \le cu$ for all $k \in \mathbb{N}$ (with some $c > 0$ independent of k) and $\|x_k - x\| \to 0$, we have $\sharp B(x_k) - B(x)\sharp \to 0$.

Here ψ is the monotone companion half-norm, and $\sharp \cdot \sharp$ is the monotone companion norm.

Recall the definition of an inf-semilattice in Section 1.1 ($x \wedge y = \inf\{x, y\}$ exists for all $x, y \in X_+$) and the upper local Collatz–Wielandt spectral radius of B at $u \in X_+$ in Section 1.2.5,

$$\eta^u(B) = \inf_{n \in \mathbb{N}} \|B^n\|_u^{1/n}.$$

Theorem 13.3. *Let X_+ be an* inf*-semilattice and $B : X_+ \to X_+$ be order-preserving and homogeneous. Let $u \in \dot{X}_+$, and $B(u)$ be u-bounded. Assume that B is antitonically u-compact and antitonically continuous. Finally assume that $\eta^u(B) > 0$ and $\|B(y_n)\|_u \to 0$ for any decreasing sequence (y_n) in $X_u \cap X_+$ with $\|y_n\| \to 0$.*

Then there exists some $x \in \dot{X}_+$, such that $B(x) \ge \eta^u(B)x$ and $\eta^u(B) \le cw(B)$.

The first part of the proof has been adapted from [25, L.9.5] where B is assumed to be a linear operator on the ordered Banach space X and the cone X_+ to be normal. Use of the monotone companion metric allows to drop normality as assumption. However, without u being a normal point of the cone, compactness of B may not imply monotonic compactness.

Proof of Theorem 13.3. Let $u \in X_+$ such that $B(u)$ is u-bounded. Since B is homogeneous, we can assume that $\eta^u(B) = 1$. Otherwise, we consider $\frac{1}{\eta^u(B)}B$. We define

$$x_0 = u, \qquad x_k = y_k \wedge u, \quad y_k = B(x_{k-1}) + 2^{-k}u, \quad k \in \mathbb{N}. \tag{13.1}$$

Then $x_k \le u = x_0$ for all $n \in \mathbb{N}$. By induction, since B is order-preserving, $x_{k+1} \le x_k$ for all $k \in \mathbb{N}$. We apply the convergence principle in Proposition 2.12 with S being the set of sequences $(B(v_n))$ with (v_n) being decreasing and $v_1 \le cu$ for some $c > 0$. S has the properties requested in Proposition 2.12. Since B is antitonically u-compact, every sequence in S has a convergent subsequence. So every sequence in S converges.

Since $(B(x_n)) \in S$, there exists some $z \in X_+$ such that $(B(x_k))$ converges to z as $k \to \infty$ and $B(x_k) \ge z$ for all $k \in \mathbb{N}$. By (13.1), $y_k \to z$. Further

$$z \le B(x_{k-1}) \le y_k \le B(u) + 2^{-k}u.$$

By (13.1),

$$x_k = y_k \wedge u \ge z \wedge u =: x.$$

Notice that $\quad y_k \wedge u + z - y_k \le z \wedge u = x. \quad$ So

$$0 \le x_k - x \le y_k - z, \qquad k \in \mathbb{N}. \tag{13.2}$$

Further

$$\sharp B(x_k) - B(x)\sharp = \psi(B(x_k) - B(x)) \le \psi(B(y_k - z + x) - B(x))$$
$$\le \|B(y_k - z + x) - B(x)\|.$$

Also $x \le y_k - z + x \le B(u) + u + x \le (c\|u\| + 2)u$. Recall that $(y_k - z + x)$ converges to x with respect to the original norm.

Since B is antitonically continuous, $(B(x_k))$ converges to $B(x)$ with respect to the monotone companion norm. Since $B(x_k) \to z$ with respect to the monotone companion norm, we have $B(x) = z \ge x$.

Moreover, $x = z \wedge u = B(x) \wedge u$.

It remains to show that $x \ne 0$. Suppose that $x = 0$. Then $z = B(x) = 0$. Recall that $x_k = y_k \wedge u$ and $\|y_k\| \to 0$. Since B is order-preserving, $\|B(x_k)\|_u \le \|B(y_k)\|_u \to 0$ with the latter holding by assumption.

So there exists some $m \in \mathbb{N}$ such that $B(x_{k-1}) + 2^{-k}u \le u$ for all $k \ge m$. Hence

$$x_k = B(x_{k-1}) + 2^{-k}u = y_k, \qquad k \ge m.$$

In particular, $2^m x_m \ge u$ and $x_k \ge B(x_{k-1})$ for all $k \ge m$. Since B is order-preserving and homogeneous,

$$2^m x_{m+n} \ge B^n(2^m x_m) \ge B^n(u)$$

and

$$2^m B(x_{m+n}) \ge B^{n+1}(u).$$

Now $2^m B(x_{m+n}) \le (1/2)u$ for sufficiently large n. This shows that, for some $n \in \mathbb{N}$, $B^{n+1}(u) \le (1/2)u$ and $\|B^{n+1}\|_u \le 1/2$. By Lemma 12.2, $\eta^u(B) = \inf_{n\in\mathbb{N}} \|B^n\|_u^{1/n} < 1$, a contradiction.

This shows that $x \ne 0$ and $B(x) \ge x$. Then $B^n(x) \ge x$ for all $n \in \mathbb{N}$. By (1.14), $[B^n]_x \ge 1$ and, by (1.19), $cw(B) \ge 1 = \eta^u(B)$. □

Theorem 13.4. *Let the cone X_+ be a lattice and $B : X_+ \to X_+$ be order-preserving and homogeneous. Further let $u \in \dot{X}_+$ and some power of B be monotonically u-compact and continuous and some power of B be uniformly u-bounded. Finally assume that $r = \eta^u(B) > 0$.*

Then there exists some $x \in \dot{X}_+$ such that $B(x) \ge rx$ and $\eta^u(B) = \mathbf{r}_o^\sharp(B) = \mathbf{r}_{cw}(B) = cw(B)$.

Proof. Replacing B by $\frac{1}{\eta^u(B)}B$, we can assume that $\eta^u(B) = 1$. Let B^m be monotonically u-compact and continuous and B^ℓ be uniformly u-bounded. Set $p = m+\ell$. Then B^p is monotonically u-compact and uniformly u-bounded and continuous. By Theorem 13.3, there exists some $w \in \dot{X}_+$ such that $B^p(w) \ge w$. By Proposition 1.11, there exists some $v \in \dot{X}_+$ such that $B(v) \ge v$. By (1.19), $cw(B) \ge 1 = \eta^u(B)$. By Theorem 12.9, $\eta^u(B) \ge \mathbf{r}_o^\sharp(B) \ge \mathbf{r}_{cw}(B) \ge cw(B)$ and equality follows. □

Recall the definition of B being strictly increasing in Definition 9.1.

Theorem 13.5. *Let X_+ be a lattice and $B : X_+ \to X_+$ be monotonically continuous, strictly increasing, and homogeneous. Further assume that a power of B is monotonically u-compact and some power is uniformly u-bounded for some $u \in \dot{X}_+$. Finally assume that $r = \eta^u(B) > 0$.*

Then there exists some $x \in \dot{X}_+$ such that $B(x) = rx$.

Proof. We can assume that $\eta_u(B) = 1$. By Theorem 13.3, there exists some $x \in X_+$, $\|x\| = 1$ such that $B(x) \geq x$. Then the sequence $(x_n)_{n\in\mathbb{Z}_+}$ in X_+ defined by $x_n = B^n(x)$ is increasing. The same proof as for Theorem 9.2 implies that $\{\|x_n\|; n \in \mathbb{N}\}$ is a bounded set in $\mathbb{R}$.

Set $x_0 = x$. Then $x_n = B(x_{n-1})$ for $n \in \mathbb{N}$. Since some power of B is uniformly u-bounded, there exists some $c \geq 0$ and $m \in \mathbb{N}$ such that $x_n \leq c\|x_{n-m}\|u = cu$ for $n \geq m$. Since some power of B is monotonically u-compact and $(x_n) = (B^n(x))$ is increasing, a similar application of the convergence principle in Proposition 2.12 provides that $x_n \to y$ for some $y \in X_+$ with $\|y\| = 1$. Since B is monotonically continuous, $\psi(B(y) - B(x_n)) \to 0$. So $\psi(B(y) - x_{n+1}) \to 0$. Then $\sharp y - B(y)\sharp \leq \sharp y - x_{n+1}\sharp + \sharp x_{n+1} - B(y)\sharp = \psi(y - x_{n+1}) + \psi(B(y) - x_{n+1}) \to 0$. Thus $y = B(y)$. □

Proposition 13.6. *Let $u \in \dot{X}_+$ and $B : X_+ \to X_+ \cap X_u$ be homogeneous and order-preserving. Further let B be monotonically u-compact, uniformly u-bounded and monotonically continuous.*

Let $\epsilon > 0$ and ψ the companion half-norm. Set $B_\epsilon(x) = B(x) + \epsilon\psi(x)u$.

Then there exists some $v \in \dot{X}_+$ such that $B_\epsilon(v) = r_\epsilon v$ with $r_\epsilon = \eta^u(B_\epsilon) = CW_u(B_\epsilon) = \mathbf{r}_{cw}(B_\epsilon) = cw(B_\epsilon) > 0$.

Proof. One readily checks that B_ϵ satisfies the assumptions of Theorem 13.3 and $r_\epsilon = \eta^u(B_\epsilon) \geq \epsilon > 0$.

We can assume that $r_\epsilon = 1$. By Theorem 13.3, there exists some $w \in X_+$, $\psi(w) = 1$, such that $B_\epsilon w \geq w$. Let $w_n = B_\epsilon^n(w)$ for $n \in \mathbb{Z}_+$. Then (w_n) is an increasing sequence in $X_+ \cap X_u$. We claim that $\psi(w_n) = 1$ for all $n \in \mathbb{N}$. Suppose not. Then there exists some $n \in \mathbb{N}$ such that $\psi(w_n) > \psi(w_{n-1})$. Then there exists some $\delta > 0$ such that

$$B_\epsilon(w_n) \geq B_\epsilon(w_{n-1}) + \delta u.$$

Since $w_n = B_\epsilon(w_{n-1}) \in X_u$, there exists some $\tilde{\delta} > 0$ such that

$$B_\epsilon(w_n) \geq (1 + \tilde{\delta})w_n.$$

This implies $\eta^u(B_\epsilon) \geq 1 + \tilde{\delta}$, a contradiction. Since B is uniformly u-bounded, it is also uniformly u-bounded with respect to the companion half-norm ψ by Proposition 10.2. So there exists some $c > 0$ such that $w_n = B(w_{n-1}) + \epsilon u \leq cu$. Since (w_n) is increasing and B_ϵ is monotonically u-compact, $w_{n+1} = B_\epsilon(w_n) \to v$ for some $v \in X_+$, $\psi(v) = 1$. Since B_ϵ is monotonically continuous, $v = B_\epsilon(v)$ and $1 \leq cw(B_\epsilon) \leq \mathbf{r}_{cw}(B_\epsilon) \leq \eta^u(B_\epsilon)$. Since $v \geq \epsilon u$ and B is uniformly u-bounded, v is u-comparable. This implies $CW_u(B_\epsilon) \leq 1 = \eta^u(B)$. Since $CW_u(B_\epsilon) \geq \eta^u(B_\epsilon)$, we have equality. □

Theorem 13.7. *Let $u \in \dot{X}_+$ and $B : X_+ \to X_+$ be homogeneous, order-preserving and continuous. Assume that B is uniformly u-bounded and monotonically u-compact.*

Then $\mathbf{r}_+(B) \geq CW_u(B) = \eta^u(B)$ with equality holding if u is a normal point of X_+ or a power of B has the lower KR property. If $r := CW_u(B) > 0$, there exists some $v \in \dot{X}_+$ such that $B(v) \geq rv$.

Proof. By Theorem 12.10, $\mathbf{r}_+(B) \geq \gamma_B(u) \geq \eta^u(B)$ with equality holding if u is a normal point of X_+ or some power of B has the lower KR property. By (12.8), $\eta^u(B) \leq CW_u(B)$.

If $CW_u(B) = 0$, the assertion holds; so we assume that $CW_u(B) > 0$.

Choose a sequence (ϵ_n) in $(0,1)$ with $\epsilon_n \to 0$. Let $B_n : X_+ \to X_+$ be given by $B_n(x) = B(x) + \epsilon_n \psi(x)u$. We combine Proposition 13.6 and Theorem 12.17 and obtain $\eta^u(B) = CW_u(B) \leq \mathbf{r}_+(B)$.

If u is a normal point of X_+ or some power of B has the lower KR property, $\eta_u(B) = \mathbf{r}_+(B)$ which implies $CW_u(B) = \mathbf{r}_+(B)$.

Assume that $r := CW_u(B) > 0$. Then $\eta^u(B) = r > 0$ and there exists some $v \in \dot{X}_+$ with $B(v) \geq rv$ by Theorem 13.3. □

14. A rank-structured population model with mating

Let $X \subseteq \mathbb{R}^{\mathbb{N}}$ be an ordered normed vector space with cone $X_+ = X \cap \mathbb{R}^{\mathbb{N}}_+$. Assume that the norm has the property that $x_j \leq \|x\|$ for all $x = (x_j) \in X_+$ and all $j \in \mathbb{N}$. This implies that $X \subseteq \ell^\infty$ and $\|x\|_\infty \leq \|x\|$ for all $x \in X$.

Define a map $B : X_+ \to \mathbb{R}^{\mathbb{N}}_+$, $B(x) = (B_j(x))$, by

$$\left.\begin{aligned} B_1(x) &= \quad q_1 x_1 + \sum_{j,k=1}^{\infty} \beta_{jk} \min\{x_j, x_k\} \\ B_j(x) &= \quad \max\{p_{j-1}x_{j-1}, q_j x_j\}, \qquad j \geq 2 \end{aligned}\right\} x = (x_j) \in X_+. \tag{14.1}$$

Here $p_j, q_j \geq 0$, $\beta_{j,k} \geq 0$ for all $j,k \in \mathbb{N}$. $\sum_{j,k=1}^{\infty}$ is to be understood as $\lim_{n\to\infty} \sum_{j,k=1}^{n}$.

The dynamical system $(B^n)_{n\in\mathbb{N}}$ can be interpreted as the dynamics of a rank-structured population and, in a way, is a discrete version (in a double sense) of the two-sex models with continuous age-structure in [18, 19]. $B_1(x)$ is the number of newborn individuals who all have the lowest rank 1. Procreation is assumed to require some mating. Mating is assumed to be rank-selective and is described by taking the minimum of individuals in two ranks. The numbers β_{jk} represent the probabilities that females of rank j and males of rank k mate and the per pair fertilities, where a 1:1 sex ratio is assumed at each rank. The maps B_j, $j \geq 2$, describe how individuals survive and move upwards in the ranks from year to year where one cannot move by more than one rank within a year. We assume that

$$p_j \leq 1, q_j \leq 1, \qquad j \in \mathbb{N}, \tag{14.2}$$

which is reasonable if x_k is interpreted as the number of individuals at rank k. If x_k were the biomass of individuals at rank k, such an assumption would make less sense. An individuals at rank j is at least $j-1$ years old, and so mortality eventually wins the upper hand such that the assumption $p_j \to 0$ and $q_j \to 0$ is natural though we will not always assume this.

Since $B_j(x) \le (p_{j-1}+q_j)\|x\|$ for $j \ge 2$, B is u-bounded with respect to $u=(u_j)$ with

$$u_1 = 1, \qquad u_j = p_{j-1} + q_j, \quad j \ge 2, \tag{14.3}$$

provided that $u \in X$.

If we choose $X = \ell^1$, it is sufficient to assume that

$$\sup_{k\in\mathbb{N}} \sum_{j=1}^{\infty} \beta_{jk} < \infty \qquad \text{or} \qquad \sup_{j\in\mathbb{N}} \sum_{k=1}^{\infty} \beta_{jk} < \infty. \tag{14.4}$$

If $X = \ell^\infty$, it is sufficient to assume that

$$\sum_{j,k=1}^{\infty} \beta_{jk} < \infty. \tag{14.5}$$

It does not appear that (14.5) can be improved for $X = c_0$ or $X = bv$. To make B map bv into itself, it is sufficient to assume that (p_j), $(q_j) \in bv$. If $X = c, c_0, \ell^p$ $(1 \le p \le \infty)$, no assumptions other than (14.2) are needed for B to map X into X.

To derive estimates for the cone spectral radius of B, one runs into algebraic difficulties very soon except when attempting $cw(B)$. Let x^n be the sequence where the first n terms are 1 and all others zero. Then, for $m \ge 2$,

$$cw(B) \ge [B]_{x^m} = \min\Big\{ q_1 + \sum_{j,k=1}^{m} \beta_{j,k},\ \inf_{j=2}^{m} \max\{p_{j-1}, q_j\} \Big\}. \tag{14.6}$$

Let e^m be the sequence where the mth term is one and all other terms zero. Then

$$[B]_{e^1} = q_1 + \beta_{11}, \qquad [B]_{e^m} = q_m, \qquad m \ge 2.$$

Since $cw(B)$ is an upper bound for all these numbers, we obtain that

$$cw(B) \ge \max\{q_1 + \beta_{11}, \sup_{j\ge 2} q_j\}. \tag{14.7}$$

Lemma 14.1. *Let $r > q_j$ for all $j \in \mathbb{N}$, $j \ge 2$, and $v \in \dot{X}_+$ with $B(v) \ge rv$. Then*

$$1 \le r^{-1} q_1 + \sum_{j,k=1}^{\infty} \beta_{jk} \min\{r^{-j} P_j, r^{-k} P_k\}. \tag{14.8}$$

Here, the right-hand side of the equality may be infinite.

Proof. From $rv \le B(v)$, we obtain the inequalities,

$$
\begin{aligned}
rv_1 &\le q_1 v_1 + \sum_{j,k=1}^{\infty} \beta_{jk} \min\{v_j, v_k\}, \\
rv_j &\le \max\{p_{j-1} v_{j-1},\ q_j v_j\}, \qquad j \ge 2.
\end{aligned}
\tag{14.9}
$$

We claim that

$$
v_j \le \frac{p_{j-1}}{r} v_{j-1}, \qquad j \ge 2. \tag{14.10}
$$

Suppose that $j \ge 2$ and $rv_j > p_{j-1} v_{j-1}$. Then $rv_j \le q_j v_j$. So $v_j = 0$ because $r > q_j$, a contradiction.

By iteration of (14.10), for all $j \in \mathbb{N}$,

$$
v_j \le r^{1-j} P_j v_1, \qquad P_j = \prod_{i=1}^{j-1} p_i, \quad j \ge 2, \quad P_1 = 1. \tag{14.11}
$$

We notice that $v_1 = 0$ implies $v = 0$; so we can assume $v_1 > 0$. We substitute this formula into the one for v_1 and divide by rv_1; this yields (14.8). □

The following conditions are necessary and sufficient for $cw(B) > 0$.

Theorem 14.2. *$cw(B) > 0$ if and only if at least one of the subsequent three conditions hold:*

(i) *$q_j > 0$ for some $j \in \mathbb{N}$;* (ii) *$\beta_{11} > 0$;*
(iii) *for some $j, k \in \mathbb{N}$ with $m = \max\{j, k\} \ge 2$, we have $\beta_{jk} > 0$ and $p_i > 0$ for $i = 1, \dots, m-1$.*

Proof. That each of the conditions is sufficient follows from (14.6) and (14.7). To see necessity, assume that $q_j = 0$ for all $j \in \mathbb{N}$ and $cw(B) > 0$. Then there exists some $r > 0$ and $v \in \dot{X}_+$ such that $B(v) \ge rv$. By Lemma 14.1,

$$
1 \le \sum_{j,k=1}^{\infty} \beta_{jk} \min\{r^{-j} P_j, r^{-k} P_k\},
$$

with the right-hand side possibly being infinity. Then $\beta_{11} > 0$ or there exists $i, k \in \mathbb{N}$ with $m = \max\{j, k\} > 1$ and $P_m > 0$. □

Inspired by (14.11), for $m \in \mathbb{N}$ and $r > 0$, we define

$$
w_1 = 1, \qquad w_j = r^{1-j} P_j \ \text{ for } \ j = 2, \dots, m, \qquad w_j = 0 \text{ for } j > m. \tag{14.12}
$$

Lemma 14.3. *$B(w) \ge s(r, m) w$ with $s(r, m)$*

$$
= \min\left\{ q_1 + \sum_{j,k=1}^{m} \beta_{jk} \min\{r^{1-j} P_j, r^{1-k} P_k\}, \max\{r, q_2\}, \dots, \max\{r, q_m\} \right\}.
$$

So

$$
cw(B) \ge \sup\{s(r, m); m \in \mathbb{N}, r > 0\}.
$$

Proof. For $j \geq 2, \dots, m$,

$$B_j(w) = \max\{p_{j-1}w_{j-1}, q_j w_j\} = \max\{r^{2-j}P_j, q_j r^{1-j}P_j\} = \max\{r, q_j\}w_j.$$

For $j > m$, $B_j(w) \geq 0 = \max\{r, q_j\}w_j$. Finally,

$$B_1(w) \geq \Big(q_1 + \sum_{j,k=1}^{m} \beta_{jk} \min\{r^{1-j}P_j, r^{1-k}P_k\}\Big)w_1.$$

So $B(w) \geq s(r,m)w$. The last inequality follows from the definition of $cw(B)$ in Section 1.2.4. □

Motivated by these results, we set

$$\mathcal{R}_0 := q_1 + \sum_{j,k=1}^{\infty} \beta_{jk} \min\{P_j, P_k\}. \tag{14.13}$$

Notice that P_j is the probability of reaching rank j within j years after birth. So, with a grain of salt, $\mathcal{R}_0$ is the average number of rank 1 individuals one typical rank 1 individual can produce during its lifetime, and $\mathcal{R}_0$ can be interpreted as a *basic reproduction number.*

Theorem 14.4. *Assume that $\mathcal{R}_0 < \infty$. Then $\mathcal{R}_0 > 1$ if and only if $cw(B) > 1$. If $\mathcal{R}_0 > 1$, then $1 < cw(B) < \mathcal{R}_0$ and $cw(B)$ is the unique solution $r > 1$ of the equation*

$$1 = r^{-1}q_1 + \sum_{j,k=1}^{\infty} \beta_{jk} \min\{r^{-j}P_j, r^{-k}P_k\}. \tag{14.14}$$

Moreover $B(v) = rv$ for some $v \in \dot{X}_+$. Finally, there exist some $s > 1$ and some $w \in X_+$ with $w_1 = 1$ and $w_j = 0$ for all but finitely many j such that $B(w) \geq sw$.

Proof. Assume $\infty > \mathcal{R}_0 > 1$. Then the right-hand side of (14.14) is a strictly decreasing continuous function of $r \geq 1$ that converges to 0 as $r \to \infty$. By the intermediate value theorem, there exists a unique solution $r = r_0 > 1$ of (14.14). Let $1 < t < r_0$ be arbitrary. Then, for sufficiently large $m \in \mathbb{N}$,

$$1 < t^{-1}q_1 + \sum_{j,k=1}^{m} \beta_{jk} \min\{t^{-j}P_j, t^{-k}P_k\}.$$

By Lemma 14.3, $s(t,m) = t$ and $cw(B) \geq t$. Since this holds for any $t \in (1, r_0)$, $cw(B) \geq r_0 > 1$.

Now assume that $\mathcal{R}_0 < \infty$ and $cw(B) > 1$. Choose an arbitrary $s \in (1, cw(B))$. Then $B(v) \geq sv$ for some $v \in \dot{X}_+$. By Lemma 14.1,

$$1 \leq s^{-1}q_1 + \sum_{j,k=1}^{\infty} \beta_{jk} \min\{s^{-j}P_j, s^{-k}P_k\} < \mathcal{R}_0.$$

This also implies $s \leq r_0$ where again r_0 is the unique solution of (14.14). Since $s \in (1, cw(B))$ can be arbitrarily chosen, $cw(B) \leq r_0$.

Since $r_0 = cw(B) > 1$ solves (14.14), $r_0^{-1}\mathcal{R}_0 > 1$ and so $cw(B) < \mathcal{R}_0$.

Similarly as in the proof of Lemma 14.3, one shows that $v = (v_j)$ with $v_j = r_0^{1-j} P_j$ satisfies $B(v) = r_0 v$.

Recall Lemma 14.3. Since $\mathcal{R}_0 > 1$, by choosing $m \in \mathbb{N}$ large enough and $r \in (1, \mathcal{R}_0)$ close enough to 1, one can achieve that $s(r,m) > r > 1$ and $B(w) \geq s(r,m)w$ for some $w \in \mathbb{R}_+^{\mathbb{N}}$, $w_1 = 1$ and $w_j = 0$ for $j > m$. □

Theorem 14.5. *Let $\mathcal{R}_0 < \infty$ and $q^\diamond := \sup_{j \geq 2} q_j$. Then the following two equivalences hold:*

- *$cw(B) < 1$ if and only if $\max\{\mathcal{R}_0, q^\diamond\} < 1$.*
- *$cw(B) = 1$ if and only if $\max\{\mathcal{R}_0, q^\diamond\} = 1$.*

Further, the following hold:

If $\mathcal{R}_0 \leq 1$, then $cw(B) \geq \max\{\mathcal{R}_0, q^\diamond\}$.

If $\limsup_{j \to \infty} p_j < 1$ and $\mathcal{R}_0 < 1$, then $\mathcal{R}_0 < cw(B) < 1$.

Proof. By (14.7), $cw(B) \geq q^\diamond$.

By contraposition of Theorem 14.4, $\mathcal{R}_0 \leq 1$ if and only if $cw(B) \leq 1$.

Let $\mathcal{R}_0 \leq 1$. By Lemma 14.3, $s(1,m) = q_1 + \sum_{j,k}^{m} \beta_{jk} \min\{P_j, P_k\}$ and $cw(B) \geq s(1,m)$ for all $m \in \mathbb{N}$. By taking the limit as $m \to \infty$, $cw(B) \geq \mathcal{R}_0$.

In particular, if $\mathcal{R}_0 = 1$, then $cw(B) = 1$.

Suppose that $1 \leq cw(B)$ and $q^\diamond < 1$. By Lemma 14.1, for all $r \in (q^\diamond, 1)$ there exists some $v \in \dot{X}_+$ such that $B(v) \geq rv$ and (14.8) holds. Taking the limit $r \to 1$ yields $1 \leq \mathcal{R}_0$.

By contraposition of this result and of Theorem 14.4, $\mathcal{R}_0 \leq 1$ implies $cw(B) \leq 1$, and $\mathcal{R}_0 < 1$ and $q^\diamond < 1$ imply $cw(B) < 1$.

Suppose $\limsup_{j \to \infty} p_j < 1$ and $\mathcal{R}_0 < 1$. We can assume that $q^\diamond \leq \mathcal{R}_0 < 1$; otherwise $cw(B) \geq q^\diamond > \mathcal{R}_0$.

Choose α strictly between $\limsup_{j \to \infty} p_j$ and 1. Then there exists some $c > 0$ such that $P_k \leq c\alpha^k$ for all $k \in \mathbb{N}$. So there exists some $r \in (\alpha, 1)$, $r > q^\diamond$, such that $\sum_{j,k}^{\infty} \beta_{jk} \min\{r^{1-j} P_j, r^{1-k} P_k\} < r$. By Lemma 14.3, for all $m \in \mathbb{N}$, $s(r,m) \geq \sum_{j,k=1}^{m} \beta_{jk} \min\{r^{1-j} P_j, r^{1-k} P_k\}$. So

$$cw(B) \geq \sum_{j,k=1}^{\infty} \beta_{jk} \min\{r^{1-j} P_j, r^{1-k} P_k\} > \mathcal{R}_0. \qquad \square$$

The following trichotomy holds.

Corollary 14.6. *Let $\mathcal{R}_0 < \infty$ and $q^\diamond := \sup_{j \geq 2} q_j < 1$. Then one and only of the following three possibilities hold:*

Either $\mathcal{R}_0 \leq cw(B) < 1$, or $cw(B) = \mathcal{R}_0 = 1$, or $\mathcal{R}_0 > cw(B) > 1$. In particular $\mathcal{R}_0 - 1$ and $cw(B) - 1$ have the same sign.

We now make the connection to the cone spectral radius $\mathbf{r}_+(B)$.

Theorem 14.7. *Let* $X = \ell^\infty$ *and* $\sum_{j,k=1}^\infty \beta_{jk} < \infty$. *Further assume*

$$\limsup_{j\to\infty} p_j < 1 \quad \textit{and} \quad \sup_{j\geq 2} q_j < 1.$$

Then $\mathcal{R}_0 > \mathbf{r}_+(B) > 1$, *or* $\mathcal{R}_0 = \mathbf{r}_+(B) = 1$, *or* $\mathcal{R}_0 < \mathbf{r}_+(B) < 1$.

If, in addition, $\limsup_{j\to\infty} p_j \leq \mathcal{R}_0$ *and* $\limsup_{j\to\infty} q_j \leq \mathcal{R}_0$, $cw(B) = \mathbf{r}_+(B) = CW(B) := \inf_{x\in\check{X}_+} \|B\|_x$.

Proof. We first show that $B = K + H$ where $K, H : X_+ \to X_+$ are continuous, homogeneous, and order-preserving, K compact and H a strict contraction. Let $\alpha > p^\infty = \limsup_{j\to\infty} p_j$ and $\alpha < q^\infty = \limsup_{j\to\infty} q_j$.

Choose some $m \in \mathbb{N}$ and $\alpha < 1$ such that $p_j \leq \alpha$ and $q_j \leq \alpha$ for $j \geq m$. Define $H(x) = (H_j(x))$ with $H_j(x) = 0$ for $j = 1, \dots, m$ and $H_j(x) = B_j(x) = \max\{p_{j-1}x_{j-1}, q_j x_j\}$ for $j \geq m+1$. Set $K = B - H$. Then $K_j(x) = 0$ for $j \geq m+1$ and K is compact. An elementary but tedious calculation with many cases shows that $|\max\{s,t\} - \max\{u,v\}| \leq \max\{|s-u|, |t-v|\}$ for $s,t,u,v \in \mathbb{R}$. So

$$\begin{aligned} |H_j(x) - H_j(y)| &\leq \max\{p_{j-1}|x_{j-1} - y_{j-1}|, q_j|x_j - y_j|\} \\ &\leq \max\{p_{j-1}, q_j\}\|x-y\|_\infty, \qquad j \in \mathbb{N}, x = (x_j), y = (y_j) \in \ell^\infty. \end{aligned}$$

This implies $\|H(x) - H(y)\|_\infty \leq \alpha < 1$. Assume that $\mathcal{R}_0 \geq p^\infty$ and $\mathcal{R}_0 \geq q^\infty$. By Theorem 14.5, $\mathbf{r}_+(B) \geq cw(B) > \max\{p^\infty, q^\infty\}$. Choose α strictly between $\mathbf{r}_+(B)$ and $\max\{p^\infty, q^\infty\}$. By Theorem 1.9 (a), B has the KR property, and $\mathbf{r}_+(B) = cw(B)$ by Theorem 1.15. Further, in Theorem 12.17, the maps B_ϵ have the KR property and so $\mathbf{r}_+(B) = CW(B)$. The remaining statements now follow from Theorem 14.5. □

B is compact on X_+ and uniformly u-bounded with u in (14.3) if $B_1 : X_+ \to \mathbb{R}_+$ is bounded and

$$\begin{cases} u \in \ell^p, & X = \ell^p, p \in [1,\infty) \\ u \in c_0, & X = c_0, c, \ell^\infty \\ (p_j), (q_j) \in bv \cap c_0, & X = bv, bv \cap c_0. \end{cases} \tag{14.15}$$

Notice that, if $x^n = (x_j^n)$ is a bounded sequence in X, then it is a bounded sequence in ℓ^∞ and, after a diagonalization procedure, has a subsequence (y^m) such that $(y_j^m)_{m\in\mathbb{N}}$ converges for each $j \in \mathbb{N}$.

Apparently, $X = \ell^\infty, c, c_0$ require the weakest assumption for compactness of B in terms of u. It is not clear, however, whether an eigenvector in any of these three spaces would also be in bv if $(q_j), (p_j) \in bv \cap c_0$. If $u \in \ell^1$, then u is also in all the other spaces we have considered, and $cw(B) = \mathbf{r}_{cw}(B) = \eta^u(B) = CW_u(B)$ in all spaces with the radii and bounds not depending on the space. If X is normal, then also $\mathbf{r}_+(B) = cw(B)$ does not depend on the space. For $X = bv$, we have $\mathbf{r}_+(B) = cw(B)$ as well provided that $(p_j), (q_j) \in bv \cap c_0$ such that B is compact also in bv or in $bv \cap c_0$.

Corollary 14.8. *Let X be one of the spaces in* (14.15) *and* (14.15) *hold and* $B_1 : X_+ \to \mathbb{R}_+$ *be bounded. Let* $q_j < 1$ *for all* $j \in \mathbb{N}$. *Then either* $\mathcal{R}_0 > \mathbf{r}_+(B) > 1$, *or* $\mathcal{R}_0 = \mathbf{r}_+(B) = 1$, *or* $\mathcal{R}_0 < \mathbf{r}_+(B) < 1$. *Further* $\mathbf{r}_+(B) = CW_u(B) = cw(B)$.

These results will be used in [23] to establish $\mathbf{r}_+(B)$ as a sharp threshold parameter deciding about local stability of the extinction equilibrium versus population persistence in a nonlinear version of this population model.

Proof. If (14.15) holds and B_1 is bounded, then B is compact and $cw(B) = \mathbf{r}_+(B) = CW_u(B)$ by Theorem 12.18. The trichotomy now follows in the same way as in the proof of Theorem 14.7. □

We have not succeeded in obtaining further estimates of $\mathbf{r}_+(B)$ using $CW(B)$. To find estimates of $\mathbf{r}_+(B)$ from above, we can use Corollary 12.11.

Assume $\sum_{j,k=1}^{\infty} \beta_{jk} < \infty$. Take the sequence e where all terms are 1 for $X = \ell^\infty$ or for $X = bv$ provided that $u \in bv \cap c_0$. Then

$$\mathbf{r}_+(B) \le \eta^e(B) \le \|B(e)\|_e = \max\Big\{q_1 + \sum_{j,k=1}^{\infty} \beta_{jk}, \sup_{m=1}^{\infty} p_m, \sup_{m=2}^{\infty} q_m\Big\}.$$

Notice that the estimate is only of interest if $\mathcal{R}_0 < 1$. If $\mathcal{R}_0 \ge 1$, the estimate $\mathbf{r}_+(B) \le \mathcal{R}_0$ is better. We obtain this estimate also for $X = c_0, bv \cap c_0, \ell^1$ if $u \in X$ and there exist $\theta \in (0,1)$ and $c > 0$ such that $p_{m-1} + q_m \le c\theta^m$ for all $m \ge 2$. Then B is uniformly w-bounded for $w = (\theta^n)$ and

$$\mathbf{r}_+(B) \le \|B\|_w = \max\Big\{q_1 + \sum_{j,k=1}^{\infty} \beta_{jk}\theta^{j+k-1}, \sup_{m=1}^{\infty} p_m/\theta, \sup_{m=2}^{\infty} q_m\Big\}.$$

This estimate also holds for all $\tilde{\theta} \in (\theta, 1)$, and so we can take the limit for $\theta \to 1$ and obtain the previous estimate.

We obtain another estimate if there exist $\alpha > 1$ and $c > 1$ such that $p_{m-1} + q_m \le cm^{-\alpha}$ for all $m \ge 2$. Then B is uniformly w-bounded for $w = (n^{-\alpha})_{n\in\mathbb{N}}$ and

$$\mathbf{r}_+(B) \le \|B\|_w = \max\Big\{q_1 + \sum_{j,k=1}^{\infty} \beta_{jk}j^{-\alpha}k^{-\alpha}, \sup_{m=1}^{\infty} p_m\Big(\frac{m+1}{m}\Big)^\alpha, \sup_{m=2}^{\infty} q_m\Big\}.$$

This estimate also holds for all $\tilde{\alpha} \in (1, \alpha)$, so we can take the limit $\alpha \to 1$.

Acknowledgment

I thank Wolfgang Arendt, Gustav Gripenberg, Karl-Peter Hadeler, Roger Nussbaum, and an anonymous referee for useful hints and comments and Onno van Gaans and Marcel de Jeu for their support as editors.

References

[1] M. Akian, S. Gaubert, R.D. Nussbaum, *A Collatz–Wielandt characterization of the spectral radius of order-preserving homogeneous maps on cones*, arXiv:1112.5968v1 [math.FA], 2011

[2] W. Arendt, P.R. Chernoff, T. Kato, *A generalization of dissipativity and positive semigroups*, J. Operator Theory **8** (1982), 167–180

[3] N. Bacaër, N., E.H. Ait Dads, *Genealogy with seasonality, the basic reproduction number, and the influenza pandemic*, J. Math. Biol. **62** (2011), 741–762

[4] F.L. Bauer, *An elementary proof of the Hopf inequality for positive operators*, Numer. Math. **7** (1965), 331–337

[5] G. Birkhoff, *Extensions of Jentzsch's Theorem*, Trans. Amer. Math. Soc. **85** (1957), 219–226

[6] E. Bohl, *Monotonie: Lösbarkeit und Numerik bei Operatorgleichungen*, Springer, Berlin Heidelberg 1974

[7] F.F. Bonsall, *Endomorphisms of a partially ordered vector space without order unit*, J. London Math. Soc. **30** (1955), 133–144

[8] F.F. Bonsall, *Linear operators in complete positive cones*, Proc. London Math. Soc. **8** (1958), 53–75

[9] F.F. Bonsall, *Lectures on Some Fixed Point Theorems of Functional Analysis*, Tata Institute of Fundamental Research, Bombay 1962

[10] L. Collatz, *Einschließungssatz für die Eigenwerte von Integralgleichungen*, Math. Z. **47** (1942), 395–398

[11] L. Collatz, *Einschließungssatz für die charakteristischen Zahlen von Matrizen*, Math. Z. **48** (1942), 221–226

[12] J.M. Cushing, Y. Zhou, *The net reproductive value and stability in matrix population models*, Nat. Res. Mod. **8** (1994), 297–333

[13] O. Diekmann, J.A.P. Heesterbeek, J.A.J. Metz, *On the definition and the computation of the basic reproduction ratio R_0 in models for infectious diseases in heterogeneous populations*, J. Math. Biol. **28** (1990), 365–382

[14] N. Dunford, J.T. Schwartz, *Linear Operators. Part I. General Theory*, John Wiley, Classics Library Edition, New York, 1988

[15] S.P. Eveson, R.D. Nussbaum, *Applications of the Birkhoff–Hopf theorem to the spectral theory of positive linear operators*, Math. Proc. Camb. Phil. Soc. **117** (1995), 491–512

[16] K.-H. Förster, B. Nagy, *On the Collatz–Wielandt numbers and the local spectral radius of a nonnegative operator*, Linear Algebra and its Applications **120** (1980), 193–205

[17] G. Gripenberg, *On the definition of the cone spectral radius*, Proc. Amer. Math. Soc (to appear)

[18] K.P. Hadeler, *Pair formation in age-structured populations*, Acta Appl. Math. **14** (1989), 91–102

[19] M. Iannelli, M. Martcheva, F.A. Milner, *Gender-Structured Population Models: Mathematical Methods, Numerics, and Simulations*, SIAM, Philadelphia 2005

[20] W. Jin, *Persistence of discrete dynamical systems in infinite dimensional state spaces*, dissertation, Arizona State University, May 2014

[21] W. Jin, H.L. Smith, H.R. Thieme, *Persistence and critical domain size for diffusing populations with two sexes and short reproductive season*, J. Dynamics and Differential Equations (to appear)

[22] W. Jin, H.R. Thieme, *Persistence and extinction of diffusing populations with two sexes and short reproductive season*, Discrete and Continuous Dynamical System – B 19 (2014), 3209–3218

[23] W. Jin, H.R. Thieme, *An extinction/persistence threshold for sexually reproducing populations: the cone spectral radius*, Disc. Cont. Dyn. Systems B 21 (2016), 447–470

[24] M.A. Krasnosel'skij, *Positive Solutions of Operator Equations*, Noordhoff, Groningen 1964

[25] M.A. Krasnosel'skij, Je.A. Lifshits, A.V. Sobolev, *Positive Linear Systems: The Method of Positive Operators*, Heldermann Verlag, Berlin 1989

[26] Krause, U., *Positive Dynamical Systems in Discrete Time. Theory, Models and Applications*, De Gruyter, Berlin 2015

[27] M.G. Krein, M.A. Rutman, *Linear operators leaving invariant a cone in a Banach space* (Russian), Uspehi Mat. Nauk (N.S.) **3** (1948), 3–95, English Translation, AMS Translation 1950 (1950), No. 26

[28] B. Lemmens, B. Lins, R.D. Nussbaum, M. Wortel, *Denjoy–Wolff theorems for Hilbert's and Thompson's metric spaces*, *J. dAnalyse Math.* (to appear), arXiv: 1410.1056v2[math.DS]

[29] B. Lemmens, R.D. Nussbaum, *Nonlinear Perron–Frobenius Theory*, Cambridge University Press, Cambridge 2012.

[30] B. Lemmens, R.D. Nussbaum, *Continuity of the cone spectral radius*, Proc. Amer. Math. Soc. **141** (2013), 2741–2754

[31] J. Mallet-Paret, R.D. Nussbaum, *Eigenvalues for a class of homogeneous cone maps arising from max-plus operators*, Discr. Cont. Dyn. Sys. (DCDS-A) **8** (2002), 519–562

[32] J. Mallet-Paret, R.D. Nussbaum, *Generalizing the Krein–Rutman theorem, measures of noncompactness and the fixed point index*, J. Fixed Point Theory and Appl. **7** (2010), 103–143

[33] J. Mallet-Paret, R.D. Nussbaum, *Asymptotic fixed point theory and the beer barrel theorem*, J. Fixed Point Theory Appl. **4** (2008), 203–245

[34] R.D. Nussbaum, *Eigenvectors of nonlinear positive operators and the linear Krein–Rutman theorem*, Fixed Point Theory (E. Fadell and G. Fournier, eds.), 309–331, Springer, Berlin New York 1981

[35] R.D. Nussbaum, *Eigenvectors of order-preserving linear operators*, J. London Math. Soc. **2** (1998), 480–496

[36] H.H. Schaefer, *Positive Transformationen in lokalkonvexen halbgeordneten Vektorräumen*, Math. Ann. **129**, 1955, 323–329

[37] H.H. Schaefer, *Halbgeordnete lokalkonvexe Vektorräume. II*, Math. Ann. **138** (1959), 259–286

[38] H.H. Schaefer, *Topological Vector Spaces*, Macmillan, New York 1966

[39] H.H. Schaefer, *Banach Lattices and Positive Operators*, Springer, Berlin Heidelberg 1974

[40] Smith, H.L., H.R. Thieme, *Dynamical Systems and Population Persistence*, Amer. Math. Soc., Providence 2011

[41] H.R. Thieme, *Spectral bound and reproduction number for infinite dimensional population structure and time-heterogeneity*, SIAM J. Appl. Math. **70** (2009), 188–211

[42] H.R. Thieme, *Eigenvectors and eigenfunctionals of homogeneous order-preserving maps*, arXiv:1302.3905v1 [math.FA], 2013

[43] H.R. Thieme, *Comparison of spectral radii and Collatz–Wielandt numbers for homogeneous maps, and other applications of the monotone companion norm on ordered normed vector spaces*, arXiv:1406.6657v2 [math.FA], 2016

[44] H.R. Thieme, *Eigenfunctionals of homogeneous order-preserving maps with applications to sexually reproducing populations*, J. Dynamics and Differential Equations (to appear)

[45] A.C. Thompson, *On certain contraction mappings in a partially ordered vector space*, Proc. AMS **14** (1963), 438–443

[46] A.C. Thompson, *On the eigenvectors of some not-necessarily-linear transformations*, Proc. London Math. Soc. **15** (1965), 577–598

[47] A.J. Tromba, *The beer barrel theorem, a new proof of the asymptotic conjecture in fixed point theory*, Functional Differential Equations and Approximations of Fixed Points (H.-O. Peitgen, H.-O. Walther, eds.), 484–488, Lecture Notes in Math. 730, Springer, Berlin Heidelberg 1979

[48] P. van den Driessche, J. Watmough, *Reproduction numbers and sub-threshold endemic equilibria for compartmental models of disease transmission*, Math. Biosci. **180** (2002), 29–48

[49] H. Wielandt, *Unzerlegbare, nicht negative Matrizen*, Math. Z. **52** (1950), 642–648

[50] K. Yosida, *Functional Analysis*, sec. ed., Springer, Berlin Heidelberg 1965–1968

Horst R. Thieme
School of Mathematical and Statistical Sciences
Arizona State University
Tempe, AZ 85287-1804, USA
e-mail: `hthieme@asu.edu`

Ordered Structures and Applications: Positivity VII
Trends in Mathematics, 469–480

Piling Structure of Families of Matrix Monotone Functions and of Matrix Convex Functions

Jun Tomiyama

Abstract. Given a nontrivial open interval I on the real line, we consider the set $P_n(I)$ (resp. $K_n(I)$) of all n-monotone (resp. n-convex) functions defined on I. Then $\{P_n(I)\}_{n=1}^{\infty}$ (resp. $\{K_n(I)\}_{n=1}^{\infty}$) is a decreasing sequence of sets "piled" on their intersection, the set of all operator monotone functions $P_\infty(I)$ (resp. the set of all operator convex functions $K_\infty(I)$) on I. In this article we give criteria for a function to belong to such a set, and we describe the gap between the sets for n and $n+1$. In fact, for every n we provide abundant examples of n-monotone (resp. n-convex) functions in the gap between $P_n(I)$ and $P_{n+1}(I)$ (resp. $K_n(I)$ and $K_{n+1}(I)$). When I is finite we show that this gap contains polynomials of degree $2n-1$ and $2n$ (resp. $2n$ and $2n+1$).

Mathematics Subject Classification (2010). Primary 47A53; Secondary 26A48, 26A51.

Keywords. Matrix monotone functions, matrix convex functions, gaps.

1. Introduction

Let I be an interval (open, closed, any type of an interval on the real line $\mathbb{R}$ is admissible) and let $f(t)$ be a real-valued continuous function on I. Let M_n be the matrix algebra of the degree n. Throughout this article then every matrix a treated here is assumed to be selfadjoint and its spectrum $\sigma(a)$ is contained in I. We call the function f n-monotone if for any pair of matrices $\{a, b\}$ in M_n we have the order $f(a) \leq f(b)$ whenever $a \leq b$. Similarly call f n-convex if for such a pair we have

$$f(\lambda a + (1-\lambda)b) \leq \lambda f(a) + (1-\lambda)f(b),$$

where $0 \leq \lambda \leq 1$. The sequence $\{P_n(I)\}_{n=1}^{\infty}$ of sets of all n-monotone functions, and likewise the sequence $\{K_n(I)\}_{n=1}^{\infty}$ of sets of all n-convex functions, is decreasing with intersections $P_\infty(I)$ resp. $K_\infty(I)$. Members of these classes are said to be operator monotone and operator convex. These notions were introduced and developed by K. Loewner with his two students O. Dobsch and U. Kraus (1934–1936). Since then the theory has been developed with a large variety of applications to many fields of both pure and applied mathematics and quite recently to quantum information theory.

As to the structure of the piles, it has been suggested in the literature that the inclusion $P_{n+1}(I) \subset P_n(I)$ must be proper for all n (and likewise for the convex case). Yet concrete examples of functions in the gap between $P_n(I)$ and $P_{n+1}(I)$ are surprisingly lacking; even for $n = 2$ only one example was known in [17]. In this article, mainly based on the joint works [7] and [8] together with other joint works [4] and [13], we discuss the existence of gaps for both sequences $\{P_n(I)\}$ and $\{K_n(I)\}$ providing abundance examples in those gaps for all n.

2. Criteria for n-monotonicity and n-convexity and the Local Property Theorem

In case of piling structure of classical calculus, we have a gap between the set of all C^∞-functions and that of analytic functions on the interval I. A great contribution by Loewner [10] was to show that we do not have such gap for operator monotone functions through the following characterization of an operator monotone function.

Theorem 2.1. *A function f defined on an open interval I becomes operator monotone if and only if it has an analytic continuation to the upper half-plane as a Pick function, that is, keeping this half-plane as its range.*

Now in order to look for the piling structure for $\{P_n(I)\}$ and $\{K_n(I)\}$ we need criteria for n-monotone functions and n-convex functions.

To begin with we first introduce the notion of divided differences and regularization process. Let $t_1, t_2, t_3, \ldots$ be a sequence of distinct points. We write those divided differences with respect to a function f as

$$[t_1, t_2]_f = \frac{f(t_1) - f(t_2)}{t_1 - t_2}$$

and inductively,

$$[t_1, t_2, \ldots, t_{n+1}]_f = \frac{[t_1, t_2, \ldots, t_n] - [t_2, t_3, \ldots, t_{n+1}]}{t_1 - t_{n+1}}.$$

When f is sufficiently smooth, we can define

$$[t_1, t_1]_f = f'(t_1),$$

and then inductively such as

$$[t_1, t_1, t_2]_f = \frac{f'(t_1) - [t_1, t_2]}{t_1 - t_2}, \quad [t_1, t_1, t_1] = f"(t_1)/2.$$

When there appears no confusion we omit the index f. In this way we see that $(n+1)$th divided difference $[t_0, t_0, \ldots, t_0]$ is $f^{(n)}(t_0)/n!$, which is nothing but the

nth coefficient of the Taylor expansion of $f(t)$ at the point t_0. An important property of divided differences is that they are permutation free so that one may find another forms of the definition for divided differences in different orders of $t_i's$.

In the following calculation we often assume that relevant functions are smooth enough. This is allowed because of the following so-called regularization process of those functions. Let $\varphi(t)$ be an even C^∞-function defined on R. It is required to be positive, supported on the interval $[-1,1]$, and with the integral being one, that is, a mollifier. Let $f(t)$ be a continuous function on (α,β), then we form its regularization $f_\varepsilon(t)$ for a small positive ε by

$$f_\varepsilon(t) = 1/\varepsilon \int \varphi(\frac{t-s}{\varepsilon})f(s)ds = \int \varphi(s)f(t-\varepsilon s)ds.$$

This regularization converges to f uniformly on any closed subinterval when ε goes to zero. Moreover $f_\varepsilon(t)$ becomes a C^∞-function, and important points are the facts that when f is monotone or convex at some level (such as n-monotone or n-convex) on (α,β) f_ε becomes monotone or convex at the same level on the interval $(\alpha+\varepsilon,\beta-\varepsilon)$. Therefore, we may prove the required property for the smooth function f_ε and come back to the original function f keeping that property. Here it is also to be noticed that the regulation of a derivative is the derivative of the regulation, that is, $(f')_\varepsilon(t) = (f_\varepsilon)'(t)$.

Now we state the criteria of n-monotone functions. There are two criteria; one global (combinatorial) and the other local. Given a function f on the interval I and an n-tuple $\{t_1,t_2,\ldots,t_n\}$ (not necessarily assumed to be distinct) from I the following matrix

$$L_n^f(t_1,t_2,\ldots,t_n) = ([t_i,t_j]_f)_{i,j=1}^n$$

is called the Loewner matrix for a function f. In the following we often write L_n^f instead of $L_n^f(t_1,t_2,\ldots,t_n)$.

Criterion $\mathbf{I_a}$. *Let f be a class C^1-function on I. Then f is n-monotone if and only if for an arbitrary n-tuple $\{t_1,t_2,\ldots,t_n\}$ in I its Loewner matrix is positive semidefinite.*

For the proof of this result we just refer to [1, Theorem V.3.4] or [5, Theorem 6.6.36].

Comparing with this criterion the next local criterion for n-monotonicity is quite useful although in the form we should know the differentiated forms in high differentiation.

Criterion $\mathbf{I_b}$. *Let f be a function in C^{2n-1} on the above interval I. Then f is n-monotone if and only if the following $n\times n$ Hankel matrix*

$$M_n(f;t) = \left(\frac{f^{(i+j-1)}(t)}{(i+j-1)!}\right)$$

is positive semidefinite for every $t \in I$.

In fact, with this criterion we can show that the function $\log t$ on the positive half-line is operator monotone whereas the exponential function e^t is not even 2-monotone (a drastic appearance of noncommutativity). These two criteria are now established facts. There were however serious troubles to establish Criterion $\mathrm{I_b}$. Assuming enough smoothness as mentioned above, for the implication from $\mathrm{I_a}$ to $\mathrm{I_b}$ we make use of the method using the so-called extended Loewner matrix $L_n^{ef}(t_1, t_2, \ldots, t_n)$ defined for an n-tuple $\{t_1, t_2, \ldots, t_n\}$ in I as

$$L_n^{ef}(t_1, t_2 \ldots, t_n) = ([t_1, t_2, \ldots, t_i, t_1, t_2, \ldots, t_j])_{i,j=1}^n.$$

As in the case of the Loewner matrix we write often L_n^{ef} as a shorthand notation for $L_n^{ef}(t_1, t_2, \ldots .t_n)$.

For the implication $I_a \to I_b$ we leave its details to the comprehensive book [3].

The conclusion of I_b is obtained by the semidefiniteness of the extended Loewner matrix L_n^{ef} and then by considering the limiting case where all $\{t_i\}$ coincide, The general relation between determinant of the Loewner matrix L_r^f of size r and that of its extended form L_r^{ef} for $\{t_1, t_2, \ldots, t_r\}$ is

$$\det L_r^f = \prod_{i>j} (t_i - t_j)^2 \det L_r^{ef}.$$

Hence they have the same sign provided that all t_k's are distinct. Thus, considering this situation positive semidefiniteness of $M_n(f;t)$ is obtained through determinants of principal submatrices of L_n^{ef}.

A starting point for the converse assertion is the next proposition. We state here its proof because of its importance in our discussion although such proof is found in the book [3].

Proposition 2.2. *Let f be a function in the class $C^{2n-1}(I)$. Suppose there exist an interior point t_0 such that $M_n(f;t_0) > 0$. Then there exists a positive number δ such that f is n-monotone in the subinterval $(t_0 - \delta, t_0 + \delta)$.*

Proof. Note first that, by the standard result in linear algebra, determinant of each leading principal submatrix of $M_n(t_0; f)$ is positive. It follows by the continuous dependence of matrix entries on points that we can find a small positive δ such that determinants of all leading principal submatrices of the extended Loewner matrix L_n^{ef} are positive in the open interval $(t_0 - \delta, t_0 + \delta)$ inside I. Thus from the above relations between determinants of principal submatrices of L_n^f and those of the extended Loewner matrix L_n^{ef} we see that those of leading principal submatrices of L_n^f are positive provided that given n-tuple $\{t_k\}$ consists of distinct points. Therefore, here the corresponding Loewner matrix becomes positive definite by the standard result in linear algebra. Since the set of such n-tuples is dense in the set of all n-tuples without restrictions, the matrix L_n^f becomes positive semidefinite in this open interval and then by I_a the function f becomes n-monotone in the interval. □

Thus for a closed subinterval inside I we have a covering of open intervals and hence a finite numbers of covering in each of which f is n-monotone. We have to however connect these facts to assert that f is n-monotone on that closed subinterval of I and then shift to the whole I. Therefore for these steps we need the following Local Property Theorem.

Theorem (Local Property Theorem). *Let (α, β) and (γ, δ) be two overlapping open intervals, where $\alpha < \gamma < \beta < \delta$. Suppose a function f is n-monotone on these intervals, then f is n-monotone on the larger interval (α, δ)*

Though its formulation looks very simple, this theorem is very deep and its proof is extremely hard. Nevertheless, to our surprise, Loewner himself wrote in his paper [10, p. 212, Theorem 5.6] that "the proof of this theorem is very easy, hence we leave its proof to the reader". Furthermore, when his student Dobsch used this result in [2], he just cited the result as already proved one by Loewner. Fortunately, forty years later Donoghue gave a comprehensive proof in his book [3], which amounts almost fifty (!) pages (together with the theory of interpolation functions of complex variable), and Criterion $\mathrm{I_b}$ is now assumed to be an established one. Since, however, Donoghue's proof is too long (as a whole) we are still looking for a simple minded proof of this Local Property Theorem for matrix monotone functions. On the other hand, the Local Property Theorem for matrix convex functions is still far beyond our scope as we shall explain later.

As for the criteria of n-convexity of functions we are in a similar situation having a serious trouble for the second criterion lacking in the Local Property Theorem of convexity! In fact, we are at present in the situation as follows.

Criterion $\mathrm{II_a}$. *Let f be a function in C^2 on I. Then f is n-convex if and only if for an arbitrary n-tuple $\{t_1, t_2, \dots, t_n\}$ in I the Kraus matrix of size n,*

$$K_n^f(s) = ([t_i, t_j, s])_{i,j=1}^n = ([t_i, s, t_j])_{i,j=1}^n$$

is positive semidefinite. Here s is fixed among $\{t_1, t_2, \dots, t_n\}$.

An expected local criterion would be

Criterion $\mathrm{II_b}$. *Let f be a function in C^{2n} in the interval I, then f is n-convex if and only if the following Hankel matrix*

$$K_n(f; t) = \left(\frac{f^{(i+j)}(t)}{(i+j)!} \right)$$

is positive semidefinite for every $t \in I$

For the proof of Criterion $\mathrm{II_a}$ we refer to [5, Theorem 6.6.52 (1)]. The implication from $\mathrm{II_a}$ to $\mathrm{II_b}$ has been proved in a similar way as in the case of Criteria $\mathrm{I_a}$ and $\mathrm{I_b}$ in [7] and [8]. Here however because of the difference of the order of relevant divided differences computations become much more complicated to paraphrase the original determinant into the determinant of the extended Kraus matrix, $K_n^{ef}(s) = ([t_1, \dots, t_i, s, t_1, \dots, t_j])_{i,j=1}^n$ similar to L_n^{ef}.

On the other hand, the Local Property Theorem for n-convex functions is proved only in the case $n = 2$ in [7] and at present we have been unable to prove the theorem even for 3-convex functions. For the moment, all we can say now is the following fact that corresponds to the previous Proposition 2.2 for n-monotone functions.

Proposition 2.3. *Suppose that $K_n(f;t_0) > 0$ at some interior point t_0 in I, then there exists a small open subinterval J in I containing t_0 on which the function f is n-convex.*

This is proved along the similar way as in the above-mentioned proposition for monotone functions through the relations between leading determinants of the Kraus matrix and those of the extended Kraus matrix $K_n^{ef}(s)$ defined above.

A general relation between determinant of the Kraus matrix $K_r^f(s)$ of size r and determinant of the extended Kraus matrix $K_r^{ef}(s)$ of the same size r is given in the following form (cf. [7]),

$$\det K_r^f(s) = \prod_{k=1}^{r-1}\prod_{l=1}^{r-k}(t_{k+l} - t_l)^2 \det K_r^{ef}(s).$$

Hence they have the same sign for $r = 2, 3, \ldots, n$ provided that those r-tuples consist of distinct points.

We skip the proof of the local property for 2-convex functions (cf. [8]).

The following observation is useful through this note.

Proposition 2.4.

(1) *Let f be a function in C^1 and 2-monotone on the interval I. If the derivative f' vanishes at some point t_0, then f becomes a constant function.*

(2) *Let f be a function in C^2 and 2-convex on the interval I. If the second derivative f'' vanishes at some point t_0, then f is at most a linear function.*

We leave their proofs by using Ia and IIa as an exercise for the reader.

3. Existence of proper examples of n-monotone functions and n-convex functions with gaps for them

For many years, as examples of n-monotone functions and n-convex functions for an arbitrary n most literatures were used to present operator monotone functions and operator convex functions. This is right anyway but sounds somewhat strange in true mathematical sense because they are somewhat out of proper examples. Both classes of matrix monotone functions $\{P_n(I)\}$ and matrix convex functions $\{K_n(I)\}$ form decreasing sequences down to the classes $P_\infty(I)$ and $K_\infty(I)$. Thus the natural question about the piling structure of these sequences is the existence of gaps for each inclusion for $P_n(I)$ and $K_n(I)$.

In this situation, so many papers on monotone operator functions have been published since the introduction of this concept by Loewner, and most papers (notably Donoghue's book [3, p. 84])had been asserting the existence of gaps for the sequence $\{P_n(I)\}$ for arbitrary n, but no explicit examples were given for $n \geq 3$

until we provided first such examples in [4]. Before this article, only one example was known by [17] for the gap between $P_2(I)$ and $P_3(I)$.

Here we shall provide abundance of proper such examples with explicit contents of the gaps for arbitrary n. Moreover, in case of finite intervals we can provide such examples even as polynomials. Before going into our discussions we review general aspect of the existence problem for gaps depending on intervals. Let I and J be finite interval in the same types (open, closed etc). There is then a linear transition function with a positive coefficient for t from I to J and for the converse too. Since this function together with its inverse are both operator monotone and operator convex, once we find functions belonging to the gap $P_n(I) \setminus P_{n+1}(I)$ (likewise $K_n(I) \setminus K_{n+1}(I)$) for any n those transposed functions on J belong to the gap on J in the same order.Therefore so far finite intervals are concerned we may choose any convenient interval for which we usually employ the interval of the form $[0, \alpha)$. Relations between two (nontrivial) infinite intervals are more or less the same. In fact, if they are in the same direction the transferring function is just a shift. When they are in the opposite direction it becomes a combination of a shift and the reflection. Anyway in both cases we can easily transfer gaps for the one interval to those of the other one. Therefore the rest is the case where the one is a finite interval, say $[0,1)$, and the other is an infinite one, say $[0,\infty)$. For this relation we notice first that the function $1/t$ is known to be operator convex in the interval $(0,\infty)$. Hence the function $h(t) = \frac{t}{1-t} : [0,1) \to [0,\infty)$ is operator monotone and operator convex. The inverse of this function, $h^{-1}(t) = \frac{t}{1+t} : [0,\infty) \to [0,1)$ is also operator monotone but operator concave.It follows that although we can freely transfer gaps for matrix monotone functions each other between arbitrary intervals, we can not treat the case of matrix convex functions in the same way.

Anyway, however, the following result solves the problem of the existence of proper examples with gaps for finite intervals.

Theorem 3.1 ([7, 13]). *Let I be a finite interval and let n and m be natural numbers with $n \geq 2$.*

(1) *If $m \geq 2n-1$, there exists an n-monotone polynomial p_m on I of degree m.*
(2) *If $m \geq 2n$ there exists an n-convex and n-monotone polynomials p_m on I of degree m. Likewise there exists an n-concave and n-monotone polynomial q_m on I of degree m.*
(3) *If $m = 2, 3, \ldots, 2n-2$, there are no n-monotone polynomials of degree m on I.*
(4) *If $m = 3, 4, \ldots, 2n-1$, there are no n-convex polynomials of degree m on I.*

Sketch of the proof. We first introduce the polynomial p_m of degree m given by

$$p_m(t) = b_1 t + b_2 t^2 + \cdots + b_m t^m,$$

where

$$b_k = \int_0^1 t^{k-1} d\mu \text{ with } supp(\mu) = [0,1]$$

Then the ℓth derivative $p_m^{(\ell)}(0) = \ell! b_\ell$ for $\ell = 1, 2, \ldots, 2n-1$, and consequently

$$M_n(p_m; 0) = \left(\frac{p_m^{(i+j-1)}(0)}{(i+j-1)!} \right)_{i,j=1}^n = (b_{i+j-1})_{i,j=1}^n .$$

Now take a vector $c = (c_1, c_2, \ldots, c_n)$ in an n-dimensional space, then

$$(M_n(p_m; 0)c|c) = \sum_{i,j=1}^n b_{i+j-1} c_j \bar{c}_i = \int_0^1 \left| \sum_{i=1}^n c_i t^{i-1} \right|^2 d\mu .$$

From this we can say that the matrix $M_n(p_m; 0)$ is positive definite, and then by Proposition 2.2 we can find a positive number α such that $M_n(p_m; t)$ is positive in the interval $[0, \alpha)$. Hence by Criterion $\mathrm{I_b}$ the polynomial $p_m(t)$ becomes n-monotone here. This shows the assertion (1).

The first half of the proof of (2) goes in a similar way but use both matrices $M_n(p_m; 0)$ and $K_n(p_m; 0)$. Here besides the calculation for $M_n(p_m; 0)$ as above we have

$$K_n(p_m; 0) = \left(\frac{p_m^{(i+j)}(0)}{(i+j)!} \right)_{i,j=1}^n = (b_{i+j})_{i,j=1}^n$$

and

$$(K_n(p_m; 0)c|c) = \sum_{i,j=1}^n b_{i+j} c_j \bar{c}_i = \int_0^1 t \left| \sum_{i=1}^n c_i t^{i-1} \right|^2 d\mu .$$

Thus, both matrices are positive definite. Hence by Proposition 2.2 and 2.3 we can find a positive number α such that p_m becomes both n-monotone and n-convex in the interval $[0, \alpha)$.

For the second assertion we consider the polynomial $q_m(t)$ of degree m whose coefficients $\{b_k\}$ are defined as

$$b_k = \int_{-1}^0 t^{k-1} d\nu \text{ with } supp(\nu) = [-1, 0] .$$

The corresponding computation for $M_n(q_m; 0)$ shows that it is still positive definite whereas $K_n(q_m; 0)$ becomes negative definite because of the range of the integration. Therefore, by the same reason as above there exists a positive number α such that q_m becomes n-monotone and n-concave in the interval $[0, \alpha)$.

Proof of (3). Let f_m be an n-monotone polynomial of degree m on I with $2 \le m \le 2\mathrm{n}-2$. We may assume as above that I contains 0. Write

$$f_m(t) = b_0 + b_1 t + \cdots + b_m t^m \quad \text{where } b_m \neq 0.$$

We have then

$$f_m^{(m-1)}(0) = (m-1)! b_{m-1}, \quad f_m^{(m)}(0) = m! b_m, \quad f_m^{(m+1)}(0) = 0.$$

Consider the matrix $M_n(f_m; 0)$. We have to check two cases where $m = 2k$, even and $m = 2k-1$, odd. Note first that in both cases $k+1 \le n$. In the first case, the principal submatrix of $M_n(f_m;0)$ consisting of the rows and columns with numbers k and $k+1$ is given by

$$\begin{pmatrix} b_{m-1} & b_m \\ b_m & 0 \end{pmatrix}$$

and it has determinant $-b_m^2 < 0$. In the latter case, we consider the principal submatrix consisting of rows and columns with numbers $k-1$ and $k+1$ given by

$$\begin{pmatrix} b_{m-2} & b_m \\ b_m & 0 \end{pmatrix}$$

and this matrix also has determinant $-b_m^2 < 0$. Since $M_n(f_m;0)$ is supposed to be positive semidefinite by I_b we have in both cases contradictions.

The assertion (4) is proved in a similar way using the matrix $K_n(f_m;0)$ since we have now the implication $II_a \to II_b$. □

The above theorem provides for a finite interval I abundance of examples of polynomials of proper examples of n-monotone functions and n-convex functions according to the choice of those probability measures μ and ν. Moreover the assertions (3) and (4) assure the contents of gaps $P_n(I)\setminus P_{n+1}(I)$ and $K_n(I)\setminus K_{n+1}(I)$ for any natural number n. In fact, those polynomials of degrees $2n-1$ and $2n$ constructed in (1) (resp. of degrees $2n$ and $2n+1$ constructed in (2)) are belonging to the gap for monotone functions (resp. for convex functions).

Here the author is wondering how fat are those sets of polynomials in the set of $P_n(I)$ and $K_n(I)$ for a finite interval I, for instance dense?

Now as an immediate consequence of the theorem we have as we noticed above

Corollary 3.2. *Let I be a nontrivial infinite interval. Then for any natural number n there exists a gap between $P_n(I)$ and $P_{n+1}(I)$.*

We remark however that those transferred functions are no longer polynomials but rational functions instead.

For gaps of matrix convex functions we need further arguments but finally obtain the same result.

Proposition 3.3. *Let I be a nontrivial infinite interval. Then for any natural number n the gap between $K_n(I)$ and $K_{n+1}(I)$ is not empty.*

For this result we need the following.

Lemma 3.4. *A nonnegative n-concave function f defined in the interval $[0,\infty)$ is necessarily n-monotone.*

Proof. Take a pair of $n\times n$ matrices a,b such that $0\le a\le b$. Then for $0<\lambda<1$ we can write as

$$\lambda b = \lambda a + (1-\lambda)\lambda(1-\lambda)^{-1}(b-a).$$

Hence by assumptions,

$$f(\lambda b) \geq \lambda f(a) + (1-\lambda) f(\lambda(1-\lambda)^{-1}(b-a)) \geq \lambda f(a).$$

Taking λ to go to 1, we have that $f(a) \leq f(b)$. □

Proof of Proposition 3.3.. Assuming that $I = [0, \infty)$ we prove the result in a concave version. Let f be an n-monotone and n-concave polynomial in $[0, 1)$ of degree $2n$. By adding a suitable constant we may assume that f is nonnegative. The composition function

$$g(t) = f\left(\frac{t}{1+t}\right), \qquad t \geq 0$$

is n-concave. Note that by (3) of the theorem f can not be $(n+1)$-monotone and so g can not be $(n+1)$-monotone either. Now suppose g to be $(n+1)$-concave, then by the above lemma it becomes $(n+1)$-monotone, a contradiction. □

In connection with the above fact the following result shows that on an (nontrivial) infinite interval we seldom have matrix monotone (resp. convex) polynomials.

Proposition 3.5. *Let I be an infinite interval and n a natural number with $n \geq 2$.*

(1) *An n-monotone polynomial on I is at most a linear function.*
(2) *An n-convex polynomial on I is at most a quadratic function.*

We skip proofs of these standard results but only mention that for proofs we make use of the following result, that is, for a natural integer $p \geq 2$ the function t^p is not 2-monotone and for $p \geq 3$ the function t^p is not 2-convex (use Criterion $\mathrm{I_b}$ and the first part of $\mathrm{II_b}$).

4. Concluding remark

We could regard the theory of matrix functions as a noncommutative calculus comparing with usual calculus based on numbers, and then meet the first appearance of noncommutativity in the theory when we begin to treat matrices in M_2. In this aspect it is to be particularly noticed and interesting to see that many results known for operator monotone functions (likewise for operator convex functions) are already found at the step of 2-monotonicity (2-convexity as well). For instance, the well-known Loewner–Heinz theorem in the operator theory states that (in terms of operator monotone functions) the function $f(t) = t^p$ on the positive half-line becomes operator monotone if and only if $0 \leq p \leq 1$. As proved in [8, Prop. 3.1], however,this restriction of p is already obtained at the stage of 2-monotonicity, that is, the function t^p becomes 2-monotone if and only if $0 \leq p \leq 1$, and no further obstruction appears through 2 to the infinity (similar fact holds for operator convexity of this function too). It should be however noticed that this does not mean the whole proof itself is finished at the stage of 2-monotonicity. We have to show the rest that in this restriction the function keeps the monotonicity for all degrees. In case of Proposition 2.4 aspects of those results are different because assertions there hold already at the step of 2-monotonicity, whereas results were known before as results for operator monotone functions and operator convex functions.

So far we have clarified the basic structure of single piling structure providing proper examples of n-monotone functions and n-convex functions with the existence of gaps for arbitrary n. There are however problems about the interplay between two kinds of piling structure, $\{P_n(I)\}$ and $\{K_n(I)\}$ in the following meaning. Actually there are many important equivalent relations between operator monotone functions and operator convex function, notably the things surrounding around operator Jensen's inequality discussed in [6]. The results are proved as consequences of seesaw games in such a way that if one assertion holds at the degree $2n$ for one piling then the counter assertion holds at the degree n for another piling, and finally they become an equivalent version at the top. We regard the thing about the classes of n-monotone functions and n-convex functions as problems of double piling structure, which are extensively discussed in [14] and [16]. These works are however only the beginning of the double piling problems, and there remain still many problems. For instance, suppose that if the one assertion holds at the degree $m(n)$, then the other assertion holds at the degree n. We could call the relation is bounded if the set $\{m(n) - n\}$ becomes bounded and say it an unbounded relation if the set is unbounded. Very few relations however we have proved them bounded (cf. [14]), and for many other relations known in literature we do not know whether they would be bounded or unbounded relations. Moreover, we could say that no truly unbounded relation has ever been found yet, although we see many unbounded relations as a glance.

References

[1] R. Bhatia, *Matrix analysis*, Springer, 1997.

[2] O. Dobsch, *Matrixfunktionen beschränkter Schwankung*, Math. Z., **43** (1937), 373–388.

[3] W.F. Donoghue, *Monotone matrix functions and analytic continuation*, Springer, 1974.

[4] F. Hansen, G. Ji and J. Tomiyama, *Gaps between classes of matrix monotone functions*, Bull. London Math. Soc., **36** (2004), 53–58.

[5] R. Horn and C. Johnson, *Topics in matrix algebras*, Cambridge University Press, 1991.

[6] F. Hansen and G.K. Pedersen, *Jensen's inequality for operators and Loewner's theorem*, Math. Ann., **258** (1982), 229–241.

[7] F. Hansen and J. Tomiyama, *Differential analysis of matrix convex functions*, Linear Alg. and its Appl., **42** (2007), 102–116.

[8] F. Hansen and J. Tomiyama, *Differential analysis of matrix convex functions II*, J. Inequal. Pure Appl. Math., **10** (2009), no. 2, Article 32, 5 pp.

[9] F. Kraus, *Über konvexse Matrixfunktionen*, Math. Z., **41** (1936), 18–42.

[10] K. Loewner, *Über monotone Matrixfunktionen*, Math. Z., **38** (1934), 177–216.

[11] K. Loewner, *Advanced matrix theory, Lecture notes*, Stanford University, 1957.

[12] R. Mathias, *Concavity of monotone matrix functions of finite order*, Linear and Multilinear Alg., **27** (1990), 129–138.

[13] H. Osaka, S. Silvestrov and J. Tomiyama, *Monotone operator functions, gaps and power moment problem*, Math. Scand., **100** (2007), 161–183.

[14] H. Osaka and J. Tomiyama, *Double piling structure of matrix monotone functions and of matrix convex functions*, Linear Alg. and its Appl. **431** (2009), 1825–1832.

[15] H. Osaka and J. Tomiyama, *Note on the structure of the space of matrix functions*, Analysis for Sciences, Engineering and Beyond, The tribute workshop in honor of Gunnar Sparr held in Lund, May 8–9, 2008, Spring Proceedings in Mathematics, vol 6, 319–324.

[16] H. Osaka and J. Tomiyama, *Double piling structure of matrix monotone functions and of matrix convex functions II*, Linear Alg. and its Appl. **437** (2012), 735–748.

[17] G. Sparr, *A new proof of Loewner's theorem on monotone matrix functions*, Math. Scand., **47** (1980), 266–274.

Jun Tomiyama
Department of Mathematics
Tokyo Metropolitan University
Minami-Osawa
Hachioji City, Japan
e-mail: `juntomi@med.email.ne.jp`

Ordered Structures and Applications: Positivity VII
Trends in Mathematics, 481–502

The Riesz Space of Minimal Usco Maps

Jan Harm van der Walt

Abstract. We consider upper semi-continuous compact-valued (usco) maps with values in a Banach lattice. Recently, it was shown that the space $\mathcal{M}(X,Y)$ of minimal upper semi-continuous compact-valued maps from a topological space X into a metrizable topological vector space Y is a vector space which contains the space $\mathcal{C}(X,Y)$ of continuous functions from X into Y as a linear subspace. In this paper, we consider the situation when the range space is a Banach lattice E. In this case, $\mathcal{C}(X,E)$ is a Riesz space with respect to the usual pointwise ordering. We show that $\mathcal{M}(X,E)$ is equipped in a natural way with a partial order that extends the order on $\mathcal{C}(X,E)$. With respect to this order, $\mathcal{M}(X,E)$ is an Archimedean Riesz space. Moreover, if E has compact order intervals, then $\mathcal{M}(X,E)$ is Dedekind complete. An application is made to the characterisation of the Dedekind completion of $\mathcal{C}(X,E)$.

Mathematics Subject Classification (2010). Primary 54C60, 54C40, 58C06, 46A40; Secondary 46B42.

Keywords. Set-valued map, Banach lattice, Riesz space.

1. Introduction

Set-valued maps appear naturally in many areas of mathematics and its applications including functional analysis [25], optimisation and control theory [9, 19], mathematical economics [18] and biomathematics [22], but is also of intrinsic interest. However, the study of the structure and properties of *spaces* of set-valued maps is fairly recent. Here we may mention the early work of Aseev [4] and the more recent contributions of Anguelov and Kalenda [2], Hammer and McCoy [13], McCoy [23, 24] and Holá [14, 15, 16]. The results presented in the current paper is a contribution to this direction of inquiry. In particular, we consider here the order structure of spaces of set-valued maps, specifically of the space $\mathcal{M}(X,E)$ of minimal upper semi-continuous compact-valued (musco) maps with values in a Banach lattice E. It will be shown that $\mathcal{M}(X,E)$ is in a natural way an Archimedean Riesz space, containing the Riesz space $\mathcal{C}(X,E)$ as a Riesz subspace.

The paper is organized as follows. In Section 3, for the convenience of the reader unfamiliar with set-valued analysis, we recall some basic notions and results related to this field. Section 4 deals with the relationship between upper semi-continuous compact-valued (usco) maps and semi-continuous (point-valued) functions in a Banach lattice. These results are extensions of known results in the case when the range space is $\mathbb{R}$, see for instance [2]. The Riesz space structure of $\mathcal{M}(X,E)$ is introduced in Section 5. As an application we show, in Section 6, that the Dedekind completion of $\mathcal{C}(X,E)$ is Riesz isomorphic to $\mathcal{M}(X,E)$ whenever X is compact and E is an AM-space with compact order intervals.

Finally, let us fix some notation. We will denote by X, Y and Z topological spaces with no additional structure, unless otherwise stated. $\mathcal{K}(Y)$ denotes the set of non-empty compact subsets of Y, while $\mathcal{F}(Y)$ is the set of nonempty closed subsets of Y. The set of open neighbourhoods at $x \in X$ is denoted by $\mathcal{V}_x$. E is always a Banach lattice. For those readers who are less familiar with the theory of Riesz spaces, and Banach lattices in particular, we have attempted to include references to all results from this field that are used in the paper.

2. Set-valued maps

A set-valued map $f : X \rightrightarrows Y$ is a function from X into 2^Y. That is, $f(x)$ is a subset of Y for each $x \in X$. We will always assume that $f(x) \neq \emptyset$ for all $x \in X$. A map $f : X \rightrightarrows Y$ may be identified with its graph

$$Gr(f) = \{(x,y) \in X \times Y \mid y \in f(x)\}.$$

This enables us to define set-theoretic notions such as inclusion, intersections and unions of set-valued maps in terms of the graphs of such maps. In particular, $f \subseteq g$ means that $Gr(f) \subseteq Gr(g)$, while $f \cap g$ denotes the map with graph $Gr(f \cap g) = Gr(f) \cap Gr(g)$, that is, $(f \cap g)(x) = f(x) \cap g(x)$ for every $x \in X$. In the same way, $(f \cup g)(x) = f(x) \cup g(x)$.

Next we recall three concepts of continuity for set-valued maps.

Definition 2.1. $f : X \rightrightarrows Y$ is *upper semi-continuous* at $x_0 \in X$ if for every open set $U \supseteq f(x_0)$ there exists $V \in \mathcal{V}_{x_0}$ so that $f(x) \subseteq U$ for every $x \in V$. If f is upper semi-continuous at every $x \in X$, then we say that f is upper semi-continuous.

Definition 2.2. $f : X \rightrightarrows Y$ is *lower semi-continuous* at $x_0 \in X$ if for every open set $U \subseteq Y$ so that $f(x_0) \cap U \neq \emptyset$ there exists $V \in \mathcal{V}_{x_0}$ so that $f(x) \cap U \neq \emptyset$ for every $x \in V$. If f is lower semi-continuous at every $x \in X$, then we say that f is lower semi-continuous.

Definition 2.3. $f : X \rightrightarrows Y$ is *continuous* at $x \in X$ if f is both upper semi-continuous and lower semi-continuous at $x \in X$. If f is continuous at every $x \in X$, then f is said to be continuous.

In this paper we are concerned with compact-valued maps only. That is, maps $f : X \rightrightarrows Y$ such that $f(x)$ is a compact subset of Y for every $x \in X$. For this reason, all results in this section will be formulated for compact-valued maps, even in cases when the result is known in a more general form.

An *usco* map $f : X \rightrightarrows Y$ is one which is upper semi-continuous and compact valued. An usco map f is called *minimal* (*musco*) if $g = f$ for every usco map g satisfying $g \subseteq f$, while f is *quasi-minimal* if it contains exactly one usco map. If f is quasi-minimal, then we denote by $\langle f \rangle$ the unique usco map contained in f. The set of all musco maps is denoted by $\mathcal{M}(X, Y)$. Clearly every continuous function $f : X \to Y$ may be identified, in a canonical way, with a musco map $S(f)$ by setting $S(f)(x) = \{f(x)\}$ for every $x \in X$. That is, we have an injective function

$$S : \mathcal{C}(X, Y) \to \mathcal{M}(X, Y). \tag{2.1}$$

In general, equality does not hold in the inclusion $\mathcal{C}(X, Y) \subseteq \mathcal{M}(X, Y)$, as can be seen at the hand of elementary examples. However, if X is a Baire space and Y is a metric space, then every usco map $f : X \rightrightarrows Y$ is nearly everywhere continuous, see [12].

Proposition 2.4. *Assume that X is a Baire space and Y is a metric space. If $f : X \rightrightarrows Y$ is usco, then there exists a dense $G - \delta$ set $D \subseteq X$ so that f is continuous at every point in D. Furthermore, if f is musco, then f is point valued at every point in D.*

Let us now recall some further results on usco maps that will be used in subsequent sections. The first result shows that musco maps are abundant, at least relative to the usco maps. It is worth noting that this result is equivalent to the axiom of choice, see for instance [5].

Proposition 2.5. *If $f : X \rightrightarrows Y$ is usco, then there is a musco map $g : X \rightrightarrows Y$ so that $g \subseteq f$.*

Proposition 2.6. *Consider a map $f : X \rightrightarrows Y$ and a point $x \in X$. Then the following statements are equivalent.*

(i) *f is usco at x.*
(ii) *If a net (x_α) in X converges to x and $y_\alpha \in f(x_\alpha)$, then the net (y_α) contains a subnet that converges to some $y \in f(x)$.*

The following well-known result may be found in [8].

Proposition 2.7. *Consider two maps $f, g : X \rightrightarrows Y$ with f usco. Then the following statements are true.*

(i) *The graph $Gr(f)$ of f is a closed subset of $X \times Y$.*
(ii) *If the graph $Gr(g)$ of g is closed and $g \subseteq f$, then g is usco.*

Given a map $f : X \rightrightarrows Y$, one can generate an usco map containing f, provided that $Gr(f)$ is contained in the graph of some usco map, see [6].

Proposition 2.8. *Consider a dense subset D of X and a map $f : D \rightrightarrows Y$. If Y is Hausdorff and the graph of f is contained in the graph of an usco map $g : X \rightrightarrows Y$,* *then there exists a unique usco map* $USC(f) : X \rightrightarrows Y$ *such that* $f \subseteq USC(f) \subseteq h$ *for every usco map* $h : X \rightrightarrows Y$ *that satisfies* $h \supseteq f$*. In particular,*

$$USC(f)(x) = \bigcap\{\overline{f(V)} \mid V \in \mathcal{V}_x\},\ x \in X.$$

New usco maps may be generated from a given usco map, or a family of usco maps, in the following ways, see for instance [2].

Proposition 2.9. *Let* $f : X \rightrightarrows Y$ *be usco,* $U \subseteq X$ *an open set and* $C \subseteq Y$ *a closed set. If* $f(x) \cap C \neq \emptyset$ *for every* $x \in U$*, then the map* $g : X \rightrightarrows Y$ *defined by*

$$g(x) = \begin{cases} f(x) \cap C & \text{if } x \in U \\ f(x) & \text{if } x \notin U \end{cases}$$

is usco.

Proposition 2.10. *Let* $f_i : X \rightrightarrows Y$ *be usco for every* $i \in I$*. The following statements are true:*

(i) $\bigcup_{i \in I} f_i$ *is usco if* I *is finite.*

(ii) *If* $\bigcap_{i \in J} f_i(x) \neq \emptyset$ *for every* $x \in X$ *and every finite set* $J \subseteq I$*, then* $\bigcap_{i \in I} f_i$ *is usco.*

Proposition 2.11. *Suppose that* $f : X \rightrightarrows Y_1$ *and* $g : X \rightrightarrows Y_2$ *are usco. Then the map* $f \times g : X \rightrightarrows Y_1 \times Y_2$ *given by* $(f \times g)(x) = f(x) \times g(x)$*,* $x \in X$*, is usco.*

The next result may be found in [6].

Proposition 2.12. *Assume that* Y *and* Z *are Hausdorff. If* $f : X \rightrightarrows Y$ *is usco and* $\varphi : Y \to Z$ *is continuous, then the map* $g : X \rightrightarrows Z$ *defined by* $g(x) = \varphi(f(x))$*,* $x \in X$*, is usco. If* f *is musco, then so is* g*.*

Lastly, we give a characterisation of quasi-minimal usco maps, see [27].

Proposition 2.13. *Let* Y *be a metric space. An usco map* $f : X \rightrightarrows Y$ *is quasi-minimal if and only if there exists for each* $\epsilon > 0$ *an open and dense subset* D_ϵ *of* X *so that* $\mathrm{diam}(f(x)) < \epsilon$ *for every* $x \in D_\epsilon$*.*

3. The linear space $\mathcal{M}(X, Y)$

We now recall the way in which the linear structure is defined on $\mathcal{M}(X, Y)$, with Y a metrizable (real)[1] topological vector space, see for instance [27]. For all $f, g : X \rightrightarrows Y$ and any $\alpha \in \mathbb{R}$ the Minkowski operations, see for instance [3], are defined as

$$(f \oplus g)(x) = \{y + z \mid y \in f(x),\ z \in g(x)\},$$
$$(f \ominus g)(x) = \{y - z \mid y \in f(x),\ z \in g(x)\}$$

[1]The results concerning the linear structure of $\mathcal{M}(X, Y)$ hold also for vector spaces Y over the complex numbers. However, since we are interested here only in real Banach lattices, we restrict ourselves to this more particular setting.

and

$$(\alpha \odot f)(x) = \{\alpha y \mid y \in f(x)\}.$$

Proposition 3.1. *Let Y be a metrizable topological vector space, $f, g : X \rightrightarrows Y$ quasi-minimal usco maps and α a real number. Then $f \oplus g$, $f \ominus g$ and $\alpha \odot f$ are quasi-minimal usco maps.*

Proposition 3.1 allows us to formulate the following.

Definition 3.2. Let Y be a metrizable topological vector space. For $f, g \in \mathcal{M}(X,Y)$ and $\alpha \in \mathbb{R}$, set $f + g = \langle f \oplus g \rangle$, $f - g = \langle f \ominus g \rangle$ and $\alpha f = \langle \alpha \odot f \rangle$.

Theorem 3.3. *Let Y be a metrizable topological vector space. Then $\mathcal{M}(X,Y)$ is a vector space over $\mathbb{R}$ with respect to the operations given in Definition 3.2. Furthermore, $\mathcal{C}(X,Y)$, equipped with the usual pointwise operations, is a linear subspace of $\mathcal{M}(X,Y)$.*

Note that Definition 3.2 implies that

$$f + g \subseteq f \oplus g,\ f - g \subseteq f \ominus g,\ \alpha f \subseteq \alpha \odot f \tag{3.1}$$

for all $f, g \in \mathcal{M}(X,Y)$ and $\alpha \in \mathbb{R}$. In fact, $\alpha \odot f$ is musco for every $\alpha \in \mathbb{R}$ and $f \in \mathcal{M}(X,Y)$ so that $(\alpha f)(x) = (\alpha \odot f)(x)$, $x \in X$.

4. Usco maps with values in a Banach lattice

We now consider usco maps with values in a Banach lattice E. We examine the relationship between such maps and point-valued upper and lower semi-continuous functions taking values in E, as introduced by Ercan and Wickstead [11].

Definition 4.1. A function $f : X \to E$ is *lower semi-continuous* if $f^{-1}(U + E^+)$ is open in X for every open subset U of E, while f is *upper semi-continuous* if $f^{-1}(U - E^+)$ is open in X for every open subset U of E.

When $E = \mathbb{R}$, Definition 4.1 reduces to the usual ones for lower and upper semi-continuity of real-valued functions.

In the case of an usco map $f : X \rightrightarrows \mathbb{R}$ it is known [2] that the functions $L_f : X \ni x \mapsto \inf f(x) \in \mathbb{R}$ and $U_f : X \ni x \mapsto \sup f(x) \in \mathbb{R}$ are lower and upper semi-continuous, respectively. Conversely, it is easy to see that if $f, g : X \to \mathbb{R}$ are lower semi-continuous and upper semi-continuous, respectively, and $f \leq g$, then the map $h : X \rightrightarrows \mathbb{R}$ given by $h(x) = [f(x), g(x)]$, $x \in X$, is usco. Combining these results, we see that for an usco map $f : X \rightrightarrows \mathbb{R}$, the map $\tilde{f} : X \rightrightarrows \mathbb{R}$ defined by $\tilde{f}(x) = [L_f(x), U_f(x)]$, $x \in X$, is also usco. We now provide generalisations of these results to functions with values in a Banach lattice.

Proposition 4.2. *Suppose that $f : X \rightrightarrows E$ is usco and E is an AM space. Then $L_f : X \ni x \mapsto \inf f(x) \in E$ is lower semi-continuous and $U_f : X \ni x \mapsto \sup f(x) \in E$ is upper semi-continuous in each of the following cases:*

(i) *E has a strong order unit e.*
(ii) *E has an order-continuous norm.*

Proof. Note that every relatively compact subset of E has a supremum and an infimum, see [26, Theorem 2.1.12], so that the functions L_f and U_f are well defined.

(i) Suppose that E has a strong order unit. Let $U = B_\delta(z)$ for some $\delta > 0$ and $z \in E$, and suppose that $x_0 \in L_f^{-1}(U + E^+)$. Without loss of generality, we may assume that $L_f(x_0) \geq z$ so that $L_f(x_0) \in W = B_{\delta/2}(z) + E^+$. Therefore $f(x_0) \subset W$, and since W is open in E and f is usco, there exists $V \in \mathcal{V}_{x_0}$ so that $f(x) \subset W$ for every $x \in V$. But, since E is an AM-space with unit, W is bounded from below by some $y \in U$. Therefore $f(x) \subset y + E^+$ for every $x \in V$ so that $x_0 \in V \subseteq L_f^{-1}(U + E^+)$. Thus $L_f^{-1}(U + E^+)$ is open so that L_f is lower semi-continuous. The proof that U_f is upper semi-continuous follows by essentially similar arguments.

(ii) Assume that E has order-continuous norm. Note that for all $z_1, \ldots, z_k$ and $y_1, \ldots, y_k$ in E we have

$$\left| \bigwedge_{i=1}^{k} z_i - \bigwedge_{i=1}^{k} y_i \right| \leq \bigvee_{i=1}^{k} |z_i - y_i|.$$

Therefore, since E is an AM-space, it follows that

$$\left\| \bigwedge_{i=1}^{k} z_i - \bigwedge_{i=1}^{k} y_i \right\| \leq \bigvee_{i=1}^{k} \|z_i - y_i\|. \tag{4.1}$$

Since $f(x)$ is compact for every $x \in X$, there exists a countable set $C(x) = \{z_n(x) \mid n \in \mathbb{N}\} \subseteq f(x)$ which is norm dense in $f(x)$ and has the same infimum as $f(x)$, that is, $L_f(x) = \inf C(x)$, see for instance [26, proof of Theorem 2.1.12]. Fix $x_0 \in X$ and $\epsilon > 0$. Since f is usco, there exists $V \in \mathcal{V}_{x_0}$ so that $f(x) \subset U_\epsilon = \bigcup_{n=1}^{\infty} B_{\epsilon/2}(z_n(x_0)$ for every $x \in V$. Now consider some arbitrary but fixed $x \in V$. For every $n \in \mathbb{N}$ there exists $m_n \in \mathbb{N}$ so that $\|z_n(x) - z_{m_n}(x_0)\| < \frac{\epsilon}{2}$. Therefore (4.1) implies that

$$\left\| \bigwedge_{n=1}^{k} z_n(x) - \bigwedge_{n=1}^{k} z_{m_n}(x_0) \right\| < \frac{\epsilon}{2} \tag{4.2}$$

for every $k \in \mathbb{N}$. But, since E has an order-continuous norm, it follows that the sequence $\left(\bigwedge_{n=1}^{k} z_n(x)\right)$ converges in norm to $L_f(x)$ while $\left(\bigwedge_{n=1}^{k} z_{m_n}(x_0)\right)$ converges in norm to some $z \geq L_f(x_0)$. Setting $w = L_f(x_0) - z + L_f(x)$, it is clear that $w \leq L_f(x)$. It follows from (4.2) that $\|L_f(x) - z\| < \epsilon$ so that $w \in B_\epsilon(L_f(x_0))$. Therefore $V \subseteq L_f^{-1}(B_\epsilon(L_f(x_0)) + E^+)$ so that L_f is lower semi-continuous. That U_f is upper semi-continuous follows in the same way. □

Proposition 4.3. *Consider two functions $f, g : X \to E$ so that $f(x) \leq g(x)$ for every $x \in X$, and let $h : X \rightrightarrows E$ be given by $h(x) = [f(x), g(x)]$, $x \in X$. The following statements are true:*

(i) *If h is upper semi-continuous, then f is lower semi-continuous and g is upper semi-continuous.*

(ii) *If f is lower semi-continuous, g is upper semi-continuous and E has compact order intervals, then h is usco.*

Proof. (i) Let $U \subseteq E$ be an open set. If $x_0 \in f^{-1}(U+E^+)$, then $h(x_0) \subset U+E^+$. Since h is upper semi-continuous, it follows that there exists $V \in \mathcal{V}_{x_0}$ so that $f(x) \in h(x) \subset U+E^+$ whenever $x \in V$. Hence $V \subseteq f^{-1}(U+E^+)$ so that f is lower semi-continuous. That g is upper semi-continuous follows in the same way. (ii) Consider $x_0 \in X$ and an open subset U of E containing $h(x_0)$. We claim that there exists $\epsilon > 0$ so that $V_\epsilon = (B_\epsilon(f(x_0)) + E^+) \cap (B_\epsilon(g(x_0)) - E^+) \subseteq U$. Since $h(x_0)$ is compact, it follows from the Lebesgue Covering Lemma [17, Theorem 26] that there exists $\delta > 0$ so that $U_\delta = \bigcup_{y \in h(x_0)} B_\delta(y) \subseteq U$. Let $\epsilon < \frac{\delta}{2}$, and fix some $y' \in V_\epsilon$. Then there exists $z, w \in E$ so that $z \leq y \leq w$, and $\|f(x_0) - z\| < \epsilon$ and $\|g(x_0) - w\| < \epsilon$. Without loss of generality, we may assume that $z \leq f(x_0) \leq g(x_0) \leq w$. Let $y = \inf\{\sup\{y', f(x_0)\}, g(x_0)\}$ so that $y \in [f(x_0), g(x_0)] = h(x_0)$. According to [20, Theorem 12.1 and Theorem 12.4 (ii)] we have

$$\begin{aligned}
|y - y'| &= |\inf\{\sup\{y', f(x_0)\}, g(x_0)\} - \inf\{y', w\}| \\
&\leq |\sup\{y', f(x_0)\} - y'| + |g(x_0) - w| \\
&= |\sup\{y', f(x_0)\} - \sup\{y', z\}| + |g(x_0) - w| \\
&\leq |f(x_0) - z| + |g(x_0) - w|.
\end{aligned}$$

Therefore $\|y - y'\| \leq \|f(x_0) - z\| + \|g(x_0) - w\| < 2\epsilon < \delta$ so that $y \in U_\delta$. Consequently $V_\epsilon \subseteq U_\delta \subseteq U$, as desired. Since f is lower semi-continuous and g is upper semi-continuous there exists $W \in \mathcal{V}_{x_0}$ so that $f(x) \in B_\epsilon(f(x_0)) + E^+$ and $g(x) \in B_\epsilon(g(x_0)) - E^+$ for all $x \in W$. Therefore $h(x) = [f(x), g(x)] \subset V_\epsilon \subseteq U$ whenever $x \in W$ so that h is upper semi-continuous at x_0. Since $x_0 \in X$ was arbitrary, it follows that h is upper semi-continuous on X, and since $h(x)$ is compact for every $x \in X$ by assumption, it follows that h is usco. □

Before stating the final result of this section, let us recall some terminology and notation. A subset A of a Riesz space L is *order convex* if $w \in A$ whenever there exist $y, z \in A$ so that $y \leq w \leq z$. The order-convex cover of A is the smallest order-convex subset of L containing A, and is denoted by $[A]$. One may note that

$$[A] = \left\{ w \in L \;\middle|\; \begin{array}{c} \exists \quad y, z \in A : \\ y \leq w \leq z \end{array} \right\}.$$

Proposition 4.4. *If $f : X \rightrightarrows E$ is usco and E has compact order intervals, then the map $[f] : X \rightrightarrows E$ given by $[f](x) = [f(x)]$, $x \in X$, is usco.*

Proof. Since E has compact order intervals, it follows from [7, Theorem 1] that $[f](x)$ is compact for every $x \in X$. Fix $x_0 \in X$ and a net (x_α) in X converging to x_0. Consider a net $(y_\alpha)_{\alpha \in J}$ in Y so that $y_\alpha \in [f](x_\alpha)$ for each $\alpha \in J$. For every $\alpha \in J$ there exist $z_\alpha, w_\alpha \in f(x_\alpha)$ so that $z_\alpha \leq y_\alpha \leq w_\alpha$. In view of Proposition 2.6 we may assume, without loss of generality, that there exist $z, w \in f(x)$ so that (z_α) converges to z and (w_α) converges to w. It follows from [31, Theorem 100.2 (i)] that

$z \leq w$. For every $\epsilon > 0$, let $U_{\frac{\epsilon}{2}} = \bigcup_{y\in[z,w]} B_{\frac{\epsilon}{2}}(y)$. As in the proof of Proposition 4.3 (ii), we find a number $\delta > 0$ so that $V_\delta = (B_\delta(z) + E^+) \cap (B_\delta(w) - E^+) \subseteq U_{\frac{\epsilon}{2}}$. Hence there exists $\alpha_\epsilon \in J$ so that $y_\alpha \in U_{\frac{\epsilon}{2}}$ for every $\alpha \geq \alpha_{\frac{\epsilon}{2}}$. Let (ϵ_n) be a sequence of real numbers that decreases to 0, and let (α_n) be a strictly increasing sequence in J so that, for each $n \in \mathbb{N}$, $y_\alpha \in U_{\frac{\epsilon_n}{2}}$ for all $\alpha \geq \alpha_n$. For each $\alpha \in J$ so that $\alpha \geq \alpha_1$ and $\alpha \not\geq \alpha_2$, select $u_\alpha \in [z,w]$ so that $\|y_\alpha - u_\alpha\| < \frac{\epsilon_1}{2}$. In general, if $\alpha \geq \alpha_n$ and $\alpha \not\geq \alpha_{n+1}$, select $u_\alpha \in [z,w]$ so that $\|y_\alpha - u_\alpha\| < \frac{\epsilon_n}{2}$. Since $[z,w]$ is compact, the net $(u_\alpha)_{\alpha\geq\alpha_1}$ has a subnet $(u_\alpha)_{\alpha\in K}$, with K cofinal in $\{\alpha \in J \mid \alpha \geq \alpha_1\}$, that converges to some $y \in [z,w]$. The subnet $(y_\alpha)_{\alpha\in K}$ also converges to y. Indeed, for every $n \in \mathbb{N}$ there exists $\alpha'_n \in K$ so that $\|u_\alpha - y\| < \frac{\epsilon_n}{2}$ whenever $\alpha \in K$ and $\alpha \geq \alpha'_n$. If $\alpha \in K$ satisfies $\alpha \geq \alpha'_n$ and $\alpha \geq \alpha_n$, then $\|y_\alpha - y\| \leq \|y_\alpha - u_\alpha\| + \|y - u_\alpha\| < \epsilon_n$. Since (ϵ_n) decreases to 0 it follows that $(y_\alpha)_{\alpha\in K}$ converges to y. But $z_\alpha \leq y_\alpha \leq w_\alpha$ for all $\alpha \in J$, so $z \leq y \leq w$ by [31, Theorem 100.2 (i)]. Since $z, w \in f(x)$ it follows that $y \in [f](x)$ so that $[f]$ is usco by Proposition 2.6. □

5. The Riesz space $\mathcal{M}(X, E)$

We now introduce the structure of a Riesz space on the set $\mathcal{M}(X, E)$ of minimal usco maps. In this regard, we define the positive cone of $\mathcal{M}(X, E)$ in a rather obvious way. Let $\mathcal{M}(X, E)^+$ denote the subset of $\mathcal{M}(X, E)$ consisting of those maps f that satisfy the inclusions

$$f(x) \subset E^+, \ x \in X. \tag{5.1}$$

Theorem 5.1. *The set $\mathcal{M}(X, E)^+$ is a cone in $\mathcal{M}(X, E)$. Furthermore, $\mathcal{M}(X, E)$ is an Archimedean Riesz space with respect to the order induced by $\mathcal{M}(X, E)^+$.*

Proof. The cone axioms are trivially satisfied due to the inclusions given in (3.1). To see that $\mathcal{M}(X, E)$ is a Riesz space with respect to the order induced by $\mathcal{M}(X, E)^+$, consider $f \in \mathcal{M}(X, E)$. Denote by $f^+ : X \rightrightarrows E$ the map defined by

$$f^+(x) = \{z^+ \mid z \in f(x)\}, \ x \in X. \tag{5.2}$$

Since the function $E \ni z \mapsto z^+$ is continuous [31, Theorem 100.1], it follows from Proposition 2.12 that $f^+ : X \rightrightarrows E$ is musco. Since $f^+(x) \subseteq E^+$ for every $x \in X$, it follows that $f^+ \geq 0$. Furthermore, $(f^+ \ominus f)(x) \cap E^+ \neq \emptyset$ for every $x \in X$. Since E^+ is closed in E, it follows from Proposition 2.9 that $g : X \rightrightarrows E$ given by $g(x) = (f^+ \ominus f)(x) \cap E^+$, $x \in X$, is usco. Because $f^+ \ominus f$ is quasi-minimal by Proposition 3.1, it now follows from (3.1) and Proposition 2.5 that $(f^+ - f)(x) \subseteq g(x) \subset E^+$ for every $x \in X$. Thus $f \leq f^+$. In order to show that $f^+ = f \vee 0$, we consider any $h \in \mathcal{M}(X, E)^+$ so that $f \leq h$. For every $x \in X$, we have $(h - f)(x) \subseteq E^+$ so that $(h \ominus f)(x) \cap E^+ \neq \emptyset$ by (3.1). Since $h(x) \subset E^+$, it therefore follows by (5.2) that $(h \ominus f^+)(x) \cap E^+ \neq \emptyset$ for every $x \in X$. Thus by Propositions 2.5 and 2.9 there is a musco map $p : X \rightrightarrows E$ such that $p(x) \subseteq (h \ominus f^+)(x)$ and $p(x) \subset E^+$ for every $x \in X$. But $h - f^+ \subseteq h \ominus f^+$ by

(3.1). Since f^+ and h are both minimal, and hence quasi-minimal, it follows from Proposition 3.1 that $h \ominus f^+$ is quasi-minimal. Definition 3.2 therefore implies that $h - f^+ = p \geq 0$ so that $f^+ \leq h$, which shows that $f^+ = f \vee 0$. Since this is true for any $f \in \mathcal{M}(X,E)$, it follows from [20, Theorem 11.5 (v)] that $\mathcal{M}(X,E)$ is a Riesz space with respect to the order induced by the cone $\mathcal{M}(X,E)^+$.

Lastly we verify that $\mathcal{M}(X,E)$ is Archimedean. To this end, consider $f, g \in \mathcal{M}(X,E)^+$ so that $nf \leq g$ for every $n \in \mathbb{N}$. Then $(g - nf)(x) \subseteq E^+$ for every $x \in X$ and $n \in \mathbb{N}$ so that, by (3.1), we have $(g \ominus n \odot f)(x) \cap E^+ \neq \emptyset$ for $x \in X$ and $n \in \mathbb{N}$. Fix $x \in X$. It follows that for each $n \in \mathbb{N}$ there exist $y_n \in f(x)$ and $z_n \in g(x)$ so that $ny_n \leq z_n$. Since $f(x)$ and $g(x)$ are compact we may assume, without loss of generality, that the sequences (y_n) and (z_n) converge to $y \in f(x)$ and $z \in g(x)$, respectively. Since $0 \leq ny_n \leq z_n$ it follows that $\|y_n\| \leq \frac{1}{n}\|z_n\|$ for every $n \in \mathbb{N}$. Hence $y = 0$ so that the musco map $g : X \ni x \mapsto \{0\} \subseteq E$ is contained in f. Since f is musco, it follows that $f = 0$. Consequently $\mathcal{M}(X,E)$ is Archimedean by [20, Theorem 22.2 (ii)]. □

It follows immediately from the definition of the cone $\mathcal{M}(X,E)^+$ of $\mathcal{M}(X,E)$ that the order on $\mathcal{M}(X,E)$ may be characterised in terms of the order on E as follows: For all $f, g \in \mathcal{M}(X,E)$, $f \leq g$ if and only if, for all $x \in X$, $z \leq y$ for some $z \in f(x)$ and $y \in g(x)$. However, more can be said.

Proposition 5.2. *For all $f, g \in \mathcal{M}(X,E)$, the following statements are true:*

(i) *$f \leq g$ if and only if for every $x \in X$ there exist $y \in f(x)$ and $z \in g(x)$ so that $y \leq z$.*

(ii) *$f \leq g$ if and only if the following two conditions are satisfied for every $x \in X$:*

  (a) *For every $y \in f(x)$ there exists $z \in g(x)$ so that $y \leq z$.*

  (b) *For every $z \in g(x)$ there exists $y \in f(x)$ so that $y \leq z$.*

(iii) *$f \vee g = \langle f \overline{\vee} g \rangle$ where $f \overline{\vee} g : X \rightrightarrows E$ is defined by $(f \overline{\vee} g)(x) = \{y \vee z \mid y \in f(x),\ z \in g(x)\}$, $x \in X$.*

(iv) *$f \wedge g = \langle f \overline{\wedge} g \rangle\rangle$ where $f \overline{\wedge} g : X \rightrightarrows E$ is defined by $(f \overline{\wedge} g)(x) = \{y \wedge z \mid y \in f(x),\ z \in g(x)\}$, $x \in X$.*

Proof. (i) Suppose that $f \leq g$. Then $(g - f)(x) \subseteq E^+$ for every $x \in X$. But according to (3.1), $(g - f)(x) \subseteq (g \ominus f)(x) = \{z - y \mid y \in f(x), z \in g(x)\}$, $x \in X$. Thus $(g \ominus f)(x) \cap E^+ \neq \emptyset$ for each $x \in X$. That is, for each $x \in X$ there exist $y \in f(x)$ and $z \in g(x)$ so that $y \leq z$. Now assume that, for each $x \in X$, there is $y \in f(x)$ and $z \in g(x)$ so that $y \leq z$. Then $(g \ominus f)(x) \cap E^+ \neq \emptyset$ for each $x \in X$. By Proposition 2.9 the map $h : X \rightrightarrows E$ given by $h(x) = (g \ominus f)(x) \cap E^+$, $x \in X$, is usco so that, by Proposition 2.5, there is a musco map $k : X \rightrightarrows E$ contained in h. Since $k \subseteq h \subseteq g \ominus f$ and $g \ominus f$ is quasi-minimal usco by Proposition 3.1, it follows by Definition 3.2 that $g - f = k$. But $k \geq 0$ by definition, so $f \leq g$.

(ii) Suppose that $f \leq g$. Define $M_f : X \rightrightarrows E$ and $M_g : X \rightrightarrows E$ as

$$M_f(x) = \left\{ y \in f(x) \;\middle|\; \exists \begin{array}{c} z \in g(x) \ : \\ y \leq z \end{array} \right\}, \quad x \in X \tag{5.3}$$

and

$$M_g(x) = \left\{ z \in g(x) \;\middle|\; \begin{array}{c} \exists \quad y \in f(x) \;: \\ y \leq z \end{array} \right\},\; x \in X.$$

By (i), $M_f(x), M_g(x) \neq \emptyset$ for each $x \in X$. Next we show that $M_f(x)$ is compact for every $x \in X$. Consider some y_0 in the closure of $M_f(x)$. Then there exists a sequence $(y_n) \subseteq M_f(x)$ which converges to y_0. According to the definition (5.3) of M_f, there is a sequence $(z_n) \subseteq g(x)$ so that $y_n \leq z_n$, $n \in \mathbb{N}$. Since $g(x)$ is compact, there is a subsequence (z_{n_k}) of (z_n) which converges to some $z_0 \in g(x)$. Since $y_{n_k} \leq z_{n_k}$ for each $k \in \mathbb{N}$, it follows from [31, Theorem 100.2 (i)] that $y_0 \leq z_0$. Furthermore, $(y_n) \subseteq f(x)$ so that, by compactness of $f(x)$, $y_0 \in f(x)$. Therefore $y_0 \in M_f(x)$ by the definition of M_f. We have therefore established that $M_f(x)$ is closed for each $x \in X$. Since $M_f(x) \subseteq f(x)$, it follows from the compactness of $f(x)$ that, for each $x \in X$, $M_f(x)$ is compact. In exactly the same way it follows that $M_g(x)$ is compact for every $x \in X$. We now show that the graph $Gr(M_f)$ of M_f is a closed subset of $X \times E$. Let (x_0, y_0) be a limit point of $Gr(M_f)$. For each $n \in \mathbb{N}$, let

$$U_n = \left\{ y \in E \;\middle|\; \begin{array}{c} \exists \quad z \in g(x_0) \;: \\ \|y - z\| < \frac{1}{2n} \end{array} \right\}.$$

Clearly U_n is open in E and $g(x_0) \subseteq U_n$. Since g is usco, there exists $V_n \in \mathcal{V}_{x_0}$ so that $g(x) \subset U_n$ for all $x \in V_n$. Since (x_0, y_0) is a limit point of $Gr(M_f)$, there exist $x_n \in V_n$ and $y_n \in M_f(x_n)$ so that $\|y_n - y_0\| < 1/2n$. According to the definition of M_f there exists $z_n \in g(x_n)$ so that $y_n \leq z_n$. Since $g(x_n) \subseteq U_n$ there exists $w_n \in g(x_0)$ so that $\|z_n - w_n\| < 1/2n$. Since $(w_n) \subset g(x_0)$ and $g(x_0)$ is compact, there exist a subsequence (w_{n_k}) of (w_n) and $w_0 \in g(x_0)$ so that (w_{n_k}) converges to w_0. Since $\|z_n - w_n\| < 1/2n$ for every $n \in \mathbb{N}$ it follows that (z_{n_k}) converges to w_0 as well. Furthermore, the sequence (y_n) converges to y_0 so that the subsequence (y_{n_k}) also converges to y_0. Therefore $y_0 \leq w_0$ by [31, Theorem 100.2(i)]. But (x_0, y_0), being an accumulation point of $Gr(M_f) \subseteq Gr(f)$, is an element of $Gr(f)$ by Proposition 2.7 (i), that is, $y_0 \in f(x_0)$. It therefore follows that $y_0 \in M_f(x_0)$ so that $(x_0, y_0) \in Gr(M_f)$, hence $Gr(M_f)$ is closed in $X \times E$. Thus M_f is usco by Proposition 2.7 (ii). In exactly the same way, it follows that M_g is usco. But $M_f \subseteq f$ and $M_g \subseteq g$ with f and g both musco. Hence $M_f = f$ and $M_g = g$ which proves that (a) and (b) hold. The converse follows by (i) of this proposition.

(iii) Since $E \times E : (a, b) \mapsto a \vee b \in E$ is continuous, it follows from Propositions 2.11 and 2.12 that $f \nabla g$ is usco. Fix $\epsilon > 0$. Since f and g are minimal, and hence quasi-minimal usco maps, it follows by Proposition 2.13 that there exists an open and dense subset D_ϵ of X so that $\mathrm{diam}(f(x)) < \frac{\epsilon}{2}$ and $\mathrm{diam}(g(x)) < \frac{\epsilon}{2}$ for every $x \in D_\epsilon$. Fix $x \in D_\epsilon$ and $w_0 = y_0 \vee z_0, w_1 = y_1 \vee z_1 \in f \nabla g(x)$ where $y_0, y_1 \in f(x)$ and $z_0, z_1 \in g(x)$. Using [20, Theorem 11.18 (v)] and the triangle inequality, it follows that $|w_0 - w_1| \leq |y_0 - y_1| + |z_0 - z_1|$. Therefore $\|w_0 - w_1\| < \epsilon$ so that $\mathrm{diam}(f \nabla g(x)) < \epsilon$ for every $x \in D_\epsilon$. Proposition 2.13 now implies that $f \nabla g$ is quasi-minimal. Since $\langle f \nabla g \rangle \subseteq f \nabla g$, it follows by part (i) of this proposition that $f \leq \langle f \nabla g \rangle$ and $g \leq \langle f \nabla g \rangle$. According to [20, Theorem 11.8 (i)], $f \vee g = (f - g)^+ + g$

so that, for each $x \in X$, $(f \vee g)(x) \subseteq h(x)$ where $h : X \rightrightarrows E$ is given by $h(x) = \{(y_0 - z_0)^+ + z_1 \mid y_0 \in f(x),\ z_0, z_1 \in g(x)\}$, $x \in X$. Using arguments similar to those used to show that $f \nabla g$ is quasi-minimal, as well as Proposition 3.1, it follows that h is a quasi-minimal usco map. By [20, Theorem 11.8 (i)] it follows that $(f \nabla g)(x) = \{(y - z)^+ + z \mid y \in f(x), z \in g(x)\} \subseteq h(x)$ for every $x \in X$. Therefore $\langle f \nabla g \rangle \subseteq h$ and $f \vee g \subseteq h$. Since h is quasi-minimal, it therefore follows that $\langle f \nabla g \rangle = f \vee g$, as required.

(iv) The proof is similar to that of (iii). □

The following consequence of Proposition 5.2 is an extension of an elementary property of continuous functions to musco maps.

Corollary 5.3. *The following are equivalent for all $f, g \in \mathcal{M}(X, E)$.*

(i) *$f \leq g$.*
(ii) *There exists a dense subset D of X so that for each $x \in D$, there exist $y \in f(x)$ and $z \in g(x)$ so that $y \leq z$.*

Proof. That (i) implies (ii) follows trivially from Proposition 5.2 (i). Now assume that (ii) is true. Suppose that $(g \ominus f)(x_0) \cap E^+ = \emptyset$ for some $x_0 \in X$. Since $g \ominus f$ is usco by Proposition 3.1 and E^+ is closed in E, there exists $V \in \mathcal{V}_{x_0}$ so that $(g \ominus f)(x) \cap E^+ = \emptyset$ for every $x \in V$. But $(g \ominus f)(x) \cap E^+ \neq \emptyset$ for every x in the dense set D, and hence for some $x \in V$, contradicting the assertion that $(g \ominus f)(x) \cap E^+ = \emptyset$ for all $x \in V$. Thus $(g \ominus f)(x) \cap E^+ \neq \emptyset$ for all $x \in X$. It now follows from Proposition 5.2 (i) that $f \leq g$. □

The remainder of this section is concerned with infima and suprema of arbitrary subsets of $\mathcal{M}(X, E)$. Three characterisations of suprema of upward directed sets will be considered. In two cases, Corollaries 5.10 and 5.13, we first derive characterisations of sets downward directed towards 0 and then obtain the results for upward directed sets in the obvious way. Before we present these results, we note the following technical result.

Proposition 5.4. *Assume that $f, g \in \mathcal{M}(X, E)$ and $f \leq g$. Let $h : X \rightrightarrows E$ be defined by*

$$h(x) = \left\{ w \in E \,\middle|\, \begin{array}{c} \exists \quad y \in f(x), z \in g(x) \ : \\ y \leq w \leq z \end{array} \right\}.$$

Then the following statements are true:

(i) *$h(x)$ is closed and order convex for each $x \in X$.*
(ii) *The graph of h is closed in $X \times E$.*
(iii) *If E has compact order intervals, then h is usco.*

Proof. (i) Consider a sequence (y_n) in $h(x)$ that converges to some $y \in E$. According to the definition of h, there exist two sequences (z_n) and (w_n) in $f(x)$ and $g(x)$, respectively, so that $z_n \leq x_n \leq w_n$ for each $n \in \mathbb{N}$. Since $f(x)$ and $g(x)$ are compact, there exist subsequences (z_{n_k}) and (w_{n_k}) of (z_n) and (w_n) converging to $z \in f(x)$ and $w \in g(x)$, respectively. Since (y_{n_k}) converges to y and

$z_{n_k} \le y_{n_k} \le w_{n_k}$ for each $k \in \mathbb{N}$ it follows that $z \le y \le w$, see [31, Theorem 100.2 (ii)], so that $y \in h(x)$. Hence $h(x)$ is closed in E. To see that $h(x)$ is order convex, consider $y_0, y_1 \in h(x)$ and $y \in E$ so that $y_0 \le y \le y_1$. It follows from the definition of h that there exist $z_0 \in f(x)$ and $z_1 \in g(x)$ so that $z_0 \le y_0 \le y \le y_1 \le z_1$ so that $y \in h(x)$.

(ii) Suppose that (x_0, y_0) is a limit point of $Gr(h)$. Then for every $\epsilon > 0$ and $V \in \mathcal{V}_{x_0}$ there exist $x_V \in V$ and $y_\epsilon \in h(x_V)$ so that $\|y_0 - y_\epsilon\| < \frac{\epsilon}{2}$. For $\epsilon > 0$, let $U_\epsilon = \{y \in E \mid \inf\{\|y - z\| \mid z \in f(x_0)\} < \frac{\epsilon}{2}\}$, $W_\epsilon = \{y \in E \mid \inf\{\|y - z\| \mid z \in g(x_0)\} < \frac{\epsilon}{2}\}$. Since f and g are upper semi-continuous, there exists $V_\epsilon \in \mathcal{V}_{x_0}$ so that $f(x) \subset U_\epsilon$ and $g(x) \subseteq W_\epsilon$ whenever $x \in V_\epsilon$. According to the definition of h, for every $\epsilon > 0$ there exist $z_\epsilon \in f(x_{V_\epsilon})$ and $w_\epsilon \in g(x_{V_\epsilon})$ so that $z_\epsilon \le y_\epsilon \le w_\epsilon$. Since $z_\epsilon \in U_\epsilon$ and $w_\epsilon \in W_\epsilon$, there exist $z'_\epsilon \in f(x_0)$ and $w'_\epsilon \in g(x_0)$ so that $\|z_\epsilon - z'_\epsilon\| < \frac{\epsilon}{2}$ and $\|w_\epsilon - w'_\epsilon\| < \frac{\epsilon}{2}$. Since $f(x_0)$ and $g(x_0)$ are compact there exist a sequence $\epsilon_n \downarrow 0$ and $z_0 \in f(x_0)$, $w_0 \in g(x_0)$ so that (z'_{ϵ_n}) converges to z_0 and (w'_{ϵ_n}) converges to w_0. It follows that the sequences (y_{ϵ_n}), (z_{ϵ_n}) and (w_{ϵ_n}) converge to y_0, z_0 and w_0, respectively. Since $z_{\epsilon_n} \le y_{\epsilon_n} \le w_{\epsilon_n}$ for all $n \in \mathbb{N}$, it follows from [31, Theorem 100.2 (i)] that $z_0 \le y_0 \le w_0$ so that $y_0 \in h(x_0)$, hence $Gr(h)$ is closed in $X \times E$.
(iii) Suppose that E has compact order intervals. According to Proposition 2.10 (i), $f \cup g$ is usco so that $[f \cup g]$ is usco by Proposition 4.4. But h is contained in $[f \cup g]$ and has a closed graph. Therefore h is usco by Proposition 2.7. □

Consider an upward directed subset $\{f_\lambda\} = \{f_\lambda \mid \lambda \in \Lambda\}$ of $\mathcal{M}(X, E)^+$ which is bounded from above. Let $\{g_\sigma\} = \{g_\sigma \mid \sigma \in \Sigma\}$ be the set of upper bounds of $\{f_\lambda\}$. Since $\mathcal{M}(X, E)$ is Archimedean by Theorem 5.1, the set $\{g_\sigma - f_\lambda\}$ is downward directed towards 0, that is, $\inf\{g_\sigma - f_\lambda \mid \lambda \in \Lambda,\ \sigma \in \Sigma\} = 0$, see [20, Theorem 22.5]. For each $\lambda \in \Lambda$ and $\sigma \in \Sigma$, let

$$h_\sigma^\lambda(x) = \left\{ w \in E \,\middle|\, \begin{array}{l} \exists \quad y \in f_\lambda(x),\ z \in g_\sigma(x)\ : \\ \qquad y \le w \le z \end{array} \right\},\ x \in X. \tag{5.4}$$

Define the map $h : X \rightrightarrows E$ by setting

$$h(x) = \bigcap_{\sigma \in \Sigma, \lambda \in \Lambda} h_\sigma^\lambda(x),\ x \in X. \tag{5.5}$$

The supremum of $\{f_\lambda\}$, if it exists, can now be characterised as follows.

Theorem 5.5. *Let $\{f_\lambda\} \subseteq \mathcal{M}(X, E)$ be upward directed and bounded from above. The map $h : X \rightrightarrows E$ defined through (5.4)–(5.5) satisfies the following:*

(i) *If $f_0 \subseteq h$ is musco, then $f_0 = \sup f_\lambda$.*
(ii) *If $f_0 = \sup f_\lambda \in \mathcal{M}(X, E)$, then $f_0 \subseteq h$.*

Proof. (i) Suppose that $f_0 \subseteq h$ is musco. Then $f_0(x) \subseteq h_\sigma^\lambda(x)$ for all f_λ and g_σ and every $x \in X$. That is, for every $y \in f_0(x)$ there exist $z \in f_\lambda(x)$ and $w \in g_\sigma(x)$ so that $z \le y \le w$. Therefore, by Proposition 5.2 (i), $f_\lambda \le f_0 \le g_\sigma$ for all f_λ and g_σ so that $f_0 = \sup f_\lambda$.

(ii) Suppose that $f_0 = \sup f_\lambda$ so that $f_\lambda \le f_0 \le g_\sigma$ for all λ and σ. Applying Proposition 5.2 (ii) we find that for each $x \in X$, $y \in f_0(x)$ and all f_λ and g_σ there exist $z_\lambda \in f_\lambda(x)$ and $w_\sigma \in g_\sigma(x)$ so that $z_\lambda \le y \le w_\sigma$. Therefore $f_0 \subseteq h_\sigma^\lambda$ for every f_λ and each g_σ. Hence $f_0 \subseteq h$, as desired. □

Corollary 5.6. *If E has compact order intervals, then $\mathcal{M}(X,E)$ is Dedekind complete.*

Proof. Let $\{f_\lambda \mid \lambda \in \Lambda\}$ be upward directed and bounded from above, and denote by $\{g_\sigma \mid \sigma \in \Sigma\}$ the set of upper bounds of $\{f_\lambda\}$. Since E has compact order intervals, it follows from Proposition 5.4 (iii) that the map h_σ^λ associated with each f_λ and g_σ through (5.4) is usco. Consider a finite subset $\{h_{\sigma_1}^{\lambda_1}, \ldots, h_{\sigma_1}^{\lambda_1}\}$ of $\{h_\sigma^\lambda \mid \lambda \in \Lambda,\ \sigma \in \Sigma\}$. Since $\{f_\lambda\}$ is upward directed and $\{g_\sigma\}$ is downward directed, it follows that there exist $f_{\lambda'}$ and $g_{\sigma'}$ so that $f_{\lambda_i} \le f_{\lambda'} \le g_{\sigma'} \le g_{\sigma_i}$ for all $i = 1, \ldots, n$. Therefore $h_{\sigma'}^{\lambda'} \subseteq h_{\sigma_i}^{\lambda_i}$ for every $i = 1, \ldots, n$ so that $\bigcap_{i=1}^n h_{\sigma_i}^{\lambda_i}(x) \neq \emptyset$ for every $x \in X$. Thus Proposition 2.10 (ii) implies that h is usco. It now follows from Proposition 2.5 and Theorem 5.5 (i) that $\{f_\lambda\}$ has supremum in $\mathcal{M}(X,E)$, and the proof is complete. □

In order to formulate the remaining results of this section, we introduce the following.

Definition 5.7. A Riesz space L has *property* (C) if there exists a countable subset A of $L^+ \setminus \{0\}$ so that, for every $z \in L$, $z = 0$ if $|z| \not\geq y$ for all $y \in A$.

Before proceeding to formulate our next result, it is worth noting some examples of Riesz spaces with property (C).

Example. The following Riesz spaces have property (C):

(i) The Banach lattices c_0 and ℓ^p, with $1 \le p \le \infty$. In each of these spaces the set $\{\alpha e_n \mid \alpha \in \mathbb{Q} \cap (0,\infty),\ n \in \mathbb{N}\}$, where e_n is the sequence with all terms equal to 0 except for a 1 in the nth entry, serves to verify Definition 5.7.

(ii) Every finite-dimensional Banach lattice E. If E has dimension k, let $\{e_1, \ldots, e_k\}$ be an algebraic basis for E consisting of positive elements. The set $\{\alpha e_i \mid \alpha \in \mathbb{Q} \cap (0,\infty),\ i = 1, \ldots, k\}$ satisfies the condition in Definition 5.7.

(iii) The space $C([0,1])$. Let $A = \{(p_n, q_n) \in \mathbb{Q}^2 \cap [0,1]^2 \mid p_n < q_n\}$, and set $I_n = [p_n, q_n]$ for each $(p_n, q_n) \in A$. For $n, k \in \mathbb{N}$, let

$$f_{n,k}(t) = \begin{cases} 1 & \text{if } t \in I_n \\ kt - kp_n + 1 & \text{if } p_n - \frac{1}{k} \le t < p_n \\ -kt + kq_n + 1 & \text{if } q_n < t \le q_n + \frac{1}{k} \\ 0 & \text{if } t < p_n - \frac{1}{k} \text{ or } t > q_n + \frac{1}{k}. \end{cases}$$

The set $\{\alpha f_{n,k} \mid \alpha \in \mathbb{Q} \cap (0,\infty),\ n, k \in \mathbb{N}\}$ fulfils the conditions on Definition 5.7. More generally, $\mathcal{C}(X)$ has property (C) whenever X is completely regular and separable.

We now proceed to establish two further characterisations of suprema in $\mathcal{M}(X,E)$, one in terms of the norm on E and one in terms of the order on E. As we will see, neither of these results are entirely general as each requires some conditions, either on E or on X.

Theorem 5.8. *Consider a downward directed set $\{f_\lambda\}_{\lambda\in\Lambda}$ in $\mathcal{M}(X,E)^+$. If X is completely regular and $f_\lambda \downarrow 0$ then for every $z \in E^+ \setminus \{0\}$, there exists an open and dense subset D_z of X so that*

$$\begin{array}{ll} \forall & x \in D_z \ : \\ \exists & \lambda_x \in \Lambda \ : \\ \forall & \lambda \in \Lambda,\ f_\lambda \le f_{\lambda_x} \ : \\ & f_\lambda(x) \subset \{w \in E^+ \mid w \not\geq z\}. \end{array} \tag{5.6}$$

If E has property (C) *and X is a Baire space, then* (5.6) *implies that $f_\lambda \downarrow 0$.*

The proof of Theorem 5.8 relies on the following.

Lemma 5.9. *Let X be completely regular. Suppose that $\{f_\lambda \mid \lambda \in \Lambda\}$ decreases to 0 in $\mathcal{M}(X,E)$. If $U \subseteq X$ is nonempty and open, then $\{f_\lambda|U\}$ decreases to 0 in $\mathcal{M}(U,E)$.*

Proof. Suppose that, for some nonempty and open subset U of X, $\{f_\lambda|U\}$ does not decrease to 0 in $\mathcal{M}(X,U)$. Then there exists $g \in \mathcal{M}(U,E)^+$ so that $g \neq 0$ and $g \le f_\lambda|U$ for every $\lambda \in \Lambda$. Without loss of generality, we may assume that $0 \notin g(x)$ for every $x \in U$. Consider an open subset V of U so that $\overline{V} \subset U$ and a point $x_0 \in V$. Let $\varphi : X \to [0,1]$ be a continuous function so that $\varphi(x_0) = 1$ and $\varphi(X \setminus V) = \{0\}$. Let

$$h(x) = \begin{cases} \varphi(x)g(x) & \text{if } x \in U \\ \{0\} & \text{if } x \notin U. \end{cases}$$

Clearly h is compact valued on X and upper semi-continuous on $X \setminus \overline{V}$. To see that h is usco on U, we apply Proposition 2.11 and Proposition 2.12 to the set-valued map $U \ni x \mapsto h(x) \times \{\varphi(x)\} \subseteq E \times [0,1]$ and the continuous function $E \times [0,1] \ni (y,t) \mapsto ty \in E$. Thus h is usco on X. By Proposition 2.5 there exists $f \in \mathcal{M}(X,E)$ so that $f \subseteq h$. It is obvious that $f \ge 0$. But $0 \notin \{\varphi(x_0)z \mid z \in g(x_0)\} \supseteq f(x_0)$ so that $f \neq 0$. Fix $x_1 \in U$ and $z \in f(x_1)$. Since $f \subseteq h$ there exists $z_0 \in g(x_1)$ so that $z = \varphi(x_1)z_0 \le z_0$. But $g \le f_\lambda|U$ for every $\lambda \in \Lambda$. Thus, for each $\lambda \in \Lambda$, there exists $y_\lambda \in f_\lambda(x_1)$ so that $z \le y_\lambda$. For $x \notin U$, $f(x) = \{0\}$ and $0 \le y_\lambda$ for every $\lambda \in \Lambda$ and $y_\lambda \in f_\lambda(x)$. Thus, by Proposition 5.2 (i), f is a lower bound for $\{f_\lambda \mid \lambda \in \Lambda\}$. Hence $\{f_\lambda \mid \lambda \in \Lambda\}$ does not decrease to 0 in $\mathcal{M}(X,E)$. □

Proof of Theorem 5.8. Suppose that $f_\lambda \downarrow 0$. Fix $z \in E^+ \setminus \{0\}$ and let $f_z(x) = \{z\}$ for every $x \in X$. Since $f_z \in \mathcal{M}(X,E)^+ \setminus \{0\}$ and $f_\lambda \downarrow 0$ it follows that there exists $\lambda_z \in \Lambda$ so that $f_\lambda \not\geq f_z$ for all $f_\lambda \le f_{\lambda_z}$. Therefore Proposition 5.2 (i) implies that the set

$$D_z = \left\{ x \in X \;\middle|\; \begin{array}{ll} \exists & \lambda_x \in \Lambda \ : \\ \forall & \lambda \in \Lambda,\ f_\lambda \le f_{\lambda_x} \ : \\ & f_\lambda(x) \subseteq \{w \in E^+ \mid w \not\geq z\} \end{array} \right\}$$

is nonempty. Since $\{f_\lambda\}$ is downward directed, it follows that if $f_{\lambda_0}(x) \subset \{w \in E^+ \mid w \ngeq z\}$ and $f_\lambda \leq f_{\lambda_0}$, then $f_\lambda(x) \subset \{w \in E^+ \mid w \ngeq z\}$. Hence

$$D_z = \left\{ x \in X \;\middle|\; \begin{array}{ll} \exists & \lambda_x \in \Lambda : \\ & f_{\lambda_x}(x) \subseteq \{w \in E^+ \mid w \ngeq z\} \end{array} \right\}.$$

Since $\{w \in E^+ \mid w \ngeq z\}$ is open in E^+ and $f_{\lambda_x} : X \rightrightarrows E^+$ is usco, the set D_z is open. D_z is also dense in X. Indeed, suppose that D_z is not dense in X. Then there exists a nonempty open set U of X so that $U \subseteq X \setminus D_z$. For every $x \in U$ and $\lambda \in \Lambda$, there exists $y \in f_\lambda(x)$ so that $y \geq z$. It follows from Proposition 5.2 (i) that $f_z|U \leq f_\lambda|U$ in $\mathcal{M}(U, E)$ for all $\lambda \in \Lambda$. But X is completely regular, so by Lemma 5.9 $f_\lambda|U \downarrow 0$ in $\mathcal{M}(U, E)$. Thus $f_z|U = 0$ so that $z = 0$, contrary to our assumption that $z \in E^+ \setminus \{0\}$.

Now assume that E has property (C) and that X is a Baire space. Let

$$A = \{z_n \mid n \in \mathbb{N}\} \subset E^+ \setminus \{0\}$$

be a set that satisfies the conditions in Definition 5.7. Suppose that D_z is open and dense in X for every $z \in E^+ \setminus \{0\}$. In particular, D_{z_n} is open and dense in X for every $n \in \mathbb{N}$. Since X is a Baire space, the set $D = \bigcap_{n \in \mathbb{N}} D_{z_n}$ is dense in X. It follows from (5.6) that

$$\begin{array}{ll} \forall & x \in D,\ n \in \mathbb{N} : \\ \exists & \lambda_x^n \in \Lambda : \\ \forall & \lambda \in \Lambda,\ f_\lambda \leq f_{\lambda_x^n} : \\ & f_\lambda(x) \subseteq \{w \in E^+ \mid w \ngeq z_n\} \end{array}$$

Let $f \in \mathcal{M}(X, E)^+$ be a lower bound for $\{f_\lambda\}$. For every $x \in D$ and $w \in f(x)$ we have $w \ngeq z_n$, $n \in \mathbb{N}$. Therefore, according to Definition 5.7, $f(x) = \{0\}$ for every $x \in D$. Since D is dense in X it follows from Corollary 5.3 that $f = 0$ so that $f_\lambda \downarrow 0$. □

Corollary 5.10. *Consider an upward directed set $\{f_\lambda\}_{\lambda \in \Lambda}$ in $\mathcal{M}(X, E)^+$, bounded from above by $f \in \mathcal{M}(X, E)$. If X is completely regular and $f_\lambda \uparrow f$, then for every $z \in E^+ \setminus \{0\}$ there exists an open and dense subset D_z of X so that*

$$\begin{array}{ll} \forall & x \in D_z : \\ \exists & \lambda_x \in \Lambda : \\ \forall & \lambda \in \Lambda,\ f_\lambda \geq f_{\lambda_x} : \\ & (f - f_\lambda)(x) \subseteq \{w \in E^+ \mid w \ngeq z\}. \end{array} \tag{5.7}$$

If E has property (C) and X is a Baire space, then (5.7) *implies that $f_\lambda \uparrow f$.*

Proof. Since $f_\lambda \uparrow f$ if and only if $f - f_\lambda \downarrow 0$, the result follows directly from Theorem 5.8. □

The final result of this section is a metric characterisation of the supremum of an upward directed set. This result may be compared to [28, Theorems 5 and

6] where order-convergence of sequences of nearly finite real normal lower semi-continuous functions is characterised in terms of pointwise convergence on a residual set.

Theorem 5.11. *Suppose that* $\{f_\lambda\}_{\lambda\in\Lambda}$ *is a downward directed subset of* $\mathcal{M}(X,E)^+$. *If, for every* $\epsilon > 0$, *there exists an open and dense set* $D_\epsilon \subseteq X$ *so that*

$$\begin{array}{ll} \forall & x \in D_\epsilon \ : \\ \exists & V \in \mathcal{V}_x,\ \lambda_x \in \Lambda \ : \\ \forall & y \in V,\ f_\lambda \leq f_{\lambda_x},\ z \in f_\lambda(y) \ : \\ & \|z\| < \epsilon, \end{array} \tag{5.8}$$

then $f_\lambda \downarrow 0$ *in* $\mathcal{M}(X,E)$. *The converse is true if* X *is a completely regular Baire space, the norm on* E *is order continuous and* E *has property* (C).

The proof of Theorem 5.11 relies on the following.

Lemma 5.12. *If* $f : X \rightrightarrows E$ *is usco, then the functions*

$$L_f : X \ni x \mapsto \inf\{\|y\| \mid y \in f(x)\} \in \mathbb{R}$$

and

$$U_f : X \ni x \mapsto \sup\{\|y\| \mid y \in f(x)\} \in \mathbb{R}$$

are lower semi-continuous and upper semi-continuous, respectively.

Proof. Fix $x_0 \in X$ and $m < L_f(x_0)$ so that, for some $\epsilon > 0$, $\|z\| > m + \epsilon$ for all $z \in f(x_0)$. Therefore $f(x_0) \subset E \setminus \overline{B}_{m+\epsilon}(0)$. Since f is upper semi-continuous, it follows that there exists $V \in \mathcal{V}_{x_0}$ so that $f(x) \subset E \setminus \overline{B}_{m+\epsilon}(0)$ for every $x \in V$. Thus $L_f(x) \geq m + \epsilon > m$ for all $x \in V$ so that L_f is lower semi-continuous at x_0. Since $x_0 \in X$ is arbitrary, L_f is lower semi-continuous on X.

That U_f is upper semi-continuous follows by an essentially similar argument. □

Proof of Theorem 5.11. Suppose that $\{f_\lambda\}$ does not decrease to 0 in $\mathcal{M}(X,E)$. That is, there exists $f_0 \in \mathcal{M}(X,E)^+ \setminus \{0\}$ so that $f_0 \leq f_\lambda$ for all $\lambda \in \Lambda$. Consider the function

$$L_{f_0} : X \ni x \mapsto \inf\{\|y\| \mid y \in f(x)\} \in \mathbb{R}.$$

Since $f_0 \neq 0$ is musco, it follows from Proposition 2.8 that there exists a nonempty, open subset U of X so that $0 \notin f_0(x)$ for every $x \in U$. Therefore, as $f_0(x)$ is compact, $L_{f_0}(x) > 0$ for every $x \in U$. Since L_{f_0} is lower semi-continuous by Lemma 5.12, there exist an $\epsilon > 0$ and a nonempty, open subset V of U so that $L_{f_0}(x) > \epsilon$ for every $x \in V$. Since $f_0 \leq f_\lambda$ for each $\lambda \in \Lambda$, it follows from Proposition 5.2 (ii) that $\|z\| \geq L_{f_0}(x)$ for every $x \in X$, $\lambda \in \Lambda$ and $z \in f_\lambda(x)$. Therefore $\|z\| > \epsilon$ for every $x \in V$, $\lambda \in \Lambda$ and $z \in f_\lambda(x)$ so that (5.8) does not hold.

Now assume that E has order-continuous norm, satisfies property (C) and X is a completely regular Baire space. Suppose that $f_\lambda \downarrow 0$ but (5.8) does not hold for some fixed $\epsilon > 0$. Note that, due to the upper semi-continuity of the f_λ, the set

$$D_\epsilon = \left\{ x \in X \,\middle|\, \begin{array}{ll} \exists & \lambda_0 \in \Lambda \ : \\ \forall & f_\lambda \leq f_{\lambda_0},\ z \in f_\lambda(x) \ : \\ & \|z\| < \epsilon \end{array} \right\} = \left\{ x \in X \,\middle|\, \begin{array}{ll} \exists & \lambda \in \Lambda \ : \\ \forall & z \in f_\lambda(x) \ : \\ & \|z\| < \epsilon \end{array} \right\}$$

is open in X. Since (5.8) does not hold, D_ϵ is not dense in X. Therefore $X \setminus D_\epsilon$ has nonempty interior. Then there exists a nonempty, open subset U of X so that for every $\lambda \in \Lambda$ and $x \in U$, there exists $z \in f_\lambda(x)$ so that $\|z\| \geq \epsilon$. According to Proposition 2.9, the map $g_\lambda : X \rightrightarrows E$ defined by

$$g_\lambda(x) = \begin{cases} f_\lambda(x) \cap (E \setminus B_\epsilon(0)) & \text{if} \quad x \in U \\ f_\lambda(x) & \text{if} \quad x \notin U \end{cases}$$

is usco for every $\lambda \in \Lambda$. But $g_\lambda \subseteq f_\lambda$ and f_λ is musco, so that $g_\lambda = f_\lambda$ for each $\lambda \in \Lambda$. Hence $\|z\| \geq \epsilon$ for all $z \in f_\lambda(x)$ with $\lambda \in \Lambda$ and $x \in U$. Let $A = \{w_n \mid n \in \mathbb{N}\}$ be a countable subset of $E^+ \setminus \{0\}$ satisfying the conditions of Definition 5.7. According to Theorem 5.8 each of the sets

$$D_n = \left\{ x \in X \,\middle|\, \begin{array}{ll} \exists & \lambda_x \in \Lambda \ : \\ \forall & \lambda \in \Lambda,\ f_\lambda \leq f_{\lambda_x} \ : \\ & f_\lambda(x) \subseteq \{w \in E^+ \mid w \not\geq w_n\} \end{array} \right\}$$

is open and dense in X. Let D denote the intersection of the D_n. Since X is a Baire space, it follows that D is dense in X. Fix $x_0 \in D \cap U$. Since $x_0 \in D$, there exists for each $n \in \mathbb{N}$ a $\lambda_n \in \Lambda$ so that $f_\lambda(x_0) \subset \{w \in E^+ \mid w \not\geq w_n\}$ for all $f_\lambda \leq f_{\lambda_n}$. The sequence (λ_n) may be chosen in such a way that (f_{λ_n}) is a decreasing sequence. For each $n \in \mathbb{N}$, select $y_n \in f_{\lambda_n}(x_0)$ so that (y_n) is a decreasing sequence. Note that, since $y_n \in f_{\lambda_n}(x_0)$ and $x_0 \in U$, it follows that $\|y_n\| \geq \epsilon$ for all $n \in \mathbb{N}$. E is super Dedekind complete since it has order-continuous norm, see [31, Theorem 103.9]. Therefore $\inf\{y_n \mid n \in \mathbb{N}\}$ exists in E. Suppose that $\inf\{y_n \mid n \in \mathbb{N}\} > 0$. Then there exists $k \in \mathbb{N}$ so that $\inf\{y_n \mid n \in \mathbb{N}\} \geq w_k$ so that $y_n \geq w_k$ for all $n \in \mathbb{N}$. Since this is clearly not possible, it follows that $\inf\{y_n \mid n \in \mathbb{N}\} = 0$. It now follows from the order-continuity of the norm on E that $\|y_n\| \downarrow 0$, contrary to the fact that $\|y_n\| \geq \epsilon > 0$ for all $n \in \mathbb{N}$. Hence our assumption that (5.8) does not hold for some fixed $\epsilon > 0$ is false. This completes the proof. □

Corollary 5.13. *Suppose that $\{f_\lambda\}$ is an upward directed subset of $\mathcal{M}(X,E)$, and $f \in \mathcal{M}(X,E)$ is an upper bound for $\{f_\lambda\}$. If, for every $\epsilon > 0$, there exists an open and dense set $D_\epsilon \subseteq X$ so that*

$$\begin{array}{ll} \forall & x \in D_\epsilon \ : \\ \exists & V \in \mathcal{V}_x,\ \lambda_x \in \Lambda \ : \\ \forall & y \in V,\ f_\lambda \geq f_{\lambda_x},\ z \in f_\lambda(y),\ w \in f(y) \ : \\ & \|w - z\| < \epsilon, \end{array} \tag{5.9}$$

then $f_\lambda \uparrow f$ in $\mathcal{M}(X,E)$. The converse is true if X is a completely regular Baire space, the norm on E is order continuous and E has property (C).

Proof. Suppose that (5.9) holds. Since $\{f - f_\lambda\} \subset \mathcal{M}(X,E)^+$ and $(f - f_\lambda)(x) \subseteq \{w - z \mid w \in f(x),\ z \in f_\lambda(x)\}$ by (3.1), it follows by Theorem 5.11 that $f - f_\lambda \downarrow 0$, hence $f_\lambda \uparrow f$.

Now assume that X is a completely regular Baire space, the norm on E is order continuous and E has property (C). Suppose that $f_\lambda \uparrow f$ so that $f - f_\lambda \downarrow 0$. It follows from (3.1) and Theorem 5.11 that for every $\epsilon > 0$ there exists an open and dense set $D'_\epsilon \subseteq X$ so that

$$\begin{array}{ll} \forall & x \in D'_\epsilon \ : \\ \exists & V \in \mathcal{V}_x,\ \lambda_x \in \Lambda \ : \\ \forall & y \in V,\ f_\lambda \geq f_{\lambda_x} \ : \\ \exists & z \in f_\lambda(y),\ w \in f(y) \ : \\ & \|w - z\| < \epsilon. \end{array}$$

Since $f \ominus f_\lambda$ is quasi-minimal usco by Proposition 3.1, the result follows from Proposition 2.13 and the monotonicity of the norm. □

We may note that the requirement on the order-continuity of the norm in Theorem 5.11 and Corollary 5.13 cannot be omitted, as is shown in the following.

Example. Let $E = \mathcal{C}([-1,1])$. Consider the sequence $(f_n) \subset \mathcal{M}(\mathbb{R},E)^+$ where $f_n(x) = \{w_n\}$ for each $x \in \mathbb{R}$, with

$$w_n(t) = \begin{cases} 1 - n|t| & \text{if} \quad |x| \leq \frac{1}{n} \\ 0 & \text{if} \quad 1 \geq |x| > \frac{1}{n} \end{cases}$$

for each $n \in \mathbb{N}$. For every $x \in \mathbb{R}$, $n \in \mathbb{N}$ and $w \in f_n(x)$, $\|w\| = 1$ so that the decreasing sequence (f_n) does not satisfy (5.8). However, if $f \in \mathcal{M}(\mathbb{R},E)^+$ is a lower bound for (f_n), then for each $x \in \mathbb{R}$ and $w \in f(x)$, $w(t) = 0$ if $t \neq 0$. Thus $f(x) = \{0\}$ for each $x \in \mathbb{R}$ so that $f = 0$. Hence (f_n) decreases to 0 in $\mathcal{M}(\mathbb{R},E)$.

6. An application: Dedekind completion of $\mathcal{C}(X, E)$

In this final section we apply the results obtained so far to the problem of characterising the Dedekind completion of the Riesz space $\mathcal{C}(X,E)$ of continuous functions from X into E in terms of functions defined on the same topological space X. In most cases, our results apply only to rather particular situations involving restrictions both on the topological space X and the Banach lattice E. However, the following is true in the most general setting considered here.

Proposition 6.1. *$\mathcal{C}(X,E)$ is a Riesz subspace of $\mathcal{M}(X,E)$.*

Proof. $\mathcal{C}(X,E)$ is a linear subspace of $\mathcal{M}(X,E)$ by Theorem 3.3. In particular, the function $S : \mathcal{C}(X,E) \to \mathcal{M}(X,E)$ defined by $S(f)(x) = \{f(x)\}$, $x \in X$, is a linear injection. To see that S is a Riesz homomorphism, so that $\mathcal{C}(X,E)$ is a Riesz subspace of $\mathcal{M}(X,E)$, consider $f, g \in \mathcal{C}(X,E)$. The continuous function $h(x) = f(x) \vee g(x)$ is the supremum of f and g in $\mathcal{C}(X,E)$. However, $(S(f)\nabla S(g))(x) = \{h(x)\}$ for every $x \in X$ so that $S(f) \vee S(g) = S(f \vee g)$ in $\mathcal{M}(X,E)$. □

We now proceed with the main result of this section. We will show that the Dedekind completion $\mathcal{C}(X,E)^\sharp$ of $\mathcal{C}(X,E)$ is Riesz isomorphic to $\mathcal{M}(X,E)$ whenever X is a compact Hausdorff space and E is an AM-space with compact order intervals. Recall [20, Definition 32.1] that a Dedekind complete Riesz space K is the Dedekind completion of the Riesz space L whenever L is Riesz isomorphic to some Riesz subspace $\hat{L}$ of K, and for every $f \in K$,

$$\sup\{\hat{g} \in \hat{L} \mid \hat{g} \leq f\} = f = \inf\{\hat{g} \in \hat{L} \mid f \leq \hat{g}\}.$$

An immediate application of a result of Veksler [29], see also [31, Theorem 83.18], yields the following.

Theorem 6.2. *Suppose that E has compact order intervals. Then there exists a Riesz subspace L of $\mathcal{M}(X,E)$ that is Riesz isomorphic to the Dedekind completion $\mathcal{C}(X,E)^\sharp$ of $\mathcal{C}(X,E)$ under an isomorphism that leaves $\mathcal{C}(X,E)$ invariant.*

Proof. The result follows immediately from Corollary 5.6, Proposition 6.1 and [31, Theorem 83.18]. □

We should note that, in general, it is not known whether or not the embedding of $\mathcal{C}(X,E)^\sharp$ into $\mathcal{M}(X,E)$ is unique. However, in a very special case, more can be said.

Theorem 6.3. *Suppose that X is a compact Hausdorff space and E is an AM-space with compact order intervals. Then $\mathcal{C}(X,E)^\sharp$ is Riesz isomorphic to $\mathcal{M}(X,E)$.*

Proof. Since E has compact order intervals, it follows from [30, Theorem 5] that E has order-continuous norm and is therefore Dedekind complete, see [31, Theorem 103.9]. Consider any $f \in \mathcal{M}(X,E)$. Since $L_f : X \ni \mapsto \inf f(x) \in E$ is lower semi-continuous by Proposition 4.2 and X is compact and Hausdorff, it follows from [11, Proposition 5.11] that L_f is the pointwise supremum of the set

$$\mathcal{A} = \{g \in \mathcal{C}(X,E) \mid g(x) \leq L_f(x),\ x \in X\}.$$

If $g \in \mathcal{A}$, then $g \leq f$. Conversely, if $g \leq f$ and $g \in \mathcal{C}(X,E)$ then, for every $x \in X$, $g(x) \leq y$ for each $y \in f(x)$. Thus $g(x) \leq L_f(x)$ for every $x \in X$ so that $\mathcal{A} = \{g \in \mathcal{C}(X,E) \mid g \leq f\}$. Let $h = \sup \mathcal{A}$ so that $g \leq h \leq f$ for every $g \in \mathcal{A}$. Since X is compact, and hence a Baire space, it follows from Proposition 2.4 that there is a dense $G-\delta$ set $D \subseteq X$ so that both f and h are point valued on D. For every $x \in D$ and $g \in \mathcal{A}$ we have $g(x) \leq h(x) \leq f(x) = L_f(x)$ so that $h(x) = f(x)$. It now follows from Corollary 5.3 that $f = h = \sup \mathcal{A}$. In exactly the same way, it follows that $f = \inf\{g \in \mathcal{C}(X,E) \mid f \leq g\}$. Since $\mathcal{M}(X,E)$ is Dedekind complete by Corollary 5.6, the proof is complete. □

Theorem 6.3 has a rather limited scope. Indeed, the assumption that E has both compact order intervals and an M-norm is highly restrictive. In fact, Wickstead [30, Theorem 8] showed that relatively compact sets and order-bounded sets in E coincide if and only if E is linearly order isomorphic and homeomorphic to the space $\mathcal{C}_0(T)$ of continuous functions vanishing at infinity on some discrete space

T. On the other hand, E is an AM-space if and only if every relatively compact subset of E has a supremum and an infimum, see for instance [26, Theorem 2.1.12]. Therefore Theorem 6.3 applies only to spaces of the form $E = \mathcal{C}_0(T)$, as described above.

We may note that Theorem 6.3 applies to the case when E is finite dimensional. However, in this case we may relax the assumptions on X – it is sufficient to assume that X is a completely regular cb-space. Recall [21] that a topological space is a cb-space if every real-valued, locally bounded lower semi-continuous function on X is bounded from above by a continuous function.

Theorem 6.4. *If E is finite dimensional and X is a completely regular weak cb-space, then $\mathcal{M}(X,E)$ is Riesz isomorphic to the Dedekind completion of $\mathcal{C}(X,E)$.*

Proof. Let E have dimension n with algebraic basis $B = \{e_1, \ldots, e_n\} \subset E^+$. Consider any $f \in \mathcal{M}(X,E)$. Since $L_f : X \ni \mapsto \inf f(x) \in E$ is lower semi-continuous by Proposition 4.2, it follows that each of the coordinate functions

$$L_f^i(x) = \pi_i(L_f(x) = \inf \pi_i(f(x)),$$

with $\pi_i : E \to \mathbb{R}$ the canonical projection associated with the basis B, is lower semi-continuous. Hence, X being completely regular, each L_f^i is the poinwise supremum of the set $\mathcal{A}_i = \{g \in \mathcal{C}(X) \mid g(x) \leq L_f^i(x),\ x \in X\}$. Thus L_f is the pointwise supremum of $\mathcal{A} = \prod_{i=1}^n \mathcal{A}_i = \{g \in \mathcal{C}(X,E) \mid g \leq L_f\}$. The arguments used in the proof of Theorem 6.3, applied to the specific case under consideration here, lead to the conclusion that $f = \sup\{g \in \mathcal{C}(X,E) \mid g \leq f\}$. That $f = \inf\{g \in \mathcal{C}(X,E) \mid f \leq g\}$ follows in the same way. Hence the proof is complete. □

Theorems 6.3 and 6.4 are generalizations of Anguelov's characterisation [1] of the Dedekind completion of $\mathcal{C}(X)$ in terms of Hausdorff-continuous interval functions. In [1] it is shown that for a completely regular space X, the Dedekind completion of $\mathcal{C}(X)$ is order isomorphic to the ideal generated by $\mathcal{C}(X)$ in the space $\mathbb{H}_{ft}(X)$ of finite Hausdorff continuous functions. Dăneţ [10] showed that the ideal generated by $\mathcal{C}(X)$ in $\mathbb{H}_{ft}(X)$ is all of $\mathbb{H}_{ft}(X)$ if and only if X is a weak cb-space, in particular when X is a cb-space. Anguelov and Kalenda [2] showed that $\mathbb{H}_{ft}(X)$ may be identified in a natural way with $\mathcal{M}(X,\mathbb{R})$. Therefore Anguelov and Dăneţ's results on the Dedekind completion of $\mathcal{C}(X)$ may be rephrased as follows: For a completely regular topological space X, the Dedekind completion of $\mathcal{C}(X)$ is the ideal generated by $\mathcal{C}(X)$ in $\mathcal{M}(X,\mathbb{R})$. This ideal is equal to $\mathcal{M}(X,\mathbb{R})$ if and only if X is a weak cb-space.

References

[1] R. Anguelov, *Dedekind order completion of $C(X)$ by Hausdorff continuous functions*, Quaestiones Mathematicae, **27** (2004), 153–170.

[2] R. Anguelov and O.F.K. Kalenda, *The convergence space of minimal usco mappings*, Czechoslovak Mathematical Journal, **59** (2009), 101–128.

[3] R. Anguelov and J.H. van der Walt, *Algebraic and Topological Structure of Some Spaces of Set-valued Maps*, Computers and Mathematics with Applications, **66** (2013), 1643–1654.

[4] S.M. Aseev, *Quasilinear operators and their application in the theory of multivalued mappings*, Proceedings of the Steklov Institute of Mathematics, **167** (1986), 23–52.

[5] J.M. Borwein and I. Kortezov, *Constructive minimal uscos*, Comptes Rendus de l'Académie Bulgare des Sciences, **57** (2004), 9–12.

[6] J.M. Borwein and W.B. Moors, *Essentially smooth Lipschitz functions*, Journal of Functional Analysis, **149** (1997), 305–351.

[7] M.N. Chaudhary and H.R. Atkinson, *Ordered Banach spaces with compact order intervals*, Journal of the London Mathematical Society, **17** (1977), 107–109.

[8] J.P.R. Christensen, *Theorems of Namioka and R.E. Johnson type for upper semi-continuous and compact valued set-valued mappings*, Proceedings of the American Mathematical Society, **86** (1982), 649–655.

[9] F.H. Clarke, *Optimization and nonsmooth analysis*, John Wiley & Sons, New York, 1983.

[10] N. Dăneţ, *When is every Hausdorff continuous interval-valued function bounded by continuous functions?*, Quaestiones Mathematicae **32** (2009), 363–369.

[11] Z. Ercan and A.W. Wickstead, *Banach lattices of continuous Banach lattice-valued functions*, Journal of Mathematical Analysis and Applications, **198** (1996), 121–136.

[12] M.K. Fort, *Points of continuity of semi-continuous functions*, Publicationes Mathematicae Debrecen, **2** (1951), 100–102.

[13] S.T. Hammer and R.A. McCoy, *Spaces of densely continuous forms*, Set-Valued Analysis, **5** (1997), 247–266.

[14] L'. Holá, *Complete metrizability of generalized compact-open topology*, Topology and its Applications, **91** (1999), 159–167.

[15] L'. Holá, *Spaces of dense continuous forms, usco and minimal usco maps*, Set-Valued Analysis, **11** (2003), 133–151.

[16] L'. Holá, *Spaces of densely continuous forms, Topology Proceedings*, **31** (2007), 145–150.

[17] J.L. Kelley, *General topology*, Van Nostrand, Toronto, New York, London, 1955.

[18] E. Klein and A.C. Thompson, *Theory of correspondences, including applications to mathematical economics*, Wiley, New York, 1984.

[19] F. Lempio and V. Veliov, *Discrete Approximation of Differential Inclusions*, Bayreuther Mathematische Schriften, **54** (1998), 149–232.

[20] W.A.J. Luxemburg and A.C. Zaanen, *Riesz spaces I*, North-Holland, Amsterdam 1971.

[21] J.E. Mack, *On a class of countably paracompact spaces*, Proceedings of the American Mathematical Society, **16** (1965), 467–472.

[22] S. Markov, *Biomathematics and interval analysis: a prosperous marriage*, In: M.D. Todorov, Ch. I. Christov, Eds., AIP Conference Proceedings, Vol. 1301, Amitans'2010, American Institute of Physics, Melville, New York, 2010, pp. 26–36.

[23] R.A. McCoy, *Fell topology and uniform topology on compacta on spaces of multifunctions*, Rostocker Mathematisches Kolloqium, **51** (1997), 127–136.

[24] R.A. McCoy, *Spaces of semi-continuous forms*, Topology Proceedings, **23** (1998), 249–275.

[25] H.J.M. Messerschmidt and M.F.E. de Jeu, *A strong open mapping theorem for surjections from cones onto Banach spaces*, Advances in Mathematics, **259** (2014), 43–66.

[26] P. Meyer-Nieber, *Banach lattices*, Springer-Verlag, Berlin 1991.

[27] J.H. van der Walt, *Linear and quasi-linear spaces of set-valued maps*, Computers and Mathematics with Applications, **68** (2014), 1006–1015.

[28] J.H. van der Walt, *Convergence of sequences of semi-continuous functions*, Journal of Mathematical Analysis and Applications, **388** (2012), 739–752.

[29] A.I. Veksler, *A new construction of the Dedekind completion of vector lattices and l-groups with division*, Siberian Mathematical Journal, **10** (1969), 891–896.

[30] A.W. Wickstead, *Compact subsets of partially ordered Banach spaces*, Mathematische Annalen, **212** (1975), 271–284.

[31] A.C. Zaanen, *Riesz spaces II*, North-Holland, Amsterdam, 1983.

Jan Harm van der Walt
Department of Mathematics and Applied Mathematics
University of Pretoria
Hatfield 0083, South Africa
e-mail: `janharm.vanderwalt@up.ac.za`

Ordered Structures and Applications: Positivity VII
Trends in Mathematics, 503–507

List of Participants

1. **Aksoy, Asuman**, Claremont McKenna College, Claremont, CA, USA
2. **Alekhno, Egor**, Belarussian State University, Minsk, Belarus
3. **Alkurdi, Taleb**, Leiden University, Leiden, the Netherlands
4. **Alpay, Safak**, Middle East Technical University, Ankara, Turkey
5. **Altomare, Francesco**, University of Bari, Bari, Italy
6. **Arendt, Wolfgang**, University of Ulm, Ulm, Germany
7. **Astashkin, Sergey**, Samara State University, Samara, Russia
8. **Athanassiadou, Evaggelia**, University of Athens, Athens, Greece
9. **Avgerinos, Evgenios**, Univ of the Aegean, Rhodes, Greece
10. **Banerjee, Bikram**, Ranaghat College, Kolkata, India
11. **Barret, Martine**, University of La Réunion, Saint Denis, France
12. **Becker, Richard**, Université Paris VI, Paris, France
13. **Ben Amor, Mohamed Amine**, IPEST-University of Carthage, La Marsa, Tunisia
14. **Boulabiar, Karim**, University of Tunis, Tunis, Tunisia
15. **Boza, Santiago**, Polytechnical University of Catalonia, Barcelona, Spain
16. **Brandani da Silva, Eduardo**, Maringa State University, Maringa, Brazil
17. **Bu, Qingying**, University of Mississippi, Mississippi, USA
18. **Buskes, Gerard**, University of Mississippi, University, USA
19. **Çağlar, Mert**, İstanbul Kültür University, İstanbul, Turkey
20. **Caponetti, Diana**, University of Palermo, Palermo, Italy
21. **Cappelletti Montano, Mirella**, University of Bari, Bari, Italy
22. **Cavaliere, Paola**, University of Salerno, Fisciano (Salerno), Italy
23. **Çavuşoğlu, Özlem**, Delft University of Technology, Delft, the Netherlands
24. **Chen, Guiling**, Leiden University, Leiden, the Netherlands
25. **Chen, Jin Xi**, Southwest Jiaotong University, Chengdu, P.R. China
26. **Chen, Zi Li**, Southwest Jiaotong University, Chengdu, P.R. China
27. **Chil, Elmiloud**, Tunis, Monflery, Tunisia
28. **Choi, Yun Sung**, Postech, Pohang, South Korea
29. **Ciesielski, Maciej**, Poznań University of Technology, Poznań, Poland
30. **Conradie, Jurie**, University of Cape Town, Cape Town, South Africa

31. **Corduneanu, Silvia-Otilia**, Gh. Asachi Technical University of Iaşi, Iaşi, Romania
32. **Curbera, Guillermo**, University of Sevilla, Sevilla, Spain
33. **Dăneţ, Nicolae**, Technical Univ. of Civil Engin. of Bucharest, Bucharest, Romania
34. **Dăneţ, Rodica Mihaela**, Technical Univ. of Civil Engin. of Bucharest, Bucharest, Romania
35. **de Jeu, Marcel**, Leiden University, Leiden, the Netherlands
36. **de Pagter, Ben**, Delft University of Technology, Delft, the Netherlands
37. **de Rijk, Björn**, Leiden University, Leiden, the Netherlands
38. **del Campo, Ricardo**, Universidad de Sevilla, Sevilla, Spain
39. **Dirksen, Sjoerd**, Universität Bonn, Bonn, Germany
40. **Dodds, Peter**, Flinders University, Bedford park, Australia
41. **Dodds, Theresa**, Flinders University, Bedford Park, Australia
42. **Drnovšek, Roman**, University of Ljubljana, Ljubljana, Slovenia
43. **Duru, Hülya**, İstanbul University, İstanbul, Turkey
44. **Ercan, Zafer**, Abant Izzet Baysal University, Bolu, Turkey
45. **Erkursun, Nazife**, University of Tübingen, Tübingen / Konya, Germany / Turkey
46. **Estaremi, Yousef**, Payame Noor University, Tehran, Iran
47. **Fechner, Włodzimierz**, University of Silesia, Katowice, Poland
48. **Fernández, Antonio**, Universidad de Sevilla, Sevilla, Spain
49. **Flores, Julio**, Rey Juan Carlos University, Madrid, Spain
50. **Foralewski, Paweł**, Adam Mickiewicz University, Poznań, Poland
51. **Fourie, Jan**, North-West University, Potchefstroom, South Africa
52. **Galé, José E.**, Universidad de Zaragoza, Zaragoza, Spain
53. **Ganiev, Inomjon**, International Islamic Univ. Malaysia, Kuala Lumpur, Malaysia
54. **Gao, Niushan**, University of Alberta, Edmonton, Canada
55. **Ger, Roman**, Silesian University, Katowice, Poland
56. **Gerlach, Moritz**, University of Ulm, Ulm, Germany
57. **Glueck, Jochen**, University of Ulm, Ulm, Germany
58. **Goldberg, Moshe**, Technion-Israel Institute of Technology, Haifa, Israel
59. **Gordon, Yehoram**, Technion-Israel Institute of Technology, Haifa, Israel
60. **Grobler, Jacobus**, North-West University, Potchefstroom, South Africa
61. **Hempfling, Thomas**, Springer Basel AG / Birkhäuser, Basel, Switzerland
62. **Hernandez, Francisco L.**, Universidad Complutense de Madrid, Madrid, Spain
63. **Heymann, Retha**, University of Tübingen, Tübingen, Germany
64. **Himmel, Martin**, Johannes Gutenberg University Mainz, Mainz, Germany
65. **Hu, Huiying**, Shanghai Normal University, Shanghai, China

66. **Hudzik, Henryk**, Adam Mickiewicz University, Poznań, Poland
67. **Kaashoek, Rien**, VU Amsterdam, Amsterdam, the Netherlands
68. **Kalauch, Anke**, TU Dresden, Dresden, Germany
69. **Kandić, Marko**, University of Ljubljana, Ljubljana, Slovenia
70. **Kawabe, Jun**, Shinshu University, Nagano, Japan
71. **Kikianty, Eder**, University of the Witwatersrand, Johannesburg, South Africa
72. **Kitover, Arkady**, Rider University, Lawrenceville, USA
73. **Koelink, Erik**, Radboud Universiteit, Nijmegen, the Netherlands
74. **Kolwicz, Paweł**, Poznań University of Technology, Poznań, Poland
75. **Kunze, Markus**, University of Ulm, Ulm, Germany
76. **Kusraeva, Zalina**, South Math. Inst. of the Russian Acad. of Sciences, Vladikavkaz, Russia
77. **Labuschagne, Coenraad**, Univ. of Johannesburg, Johannesburg, South Africa
78. **Lajara, Sebastian**, Universidad de Castilla-La Mancha, Albacete, Spain
79. **Lee, Han Ju**, Dongguk University, Seoul, South Korea
80. **Lemmens, Bas**, University of Kent, Canterbury, United Kingdom
81. **Leonessa, Vita**, University of Basilicata, Potenza, Italy
82. **Leśnik, Karol**, Poznań University of Technology, Poznań, Poland
83. **Leung, Denny**, National University of Singapore, Singapore, Singapore
84. **Lindemulder, Nick**, Universiteit Utrecht, Utrecht, the Netherlands
85. **Luca-Tudorache, Rodica**, Gheorghe Asachi Technical University, Iasi, Romania
86. **Luxemburg, Wim**, California Institute of Technology, Pasadena, Ca, USA
87. **Malinowski, Helena**, TU Dresden, Dresden, Germany
88. **Marraffa, Valeria**, University of Palermo, Palermo, Italy
89. **Mastyło, Mieczysław**, Adam Mickiewicz University, Poznań, Poland
90. **McKenzie, Stuart**, Queen's University Belfast, Belfast, Northern Ireland
91. **Mekler, Alexander**, University of Bremen, Dresden, Germany
92. **Messerschmidt, Miek**, Leiden University, Leiden, the Netherlands
93. **Mısırlıoğlu, Tunç**, İstanbul Kültür University, İstanbul, Turkey
94. **Mordukhovich, Boris**, Wayne State University, Detroit, USA
95. **Mouton, Sonja**, Stellenbosch University, Stellenbosch, South Africa
96. **Muzundu, Kelvin**, University of Zambia, Lusaka, Zambia
97. **Naranjo, Francisco**, Universidad de Sevilla, Sevilla, Spain
98. **Nussbaum, Roger**, Rutgers University, New Brunswick, USA
99. **Oikhberg, Timur**, University of Illinois, Urbana, USA
100. **Orhon, Mehmet**, University of New Hampshire, Durham, USA
101. **Osaka, Hiroyuki**, Ritsumeikan University, Kusatsu, Japan
102. **Ostaszewska, Urszula**, University of Białystok, Białystok, Poland

103. Panfil, Agata, Adam Mickiewicz University, Poznań, Poland
104. Papachristodoulos, Christos, University of Athens, Athens, Greece
105. Papanastassiou, Nikolaos, University of Athens, Athens, Greece
106. Pedersen, Henrik Laurberg, University of Copenhagen, Copenhagen, Denmark
107. Pinchuck, Andrew, Rhodes University, Grahamstown, South Africa
108. Pliev, Marat, South Math. Institute of the Russian Academy of Science, Vladikavkaz, Russia
109. Płuciennik, Ryszard, Poznań University of Technology, Poznań, Poland
110. Polyrakis, Ioannis, National Technical University of Athens, Athens, Greece
111. Porcello, Giovanni, University of Palermo, Palermo, Italy
112. Ranjbari, Asghar, University of Tabriz, Tabriz, Iran
113. Raynaud, Yves, UPMC, Paris, France
114. Ressel, Paul, Katholische Universität Eichstätt-Ingolstadt, Eichstätt, Germany
115. Reynov, Oleg, Saint Petersburg State University, Saint Petersburg, Russia
116. Rhandi, Abdelaziz, University of Salerno, Salerno, Italy
117. Riihentaus, Juhani, University of Oulu, Oulu, Finland
118. Roberts, Stephan, University of Mississippi, Oxford, USA
119. Roelands, Mark, University of Kent, Canterbury, United Kingdom
120. Rozendaal, Jan, Delft University of Technology, Delft, the Netherlands
121. Sablik, Maciej, Silesian University, Katowice, Poland
122. Sakhnovych, Oleksandr, University of Vienna, Vienna, Austria
123. Schep, Anton, University of South Carolina, Columbia, United States
124. Schwanke, Christopher, University of Mississippi, Oxford, USA
125. Semenov, Evgeny, University of Voronezh, Voronezh, Russia
126. Shahidi, Farruh, International Islamic Univ. Malaysia, Kuala Lumpur, Malaysia
127. Shamsi Gamchi, Saeedeh, Ferdowsi university of Mashhad, Mashhad, Iran
128. Skvortsov, Valentin, Casimir the Great University and Moscow State University, Bydgoszcz/Moscow, Poland/Russia
129. Smela, Krzysztof, Technical University of Rzeszów, Rzeszów, Poland
130. Sofi, Mohd Amin, Kashmir University, Srinagar, India
131. Spijker, Marc, Leiden University, Leiden, the Netherlands
132. Stovall, Jessica, University of North Alabama, Florence, USA
133. Sukochev, Fedor, University of New South Wales, Sydney, Australia
134. Sun, Leping, Shanghai Normal University, Shanghai, P.R. China
135. Sun, Yan, Shanghai Normal University, Shanghai, P.R. China
136. Szwedek, Radosław, Adam Mickiewicz University, Poznań, Poland
137. Tan, Neşet Özkan, Delft University of Technology, Delft, the Netherlands
138. Thieme, Horst, Arizona State University, Tempe, USA

139. **Tomiyama, Jun**, Tokyo Metropolitan University, Tokyo, Japan
140. **Tradacete, Pedro**, Universidad Carlos III de Madrid, Leganes, Spain
141. **Troitsky, Vladimir**, University of Alberta, Edmonton, Canada
142. **Tulone, Francesco**, Palermo University, Palermo, Italy
143. **Uchiyama, Mitsuru**, Shimane University, Matsue, Japan
144. **van der Walt, Jan Harm**, University of Pretoria, Pretoria, South Africa
145. **van Gaans, Onno**, Leiden University, Leiden, the Netherlands
146. **van Neerven, Jan**, Delft University of Technology, Delft, the Netherlands
147. **van Tubergen, Koos**, Hogeschool Inholland, Diemen, the Netherlands
148. **van Waaij, Jan**, Leiden University, Leiden, the Netherlands
149. **van Zuijlen, Willem**, Leiden University, Leiden, the Netherlands
150. **Vardy, Jessica**, University of the Witwatersrand, Johannesburg, South Africa
151. **Väth, Martin**, Free University of Berlin, Berlin, Germany
152. **Veraar, Mark**, Delft University of Technology, Delft, the Netherlands
153. **Watson, Bruce**, University of the Witwatersrand, Johannesburg, South Africa
154. **Westerbaan, Bram**, Radboud University, Nijmegen, the Netherlands
155. **Wickstead, Anthony**, Queen's University Belfast, Belfast, U.K.
156. **Wisła, Marek**, Adam Mickiewicz University, Poznań, Poland
157. **Wnuk, Witold**, Adam Mickiewicz University, Poznań, Poland
158. **Wójtowicz, Marek**, Casimir the Great University, Bydgoszcz, Poland
159. **Wortel, Marten**, University of Kent, Canterbury, UK
160. **Xanthos, Foivos**, University of Alberta, Edmonton, Canada
161. **Zajkowski, Krzysztof**, University of Bialystok, Bialystok, Poland
162. **Zeng, Liuchuan**, Shanghai Normal University, Shanghai, China

MIX
Papier aus verantwortungsvollen Quellen
Paper from responsible sources
FSC® C105338

If you have any concerns about our products,
you can contact us on
ProductSafety@springernature.com

In case Publisher is established outside the EU,
the EU authorized representative is:
Springer Nature Customer Service Center GmbH
Europaplatz 3, 69115 Heidelberg, Germany

Printed by Libri Plureos GmbH
in Hamburg, Germany